Function	Equation	Behavior	Graph
Quadratic (polynomial of degree 2)	$y = ax^2 + bx + c$	One turning point (the vertex) Concave up when $a > 0$ Concave down when $a < 0$	

Cubic (polynomial of degree 3)	$y = ax^3 + bx^2 + cx + d$	Typically, two turning points
		One inflection point
		Rises toward $+\infty$ as $x \to \infty$
		   when $a > 0$
		Falls toward $-\infty$ as $x \to \infty$
		   when $a < 0$	

Quartic (polynomial of degree 4)	$y = ax^4 + bx^3 + cx^2 + dx + e$	Typically, three turning points
		Typically, two inflection points
		Rises toward $+\infty$ as $x \to \infty$
		   when $a > 0$
		Falls toward $-\infty$ as $x \to \infty$
		   when $a < 0$	

Functioning in the Real World

A Precalculus Experience

Second Edition

Sheldon P. Gordon
Farmingdale State University of
New York

Florence S. Gordon
New York Institute of Technology

Alan C. Tucker
SUNY at Stony Brook

Martha J. Siegel
Towson State University

PEARSON
Addison
Wesley

Boston San Francisco New York
London Toronto Sydney Tokyo Singapore Madrid
Mexico City Munich Paris Cape Town Hong Kong Montreal

Publisher: Greg Tobin
Managing Editor: Karen Guardino
Sponsoring Editor: Anne Kelly
Project Editor: Rachel Reeve
Production Supervisor: Cindy Cody
Marketing Manager: Becky Anderson
Marketing Coordinator: Julia Coen
Media Producer: Sharon Smith
Prepress Supervisor: Caroline Fell
Senior Designer: Barbara Atkinson
Cover and Interior Designer: Andrea Menza
Production Services: Jenny Bagdigian
Compositor: WestWords Inc.
Art Creation: Techsetters
Cover Photograph: Eyewire/Getty

Dedicated to:
Craig Eric Gordon and Kenneth Scott Gordon,
Our love, our pride, our hope, our joy.

Library of Congress Cataloging-in-Publication Data
  Gordon, Sheldon P.
    Functioning in the real world : a precalculus experience.— 2nd ed. / Sheldon P.
Gordon, Florence S. Gordon, Alan C. Tucker.
       p. cm.
    Rev. ed. of: Functioning in the real world / Sheldon P. Gordon . . . [et al.]. c1997.
    Includes bibliographical references and index.
    ISBN 0-201-38389-6
    1. Functions. I. Gordon, Florence S. II. Tucker, Alan, 1943 July 6-
    III. Functioning in the real world. IV. Title.

QA331.3.G67 2004
515—dc21

2001056076

This book is based on a portion of the materials developed with the support of the Division of Undergraduate Education of the National Science Foundation under grants #USE-91-50440 and #DUE-9254085 for the Math Modeling/PreCalculus Reform Project. However, any views expressed are not necessarily those of the Foundation.

Contents

Preface vii

CHAPTER 1 **Functions in the Real World**

1.1 Functions Are All Around Us 1
1.2 Describing the Behavior of Functions 8
1.3 Representing Functions Symbolically 18
1.4 Connecting Geometric and Symbolic Representations 25
1.5 Mathematical Models 34
Chapter Summary and Review Problems 39

CHAPTER 2 **Families of Functions**

2.1 Introduction 43
2.2 Linear Functions 44
2.3 Linear Functions and Data 59
2.4 Exponential Growth Functions 70
2.5 Exponential Decay Functions 87
2.6 Logarithmic Functions 99
2.7 Power Functions 113
2.8 Comparing Rates of Growth and Decay 131
2.9 Inverse Functions 140
Chapter Summary and Review Problems 154

CHAPTER 3 **Fitting Functions to Data**

3.1 Introduction to Data Analysis 161
3.2 Linear Regression Analysis 164
3.3 Fitting Nonlinear Functions to Data 181
3.4 How to Fit Exponential and Logarithmic Functions to Data 196
3.5 How to Fit Power Functions to Data 208
3.6 How Good is the Fit? 221
3.7 Linear Models with Several Variables 237
Chapter Summary and Review Problems 243

CHAPTER 4 **Extended Families of Functions**

4.1 Introduction to Polynomial Functions 249
4.2 The Behavior of Polynomial Functions 255
4.3 Modeling with Polynomial Functions 273

4.4   The Roots of Polynomial Equations: Real or Complex?   285

4.5   Finding Polynomial Patterns   292

4.6   Building New Functions from Old: Operations on Functions   304

4.7   Building New Functions from Old: Shifting, Stretching, and Shrinking   320

4.8   Using Shifting and Stretching to Analyze Data   331

4.9   The Logistic and Surge Functions   346

Chapter Summary and Review Problems   352

CHAPTER ◆**5**···· **Modeling with Difference Equations**

5.1   Eliminating Drugs from the Body   357

5.2   Modeling with Difference Equations   371

5.3   The Logistic or Inhibited Growth Model   385

5.4   Newton's Laws of Cooling and Heating   407

5.5   Geometric Sequences and their Sums   416

Chapter Summary and Review Problems   425

CHAPTER ◆**6**···· **Introduction to Trigonometry**

6.1   The Tangent of an Angle   429

6.2   The Sine and Cosine of an Angle   440

6.3   The Sine, Cosine and Tangent in General   456

6.4   Relationships Among Trigonometric Functions   463

6.5   The Law of Sines and the Law of Cosines   469

Chapter Summary and Review Problems   479

CHAPTER ◆**7**···· **Modeling Periodic Behavior**

7.1   Introduction to the Sine and Cosine Functions   483

7.2   Modeling Periodic Behavior with Sine and Cosine   494

7.3   Solving Equations with Sine and Cosine: The Inverse Functions   515

7.4   The Tangent Function   527

Chapter Summary and Review Problems   533

CHAPTER ◆**8**···· **More About the Trigonometric Functions**

8.1   Relationships Among Trigonometric Functions   539

8.2   Approximating Sine and Cosine with Polynomials   550

8.3   Properties of Complex Numbers   566

8.4   The Road to Chaos   574

Chapter Summary and Review Problems   581

CHAPTER **Geometric Models**

9.1   Introduction to Coordinate Systems   583

9.2   Analytic Geometry   585

9.3   Conic Sections: The Ellipse   595

9.4   Conic Sections: The Hyperbola and the Parabola   606

9.5   Parametric Curves   619

9.6   The Polar Coordinate System   630

9.7   Families of Curves in Polar Coordinates   636

Chapter Summary and Review Problems   644

CHAPTER **Matrix Algebra and Its Applications**

10.1   Geometric Vectors   649

10.2   Linear Models   659

10.3   Scalar Products   670

10.4   Matrix Multiplication   686

10.5   Gaussian Elimination   700

Chapter Summary and Review Problems   718

CHAPTER **Probability Models (Available on website only)**

11.1   Introduction to Probability Models

11.2   Binomial Probability and the Binomial Formula

11.3   Using Probability to Investigate Polynomials

11.4   Geometric Probability

11.5   Estimating Areas of Plane Regions

11.6   The Normal Distribution Function

11.7   Waiting Time at a Red Light

11.8   Random Patterns in Chaos

Chapter Summary and Review Problems

CHAPTER **More About Difference Equations (Available on website only)**

12.1   Solutions of Difference Equations

12.2   Constructing Solutions of First Order Difference Equations

12.3   Modeling with First Order Non-Homogeneous, Difference Equations

12.4   Financing a Car or a Home

12.5   Solving the Fibonacci and Other Second Order Difference Equations

12.6   The Predator-Prey Model

12.7   Competing Species

12.8   Iteration and Chaos

Chapter Summary and Review Problems

APPENDIX **A** Some Mathematical Moments to Remember   721

APPENDIX **B** Solving Systems of Linear Equations Algebraically   727

APPENDIX **C** Solving Systems of Linear Equations using Matrices   731

APPENDIX **D** Symmetry   732

APPENDIX **E** Arithmetic of Complex Numbers   734

APPENDIX **F** Introduction to Data Analysis   737

APPENDIX **G** 2002 World Population Data   740

ANSWERS Selected Answers   746

Index   775

Preface

To the Student

Picture yourself as a homeowner whose only tools are a set of screwdrivers. You are perfectly capable of driving screws into or out of wood. But what about other jobs around the house? A screwdriver is not useful for banging a nail into the wall or cutting a board in two. Other tools are needed, and for larger jobs, power tools are essential.

In many ways this situation is analogous to standard mathematics courses in which the emphasis has been almost completely on algebraic methods. These methods give you a powerful set of tools—you can collect like terms, factor various expressions, cancel common factors, expand powers of binomial terms, multiply out polynomials, and so forth. But there are many jobs requiring mathematics that cannot be solved at all using only algebraic methods. For such problems other mathematical tools, including graphical and numerical methods, are far more useful.

For instance, suppose a doctor wants to study a patient's heartbeat using an EKG or a patient's brainwaves using an EEG. There are no known formulas to express these quantities algebraically, but the doctor certainly can get critical information about a patient by interpreting the graphs produced by these devices. Suppose an engineer develops a new tread design for automobile tires and wants to test its braking effectiveness for a car going 20, 30, 40, 50, and 60 miles per hour. There is no exact formula for either the braking distance or the time until the car comes to rest—too many unpredictable factors are involved. All the engineer has is a set of measurements from the experimental runs, and he or she must make decisions based on an understanding of what information the data provides.

Both cases illustrate themes that run through this book. We focus on the applications of mathematics to situations all around us and on the *function* concept that allows us to study these phenomena. That's why this book is titled *Functioning in the Real World*. In this course you will learn to use a combination of algebraic, graphical, and numerical methods, depending on which is the most helpful tool in any given context. You will develop an understanding of the mathematical concepts and learn how to apply them to realistic problems, and not merely perform operations mechanically. You will learn to interpret results, not just obtain answers. You will use technology as a tool for answering the kinds of questions that arise naturally. But this tool is not intended merely to give you answers; technology will help you learn the mathematics. Above all, you will increase your ability to think mathematically so that you can apply mathematics in many other arenas—in other math courses, in courses in other disciplines, in your eventual careers, and in all other aspects of your life.

To do all this, you will look at a much wider variety of topics than you probably have seen in previous math courses. You also will face much more varied types of problems than you might have encountered before. Many of these problems will require you to *think* about the mathematics, not just redo a worked-out example from the text with the numbers changed slightly. To do such problems, you will have to *understand* the mathematical ideas, not merely memorize solutions. To gain that understanding, you will have to pay careful attention in class and read the text thoroughly.

Real-world problems, such as the realistic applications in this book, involve far more than solving an equation that someone hands you. Instead, you face a situation from which you have to identify and extract the mathematical component (which usually means creating an equation), you then have to solve that equation (perhaps using pencil and paper or more likely some kind of technology), you have to interpret the results to make sure that they make sense and are realistic, and finally you have to communicate your answers to other people.

In some ways, this course will be more challenging than others you have taken, but it certainly will be much more rewarding because you will see the value of mathematics all around you. We'd like to give you some suggestions that will make things easier for you in this course:

- ◆ Read the book. It was written for you and is very readable.
- ◆ Work in teams outside of class. Students are very good at explaining mathematical ideas to one another in terms that they understand.
- ◆ Attacking problems in teams is not only a good learning strategy but also the way people in science, engineering, business, and other fields function in the real world.
- ◆ Ask questions in class. Many of the problems are ideal for class discussions, but it is very hard for a teacher to answer questions no one asks.
- ◆ Feel free to suggest your own interpretations. Many of the problems can be approached in very different ways, and different students (and instructors) are likely to come up with different solutions depending on their viewpoint.
- ◆ Talk to your instructor during office hours if you need help. Your professor understands that this is a demanding course for some students and will be happy to work with you. But you must seek out the help.
- ◆ If you use a graphing calculator, carry it to class and use it often to add a graphical dimension to whatever you are studying.
- ◆ Some of you already may use more sophisticated computer programs, the mathematical power tools that are widely available today. These might include software packages, such as Derive^{TM}, Mathematica^{TM}, or Maple^{TM}, that can perform virtually any algebraic operation or a spreadsheet, such as Excel^{TM}, that organizes data and produces sophisticated graphs quickly and easily. Become comfortable with this software as soon as you can so that they are familiar tools throughout the course.
- ◆ Realize that the most powerful and effective tool you have is your mind. It does things no machine is capable of doing—thinking, understanding, creating, and interpreting.

Remember, a carpenter equipped with *all* the right tools is able to build almost anything. Likewise, a student equipped with all the right mathematical tools and the knowledge and judgment to select the right one is prepared for almost anything. Above all, we hope that you will have a very exciting and rewarding experience as you are *Functioning in the Real World*.

To the Instructor

The mathematics curriculum is in the process of change to establish a better balance among geometric, numerical, symbolic, and verbal approaches. There is a much greater emphasis on understanding fundamental mathematical concepts, on realistic applications, on the use of technology, on student projects, and on more active learning environments. These approaches tend to make greater intellectual demands on the students compared to traditional courses that place heavy emphasis on rote memorization and manipulation of formulas.

With support from the National Science Foundation, the Math Modeling/ PreCalculus Reform Project developed a new precalculus or college algebra/ trigonometry experience with the following goals:

- ◆ Extend the common themes in most of the calculus reform projects.
- ◆ Reflect the common themes for new curricula and pedagogy as called for in the MAA/CUPM recommendations, the AMATYC *Crossroads* standards recommendations, and the NCTM *Standards* recommendations.
- ◆ Focus more on mathematical concepts and mathematical thinking by achieving a balance among geometric, numerical, symbolic, and verbal approaches rather than focusing almost exclusively on developing algebraic skills.
- ◆ Provide students with an appreciation of the importance of mathematics in a scientifically oriented society by emphasizing mathematical applications and models.
- ◆ Introduce some modern mathematical ideas and applications that usually are not encountered in traditional courses at this level.
- ◆ Provide students with the skills and knowledge they will need for subsequent mathematics courses.
- ◆ Reflect the major changes in the mathematical needs of other disciplines and so provide students with the skills and knowledge that they will need for courses in the disciplines.
- ◆ Make appropriate use of technology without becoming dominated by the technology to the exclusion of the mathematics.

Philosophy of the Project

To accomplish these goals, we have adopted several basic principles advocated by most leading mathematics educators.

Students should see the power of mathematics.

Most students take precalculus or college algebra/trigonometry courses at the college level because they are prerequisites for future math courses or for courses in one of the client disciplines or as general education requirement, not because they are turned on by the elegance of mathematics. When students see interesting and significant applications, they see the power of mathematics and so are willing to put in the effort needed to learn the subject.

Such applications are the primary focus of the *Functioning in the Real World* course. For instance, in Chapter 2, we build mathematical models for the growth of populations and for the decay in the level of a medication in the blood. In Chapter 3 we analyze the data related to the *Challenger* disaster showing how mathematics could have been used to decide against the launch that day. In Chapter 4 we analyze the spread of AIDS to determine what kind of mathematical model best describes its growth. In Chapter 7 we see how periodic phenomena in nature, such as temperatures, tides, or hours of daylight over the course of a year, can be modeled mathematically.

Students should focus on mathematical ideas, not mathematical calculations.

Our goal is to achieve mathematical understanding, which too often is lost when students concentrate on routine computations that can, should, and in practice will be done by machine. Computer Algebra Systems (CAS) now perform virtually any type of manipulation we could ever expect a student to do. Even if such systems are not necessarily incorporated into the new calculus courses, their existence has major implications about what is important to teach. Further, graphing calculators allow us to study considerably more complicated problems from a geometric or numeric perspective than conventional problems that require factoring artificially constructed functions. Consequently, there is less need to develop as broad a level of manipulative skills in precalculus courses as in the past. However, the algebra that arises in contexts and modeling throughout the book is very substantial, especially the properties of exponential and logarithmic functions. Rather than focusing on producing students who are little better than imperfect organic clones of CAS systems, we emphasize the power of the human mind to inquire, explore, analyze, and interpret.

Technology certainly has a role in the *Functioning* course. We use graphing technology to compare the growth behavior of different members within a family of functions or to compare the behavior among different families of functions. This theme comes up repeatedly throughout the book. Another technological theme is to find the equation of the function (linear, exponential, power, logarithmic, trigonometric, logistic, and so on) that best fits real sets of data. In Chapter 5 we use technology to generate and display solutions of difference equations and investigate their dependence on initial conditions.

Whether you choose to have your students use graphing calculators, CAS, spreadsheets, or other computer packages, technology should be used to motivate and explain the mathematical ideas, not just to produce answers. Whatever technology you and your students use, we firmly believe that the overriding focus should remain on the mathematics and not on the machinery.

Students should DO mathematics, not just passively watch mathematics.

The typical mathematics course involves classroom lectures and homework. But mathematics is all around us, and the best way to appreciate its power and usefulness is to apply it directly. To achieve this, we suggest having students do several mathematical projects either individually or in small groups. Such projects give the students the opportunity to "get their hands dirty" by:

- ◆ formulating mathematical questions
- ◆ collecting appropriate sample data
- ◆ analyzing that data
- ◆ drawing conclusions based on the analysis
- ◆ preparing reports, which promote the organization of ideas as well as their communication skills

For instance, when we look at linear functions early in Chapter 2, students can be asked to select a set of data of interest to them and estimate, by eye, the equation of the line that best fits the data. In Chapter 3, students can be asked to fit a variety of functions (linear, exponential, power, and logarithmic) to a set of data and make predictions based on the results. (See the *Instructor's Resource Manual* for additional suggestions related to each chapter.) The results of doing projects increase the students' level of enthusiasm for the subject matter and their understanding of the mathematical ideas.

Students should be exposed to a broad view of mathematics.

Traditional precalculus courses focus exclusively on preparing students for calculus, particularly in terms of the algebraic skills they may need. The *Functioning* course is intended to prepare students for mathematics in a more general sense. We have incorporated a variety of nontraditional topics in the course that simultaneously introduce new mathematical ideas while advancing the students toward a calculus experience.

- ◆ The laboratory sciences routinely use mathematics to analyze laboratory data, but rarely explain the underlying mathematical foundations—students may use semi–log paper or log–log paper to produce the appropriate results without understanding why. Throughout the book, we emphasize the concept of fitting functions to data using real-life situations, which motivates the mathematical ideas as well as providing contexts in which to develop important algebraic skills at a high level. For example, in Chapter 3 we analyze the growth of the U.S. population via an exponential function; in Chapter 5 we return to the U.S. population and attempt to fit it with a logistic, or inhibited growth, model.

- ◆ In most calculus reform courses the emphasis on differential equations has greatly increased. We extensively discuss difference equations in Chapter 5 to introduce most of the comparable ideas and models in a discrete setting. This discussion reinforces critical ideas on functional behavior and the modeling of real-world phenomena while providing opportunities for honing important algebraic skills. We return to the study of difference equations in supplementary Chapter 12 (available on the web at www.aw.com/ggts).

- ◆ Probabilistic reasoning has become increasingly important in recent decades, and we use it in supplementary Chapter 11 (also available on the web) and we use it in the service of reinforcing other previous ideas and methods from the study of functions.

What's New in the Second Edition

In the second edition, we have tried to build on the strengths of the first edition. The new edition contains a wealth of new applications, examples and problems, and all real-world data sets have been updated. All concepts and methods are approached using the Rule of Three: graphically, symbolically, and numerically.

The new edition has been reorganized and completely rewritten to provide a slower pace through topics that some students find challenging. It also contains a more prominent role for algebraic topics, where the algebraic steps involved in derivations are now highlighted to assist students who may have forgotten some of the algebra they learned in prior courses. Many of the problem sets now include collections of problems, called *Exercising Your Algebra Skills,* to give those students who need it some practice with routine algebra. The book also includes a considerably expanded treatment of trigonometry and the use of the trig functions as models of periodic behavior; there are now three chapters devoted to these ideas and methods.

In particular, some of the major changes in the second edition are:

Chapter 2: Families of Functions　The long section on linear function has been split into two shorter sections to slow the pace. Similarly, the treatment of exponential functions has been slowed by presenting the material in two sections, one on exponential growth functions where the motivating illustration remains population growth, and the other on exponential decay functions where the unifying application is the level of a medication in the bloodstream. Both of these applications then serve as the key themes in later chapters on difference equation models.

Chapter 3: Fitting Functions to Data　Many new examples, particularly relating to power functions, have been added and considerably more emphasis has been placed on the judgmental issues and mathematical reasoning involved in deciding which function is the best fit to a set of data. A new optional section on multivariable linear regression has been added.

Chapter 4: Extended Families of Functions　The material has been reorganized to bring together all the discussions related to polynomial functions. New sections have been added that relate the ideas on shifting and stretching functions to operations on tables of data. A new section has been added on the logistic and surge functions as applications of the material on building new functions from old.

Chapter 5: Modeling with Difference Equations　The introduction to difference equation models has been totally rewritten and reorganized. The two unifying themes are models for the level of a medication in the blood and population growth. The more challenging material, particularly the heavy manipulative topics involving methods for determining closed form solutions, has been moved to Chapter 12.

Chapter 6: Introduction to Trigonometry　A new chapter on right angle trigonometry has been written for those students who need an exposure to this material. The development starts with the tangent ratio, because it makes more sense in most applications to consider the adjacent and opposite sides. The chapter includes the law of sines and the law of cosines.

Chapter 7: Modeling Periodic Behavior　This chapter presents the use of the trigonometric functions as models for periodic behavior.

Chapter 8: More About the Trigonometric Functions　This chapter presents more advanced ideas on the trigonometric functions, especially trigonometric identities, complex numbers and DeMoivre's theorem, and the use of DeMoivre's theorem in understanding some chaotic phenomena.

Chapter 9: Geometric Models　The material on the conic sections has been split into several sections, one on the ellipse and another on the hyperbola and parabola. Additional applications of the hyperbola have been added.

Chapter 10: Matrix Algebra and its Applications   A new section introducing geometric and physical vectors has been added for those instructors who only have time to develop these ideas. Additional examples and problems have been added that link matrix methods more directly to previous topics in the book, including functions, trigonometry, and finding equations of conic sections.

Chapter 11: Probability Models   In the first edition, we had integrated some ideas on probability throughout the text. However, all reports from users that we received indicated that, unfortunately, very few people were able to take advantage of this material because of time pressures. Accordingly, we collected all of this material into a chapter on probability models. We have also added a new section introducing the normal distribution and its uses. Because relatively few instructors have the time to get to this chapter, it is being posted on the web at www.aw.com/ggts, where it is available for downloading. Wherever appropriate, we refer to this chapter, or individual sections, as a supplementary chapter or supplementary sections.

Chapter 12: More About Difference Equations   As mentioned, the more sophisticated ideas and methods on difference equations have been combined into the new Chapter 12. Moreover, additional sections have been added introducing several models based on systems of difference equations; these include the predator-prey model and a model for competitive species. This supplementary chapter is also available for downloading from the web at www.aw.com/ggts.

The Intended Audiences

The materials in this book were developed with different courses in mind:

1. An alternative to the usual one- or two-semester precalculus course designed to prepare students for calculus, one that is in the spirit of the MAA's CUPM recommendations and AMATYC's *Crossroads* Standards.

2. An alternative (perhaps as a course in mathematical modeling) to a one- or two-semester course in college algebra and trigonometry.

3. An alternative to traditional high-school precalculus courses, one that is in the spirit of the NCTM *Standards* and the recommendations of the Pacesetter curriculum project.

4. An alternative to related courses that often are used as a terminal or capstone mathematics course.

5. An alternative to a precalculus-level course for education majors.

We presume that the students taking this course have had a reasonably good mathematical background at the level of intermediate algebra and a previous exposure to some right-angle trigonometry and logarithms (but likely remember very little of it).

As an alternative to traditional precalculus and college algebra/trig courses, the *Functioning* course certainly provides a strong preparation for a standard calculus course. However, the primary emphases we have adopted make the approach particularly well-suited as preparation for virtually any calculus course. We put a strong emphasis on:

♦ the applications of the mathematics

♦ mathematical reasoning and understanding of mathematical concepts, not just symbol manipulation

◆ the use of the Rule of Three: Topics are approached geometrically, numerically, and symbolically wherever possible

◆ the use of verbal reasoning and communication skills

Overall, we believe that it is very important to emphasize repeatedly why you are teaching a course with a very different approach. Students need to be reminded of the reasons why you are expecting more of them and asking them to do different things. Pointing out the limitations of purely algebraic methods or the power of modern technology to solve problems that could not be touched just a few years ago helps. Pointing out the type of traditional manipulative operations that now can be done easily by machine also helps. Most importantly, remind students that they now need to develop the thinking skills to know the questions to ask, to decide which tools to use for answering those questions, and to develop the ability to interpret and communicate these solutions.

Suggested Time Frame

The suggested pacing below is for a one-semester course that meets four hours per week. By including the optional sections and Chapters 10–12, this text can give a two-semester sequence that provides a strong foundation in precalculus ideas and in applied mathematics, particularly discrete mathematics. The text lends itself to many other possible courses and the suggested timetable can provide instructors with some guidance of the time needed for various topics.

Chapter 1 Functions in the Real World **1–1.5 Weeks**	Much of this chapter can be assigned as independent reading. However, we suggest spending some of the first week talking about and developing the critical ideas on the behavior of functions and introducing students to the function concept from geometric, numerical, symbolic, and verbal points of view.

Chapter 2   Families of Functions
             3 weeks
             Section: 2.2 and 2.3   Class hours: 3
                     2.4 and 2.5   3
                     2.6   1
                     2.7   2
                     2.8 (optional)   0.5–1
                     2.9   1
	This chapter is critical in helping students to reach the same plateau of mathematical background and to make the transition to a new way of looking at and thinking about mathematics. You should give students time to reorient themselves.

Chapter 3   Fitting Functions to Data
             2–2.5 Weeks
             Section: 3.1 and 3.2   Class Hours: 2
                     3.3   1.5–2
                     3.4   1.5–2
                     3.5   1.5–2
                     3.6 (optional)   1
                     3.7 (optional)   1
	This chapter provides the link between mathematics and the real world. It shows where functions come from, reinforces ideas about the behavior of families of functions, and provides the opportunities to develop important algebraic skills. Section 3.6 may be assigned as reading.

Chapter 4	**Extended Families of Functions**			This chapter extends the idea of families of functions to polynomials. It also introduces the idea of constructing new functions from old, including operations on functions and shifting and stretching functions.
	2.5–3 Weeks			
	Section: 4.1	Class hours:	1.5	
	4.2		1.5	
	4.3		0.5–1	
	4.4 (optional)		0.5–1	
	4.5 (optional)		1	
	4.6		1.5	
	4.7		1.5	
	4.8 (optional)		1	
	4.9 (optional)		1	
Chapter 5	**Modeling with Difference Equations**			This chapter develops and analyzes models for describing: • population growth • logistic growth • eliminating drugs from the body • radioactive decay • Newton's laws of heating and cooling • geometric sequences and their sums. If pressed for time, you may wish to select some, but not all, of the models discussed.
	1.5–2.5 Weeks			
	Section: 5.1	Class hours:	1.5	
	5.2		1.5–2	
	5.3		1.5–2	
	5.4		1.5–2	
	5.5		1.5	
Chapter 6	**Introduction to Trigonometry**			This chapter presents a brief introduction to right angle trigonometry, including some of the simplest and most common trig identities.
	1.5 Weeks			
	Section: 6.1	Class hours:	1	
	6.2		1–1.5	
	6.3		1	
	6.4		1.5	
	6.5		1.5	
Chapter 7	**Modeling Periodic Behavior**			This chapter uses trigonometric functions to model periodic phenomena. The four parameters in sinusoidal functions are introduced and developed in contexts involving periodic phenomena.
	2–2.5 Weeks			
	Section: 7.1	Class hours:	2	
	7.2		3–4	
	7.3		1.5	
	7.4		1–1.5	
Chapter 8	**More About the Trigonometric Functions**			This chapter • examines the relationship between the trigonometric functions • approximates sine and cosine functions with polynomials • examines the properties of complex numbers • explains chaotic phenomena. Coverage of Section 8.4 may take several class hours, particularly if you wish to include live computer graphics demonstrations.
	2 Weeks			
	Section: 8.1	Class hours:	2–2.5	
	8.2		2	
	8.3		1.5	
	8.4 (optional)		1.5	
Chapter 9	**Geometric Models**			This chapter includes analytic geometry, the conic sections, parametric curves, and curves in the polar coordinate system. The two parts of this chapter, Sections 9.1–9.4 (analytic geometry) and Sections 9.5–9.7 (polar coordinates and parametric curves) can be considered as minichapters that could be covered independently, if so desired.
	2.5–3 Weeks			
	Section: 9.1	Class hours:	0.5	
	9.2		1.5	
	9.3		2	
	9.4		2	
	9.5 (optional)		1.5	
	9.6 (optional)		1	
	9.7 (optional)		1.5	

Detailed suggestions can be found in the *Instructors Resource Manual*.

Supplements

- ◆ The **Instructor's Solutions Manual** (available through http://suppscentral. aw.com) contains complete solutions to all problems for instructors.
- ◆ The **Student Solutions Manual** (ISBN 0-201-61137-6) contains complete solutions to selected problems, particularly nonroutine problems.
- ◆ The **Instructor's Resource Manual** (available through http://suppscentral. aw.com) contains
  - — detailed suggestions for topic selection and pacing for the course,
  - — additional problems and examples,
  - — ideas for classroom activities,
  - — suggestions for student projects related to each chapter,
  - — suggestions for computer laboratory exercises and assignments, and
  - — samples of tests, exams, and project assignments.

Acknowledgments

We gratefully acknowledge the contributions of many people whose advice and assistance immeasurably helped the project and the development of these materials. Judy Broadwin (Jericho High School, NY) brought the secondary school perspective into the project, contributed many innovative suggestions for graphing calculator usage, and assisted in project dissemination activities; we greatly appreciate her ongoing efforts. We also want to thank Ben Fusaro (Florida State University), Joe Fiedler (California State University-Bakersfield), Ignacio Alarcón (Santa Barbara College), Jim Sandefur (Georgetown University), Walter Meyer (Adelphi University), Arlene Kleinstein (Farmingdale State University), Yajun Yang (Farmingdale State University), Ray Bigliani (Farmingdale State University), Anna Silverstein (New York Institute of Technology), Bob Feldman (New York Institute of Technology), Walter Yurek (WorWic College), Tony Peressini (University of Illinois at Champaign-Urbana), Paula Maida (Western Connecticut State University), I-Lok Chang (American University), Harry Hauser (Suffolk Community College), William Abrams (Longwood College), Anne Landry (Dutchess Community College), William Steger (Essex Community College), Sylvia Sorkin (Essex Community College), and Chuck Laufman (Westbury High School) for their valuable suggestions and contributions to the project.

We are indebted to Elka Block and Frank Purcell for the lovely and thorough job they did in producing both the *Instructor's Solutions Manual* and the *Student's Solutions Manual* that accompany the book. We want to thank Paula Maida for her creative contributions to the *Instructor's Resource Manual*. We also acknowledge the careful and intensive work that Paul Lorczak (MathSoft, Inc.) has done in checking the entire manuscript.

We want to thank the National Science Foundation for their support and for making this project possible. In particular, we are indebted to James Lightbourne, William Haver, Elizabeth Teles, and Lee Zia at the NSF Division of Undergraduate Education for their ongoing encouragement and assistance. It is truly appreciated.

We want to express our appreciation to the reviewers of the first edition for their many contributions to the development of the book. For this edition, it is our pleasure to thank the following reviewers, whose comments and suggestions resulted in significant improvements.

Vic Akatsa	Chicago State University
Kathleen Bavelas	Manchester Community College
Therese Bennett	Southern Connecticut State University
Laurie Burton	Western Oregon University
Bob Dobrow	Carleton College
William Fox	Francis Marion University
Debra L. Hydorn	Mary Washington College
Matthew Isom	Arizona State University
Sandra Lofstock	California Lutheran University
Mohammad Moazzam	Salisbury State University
Bernd Rossa	Xavier University
Marvin Stick	University of Massachusetts–Lowell
Richard West	Francis Marion University

We are indebted to the wonderful team at Addison-Wesley, who share our vision for the course and helped bring this book to fruition. We appreciate all of the outstanding contributions of Anne Kelly, Rachel S. Reeve, Cindy Cody, Jenny Bagdigian, Becky Anderson, Karen Guardino, Greg Tobin, Julia Coen, and Sharon Smith. They are a great team, and it has been a true pleasure working with each and every one of these professionals. We also appreciate the wonderful efforts of Beverly Fusfield at Techsetters and Pat McCutcheon at WestWords.

Sheldon P. Gordon
E. Northport, NY

Florence S. Gordon
E. Northport, NY

Alan C. Tucker
Cold Spring Harbor, NY

Martha J. Siegel
Baltimore, MD

Functions in the Real World

1.1 Functions Are All Around Us

The notion of *function* is a fundamental idea in mathematics. Functions are the basis of most mathematical applications in nearly all areas of human endeavor. To see how functions can arise in unexpected places, look at the graph shown in Figure 1.1.

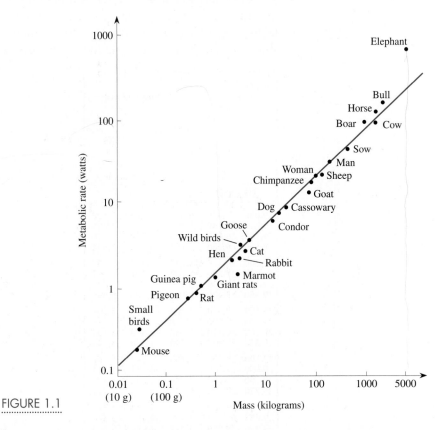

FIGURE 1.1

This graph appears in many introductory biology textbooks. It shows the results of a study comparing the masses of various mammals and birds with their metabolic rates. The biologist who conducted the study first plotted the *data*—the raw measurements on body mass (measured in kilograms) and metabolic rate (measured in watts)—on a graph and then drew a line that passes very close to most of the points. What does this graph show? It is clear from the pattern of the data points that there must be some relationship between the body mass and the metabolic rate of mammals and birds. If there were no relationship, the points would not fall into such a clear pattern. Thus we conclude that, in some way, the metabolic rate of an organism depends on the mass of that organism. Such a relationship is a *function,* and we say that metabolic rate R is a function of body mass W.

Informally, a function is a rule that associates a set of values of one quantity with a set of values of another quantity. Functions are usually represented in four different ways:

1. by formulas or equations,
2. by graphs,
3. by tables, and
4. in words.

For instance, a function might be expressed as a mathematical formula such as $A = s^2$, which gives the area A of a square in terms of its side s. The equation might be $D = 50t$, which gives the distance D you travel at a constant rate of 50 mph in terms of the time t that you drive. A function might be given as a graph, as in the relationship between metabolic rate R as a function of body mass W of various organisms illustrated in Figure 1.1. A function might be given as a table of data. For instance, you compute your income tax for the Internal Revenue Service by using a table—for each level of taxable income, there is a corresponding tax levied, as shown in Figure 1.2. The rule for a function might be expressed in words, as in

If line 37 (taxable income) is—		And you are—		
At least	But less than	Single	Married filing jointly	Married filing sepa-rately
				Your tax is—
29,000				
29,000	29,050	5,092	4,354	5,592
29,050	29,100	5,106	4,361	5,606
29,100	29,150	5,120	4,369	5,620
29,150	29,200	5,134	4,376	5,634
29,200	29,250	5,148	4,384	5,648
29,250	29,300	5,162	4,391	5,662
29,300	29,350	5,176	4,399	5,676
29,350	29,400	5,190	4,406	5,690
29,400	29,450	5,204	4,414	5,704
29,450	29,500	5,218	4,421	5,718
29,500	29,550	5,232	4,429	5,732
29,550	29,600	5,246	4,436	5,746
29,600	29,650	5,260	4,444	5,760
29,650	29,700	5,274	4,451	5,774
29,700	29,750	5,288	4,459	5,788
29,750	29,800	5,302	4,466	5,802
29,800	29,850	5,316	4,474	5,816
29,850	29,900	5,330	4,481	5,830
29,900	29,950	5,344	4,489	5,844
29,950	30,000	5,358	4,496	5,858

FIGURE 1.2

"The cost of postage is 37 cents for the first ounce and 23 cents for each additional ounce."

Typically, when a functional relationship exists between two quantities, the values of one of the quantities depends on the values of the other quantity. That is:

> A **function** is a rule that assigns to each value of one quantity precisely one related value of another quantity.

But, if one value of a quantity leads to two or more values of the other quantity, the relationship between them is *not* a function. For instance, consider the relationship between the number of home runs that a batter has hit by the end of the baseball season and the number of runs he has batted in (RBIs). How many RBIs are associated with 10 home runs? Many different players hit 10 home runs say, but each likely had a different number of RBIs, so this relationship is not a function.

Representing Functions with Formulas and Equations

When you think of functions, the first thing you probably think of is a relationship between two quantities that is given by a formula, such as $A = \pi r^2$, which gives the area A of a circle in terms of its radius r. Similarly, the ideal gas law from chemistry, which says that $P = kT/V$, expresses the pressure P of a gas as a function of its temperature T, where V is the volume of the container that holds the gas and k is a constant. The conversion between Fahrenheit and Celsius temperature readings, $F = \frac{9}{5}C + 32$, expresses the functional relationship between the two temperature scales.

Frequently, when we observe that one quantity is a function of another, we would like to determine an appropriate formula that expresses this relationship. For example, throughout most of human history, people believed that objects fall at a constant speed. Then, in about 1590, Galileo realized that this belief might not necessarily be true. He also had the insight to realize that this conjecture could be tested experimentally. Galileo conducted his now-famous experiments of dropping objects from the top of the Tower of Pisa and found that they fell at ever-increasing rates and that the weight of the objects didn't affect how fast they fell. Galileo's study of the relationship between the distance that an object falls and the time it takes to fall was the key connection for Newton that enabled him to develop his theories of motion that transformed the physical sciences. Based on either Newton's laws of motion or the analysis of data from such an experiment, a formula for the height y of an object dropped from the top of the 180-foot-high Tower of Pisa is $y = 180 - 16t^2$, where t is the number of seconds since the object was dropped. We show where this formula comes from later in the book.

Representing Functions with Graphs

Many effective ways are used to display functions graphically in everyday life—in newspapers, magazines, and scientific, business, and government reports. The graph of a function is valuable because it displays accurate information about a quantity while simultaneously giving an overview of the behavior of that quantity. In particular, a graph can show any trends or patterns in the process being studied.

The graph shown in Figure 1.3 shows the increase in life expectancy in the United States in years since the beginning of the twentieth century. This graph is a function of time t because, for any given year, there is a single value for the life

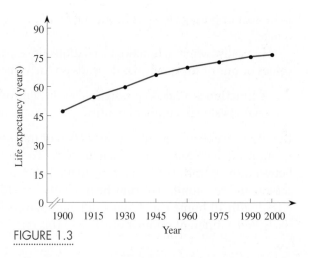

FIGURE 1.3

expectancy of a child born in that year. From the graph, for example, we can esti-mate that a child born in 1900 would have had, on average, a life expectancy of about 47 years, and that a child born in 1990 would have a life expectancy of about 75 years. The rise in life expectancy is a remarkable achievement due to advances in science and medicine and improvements in lifestyles. However, there are also some unfortunate aspects connected with living longer. Can you think of any?

From this graph, not only can you observe the rising trend, but you can also look ahead to predict life expectancies in the not-too-distant future. Note that life expectancy is not merely increasing, but it is actually increasing more slowly as time goes by.

Think About This   What is the significance of this growth pattern for life expectancy if it continues? ␥

Functions are displayed graphically in many ways in newspapers and maga-zines. Keep an eye out for them in your daily activities.

Representing Functions with Tables

Consider the following table of values, which shows the acceleration of a Pontiac Trans Am. The table gives the time in seconds needed to reach different speeds.

Final speed, v (mph)	30	40	50	60	70
Time, t (sec)	3.00	4.29	5.52	7.38	9.81

Although you may not have thought of something such as this as being a func-tion, the time t needed for a Trans Am to achieve a certain speed v is a function of the speed. There may not be an explicit formula for this time as a function of the final speed, but it nevertheless satisfies the definition of a function: For each final speed v, there is a unique time t needed for a Trans Am to accelerate to that speed.

We can plot these points and connect them with a series of straight line seg-ments or even by a smooth curve, as shown in Figure 1.4. Note that the times de-pend on the speeds. Thus we plot the speeds on the horizontal axis and the times needed to achieve those speeds on the vertical axis. Also, you should realize that the values in the table represent only the actual measured points. Drawing a smooth curve through the points requires making assumptions about what happens be-

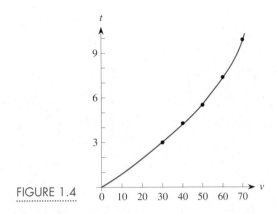

FIGURE 1.4

tween those points. The curve is just an artist's rendition of what the pattern could be; the actual pattern might have some minor variations. Note that we have now represented the same function by both a table and a graph.

Think About This   Estimate the time needed for a Trans Am to accelerate from 0 to 45 mph and from 0 to 75 mph. Which estimate do you think is more accurate? Why? ⌐

Now consider the following daily high temperatures in Phoenix during a severe heat wave in June 1990.

Date	19	20	21	22	23	24	25	26	27	28	29
Temperature (°F)	109	113	114	113	113	113	120	122	118	118	108

Note that a single high-temperature reading is associated with each day, so high temperature is a function of the day. This function makes sense only for the 11 days—June 19 through June 29—and its values consist of the high-temperature readings 108, 109, 113, 114, 118, 120, and 122.

However, the date is *not* a function of the high temperature because a given temperature (say, 113°) was reached on *more than one* date (in this case the 20th, the 22nd, the 23rd, and the 24th).

The function that associates the high temperature in Phoenix with the corresponding day of the month can be depicted graphically by plotting the individual points, as shown in Figure 1.5, so again we have represented the same function by both a table and a graph. The points in the figure can be joined by a series of line segments or by a smooth curve to give a sense of an overall trend or pattern, as shown in Figure 1.6. However, doing so requires some careful thought. When we connect the points, we are *not* indicating that this graph represents temperature as a function of time; we are just connecting the maximum temperatures recorded each day, and the curve shown gives absolutely no information about the temperature at any intermediate time. In fact, the actual graph of temperature versus time would typically show the type of oscillatory effect depicted in Figure 1.7.

Representing Functions with Words

Functions are expressed verbally in many different ways. The maximum load that a jet plane can lift is *related* to its wingspan. The number of different species that can live on an island *depends* on the size of the island. The population of the world *over*

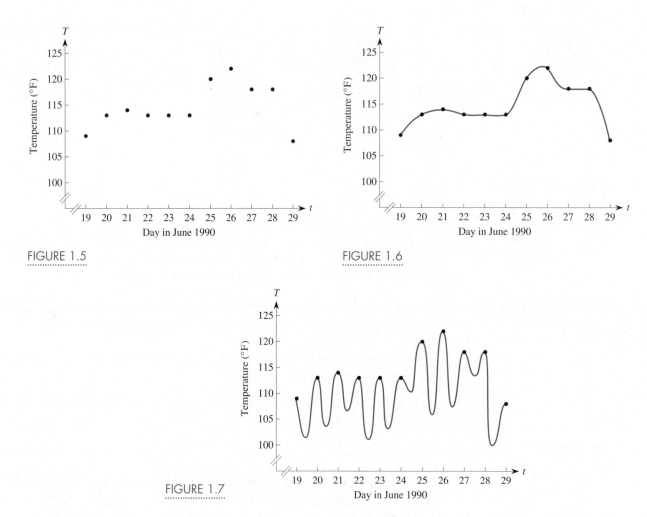

FIGURE 1.5

FIGURE 1.6

FIGURE 1.7

time is a *function of* time. If money is borrowed at simple interest (no compounding), the amount of interest earned is *a multiple of* the amount borrowed.

Why Study Functions?

Any situation involving two quantities usually raises several questions:

1. Is there a functional relationship between the two quantities?
2. If there is a relationship, can we find a formula for it?
3. Can we construct a table or graph relating the two quantities, especially if we can't find a formula?
4. If we can find a formula, or if we have a graph of the relationship, or if we have a table of values relating the two quantities, how do we use it? That is, how can knowledge of the function aid in understanding the relationship between the two quantities or allow us to make predictions or informed decisions about one of the variables based on the other?

Figure 1.1 clearly suggests a relationship between metabolic rates R in mammals and birds and their body mass W. This same relationship can then be used to predict the metabolic rate of other species—say, lions or Kodiak bears—based on knowledge of their mass. This relationship could even be used to predict the metabolic rate of an extinct pterodactyl from estimates of its body mass made from its skeletal remains.

However, it could not be used to predict the metabolic rate for a crocodile because a crocodile is neither a mammal nor a bird; the relationship observed in Figure 1.1 for mammals and birds may not apply to reptiles.

Think About This   Based on the graph shown in Figure 1.1, would you use the relationship to predict the metabolic rate for extinct mammoths, which were slightly larger than today's elephant? ▭

Problems

1. Which of the following relationships are functions and which are not? Explain your reasoning. For those that are functions, identify which of the two quantities depends on the other. Again, explain your reasoning.

   a. The number of miles driven in a car versus the number of gallons of gas used.

   b. The price of a diamond versus the number of carats.

   c. The major league baseball player who has a certain number of home runs at the end of the season.

   d. The student who has a specific score on the SAT in a particular year.

   e. The amount of rain that falls on any particular day of the year in Seattle.

   f. The day of the year on which given amounts of snow, in inches, fall in Buffalo.

2. Match each of the following functions with a corresponding graph. Explain your reasoning.

   a. The population of a country as a function of time.

   b. The path of a thrown football as a function of time.

   c. The distance driven at a constant speed as a function of time.

   d. The daily high temperature in a city as a function of time over several years.

   e. The number of cases of a disease as a function of time.

   f. The percentage of families owning VCRs as a function of time.

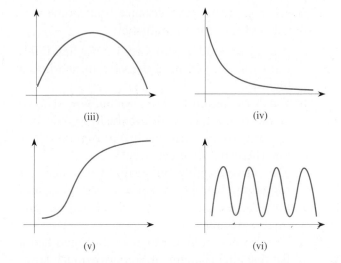

(iii)   (iv)

(v)   (vi)

3. The following graphs show the noise level of a crowd of college students watching their school's basketball team playing at home in the championship finals for the league title. Match the three graphs with the corresponding scenarios (reactions) and then draw a graph for the remaining scenario.

   a. Our team started slowly but eventually began to pull away.

   b. It was a disaster from start to finish.

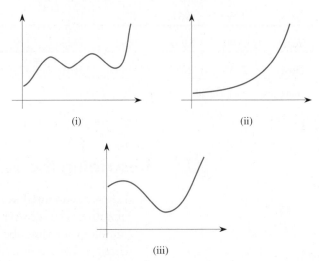

(i)   (ii)

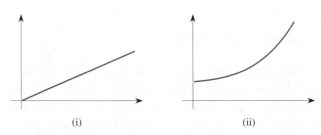

(i)   (ii)   (iii)

c. The score kept seesawing back and forth, but we finally won on a three-point shot at the buzzer.

d. Our team started well, then the opposition took the lead, but we finally won.

4. Consider the scenario: "You left home to run to the local gym. You started at a constant rate of speed but sped up when you realized how energetic you felt. About halfway there, you began to tire, so you started slowing down." Sketch a graph of your distance from home as a function of time.

5. Sketch a graph of your distance from home as a function of time for each situation.

a. You drove steadily across town, speeding up as traffic diminished until the road turned into a highway.

b. You drove steadily toward town but slowed down as the traffic increased. Eventually you inched forward around a car that had broken down before you could resume normal speed.

c. You drove steadily but realized you had left something behind, so you returned home and then drove all the way to school without any further trouble.

d. You drove steadily across town but then had a flat tire; after changing it, you drove much faster so that you wouldn't be too late for class.

6. For each of the scenarios in Problem 5, sketch a graph of the *total distance* you've traveled as a function of time.

7. Consider again the graph in Figure 1.3. Write a paragraph or two interpreting what the increase in life expectancy over the past century means. For example, you might consider it in terms of your own expected life span compared to those of your children and grandchildren. Alternatively, you might consider the effects on the overall distribution of people of different ages in the population at large, or you might discuss the question of whether there is a natural limit to how long the human life span can be extended in the future. Compare the values for life expectancies in the United States in Figure 1.3 with the values for life expectancies of other nations given in Appendix G.

8. Which table of values represents a function and which doesn't? Explain your reasoning.

a.

x	0	3	6	1	5	2	4
y	8	6	2	2	4	5	3

b.

x	0	2	3	4	1	3	5
y	8	4	7	2	6	10	9

9. The Dow-Jones average of 30 industrial stocks is probably the most closely watched measure of stock market performance. Below are the Dow values at the beginning of each year from 1980 to 2000.

Write a short paragraph describing the behavior of the stock market over this period of time. When did it rise? When did it fall? Which years would have been the best times to buy stocks? Which would have been the worst times to do so?

Year	1980	1981	1982	1983	1984	1985	1986	1987	1988 -	1989	
Dow	839	964	875	1047	1259	1212	1547	1896	1939	2169	
Year	1990	1991	1992	1993	1994	1995	1996	1997	1998	1999	2000
Dow	2753	2634	3169	3301	3758	3834	5177	6447	7965	9184	11358

Source: *Wall Street Journal.*

1.2   Describing the Behavior of Functions

Functions are used to represent quantities in the real world. Because most of these quantities change over time or depend on some other quantity, we need some terminology to describe the *behavior of the function*—that is, how the function changes. We can describe the behavior of a function in two different ways.

Increasing and Decreasing Functions

The first and most immediate aspect of behavior is whether the function is increasing or decreasing. The graph of a function is *increasing* if, as you look from left to right, the vertical values get larger; that is, the graph rises. We also describe this behavior as *growth*. Similarly, the graph of a function is *decreasing* if, as you look from left to right, the vertical values get smaller; that is, the graph falls. We also describe this behavior as *decay*. Figure 1.8 illustrates these characteristics.

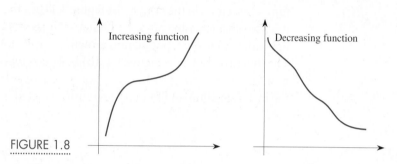

FIGURE 1.8

For instance, the world's population is growing. Therefore the function that expresses the population over time is an increasing function. Also, the heavier a car is, the lower its gas mileage will be. Therefore the function that relates gas mileage to the weight of a car is a decreasing function.

Of course, not every quantity merely increases or decreases. Often, a quantity will rise some of the time and fall some of the time, such as the height of a bouncing ball, the value of the Dow-Jones average, or the high temperature recorded in a particular location each day of the year. Thus a function whose graph looks like the one shown in Figure 1.9 increases for some values of the variable and decreases for others. Here the function rises (increases) to a maximum or largest value compared to nearby points, then falls (decreases) to a minimum value compared to nearby points, and then rises again. We call any point where the behavior of the function changes from increasing to decreasing or from decreasing to increasing a **turning point** of the function. Turning points occur at points where a function reaches a *local maximum* (the value where the function is larger than any nearby value) or a *local minimum* (the value where the function is smaller than any nearby value).

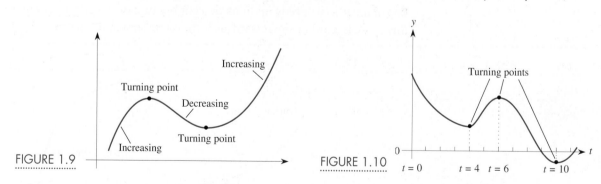

FIGURE 1.9

FIGURE 1.10

Note that a function increases or decreases over an interval of values on the horizontal axis; it has a turning point at a particular point corresponding to a single value along the horizontal axis. Figure 1.10 shows a function of t decreasing from $t = 0$ to $t = 4$, increasing from $t = 4$ to $t = 6$, decreasing from $t = 6$ to $t = 10$, and then increasing after $t = 10$. This function has turning points at $t = 4$, $t = 6$, and $t = 10$.

Sketch the graph of a different function having the same behavior. ▭

We describe a function that is always increasing or always decreasing (it has no turning points) as being *strictly increasing* or *strictly decreasing*.

Concavity: How A Function Bends

There is a second aspect to a function's behavior. Figure 1.11 shows two increasing functions. How do they differ? In Figure 1.11(a), the function isn't merely increasing; it is actually increasing faster and faster as time goes by. Think of the curve as bending upward. For instance, population growth typically follows this type of growth pattern. The function shown in Figure 1.11(b) is also increasing, but it is increasing more and more slowly as time goes by. Think of the curve as bending downward. For instance, the increasing human lifespan previously depicted in Figure 1.3 grows in this way.

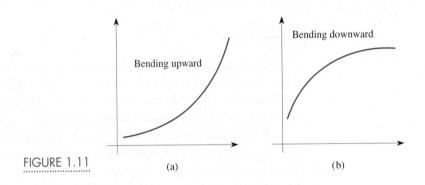

FIGURE 1.11   (a)   (b)

Now look at the two decreasing functions in Figure 1.12. The function shown in Figure 1.12(a) decreases very rapidly at first and then more slowly as time passes—it is decreasing at a decreasing rate. For instance, if a pollutant is released into a lake, the level of pollution in the lake will decrease ever more slowly as time goes by. Like the function shown in Figure 1.11(a), this curve is also bending upward. The graph in Figure 1.12(b) also decreases, but it is decreasing slowly at first and then more and more rapidly—it is decreasing at an increasing rate. For instance, if an object is tossed off the roof of a tall building, its height above ground will decrease in this manner as it speeds up in its descent because of the effects of gravity. Note that this curve is bending downward, as is the curve shown in Figure 1.11(b).

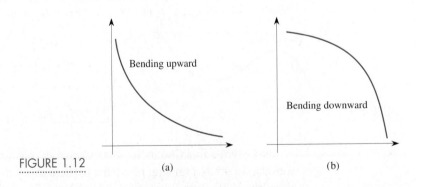

FIGURE 1.12   (a)   (b)

We use the term **concavity** to describe the way a function *bends*. Curves that bend upward, such as those shown in Figures 1.11(a) and 1.12(a), are **concave up.**

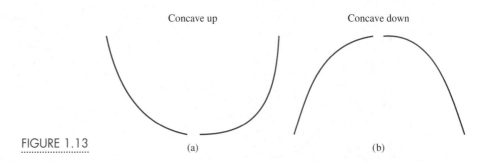

Concave up   Concave down

FIGURE 1.13   (a)   (b)

Note that one curve is increasing and that the other is decreasing, so concavity is a completely different concept from increasing/decreasing. Similarly, curves that bend downward, such as those shown in Figures 1.11(b) and 1.12(b), are **concave down.** Again, note that one is increasing and the other is decreasing. Figure 1.13(a) illustrates the two types of concave up behavior, and Figure 1.13(b) illustrates the two types of concave down behavior.

<u>Think About This</u>

Imagine a ball bouncing up and down across the floor in front of you. Is the path of the ball concave up or concave down? ⬚

Just as a function can be increasing over one interval and decreasing over another, a function can be concave up over one interval and concave down over another. For instance, think of the behavior of the Dow-Jones average. This function is increasing during some time intervals and is decreasing during other time intervals, as shown in Figure 1.14. It is also concave up over some time intervals and is concave down over other time intervals.

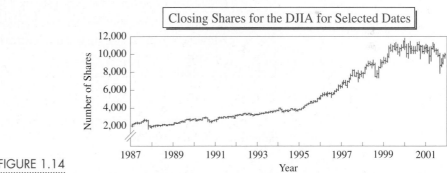

FIGURE 1.14

A point on a graph where the concavity changes from concave up to concave down or vice versa is called a **point of inflection** or an **inflection point.** In Figure 1.15, we show two curves, one having a point of inflection where the curve changes from concave up to concave down and the other where the curve changes from concave down to concave up. Observe that neither point of inflection occurs at the turning points where the curve reaches a local maximum or a local minimum, so turning points are not the same as inflection points.

Note that the function on the left in Figure 1.15 grows faster and faster to the left of the inflection point and then grows slower and slower to the right of the inflection point. As a result, the function is growing most rapidly *at* the inflection point.

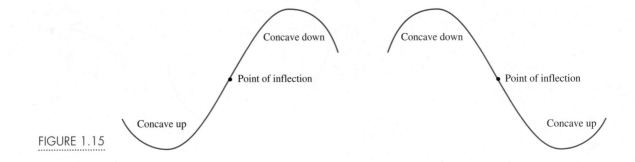

FIGURE 1.15

Think About This   What happens at the inflection point of the function on the right in Figure 1.15? ▭

We summarize the preceding information about functions as follows:

> A function of *x* is *increasing* if the values of the function increase as *x* increases.
>
> A function of *x* is *decreasing* if the values of the function decrease as *x* increases.
>
> The graph of an *increasing* function *rises* from left to right.
>
> The graph of a *decreasing* function *falls* from left to right.
>
> The points where a function changes from increasing to decreasing or from decreasing to increasing are the *turning points*.
>
> The graph of a function is *concave up* if it bends upward.
>
> The graph of a function is *concave down* if it bends downward.
>
> The points where the concavity changes from concave up to concave down or from concave down to concave up are the *points of inflection* or *inflection points*.

In addition, the rate of change and concavity of the graph of a function are related in the following ways.

> If the graph of a function is increasing and concave up, it is increasing at an increasing rate.
>
> If the graph of a function is increasing and concave down, it is increasing at a decreasing rate.
>
> If the graph of a function is decreasing and concave up, it is decreasing at a decreasing rate.
>
> If the graph of a function is decreasing and concave down, it is decreasing at an increasing rate.
>
> A function grows fastest or decays fastest *at* a point of inflection.

EXAMPLE 1

Identify all intervals where the function f shown in Figure 1.16 is

a. increasing;

b. decreasing;

c. concave up;

d. concave down.

Then indicate all points where the function has a

e. turning point;

f. local maximum;

g. local minimum;

h. point of inflection.

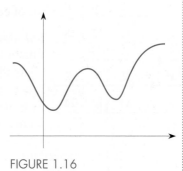

FIGURE 1.16

Solution   For a–d, the task is to find the *intervals* of x-values where the different types of behavior occur. We begin by redrawing the graph and introducing all the points x_1, $x_2, \ldots, x_9$ where the behavior of the function changes, as shown in Figure 1.17.

a. The function is increasing for values of x between x_3 and x_5 and again between x_7 and x_9, as shown in Figure 1.17(a).

b. The function is decreasing between x_1 and x_3 and again between x_5 and x_7, as shown in Figure 1.17(a).

c. The curve is concave up between x_2 and x_4 and again between x_6 and x_8, as shown in Figure 1.17(b).

d. The function is concave down between x_1 and x_2, between x_4 and x_6, and again from x_8 to x_9, as shown in Figure 1.17(b).

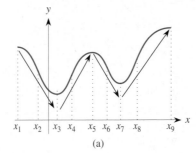
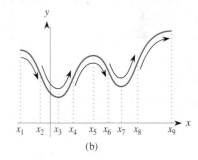

FIGURE 1.17   (a)   (b)

Next, we look for particular points on the curve.

e. The turning points for this function are at $x = x_3$, at $x = x_5$, and at $x = x_7$.

f. The function has a local maximum at $x = x_1$ (when compared to other nearby points); the function also has a local maximum at $x = x_5$ and again at $x = x_9$.

g. Similarly, the function reaches a local minimum at $x = x_3$ (when compared to other nearby points) and again at $x = x_7$.

h. The points of inflection occur where the concavity changes, which happens at $x = x_2$, at $x = x_4$, at $x = x_6$, and at $x = x_8$.

EXAMPLE 2

x	1	2	3	4
y	6	11	15	18

x	1	2	3	4
y	6	11	17	24

Two functions are defined in the accompanying tables of values. Describe the behavior of each function.

Solution   Both functions are obviously increasing as x increases. Note that the first function grows first by 5 (from 6 to 11), then by 4 (from 11 to 15), and then by 3 (from 15 to 18), so it is growing at a decreasing rate. Therefore it is concave down. The second function, however, grows by larger and larger amounts—first by 5 (from 6 to 11), then by 6 (from 11 to 17), and then by 7 (from 17 to 24)—so the function is increasing at an increasing rate and thus is concave up. Plot the points to verify both behaviors.   ◆

Periodic Behavior

Another behavior pattern for functions is extremely common in real life. Many natural processes have the property of being *periodic*—that is, the pattern repeats over and over. We see this in the height of tides that rise and fall in the same pattern roughly every 12 hours in most coastal locations. It also occurs in the pattern of temperature readings in any location from one year to the next. Spotting a periodic function from its graph is easy: The identical pattern appears repeatedly. For instance, consider the following data based on historical records giving the average number of tornados reported in the United States, per month, in a typical year.

Month	Jan	Feb	Mar	Apr	May	Jun	Jul	Aug	Sep	Oct	Nov	Dec
Tornados	16	24	60	111	191	179	96	66	41	26	31	22

Source: National Oceanic and Atmospheric Administration.

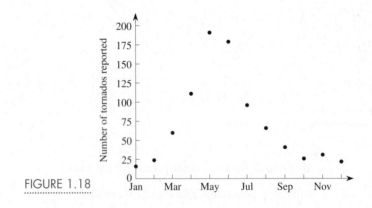

FIGURE 1.18

Figure 1.18 shows a graph of these points. Note, either from the table or the graph, how the values increase from a minimum level of tornado sightings in January to a maximum number in May and then decrease toward the minimum as the year ends. Because these values are based on historical averages, this cycle will likely repeat yearly with little change from one year to the next. It is therefore a roughly periodic phenomenon. Figure 1.19 shows a smooth curve that captures the longer term behavior of this roughly periodic function.

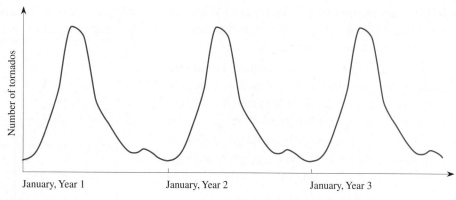

FIGURE 1.19

Problems

1. Which of the functions are strictly increasing, strictly decreasing, or neither?

   a. The cost of first-class postage on January first of each year.

   b. The time of sunrise associated with each day of the year.

   c. The high temperature associated with each day of the year.

   d. The closing price of one share of IBM stock for each trading day on the stock exchange.

   e. The area of an equilateral triangle in terms of its base b.

   f. The height of a bungee jumper t seconds after leaping off a bridge.

   g. The height of liquid in a 55-gallon tank h hours after a leak develops.

   h. The daily cost of heating a home as a function of the day's average temperature.

   i. The world record times for running the 100 meter dash.

2. Consider the function shown in the accompanying graph. Use two different colored pens or pencils. With one, trace all parts of the curve where the function is increasing. With the other, trace all parts of the curve where the function is decreasing. Then mark all turning points on the curve.

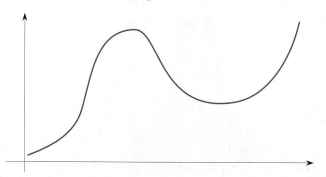

3. Consider the function shown in the accompanying graph. Use two different colored pens or pencils. With one, trace all parts of the curve where the function is concave up. With the other, trace all parts of the curve where the function is concave down. Then mark all points of inflection on the curve.

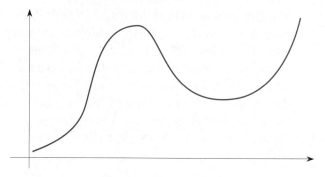

4. Sketch the graph of a single smooth curve that is first increasing and concave up, then increasing and concave down, and finally decreasing and concave down. Mark all turning points and points of inflection on your curve.

5. Sketch the graph of a single smooth curve that is first decreasing and concave up, then increasing and concave up, and finally increasing and concave down. Mark all turning points and points of inflection on your curve.

6. Sketch a possible graph of the temperature in your hometown over an entire week as a function of time. On the graph indicate all the turning points. Where

is the temperature function increasing? Where is it decreasing? Where is the temperature function concave down? Where is it concave up?

7. Each year the world's annual consumption of water rises, as does the amount of increase in water consumption. Sketch a graph of the annual world consumption of water as a function of time.

8. A human fetus grows rapidly at first and then grows with decreasing rapidity. Draw a graph showing the size of a fetus as a function of time.

9. Sales of microwave ovens grew slowly when they were first introduced and then increased dramatically as more people appreciated their usefulness. Eventually, sales began to slow as most households already owned one. Sketch the graph of microwave oven sales as a function of time. Indicate the location of the point of inflection.

10. Sales of VCRs grew slowly at first and then increased tremendously as people came to accept them widely. Eventually new sales began to level off as market saturation neared. Sketch a possible graph of the percentage of U.S. homes owning a VCR as a function of time, paying careful attention to the behavior of the function. Indicate any turning points and points of inflection.

11. The Environmental Protection Agency (EPA) monitors the levels of industrial pollutants in many lakes and rivers. The following graphs show the level of pollutants L in four different lakes as a function of time t. For each, write a short paragraph either from the point of view of the EPA bringing charges against a company for polluting or from the point of view of a company defending itself against such charges.

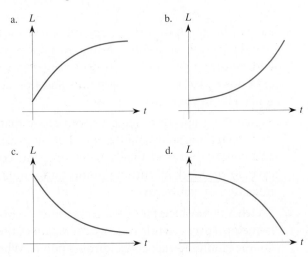

12. Each part of the table of values below defines a function. Determine the concavity of each function.

x	y	x	y	x	y
1	36	10	160	3	84
2	31	15	172	7	74
3	27	20	189	11	61
4	24	25	209	15	45
5	22	30	243	19	22

13. The Apollo-12 mission involved a flight to the moon (250,000 miles from Earth), five circular orbits about the moon, and a return to Earth.

   a. Assume (incorrectly) that the spacecraft traveled at a constant speed between the Earth and the moon. Sketch a rough graph of the distance from Earth as a function of time.

   b. Assume (correctly) that the spacecraft's speed diminished the farther it got from Earth's gravity until it neared the moon and then increased due to the moon's gravitational force. The behavior of the spacecraft's speed reversed on the return trip. Sketch a rough graph of the spacecraft's distance from the Earth as a function of time. (Think concavity!)

14. Water is being poured, at a constant rate, into vases having the shapes shown. Sketch a graph showing

a.

b.

the level of the water as a function of time, paying careful attention to concavity. What is the significance of each of the points of inflection, if any?

c.

d.

15. Decide which functions in a–j are periodic. (Assume that the graphs continue indefinitely to the left and right in the same pattern.)

a.

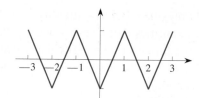

b.

c.

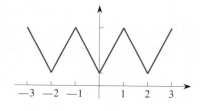

d.

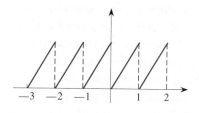

e.

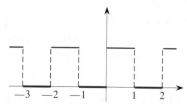

f.

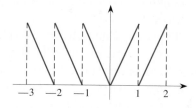

g.

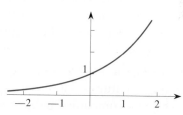

h.

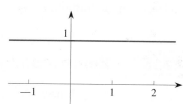

i.

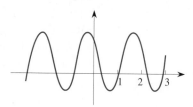

j.

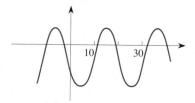

16. Janis trims her fingernails every Saturday morning. Sketch the graph of the length of her nails as a function of time. Is this process periodic?

17. Craig is a perfectly normal individual with a pulse rate of 60 beats per minute and a blood pressure of 120 over 80. Thus his heart is beating 60 times/minute and his blood pressure is oscillating between a low (diastolic) reading of 80 and a high (systolic) reading of 120. Sketch the graph of his blood pressure as a function of time. Be sure to indicate appropriate scales on each axis.

18. **a.** The thermostat in Sylvia's home in Baltimore is set at 66°F during the winter. Whenever the temperature drops to 66° (roughly every half-hour), the furnace comes on and stays on until the temperature reaches 70°. Sketch the graph of the temperature in her house as a function of time. Be sure to indicate appropriate scales on each axis.

    b. Gary, who lives in upstate New York, also has his thermostat set to come on at 66°F. How will a sketch of the temperature in his house differ from the one you drew in part (a) for Sylvia's house?

    c. Jodi, who lives in central Florida, likewise has her thermostat set to come on at 66°. How will a sketch of the temperature in her house differ from the other two?

19. Astronomers have been observing sunspots on the face of the sun for centuries. These dark spots on the sun, which are accompanied by the release of bursts of electromagnetic radiation that disrupt radio and TV signals, occur in periodic cycles. The accompanying figure is a graph of the number of sunspots observed each year.

    a. Estimate the period of the sunspot cycle.
    b. Estimate when the next two peaks will occur in the cycle.
    c. Suppose that you were required to come up with a reasonable estimate for the maximum number of sunspots that will occur during the next peak in the cycle. How might you create such an estimate based on the information given in the figure?

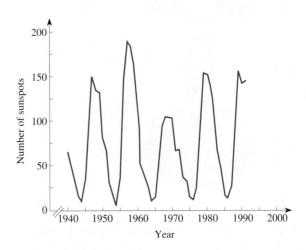

1.3 ····· Representing Functions Symbolically

A function is a rule that associates one and only one value of a quantity (say, y) with each value of another quantity (say, x). The quantities x and y are variables. We use a single letter, such as f, g, or h, as the name of a function. The particular formula for the function, if it is known, is usually written as $y = f(x)$. It is read as "y equals f of x" or possibly "y is a function of x." For instance, the function f that takes any real number x and squares it can be written as

$$y = f(x) = x^2.$$

Some particular values of this function are

$$f(3) = 3^2 = 9; \qquad\qquad f(\tfrac{1}{3}) = (\tfrac{1}{3})^2 = \tfrac{1}{9};$$
$$f(-5) = (-5)^2 = 25; \qquad\qquad f(0.02) = (0.02)^2 = 0.0004;$$
$$f(-0.01) = (-0.01)^2 = 0.0001; \qquad\qquad f(\pi) = \pi^2 \approx 9.8696.$$

In each case we replaced the variable x in $f(x) = x^2$ with the indicated value, $x = 3$, $x = \tfrac{1}{3}$, and so on, and then evaluated the expression 3^2, $(\tfrac{1}{3})^2$, and so on. Similarly, the function g that takes the square root of any nonnegative real number x can be written as

$$y = g(x) = \sqrt{x}.$$

For instance, some values of this function g are

$$g(16) = \sqrt{16} = 4;$$
$$g(0.04) = \sqrt{0.04} = 0.2;$$
$$g\left(\tfrac{1}{4}\right) = \sqrt{\tfrac{1}{4}} = \tfrac{1}{2}.$$

But $g(-25) = \sqrt{-25}$ does not make sense because it isn't possible to take the square root of a negative number in the real number system. That is, the function g is *not defined* for $x = -25$.

The function h that gives the reciprocal of any nonzero number x can be written as

$$y = h(x) = \tfrac{1}{x}.$$

Some values of this function h are

$$h(5) = \tfrac{1}{5} = 0.2;$$
$$h(200) = \tfrac{1}{200} = 0.005;$$
$$h(-0.125) = -\tfrac{1}{0.125} = -8.$$

But $h(0)$ does not make sense because division by 0 is not possible. That is, the function h is *not defined* at $x = 0$.

To work with functions requires some terminology. In the form $y = f(x)$, we call x the **independent variable** because it can take on any appropriate value. We call y the **dependent variable** because its value depends on the choice of x.

You can use letters other than f, g, or h to represent functions; other common choices are F, G, or f_1, f_2, f_3, and so on. You can use letters other than x to represent the independent variable; other common choices are t (for time), θ (for an angle), and r (for radius). Similarly, you can use letters other than y to represent the dependent variable; for instance, you can use A for area, D for distance, P for population, and C for cost.

The area A of a circle is a function of its radius r—for each radius r, there is one and only one area A. We write this function as $A = f(r) = \pi r^2$. Here r is the independent variable, A is the dependent variable because the area depends on the choice of r, and f is the function. The distance D that a car moves in t hours at a steady speed of 50 miles per hour (mph) is given by $D = g(t) = 50t$. Here t is the independent variable, D is the dependent variable, and g is the function.

Suppose that you toss a ball straight up with an initial velocity of 64 feet per second. The function

$$y = f(t) = 64t - 16t^2$$

gives the height in feet of the ball above ground level after t seconds, until the instant that the ball hits the ground. Picture what happens. As the ball rises, it slows due to the effect of gravity. Eventually it reaches a maximum height and then begins to fall back to the ground. As the ball falls, its speed increases, again because of gravity.

Now let's see how the function f gives the height of the ball above ground at any time t. After half a second, when $t = \tfrac{1}{2}$, the ball is 28 feet above the ground because

$$y = f\left(\tfrac{1}{2}\right) = 64\left(\tfrac{1}{2}\right) - 16\left(\tfrac{1}{2}\right)^2 = 32 - 4 = 28 \text{ feet.}$$

After 1 second, it is at a height of

$$y = f(1) = 64(1) - 16(1)^2 = 48 \text{ feet.}$$

After 2 seconds, it is at a height of

$$y = f(2) = 64(2) - 16(2)^2 = 64 \text{ feet},$$

which happens to be the maximum height the ball reaches. After 3 seconds, the height is

$$y = f(3) = 64(3) - 16(3)^2 = 48 \text{ feet},$$

and the ball is on its way down. After 4 seconds, the ball is back at ground level because

$$f(4) = 64(4) - 16(4)^2 = 0.$$

Domain and Range of a Function

In each of the preceding functions, there were some natural limitations on the possible values for both the independent variable and the dependent variable. The ball is released at time $t = 0$ and returns to the ground at $t = 4$ seconds. It therefore makes no sense in this problem to think about what happens before time $t = 0$ or after time $t = 4$. Thus the permissible values for t are between 0 and 4 seconds. Furthermore, the ball rises to its maximum height of 64 feet and then falls back to the ground. Therefore the only meaningful values for the height of the ball are between $y = 0$ and $y = 64$ feet. (Of course, it is more realistic to think of throwing a ball upward from about 4 or 5 feet above the ground rather from ground level—we used ground level here just to simplify the mathematics.)

Similarly, the function $y = g(x) = \sqrt{x}$ makes sense only if the independent variable x is not negative. The possible corresponding values for y must be positive or zero. The function $y = h(x) = 1/x$ makes sense only if x is not zero. The possible corresponding y-values of this function can be any number other than 0 because there is no value of x such that $y = 1/x = 0$. Finally, for the function $y = f(x) = x^2$, there is no limitation on the possible values of x, but there certainly is a limitation on the corresponding values for $y = x^2$ because they can never be negative.

For any function f, the set of *all possible values for the independent variable* is called the **domain** of f; the set of *all possible values for the dependent variable* is called the **range** of f.

Typically, the domain and range consist of intervals of values for the independent variable and the dependent variable, respectively. For instance, with the function representing the height of the ball, the domain consists of the interval from $t = 0$ to $t = 4$ and the range consists of the interval from $y = 0$ to $y = 64$. We can also use inequalities to write these intervals expressing the domain as $0 \le t \le 4$ and the range as $0 \le y \le 64$.

Because each of these intervals contains endpoints ($t = 0$ and $t = 4$ for the domain and $y = 0$ and $y = 64$ for the range), they are called *closed intervals*. We can also write these intervals using *interval notation*, so that the domain is the closed interval $[0, 4]$ and the range is the closed interval $[0, 64]$. We use square brackets to indicate that the endpoint value is included in the interval.

If one or both endpoint values is *not* included in an interval, we use parentheses instead of square brackets. For instance, if an interval is $3 < x < 6$, where both endpoints are not included, we write it in interval notation as the *open interval* $(3, 6)$. (Caution: Don't misinterpret this notation as the coordinates of the point with $x = 3$ and $y = 6$. The symbols are identical, but the meaning should be clear from the context.)

An interval can also contain one endpoint, but not the other, as in $-2 < x \leq 8$. We write this in interval notation as $(-2, 8]$, where the square bracket on the right indicates that $x = 8$ is included and the left parenthesis indicates that $x = -2$ is not included. For instance, the domain of the function $y = g(x) = \sqrt{x}$ is the interval $[0, \infty)$, which indicates that $x \geq 0$; it is closed on the left, and extends toward ∞, but never reaches ∞, and so is open on the right.

EXAMPLE 1

Find the values of the function $y = g(x) = \sqrt{x}$ at $x = 0, \frac{1}{4}, 1, 2, \pi$, and 4. What is the domain and range of this function?

Solution   For each of the given values of x in the domain of this function, the corresponding y-values in its range are

$$y = g(0) = \sqrt{0} = 0;$$
$$y = g(\tfrac{1}{4}) = \sqrt{\tfrac{1}{4}} = \tfrac{1}{2};$$
$$y = g(1) = \sqrt{1} = 1;$$
$$y = g(2) = \sqrt{2} \approx 1.41421 \ldots;$$
$$y = g(\pi) = \sqrt{\pi} \approx \sqrt{3.14159 \ldots} \approx 1.77245;$$
$$y = g(4) = \sqrt{4} = 2.$$

Note that you can take the square root of any positive value of x or of 0, so the domain of the function g consists of all nonnegative numbers. Similarly, the square root of any such number is positive or 0, so the range of g also consists of all nonnegative numbers.

We can use inequality notation to write $x \geq 0$ for the domain and $y \geq 0$ for the range. Alternatively, using interval notation, we have $[0, \infty)$ for the domain and $[0, \infty)$ for the range.

EXAMPLE 2

Discuss the range of the function

$$y = F(x) = x + \frac{1}{x}$$

when the domain for F is restricted to the set of all positive numbers.

Solution   We start by looking at the graph of the function F, as shown in Figure 1.20. Note that the function is decreasing rapidly to the right of $x = 0$. It has a turning point at

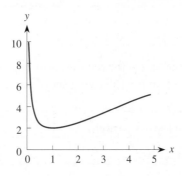

FIGURE 1.20

about $x = 1$, and then it increases slowly thereafter. Try different values for x with your calculator to verify that this result is indeed the case numerically. Also, extend the graph farther to the right with your function grapher to see that this pattern continues indefinitely. It turns out that the smallest possible value for y, which is $y = 2$ exactly, corresponds to $x = 1$ at the turning point. For any other value of x, the value for y is larger. Therefore the range of F is all values $y \geq 2$.

◆

You can visualize a function f as an operation that *transforms* each value x from its domain into the corresponding value y in its range. Figure 1.21 illustrates how each point x in the domain is transformed into a single point in the range. Thus x_1 is transformed into y_1. We also can say that x_1 is carried into y_1 or that x_1 is *mapped* into y_1. Similarly, x_2 is carried into y_2 and x_3 is mapped into y_3. Note that x_4 and x_5 are both transformed into y_4, which is perfectly legitimate for a function. Each x-value must be mapped into a single y-value, although it is certainly possible for several different x's to be mapped into the same y. Think about the function $y = f(x) = x^2$, where both $x = 2$ and $x = -2$ are transformed into $y = 4$.

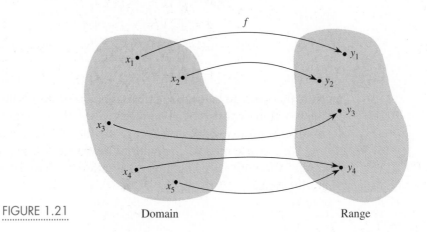

FIGURE 1.21   Domain   Range

We now summarize the preceding ideas in a formal definition of a function.

Definition of a Function

A **function** f is a rule that assigns to each permissible value of the independent variable x one and only one value of the dependent variable y.

The **domain** of f is the set of all possible values for the independent variable.

The **range** of f is the set of all possible values for the dependent variable.

EXAMPLE 3

Discuss the domain and range for the function relating acceleration time t to final speed v for a Trans Am, based on the following set of data.

Final speed, v (mph)	30	40	50	60	70
Time, t (sec)	3.00	4.29	5.52	7.38	9.81

Solution   The independent variable is the final speed v, so the domain of this function consists of all possible speeds. We therefore might conclude that the domain would be 0 to 70 mph; however, if we want to use the function to predict the time needed to reach a higher speed, we would need a somewhat larger domain—say, 0 to 100 mph. It probably isn't reasonable to think of speeds any faster than that. The dependent variable is the time t needed for a Trans Am to accelerate to a given speed, v. If we use only speeds between 0 and 70 mph, the associated range would be 0 to 9.81 seconds. If we use the extended domain of v of 0 to 100 mph, however, the associated range might be more like 0 to 20 seconds. It takes about 2.5 seconds to accelerate from 60 to 70 mph. The pattern suggests that it will take even more time to accelerate from 80 to 90 mph and still more time to go from 90 to 100 mph. Thus an estimate of 20 seconds to accelerate from 0 to 100 mph is reasonable.

◆

Consider the relationship between people and their telephone numbers. Is this relationship a function? If there is even one person who has two different telephone numbers, the relationship does not satisfy the definition and so *is not* a function. But a person's height *is* a function of the person—each individual has one and only one height at any particular time.

Often a verbal description of a function includes the idea of *proportionality* from elementary algebra. Recall that y is proportional to x means that $y = k \cdot x$, for some constant of proportionality k. For instance, the area of a circle is proportional to the square of the radius because $A = \pi r^2$ and π is a constant (it is the constant of proportionality). Similarly, y is inversely proportional to x if $y = k \cdot 1/x = k/x$, where k is a constant of proportionality.

Throughout this book, unless some restriction is indicated, we assume that all functions discussed are defined (either mathematically or practically) on the largest possible domain that makes sense.

Problems

1. Which of the relationships are functions and which are not? For those that are not functions, explain why. For those that are functions, identify the independent and dependent variables and give a reasonable domain.

   a. The cost of first-class postage on January first of each year since 1900.

   b. The weight of letters you can mail with $n = 1, 2, 3, \ldots$ postage stamps.

   c. The time of sunrise associated with each day of the year.

   d. The time of high tide associated with each day of the year.

   e. The high temperature associated with each day of the year.

   f. The closing price of one share of IBM stock each trading day on the stock exchange.

   g. The area of a rectangle whose base is b.

   h. The area of an equilateral triangle whose base is b.

   i. The height of a bungee jumper t seconds after leaping off a bridge.

   j. The time it takes the bungee jumper to reach a height H above the ground.

   k. The number of baseball players who have n home runs in a full season.

   l. The height of liquid in a 55-gallon tank h hours after a leak develops.

   m. The daily cost to a family of heating their home versus the average temperature that day.

2. The balance B, in thousands of dollars, in a CD account at a bank is a function of time t, in years, since you opened the account, so $B = f(t)$.

   a. What does $f(4) = 2$ tell you? What are appropriate units?

   b. Is f an increasing or decreasing function of t?

   c. Discuss the concavity of f.

3. The height H in inches of a child is a function of the child's age a, so $H = f(a)$.

   a. What does $f(10) = 50$ tell you? What are appropriate units?

   b. Is f an increasing or decreasing function of a?

   c. Discuss the concavity of f.

4. The surface area S of a sphere of radius r is 4π times the square of the radius. Write a formula for S as a function of r.

5. The pressure P of a gas in a container of fixed size is proportional to the temperature T of the gas. Write a formula for the pressure as a function of temperature.

6. The pressure P of a gas held at a constant temperature in a container is inversely proportional to the volume V of the container. Write a formula for the pressure as a function of volume.

7. The force of gravity F between two objects is inversely proportional to the square of the distance d between the objects. Write a formula for F as a function of d.

8. When a cup of hot coffee is left to cool on the table where the air temperature is 70°F, the change ΔT in the temperature T of the coffee is proportional to the difference between the temperature of the coffee and the room temperature. Write a formula for ΔT as a function of T.

9. Kim has a peanut butter sandwich on white bread each day. The number of calories C in the sandwich, as a function of the number of grams P of peanut butter, is $C = f(P) = 150 + 6P$.

   a. What is $f(1)$? What does it mean?

   b. What is $f(10)$? $f(15.5)$? $f(20)$? $f(30)$?

   c. How many calories come from the bread alone?

   d. Explain why using $P = -1$ makes no sense.

   e. What is a reasonable domain and range for this function?

10. Suppose that Jim wants his peanut butter sandwich on rye bread instead of white bread. Rye bread contains 85 calories per slice. What would be the corresponding formula for the number of calories in Jim's sandwich?

11. The number of calories in a peanut butter and jelly sandwich on white bread is $C = 150 + 6P + 2.7J$, where P and J are the number of grams of peanut butter and jelly, respectively.

   a. How many calories are in a sandwich with 24 g of peanut butter and 20 g of jelly?

   b. Suppose that Adam is on a diet and wants to limit his calorie intake from a peanut butter and jelly sandwich to a maximum of 300 calories. Find two reasonable combinations of amounts of peanut butter and jelly that produce a sandwich with exactly 300 calories.

   c. Which is more caloric, a gram of peanut butter or a gram of jelly? Explain how you know.

12. A car rental company charges a fixed daily rate for a midsize car plus a charge for each mile more than 100 miles that the car is driven per day. A formula for the cost of a rental car driven more than 100 miles is $c = f(m) = 35 + 0.25(m - 100)$, where m is the number of miles that the car is driven.

   a. Find $f(100)$. What does it mean?

   b. Find $f(150)$, $f(200)$, and $f(500)$.

   c. What is a reasonable domain and range for this function?

13. Suppose that you throw a ball upward, with an initial velocity of 60 ft/sec, from the roof of a 120-ft-high building.

   a. Sketch a possible graph of the height of the ball as a function of time, as you visualize it.

   b. Suppose that the height of the ball as a function of time is given by
   $$H(t) = 120 + 60t - 16t^2.$$
   Find the height of the ball when $t = 1$; when $t = 4$.

   c. Find $H(2)$ and $H(3)$. What do they represent?

   d. Use your function grapher to estimate how long it takes for the ball to reach its maximum height. What is the maximum height?

   e. How long does it take until the ball first hits the street below?

   f. What are the domain and range for this function?

14. For the function $f(t) = t^2 - 5$, find the values corresponding to $t = 2, 4, 6, 10$.

15. For the function[1]
   $$F(x) = \frac{1}{x^2 - 4},$$
   find $F(0)$, $F(1)$, $F(3)$, $F(4)$, $F(5)$. Why did we skip $x = 2$? Are there any other values of x that should be skipped? What is the domain of this function?

---

[1] Note that when you enter this expression in a calculator or most computer programs, you must key the expression in as $1/(x\char94 2 - 4)$. Pay careful attention to when you need to use parentheses in any such expression.

16. For the function

$$g(x) = \frac{x^2 + 4}{x^2 - 9},$$

find $g(0), g(1), g(2), g(4), g(-1)$. Why did we skip $x = 3$? Are there any other values of x that should be skipped? What is the domain of this function?

17. For the function $g(s) = s + \sqrt{s}$, find the values corresponding to $s = 4, 16, 25, 100$. Are there any values for s that will make the function come out negative? What does that tell you about the range of g? What is its domain?

18. For the function $z = f(q) = q^3 + 5$, find the value of the dependent variable that corresponds to a value of the independent variable of 4. Find the value of the independent variable that corresponds to a value of the dependent variable of 6.

19. For the function $f(x) = x^3 - 8x^2 + 15x - 1$, find three different values of x between 1 and 8 for which $f(x) < 0$. Then find at least two noninteger values of x for which $f(x) < 0$.

20. A simple substitution code in which each letter is replaced by a different letter can be thought of as a function f whose domain is the letters of the alphabet $A, B, \ldots, Z$. Suppose that $f(A) = M$, $f(B) = D$, $f(C) = K$, $f(D) = V$, $f(E) = X$, $f(F) = B$, $f(G) = P$, $f(H) = T$, $f(I) = J$, $f(J) = S$, $f(K) = Z$, $f(L) = Q$, $f(M) = H$, $f(N) = O$, $f(O) = A$, $f(P) = L$, $f(Q) = W$, $f(R) = C$, $f(S) = F$, $f(T) = Y$, $f(U) = R$, $f(V) = G$, $f(W) = I$, $f(X) = U$, and $f(Y) = N$.

a. What is $f(Z)$?

b. What is the solution to the equation $f(x) = R$?

1.4 Connecting Geometric and Symbolic Representations

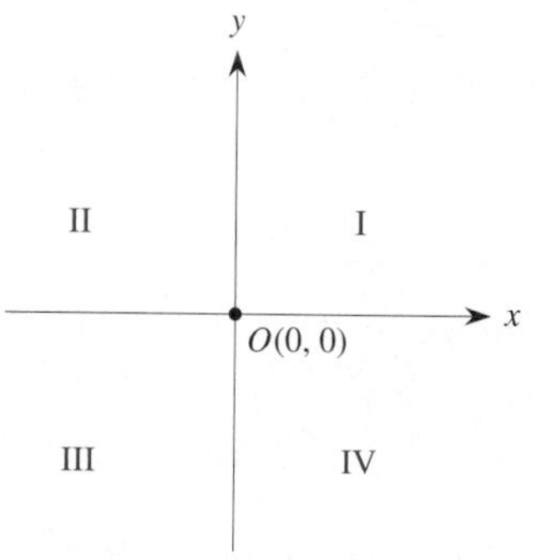

FIGURE 1.22

One of the most significant advances in mathematics is based on the idea of connecting the geometric and symbolic representations of functions. It allows you to think of functions from a visual rather than an exclusively symbolic perspective.

Begin by drawing two perpendicular *axes,* as shown in Figure 1.22, whose point of intersection O is the *origin.* The *horizontal axis* represents values of the *independent variable* (in this case, x); by convention, these values increase from left to right, as indicated by the arrow. The *vertical axis* represents values of the *dependent variable* (in this case, y); by convention these values increase upward, as indicated by the vertical arrow. The two axes divide the plane into four *quadrants:* the first quadrant (I), the second quadrant (II), the third quadrant (III), and the fourth quadrant (IV). Whenever appropriate, indicate the units used for each variable and label the axes accordingly.

This representation is called a *rectangular* or *Cartesian coordinate system,* and it is a way of associating points in the plane with ordered pairs of numbers. Every point P in the plane can be represented by an ordered pair of numbers, (x, y). Alternatively, every ordered pair, such as $(2, 5)$, $(27, 1)$, or $(23.84, 21.02)$, represents a point in the plane. We call (x, y) the **coordinates** of the point P. In mathematics, the letters x and y generally represent the independent and dependent variables, respectively. However, in any given context, you should use letters that suggest the quantities being studied.

Consider again the function

$$y = f(t) = 64t - 16t^2,$$

which represents the height, at any time t, of a ball tossed vertically upward with initial velocity of 64 feet per second. The formula for the function f gives the vertical height y of the ball at any instant t. When $t = 0$, $y = 0$ and the corresponding point $(0, 0)$ is the origin. Further, as we calculated before, $f(\frac{1}{2}) = 28$, $f(1) = 48$,

$f(2) = 64$, $f(3) = 48$, and $f(4) = 0$, which give rise to the points $(\frac{1}{2}, 28)$, $(1, 48)$, $(2, 64)$, $(3, 48)$, and $(4, 0)$ in the (t, y) coordinate system. These six points are plotted in Figure 1.23.

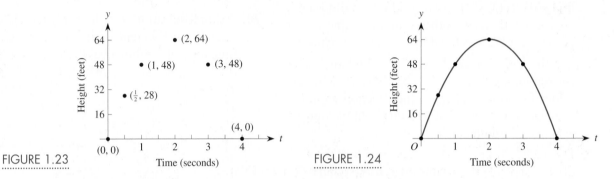

FIGURE 1.23

FIGURE 1.24

We can determine many other ordered pairs (t, y) satisfying the equation $y = 64t - 16t^2$. (Simply pick any other value for t between 0 and 4 and calculate the associated value of y by using the equation.) Each such ordered pair can be plotted as a point in the coordinate system. When all possible points are plotted, they form the curve shown in Figure 1.24. This curve is the *graph of the function f.* It consists of *all* points in the plane whose coordinates (t, y) satisfy the given equation. Thus we have a direct connection between the graph of a function and its algebraic equation. The graph of a function is therefore another representation of the same function. Note that the graph shown in Figure 1.24 represents the *height* of the ball at any time t; it doesn't show the *path* of the ball, which goes straight up and then down.

> The **graph** of a function $y = f(x)$ consists of *all* points (x, y) in the plane whose coordinates satisfy the equation of the function.

A table of values for a function is also useful when you're creating a hand-drawn graph of the function f from the formula $y = f(x)$. It provides a simple method of organizing the values of the independent variable x and the associated values of the dependent variable y that produce each point to be plotted. The number of points that you need to calculate for a table to draw a reasonable graph of a function depends on how complicated the behavior of the function is. For a line, all you need is two points because two points completely determine a line. We used six points to produce the graph of the height of the thrown ball shown in Figure 1.24. For comparison, a graphing calculator uses about 100 points to construct a curve.

When drawing the graph of a function, you should determine several key points. One point is where the graph crosses the vertical axis. You can easily find this point if you have a formula for the function: Just set the independent variable x equal to zero in the algebraic formula for the function and calculate the corresponding y-value. Although often desirable, finding the point(s) where the curve crosses the horizontal axis is usually more complicated. To find them, set the dependent variable y equal to zero and then solve the resulting equation. For the function representing the height of the ball, we can factor the expression for y and then set $y = 0$:

$$y = 64t - 16t^2 = 16t(4 - t) = 0.$$

When you solve this equation for t, you get either $t = 0$ or $t = 4$. The time $t = 0$ is the instant when the ball is first released, so $y = 0$. At the instant when $t = 4$, the corresponding value for y, which represents the height of the ball, is also zero. That is, at time $t = 4$, the ball has come back to the ground. You can see the pattern for the values of this function (and thus the pattern for the height of the ball) in the following table.

Time t (sec)	0	0.5	1.0	1.5	2.0	2.5	3.0	3.5	4.0
Height y (ft)	0	28	48	60	64	60	48	28	0

EXAMPLE 1

Determine the domain and range of the function shown in Figure 1.25.

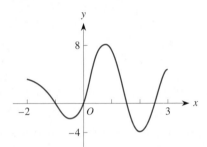

FIGURE 1.25

Solution   Note that the axes shown are labeled x and y; x is the independent variable, and y is the dependent variable. Further, observe that the graph extends from $x = -2$ at the left to $x = 3$ at the right, so the domain of this function is from -2 to 3. We can write this domain in terms of inequalities as $-2 \le x \le 3$. Similarly, the graph extends vertically from a low of $y = -4$ to a high of $y = 8$, so the range is $-4 \le y \le 8$.   ◆

In many situations, we typically start with a set of data collected from some experiment or from measurements taken on some process. We then graph the data to get a feel for the behavior of the quantity. Often, we try to connect the points on the graph with a smooth curve to get a better indication of the behavior of the quantity. Finally, we would like to obtain an equation for a function that fits these data points because many questions can be answered far more easily and accurately when an equation is available. We illustrate this methodology in Examples 2–4.

EXAMPLE 2

The snow tree cricket, which lives in the Colorado Rockies, has been studied by field biologists who have gathered the following measurements on how the chirp rate depends on the air temperature.

Temperature T (°F)	50	55	60	65	70	75	80
Rate R (chirps/min)	40	60	80	100	120	140	160

Plot the points to determine the kind of trend in the data.

Solution   Plotting these data points gives a visual dimension, as shown in Figure 1.26. Note that the chirp rate is growing at a constant rate as the temperature increases. Moreover, the corresponding points in the figure seem to fall into a straight line pattern, as indicated by the line drawn through them. In Chapter 2, we discuss how to find the equation of this line and how to predict the chirp rate R of the cricket based on the temperature T, or vice versa.

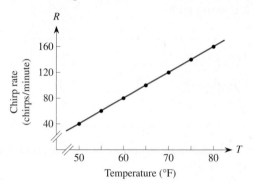

FIGURE 1.26

EXAMPLE 3

The following table of values gives the population, in millions, of the state of Florida since 1990.

a. Plot the data points and describe the behavior pattern.

b. If this trend continues, estimate the population in the year 2000.

Year	1990	1991	1992	1993	1994	1995	1996	1997	1998	1999	2000
Population	12.94	13.32	13.70	14.10	14.51	14.93	15.36	15.81	16.27	16.74	?

Solution

a. The graph of this set of data is shown in Figure 1.27. The growth pattern clearly is not a straight line pattern; rather, the population grows ever faster. The function is both increasing and concave up.

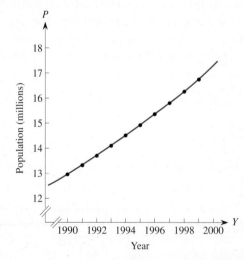

FIGURE 1.27

b. The increase from 1997 to 1998 was $16.27 - 15.81 = 0.46$ million, and the increase from 1998 to 1999 was $16.74 - 16.27 = 0.47$ million. As a result, we could estimate that the increase from 1999 to 2000 might be about 0.48 million, so our prediction for the year 2000 is about $16.74 + 0.48 = 17.22$ million people. We determine a formula for this function in Chapter 2 so that we can make such a prediction in a much simpler and more confident way.

EXAMPLE 4

The following table of values shows measurements, at different times, of the height of an object dropped from the top of the 1250-foot-high Empire State Building. Construct a graph of the height as a function of time and describe its behavior.

Time (sec)	0	1	2	3	4	5	6	7	8
Height (ft)	1250	1234	1186	1106	994	850	674	466	226

Solution The graph of the height of the object versus time in Figure 1.28 shows that the object is falling ever faster as time goes by. The function is decreasing and concave down. Again note that the graph represents the height of the object, not its path, which is straight down.

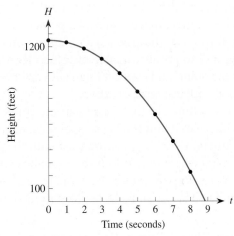

FIGURE 1.28

Although we could estimate from either the table or the graph how long it takes the object to hit the ground or to pass, say, the 30th floor, we could answer such questions more precisely if we knew the formula for the function.

In Examples 2–4, we simply connected the points to construct a smooth curve that seemed to fit the pattern. Doing so, however, can sometimes lead to serious errors. Suppose that we had some data on the turkey population of the United States taken on January 1 each year. It would likely show a growth trend similar to that in Example 3 on the population of Florida. However, a little thought will convince you that this population will change quite drastically about the middle of November each year. The smooth curve drawn using the January 1 turkey census data would therefore be a rather poor description of the actual population over all intermediate times.

Nevertheless, the idea of connecting a series of points to form a curve is precisely how a computer or graphing calculator produces the graph of a function. We strongly urge you to become comfortable with using a graphing calculator or a computer program to investigate the graph of any desired function given by a formula. The visual dimension invariably provides a wealth of information about the behavior of the function, and we continually turn to graphical images throughout this book.

Connections Between Geometric, Numerical, and Symbolic Representations

We have shown that a function can be represented in a variety of ways—as a formula giving a symbolic representation, as a graph giving a geometric representation, as a table giving a numerical representation, or in words giving a verbal representation. The first three ways (formula, graph, or table) are the most useful, but each approach provides a very different perspective.

The problem is: Can you always move back and forth between these different representations? Figure 1.29 illustrates schematically the interrelationships between the three most useful representations—symbolic, geometric, and numeric—by arrows. Ideally, you should be able to start with any one of these representations for a function and shift to the other two representations. Some of these shifts are very simple. If you know a formula for a function, you can create a table of numerical values. Similarly, if you know a formula, you can create its graph at the push of a button, using a graphing calculator or computer graphics program or even by hand, as the graph of a function consists of all points (x, y) that satisfy the equation of the function. If you have the graph of a function, you can easily read off a set of points on the curve and so produce a table of values. If you have a table of values, you can easily plot the points and connect them with an appropriate, usually smooth, curve to generate a graphical representation.

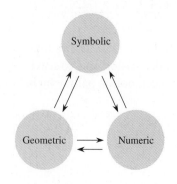

FIGURE 1.29

Unfortunately, the two remaining shifts in perspective are considerably more complicated. If you start with a table of values, how do you produce a formula for the function? Similarly, if you start with the graph of a function, how do you construct a formula for it? Both shifts can be extremely difficult, but fortunately modern technology provides the tools by which you can create reasonably accurate formulas. That often is the key step in most real-life applications of mathematics.

Does Every Curve Represent a Function?

Let's consider one last question: Is every curve the graph of some function $y = f(x)$? Consider the five curves shown in Figure 1.30. Are they all the graphs of functions? That is, does each value of the independent variable x correspond to one and only one value of the dependent variable y? We can test a curve in the following way. Imagine a vertical line moving across the curve from left to right so that it passes through every possible value of x in the domain. If, for each x, the line crosses the curve at only one point, there is exactly one y-value for that x and so the curve represents a function. If the vertical line crosses the curve at more than one point for *any* value of x, the curve does not represent a function. This criterion, called the *vertical line test,* shows that the curves (a), (b), and (c) are all graphs of functions. However, when the vertical line test is applied to curves (d) and (e), the line crosses both curves at more than one y-value and thus neither are graphs of functions.

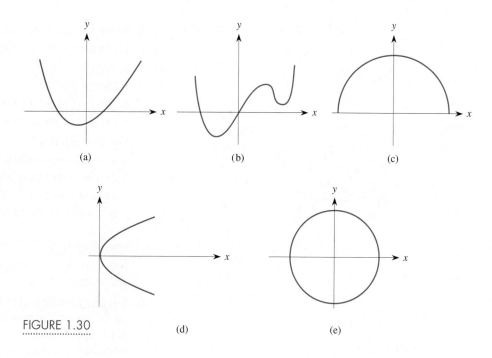

FIGURE 1.30

Problems

1. Which of the following graphs are functions? For each function, (**a**) give its domain and range, (**b**) identify where it is increasing or decreasing, and (**c**) identify where it is concave up or concave down.

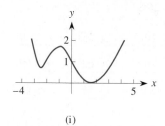

(i)

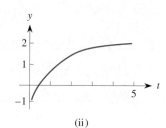

(ii)

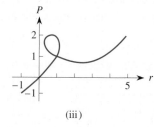

(iii)

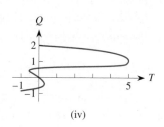

(iv)

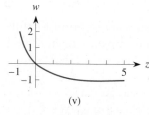

(v)

2. The following questions all relate to the accompanying graph of a function $y = f(x)$.

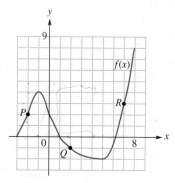

a. What are the coordinates of the points P, Q, and R?
b. Is the point $(-2, 5)$ on the curve?
c. Is the point $(5, -2)$ on the curve?
d. What is $f(5)$?
e. Find x if $f(x) = -1$.
f. Is $f(4) = -1$ true or false?
g. Find y when $x = 2$.
h. Find x when $y = 2$.
i. Solve $f(x) = 0$ for x.
j. Solve $f(0) = y$ for y.

t	1970	1980	1990	1991	1992	1993	1994	1995
$f(t)$	400	300	210	190	175	162	150	135

3. The accompanying graphs are based on the set of data above, but something is wrong with each graph. What was done incorrectly in each instance?

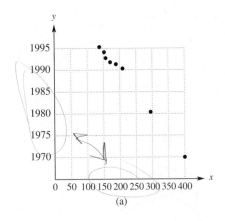
(a)

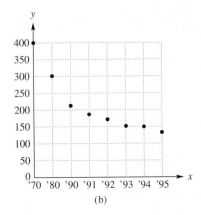
(b)

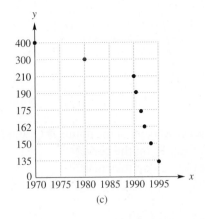
(c)

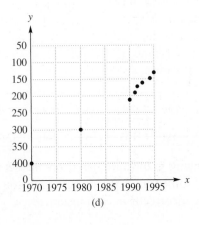
(d)

4. Refer to the function f relating the snow cricket's chirp rate to the air temperature in Example 2.
   a. What is $f(60)$?
   b. Solve $f(x) = 120$.
   c. In a complete sentence, tell what the equation $f(62) = 88$ means.

5. For the function $f(x) = x^2 - 3x + 2$, find the values of y corresponding to $x = -3, -2, -1, \ldots, 4, 5$. Plot the corresponding points and connect them with a smooth curve. Then use your function grapher to graph the function. How do the two graphs compare? Find the values of the function corresponding to $x = \frac{1}{2}$, $x = \frac{3}{2}$, $x = -\frac{5}{2}$ and indicate the location of the corresponding points on the curve you drew.

6. For the function $g(t) = 9 - t^2$, use an appropriate set of values for t and the corresponding y values to get a feel for the behavior of the curve when you draw and connect the points. How does your sketch compare to what you see when you use your function grapher?

7. Repeat Problem 6 for $h(s) = s^3 - 7s + 5$.

8. One of the functions f or g in the following table of values is concave up and the other is concave down. Which is which? Explain how you know.

x	5	10	15	20
$f(x)$	80	70	62	56
$g(x)$	80	70	58	43

9. A function $f(x)$ whose values are given in the following table is increasing and concave up. Give a possible value for $f(5)$.

x	4	5	6
$f(x)$	10	??	20

10. A function $f(x)$ whose values are given in the following table is increasing and concave down. Give a possible value for $f(50)$.

x	30	40	50
$f(x)$	12	20	??

11. For the function shown in the accompanying figure, indicate

 a. the intervals of *x*-values where the function is increasing.
 b. the intervals where the function is decreasing.
 c. all points *x* where the function has a turning point.
 d. all points *x* where the function has a local maximum.
 e. all points *x* where the function has a local minimum.
 f. all points *x* where the function has points of inflection.
 g. the intervals of *x*-values where the function is concave up.
 h. the intervals where the function is concave down.
 i. approximately where the function is increasing most rapidly.
 j. approximately where the function is decreasing most rapidly.
 k. the location of any *zeros* of the function (points where the curve crosses the *x*-axis).
 l. For any of parts (a)–(k) that asks for intervals, write the interval both in terms of inequalities and interval notation.

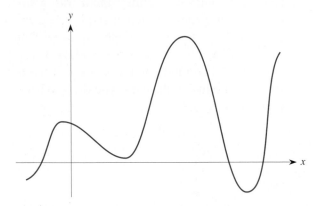

12. Consider the data below. Assume that these values represent a sample of values for a smooth, or continuous, function.

 a. Over what intervals of *x*-values is the function increasing?
 b. Over what intervals is the function decreasing?
 c. Near what *x*-values is the function at a local maximum?

 d. Near what *x*-values is the function at a local minimum?
 e. Between what pair of successive *x*-values is the function increasing most rapidly?
 f. Between what pair of successive *x*-values is the function decreasing most rapidly?
 g. Over what intervals is the function concave up?
 h. Over what intervals is the function concave down?
 i. Near what *x*-values does the function have points of inflection?
 j. Estimate the location of any *zeros* of the function.

13. Functions *f*, *g*, and *h* in the following table are increasing functions of *x*, but each function increases according to a different behavior pattern. Which of the accompanying graphs best fits each function?

x	*f(x)*	*g(x)*	*h(x)*
1	11	30	5.4
2	12	40	5.8
3	14	49	6.2
4	17	57	6.6
5	21	64	7.0
6	26	70	7.4

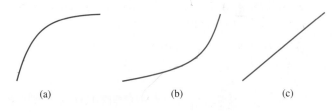

(a)   (b)   (c)

14. Functions *f*, *g*, and *h* in the following table are decreasing functions of *t*, but each function decreases according to a different behavior pattern. Which of the accompanying graphs best fits each function?

x	−2.5	−2.0	−1.5	−1.0	−0.5	0	0.5	1.0	1.5	2.0	2.5	3.0
f(x)	62.3	28.4	6.8	4.3	11.9	33.2	14.7	2.3	−12.5	−38.8	−5.2	11.7

t	f(t)	g(t)	h(t)
1	200	30	5.4
2	180	27.6	5.2
3	164	25.2	4.8
4	151	22.8	4.1
5	139	20.4	3.1
6	129	18.0	1.8

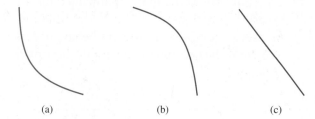

(a)　　　　　　　(b)　　　　　　　(c)

15. Sketch the graph of a function that passes through the point $(0, 1)$ and is

a. increasing and concave up for $x < 0$ and increasing and concave down for $x > 0$.

b. increasing and concave up for $x < 0$ and decreasing and concave up for $x > 0$.

c. decreasing and concave up for $x < 0$ and increasing and concave up for $x > 0$.

d. decreasing and concave up for $x < 0$ and decreasing and concave down for $x > 0$.

16. Sketch the graph of a single smooth curve that satisfies all the following conditions.

a. $f(0) = 4$　　　　**b.** $f(5) = -2$

c. f has a turning point at $x = 3$.

d. f is decreasing from 3 to 5.

e. f is increasing for $x > 5$.

f. f is concave down from 0 to 4.

g. f has a point of inflection at $x = 7$.

17. Consider the function $f(x) = x^2 - 4$ for $x \geq 3$.

a. Is f increasing or decreasing?

b. Is f concave up or concave down?

18. Consider the function $f(x) = x^3 + 7$ for $x \leq -3$.

a. Is f increasing or decreasing?

b. Is f concave up or concave down?

19. For the function $f(x) = \sqrt{x}$, find the values of y corresponding to $x = 0, 1, 2, \ldots, 6$. Plot the corresponding points and connect them with a smooth curve. Then use your function grapher to graph the function. How do the two graphs compare? Find the values of the function corresponding to $x = \frac{1}{2}$, $x = \frac{3}{2}$, $x = \frac{5}{2}$ and indicate the location of the corresponding points on the curve you drew. What is the domain?

20. For the function

$$f(x) = \frac{x}{x + 1}$$

calculate $f(0), f(1), f(2), \ldots, f(8)$. Plot the corresponding points and connect them with a smooth curve. Then use your function grapher to graph the function. How do the two graphs compare? Find the values of the function corresponding to $x = \frac{1}{2}$, $x = \frac{3}{2}$, $x = \frac{5}{2}$ and indicate the location of the corresponding points on the curve you drew. What is the domain?

1.5 Mathematical Models

A *model* is an image or representation of an object or process. A diagram of the human circulatory system is a model of the veins and arteries in the human body; the picture can be used to help us understand how blood circulates throughout the body. Similarly, an architect's sketch of a proposed shopping center is a model of the actual center; a wooden model built to scale is a still more realistic representation for that shopping center.

> A **model** is a representation that highlights the most important characteristics of an object or process.

Models can be found everywhere: the tide tables used by fishermen; a computer scientist's flowchart for a new program; plastic replicas of jet fighters; and many, many more. Because our focus here is on mathematics, the models we present are *mathematical models.* A mathematical model is a representation of a process expressed by a formula, an equation, a graph, a sequence of numbers, or a table of values. Once we have developed any such mathematical representation, we can use it to examine the behavior of the actual process.

For example, the equation for the motion of the ball thrown vertically upward from ground level with an initial velocity of 64 ft/sec,

$$y = f(t) = 64t - 16t^2,$$

is a mathematical model that describes one important aspect of the motion of that ball—its height at any time t. There may be other aspects of the motion that may not be captured in this mathematical model (such as releasing it from some initial height above the ground or the effects of air resistance). These other factors may likely require development of a more sophisticated model, but often a simple model gives a reasonably accurate first approximation.

How can we use mathematics to describe the real world via a mathematical model? We begin by looking at some process in the real world, such as the motion of a ball thrown upward, the growth of a population, or a person's reaction to a drug. Typically, the process of trying to explain what is happening requires some simplifying assumptions. For instance, in modeling the motion of a ball, we assumed that the only force acting on the ball is the force of gravity and ignored the negligible effects of air resistance. (Of course, if the ball were replaced with a balloon, a feather, or a piece of paper, this assumption would be invalid.)

After making reasonable assumptions, we then express the process in a mathematical form, which leads to a mathematical representation of the process in terms of a formula, an equation, a graph, or a table. This mathematical model then has to be interpreted. Does it truly seem to reflect what happens in the real world? Does the behavior of the function mirror the behavior of the process? If so, we can use the mathematical model to describe of the process under study and as the basis for predictions about the process. If the model doesn't adequately reflect the actual process, we may have done something wrong—overlooked some important aspect of the situation or ignored some critical factor; made some erroneous assumptions; or made an error in our work. We illustrate this interplay between mathematics and the real world via mathematical modeling schematically in Figure 1.31.

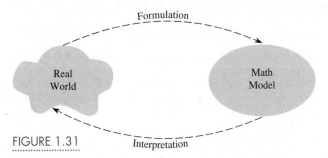

FIGURE 1.31

The concept of function is closely connected to the idea of a mathematical model, and most mathematical models are expressed as functions. Let's look at an example of this process. Researchers have studied the relationship between the level of animal fat in women's diets and the death rate from breast cancer in different countries. Some of their data are shown in the following table, which gives the average daily intake of animal fats in grams per day and the age-adjusted death rate from breast cancer per 100,000 women. We begin by looking at these data, first as a set of numerical values in the table and then visually on a graph, as shown in

Figure 1.32. There is obviously a relationship between the death rate and the daily fat intake. Clearly the death rate D increases as the daily fat intake F increases, so D is an increasing function of F. Moreover, the points fall into a straight line pattern.

Country	Japan	Spain	Austria	U.S.	U.K.
Daily fat intake(grams)	20	40	90	100	120
Death rate per 100,000	3	7	17	19	23

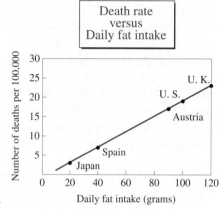

FIGURE 1.32

It turns out that the equation for this line is

$$D = f(F) = 0.2F - 1,$$

and we use this equation as our mathematical model in the following example.

EXAMPLE

The average daily animal fat intake in Mexico is $F = 23$ grams and the average daily animal fat intake in Denmark is $F = 135$ grams. Predict the death rate from breast cancer per hundred thousand women in Mexico and Denmark.

Solution   We use the mathematical model to predict that the death rate from breast cancer in Mexico for the average daily fat intake $F = 23$ will be

$$D = 0.2(23) - 1 = 3.6 \text{ per } 100{,}00 \text{ Mexican women.}$$

Similarly, for the average daily fat intake $F = 135$ grams in Denmark, the equation predicts a death rate of

$$D = 0.2(135) - 1 = 26 \text{ per } 100{,}000 \text{ Danish women.}$$

The type of prediction for the death rate for breast cancer in Mexico is called *interpolation* because we are predicting the value of a quantity using a measurement *within* the set of data. The type of prediction for the death rate in Denmark is called *extrapolation* because we are predicting the value of a quantity *beyond* the set of data.

In Section 2.2, we show how to find an equation such as the one relating the death rate to the daily fat intake. Once we have such an equation as a mathematical model, we can base some informed judgments on it. This model is based on the *average* daily intake in each country, which can vary tremendously among individuals. Even so, there is obviously a link between consumption of animal fat and the incidence of breast cancer. Thus the mathematical relationship indicates that women should drastically reduce their daily animal fat intake to reduce their chance of breast cancer. Furthermore, knowing that such a link exists, researchers have since been conducting follow-up studies to determine why the link exists. They have also found links to other items in the diet as well as in the environment. Thus, the incidence of breast cancer depends not just on a single variable but on a number of different variables, so it is a function of several independent variables. Many situations that you encounter in real life are examples of functions of more than one variable. Although the study of such functions is somewhat beyond the scope of this book, we introduce and briefly discuss functions of several variables in Section 3.8.

Parameters and Mathematical Models

Consider again the formula for the height y at any time t of an object dropped from the top of the 180-foot-tall Tower of Pisa: $y = 180 - 16t^2$. In comparison, the comparable formula for the height of an object dropped from the top of the 555-foot-high Washington Monument at any time t is

$$y = 555 - 16t^2,$$

and the height of an object dropped from the top of the 1821-foot-high CN tower in Toronto at any time t is

$$y = 1821 - 16t^2.$$

Each of these functions has the same structure mathematically; what differs among them is the leading number that represents the initial height from which the object is dropped. Based on these specific functions, we can hypothesize a general formula for the height y at any time t of an object dropped from any initial height y_0:

$$y = y_0 - 16t^2.$$

In this formula the height y clearly is a function of time t— they are the dependent and independent variables, respectively. The quantity y_0 can also take on different values, but it isn't a variable in the same sense that t and y are. Although y_0 can take on different values, in any particular situation it has just one value—in this case the initial height of the object. That value doesn't change even though the variables t and y change during the event. A quantity such as y_0 is called a **parameter**. Note that each value of y_0 yields a different function, although each function has the same form.

Now suppose that you're driving steadily at a rate of 40 mph; the relationship between the distance D you travel and the time you drive is $D = 40t$. If you drive at a steady 50 mph, the relationship is $D = 50t$, and if you drive at a steady 65 mph, the relationship is $D = 65t$. Obviously, distance is a function of time. The independent variable is time t, the dependent variable is distance D, and we write the function as $D = r \cdot t$. This relationship holds for any choice of speed r and gives a slightly different function, but one having the identical form, for each possible value of r. The quantity r is a parameter in the formula for this distance function.

How Accurate is a Mathematical Model?

Recall that a mathematical model is only a mathematical description of a process, not the process itself. So, every model carries with it some degree of inaccuracy. For instance, we used the formula $y = 64t - 16t^2$ as a mathematical model for the height at any time t of an object thrown vertically upward with an initial velocity of 64 ft/sec. With this model, we can find how long it takes for the object to come back to the ground. That occurs when $y = 0$, so we solve the equation

$$y = 64t - 16t^2 = 0,$$

We factor out the common factor of $16t$ to get

$$16t(4 - t) = 0$$

so that the object is at ground level when $t = 0$ (when the object is initially released) or $t = 4$ (when it has come back to the ground).

However, according to the laws of physics, the coefficient of t^2 is actually one half of the Earth's gravitational constant. Its value is not precisely -16, but rather more like -16.1, so the solution $t = 4$ is not quite accurate. Instead, we really should say that t is *about* 4 or that it is *approximately equal* to 4, which we write symbolically as $t \approx 4$. We can improve on this *estimate* by using the more accurate value of -16.1 for the coefficient of t^2 and then solving the equation

$$64t - 16.1t^2 = 0.$$

The only common factor now is t, so that

$$t(64 - 16.1t) = 0,$$

giving either $t = 0$ or $t = 64/16.1 \approx 3.975$, which is correct to three decimal places. How many decimal places are reasonable for this answer? We could use more decimal places when we divide out the fraction—say $t \approx 3.97516$ or even $t \approx 3.97515528$—but when we are measuring time in seconds, both results are unrealistic levels of accuracy and should be avoided. Even using the three decimal places in $t \approx 3.975$ may be too many, both from a practical point of view—think about timing in Olympic events where time is usually measured to the hundredth of a second—and from a mathematical point of view—we used only one decimal place in the coefficient, -16.1. In any context, you should determine a reasonable number of decimal places for your final answer, both practically and in terms of the number of digits used.

In fact, rarely in applied situations do you get an "exact" answer such as $x = 5$ or $x = \sqrt{8}$. Even when you do get an exact answer involving a radical or a fraction, you should usually convert it to a decimal, which automatically introduces another level of inaccuracy. Thus $\sqrt{8} \approx 2.828$ or $\sqrt{8} \approx 2.82843$ or $\sqrt{8} \approx 2.82842712$. But $\sqrt{8}$ is an irrational number and its decimal equivalent is an infinite, nonrepeating decimal. Just because your calculator displays 10 or 12 decimal places does not necessarily mean that the result is exactly that number.

Problems

1. An uncooked chicken (temperature of 70° F) is placed in a hot oven at a temperature of 350° to cook. The chicken is removed when its internal temperature reaches 180°. Sketch a possible graph for the temperature T of the chicken as a function of time t. What would be appropriate values for the domain and range of this function? Describe the behavior (increasing/decreasing, concavity) for the graph.

2. A warm can of soda (80° F) is placed in a refrigerator at a temperature of 36° and left there to cool. Sketch a graph of the temperature T of the soda as a function of time t. Identify appropriate intervals for the domain and range of this temperature function. Describe the behavior (increasing/decreasing, concavity) of this function.

3. An Olympic diver dives off the 10 meter platform, enters the water cleanly, and rises slowly to the surface. Sketch a possible graph for the height of the diver above water level as a function of time. What might be appropriate values for the domain and range of this function? (Estimate how long it will

probably take the diver to reach the water from the platform.) Describe the behavior of this function.

4. Repeat Problem 3 by sketching the graph of the diver's height above the diving platform as a function of time. How does the shape of this graph compare to the one you drew in Problem 3?

5. Police sometimes use the formula $s = f(d) = \sqrt{24d}$ as a model to estimate the speed s in miles per hour that a car was going on dry concrete pavement if it left a set of skid marks d feet long. Using this model, estimate the speed of a car whose skid marks stretched

   a. 60 ft.   **b.** 100 ft.
   c. 140 ft.   **d.** 200 ft.
   e. Suppose that you're driving at 60 mph on dry concrete pavement and slam on your brakes. How long will your skid marks be, according to this model?

6. When a basketball player takes a long shot, the height H of the ball above the floor can be modeled by the equation $H(t) = -16t^2 + 24t + 7$, where t is the number of seconds since the ball was released.

   a. Use your calculator to estimate the maximum height that the ball reaches, correct to two decimal places.

   b. The rim of the basket is 10 feet above the floor. Use your calculator to estimate all times t when the ball is at the height of the rim.

7. At the beginning of this section, we gave the equation for the height y of a ball thrown vertically upward from ground level with initial velocity 64 feet per second,

$$y = f(t) = 64t - 16t^2,$$

as a function of time t.

   a. Suppose that the initial velocity of the ball is 80 feet per second, Write a comparable formula for the height y as a function of time t.

   b. If you think of the initial velocity v_0 of the ball as a parameter, write a formula for the height y as a function of time t with any initial velocity v_0.

8. In each expression for a function, identify which letters represent variables, which letters represent functions, and which letters therefore represent parameters.

   a. $y = f(x) = ax^3 + bx^2 + cx + d$
   b. $z = g(t) = at^b$
   c. $z = h(t) = ab^t$
   d. $Q = k(m) = \dfrac{am}{bm^2 - c}$

Chapter Summary

In this chapter we introduced you to functions, their importance, and some of their uses. Specifically we showed you how to work with functions in the following ways.

- ◆ How functions arise in the real world.
- ◆ How functions can be represented in different ways—by graphs, by tables, by formulas, and in words—and how to move from one representation to another.
- ◆ How to identify whether a relationship between two variables given by an equation, a graph, or a table is or is not a function.
- ◆ How to decide which is the independent variable and which is the dependent variable.
- ◆ What the domain and range of a function are.
- ◆ The important characteristics about the behavior of functions—where they increase and decrease, where their turning points are, where they are concave up and concave down, where their points of inflection are, and whether they are periodic.
- ◆ How to interpret concavity—whether the growth (increase) or decay (decrease) in a function is speeding up or slowing down.
- ◆ How mathematics is used to model phenomena in the real world.

Review Problems

Budget ($ millions)	10.0	3.4	27.0	6.2	9.7
Attendance (millions)	1.0	0.5	2.0	0.6	1.3
Budget ($ millions)	7.0	4.8	18.0	6.5	13.0
Attendance (millions)	1.0	1.1	4.0	0.6	3.0
Budget ($ millions)	9.0	15.7	7.0	3.2	14.7
Attendance (millions)	0.5	1.3	1.0	0.5	2.7

1. In determining the amount of radiation to apply to a tumor site, doctors take into account the depth of the tumor within the body. What is the independent variable and what is the dependent variable in such a relationship? Give reasons for your answer.

2. The accompanying graph describes the loudness of a crowd watching a baseball game during the ninth inning. Write a scenario that might explain what was happening on the field as the inning progressed.

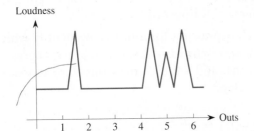

3. An experimental form of insulin is being administered every 4 hours to a person with diabetes. The body uses or excretes about 40% of the drug over the 4-hour period. Draw a graph that shows the amount of the drug in the body as a function of time over a 24-hour period.

4. Populations tend to grow steadily until there are too many members for the space and resources available. Then the population size levels off. Sketch a function that gives population size as a function of time.

5. Determine whether each table of values could represent a function. If not, explain why not.

a.

x	1	2	3	4	5	6
$f(x)$	10	10	12	14	18	25

b.

x	11	15	9	20	15	8
$g(x)$	12	13	13	15	16	17

6. The table of values shows the budget and attendance at 15 U.S. zoological parks. Write a short description of how attendance and budget are related.

7. The table of values shows the number, in millions, of prerecorded cassette tapes sold in the United States in various years between 1982 and 1998.

   a. Draw a graph of the number of cassettes sold as a function of the year since 1982.
   b. In approximately what year did the sales of cassettes reach its maximum?
   c. During which year, approximately, did the sale of cassettes change most rapidly? Most slowly?

Year	1982	1985	1990
Cassettes sold	182.3	339.1	442.2
Year	1993	1994	1995
Cassettes sold	339.5	345.4	272.6
Year	1996	1997	1998
Cassettes sold	225.3	172.6	158.5

Source: *2000 Statistical Abstract of the United States*

8. For the function $f(x) = 3x^2 - 2x + 1$, find $f(0)$, $f(1)$, $f(1.1)$, $f(1.01)$, $f(-3)$, and $f(a)$.

9. During the 1990s, the average cost of a new car bought in the United States can be approximated by the function $C = f(t) = 659.7t + 15598$, where C is the cost of the car and t is the number of years since 1990. (Source: *2000 Statistical Abstract of the United States*)

   a. Determine the average cost of a car purchased in 1995.
   b. If this trend continues, estimate the average cost of a car in 2003.
   c. According to this function, the average cost of a new car was increasing, on average, $659.70 each year after 1990. Use this information to calculate

how many years it would take for the average price of a new car to reach $20,000.

10. A study of the relationship between the average longevity (in years) and the gestation period (in days) for a sample of animals shows that the animals' average longevity L can be predicted reasonably well as a function of the gestation period t by the function $L = f(t) = 1.04t^{0.49}$, where t is the gestation period in days.

   a. Estimate the lifetime of a chipmunk whose gestation period is 31 days.

   b. The gestation times in the study extend from 15 days (opossum) to 645 days (elephant). What would the range of the average longevity be if the given function were a good predictor?

   c. Use your function grapher to graph the function. Is it increasing or decreasing? Is it concave up or concave down?

   d. Use the graph to estimate the gestation period of an animal whose average longevity is 15 years. (*Hint:* Find the point on the graph where $L = 15$.)

   e. The gestation time for humans is 9 months, or about 270 days. What does the formula predict for the average longevity of human beings? Can you think of any reasons why the value you obtained is so inaccurate?

11. The domain of the function $f(x) = 4x^2 + 3x - 5$ is all real numbers.

   a. Use your function grapher to estimate the coordinates of the turning point.

   b. What is the range of this function?

12. The domain of the function $f(x) = x^3 - 9$ is all real numbers.

   a. Use your function grapher to estimate the coordinates of the point of inflection.

   b. What is the range of this function?

13. Give the domain of each of the following functions.

   a. $f(x) = \sqrt{x + 5}$.   b. $f(x) = \sqrt{x^2 - 16}$.

   c. $g(x) = \dfrac{x^2 + 4}{x^2 - 9}$.   d. $h(x) = \dfrac{x^2 - 4}{x^2 + 9}$.

14. The U.S. Postal Service rates for first-class mail in 2003 were 37¢ for the first ounce and 23¢ for every additional ounce.

   a. Construct a table showing the cost of postage to mail a first-class letter weighing 0 to 1 oz, 1 to 2 oz, 2 to 3 oz, 3 to 4 oz, and 4 to 5 oz.

   b. Sketch a graph of this function $F(w)$, showing the cost of sending in 2003 a first-class letter as a function of the weight w of the letter in ounces.

15. Consider the roller coaster shown in the accompanying figure. The points A, B, C, D, E, and F divide the curve representing the track into portions that are increasing and concave up, increasing and concave down, decreasing and concave up, and decreasing and concave down. For each of the five portions of the track, A to B, B to C, and so on, (a) identify the mathematical behavior of the curve, and (b) describe whether the speed of the cars is increasing at an increasing rate, increasing at a decreasing rate, decreasing at a decreasing rate or decreasing at an increasing rate. In each case, explain your answer.

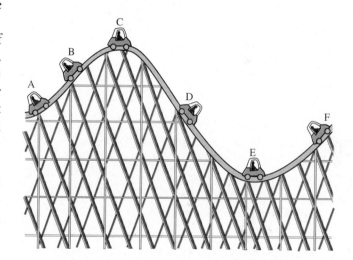

16. Picture a water slide at an amusement park. The slide starts at an initial height H_0 above the pool and smoothly drops to water level. The slide is first concave down, then concave up, then concave down, and finally concave up.

   a. Sketch a graph of the slide's height H above water level as a function of horizontal distance x.

   b. Suppose that you go down the slide in a sitting position. Sketch a graph of the height of your eye above the water as a function of horizontal distance x.

   c. Sketch the graph of the height of your eye above the water as a function of time t.

   d. Sketch the graph of your speed as a function of time t.

17. Pacific coast salmon hatch in rivers and then migrate to the ocean where they live most of their

lives. Eventually, they migrate back up the same river to spawn and then die. The accompanying graph shows the number of chinook salmon, in various years, that pass a particular point on one waterway near Seattle as they wend their way upstream to spawn. (The U.S. Department of Fisheries keeps accurate counts in its efforts to maintain a healthy salmon population.) Based on this graph, write a paragraph explaining why the number of salmon who swim upstream here is a roughly periodic function of time. What is the period? What does this graph suggest about the life span of the chinook salmon?

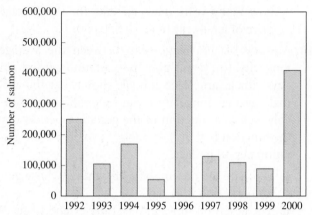

Source: Hiram M. Chittenden Locks data, U.S. Army Corps of Engineers, Seattle, Washington.

2

Families of Functions

2.1 ·········Introduction

Functions are fundamental to mathematics and its applications. Although there are many different types of functions, most of our work focuses on just a few: functions that are simple and yet sufficiently powerful to meet our needs. These types of functions can be thought of as *families of functions* because the members of each family are closely related to one another in terms of their essential properties. We have described several distinct behavior patterns already as phenomena that

1. increase at a fixed rate and so go up by the same amount each fixed time period;
2. decrease at a fixed rate and so go down by the same amount each fixed time period;
3. increase at an increasing rate and so are concave up;
4. increase at a decreasing rate and so are concave down;
5. decrease at an increasing rate and so are concave down;
6. decrease at a decreasing rate and so are concave up.

Figure 2.1 illustrates these behavior patterns.

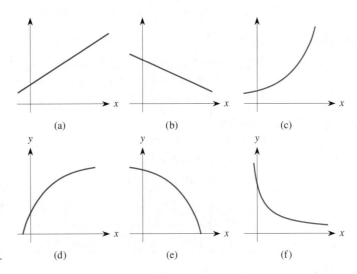

FIGURE 2.1

(a) (b) (c)
(d) (e) (f)

To model such phenomena and make predictions based on the models, you need to know families of functions that behave in each of these six ways. In this chapter, we present the families of *linear functions, exponential functions,* and *power functions,* as well as several other useful families, that possess these behavior patterns. In later chapters, we consider other families of functions, including *polynomial functions* and *trigonometric functions,* that exhibit more complex behavior patterns.

As discussed in Section 1.3, we use the letters x and y generically for the independent and dependent variables, respectively. However, in any specific context, we use letters that directly suggest the quantities under discussion.

2.2 Linear Functions

The simplest and probably most useful family of functions are the **linear functions.** These functions model any quantity that increases steadily or decreases steadily—that is, it goes up or down by a fixed amount for any fixed change in the independent variable. The graph of such a function is always a straight line.

Linear Functions That Pass Through the Origin

The simplest type of linear function is of the form $y = mx$, which can be interpreted as *y is proportional to x.* For example, suppose that you go to a deli to buy some roast beef that is selling at $5.99 per pound. If you purchase 1 pound, the roast beef costs $C = 5.99 \times 1 = \$5.99$; if you purchase 2 pounds, it costs $C = 5.99 \times 2 = \$11.98$. If you buy N pounds, the cost is $C = 5.99N$, so the cost C of the roast beef is proportional to the number of pounds N that you buy. The multiple 5.99 is the *constant of proportionality.* We also say that the cost of the roast beef is a linear function of the number of pounds of roast beef purchased.

Similarly, the distance D that a car travels at a constant speed of 50 mph is proportional to the number of hours t driven, so the number of miles traveled is

$$D = 50t.$$

As another illustration, it is reasonable to assume that the quantity G of garbage produced in a city is proportional to the number of people P living there, so that $G = kP$, for some constant multiple k. In fact, the average amount of garbage produced annually in the United States is about 1500 pounds per person, so $k \approx 1500$. A mathematical model for the quantity G of garbage produced annually in a city whose population is P is therefore $G = 1500P$ pounds.

EXAMPLE 1

Suppose that gas costs $1.50 per gallon. **(a)** Create a function, as a table, as a graph, and as a formula, to represent the cost C of G gallons of gas. **(b)** Create comparable functions if the price of gas rises to $1.75/gal and to $2.00/gal and compare them to one another.

Solution

a. If gas costs $1.50/gal, the cost for 1 gallon is $1.50, the cost for 2 gallons is $3.00, the cost for 3 gallons is $3 \times 1.50 = \$4.50$, and so on. We therefore get the following table of values for this function.

G (gal)	1	2	3	4	5	...	10	...	20	...
C ($)	1.50	3.00	4.50	6.00	7.50	...	15.00	...	30.00	...

When we plot these points, as shown in Figure 2.2, they all fall onto a line that passes through the origin (the cost of 0 gallons of gas is $0). To find an equation for this line, we note that, because each gallon of gas costs $1.50, the cost for buying G gallons must be

$$C = f(G) = 1.50 \times G = 1.50G.$$

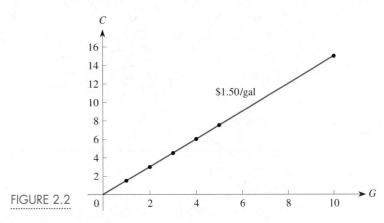

FIGURE 2.2

b. If the cost of gas rises to $1.75/gal, the corresponding function would be

$$C = g(G) = 1.75 \times G = 1.75G.$$

Similarly, if the cost of gas rises to $2.00/gal, the function would be

$$C = h(G) = 2.00 \times G = 2.00G.$$

The graphs of all three of these linear functions are shown in Figure 2.3. Note that all three lines pass through the origin, but that each is inclined at a slightly different angle. In particular, the greater cost for a gallon of gas, the steeper the line, which makes sense because filling a tank costs more.

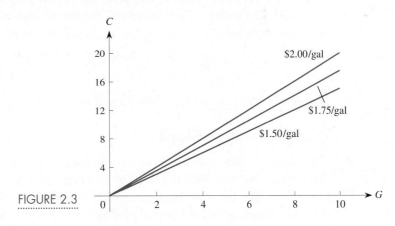

FIGURE 2.3

The Graph of a Linear Function That Passes Through the Origin

The graph of any linear function of the form $y = mx$ is a line that passes through the origin, as shown in Figure 2.4. What distinguishes one line from another is the constant m, which represents how much y changes for a given change in x. A large value for m, either positive or negative, means that y changes by a large amount for a fixed change in the variable x. A small m means that y changes relatively little for a fixed change in x. A positive value for m means that y gets larger as x gets larger. A negative value for m means that y gets smaller as x gets larger.

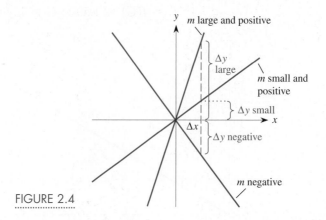

FIGURE 2.4

We use the Greek letter Δ (delta) to represent a change in any quantity. Hence Δx means the change in x, and Δy means the change in y. The quantity

$$m = \frac{\text{change in } y}{\text{change in } x} = \frac{\Delta y}{\Delta x} = \frac{\text{rise}}{\text{run}}$$

is called the **slope** of the line. More generally, the slope of a line is

$$m = \frac{\text{change in dependent variable}}{\text{change in independent variable}}.$$

The letters used for the independent and dependent variables reflect what those quantities are and often are different from x and y. For instance, based on the data in the preceding table of values for the cost of gasoline, the independent variable G is the number of gallons of gas purchased and the dependent variable C is the cost of the gas. The slope of the line, based on the first two points, is

$$m = \frac{\Delta C}{\Delta G} = \frac{3.00 - 1.50}{2 - 1} = \frac{1.50}{1} = 1.50.$$

We get the identical value for the slope if we use any two of the points.

The Meaning of Slope

The slope of a line indicates how fast the linear function is changing. For the gasoline example $C = 1.50G$, the slope of 1.50 is the cost, \$1.50, of each additional gallon of gas. For the roast beef example $C = 5.99N$, the slope, 5.99, is the cost per pound, so each additional pound of roast beef costs an additional \$5.99. If the roast beef is on sale for \$3.99, the cost of the roast beef goes up more slowly as the weight of the purchase increases, and the slope is smaller. Similarly, if the price of roast beef goes up to \$6.99/lb, the cost goes up more rapidly, and the slope is steeper.

In general, whenever a linear function (or a line) arises in some context, the slope of that line should be given in terms of units. For instance, if we use a linear function to model the growth in the U.S. prison population over time, the units for the slope of the line might be the number of new prisoners per year.

Suppose that a car gets 25 mpg. Then

Number of miles driven = 25 miles per gallon × number of gallons used, or

$$25 \text{ miles per gallon} = \frac{\text{number of miles driven}}{\text{number of gallons used}}.$$

Therefore the expression "25 miles per gallon" actually describes the slope of a linear function. The units of the slope are always a ratio: the units of the dependent variable divided by the units of the independent variable.

Because the slope indicates how fast a line rises or falls, it is also known as the *average rate of change,* or simply the **rate of change.** For a linear function, the rate of change is always constant and is equal to the slope. For a nonlinear function, which may be concave up or concave down, the rate of change is not constant, as we demonstrate later in this chapter.

Lines That Don't Pass Through the Origin

Next let's consider lines that don't pass through the origin. The equation of any such line has the form

$$y = mx + b,$$

where

m is the slope of the line and

b is the vertical intercept.

The **vertical intercept** b represents the value of y when x is zero, because $y = m \cdot 0 + b = b$. The vertical intercept is sometimes called the **y-intercept.** A vertical intercept of zero ($b = 0$) corresponds to the special case $y = mx$, which is the equation of a line that passes through the origin, as we discussed previously.

◆EXAMPLE 2

Graph the line $y = 3x - 4$ and describe it.

Solution   The line has a slope of 3, so it rises 3 units for each increase of 1 unit to the right. It has a vertical intercept of -4, so the line crosses the y-axis 4 units below the origin. The graph of this line is shown in Figure 2.5.

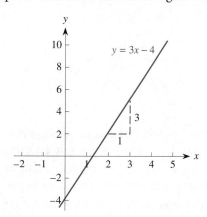

FIGURE 2.5

The graphs associated with the equations

$$y = f(x) = 2x - 1, \qquad y = g(x) = 2x + 1, \quad \text{and} \quad y = h(x) = 2x + 2$$

are shown in Figure 2.6. The three lines are parallel because they all have the same slope, $m = 2$, and so all rise 2 units for each 1 unit increase to the right. But their vertical intercepts are different: $b = -1$, $b = 1$, and $b = 2$ respectively.

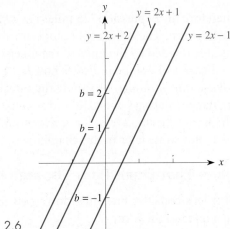

FIGURE 2.6

The graphs associated with the equations

$$y = f(x) = 2x + 1, \qquad y = g(x) = 3x + 1, \quad \text{and} \quad y = h(x) = -2x + 1$$

are shown in Figure 2.7. Note that, because all three lines cross the y-axis at the point $y = 1$, all have the same y-intercept, $b = 1$. However, the three lines have different slopes and so behave differently. The functions f and g are increasing as you move from left to right (as x increases), whereas the function h is decreasing as x increases. Moreover, the line $y = g(x) = 3x + 1$ is increasing more rapidly than the line $y = f(x) = 2x + 1$ because it has a larger slope. Again, the slope of the line determines whether a line rises or falls and how rapidly it does so.

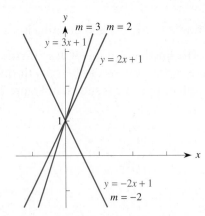

FIGURE 2.7

In summary, when the slope m is positive, the line rises as x increases from left to right and the linear function is increasing; the larger m is, the faster the line rises. When the slope m is negative, the line falls as x increases from left to right and the linear function is decreasing; the more negative the slope, the faster the line drops. (However, because a line doesn't bend, either up or down, it is neither concave up nor concave down.)

We can express the slope of a line,

$$m = \frac{\text{change in } y}{\text{change in } x} = \frac{\Delta y}{\Delta x} = \frac{\text{rise}}{\text{run}}.$$

in another way. If (x_1, y_1) and (x_2, y_2) are two points on the line, then

$$\Delta x = x_2 - x_1 \quad \text{and} \quad \Delta y = y_2 - y_1,$$

where the order of the coordinates must be the same in both Δx and Δy. Therefore the equation for the slope becomes

$$m = \frac{\Delta y}{\Delta x} = \frac{y_2 - y_1}{x_2 - x_1}.$$

Now let's explore these ideas in terms of the real world.

EXAMPLE 3

A wholesale supplier quoted the following costs, C, in dollars, for graphing calculators, depending on the number n of units ordered.

n	1	2	3	4	5	...
C	87	167	247	327	407	...

Find a linear function that models the costs and discuss the meaning of the slope and vertical intercept.

Solution   Note that the cost for each additional calculator after the first is $80. The $87 charged for the first calculator consists of the $80 for the calculator and shipping, plus an additional $7 that covers the fixed cost for processing the order. This amount remains fixed no matter how many units are purchased. Therefore the cost C of buying n calculators can be written as the linear function $C = f(n) = 80n + 7$. The slope, 80, represents the increase in the total cost for each additional unit ordered—every time n increases by 1, C increases by 80. That is, the *rate* at which the cost is increasing is $80 per calculator sold. The vertical intercept, 7, is the fixed cost for any size order. Figure 2.8 depicts the slope as the ratio

$$m = \frac{\Delta C}{\Delta n} = \frac{\text{rise}}{\text{run}} = \frac{167 - 87}{2 - 1} = 80,$$

or simply $m = \$80$ per calculator.

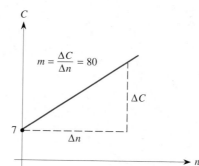

FIGURE 2.8

The value $m = 80$ was based on the two points $(1, 87)$ and $(2, 167)$. If you calculate the slope by using *any* two points on the line, you get the same value. It is this fact—that the slope, or rate of change, is the same at every point—that makes a line straight. If the rate of change varies from one point to another, then the function is not linear.

The form for the equation of the line $C = 80n + 7$ that we found in Example 3 is known as the *slope–intercept form* and usually is written as $y = mx + b$

because it highlights the *slope* $m = 80$ of the line and the *vertical intercept* $b = 7$.

The slope–intercept form is very useful for *displaying* the equation of a line. However, it is usually a poor choice for *finding* the equation of a line because the vertical intercept is often difficult to determine. Even when we can find the vertical intercept, it may have little to do with the situation we're studying. Example 4 demonstrates an easy way to apply the slope–intercept form.

EXAMPLE 4

A plumber charges $50 for a service call to come to the job and $70 per hour for labor. (a) Find a linear function for the plumber's charges for a job taking t hours (disregarding the costs for any parts). (b) What is the meaning of the slope in this function?

Solution

a. The plumber charges $50 just for coming. For each hour on the job, the charge is an additional $70, so a job lasting t hours costs an additional $70t$ dollars. Therefore the total cost is

$$C = 50 + 70t.$$

b. The slope of this line, 70, is the charge for each hour of labor and its units are dollars per hour.

Usually, a much better method for determining an equation of a line is the *point–slope form*. It is based on the idea that a line is determined by its slope m and one point (x_0, y_0) on the line. Suppose that (x, y) is any other point on the line, as shown in Figure 2.9. Because

$$m = \frac{\Delta y}{\Delta x} = \frac{y - y_0}{x - x_0},$$

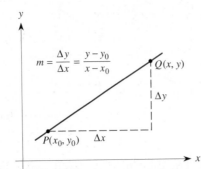

FIGURE 2.9

we can multiply both sides by $(x - x_0)$ to get

$$y - y_0 = m(x - x_0).$$

In summary we have the following formula.

> ## Point–Slope Formula for the Equation of a Line
>
> The equation of the line with slope m that passes through the point (x_0, y_0) is
>
> $$y - y_0 = m(x - x_0)$$

For instance, the line through the point $(5, 2)$ with slope 4 is

$$y - 2 = 4(x - 5).$$

You are almost always better off using the point–slope form rather than the slope–intercept form to *find* the equation of a line. In Example 5 we revisit the snow tree cricket from Example 2 in Section 1.4.

EXAMPLE 5

The following set of measurements relate the snow tree cricket's rate of chirping, in chirps per minute, to the temperature, in Fahrenheit.

Temperature, T (°F)	50	55	60	65	70	75	80
Rate, R (chirps/min)	40	60	80	100	120	140	160

a. Find a function that models the chirp-rate as a function of temperature.
b. Discuss the reasonableness of the model and give reasonable values for the domain and range.

Solution

a. The chirp rate is increasing steadily, so it is an increasing function of the temperature T. In particular, the chirp rate goes up 20 chirps/min for every 5°F increase in temperature. Equivalently, the chirp rate goes up 4 chirps/min for every 1°F increase in temperature. Figure 2.10 shows that the corresponding points clearly fall in a linear pattern. In other words the chirp rate R as a function of temperature T is a linear function.

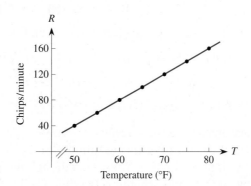

FIGURE 2.10

Because two points determine a line, we can use any two of the given points—say, $(55, 60)$ and $(75, 140)$—to find the equation for this line. Using these two points, as shown in Figure 2.11, we find that the slope of the line is

$$m = \frac{\Delta R}{\Delta T} = \frac{140 - 60}{75 - 55} = \frac{80}{20} = 4.$$

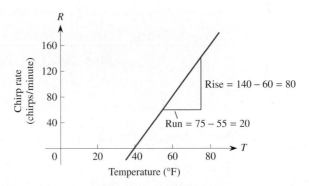

FIGURE 2.11

This value for the slope means that, for each 1°F increase in temperature, the cricket chirps 4 more times per minute. Thus, if the temperature goes up 5°F, the cricket chirps 20 more times per minute; if it goes up 10°F, the cricket chirps 40 more times per minute, and so on.

Next we apply the point–slope formula to find the equation of the line, using any point on the line. If we pick the point (55, 60) used earlier, we obtain

$$R - 60 = 4(T - 55)$$
$$= 4T - 220.$$

Adding 60 to both sides of this equation, we get

$$R = f(T) = 4T - 160.$$

This equation tells us that the vertical intercept is $R = -160$ (when $T = 0$). Of course, a chirp rate of $R = -160$ is meaningless! Was the formula wrong? No. But it makes sense to describe the snow tree cricket's chirp rate only for temperatures between, or possibly near, the given set of readings—that is, from 50°F to 80°F. It does not make real-world sense to use this linear relationship far outside of this interval, such as at 0°. The formula doesn't predict sensible chirp rates for temperatures less than 40°F, when R becomes negative. It doesn't hold at temperatures high enough to cook the cricket either. Because temperatures in the Colorado Rockies aren't likely to rise above 100°F, there is a natural domain for this function:

Domain of f = all values between 40°F and 100°F   or
$$40 \leq T \leq 100 \quad \text{or} \quad [0, 100].$$

This function is strictly increasing, so we can find the corresponding range:

$$f(40) = 4 \cdot 40 - 160 = 0 \quad \text{and} \quad f(100) = 4 \cdot 100 - 160 = 240.$$

Thus

Range of f = all values R from 0 to 240   or $0 \leq R \leq 240$ or $[0, 240].$

b. How reasonable are these results? At 100°F, the equation predicts that a snow tree cricket will chirp 240 times per minute, or 4 times per second, which we might decide is a bit unreasonable. Thus, even though the linear model predicts this value, we might want to rethink whether extending the linear model as far as $T = 100°F$ makes sense when the upper limit of the data values is $T = 80°F$. As we've said previously, it is often misleading to extrapolate too far beyond the actual data values.

◆

So far we have used the temperature to predict the chirp rate, and we thought of the temperature as the *independent variable* and the chirp rate as the *dependent variable.* However, we could reverse the role of the variables and think of temperature as a function of chirp rate. How we view a relationship determines which variable is dependent and which is independent. Thinking of temperature as a function of chirp rate would enable us to approximate temperature for given chirp rates. To do so, we again start with the formula

$$R = f(T) = 4T - 160$$

and solve it algebraically for T as a function of R. We add 160 to both sides to obtain

$$4T = R + 160$$

and then divide both sides by 4 to get

$$T = \frac{1}{4}(R + 160) = \frac{1}{4}R + 40 = g(R).$$

This linear function has slope $\frac{1}{4}$ and vertical intercept 40, except now the independent variable is R. So, if you ever encounter a snow tree cricket who is chirping merrily away, knowing this equation can help you determine the local temperature just by using your watch. Count the number of chirps in a one-minute interval and apply the formula to calculate the temperature.

We summarize the important information about linear functions as follows.

The **slope-intercept form** for the equation of a line is

$$y = mx + b,$$

where m is the **slope,** or rate of change of y with respect to x,

$$m = \frac{\Delta y}{\Delta x} = \frac{\text{rise}}{\text{run}},$$

and b is the **vertical intercept,** or value of y when $x = 0$.

The **point–slope form** for the equation of a line with slope m that passes through the point (x_0, y_0) is

$$y - y_0 = m(x - x_0).$$

Note that in the slope-intercept form for the equation of a line, there are two parameters, the slope m and the vertical intercept b. So linear functions are a *two-parameter* family of functions. We determine the equation of a particular line by finding the values of the two parameters.

EXAMPLE 6

During the early years of the Indianapolis 500 race held annually on Memorial Day, the average winning speed increased as shown in the following table. Find a formula to model these values.

Year	1919	1922	1925
Average speed (mph)	88	94.5	101

Source: *World Almanac and Book of Facts.*

Solution   The average winning speed starts at 88 mph and increases at the rate of 6.5 mph each three years. Because the average winning speed S increases consistently by 6.5 mph every three years, S is a linear function of time over the period 1919 to 1925, as shown in Figure 2.12. The slope of this line is

$$\text{Slope} = \frac{\text{rise}}{\text{run}} = \frac{\Delta S}{\Delta t} = \frac{6.5}{3} \approx 2.17,$$

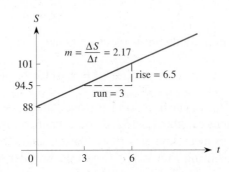

FIGURE 2.12

which shows the rate at which the winning speed increased each year. Let t be the number of years since 1919. Using the initial point $(0, 88)$ gives the equation of this line as

$$S - 88 = 2.17(t - 0) = 2.17t,$$

or

$$S = f(t) = 2.17t + 88.$$

In this formula, t represents the number of years since 1919 and S is the speed in miles per hour. Note that we would get the same result if we used the slope–intercept form. ◆

You may wonder whether this linear trend continued beyond 1925. Let's compare what it predicts with what actually happened. The fastest average winning speed in the Indy 500 was 186 mph in 1990, when $t = 71$ years after 1919. Using the linear equation $S = 2.17t + 88$, we predict an average speed of 242 mph in 1990. Clearly, although speeds have increased dramatically, they haven't kept up with the linear function we constructed based on just a few early data points. Further, this model again illustrates the danger of extrapolating too far from the given data.

Think About This   What does this information indicate about how long the 500-mile race takes? How much longer did it take the winning car to drive the 500 miles in 1919 than it took in 1990? ▭

Because the data in the table are given only at specific points (every 3 years), we say that the data are *discrete*. However, because the function $S = 2.17t + 88$ makes sense for *all* possible values of t, we treat the variable t as though it were *continuous* (or defined for all points). The graph shown in Figure 2.12 is of a continuous function because it is a solid line including infinitely many points, not just the three distinct points representing the winning speeds in the race in three particular years.

EXAMPLE 7

Search and Rescue teams are often called on to find lost hikers in remote areas in the Southwest. Members of the search team walk through the search area parallel to each other at a fixed distance d between searchers. Experience has shown that the team's chance of finding those who are lost is related to the distance of separation d. The closer together the searchers are, the better are their chances of success. Based on a number of simulated missions, the percentage of lost people who were found was used to assess the probability of finding someone based on various separation distances, as shown in the following table of values. Find a formula to model these probabilities.

Distance d (ft)	20	40	60	80	100
Probability of success P (%)	90	80	70	60	50

(These values correspond to searches conducted in the relatively open terrain of the Southwest; searchers in other regions where there is dense forest or undergrowth would have to use much narrower separation distances to achieve comparable levels of success.)

Solution   Because the value for the probability of success P decreases as distance d (the independent variable) increases, the function $P = f(d)$ is a decreasing function of d. The data indicate that each 20 foot increase in distance causes the probability of success P to decrease by 10%. Because this fact holds for any successive pair of points, P is a linear function of d, and the graph of the probability of success versus distance is a line, as shown in Figure 2.13. Based on the two data points $(20, 90)$ and $(100, 50)$, say, the slope of this line is

$$m = \frac{\Delta P}{\Delta d} = \frac{50 - 90}{100 - 20} = \frac{-40}{80} = -\frac{1}{2}.$$

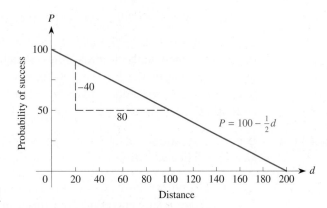

FIGURE 2.13

The negative sign reinforces the fact that P decreases as d increases. The slope is the rate at which P is decreasing as d increases.

To find the equation of the line, we use the point–slope formula. We choose any one of the given points—say, $(20, 90)$—and obtain

$$P - 90 = -\frac{1}{2}(d - 20)$$

$$= -\frac{1}{2}d + 10.$$

Adding 90 to both sides of this expression, we get

$$P = f(d) = -\frac{1}{2}d + 100.$$

Think About This   Pick any one of the other points in Example 7 and show that you get the same equation for P. ▭

What is the meaning of the vertical intercept, $P = 100$? Suppose that $d = 0$ so that the searchers are walking shoulder to shoulder; we would expect everyone to be found, or $P = 100$. What is the horizontal intercept? When $P = 0$, we have $0 = -\frac{1}{2}d + 100$, or $d = 200$. According to the model, the value $d = 200$ represents the separation distance at which no one is found. This outcome is unreasonable because, even when the searchers are far apart, the search will sometimes be successful. What this situation suggests is that, somewhere outside the data given, the linear relationship ceases to hold. As in the Indy 500 example, extrapolating too far beyond the given data may not make sense.

Some Useful Facts

Several facts about lines are useful to remember.

1. *Parallel lines* have the same slope. That is, the quantities they represent are growing at the same rate.

Think About This   The lines $y = 4x + 3$, $y - 4x = 11$, and $4x - y - 15 = 0$ are all parallel. What is their common slope? ▭

2. *Perpendicular lines* have slopes that are negative reciprocals.

The lines $y = 2x - 9$ and $y = -\frac{1}{2}x + 3$, having slopes of 2 and $-\frac{1}{2}$, respectively, are perpendicular to each other. Sketch their graphs to convince yourself of this fact. Similarly, the lines $y = 0.162x + 7.4$ and $y = -6.173x + 1.03$, which have slopes of 0.162 and $-6.173 \approx -1/0.162$, respectively, are perpendicular to each other.

Think About This   Write the equation of a line that is perpendicular to $y = \frac{5}{4}x - 7$. (Of course, your answer will likely be different from your classmates' choices.) ▭

3. The point where any two lines cross is known as their *point of intersection*. The x- and y-coordinates of this point must satisfy both equations simultaneously. You find the point of intersection by solving the system of simultaneous equations either algebraically or graphically. (See Appendix B and C for a discussion of ways to solve such systems of equations.)

Problems

1. Match each equation with its graph. (Note that the scales of the graphs are different.)

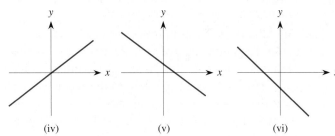

(i)   (ii)   (iii)

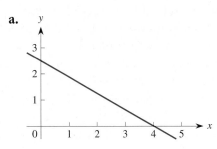

(iv)   (v)   (vi)

a. $y = x + 2$ **b.** $y = x - 3$
c. $y = -2x + 4$ **d.** $y = -3x - 4$
e. $y = \frac{1}{2}x$ **f.** $y = 3$

2. Estimate the slope of each line. Then use the slope to find an equation of the line.

a.

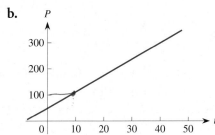

b.

3. Find the equation of the line passing through each pair of points.

   a. $(1, -2), (2, 5)$ **b.** $(1, -2), (3, -2)$
   c. $(3.52, 4.96), (-1.91, 8.36)$

4. The graph of Fahrenheit temperature F versus Celsius temperature C is a line. Water boils at 212°F and 100°C and freezes at 32°F and 0°C.

   a. Sketch the graph of the line.
   b. Find the slope of the line relating the two temperature scales.
   c. Find the equation of this line.
   d. Use the equation to find the Fahrenheit temperature that corresponds to 30°C.
   e. Use the equation to find the Celsius temperature that corresponds to 98.6°F.
   f. When is the Fahrenheit temperature the same numerical value as the Celsius temperature?

5. In 1990, 442.2 million prerecorded cassette tapes and 865.7 million CDs were sold in the United States. In 1998, 158.5 million cassettes tapes and 1,124.3 million CDs were sold. Assume (incorrectly) that the pattern of sales for both items is linear.

   a. Find the equation for the number of cassette tapes sold as a linear function of time.
   b. Find the equation for the number of CDs sold as a linear function of time.
   c. What is the practical significance of the slopes in parts (a) and (b)?
   d. If the trends in sales of both items were indeed linear, find when the number of CDs sold overtook the number of cassette tapes sold.
   e. Use the data given to find the total number of both CDs and cassette tapes sold in 1990 and 1998 and use these values to find the equation for the total number of sales of both items combined as a linear function of time.
   f. Use the fact that, in 1995, 272.6 million cassette tapes and 272.6 million CDs were sold to explain why assuming that the sales trends were linear is incorrect.

6. The charges for a taxi ride are an initial charge of $1.80 and $0.75 for each mile driven.

   a. Write a formula for the charge for a taxi ride as a linear function of the distance traveled.
   b. What is the meaning of the slope of this linear function?
   c. What is the cost of a 12-mile trip?
   d. Suppose that you have only $15. How far can you go in the taxi? (Assume that you will give a $2 tip out of the $15 you have.)

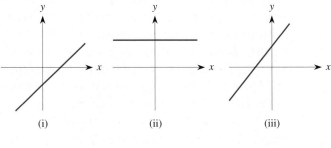

7. A long-distance telephone company charges 40¢ to place a call from Los Angeles to London and 30¢ for each minute.

   a. Write the equation of a linear function that models this situation.
   b. What is the practical significance of the slope? Of the vertical intercept?
   c. What is the cost of a 26-minute call?
   d. Suppose that there is a 30% discount on the rates for calls made in off-peak hours. Repeat parts (a)–(c).

8. (Continuation of Problem 7) A competing long-distance company claims that it is cheaper because its rates on the Los Angeles to London call are 15¢ to place the call and 36¢ for each minute.

   a. For the 26-minute call in Problem 7(c), which carrier is actually cheaper?
   b. Graph both lines. What does the point where they intersect signify?
   c. Find the length of call at which the second company becomes more expensive than the first.

9. A disk jockey (DJ) charges a flat fee of $120 per party plus $60 for each hour of the party. A second DJ charges $100 per party plus $75 for each hour.

   a. For each DJ find a formula that gives the cost of hiring the DJ as a function of the number of hours the party lasts.
   b. Sketch the graphs of both functions on the same set of axes.
   c. How do you decide which DJ costs less?

10. The net income of the Apex Company was $240 million in 1980 and has been increasing by $30 million per year since. Over the same period, the net income of its chief competitor, the Best Corporation, has been growing by $20 million per year, starting with $300 million in 1980. Which company earned more in 1990? When did Apex surpass Best?

11. According to the IRS, the formula $T = 0.15I$, which gives income tax as a function of taxable income, applies only for single taxpayers with taxable incomes up to $21,450. The IRS tax table states: "If the taxable income is over $21,450, But not over $51,900, Enter on Form 1040: $3,217.50 + 28% of the amount over 21,450."

   a. Rewrite this statement as an equation that can be used to calculate your taxes. What are the domain and range of the resulting function?

   b. What is the practical meaning of the value you get for the slope?
   c. Sketch a single graph showing both tax formulas. Is there any discrepancy?

12. When filing income tax returns, many people can claim deductions for depreciation on items such as cars and computers used for business purposes. The idea is that the value of such an asset decreases, or depreciates, over time. The simplest method used to find the depreciated value is called *straight-line depreciation*, which assumes that the item's value decreases as a linear function of time. If an $1800 computer system depreciates completely in five years, find a formula for its value as a function of time. What is it worth after three years?

13. The Athabasca glacier in southern Alberta, Canada, is part of the largest mass of ice in the Rocky Mountains. (Tourists who visit the Jasper and Banff National Parks can take a side trip out onto the actual glacier.) Over the past 120 years, the glacier has been steadily "withdrawing" at a rate of about 15 meters per year, as it slowly melts.

   a. Express the approximate position of the southernmost extent of Athabasca as a function of time, measured in years from 1900. Measure its position northward from the U.S.–Canada border, which was about 300 kilometers south of the glacier in 1900.
   b. If the current rate of withdrawal has been in effect indefinitely, how long ago did the toe of the glacier extend over the border?
   c. Can the function in part (a) continue to apply for the next million years? Why or why not?

14. Jen is typing her term paper for Psych 101. She types the body of the paper at the rate of 35 words per minute for 30 minutes, then takes a 5-minute break, and comes back to do the references at a rate of 20 words per minute for 12 minutes.

   a. Sketch the graph of Jen's typing rate as a function of time.
   b. Sketch the graph of the total number of words she types as a function of time.
   c. Find the equations of the different line segments you drew in part (b).

15. A bicyclist pedals at the rate of 1000 ft/min for 20 minutes, then slows to 500 ft/min for 6 minutes, then races at 1200 ft/min for 4 minutes, and cools down at 500 ft/min for 5 minutes.

   a. Sketch the graph of the bicyclist's rate as a function of time.

b. Use the graph from part (a) to determine the total distance biked.

c. Sketch the graph of the distance traveled as a function of time.

d. Find the equations of the different line segments you drew in part (c).

16. The points P, Q, and R lie in order from left to right on the graph of a function f that is increasing. If the slope of line segment PQ is less than that of line segment QR, is the curve concave up or concave down? Explain your reasoning.

17. Find the equation of the line that passes through the point $(6, 4)$ and is

   a. parallel to the line $y = 5x - 3$.
   b. perpendicular to this line.

18. Find the equation of the line that passes through the point $(6, 4)$ and also passes through the point of intersection of $y = -2x + 1$ and $y = 3x + 6$.

19. The algebraic method of elimination for solving a system of linear equations involves adding a multiple of one equation to another equation to eliminate one of the variables. Consider the system of two equations in two unknowns:

$$3x - 4y = 1 \qquad \textbf{(1)}$$
$$2x + y = 8. \qquad \textbf{(2)}$$

   a. Plot the two lines carefully on a sheet of graph paper and determine the point of intersection.

   b. Solve the two equations algebraically.

   c. Add two times Equation (2) to Equation (1) to get a new linear equation. Plot that line on the same graph you created in part (a). What do you observe about the three lines?

   d. Add three times Equation (2) to Equation (1) and plot that line on the same graph. What do you observe about the four lines?

   e. Add four times Equation (2) to Equation (1) and plot that line on the same graph. What do you conclude from this result?

   f. Find an appropriate multiple of Equation (2) that, when added to Equation (1), will eliminate the x-term. What will the graph of the resulting line look like when x has been eliminated?

20. The point $(3, 4)$ is on the circle $x^2 + y^2 = 25$.

   a. Find the equation of the line that is tangent to the circle at this point.

   b. Find the points where the line intersects the x and y axes. (*Hint*: The line tangent to a circle at a point is always perpendicular to the radius at that point.)

21. **a.** Of the following three linear functions, which two represent perpendicular lines?

   i. $3x - 4y = 12$ **ii.** $2x + 5y = 10$
   iii. $8x + 6y = 7$

   b. For the two lines that are perpendicular, find the point of intersection.

Exercising Your Algebra Skills

Solve each equation for the appropriate variable.

1. $5x - 7 = 12$
2. $3x + 8 = -7$
3. $4x - 3 = -5$
4. $8x + 7 = 15$
5. $18y - 7 = 22$
6. $5.4x - 7.2 = 0.8$
7. $9 - 3x = 6$
8. $5 - 4p = -7$
9. $4.7q + 5.1 = 24.5$
10. $-1.3w + 12.8 = 22.7$
11. $4k + 7 = 9k - 8$
12. $6z - 5 = 4z + 11$
13. $3(2x - 5) = 4$
14. $2(4 - 3w) = 7$
15. $3.2(t - 1980) = 1700$
16. $1.35(t - 75) = 8$

Find (**a**) the slope and (**b**) the x and y intercepts of each line in Problems 17–20.

17. $2x - 3y = 8$
18. $2x + 3y = 8$
19. $4x + 7y + 5 = 0$
20. $3y - 2x + 4 = 0$

2.3 Linear Functions and Data

$y - 4 = 5(x + 6)$

$y = 5x - 10$

Determining Whether a Set of Data Is Linear

When you encounter tables of data relating two quantities, you will often need to determine whether a linear relationship exists between the two variables. You could plot the data points to see if the points fall into a linear pattern, but this approach is imprecise. Alternatively, you could decide whether a function $y = f(x)$ given by a table of values is linear by examining the data. If the data fall into a linear pattern, you should get the same slope no matter which pair of points you use. This reasoning gives a simple criterion for determining linearity: See whether the differences in *y*-values are constant for equally spaced *x*-values.

> If the *x*-values are uniformly spaced and there is a constant difference among the *y*-values, the data fall into a linear pattern.

You can visualize this principle by thinking of a long plank of wood and a flight of stairs. If the steps all have the same height—say, 8 inches, and the same depth, you can lay the plank on the stairs and it will touch the edge of each one, as illustrated in Figure 2.14. The plank plays the role of a line. But, if the stairs have different heights or depths, the plank won't touch every one of the edges—those edges do not fall in a linear pattern.

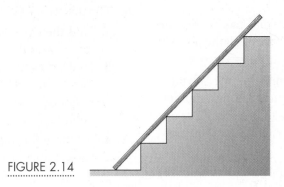

FIGURE 2.14

EXAMPLE 1

The following two sets of data represent values for a linear function and a nonlinear function. Identify which is the linear function. Find the equation of the line and the concavity of the nonlinear function.

x	*f(x)*	*x*	*g(x)*
1.0	7.0	1	2
1.2	7.8	2	3
1.4	8.6	3	6
1.6	9.4	4	11
1.8	10.2	5	18
2.0	11.0	6	27
2.2	11.8	7	38

Solution In both sets of values, the *x*-values are evenly spaced, so we can proceed to examine the successive differences in the values of the two functions, which we write as $\Delta f(x)$ and $\Delta g(x)$. For instance, for the function *f* the difference between the first two values is $7.8 - 7.0 = 0.8$. Continuing in this manner, we obtain the data on the next page.

x	$f(x)$	$\Delta f(x)$
1.0	7.0	
		$0.8 = 7.8 - 7.0$
1.2	7.8	
		$0.8 = 8.6 - 7.8$
1.4	8.6	
		0.8
1.6	9.4	
		0.8
1.8	10.2	
		0.8
2.0	11.0	
		0.8
2.2	11.8	

Because the difference is a constant 0.8 between the values of the function f, we conclude that this set of data is indeed linear. The slope of the line through these points is

$$m = \frac{\Delta y}{\Delta x} = \frac{\Delta f(x)}{\Delta x} = \frac{0.8}{0.2} = 4.$$

Further, using the first point $(1, 7)$ and the point–slope form for the equation of a line, we find that the equation of the line is

$$y - 7 = 4(x - 1)$$
$$= 4x - 4.$$

When we add 7 to both sides of this expression, we get

$$y = 4x + 3 = f(x).$$

Suppose that we try the same analysis on the values for the function g.

x	$g(x)$	$\Delta g(x)$
1	2	
		$1 = 3 - 2$
2	3	
		$3 = 6 - 3$
3	6	
		5
4	11	
		7
5	18	
		9
6	27	
		11
7	38	

The differences are not constant, so we conclude that these points don't fall into a linear pattern and hence no line passes through them. Consequently, the function g *cannot* be a linear function. In fact, because the differences are successively larger, the function is growing faster than a linear function grows. Because the function g is increasing at an increasing rate, it is concave up.

So far, we have given you information on some process or quantity that clearly is a linear function. In practice, however, you may face a situation in which you simply assume that one quantity grows or decays in a linear manner. Or you may even encounter a set of data that appears to be roughly linear in nature, but the particular data points do not precisely fall on a line. We illustrate both situations in Examples 2 through 4.

EXAMPLE 2

In 1990, the United States imported $495 billion worth of goods. In 1998, the United States imported $912 billion worth of goods. (Source: 2000 Statistical Abstract of the United States.)

a. Assuming that the growth in imports followed a linear pattern, find an equation of the linear function that models U.S. imports.

b. What is an appropriate domain for this model?

c. Use the model to predict the amount of imports in the year 2005.

d. Predict when the United States will import $1 trillion worth of foreign goods according to this model.

Solution

a. For convenience, we take the independent variable t to be the number of years since 1990 and measure imports I in billions of dollars. We therefore have two points $(0, 495)$ and $(8, 912)$ for our linear model, as shown in Figure 2.15. The slope of the line through these points is

$$m = \frac{912 - 495}{8 - 0} \approx 52.1;$$

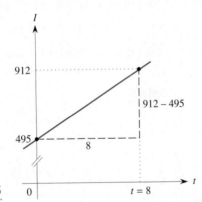

FIGURE 2.15

that is, imports have been growing at a rate of $52.1 billion per year. Using the point–slope form for a line and the point $(0, 495)$ yields

$$I - 495 = 52.1(t - 0),$$

or equivalently the slope–intercept form

$$I = 52.1t + 495.$$

b. The data extend from 1990 when $t = 0$ to 1998 when $t = 8$. A reasonable domain might be from $t = -5$ to $t = 13$, allowing us to predict 5 years before and after the data points.

c. Assuming that this linear trend continues, we predict that the value of foreign goods that will be imported in 2003, when $t = 13$ years after 1990, is

$$I = 52.1(13) + 495 = \$1172.3 \text{ billion.}$$

d. Using this linear model, we have to solve for the value of t when

$$I = 52.1t + 495 = 1000 \text{ billion} \, (=1 \text{ trillion}).$$

If we subtract 495 from both sides of this equation, we obtain

$$52.1t = 505$$

so that

$$t = 505/52.1 \approx 9.7,$$

or sometime late in 1999.

◆

In Example 2 we arbitrarily chose to count years from 1990. Let's see what happens if we choose a different baseline.

EXAMPLE 3

As in Example 2, the United States imported $495 billion worth of goods from abroad in 1990 and $912 billion in goods in 1998. Assuming that the growth in imports followed a linear pattern, find an equation of the linear function that models U.S. imports based on using the independent variable t to represent **(a)** the number of years since 1900 and **(b)** the number of years since year 0. For each model, state an appropriate domain and compare each model to the one constructed in Example 2. **(c)** Use each model to predict the amount of imports in 2003.

Solution In Example 2, we took the independent variable t to be the number of years since 1990, or equivalently used $t = 0$ in 1990, and constructed the linear model

$$I = 52.1t + 495.$$

a. Now suppose that the independent variable t is the number of years since 1900. We therefore have the two points $(90, 495)$ and $(98, 912)$. The slope of the line through these points is

$$m = \frac{912 - 495}{98 - 90} \approx 52.1,$$

which is the same value obtained before. Using the point–slope formula and the point $(90, 495)$ gives the equation of this linear function as

$$I - 495 = 52.1(t - 90)$$

or

$$I = 52.1t - 4194.$$

Although the slope remained the same, the vertical intercept changed dramatically. The reason is that we now think of the line as "starting" in 1900, not 1990, so it has been climbing for 90 years at the rate of $52.1 billion per year. A reasonable domain for this linear model might be from $t = 85$ to $t = 103$.

b. Now suppose that the independent variable t is the number of years since the year 0. Our two points are now $(1990, 495)$ and $(1998, 912)$, and the slope of the line through these points is

$$m = \frac{912 - 495}{1998 - 1990} \approx 52.1,$$

which again is the same value. Using the point–slope formula and the point $(1990, 495)$, the equation of this linear function is

$$I - 495 = 52.1(t - 1990)$$

or

$$I = 52.1t - 103{,}184.$$

There has been a huge change in the vertical intercept because this line has been climbing at a rate of \$52.1 billion per year for almost 2000 years! An appropriate domain for this linear function might be from $t = 1985$ to $t = 2003$.

c. Using any one of the three linear models, the first with $t = 13$, the second with $t = 103$, and the third with $t = 2003$, we obtain the identical prediction for the total value of imports into the United States of about $I = \$1172$ billion in 2003. ◆

So, which of these three models is correct? In one sense, all three are correct because they give the same predictions. In another sense, they are all wrong, because the equation by itself, without reference to what the variable t stands for, is incomplete—if we don't specify the meaning of the variable or its "starting" point, someone using the equation to make predictions may well use a different interpretation. We get very different answers from the first model,

$$I = 52.1t + 495,$$

if we use $t = 13$, $t = 103$, and $t = 2003$. Therefore we should write the three models as

$$I = 52.1t + 495, \qquad \text{where } t \text{ is the number of years since 1990,}$$

or

$$I = 52.1t - 4194, \qquad \text{where } t \text{ is the number of years since 1900,}$$

or

$$I = 52.1t - 103{,}184, \qquad \text{where } t \text{ is the number of years since the year 0.}$$

Capturing a Linear Pattern in Data

In most applications of linear functions in the real world, you will typically have far more than two points; in fact, you will often have a relatively large set of points that fall into a *roughly linear pattern*. In Example 4 we illustrate how to deal with this important situation.

EXAMPLE 4

The following table of values gives some measurements for the rate of chirping (in chirps/sec) of the striped ground cricket as a function of the temperature.

T (°F)	89	72	93	84	81	75	70	82	69	83	80	83	81	84	76
Chirps/sec	20	16	20	18	17	16	15	17	15	16	15	17	16	17	14

Source: Adapted from George W. Pierce, *The Songs of Insects*. Boston: Harvard University Press, 1948.

Even though the measurements presented for the snow tree cricket in Chapter 1 fell exactly onto a straight line, Figure 2.16(a) shows that comparable measurements for the striped ground cricket clearly do not. The difference may be due to errors in measurement; it may be that the striped ground cricket is less sensitive to temperature; or perhaps the snow tree cricket has more mathematical aptitude to get the situation right. Even though the points for the striped ground cricket do not fall precisely on a line, they do fall in a roughly linear pattern. Find an equation that captures this linear pattern.

Solution   Suppose that we take a piece of black thread (or a clear plastic ruler), hold it taut, and move it back and forth over the points in Figure 2.16(a). Each possible orientation for the thread represents a different line. We can then select an orientation that seems, by eye, to give the best match or *fit* to the linear pattern in the data. Usually we want roughly half the points to be above the thread and half below it, so that the line passes "midway" between the points and follows the overall trend. Such a line superimposed over the data points is shown in Figure 2.16(b). (Obviously, different people will come up with slightly different lines.) We now estimate the equation of this line that captures the overall trend of the chirp rate function.

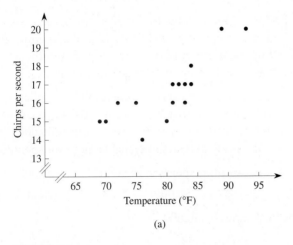
(a)

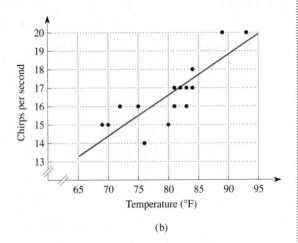
(b)

FIGURE 2.16

In Figure 2.16(b) this line seems to pass through the points (68, 14) and (90, 19). (Note that these points are *on* the line; we chose them for convenience. The points used are not necessarily actual data points. In fact, unless the line drawn happens to pass through a data point, you should not use any of the data points to estimate the slope.) Using these two points, we find that the slope of the line is approximately

$$m = \frac{19 - 14}{90 - 68} = \frac{5}{22} \approx 0.23.$$

This means that the chirp rate increases about 0.23 chirp/sec for each 1°F increase in temperature. Further, because the line apparently passes through the point (68, 14), we conclude that the equation of the line is

$$C - 14 = 0.23(T - 68),$$

or, when simplified,

$$C = 0.23T - 1.64.$$

In applying this "black thread method" to find the equation of the line, you must use two points that are *on the line* you draw. Do not use data points that are not on the line and do not force a line by drawing one that must pass through any of the data points. Note that the result you get is just an *estimate* for the equation of the line that visually best fits the linear trend in the data. In Chapter 3, we introduce methods for finding the equation of the one line that is the *best fit* to a set of data in a certain sense.

Implicit Linear Functions

We now consider a somewhat different type of situation, in which two quantities are related but in such a way that we can't necessarily identify which variable is independent and which is dependent. An ongoing debate at all levels of government concerns the allocation of money among different programs. Because typically only a fixed amount of money is available, the more that is spent on one program, the less there is to spend on other programs. Let's look at a simple case involving just two competing programs, funding road and highway repairs versus funding day-care centers.

EXAMPLE 5

Suppose that we have a total of $1,000,000 available to divide between day-care centers, which cost $200,000/center, and road repaving, which costs $50,000/mile. Find an equation of a linear function relating the number of day-care centers and the number of miles of road to be repaved. What are the domain and range of this function?

Solution Let c be the number of day-care centers and r be the number of miles of road to be repaved. Then the amount of money spent on road repaving is $50,000r$ (it costs $50,000 to repave each mile), and the amount spent on day-care centers is $200,000c$ (it costs $200,000 for each day-care center). Assuming that all the available money is spent, we get

amount spent on centers + amount spent on repaving = 1,000,000

$$200,000c \quad + \quad 50,000r \quad = 1,000,000$$

or equivalently, when we divide both sides by 50,000,

$$r + 4c = 20.$$

This equation is called the *budget constraint*. To graph this equation, we first find the points at which the graph crosses the axes (the intercepts), as shown in Figure 2.17. If $c = 0$, then $r + 4(0) = 20$, so that $r = 20$. At the other extreme, if $r = 0$, we have $0 + 4c = 20$, so that $c = 5$.

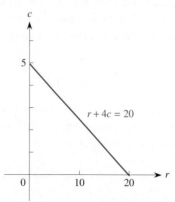

FIGURE 2.17

Because all the money not spent on road work is used for day-care centers, the number of centers funded is a function of the number of miles of roads repaved. That is, c is a function of r, and we can solve the budget constraint equation $r + 4c = 20$ for c. We subtract r from both sides of the equation and get

$$4c = 20 - r.$$

We then divide both sides by 4 and obtain

$$c = f(r) = \frac{1}{4}(20 - r) = 5 - \frac{1}{4}r.$$

Similarly, the number of miles repaved is a function of the number of centers funded, so r is a function of c. We can solve the budget constraint equation for r by subtracting $4c$ from both sides to get

$$r = g(c) = 20 - 4c.$$

To determine the applicable domain and range, we recognize that r makes sense only for values between 0 and 20, whereas c makes sense only for values between 0 and 5. Which of these is the domain and which is the range depends on which variable we think of as the independent variable and which as the dependent variable.

◆

Note that the budget constraint equation

$$r + 4c = 20$$

is an example of a third way of writing the equation of a line, called the *normal form* of a line. In general, an equation of the form $ax + by = c$ is the **normal form** of a line. It is algebraically equivalent to either the point–slope form or the slope–intercept form. For instance, if

$$3x - 5y = 15,$$

then

$$5y = 3x - 15$$

and when we divide both sides by 5, we get

$$y = \left(\frac{1}{5}\right)(3x - 15) = \left(\frac{3}{5}\right)x - 3,$$

a line that has slope $\frac{3}{5}$ and vertical intercept -3.

To graph a line given in normal form, the easiest way is to find and plot both the vertical and the horizontal intercepts and connect them with a straight line. Thus, to find the vertical intercept of $3x - 5y = 15$, we set $x = 0$ and solve $-5y = 15$ to get $y = -3$. To find the horizontal intercept, we set $y = 0$ and solve $3x = 15$ to get $x = 5$. The resulting graph is shown in Figure 2.18.

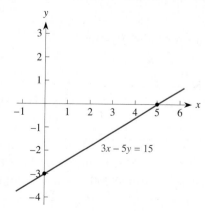

FIGURE 2.18

Incidentally, as a function, an equation such as $ax + by = c$ is called an **implicit function.** It is a function, but neither variable, x nor y, is given explicitly in terms of the other.

Problems

1. Determine which of the functions are linear. For any linear function, find the equation of the line and use it to predict the next entry to extend the table of values.

a.

x	5	6	7	8	9
y	77	71	65	59	53

b.

t	50	60	70	80	90
L(t)	23.2	23.9	24.6	25.2	25.9

c.

t	75	80	85	90	95
Q(t)	125.1	127.5	129.9	132.3	134.7

2. The data in each table of values lie along a line.

   a. For each set of data, carefully plot the points on graph paper, estimate by eye the slope and vertical intercept, and use these values to approximate the equation of the line.

   b. Then find the equation of the line algebraically. How close was your estimate?

   i.

x	1	2	3	4
y	1.81	3.34	4.87	6.40

   ii.

x	1	2	3	4
y	1.08	0.69	0.30	−0.09

3. Find the equation of a linear function that fits this set of values.

x	4	5	6	7
y	1.557	1.614	1.671	1.728

4. **a.** Explain why the equation of the line shown in the accompanying figure is *not* $y = 10x + 200$.

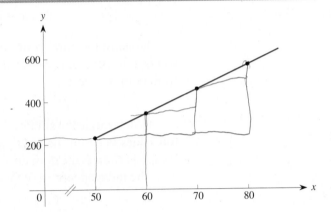

   b. Find the correct equation of the line.

5. In 1980 (when $t = 0$), \$26.5 billion were spent on water pollution prevention and cleanup in the United States. In 1990, \$33.1 billion were spent.

   a. Construct the linear function giving the amount spent on water pollution as a function of time t, where $t = 0$ in 1980.

   b. Use the linear function to estimate the amount spent in 2002.

   c. Repeat parts (a) and (b) if $t = 0$ in 1900.

6. Inspector Clueless, while investigating the murder of Mr. Jones, found the murderer's size $11\frac{1}{2}$ footprint in a flower bed. The inspector mutters something about the killer being "a man who is %#\$#\$&&# tall." If the equation of the best-fit line relating shoe size to height in inches is $S = 0.51H - 25.2$, decipher Clueless's muttering.

7. The table of values shows the total value, in billions of dollars, of electronics and electronic components produced in the United States during the 1990s.

t	1990	1993	1994	1995	1996	1997	1998
E	43.0	52.5	58.2	62.5	68.3	72.4	76.0

Source: *2000 Statistical Abstract of the United States.*

   a. Use graph paper to plot these data and use the black thread method to sketch the best-fit line.

   b. Estimate the slope of this line and tell what it means.

c. What is your best estimate for the equation of this line?

d. Use this line to estimate the total value of electronics and electronic components produced in 2004.

8. The table shows the height H in feet and the number of stories n in some notable buildings.

	H	n
Empire State Building (New York)	1250	102
John Hancock Tower (Boston)	788	61
Sears Tower (Chicago)	1450	110
NationsBank Plaza (Dallas)	921	72
TCBY Tower (Little Rock)	546	40
Peachtree Center (Atlanta)	374	31
Place Ville Marie (Montreal)	620	45

Source: *World Almanac and Book of Facts.*

a. Which variable, H or n, is the independent variable and which is the dependent variable?

b. Plot these points carefully on a sheet of graph paper and use the black thread method to locate and draw the line that seems to best fit the data points.

c. Estimate the equation of this line.

d. What is the meaning of the slope of this line?

e. Use your answer to part (c) to estimate the number of stories in a building 860 feet tall.

f. What is your best estimate of the height of a building that has 96 stories?

9. The table of values at the bottom of the page gives data relating a car's gas mileage to its weight.

a. Plot these points carefully on a sheet of graph paper and use the black thread method to locate and draw the line that seems to best fit the data points.

b. Estimate the equation of this line.

c. Use your answer to part (b) to estimate a car's gas mileage if it weighs 2350 pounds, 3100 pounds, 1950 pounds.

d. What is your best estimate of the weight of a car that gets 32 mpg?

10. A student who works as a waiter in a restaurant records the cost C of meals and the tip T left by couples. His data for one evening are as follows.

C ($)	28.55	31.04	32.76	33.38	36.10	38.54
T ($)	4.25	4.50	5.00	5.00	5.50	6.00

Source: Student project.

a. Plot these points on a sheet of graph paper and draw the best line you can to fit the points. Explain your choices of the independent and the dependent variable.

b. Suppose that the equation for this function is $T = 0.18C - 0.93$. In terms of this mathematical model, what is the increment in the tip for each $1 increment in the cost of the meal?

c. What does the slope of the line in part (b) represent? What significance does the vertical intercept have?

d. Suggest possible values for the domain and range of this function.

11. You have a fixed budget of $30 to spend on nuts and Gummi Bear™ candy for a party. The nuts cost $3 per pound, and the candy costs $2 per pound.

a. Write an equation expressing the relationship between the number of pounds of nuts and of Gummi Bears that you can buy if you spend your budget completely. This equation is your budget constraint.

b. Graph the budget constraint, assuming that you can buy any fractional amount of a pound. Label the intercepts.

c. What are the domain and range for this function?

d. Suppose that your roommate chips in an additional $30 for the party. Graph the new budget constraint on the same set of axes used for the budget constraint graphed in part (b).

Weight (lb)	2100	2200	2400	2500	2800	3000	3200
Mileage (mpg)	37	34	29	27	26	25	23

Source: Student project.

e. Keep the original budget at $30 and suppose that the Gummi Bears go on sale for half the price. Sketch the new budget constraint on the same axes used in part (d).

f. Keep the original budget at $30 and suppose that the price of nuts suddenly doubles because of a frost in the Southeast. Sketch the new budget constraint on the same axes used in part (d).

12. For the implicit equation of a line $4p - 3q = 5$, find the following.

 a. An explicit function that gives p as a function of q.
 b. The slope of the line in part (a).
 c. An explicit function that gives q as a function of p.
 d. The slope of the line in part (c).

13. a. Find the slope and vertical intercept of each line given in normal form.

 $$3y - 2x = 12 \quad \text{and} \quad 4x + 5y = 20.$$

 b. Draw the graphs of the two lines on the same axes.
 c. Find the point of intersection of the two lines
 i. graphically;
 ii. numerically by trial-and-error;
 iii. algebraically.

14. Repeat Problem 13 for the two lines

 $$3y - 2x = 12 \quad \text{and} \quad 4x + 5y = 21.$$

15. Suppose that a function f is increasing and concave up and that $f(60) = 250$ and $f(70) = 300$. Which values are possible and which are impossible? Explain.

 a. $f(65) = 270$ b. $f(65) = 275$
 c. $f(65) = 280$ d. $f(100) = 400$
 e. $f(100) = 450$ f. $f(100) = 500$
 g. $f(40) = 100$ h. $f(40) = 150$
 i. $f(40) = 200$

16. Suppose that a function f is decreasing and concave up, and that $f(10) = 80$ and $f(12) = 70$. Which values are possible and which are impossible? Explain.

 a. $f(11) = 78$ b. $f(11) = 75$
 c. $f(11) = 72$ d. $f(15) = 50$
 e. $f(15) = 55$ f. $f(15) = 60$
 g. $f(5) = 100$ h. $f(5) = 105$
 i. $f(5) = 110$

17. Draw the graph of a function f that is decreasing and concave up. Mark three points on the curve: P near the left, Q near the center, and R near the right. These points determine three line segments: PQ, QR, and PR.

 a. List the three line segments in the order of increasing slopes.
 b. List the three segments in the order of increasing steepness.

18. Repeat Problem 17 if the function is decreasing and concave down.

Exercising Your Algebra Skills

Solve each formula for the indicated variable.

1. $A = bh$, for h
2. $C = 2\pi r$, for r
3. $A = \pi r^2$, for r
4. $K = \frac{1}{2}mv^2$, for m
5. $K = \frac{1}{2}mv^2$, for v
6. $F = \frac{GmM}{d^2}$, for d
7. $T = 2\pi\sqrt{\frac{l}{g}}$, for l
8. $F = \frac{mv^2}{r}$, for r
9. $F = \frac{mv^2}{r}$, for v

Solve each equation in normal form for y in terms of x. Identify the slope in each case.

10. $4x - 5y = 20$
11. $6x + 5y = 30$
12. $5x - 4y = 10$
13. $2x + 7y = 9$

2.4 Exponential Growth Functions

The population of Florida was 12.94 million in 1990 and has been growing as shown in the following table of values. Let's see if we can find a mathematical pattern for the way in which this population is growing. If the population grows linearly, the changes or increases in population from one year to the next, ΔP, would all be the same. Let's check these differences.

Year	Population	ΔP
1990	12.94	
		$0.38 = 13.32 - 12.94$
1991	13.32	
		0.38
1992	13.70	
		0.40
1993	14.10	
		0.41
1994	14.51	
		0.42
1995	14.93	
		0.43
1996	15.36	
		0.45
1997	15.81	

Not only are the successive differences not constant, they are increasing. This makes sense because as the population grows, there are more people to have babies. Consequently, Florida's population has been growing at a faster than linear rate. We therefore need a concave up function to model this population over time.

Instead of taking differences, suppose that we take ratios of successive terms. To do so, we divide the population in any year by the population in the preceding year. This quotient gives

$$\frac{\text{Population in 1991}}{\text{Population in 1990}} = \frac{13.32 \text{ million}}{12.94 \text{ million}} \approx 1.029,$$

$$\frac{\text{Population in 1992}}{\text{Population in 1991}} = \frac{13.70 \text{ million}}{13.32 \text{ million}} \approx 1.029,$$

$$\frac{\text{Population in 1993}}{\text{Population in 1992}} = \frac{14.10 \text{ million}}{13.70 \text{ million}} \approx 1.029,$$

and so on. If you check the population figures for the subsequent years through 1997, you will find that each year the population grew by the same factor of about 1.029.

Because the ratios of successive population values are constant, we have, for any year,

$$\frac{\text{Population next year}}{\text{Population this year}} = 1.029$$

or

$$\text{Population next year} = 1.029 \cdot \text{population this year}.$$

If this trend continues, we can estimate Florida's population in 1998 as

$$\text{Population in 1998} = 1.029 \cdot \text{population in 1997} = 1.029 \cdot 15.81 = 16.27.$$

The fact that Florida's population next year is 1.029 times this year's population is equivalent to saying that

$$\text{Population next year} = 1.029 \cdot \text{population this year}$$
$$= (1 + 0.029) \cdot \text{population this year}$$
$$= \text{population this year} + 0.029 \cdot \text{population this year}.$$

In other words, *each* year between 1990 and 1997, Florida's population grew by about $0.029 = 2.9\%$ from one year to the next. The number 2.9% is called the annual growth rate for the population.

Whenever the successive ratios are constant (here they all are 1.029), the function is an **exponential function.**

We now find an equation for this exponential function $P(t)$, where t is the number of years since 1990. The starting population, when $t = 0$, is $P(0) = 12.94$, which we write as P_0. Then

when $t = 1, P(1) = 1.029 \cdot P_0,$ or 13.32;

when $t = 2, P(2) = 1.029 \cdot P(1) = 1.029 \cdot (1.029 \cdot P_0) = (1.029)^2 P_0,$ or 13.70;

when $t = 3, P(3) = 1.029 \cdot P(2) = 1.029 \cdot (1.029)^2 P_0 = (1.029)^3 P_0,$ or 14.10;

and so on. In general, after t years, the population of Florida is

$$P(t) = P_0 \cdot (1.029)^t = 12.94(1.029)^t.$$

This equation is called an *exponential growth function* with *base* 1.029. The name *exponential* is used because the independent variable (in this case, t) occurs in the exponent. The base (in this case, 1.029) is called the *growth factor*. It gives the population each year as 1.029 times the population in the preceding year. The quantity $0.029 = 2.9\%$ is the associated annual *growth rate*. Note the relationship between the growth factor and the growth rate.

> Growth factor = 1 + growth rate

In this formula, you must write the growth rate as a decimal, not as a percent. For instance, if the growth rate for a process is $4\% = 0.04$ each year, the associated growth factor is $1 + 0.04 = 1.04$.

Assuming that Florida's population continues to grow with the same exponential pattern for the next 80 years, we can graph this population function as shown in Figure 2.19. The function obviously is increasing. Moreover, the graph grows faster and faster as time goes on, so the curve is concave up. This behavior is typical of an exponential growth function. Compare this function's behavior with that of an increasing linear function. Because a linear function grows at the same rate at *every* point, its graph is a line. However, exponential growth functions such as this one are curves that may seem to climb slowly at first but eventually climb extremely rapidly. This type of behavior explains why there is widespread concern about

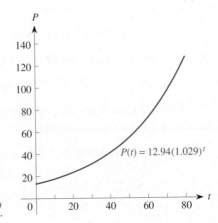

FIGURE 2.19

the exponential growth of the world's population: Eventually there won't be enough land, food, and water to sustain everyone.

The graph shown in Figure 2.19 is only an approximation to the actual graph of Florida's population. We can't have a fraction of a person, so the graph theoretically should be jagged with small steps up or down each time someone is born, dies, or moves into or out of Florida. However, on a scale of millions of people, such changes are insignificant, and our smooth curve actually is a good approximation to the population.

We summarize the formula for an exponential growth function and its parameters as follows.

Formula for an Exponential Growth Function

P is an **exponential growth function** of t with base $c > 1$, if

$$P(t) = P_0 c^t,$$

where P_0 is the initial quantity (when $t = 0$) and c is the **growth factor** by which P changes when t increases by 1 unit.

Because $c > 1$, we can write $c = 1 + a$, where a is the **growth rate** written as a decimal.

For example, if a quantity (e.g., the balance in your bank account) is growing at 5% per year, the growth rate $a = 5\% = 0.05$ and the associated growth factor is $c = 1 + a = 1.05$. The corresponding formula for the balance B in the account as a function of time t is

$$B(t) = B_0 \cdot (1.05)^t,$$

where B_0 represents the initial or starting balance.

The growth factor c in any exponential growth function $y = P_0 c^t$ plays a role similar to the slope in a linear function. The larger the growth factor c, the faster the exponential function grows. Figure 2.20 shows a series of exponential growth curves, with growth rates a of 3%, 4%, and 5% and corresponding growth factors c of 1.03, 1.04, and 1.05. All start with the same initial value at time $t = 0$, but the larger the growth factor, the faster the curve grows. A curve corresponding to a growth factor of 1.033, say, would lie between the curves for the growth factors of 1.03 and 1.04; the curve corresponding to a growth rate of 2.6% would lie below the lowest of the three curves.

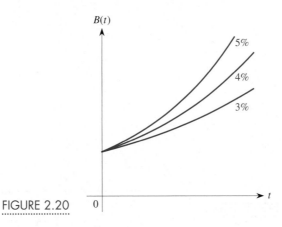

FIGURE 2.20

The coefficient P_0 in an exponential function $y = P_0c^t$ plays a role similar to the vertical intercept for a line. If we set $t = 0$, we get

$$y = P_0c^0 = P_0$$

because any number (other than 0) raised to the zero power is 1. Thus the initial or starting value P_0 of the exponential growth function represents the height at which the exponential function crosses the vertical axis (when $t = 0$). Figure 2.21 shows the graphs of three different exponential growth curves: $y = 10(1.03)^t$, $y = 30(1.03)^t$, and $y = 60(1.03)^t$. All have the same growth factor, 1.03, and so the same growth rate, $0.03 = 3\%$, but all have different initial values. Note how all three curves have similar shape, but each crosses the vertical axis (at $t = 0$) at a different height.

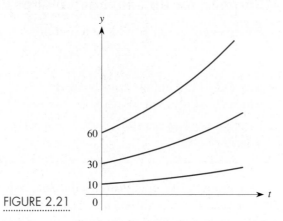

FIGURE 2.21

Comparing Linear and Exponential Growth

The key fact about linear growth is that a linear function grows at a constant rate. That is, every time the independent variable x, say, increases by 1, the linear function grows by the same amount (equal to the slope), as illustrated in Figure 2.22(a). In contrast, the key fact about exponential growth is that an exponential function grows by a fixed percentage. Suppose that the growth factor is 1.20 so that the growth rate is 20%. Then every time the independent variable x increases by 1, the exponential function grows by the same multiple, 1.2, as illustrated in Figure 2.22(b). The corresponding values are then y_0, $1.2\,y_0$, $(1.2)^2\,y_0$, $(1.2)^3\,y_0$, and so on, and these values eventually grow very rapidly as the exponent increases. As a result, eventually any exponential growth function will outstrip any linear function.

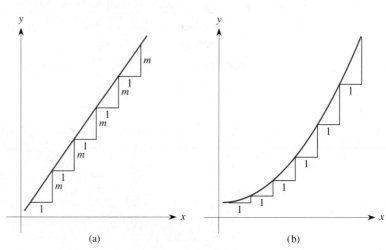

FIGURE 2.22

(a)   (b)

Applications of Exponential Growth

In Examples 1 and 2 we apply the ideas and definitions on exponential growth.

EXAMPLE 1

In 1995, the population of Mexico was 93.7 million and growing at a rate of 2.2% a year.

a. Find a formula for the population of Mexico at any time t.

b. Predict the population of Mexico in 2003.

Solution

a. Because the annual growth rate for Mexico is 2.2% = 0.022 = a, the corresponding growth factor is $c = 1 + 0.022 = 1.022$. Let t be the number of years since 1995 and $M(t)$ be the population of Mexico in millions. Then a formula for the Mexican population at any time t since 1995 is

$$M(t) = 93.7(1.022)^t.$$

b. Assuming that this exponential growth pattern continues until 2003, we have $t = 8$ years after 1995. We predict that the population of Mexico will be

$$M(8) = 93.7(1.022)^8 \approx 111.52 \text{ million people},$$

as shown in Figure 2.23.

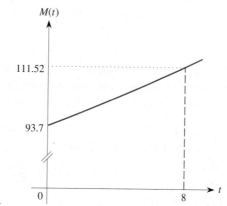

FIGURE 2.23

EXAMPLE 2

During one of New York City's recent financial crises, someone discovered a million dollar loan the city made to the U.S. Government in 1812. At first it appeared that the loan had not been repaid. For a 6% annual compound interest rate, what would this amount have become by the year 2000?

Solution   The 6% growth rate corresponds to a growth factor of $c = 1.06$ so that t years after 1812, the amount would be

$$b(t) = b_0 \cdot (1.06)^t = 1,000,000(1.06)^t.$$

For 2000, $t = 2000 - 1812 = 188$, and the resulting balance would be

$$b(188) = 1,000,000(1.06)^{188}$$
$$\approx \$57,214,047,000.$$

As depicted in Figure 2.24, that would easily have solved the municipal finance problem for many years to come. Unfortunately for New York City, the loan was later found to have been repaid, with interest, in 1815.

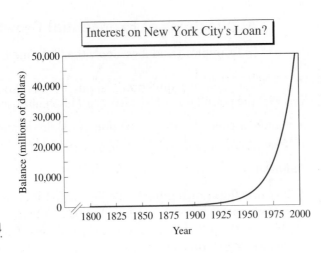

FIGURE 2.24

Think About This   How much interest did New York City receive in 1815 from this loan? ▭

Doubling Time

One of the special characteristics of any exponential growth function is that it has a unique *doubling time*—the time needed for the exponential function to double. We illustrate this concept in Example 3.

◆ **EXAMPLE 3**

Assuming that the exponential model $P(t) = 12.94(1.029)^t$ for Florida's population created at the beginning of this section continues to hold far into the future, estimate the population of Florida in **(a)** 2014, when $t = 24$; **(b)** 2038, when $t = 48$; and **(c)** 2062, when $t = 72$.

Solution   We use exponential growth model

$$P(t) = 12.94(1.029)^t$$

to predict the following values.

a. $P(24) = 12.94(1.029)^{24} \approx 25.70 \approx 2 \cdot 12.94$

b. $P(48) = 12.94(1.029)^{48} \approx 51.04 \approx 4 \cdot 12.94$

c. $P(72) = 12.94(1.029)^{72} \approx 101.35 \approx 8 \cdot 12.94$

Let's look at what these predicted population values indicate. After 24 years, Florida's population has doubled. After roughly another 24 years (i.e., $t = 48$), it has doubled again. After roughly another 24 years (i.e., $t = 72$), the population has doubled yet again. Therefore we say that the doubling time of Florida's population is about 24 years: If you take the population in any given year and compare it to the population 24 years later, you will find that it has doubled.

Think About This   To extrapolate far into the future, we must assume that the population continues to grow exponentially at the same rate of 2.9% per year. The farther we project into the future, the riskier our prediction becomes because other factors can affect the growth rate. What are some? ▭

Every population that grows exponentially has a fixed doubling time that depends only on the growth rate or the growth factor, not on the size of the population. The world's population, with an annual growth rate of about 1.5%, has a doubling time of about 38 years. (We show how to calculate doubling times later.) The current population is about 6 billion, so there will be about 12 billion people in 38 years and roughly 24 billion people in 76 years, all competing for an ever diminishing amount of resources. As another way of looking at it, if you live to be 76, the world's population will quadruple during your lifetime.

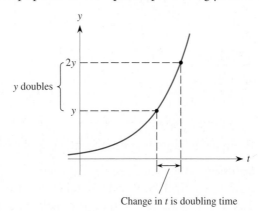

FIGURE 2.25

Change in t is doubling time

The doubling time T for any exponential growth process is the same at any point on the curve; that is, if you pick any point (t, y) on the exponential curve, the value for y will always increase to $2y$ (it has doubled) after T time units. You can visualize what this means by looking at Figure 2.25.

Predicting with Exponential Growth Functions

The purpose of creating an expression for an exponential growth function is to answer predictive questions about the quantity being modeled, as we illustrate in Examples 4 and 5.

EXAMPLE 4

Estimate when the population of Florida will reach 20 million.

Solution   Our formula for the population of Florida is $P(t) = 12.94(1.029)^t$, and we want the value of t when the curve reaches a height of 20. We therefore must solve the equation

$$12.94(1.029)^t = 20$$

for t. We can solve this equation *numerically* by using trial-and-error by substituting different values of t until we get a value for $12.94(1.029)^t$ that is very close to 20. (Your calculator may have a table feature that allows you to generate a table of values and zoom in with smaller and smaller steps to find the value of the independent variable that produces a given value—20 in this case—for the dependent variable.)

A simpler approach is to solve the equation *graphically* by drawing the graph of the function on a function grapher and tracing along the curve to determine when the function reaches a height of 20, as shown in Figure 2.26. If necessary, this approach could also involve zooming in to increase the level of accuracy. Alternatively, we could graph the two functions, $y = 12.94(1.029)^t$ and $y = 20$, and find the point of intersection using a function grapher. Whichever way we proceed, the solution is approximately $t = 15.2$ years from 1990 or early in 2005. (We develop an algebraic approach using logarithms for solving such an equation that yields an exact answer in Section 2.8.)

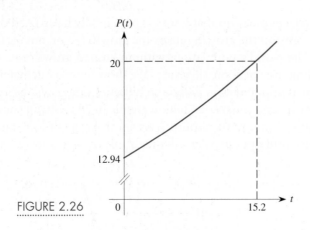

FIGURE 2.26

EXAMPLE 5

Estimate the doubling time for the population of Florida.

Solution   We again use the exponential growth model $P(t) = 12.94(1.029)^t$ and we now must find how long it takes for the population to double, that is, to reach $2 \times 12.94 = 25.88$. We therefore have to solve the equation

$$12.94(1.029)^t = 2 \times 12.94 = 25.88$$

for t. We solve this equation graphically by looking for the intersection of the two curves $y = 12.94(1.029)^t$ and $y = 25.88$, as shown in Figure 2.27. This point is at $t \approx 24.2465$, so the doubling time for Florida's population is about $24\frac{1}{4}$ years.

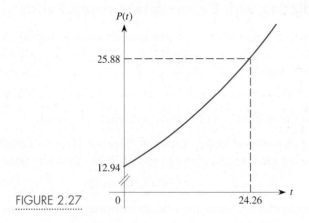

FIGURE 2.27

So far, we have thought of exponential functions as starting at time $t = 0$. In reality, the formula for any exponential function can be interpreted for negative values of the independent variable, as we demonstrate in Example 6.

EXAMPLE 6

Use the model we constructed for the population of Florida to predict what the population was in 1980.

Solution   Our formula for the population of Florida is

$$P(t) = 12.94(1.029)^t,$$

where t is the number of years since 1990. The year 1980, 10 years before 1990, therefore corresponds to $t = -10$. The formula predicts that the population in 1980 was

$$P(-10) = 12.94(1.029)^{-10} \approx 12.94(0.75135) \approx 9.72 \text{ million people,}$$

as illustrated in Figure 2.28. Note that this result is considerably below the 1990 population value of 12.94 million people, as we would expect.

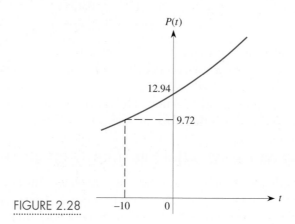

FIGURE 2.28

Just as the domain of a linear function theoretically is the set of all real numbers from $-\infty$ to $+\infty$, the domain of an exponential function is likewise theoretically from $-\infty$ to $+\infty$. Of course, in any real-world setting, there may be practical limitations to the domain. For instance, it wouldn't make sense to use the function to extrapolate the population of Florida 200 years into the past, as Florida became a state only in 1845. Moreover, as we've stated before, extrapolating far into the future or the past is risky because the trend in the data may not hold.

Also, Example 6 indicates that, when $t < 0$, the values for the exponential growth function continue to decrease from right to left. Figure 2.29 shows the typical graph of an exponential growth function $y = kc^x$, with $k > 0$. Note how it grows in the expected way toward the right and decays to 0 toward the left. The reason is that, as we move farther to the left of the vertical axis, the values of x become ever more negative. Suppose that we write $x = -z$. Recall one of the basic properties of exponents:

$$b^{-z} = \frac{1}{b^z}.$$

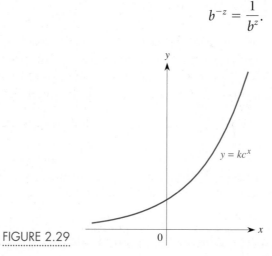

FIGURE 2.29

Consequently,

$$1.029^{-10} = \frac{1}{1.029^{10}} \approx 0.75135.$$

As the exponent becomes ever more negative,

$$1.029^{-20} \approx 0.56454,$$
$$1.029^{-30} \approx 0.42417,$$
$$1.029^{-100} \approx 0.05734,$$

the values become ever smaller and eventually approach 0. We say that the curve approaches the negative x-axis *asymptotically* because it never reaches 0 in any finite time interval. We call the horizontal axis a *horizontal asymptote* for the graph of the exponential decay function. The range of any exponential growth function $y = kc^x$, with $k > 0$, is therefore all positive values for y.

Finding an Exponential Function Through Two Points

We know that two points determine a line (because one and only one line can pass through the two points). Similarly, two points also determine an exponential function in the sense that one and only one exponential function passes through the two points, provided that the y-values for the points are either both positive or both negative. Suppose that we have any two points (x_1, y_1) and (x_2, y_2), where $x_1 < x_2$ and $0 < y_1 < y_2$, so the second point is to the right of and above the first point and both points are above the x-axis, as shown in Figure 2.30. One and only one exponential growth curve passes through the two points.

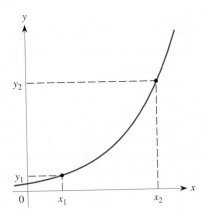

FIGURE 2.30

Think About This   By drawing several sketches, convince yourself why it is not possible to draw an exponential growth curve through two points when one is above the x-axis and the other is below the x-axis. Also, if the two points are both below the x-axis, what should you expect about the sign of the coefficient k in $y = kc^x$? ▭

We now determine a formula for the exponential function that passes through two points. Doing so also gives us a way to find the growth rate for any exponential process. For the equation of an exponential function $y = kc^x$, values for the two parameters k and c must be determined, which is why we use two points. We demonstrate how to do so in Example 7.

EXAMPLE 7

The number of cell phones in use worldwide grew from 11 million in 1990 to 319 million in 1998.

a. Assuming that the growth pattern was exponential, find the annual growth rate for the number of cell phones in use and the equation of the exponential function that models the number of cell phones in use.

b. Predict the number of cell phones in use in 2003.

Solution

a. Let t represent the number of years since 1990 and P the number of cell phones (in millions) in use. We then have the two points $(0, 11)$ and $(8, 319)$. The exponential growth function has the form

$$P(t) = P_0 c^t,$$

where the constants P_0 and c must be determined. Substituting the coordinates of the point $(0, 11)$ into the function gives

$$P(0) = P_0 c^0 = P_0 = 11,$$

because $c^0 = 1$. Thus the exponential function becomes $P(t) = 11c^t$. Using the point $(8, 319)$ gives

$$P(8) = 11c^8 = 319.$$

Solving for c^8 gives

$$c^8 = \frac{319}{11} = 29.$$

Just as we solve $x^2 = 10$ for x by taking the square root of 10 or solve $x^3 = 10$ for x by taking the cube root of 10, we solve $c^8 = 29$ for c by taking the eighth root of 29. (We discuss the details more formally in Section 2.7.) Thus

$$c = \sqrt[8]{29} \approx 1.5234.$$

(Verify that $\sqrt[8]{29} \approx 1.5234$ by taking the eighth power of 1.5234.) For the growth factor of 1.5234, the annual growth rate in the number of cell phones in use is $0.5234 = 52.34\%$. Moreover, the exponential function that models the growth in the number of cell phones is

$$P(t) = 11(1.5234)^t,$$

where t is the number of years since 1990.

b. Because 2003 is 13 years after 1990, we set $t = 13$ as shown in Figure 2.31. We then use this exponential model to predict that the number of cell phones in use in 2003 is

$$P(13) = 11(1.5234)^{13} \approx 2618.0 \text{ million},$$

or about 2.618 billion.

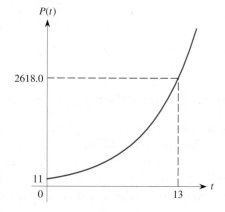

FIGURE 2.31

By letting t represent the number of years since 1990, we simplified the work in Example 7 to give the vertical intercept $(0, 11)$ as one of the points. If we can't do so, things become more complicated, as shown in Example 8.

EXAMPLE 8

Find the equation of the exponential function that passes through the points $(1, 6)$ and $(2, 9)$.

Solution   The desired exponential function has the form $f(x) = kc^x$, where we must find the correct values for the parameters k and c. Using the point $(1, 6)$, we have

$$f(1) = kc^1 = kc = 6.$$

Using the point $(2, 9)$, we have

$$f(2) = kc^2 = 9.$$

From the first of these two equations, we solve for k and get $k = 6/c$. We substitute this term into the second equation to get

$$kc^2 = \left(\frac{6}{c}\right)c^2 = 6c = 9,$$

and so

$$c = \frac{9}{6} = 1.5.$$

Therefore

$$k = \frac{6}{c} = \frac{6}{1.5} = 4,$$

and the desired exponential function is $f(x) = 4(1.5)^x$, as shown in Figure 2.32.

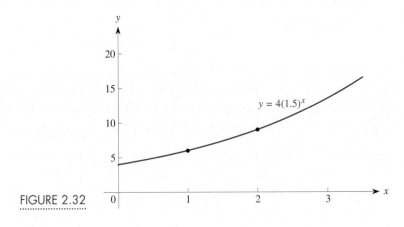

FIGURE 2.32

Determining Whether a Set of Data Is Exponential

Recall the simple criterion that determines whether a set of data follow a linear pattern: The successive differences in the dependent variable must be constant when there is a constant difference between values of the independent variable. Similarly, we can determine whether a table of data values (t, y) follows an exponential pattern by looking at the successive ratios of the y values.

> If the ratios of the successive values of the dependent variable are constant for equally spaced t values, the y values follow an exponential pattern: $y = kc^t$.

The common ratio is precisely the growth factor for the exponential growth process if the t values increase by 1 unit. For instance, with Florida's population values from one year to the next, we found that the common ratio was 1.029, which is the growth factor, and that the associated growth rate is 0.029, or 2.9% per year.

◄ **EXAMPLE 9**

One of the following functions is exponential and the other isn't. Determine which is the exponential function. The values are rounded to four decimal places.

x	y		x	y
0	20.0		0	20.0
1	21.0		1	21.0
2	22.10		2	22.05
3	23.2775		3	23.1525
4	24.6425		4	24.3101
5	26.2650		5	25.5256

Solution   We apply the criterion for an exponential pattern and examine the ratios of successive terms for each function. For the first function the ratios are

$$\frac{21.0}{20.0} = 1.05, \quad \frac{22.10}{21.0} = 1.0524, \quad \frac{23.2775}{22.10} = 1.0533,$$

$$\frac{24.6425}{23.2775} = 1.0586, \quad \text{and} \quad \frac{26.2650}{24.6425} = 1.0658.$$

The successive ratios are not constant, so this function cannot be exponential.
     For the second function the ratios are

$$\frac{21.0}{20.0} = 1.05, \quad \frac{22.05}{21.0} = 1.05, \quad \frac{23.1525}{22.05} = 1.05,$$

$$\frac{24.3101}{23.1525} = 1.04999, \quad \text{and} \quad \frac{26.5256}{24.3101} = 1.04999.$$

These ratios are essentially constant (the last two vary slightly because the entries listed in the table were rounded), so we conclude that this function is indeed exponential.   ◆

Rules for Exponents

Because exponential functions involve working with exponents, all the usual algebraic rules for manipulating exponents apply. As a reminder, we list some of the fundamental definitions and algebraic rules for exponents.

Definitions and Rules for Exponents

Property	Example
1. $a^x \cdot a^y = a^{x+y}$	$10^3 \cdot 10^2 = (10 \cdot 10 \cdot 10) \cdot (10 \cdot 10) = 10^5 = 10^{3+2}$
2. $\dfrac{a^x}{a^y} = a^{x-y}, \quad a \neq 0$	$\dfrac{10^5}{10^2} = \dfrac{10 \cdot 10 \cdot 10 \cdot 10 \cdot 10}{10 \cdot 10} = 10^3 = 10^{5-2}$
3. $(a^x)^y = a^{xy}$	$(10^3)^2 = 10^3 \cdot 10^3 = 10^6 = 10^{3 \cdot 2}$
4. $a^0 = 1$	$10^0 = 1$
5. $a^{-1} = \dfrac{1}{a}, \quad a \neq 0$	$10^{-1} = \dfrac{1}{10}$
6. $a^{-n} = \dfrac{1}{a^n}, \quad a \neq 0$	$10^{-3} = \dfrac{1}{10^3} = \dfrac{1}{1000}$
7. $a^{1/n} = \sqrt[n]{a}, \quad a \geq 0$	$10^{1/2} = \sqrt{10}, \qquad 10^{1/3} = \sqrt[3]{10}$

Problems

1. The accompanying graph shows population growth curves for four different nations. Which nation

   a. has the greatest growth rate?
   b. has the smallest growth rate?
   c. has the largest initial population?
   d. has the smallest initial population?
   e. Which nations have the same growth rate?

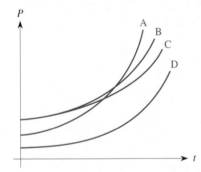

2. Determine which of the functions are exponential. For any exponential function, find the equation of the function and use it to predict the next entry to extend the table of values.

   a.

x	0	1	2	3
y	1000	1200	1440	1728

   b.

t	0	1	2	3
L(t)	300	308	320.2	335.5

   c.

t	0	10	20	30
Q(t)	200	208	216.32	224.97

3. Anne opens a bank account with $1200 at 4% annual interest. Bill opens an account with $1000 at 4.5% annual interest. Christine opens an account with $1500 at 3.8% annual interest. Doug opens an account with $1200 at 4.5% annual interest. Elka opens an account with $1300 at 4.25% annual interest. Sketch a graph showing the balances in the five accounts over time on the same set of axes. Be sure to label which account belongs to which person.

4. Use the exponential growth function $f(t) = 125(1.04)^t$ to make a prediction for 2000 if (a) t is the number of years since 1980, (b) t is the number of years since 1900, (c) t is the number of years since the year 0.

5. In 1990, the United States imported $495 billion worth of goods. In 1998, the United States imported $912 billion worth of goods. Assuming that the growth in imports has been following an exponential growth pattern, find an equation of the exponential function that models U.S. imports when

   a. the independent variable t represents the number of years since 1990.
   b. the independent variable t represents the number years since 1900.
   c. the independent variable t represents the number of years since the year 0.

d. For the three functions you created in parts (a)–(c), which parameters changed and which remained the same? Explain why the changes occurred. Explain why the parameters that stayed the same didn't change.

e. Use each model from (a)–(c) to predict the amount of imports in 2005.

Source: *2000 Statistical Abstract of the United States.*

6. Match each formula with the corresponding table of values.

a. $y = a(1.1)^s$
b. $y = b(1.05)^s$
c. $y = c(1.03)^s$

i.

s	2	3	4	5	6
$f(s)$	1.06	1.09	1.13	1.16	1.19

ii.

s	1	2	3	4	5
$g(s)$	2.20	2.42	2.66	2.93	3.22

iii.

s	3	4	5	6	7
$h(s)$	3.47	3.65	3.83	4.02	4.22

7. In 1980, a total of $119 trillion was spent on food and drinks in the United States. In 1994, the total spent was $274 trillion.

a. Find the equation of the exponential function that can be used to model the total spent on food and drinks in the United States as a function of the number of years since 1980.

b. Use your model to predict the amount spent in 1990.

c. What is your prediction for the total sales of food and drink in 2004?

d. Estimate when the total sales will reach $500 trillion if this exponential trend continues.

8. The 1990 population of Arizona was 3.7 million and growing at an annual rate of 1.7%. $P(t) = 3.7(1.7)^t$

a. Find an expression for the population at any time t.

b. What will be the population in 2005?

c. Estimate the doubling time for this population.

9. The 1995 population of Venezuela was 21.8 million and growing at an annual rate of 2.6%.

a. Find an expression for the population at any time t.

b. What will be the population be in 2005?

c. Estimate the doubling time for this population.

10. The 1995 population of France was 58.1 million and growing at an annual rate of 0.3%.

a. Find an expression for the population at any time t.

b. What will be the population in 2010?

c. Estimate the doubling time for this population.

11. In 1990, 1.36 billion metric tons of carbon dioxide were emitted into the atmosphere in the United States. In 1998, 1.595 billion metric tons were emitted.

a. Construct the exponential function giving the amount of carbon dioxide emitted into the atmosphere as a function of the number of years since 1990.

b. Use the exponential function to estimate the amount emitted in 2004.

12. The population graph shown in the accompanying figure is growing exponentially.

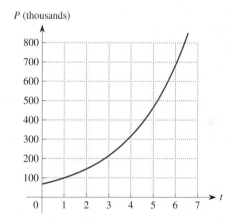

a. Use the graph to estimate the doubling time of the population.

b. Verify graphically that the doubling time does not depend on where you start on the graph.

13. The world's population passed 6 billion in late 1999 and is increasing at a rate of about 1.5% per year.

a. Find the world's population 15 years later if this trend continues.

b. Estimate how long it will take the world's population to double.

14. Find a formula for the balance in a bank account in which $100 was deposited at 6% annual interest compounded for 10 years.

15. Find the balance after 1 year if $100 is deposited at an annual rate of 6% compounded quarterly instead of yearly. What is the balance after 10 years? (*Hint:* What is the interest rate for each 3-month period?)

16. In 1998, the population of the United States was about 268.2 million with an annual growth rate of 0.7%. At the same time, the population of Mexico was about 100.1 million with an annual growth rate of 2.2%. If these growth rates continue, use either graphical or numerical methods to estimate when the population of Mexico will overtake that of the United States.

17. According to an article in the *New York Times* on May 27, 1990, a wealthy Pennsylvania merchant named Jacob DeHaven loaned $450,000 to the Continental Congress in 1776 to rescue the troops at Valley Forge. The descendants of Mr. DeHaven sued the U.S. government for what they believed they were owed. The interest rate in effect in 1776 was 6% per year. How much did the family stand to collect in 1991, assuming that interest is compounded annually?

18. The lily pads in a pond grow in such a way as to double the area of the pond that they cover daily.

    a. If the lily pads exactly cover the entire pond on the 25th day, how much of the pond do they cover on the 24th day?

    b. Write an exponential function that models the fraction of the pond covered on any particular day.

    c. If the area of the pond is 40,000 sq ft, find the area covered by the lily pads on the initial day.

    d. What area of the pond is covered by the lily pads at the end of 1 week?

19. Let $f(x)$ be an exponential function of x. If $f(7) = 25.6$ and $f(8) = 28.8$, find

    a. the growth factor;

    b. the growth rate;

    c. the value of the function when $x = 10$;

    d. a formula for $f(x)$.

20. The Dow-Jones average of 30 industrial stocks is the most famous measure of performance of the New York Stock Exchange. At the beginning of 1995 the Dow was 3834, and at the beginning of 2000 it was 11,358. Assuming (incorrectly) that the Dow increased continuously over these 5 years and that the pattern is exponential, find the exponential function that models the behavior of the Dow between 1995 and 2000. What would you predict as the value for the Dow at the beginning of 2004?

21. Repeat Problem 20, using the facts that the Dow was 964 at the beginning of 1981 and was 11,358 at the beginning of 2000.

22. a. Suppose that you're an aggressive stockbroker who is trying to convince a little old lady to invest her life savings with you. What argument would you make based on your work on either Problem 20 or 21 to convince her.

    b. Now suppose that the little old lady is your grandmother. What argument would you make based on your work on either Problem 20 or 21 to convince her to be more conservative.

23. An exponential function f is such that $f(0) = 512$ and $f(4) = 1250$. Which of the values are possible and which are impossible?

    a. $f(2) = 800$ b. $f(2) = 881$ c. $f(2) = 981$

24. The net income of the Acme Company was $240 million in 1990 and has been increasing at an annual rate of 10% per year since. Over the same period, the net income of its chief competitor, the Finest Corporation, has been growing 8% annually from an income of $300 million in 1990. Which was the richer company in 2000? Does Acme ever surpass Finest? If so, estimate when.

25. (Extension of Problem 24) Suppose that Finest grew by a fixed amount of $25 million per year since 1990 while Acme grew exponentially at an annual rate of 10%. By using trial and error, estimate when Acme surpassed Finest.

26. When Steven was 5 years old, his grandmother decided to set up a trust account to pay for his college education. She wanted the account to grow to $80,000 by Steven's 18th birthday. If she was able to invest her money at 6% per year, how much did she have to put into this trust account? (*Note:* This amount is known as the *present value* of the investment. The $80,000 is known as the *future value.*)

27. In Example 7 the number of cell phones in use increased 29-fold, from 11 million in 1990 to 319 million in 1998. This is equivalent to a 3000% increase over that 8-year period. Explain what's wrong with the reasoning that says: If the number of cell phones increased by 3000% over the 8 years, the annual growth rate is $\frac{1}{8}$ of 3000% or 375%.

28. Show that $x^{5/3} \neq \dfrac{x^5}{x^3}$

a. numerically, by finding at least one value of x for which the two expressions are different;

b. graphically, by comparing the graphs of the two functions $y = x^{5/3}$ and $y = x^5/x^3$.

Exercising Your Algebra Skills

Simplify the following.

1. $x^5 \cdot x^3$

2. $x^4 \cdot x^2$

3. $a^8 \cdot a^4$

4. $\dfrac{a^{15}}{a^6}$

5. $x^{-5} \cdot x^3$

6. $a^5 \cdot a^{-3}$

7. $\dfrac{r^8}{r^{-4}}$

8. $\dfrac{b^{15}}{b^{-6}}$

9. $\dfrac{w^{-4}}{w^{-7}}$

10. $\dfrac{w^{-7}}{w^{-4}}$

11. $x^{-1/2}x^{3/4}$

12. $y^{2/3}y^{4/3}$

13. $z^{2/3}z^{-5/3}$

14. $(x^5)^3$

15. $(x^3)^5$

16. $(a^8)^{-4}$

Perform the following operations:

17. $(a^3b^5)^4$

18. $(a^3 + b)^2$

19. $(a^3 - b)^2$

2.5 ······ Exponential Decay Functions

Prozac is one of the most widely used drugs to treat extreme depression. Once a medication such as Prozac has been absorbed into the bloodstream, it eventually is eliminated from the body by the kidneys, which purify the blood by filtering out foreign chemicals. For now, let's assume that a person takes a single dose of Prozac and that it has been completely absorbed into the blood. It is reasonable to assume that, during any fixed time period, a fixed percentage of any medication, including Prozac, is removed from the bloodstream as the kidneys process the blood. In particular, the kidneys eliminate approximately one-fourth of the Prozac in the bloodstream during any 24-hour period, so that 75% of the drug remains. (Note that this rate is specific to Prozac and that other medications are washed out of the body at different rates.)

Suppose that the original dosage of Prozac is 80 mg (milligrams). We want to develop a formula for the amount $D(t)$ present at any time t. Clearly, it must be a decreasing function because the level of the drug in the bloodstream is decaying over time.

We start with $D(0) = 80$ mg. After the first 24 hours, one quarter of 80 mg, or 20 mg, of the Prozac is eliminated, leaving three quarters of the 80 mg, or 60 mg, of the Prozac in the bloodstream. After one 24-hour period, when $t = 1$, the amount of Prozac in the system is

$$D(1) = 0.75(80) = 60 \text{ mg.}$$

After a second 24-hour period, the kidneys remove 25% of the remaining 60 mg of Prozac, so 15 mg are eliminated, leaving 75% of the remaining 60 mg of Prozac. Thus, when $t = 2$,

$$D(2) = 0.75(60) = 0.75(0.75)(80) = (0.75)^2(80) \text{ mg.}$$

After the third 24-hour period, 25% of the remaining Prozac is eliminated, leaving

$$D(3) = 0.75D(2) = 0.75(0.75)^2(80) = (0.75)^3(80).$$

Similarly,

$$D(4) = 0.75D(3) = (0.75)^4(80)$$

and

$$D(5) = 0.75D(4) = (0.75)^5(80).$$

In general, after t days the amount of Prozac in the bloodstream is given by

$$D(t) = 80(0.75)^t.$$

This function has the same form, $y = kc^t$, as the exponential growth functions presented in Section 2.4 except that the base c is 0.75, which is less than 1. It is an example of an *exponential decay function,* and its graph is shown in Figure 2.33. Note that the behavior is that of a decreasing, concave up function. Each step down is smaller than the previous one. This result makes sense because, as the amount of Prozac remaining in the bloodstream gets smaller, there is less of the drug left to eliminate, and the amount of decrease in drug strength diminishes every successive day.

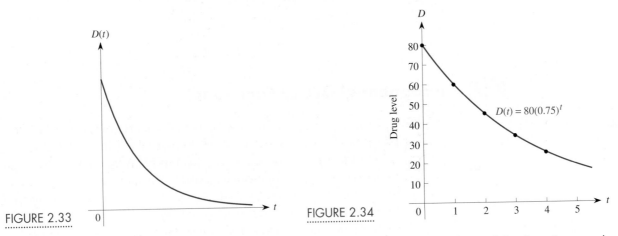

FIGURE 2.33

FIGURE 2.34

You can see this numerically by calculating the values of the function previously given:

$$D(0) = 80, \quad D(1) = 60, \quad D(2) = 45, \quad D(3) = 33.75, \quad D(4) = 25.3125, \dots,$$

which is a decreasing, concave up pattern. If you continue these calculations, you will find that the values eventually approach 0 *asymptotically;* that is, the drug level never reaches 0 in any finite time interval, as illustrated in Figure 2.34. Thus, the horizontal axis is a horizontal asymptote for the graph of the exponential decay function.

In general, the graph of any exponential decay function, $y = kc^t$, with $0 < c < 1$, is a decreasing, concave up curve that approaches 0 as t gets larger and larger. In comparison, the graph of any exponential growth function, $y = kc^t$, with $c > 1$, is an increasing, concave up curve. Because the base c for an exponential decay function is between 0 and 1, we call it the *decay factor.*

Often, we are told that a process is decaying at a given rate—say, 12% per year. The 12% = 0.12 is known as the *decay rate* and the associated decay factor c is

$$\text{Decay factor} = 1 - \text{decay rate,}$$

where the decay rate must be written as a decimal. Thus

$$c = 1 - 0.12 = 0.88,$$

because 88% (or 0.88) of the original amount is left. By comparison, for exponential growth, recall that

$$\text{Growth factor} = 1 + \text{growth rate.}$$

Note that whether we have an exponential decay function such as $y = 80(0.75)^t$ for the level of Prozac or an exponential growth function such as $y = 12.94(1.029)^t$ for the population of Florida, it is still an exponential function and the same techniques that we introduced in Section 2.4 apply. The only difference is that, for an exponential growth function, $c > 1$, whereas for an exponential decay function, $0 < c < 1$.

EXAMPLE 1

Find the amount of Prozac in the bloodstream after 1 week.

Solution   We use the formula for the exponential decay function,

$$D(t) = 80(0.75)^t,$$

we previously constructed. After 1 week, $t = 7$ days, so the level of Prozac will be

$$D(7) = 80(0.75)^7 \approx 10.679,$$

or about $10\frac{2}{3}$ mg.

EXAMPLE 2

Estimate how long it takes until the level of Prozac in the bloodstream drops to 2 mg.

Solution   Using the formula for the level of Prozac, we have to find t so that

$$D(t) = 80(0.75)^t = 2.$$

Using either numerical or graphical methods, we find that $t \approx 12.8$ days, as shown in Figure 2.35.

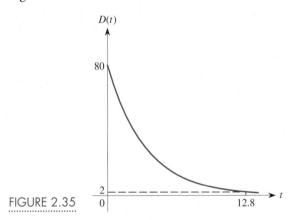

FIGURE 2.35

We summarize the formula for an exponential decay function and its parameters as follows.

Formula for an Exponential Decay Function

P is an **exponential decay function** of t with base c, $0 < c < 1$, if

$$P(t) = P_0 c^t,$$

where P_0 is the initial quantity (when $t = 0$) and c is the **decay factor** by which P changes when t increases by 1 unit. Because $0 < c < 1$, we write $c = 1 - a$, where a is the **decay rate,** written as a decimal.

The larger the decay rate a, and hence the smaller the decay factor c, the faster the exponential decay function approaches 0, as illustrated in Figure 2.36.

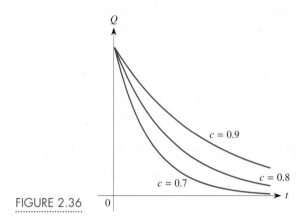

FIGURE 2.36

For example, if a quantity is decreasing at the rate of 12% per hour (e.g., the effectiveness of a medication in the body), the decay rate is $a = 0.12$ and the decay factor is $c = 1 - a = 0.88$. This reflects the fact that, if 12% of the quantity is removed each hour, then 88% of the quantity remains at the end of the hour. The corresponding formula for the exponential decay function that models the quantity Q is

$$Q(t) = Q_0 \cdot (0.88)^t,$$

where Q_0 is the initial amount of the quantity at time $t = 0$.

Half-life

Just as the doubling time for an exponential growth process is the time needed for the quantity to double, the *half-life* for an exponential decay process is the time T needed for the quantity to be reduced by half. You can visualize what this means by looking at Figure 2.37.

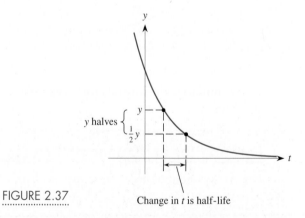

FIGURE 2.37

Change in t is half-life

Note that the half-life T for any specific process is the same at any quantity level; no matter which point (t, y) you select, the quantity will decrease to $\frac{1}{2}y$ after T time units.

EXAMPLE 3

Estimate the half-life of Prozac in the bloodstream following an 80 mg dose.

Solution   The exponential decay function that models the amount of Prozac in the bloodstream is

$$D(t) = 80(0.75)^t.$$

We want to find the time t needed for this level to drop to $\left(\frac{1}{2}\right)80 = 40$ mg, so we must solve the equation

$$80(0.75)^t = 40.$$

Using either numerical or graphical methods, as shown in Figure 2.38, we get $t \approx 2.4$ days. Therefore, no matter what level of Prozac is in the blood at any specific time, the level will be down by half about 2.4 days, or 58 hours, later.

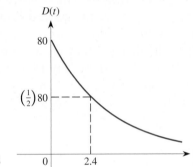

FIGURE 2.38

Radioactive Decay

One of the characteristics of any radioactive substance, such as radium or uranium, is that it transforms, or decays, to some other element, often lead, as time progresses. This decay is accompanied by the release of energy, called radioactivity, which can be detected and measured. More specifically, the rate at which an element decays is distinctive for that element. That is, during any fixed length of time, the same percentage of the mass of a radioactive element will decay. For instance, over the course of any 100-year period, approximately 4.3% of any radium present will decay to lead, leaving 95.7% of the radium at the end of 100 years, as illustrated in Figure 2.39. Thus, if someone had put aside $R_0 = 100$ grams of radium in the

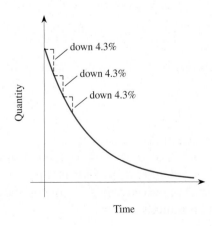

FIGURE 2.39

year 1900, we would expect to find only $R(1) = 95.7$ grams by the year 2000. By the end of a second century, the amount of radium left would be

$$R(2) = 0.957\,R(1) = (0.957)^2\,R_0$$

and by the end of a third century it would be

$$R(3) = 0.957\,R(2) = (0.957)^3\,R_0.$$

In general, the amount of radium present after t centuries is modeled by the exponential decay function

$$R(t) = (0.957)^t\,R_0$$

for any t.

Alternatively, because $4.3\% = 0.043$ is the decay rate for this exponential decay process, the decay factor is $1 - 0.043 = 0.957$. Thus, if the initial amount of radium is R_0, we can use the general formula for an exponential decay function to get $R(t) = (0.957)^t\,R_0$.

Figure 2.40 shows a graph of the amount of radium as a decaying exponential function of time. The amount of radium begins decreasing relatively rapidly, then decreases more slowly, and eventually approaches the time axis as a horizontal asymptote.

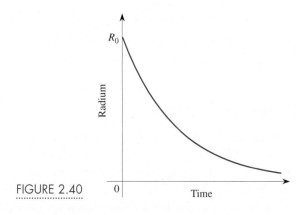

FIGURE 2.40

EXAMPLE 4

Estimate the half-life of radium.

Solution We want to determine the value of t for which

$$R(t) = (0.957)^t\,R_0 = \frac{1}{2}R_0.$$

We first divide both sides of this equation by R_0 to obtain

$$(0.957)^t = \frac{1}{2}.$$

If we now use either numerical or graphical methods, as shown in Figure 2.41, we find that $t \approx 15.77$ centuries. That is, the half-life for radium is approximately 1577 years. (The actual value for its half-life is closer to 1590 years; our calculations were based on the fact that *approximately* 4.3% of the radium decays to lead each century, and this rounding produced an error.)

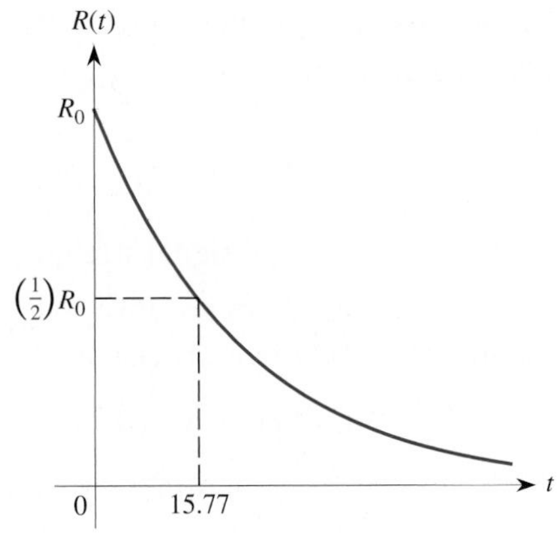

FIGURE 2.41

Think About This   What is the actual percentage of radium that decays into lead each year, based on its half-life of 1590 years?

In all of the examples so far, time is the independent variable. Example 5 illustrates a situation in which the independent variable in an exponential function may represent some other quantity.

◆ **EXAMPLE 5**

The strength of any signal in a fiber-optic cable, such as the type used for telephone and other communication lines, diminishes 15% every 10 miles.

a. Find an expression for the strength of a signal remaining after a given number of 10-mile lengths.

b. How much of the signal is left after 100 miles?

c. How far does a signal go until its strength is down to 1% of the original level?

Solution

a. If the signal diminishes by 15% every 10 miles of cable, after each 10-mile stretch, only 85% of the original signal strength remains. Let S_0 be the initial strength of some signal and let $S(n)$ be the strength of the signal remaining after n 10-mile lengths. Therefore, after the first 10-mile length of cable ($n = 1$), 85% of S_0 is left, so

$$S(1) = 0.85 S_0.$$

Similarly, after the second 10-mile length ($n = 2$), 85% of $S(1)$, the signal strength remaining after the first 10-mile length, is left. That is,

$$S(2) = 0.85 S(1).$$

Continuing this pattern, we get

$$S(0) = S_0,$$
$$S(1) = (0.85) S_0,$$
$$S(2) = (0.85) S(1) = (0.85)(0.85) S_0 = (0.85)^2 S_0,$$
$$S(3) = (0.85) S(2) = (0.85)(0.85)^2 S_0 = (0.85)^3 S_0,$$

and so on. After n 10-mile lengths of a cable,

$$S(n) = S_0 \cdot (0.85)^n,$$

which is an exponential decay function with decay factor $c = 0.85$.

b. After 100 miles, or $n = 10$ ten-mile lengths, the fraction of the original signal strength remaining is

$$S(10) = S_0 \cdot (0.85)^{10} = 0.1969\, S_0,$$

so just under 20% of the original signal strength is left.

c. To find out how far the cable can go until only 1% of the signal strength is left, we must find the value of n for which the strength remaining is $0.01\, S_0$, or

$$S(n) = S_0 \cdot (0.85)^n = 0.01 S_0.$$

If we divide both sides of this equation by the initial signal strength S_0, we get

$$(0.85)^n = 0.01.$$

Solving this equation either numerically or graphically, as shown in Figure 2.42, we find that $n \approx 28$. Therefore the signal deteriorates by 99% after about 28 ten-mile lengths, or about 280 miles.

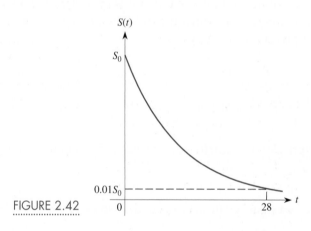

FIGURE 2.42

Think About This   In practice, this model suggests that fiber-optic signals need to be boosted if they are to go any great distance. For instance, if a booster station can clearly detect a signal at 1% of its original strength, such stations would have to be located every 280 miles. Suppose that the equipment used can clearly detect a signal at 0.1% of its original level. How far apart would the booster stations have to be?

Determining Whether a Set of Data Is Exponential Growth or Exponential Decay

In Section 2.4, we presented a simple criterion for recognizing that a set of data follows an exponential growth pattern: The successive ratios of the values of the dependent variable y are constant for equally spaced t values. The same criterion applies if the values of y are decreasing in an exponential decay pattern. In this case, the common ratio is precisely the decay factor for the process if the values of t increase by 1 unit. In general, the ratio criterion works whether the data values are increasing or decreasing. A common ratio greater than 1 gives the growth factor for an exponential growth process; a common ratio less than 1 gives the decay factor for an exponential decay process.

Finally, we consider some parallels between the family of linear functions and the family of exponential functions. The general formula for a linear function is $y = mx + b$, and the general formula for an exponential function is $y = kc^x$, so both are two-parameter families. For linear functions the more important parameter usually is the slope, and its sign determines whether the function increases or decreases. For exponential functions the more important parameter is the growth or decay factor c, and whether its value is greater than 1 or less than 1 determines whether the exponential function increases or decreases.

The following problems include both exponential growth and exponential decay situations because you need to learn to distinguish between them.

Problems

1. Determine which of the six functions could be exponential functions of the form $f(x) = kc^x$ and which cannot be exponential. Explain your reasoning.

(a)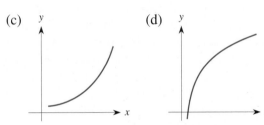

(b)

(c)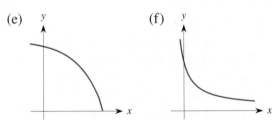

(d)

(e)

(f)

2. Determine which of the functions are exponential. For any exponential function, find the equation of the function and use it to predict the next entry to extend the table of values.

a.

x	0	1	2	3
y	2000	1800	1620	1458

b.

t	0	1	2	3
L(t)	300	240	190	150

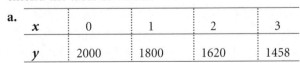

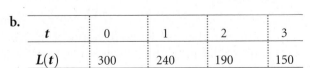

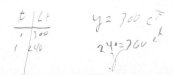

c.

t	0	10	20	30
Q(t)	400	288	207.36	149.30

3. Which of the following pairs of points can determine an exponential function of the form $y = Ac^x$ and which cannot. For those that can, sketch the graph of the exponential function and indicate the sign of A and whether the growth or decay factor c is greater than or less than 1.

(a)

(b)

(c)

(d)

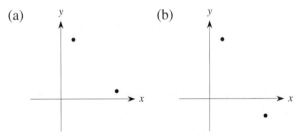

4. Decide which situations represent exponential growth, exponential decay, linear increase or decrease, or none of these patterns.

a. The value for a rare bottle of wine goes up $50 each year. *linear*

b. The value for a piece of sculpture increases 15% each year. *exponential*

c. A 3-year labor contract calls for yearly increases of $800. *linear*

d. A 3-year labor contract calls for an increase of 5% the first year, 4% the second year, and 3% the third year.

e. The value of a car drops by 40% each year.

f. The average cost of a home computer for the first-time buyer has been dropping by $300 each year.

g. The number of new cases of a disease reported over the last decade has been dropping by 12% each year.

5. The accompanying figure shows the graph of the price of each of seven collectible toys as a function of time. Match each scenario with one of the graphs and write a brief scenario for each of the remaining graphs.

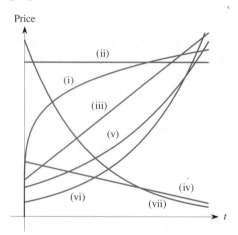

a. The price of the toy increased by 10% each year.

b. The price of the toy increased by 6% each year.

c. The price of the toy dropped by $5 each year.

d. The price of the toy remained steady.

6. Find possible equations for the exponential functions graphed in **(a)**–**(c)**.

(a)

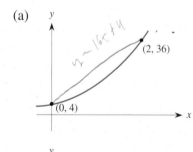

(b)

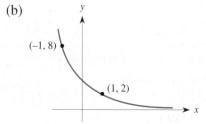

(c)

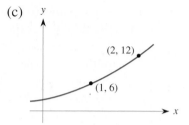

7. In 1980, about 27,700 cases of tuberculosis were reported in the United States. In 1997, there were 19,900 such cases. Source: U.S. Centers for Disease Control and Prevention.

a. Write an exponential decay function that models the number of reported cases of TB as a function of time.

b. Predict the number of cases in 2004.

c. Estimate how long it will take for the number of reported cases to drop to 10,000.

8. In 1940, there were 6,102,000 farms in the United States. By 1997, the number of farms had dropped to 1,912,000.

Source: *2000 Statistical Abstract of the United States.*

a. Assuming that the pattern of decay is exponential, find the equation of a model that can be used to predict the number of farms.

b. Use your model to predict the number of farms in 1980. How close is your prediction to the correct value of 2,440,000 farms?

c. Predict the number of farms in 2005.

d. If the trend continues, estimate when there will be 1 million farms.

e. Write a paragraph describing the long-term implications if this trend continues.

9. When a person smokes a cigarette, about 0.4 mg of nicotine is absorbed into the blood. About 35% of the nicotine is washed out of the blood every hour.

a. Find the equation of a function that models the level of nicotine in the blood after a single cigarette.

b. Use your model to estimate how long it takes for the amount of nicotine in the blood to drop to 0.005 mg.

10. The amount of the drug ampicillin (a form of penicillin) in the bloodstream decreases by about 42% every hour.

a. If the dosage of ampicillin is 250 mg, write a function that can be used to model the level of ampicillin in the blood as a function of time, if one dose is taken.

b. How much ampicillin is left in the blood after 5 hours?

c. Estimate how long it will take for the level of ampicillin to drop to 1 mg.

11. A hospital patient is administered 3 mg of morphine to control his pain. About 31% of the morphine in the blood is washed out every hour.

   a. Construct a function that models the level of morphine in the blood after one dose.

   b. How much morphine remains in the blood after 4 hours?

   c. Estimate how long it will take for the amount of morphine left to drop to 0.2 mg.

12. The level of pollution in the Great Lakes is a major concern to environmentalists.

   a. In Lake Erie, about 38% of the pollutants are washed out each year if no pollutants are added. Write a function that models the level of pollutants in the lake as a function of time.

   b. How long will it take for 90% of the pollutants to be washed out of Lake Erie if no further pollutants are added?

   c. In Lake Superior, about 0.053% of the pollutants are washed out each year if no further pollutants are added. Write a function to model the level of pollutants in Lake Superior as a function of time.

   d. How long will it take for 90% of the pollutants to be washed out of Lake Superior if no further pollutants are added?

13. One of the major concerns about above-ground nuclear testing is that it produces strontium-90, a radioactive element whose half-life is 29 years and which has worked its way into the food chain. That is, strontium-90 from fallout is deposited on grass, eaten by cows, carried into their milk, and eventually finds its way onto the kitchen table. Suppose that, as a result of a single nuclear explosion, the amount of strontium-90 in a particular valley exceeds health limits by a factor of 10. Estimate how long it will take for the strontium-90 to decay to the safety level.

14. Carbon-14, a radioactive form of carbon, is used in the carbon-dating process to measure the age of objects. About 0.012% of the carbon-14 decays into carbon-12 every century.

   a. Write a function for the amount of carbon-14 remaining in an object that originally contained C_0 grams of carbon-14.

   b. What percentage of the carbon-14 remains after a thousand years.

15. The filter in a swimming pool removes 30% of all impurities in the water every hour it operates.

   a. Find an expression for the level of impurities left in the pool after n hours, if no further impurities are added.

   b. How much is left after 5 hours?

16. Use the information in Example 5 to estimate the half-life of a signal in a fiber-optic cable. What does it mean?

17. One of the major problems associated with any organ transplant is the long-term risk of rejection, despite patients' taking anti-rejection drugs for the rest of their lives. The percentage of individuals who have not rejected a transplanted organ can be modeled by an exponential decay function as a function of time in years. According to one study, the half-life of kidney transplants done in 1988 was 9.1 years; according to another study, the half-life of kidney transplants done in 1996 was projected to be 13.3 years. Is this later result good or bad news? Explain your reasoning.

18. According to a medical study, the half-life of kidney transplants was 13.3 years.

   a. Write a formula for an exponential function that can be used to model the percentage of kidney transplant recipients who haven't rejected the kidney as a function of time.

   b. What percentage of kidney transplant recipients do you predict will still have their new kidneys functioning after 10 years?

   c. How long will it take until the percentage of kidney transplant recipients having their new kidneys will be down to 20%?

19. Treatments for different kinds of cancer are usually reported in terms of the percentage of patients who survive for 5 years after receiving the treatment, be it surgery, chemotherapy, or radiation therapy. The percentage who survive can be modeled by a exponential decay function. The 5-year survival rate for early stage malignant melanoma, a particularly severe type of skin cancer, is 80%.

   a. What percentage of patients having this treatment will survive 10 years?

   b. Use the information given to write an exponential decay function that models the percentage of patients treated for melanoma who survive any given length of time t in years.

c. What is the half-life for survival among patients having this treatment?

20. The 5-year survival rate for stage I lung cancer (the mildest and earliest form) treated by surgery is 60% to 70%.

   a. Use the middle value of 65% to write an exponential decay function that models the percentage of patients treated for stage I lung cancer who survive any given length of time t in years.

   b. What is the half-life for survival among patients having this treatment?

   c. Repeat parts (a) and (b), using the lowest survival rate of 60%.

   d. Repeat parts (a) and (b), using the highest survival rate of 70%.

21. In 1990, 442.2 million prerecorded cassette tapes and 865.7 million CDs were sold in the United States. In 1998, 158.5 million cassettes tapes and 1,124.3 million CDs were sold. Assume for now that the patterns of sales for both items are exponential functions.

   a. Find the equation for the number of cassette tapes sold as an exponential function of time.

   b. Find the equation for the number of CDs sold as an exponential function of time.

   c. What is the practical significance of the growth or decay factors and growth or decay rates in parts (a) and (b)?

   d. If the trends in sales of both items were indeed exponential functions, estimate when the number of CDs sold overtook the number of cassette tapes sold.

22. An exponential function f is such that $f(1) = 96$ and $f(5) = 6$. Which of the values are possible and which are impossible.

   a. $f(3) = 24$ **b.** $f(3) = 51$ **c.** $f(3) = 65$

23. Suppose that a scientist has some initial amount R_0 of a radioactive substance whose half-life is measured on a scale of days.

   a. Sketch the graph of the amount of this substance present as a function of time.
   Use the concavity of your graph from part (a) to answer the following questions.

   b. Suppose that you measure the amount of the substance after 10 days and find that 800 grams are left and after 11 days that 750 grams are left. Use this information to estimate the number of grams remaining after 20 days. Is the actual value higher or lower than your estimate? How do you know?

c. Suppose that you are told that the amount of the substance present after 30 days is 400 grams. Use this information and the amount left after 10 days to estimate the amount present after 20 days. Is the actual value higher or lower than your estimate? How do you know?

d. How might you use the results from (b) and (c) to come up with a better estimate of the amount of radioactive material present after 20 days?

24. A certain radioactive isotope has a half-life of 20 days. Suppose that 800 mg are present initially and consider a 60-day time period. Let r_1 represent the average daily rate of decrease of the isotope over the full 60-day period, let r_2 be the average daily rate of decrease over the first 30-day period, and let r_3 be the average daily rate of decrease over the last 30 days. List these three rates in increasing order without calculating their values.

25. The function shown in the accompanying figure is a modified exponential function of the form $y = A + B \cdot c^x$, with $c < 1$. Find appropriate values for the three constants A, B, and c.

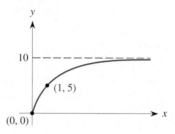

26. You have been asked to design a slide at a water amusement park that extends vertically from point A to point B. A person sliding down it will speed up due to the force of gravity. For the three possible shapes of the slide shown, along which will a person make the trip from A to B most rapidly? Give reasons for your answer. (The specific curve along which an object will slide without friction from A to B in the shortest possible time is known as the *brachistochrone* and was first solved by Jacques Bernoulli in about 1700.)

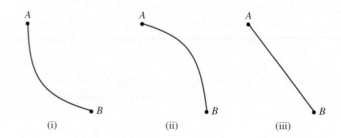

(i)   (ii)   (iii)

Exercising Your Algebra Skills

Simplify the expressions by using properties of exponents.

1. $2^m \cdot 2^n$

2. $\dfrac{1}{2^u} \cdot \dfrac{1}{2^v}$

3. $5^3 \cdot 5^x$

4. $4^{-2} \cdot 4^{3x}$

5. $3^5 \cdot 3^{-2a}$

6. $2^{-4} \cdot 2^{-3w}$

7. $\dfrac{10^{-3x}}{10^{2x}}$

8. $\dfrac{4^{3x}}{4^{-3x}}$

9. $(2^x)^5$

10. $(0.7^x)^{10}$

Write each expression as the product of two terms, one an exponential term and the other a constant.

11. 3^{x+2}

12. 5^{x-2}

13. 10^{3x+1}

14. $\left(\dfrac{1}{2}\right)^{4x+3}$

2.6 ⸺ Logarithmic Functions

In Section 2.4, we constructed an exponential function to approximate the population (in millions) of Florida as

$$P = f(t) = 12.94(1.029)^t,$$

where t is the number of years since 1990. Using this model, we can predict Florida's population at any given time, assuming that the growth rate remains 2.9% each year.

In Example 4 of Section 2.4, we estimated (using both numerical and graphical methods) that Florida's population will reach 20 million in early 2005, when $t \approx 15.2$. This problem involved finding the value of t for which

$$f(t) = 12.94(1.029)^t = 20.$$

Because this exponential function is always increasing, we know that there must be only one value of t when $P = 20$. We can always find an approximate value for t numerically or graphically, as we did in Sections 2.4 and 2.5. We now develop an algebraic approach for solving such equations exactly rather than approximately. We want a process that extracts the variable t from the exponent in $P = kc^t$. This process involves a new function called the **logarithm**. As with exponential functions, logarithms have a base b. Although logarithms can have any base b (denoted $\log_b$), we work primarily with logarithms to base 10.

Definition of Logarithms to Base 10

$$\log_{10} x = y \quad \text{means} \quad 10^y = x.$$

The logarithm to the base 10 of x is that power of 10 needed to produce x.

For instance,

$$\log_{10} 100 = \log_{10} 10^2 = 2$$

because 2 is that power of 10 needed to produce 100, or $10^2 = 100$. Also,

$$\log_{10} 1000 = \log_{10} 10^3 = 3$$

because 3 is that power of 10 needed to produce 1000, or $10^3 = 1000$. Similarly,

$$\log_{10}(0.1) = \log_{10} 10^{-1} = -1$$

because $0.1 = 1/10 = 1/10^1 = 10^{-1}$ and -1 is the power to which 10 must be raised to produce 0.1, or $10^{-1} = 0.1$.

The logarithm to the base 10 of x, $\log_{10} x$, is usually written simply as $\log x$. Because the logarithm is a function, it would be preferable to write $\log(x)$ rather than just $\log x$. But $\log(x)$ is not standard usage, so we avoid it. However, we do use parentheses for expressions such as $\log(5x)$.

The definition of the logarithm also gives two useful formulas.

Fundamental Logarithmic-Exponential Identities

$$\log(10^x) = x, \quad \text{for all real } x$$

$$10^{\log x} = x, \quad \text{for all } x > 0$$

Because these formulas hold for all appropriate values of x, they are called **identities.** Think about the two results to be sure that you understand them thoroughly. For the first identity, $\log(10^x)$ is that power of 10 needed to produce 10^x. Clearly, that power must be x itself. For instance, $\log(10^{1.234}) = 1.234$. For the second identity, the exponent in $10^{\log x}$ is $\log x$. In other words, $\log x$ is the power of 10 that gives the number x. For instance, $10^{\log 50.7} = 50.7$. The second property allows us to undo an equation involving logarithms. We discuss why the second identity is restricted to positive values of x later in this section.

Using the Logarithm

The principal reason for introducing logarithms here is to solve equations of the form

$$c^x = A$$

for the variable x in the exponent when the quantities c and A are known. For instance, we might want to solve $3^x = 8$ for the variable x in the exponent. To do so, we apply the following fundamental property of logarithms.

$$\log(c^x) = x \cdot \log c, \quad c > 0$$

A proof of this formula can be found in any algebra textbook.

EXAMPLE 1

Solve for x in the equation $3^x = 8$.

Solution To extract x from the exponent, we take the logarithm of both sides of the equation:

$$\log(3^x) = \log 8.$$

We use the above fundamental property of logarithms to get

$$\log(3^x) = x \cdot \log 3 = \log 8,$$

where log 3 and log 8 are both just numbers. Finally, we divide both sides by log 3 to obtain

$$x = \frac{\log 8}{\log 3} \approx 1.893,$$

using a calculator. The graph of the function is depicted in Figure 2.43. To verify this result, we substitute $x = 1.893$ into the original equation and get

$$3^{1.893} \approx 8.0018.$$

We would have gotten 8 exactly if we hadn't rounded $\log 8/\log 3 \approx 1.893$.

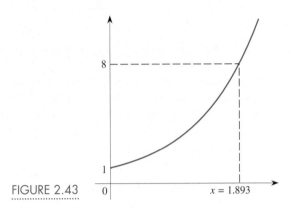

FIGURE 2.43

We now show how to obtain an exact solution to the question on the growth of Florida's population posed at the beginning of this section.

EXAMPLE 2

Determine when the population of Florida will reach 20 million.

Solution   We begin with the equation

$$P = f(t) = 12.94(1.029)^t = 20.$$

Dividing both sides of the equation by 12.94, we get

$$(1.029)^t = \frac{20}{12.94}.$$

We now take logarithms of both sides and use the fundamental property of logarithms to get

$$\log(1.029^t) = t\log(1.029) = \log\left(\frac{20}{12.94}\right).$$

Dividing both sides by $\log(1.029)$ gives

$$t = \frac{\log(20/12.94)}{\log(1.029)},$$

which is the exact solution. When we perform the actual calculations, we get $t \approx 15.23$. Thus Florida's population will reach 20 million about $15\frac{1}{4}$ years after 1990, or early in 2005, as depicted in Figure 2.44.

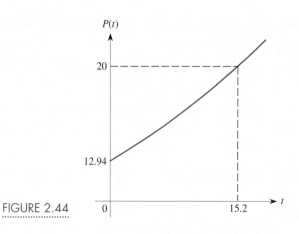

FIGURE 2.44

Think About This   In the solution to Example 2, we intentionally left the quantity in the form 20/12.94 to avoid possible rounding errors when dividing it out. What happens to the final answer if you perform the division operation early in the solution and round differently? See what happens, for instance, if you use $20/12.94 \approx 1.5$ or $20/12.94 \approx 1.55$ or $20/12.94 \approx 1.546$.

In Example 3 of Section 2.5, we estimated numerically and graphically that the half-life of Prozac is approximately 2.4 days based on the exponential decay function

$$D(t) = 80(0.75)^t$$

that models the amount of Prozac in the bloodstream following an 80 mg dose. We now show how to obtain the exact answer using logarithms.

EXAMPLE 3

Determine the half-life of Prozac in the bloodstream.

Solution   To find the half-life exactly, we must find the time t needed until the original 80 mg drug level falls to 40 mg. Therefore we must solve the equation

$$80(0.75)^t = 40.$$

If we divide both sides by 80, we get

$$(0.75)^t = \frac{40}{80} = 0.5.$$

We now take logarithms of both sides and use the fundamental property of logarithms to find

$$\log(0.75)^t = t \log(0.75) = \log(0.5).$$

When we divide both sides by $\log(0.75)$, we get

$$t = \frac{\log(0.5)}{\log(0.75)} \approx 2.4094.$$

Thus, no matter what level of Prozac is in the blood at any given time, the level will drop by half about 2.4 days later, as illustrated in Figure 2.45.

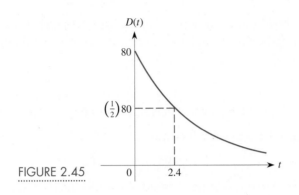

FIGURE 2.45

Properties of Logarithms

To use logarithms, you need to know their basic properties.

Properties of Logarithms

1. $\log(A^x) = x \cdot \log A$
2. $\log(A \cdot B) = \log A + \log B$
3. $\log(A/B) = \log A - \log B$

Proofs of all three of these properties can be found in any algebra textbook.

The first property is the tool we used to extract a variable from the exponent. The second property lets us simplify the *logarithm of a product* by writing it as the *sum of the individual logarithms;* for instance,

$$\log(5 \cdot 12) = \log 5 + \log 12.$$

Check this result on your calculator. Also,

$$\log(100x) = \log 100 + \log x = 2 + \log x.$$

The third property lets us simplify the *logarithm of a quotient* by writing it as the *difference of the individual logarithms;* for instance,

$$\log\left(\frac{9}{4}\right) = \log 9 - \log 4.$$

Check this result on your calculator. Also,

$$\log\left(\frac{1000}{x}\right) = \log 1000 - \log x = 3 - \log x.$$

Think About This   Is $\log(9/4)$ the same as $(\log 9)/(\log 4)$? Why or why not? Try it on your calculator to see. ⌐

Think About This   Is $\log(1000)/\log(x) = 3/\log x$ the same as $\log 1000 - \log x$? Why or why not? Graph both $y = \log(1000)/\log(x)$ and $y = 3/\log x$ to see whether it is true. ⌐

Note that all three properties of logarithms apply to logarithms with any base b, not just the base 10. Furthermore, these properties give us some alternative tools for solving some of the problems that we have already encountered. In Example 4, we show how to use the second and third properties of logarithms to determine once more when the population of Florida will reach 20 million.

EXAMPLE 4

Find when Florida's population will reach 20 million by using properties of logarithms.

Solution   We again have to solve the equation

$$12.94(1.029)^t = 20.$$

In Example 2, we began by dividing both sides by 12.94. Instead, suppose that we start by taking logarithms of both sides of the equation:

$$\log[12.94(1.029)^t] = \log 20;$$
$$\log 12.94 + \log(1.029)^t = \log 20; \qquad \log(AB) = \log A + \log B$$
$$\log 12.94 + t \cdot \log 1.029 = \log 20. \qquad \log(A^x) = x \log A$$

To solve for t, we subtract log 12.94 from both sides of the equation:

$$t \cdot \log(1.029) = \log 20 - \log 12.94 = \log\left(\frac{20}{12.94}\right). \qquad \log A - \log B = \log\left(\frac{A}{B}\right)$$

Dividing by log(1.029), we get

$$t = \frac{\log(20/12.94)}{\log(1.029)} \approx 15.23,$$

which is the same result as in Example 2.

Behavior of the Logarithmic Function

For any value of $x > 0$, there is a single corresponding value of log x, so log x is a function of x. We call this function the **logarithmic function** or simply the **log function.**

Let's now consider the behavior of the log function. Recall that the logarithm of a number x represents that power of 10 needed to produce x. Because no power of 10 ever produces 0 (10 raised to what power is 0?), log 0 is undefined. Similarly, because every power of 10 is positive, log x is not defined for negative values of x. Thus the domain of the logarithmic function is $x > 0$.

However, it is possible to have a negative power of 10, so log x can be negative. For instance, $10^{-0.25} \approx 0.56234$ means that log $0.56234 \approx -0.25$. Thus the range of the logarithmic function includes both positive and negative values. Figure 2.46 demonstrates that the logarithm of a number between 0 and 1 is negative and that

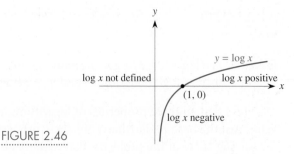

FIGURE 2.46

the logarithm of any number larger than 1 is positive. Finally, because log 1 = 0 (since $10^0 = 1$), the range of the log function consists of all real numbers.

We use these ideas to examine the behavior of the log function $f(x) = \log x$. First, let's look at some values for this function. We know that

$$\log 1 = 0,$$
$$\log 10 = 1,$$
$$\log 100 = 2,$$
$$\log 1000 = 3,$$
$$\vdots$$
$$\log 1,000,000 = 6,$$

and so on. Clearly, log x is an increasing function, at least for $x \geq 1$. Because the successive values grow more and more slowly, it is concave down. In fact, the most significant feature of the growth pattern for the logarithmic function is that it grows extremely slowly. Note what happens with the above values for the log function—to gain just one unit vertically requires going 10 times as far horizontally. Thus you need an extremely large value of x to make log x large. For instance, what value of x is needed to make log $x = 100$? By the definition, x must be 10^{100} because $\log(10^{100}) = 100$. Consequently, it takes an incredibly long time for the log curve to reach a height of 100. The log function goes to infinity as x increases, although it does so exceedingly slowly.

The log function is not defined for $x = 0$ or for negative values. But what happens for small positive values of x? Consider the values

$$\log(1) = 0,$$
$$\log(0.1) = \log(10^{-1}) = -1,$$
$$\log(0.01) = \log(10^{-2}) = -2,$$
$$\log(0.001) = \log(10^{-3}) = -3,$$
$$\vdots$$
$$\log(0.000001) = \log(10^{-6}) = -6,$$

and so on. As x gets closer and closer to 0, log x becomes more and more negative. Thus the line $x = 0$ (the y-axis) is a *vertical asymptote* of the graph of $y = \log x$ because the curve gets closer and closer to this line as x gets closer and closer to 0, but the curve never reaches it. This vertical asymptote reinforces the fact that the logarithmic function is not defined for $x = 0$, and so the graph of $y = \log x$ has no y-intercept. It does, however, have an x-intercept at $x = 1$ because $\log(1) = 0$, as illustrated in Figure 2.46.

Comparing Exponential and Logarithmic Functions

From the definition of the logarithm, it is clear that there is a close interrelationship between $y = \log x$ and the exponential function $y = 10^x$. To investigate this relationship, we start with the graphs of the two functions shown in Figure 2.47. Both are clearly increasing functions. The exponential function is concave up, whereas the log function is concave down. However, the main difference in their growth patterns is that the exponential function $y = 10^x$ grows extremely rapidly but the logarithm function $y = \log x$ grows extremely slowly.

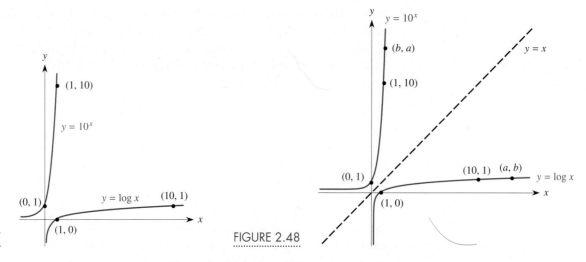

FIGURE 2.47

FIGURE 2.48

Figure 2.48 shows something striking about the graphs of the two functions $y = 10^x$ and $y = \log x$: They are reflections of each other about the diagonal line $y = x$ and thus are *symmetric* about this line (see Appendix D). Let's see why. We know that

$$\log 10 = 1,$$

so the point $(10, 1)$ is on the graph of $y = \log x$. By the definition of the logarithm,

$$\log 10 = 1 \quad \text{means} \quad 10^1 = 10.$$

But $10^1 = 10$ tells us that the point $(1, 10)$ satisfies the equation $10^x = y$, so the point $(1, 10)$ is on the exponential graph $y = 10^x$. The points $(10, 1)$ and $(1, 10)$ are reflections of each other about the line $y = x$. In general, if the point (a, b) is on the graph of $y = \log x$, then

$$\log a = b.$$

This expression is equivalent to saying that

$$10^b = a,$$

which means that the point (b, a) is on the graph of the exponential function $y = 10^x$. Hence the log graph and the exponential graph are reflections of each other about the line $y = x$. We investigate this phenomenon in more depth in Section 2.9.

Applications of Logarithmic Functions

Logarithms have many applications. For instance, chemists use a quantity known as the pH to measure how acidic a water solution is. The pH is based on the concentration of hydrogen ions (measured in moles per liter) in the solution. The hydrogen-ion concentration of pure water is 10^{-7} moles per liter. Thus the pH of pure water is

$$\text{pH} = -\log(\text{concentration}) = -\log(10^{-7}) = -(-7)\log 10 = 7 \cdot 1 = 7,$$

which is used as the reference point for a neutral solution. Water solutions whose pH values are less than 7 are said to be acidic, whereas water solutions with pH values greater than 7 are said to be basic or alkaline. The lower the pH, the more acidic the solution; the higher the pH, the more basic the solution. Orange juice, which is somewhat acidic, has a hydrogen-ion concentration of 2×10^{-4} moles per liter and so its pH is

$$
\begin{aligned}
\text{pH} = -\log(2 \times 10^{-4}) &= -[\log 2 + \log(10^{-4})] \\
&= -\log 2 - \log(10^{-4}) \\
&= -\log 2 - (-4)\log 10 \\
&\approx -0.301 + 4 \cdot 1 \approx 3.7.
\end{aligned}
$$

Hydrochloric acid, with a hydrogen-ion concentration of 10^{-1} moles per liter, has a pH of 1, which indicates that it is extremely acidic. In comparison, household ammonia, with a pH of 11.5, is extremely basic.

Think About This   Human blood has a hydrogen-ion concentration of 4×10^{-8}. What is its pH? Is blood slightly or extremely basic? ☐

Note that each one-point decrease in pH represents a tenfold increase in the hydrogen-ion concentration.

The crust of the Earth is composed of about 20 rigid plates that "float" on liquid magma (the molten material beneath the Earth's crust). The study of this phenomenon is called *plate tectonics*. A geologic fault, such as the famous San Andreas Fault in California, is the space between two plates. As plates move, they bump into one another and sometimes one plate passes slightly under another, causing the upper plate to shift and heave. The result is an earthquake on the Earth's surface. There are about a million earthquakes, mostly very minor, each year. American seismologist Charles Richter developed a way of measuring the intensity of an earthquake. The *Richter scale* is based on the idea that there is a minimum noticeable, or threshold, level of earthquake intensity, denoted by I_0. The energy involved in a threshold level earthquake is approximately equal to the energy released by 10,000 atomic bombs. Any stronger quake has an intensity denoted by I. The Richter scale relates the magnitude R of an earthquake to its intensity, or

$$
R = \log\left(\frac{I}{I_0}\right).
$$

That is, the magnitude given by the Richter scale measurement is the logarithm of the ratio of the actual intensity to the threshold level.

The largest recorded earthquake, which occurred in Japan in 1933, had magnitude $R = 8.9$ on the Richter scale. Let's see just how powerful this quake was. We have

$$
R = \log\left(\frac{I}{I_0}\right) = 8.9,
$$

so when we take powers of 10 of both sides of the equation,

$$
10^{\log(I/I_0)} = \frac{I}{I_0} = 10^{8.9} \approx 794{,}328{,}235.
$$

Therefore this quake had an intensity almost 800 million times greater than the threshold level!

How are different measurements on the Richter scale related? For instance, if the measurement for one earthquake is double that of another, how much greater is it? What does a one point change in magnitude represent?

EXAMPLE 5

How does a magnitude 5 earthquake on the Richter scale compare to a magnitude 6 quake?

Solution   If $R = 5$, we have

$$\log\left(\frac{I}{I_0}\right) = 5.$$

We undo the logarithm by taking powers of 10 of both sides of the equation and use the fundamental identity

$$10^{\log x} = x$$

to get

$$10^{\log(I/I_0)} = \frac{I}{I_0} = 10^5 = 100{,}000.$$

For $I/I_0 = 100{,}000$, we get $I = 100{,}000 I_0$, so the intensity of a magnitude 5 quake is 100,000 times the threshold level. This quake's energy is equivalent to roughly $100{,}000 \times 10{,}000 = 10^9$, or 1 billion, atomic bombs exploding simultaneously.

   Similarly, consider an earthquake measuring $R = 6$ on the Richter scale. We now get

$$I/I_0 = 10^6 = 1{,}000{,}000.$$

The intensity of this quake is 1 million times the threshold level. Thus an increase of 1 Richter unit corresponds to a tenfold increase in the intensity of the earthquake. ◆

   Suppose that one earthquake has a reading twice that of another on the Richter scale. How much stronger is it? Is it four times as strong? Is the relative intensity the same? Does it depend on the value for R? Let's compare $R = 4$ to $R = 2$ to see what happens. With $R = 4$, we have

$$\log\left(\frac{I}{I_0}\right) = 4,$$

so when we take powers of 10 of both sides of the equation,

$$10^{\log(I/I_0)} = \frac{I}{I_0} = 10^4$$

and therefore

$$I = 10^4 \cdot I_0.$$

Hence a magnitude 4 quake is 10,000 times the intensity of the threshold level. For $R = 2$, we have

$$\log\left(\frac{I}{I_0}\right) = 2$$

so that

$$10^{\log(I/I_0)} = \frac{I}{I_0} = 10^2$$

and therefore

$$I = 10^2 \cdot I_0.$$

Hence a magnitude 2 quake is 100 times the intensity of the threshold level. Consequently, a magnitude 4 quake is actually $10^4/10^2 = 100$ times stronger than a magnitude 2 quake.

Changing Bases

Throughout this book, we work with logarithms to the base 10 to undo exponential functions of the form $y = k \cdot 10^x$. However, as we stated earlier, it is possible to have bases other than 10—say, $c = 2$ or $c = 1.029$—as the base for an exponential function $y = kc^x$. Each possible base gives rise to a corresponding logarithmic function. For instance, we could work with logarithms to the base 2, written $\log_2 x$.

Definition of Logarithms to Base c

$$\log_c x = y \quad \text{means} \quad c^y = x, \qquad x > 0.$$

The logarithm to the base c of x is that power of c needed to produce x.

In practice, there is one particular base other than 10 that is widely used: the number $e = 2.71828\ldots$. The logarithm corresponding to base e is called the **natural logarithm.** It is especially important in calculus and many of the sciences. Even though we could write the natural logarithm as $\log_e x$, it is customarily written $\ln x$.

Although $\log_{10} 10 = 1$, we have $\ln 10 \approx 2.3026$ because $e^{2.3026} \approx 2.71828^{2.3026} \approx 10.0001$. Similarly, whereas $\log_{10} 100 = 2$, we have $\ln 100 \approx 4.6052$ because $e^{4.6052} \approx 2.71828^{4.6052} \approx 100.003$.

We previously said that all the properties of logarithms apply no matter what base is used. Thus the following are properties of the natural logarithm.

Properties and Identities for the Natural Logarithm

1. $\ln(A^p) = p \ln A$
2. $\ln(A \cdot B) = \ln A + \ln B$
3. $\ln(A/B) = \ln A - \ln B$
4. $\ln e^x = \log_e e^x = x$
5. $e^{\ln x} = x, \qquad \text{if } x > 0$

If different bases are used, we must be able to convert either an exponential function or a logarithm in one base into an exponential function or a logarithm in a different base. That is, for any number x, how do we convert from c^x to 10^x or convert $\log x$ to $\ln x$ and vice versa? Let's first look at converting bases of exponential functions.

EXAMPLE 6

We found that the population of Florida can be modeled by the exponential function $P(t) = 12.94(1.029)^t$. Convert this function to an equivalent expression that involves (**a**) base 10 and (**b**) base e.

Solution

a. Suppose that we try to find the appropriate power q so that

$$(1.029)^t = 10^q.$$

Using properties of logarithms, we take logs of both sides and get

$$t \log(1.029) = q \log 10 = q \cdot 1 = q.$$

The formula for the population of Florida becomes

$$P(t) = 12.94(1.029)^t = 12.94(10^q) = 12.94(10^{t \log(1.029)}) = 12.94(10^{0.0124t}),$$

since $\log(1.029) = 0.0124$. Alternatively, we might think of this result as coming from

$$10^{t \log(1.029)} = (10^{\log(1.029)})^t = (1.029)^t.$$

b. To convert the formula for the Florida population to an equivalent formula using base e, we use the property $e^{\ln x} = x$:

$$(1.029)^t = (e^{\ln 1.029})^t = (e^{0.0286})^t = e^{0.0286t}.$$

Therefore

$$P(t) = 12.94(1.029)^t = 12.94(e^{0.0286t}).$$

◆

We have three formulas for the population of Florida:

$$P(t) = 12.94(1.029)^t;$$
$$P(t) = 12.94(10^{0.0124t});$$
$$P(t) = 12.94(e^{0.0286t}).$$

These three formulas are mathematically equivalent—only the bases are different. Graph these three functions using your function grapher to convince yourself that they are identical, except perhaps for slight differences due to rounding.

In general, to convert an exponential function $y = kc^x$ from base c to base 10, where $y = k \cdot 10^{mx}$, we write $c = 10^m$ so that $m = \log c$. To convert an exponential function $y = kc^x$ from base c to base e, where $y = ke^{mx}$, we write $c = e^m$, so that $m = \ln c$.

Now let's consider how to convert a logarithm in one base to a logarithm in another base. We begin by looking at some typical values of $\log x$ and $\ln x$, rounded to four decimal places, as shown in the following table. To see if there is any clear relationship between the two sets of logarithmic values, we plot the values of $\ln x$ versus $\log x$, as shown in Figure 2.49.

The graph shows that there is a linear pattern. The line passes through the origin, so its vertical intercept is 0 and we can write

$$\ln x = m \log x$$

for some constant of proportionality m. You can also see this from the ratio of $\ln x$ and $\log x$ for any value of x—in every case, the ratio is approximately 2.3026. Check this

x	$\log x$	$\ln x$
1	0	0
2	0.3010	0.6932
3	0.4771	1.0986
4	0.6021	1.3863
5	0.6990	1.6094
6	0.7782	1.7918
7	0.8451	1.9459
8	0.9031	2.0794
9	0.9542	2.1972
10	1	2.3026

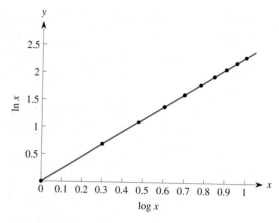

FIGURE 2.49

on your calculator by using several different values of x. Note that Figure 2.49 shows the values of $\ln x$ plotted against those of $\log x$. If we plotted either $\ln x$ against x or $\log x$ against x, we would get the usual graph of a logarithmic function—one that is increasing and concave down. Only the graph of $\ln x$ versus $\log x$ results in a line.

The value of the constant of proportionality $m = 2.3026$ is also the slope of the line through the points shown in Figure 2.49. Thus we can write

$$\ln x = 2.3026 \log x$$

for any x. Because 2.3026 appears in the last row of the preceding table as the value of $\ln 10$, we can rewrite this relationship as

$$\ln x = (\ln 10)\log x,$$

or equivalently,

$$\log x = \frac{\ln x}{\ln 10}.$$

Rewriting this equation to highlight the base of the logarithm, we get

$$\log_{10} x = \frac{\log_e x}{\log_e 10}.$$

In fact, if we perform the identical analysis with any other base (say, c instead of e), we obtain the comparable result for changing between base 10 and base c, for any c; or

$$\log_{10} x = \frac{\log_c x}{\log_c 10}.$$

Problems

1. The graphs of the following functions may surprise you. Use your function grapher to graph each function and then explain what you see and why, using the properties of logarithms.

   a. $y = \log 10^{2x}$

   b. $y = \log(2x) - \log(x)$

   c. $y = \log 10^{x^2}$

   d. $y = 10^{\log(x^2)}$

   e. $y = \log 3^x$

   f. $y = \log(10/6^x)$

2. Use your function grapher to draw simultaneously the graphs of $y = \log(2^x)$, $y = \log(3^x)$, and $y = \log(5^x)$. For each function, use the properties

of logarithms to explain why you get the graphs you do.

3. The population of Argentina was 34.6 million in 1995 and was growing exponentially at an annual rate of 1.3%.

   a. Find an expression for Argentina's population at any time t, where t is the number of years since 1995.

   b. What population would you predict for 2005 if the present trend continues?

   c. Use logarithms to find the doubling time exactly.

4. The population of Kenya is growing exponentially. Its population was 23.3 million people in 1988 and 28.3 million in 1995.

   a. Find an expression for the population at any time t, where t is the number of years since 1988.

   b. What would be the population in 2005?

   c. Use logarithms to find the doubling time.

5. Because of ardent fishermen during the summer months, the population of fish in a lake is reduced by 10% each week. Find the half-life of this dwindling fish population, using logarithms.

6. The Best Company earned $50 million in 2000, and its income is growing at a rate of 2% per year. The Acme Corporation earned $30 million that year, and its income is growing at a rate of 6.5% per year. When will Acme overtake Best in annual income?

7. a. Find the doubling time for annual growth rates of 3%, 4%, 5%, 6%, and 7%.

   b. Consider the doubling time d as a function of growth rate r. Plot your results from part (a) and decide what type of function seems to fit the behavior pattern you observe.

8. Bankers use a technique called the Rule of 70 to estimate the doubling time for money invested at different interest rates, dividing 70 by 100 times the interest rate. Thus for an interest rate of $10\% = 0.10$, bankers estimate the doubling time to be

$$\frac{70}{100 \cdot 0.10} = \frac{70}{10} = 7 \text{ years.}$$

Use your results from Problem 7 to test the accuracy of this method.

9. Determine when the cost of first-class postage for a letter will reach $1, given that first class postage rose to 29¢ in 1990 and to 37¢ in 2002.

Problems 10–13 are based on the carbon dating process to measure the age of objects. Carbon-14, a radioactive form of carbon, decays into carbon-12 with a half-life of 5730 years.

10. The famous Cro-Magnon cave paintings are found in the Lascaux Cave in France. If the level of carbon-14 radioactivity in charcoal in the cave is approximately 14% of that of living wood, estimate the date when the paintings were made.

11. The level of carbon-14 in a charred roof beam found in a 1950 excavation of an ancient Babylonian city is about 61% of the level in living wood. Estimate when the fire occurred.

12. The well-preserved body of a Stone Age man was found in melting snow in the northern Italian Alps in 1991. Examination of a tissue sample from the body indicated that 47% of the carbon-14 present in the body at the time of death had decayed. When did the man die?

13. Several groups of scientists were allowed to test the Shroud of Turin, the supposed burial cloth of Jesus, in 1991. They found that the cloth contained 91% of the amount of carbon 14 contained in newly made cloth of the same material. Based on this information, how old is the Shroud of Turin?

14. A radioactive substance decays exponentially so that after 10 years, 40% of the initial amount R_0 remains.

    a. Find an expression for the quantity remaining after t years.

    b. How much will be present after 25 years?

    c. What is the half-life of the substance?

    d. How long will it be before only 2% of the original amount is left?

15. In an effort to reduce the breeding rate of a strain of pesticide-resistant mosquitoes in the southeastern United States, a group of scientists released large numbers of sterilized male mosquitoes to mate with the fertile females who would consequently produce no offspring. Suppose that effort reduced the mosquito population by 2% per month.

    a. What percentage of the original population P_0 would remain after 1 year?

    b. How long would it take to lower the population by half?

    c. How long would it take for the population to fall to 10% of the original level?

16. In computer science, the efficiency of algorithms (methods for accomplishing a certain task) are often analyzed by how long it takes to perform the operation with n objects. Typically, as n increases, the time involved for the operation increases signif-

icantly. Two algorithms used to put a set of names in alphabetical order are compared. For one algorithm, the time needed to order n names, as a function of n, is $B(n) = \frac{1}{2}n^2$. The time for the second algorithm, as a function of n, is $S(n) = n \log n$. Which method is faster? Explain.

17. How much stronger is a magnitude 6 earthquake than a magnitude 3 earthquake?

18. How much stronger is

   a. a magnitude 7 quake than a magnitude 5 quake?
   b. a magnitude 7 quake than a magnitude 4 quake?

19. Let I_0 be the minimum (or threshold) level of sound that can be heard by human beings. If the intensity of a particular sound is I, the magnitude of the sound, measures in decibels d, is given by

$$d = 10 \log\left(\frac{I}{I_0}\right).$$

   a. Normal conversation measures about 60 decibels. How much more intense is this level than the threshold level?

b. A loud noise of about 150 decibels will cause deafness. How much more intense is this level than the threshold level?

c. An aircraft taking off has a loudness level of about 120 decibels. How much more intense is this level than the threshold level?

d. How loud (that is, how many decibels) is a sound whose intensity is 1 million times the threshold level?

e. The noise level from a rock band is about 100 billion times higher than the threshold level. What is the decibel measure of this noise level?

20. Convert the formula $D(t) = 80(0.75)^t$ for the level of Prozac in the bloodstream following an initial dose of 80 mg to equivalent formulas with base 10 and base e.

21. The population of the world can be modeled by the function $P(t) = 6(1.015)^t$, where t is the number of years since 1999. Convert this formula to equivalent formulas with base 10 and base e.

Exercising Your Algebra Skills

Simplify each expression by using the properties of logarithms.

   1. $\log x + \log x^2 + \log x^3$ **2.** $\log x + \log \sqrt{x}$
   3. $\log x^2 + \log y^3 - \log x - \log y^2$
   4. $\log(x/y) - \log(y/x)$ **5.** $\log 10^{x^2}$
   6. $10^{\log(x^2)}$

Solve each expression for x.

   7. $7^x = 11$ **8.** $1.05^x = 2$
   9. $0.4^x = 0.6$ **10.** $3(1.04)^x = 5$
   11. $12(0.86)^x = 3$ **12.** $9(0.17)^x = 0.25$

   13. $4(1.05)^x = 5(1.04)^x$ **14.** $3(0.7)^x = 6(0.5)^x$
   15. $\log x = 2$ **16.** $\log x = 0.5$

Without using a calculator, evaluate each term.

   17. $\log 1,000,000$ **18.** $\log 0.001$
   19. $\log(1/10,000)$ **20.** $\log \sqrt{10}$

Determine which of the following are true for all values of x and which are not by using algebra.

   21. $\log(x^2 - 1) = \log(x - 1) + \log(x + 1)$
   22. $\log(x^2 + 1) = \log(x^2) + \log(1)$
   23. $\log(1/x) = -\log x + \log(1)$

2.7 Power Functions

The area A of a circle with radius r is given by

$$A = f(r) = \pi r^2.$$

The surface area S of a sphere with radius r is given by

$$S = g(r) = 4\pi r^2.$$

(Picture a tennis ball whose surface is made up of four roughly circular regions, as shown in Figure 2.50). The volume V of the sphere is given by

$$V = h(r) = \frac{4}{3}\pi r^3.$$

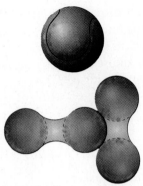

FIGURE 2.50   Tennis ball cover split apart

Similarly, the inverse square law of gravitation describes how the force of gravity of one object on any other object in the universe varies with distance. In particular, the gravitational force F on a unit mass at a distance d from the center of the Earth is given by

$$F = \frac{k}{d^2} \quad \text{or} \quad F = kd^{-2},$$

where k is a positive constant.

All four of these functions are examples of *power functions,* so called because the independent variable is raised to a constant power. In each case, the dependent variable is a constant multiple of some power of the independent variable. In general, a **power function** is any function of the form

$$y = f(x) = kx^p,$$

where k and p are any constants, positive or negative. (Compare this expression for a power function with an exponential function of the form $y = kc^x$, where the independent variable x is the exponent and the base c is a constant, as shown in Figure 2.51.) Note that the family of power functions $y = kx^p$ is a two-parameter family with parameters p and k.

Exponential function   **Power function**

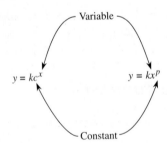

FIGURE 2.51

The definition of a power function includes the special case where $p = 1$, which gives the linear function $y = kx^1 = kx$ that passes through the origin. The definition also includes functions such as

$$y = x^2, \quad y = x^3, \quad \text{and} \quad y = x^4,$$

as well as the case where p is a positive or negative fraction or a decimal, such as

$$y = x^{1/2}, \quad y = x^{1/3}, \quad y = x^{3/2}, \quad \text{and} \quad y = x^{-2.83}.$$

In algebra, fractional exponents are usually introduced purely as a means for simplifying operations with terms involving radicals. Recall that

$$x^{1/2} = \sqrt{x}, \qquad x^{1/3} = \sqrt[3]{x}, \qquad x^{5/8} = \sqrt[8]{x^5},$$

and, in general,

$$x^{m/n} = \sqrt[n]{x^m} = \left(\sqrt[n]{x}\right)^m.$$

Power functions of the form $y = f(x) = kx^{m/n}$ arise naturally in many applications. For instance, biologists have found a relationship between the weight W of large flying birds and their wingspan S. This relationship can be modeled by the power function

$$W = f(s) = 0.15S^{9/4}.$$

This function gives the weight that can be supported by a given wingspan. For example, the wingspan S of a condor is about 10 feet. According to this model, its weight is approximately

$$W = 0.15(10)^{9/4} \approx 27 \text{ lb.}$$

To perform this calculation we must either first raise $S = 10$ to the ninth power and then take the fourth root of the result, so that

$$W = 0.15(10)^{9/4} = 0.15[(10)^9]^{1/4} = 0.15[1{,}000{,}000{,}000]^{1/4} \approx 27$$

or first take the fourth root of S and then raise the result to the ninth power, so that

$$W = 0.15(10)^{9/4} = 0.15[(10)^{1/4}]^9 = 0.15[1.778279]^9 \approx 27.$$

Symbolically, we write

$$S^{9/4} = \sqrt[4]{S^9} = \left(\sqrt[4]{S}\right)^9.$$

When you use a calculator to evaluate such an expression, be careful to use parentheses around the fractional exponent, as in

```
0.15*10^(9/4);
```

without the parentheses, the rules for the order of operations will give you a very different answer.

The graph of the power function $W = 0.15S^{9/4}$ is shown in Figure 2.52. Note that the pattern is that of an increasing, concave up function; thus, as the wingspan S of a bird increases, its weight W increases even more rapidly. Consequently, heavier birds require relatively smaller wingspans in order to fly, which is likely contrary to intuition.

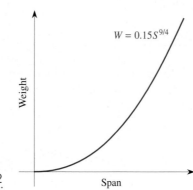

FIGURE 2.52

<u>Think About This</u> The largest known bird is the Steller's eagle, with a wingspan of 8 feet. Estimate the weight of an adult Steller's eagle. ▭

Aeronautical engineers use the same principle—based on a similar relationship between wingspan and the weight of a plane—when designing new aircraft.

Behavior of Power Functions for $x > 0$

Recall that for exponential functions the value of the base c in $y = kc^x$ leads to different behavior patterns—exponential growth when $c > 1$ and exponential decay when $0 < c < 1$. Similarly, the behavior of power functions depends on the size of the constant power p in $y = kx^p$. To simplify things initially, we let $k = 1$ so that we can consider the more basic power function $y = x^p$.

The three different behavior patterns for power functions are illustrated in Figure 2.53: the graphs of $y = x^2$, $y = x^{1/2}$, and $y = x^{-2}$ (along with the graph of $y = x$ for reference) for $x > 0$. Note how the graph of $y = x^2$ (with $p = 2$) is increasing and concave up; that of $y = x^{1/2}$ (with $p = \frac{1}{2}$) is increasing and concave down, and that of $y = x^{-2}$ (with $p = -2$) is decreasing and concave up. The specific values $p = 1$ (when we get a line through the origin) and $p = 0$ (when we get a horizontal line at height $y = 1$) are critical in separating one kind of behavior from another.

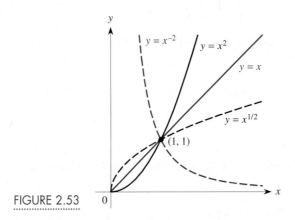

FIGURE 2.53

Let's investigate these cases in more detail by looking at a variety of different power functions of each type. Figure 2.54(a) shows the graphs of three related

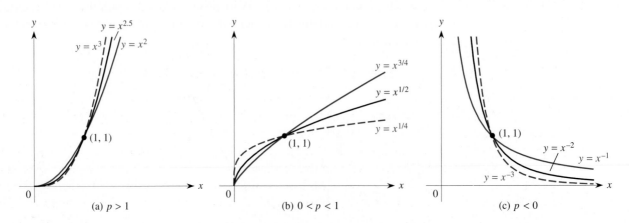

FIGURE 2.54

power functions, $y = x^2$, $y = x^{2.5}$, and $y = x^3$, all of which behave in the same manner as $y = x^2$. Similarly, Figure 2.54(b) shows the graphs of three other power functions, $y = x^{1/4}$, $y = x^{1/2}$, and $y = x^{3/4}$, all of which behave in a pattern similar to that for $y = x^{1/2}$. Finally, Figure 2.54(c) shows the graphs of $y = x^{-1}$, $y = x^{-2}$, and $y = x^{-3}$, all of which behave in a third manner, similar to $y = x^{-2}$.

These graphs suggest the following facts about power functions for $x \geq 0$:

---

1. If $p > 1$, the power function $y = x^p$ is increasing and concave up.

   If $0 < p < 1$, the power function $y = x^p$ is increasing and concave down.

   If $p < 0$, the power function $y = x^p$ is decreasing and concave up.

2. Every power function $y = x^p$ passes through the point $(1, 1)$.

3. If $p > 0$, the power function $y = x^p$ passes through the origin.

   If $p < 0$, the power function $y = x^p$ rises toward the positive y-axis as x gets closer to 0 (the y-axis is a vertical asymptote) and decays toward the positive x-axis as x approaches ∞.

Statement 2 is true because 1 raised to any power is 1 (that is, $1^p = 1$ for any p). The first part of Statement 3 is obvious because 0 raised to any positive power will be 0 (that is, $0^p = 0$ for $p > 0$). As for the second part of Statement 3, if the power p is negative, we can write it as $p = -q$ so that

$$x^p = x^{-q} = \frac{1}{x^q},$$

using one of the basic rules for exponents from algebra. Obviously, we can't have $x = 0$ because the quotient $1/x^q$ is not defined at 0. However, the closer x is to 0, the closer x^q is to 0 also, and therefore the larger $1/x^q$ is. That is, the graph of any power function of the form $y = x^p = x^{-q}$ must always rise and approach the positive y-axis as x approaches 0. The y-axis is a vertical asymptote for these curves because they approach it more and more closely but never reach it. Also, if $p < 0$, as x increases, $x^p = x^{-q} = 1/x^q$ becomes smaller and eventually approaches 0. So the x-axis is a horizontal asymptote for any power function with $p < 0$.

Behavior of Power Functions for $x \geq 1$

It is also evident from Figure 2.54 that the behavior patterns for these power functions are different when x is between $x = 0$ and $x = 1$ compared to when $x > 1$. Again, the point $(1, 1)$ serves as a demarcation between the different behaviors. To see the differences more clearly, in Figure 2.55 we zoom in on all the graphs shown in Figure 2.54 to examine what happens when $x > 1$ for different values of p. (Incidentally, in all these cases, you should think of x as the variable and p as a parameter that takes on a fixed value to produce a particular curve whose values depend on x.)

Figure 2.55 suggests the following additional fact about power functions.

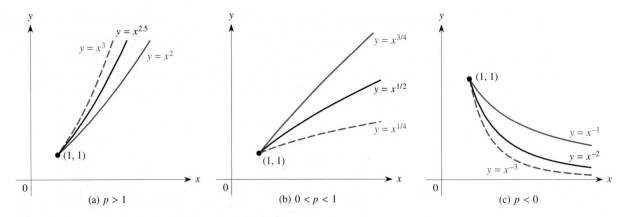

FIGURE 2.55

> 4. If $p > 0$, the larger p is, the larger x^p is when $x > 1$.
>
> If $p < 0$, the more negative p is, the more rapidly x^p dies out as x increases.

Thus, for instance, when $x > 1$,

$$x^5 > x^4 > x^3 > x^2,$$

and also

$$x^{-5} < x^{-4} < x^{-3} < x^{-2}.$$

These relationships are shown numerically in Table 2.1 for $p > 0$. For each value of $x > 1$, not only do the higher powers of x get larger, but they get *larger much faster.*

TABLE 2.1

	$x = 2$	$x = 5$	$x = 10$	$x = 20$
$y = x^2$	4	25	100	400
$y = x^3$	8	125	1000	8000
$y = x^4$	16	625	10,000	160,000
$y = x^5$	32	3125	100,000	3,200,000

We can prove this fact algebraically. For instance, if $x > 1$ and we multiply both sides of this inequality by the positive term x^3, we get

$$x^3 \cdot x > x^3 \cdot 1 \quad \text{so that} \quad x^4 > x^3.$$

This comparison gets more pronounced as x gets larger and larger. Compare the values in the table for $x = 10$ and $x = 20$. As x gets ever larger (i.e., as x approaches infinity, denoted by $x \rightarrow \infty$), any positive power completely overwhelms, or dominates, any smaller power.

Now let's look at the case when $p < 0$. For each value of $x > 1$, Table 2.2 shows that, not only do the negative powers of x get smaller as x increases, but the more negative the power, the faster the function $y = x^p$ dies out.

TABLE 2.2

	$x = 2$	$x = 5$	$x = 10$	$x = 20$
$y = x^{-2}$	0.25	0.04	0.01	0.0025
$y = x^{-3}$	0.125	0.008	0.001	0.000125
$y = x^{-4}$	0.0625	0.0016	0.0001	0.00000625
$y = x^{-5}$	0.03125	0.00032	0.00001	0.0000003125

Behavior of Power Functions for $0 < x < 1$

The preceding conclusions are based on what happens to power functions when $x > 1$. Now let's see what happens to these same power functions when $0 < x < 1$, as shown in Figure 2.56. Note that the behavior patterns are reversed from those when $x > 1$. In particular, the graphs suggest the following fact about power functions.

> 5. If $p > 1$ or $p < 0$, the larger p is, the smaller x^p is when $0 < x < 1$.
> If $0 < p < 1$, the larger p is, the larger x^p is when $0 < x < 1$.

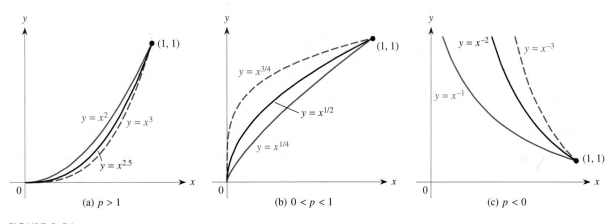

FIGURE 2.56

Thus, for instance, when x is between 0 and 1,

$$x^5 < x^4 < x^3 < x^2,$$

and

$$x^{-2} < x^{-3} < x^{-4} < x^{-5}.$$

Thus, as x approaches 0, the more positive the power p, the faster a power function dies out near the origin. Similarly, the more negative the power p, the faster a power function grows toward infinity as x approaches 0.

Think About This

What do the graphs of $y = x^{0.99} = x^{(99/100)}$ and $y = x^{1.01}$ look like? How do they behave compared to the line $y = x$? ◻

Think About This

What do the graphs of $y = x^{0.01}$, $y = x^0$, and $y = x^{-0.01}$ look like? How do they compare to one another? ◻

Applications of Power Functions

In Sections 2.2 and 2.3, we dealt with many situations that led to equations involving linear functions. In Sections 2.4–2.6, we similarly explored some situations leading to equations involving exponential functions of the form

$$y = kc^t.$$

In many of these situations, we had a value for y and had to solve for the unknown t in the exponent by using logarithms. Similarly, situations often arise that lead to equations involving power functions of the form

$$y = kx^p,$$

where a value for y is given and we have to solve for the unknown x, which in this case is raised to a constant power.

As we discussed in several simple examples in previous sections, we solve the power function equation $x^2 = 25$ by taking the square root of both sides of the equation to get $x = \pm 5$. (To envision the two solutions, imagine the graph of the function $y = x^2$: It reaches a height of 25 in two places, one for $x < 0$ and another for $x > 0$.) Similarly, we solve the equation $x^3 = 27$ by taking the cube root of both sides of the equation to get

$$x = \sqrt[3]{27} = 3.$$

Similarly, we solve the equation $x^7 = 50$ by taking the 7th root of both sides of the equation (equivalently by raising both sides to the $\frac{1}{7}$ power):

$$x = \sqrt[7]{50} = 50^{1/7} \approx 1.7487.$$

EXAMPLE 1

A bald eagle weighs about 16 pounds. Use the relationship $W = 0.15S^{9/4}$ to estimate its wingspan.

Solution We have to solve for the eagle's wingspan S corresponding to $W = 16$ pounds in the equation

$$0.15S^{9/4} = 16.$$

We first divide both sides by 0.15 to get

$$S^{9/4} = 106.667.$$

To solve for S, we have to undo the $\frac{9}{4}$ power, which we do by raising both sides of this equation to the $\frac{4}{9}$ power:

$$(S^{9/4})^{4/9} = S^1 = S = 106.667^{4/9} \approx 7.968,$$

or about 8 feet.

In Example 2 we show how these ideas about power functions arise in the context of one of the most useful applications of radioactive decay. Scientists routinely use a process known as carbon-dating to establish the age of fossils. It is based on the fact that carbon-14 decays to carbon-12 with a half-life of 5730 years.

EXAMPLE 2

Crater Lake in Oregon was formed as the result of a volcanic eruption. A charcoal sample from a tree that burned during the eruption contains about 46% of the carbon-14 found in live trees.

a. What is the decay factor for carbon-14?

b. What was the approximate date for the formation of Crater Lake?

Solution   Because the half-life of carbon-14 is 5730 years and slightly more than 50% of the radioactive carbon has disintegrated, we expect that the time involved is somewhat more than 5730 years; we might estimate, say, about 6000 years. Now let's find out more precisely.

a. The exponential decay function that models the radioactive decay process is

$$R(t) = R_0 c^t$$

for some initial quantity R_0 of the radioactive carbon and some decay factor $c < 1$. Because the half-life of carbon-14 is 5730 years, we substitute $t = 5730$ into the expression for the function to get

$$R(5730) = R_0 c^{5730} = \frac{1}{2} R_0$$

so that

$$c^{5730} = \frac{1}{2}.$$

We solve this equation for c by extracting the 5730th root of $\frac{1}{2}$, to get the decay factor:

$$c = \left(\frac{1}{2}\right)^{1/5730} \approx 0.99988.$$

b. We know that

$$R(t) = R_0 \cdot (0.99988)^t.$$

We now have to find how long it takes for the carbon-14 to decay to the point where only 46% of R_0 is present. Thus we want to find t when

$$R(t) = R_0 \cdot (0.99988)^t = 0.46 R_0.$$

Dividing through by R_0 gives

$$(0.99988)^t = 0.46.$$

To solve this exponential equation, we take logs of both sides and get

$$\log(0.99988)^t = t \log(0.99988) = \log(0.46).$$

Therefore

$$t = \log(0.46)/\log(0.99988) \approx 6470.685,$$

or about 6471 years ago, as shown in Figure 2.57. We thus conclude that Crater Lake was formed in roughly 4471 B.C.

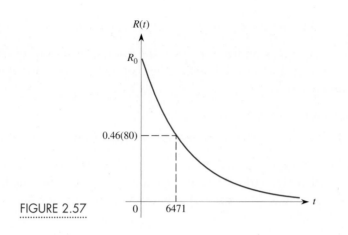

FIGURE 2.57

Note that we solved two very different mathematical problems in Example 2. In part (a), we had the power function equation $c^{5730} = \frac{1}{2}$ and solved for c by taking the 5730th root. In part (b), we had the exponential function equation $(0.99988)^t = 0.46$ and solved for t by using logarithms. In general, we take roots to undo powers and so solve equations of the form $x^n = A$ involving power functions. Similarly, we use logarithms to extract variables from an exponent and so solve equations of the form $b^x = A$ involving exponential functions.

Unfortunately, neither operation will do anything useful to solve an equation such as

$$3^x = x^4,$$

which involves both an exponential function and a power function. This equation can be solved numerically or graphically to find an approximate solution that is accurate to any desired degree of accuracy, by using a graphing calculator or a computer graphics program. (We ask you to solve it as a problem at the end of the section.) But, the equation *cannot* be solved algebraically to find an exact solution.

Fitting Power Functions to Two Points

We have seen that two points determine a line or an exponential function because only one line or one exponential function passes through those points. Similarly, two points also determine a power function, which we illustrate in Examples 3–5.

EXAMPLE 3

Find the power function that passes through the points $(1, 5)$ and $(4, 60)$.

Solution A power function is of the form $y = kx^p$, for some constants k and p. We substitute the coordinates of the first point $(1, 5)$ into this expression to obtain

$$k \cdot (1)^p = k = 5.$$

Therefore the expression for the power function reduces to $y = 5x^p$. We use the second point, $(4, 60)$, to determine the power p. When we substitute the coordinates $x = 4$ and $y = 60$, we get

$$5(4)^p = 60.$$

We divide both sides of this equation by 5 to obtain

$$4^p = 12.$$

To extract the unknown p from the exponent, we take logs of both sides to get

$$\log(4^p) = p \log 4 = \log 12,$$

so that

$$p = \frac{\log 12}{\log 4} \approx 1.792.$$

Consequently, the power function that passes through the two points $(1, 5)$ and $(4, 60)$ is

$$y = 5x^{1.792}.$$

As depicted in Figure 2.58, this function is increasing and concave up.

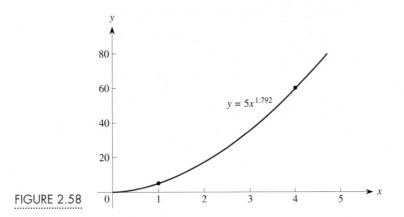

FIGURE 2.58

Example 4 illustrates a somewhat more complicated situation.

EXAMPLE 4

Find the power function that passes through the points $(2, 5)$ and $(4, 60)$.

Solution   When we substitute the coordinates of the first point $(2, 5)$ into the equation of a power function $y = kx^p$, we get

$$k \cdot 2^p = 5,$$

which involves both unknowns k and p. Similarly, when we substitute the coordinates of the second point $(4, 60)$, we get

$$k \cdot 4^p = 60.$$

To eliminate one of the unknowns, we divide the second equation by the first:

$$\frac{k \cdot 4^p}{k \cdot 2^p} = \frac{4^p}{2^p} = \frac{60}{5} = 12.$$

Because

$$\frac{4^p}{2^p} = \left(\frac{4}{2}\right)^p = 2^p,$$

the preceding equation becomes

$$2^p = 12.$$

We solve for p by using logarithms:

$$\log 2^p = p \log 2 = \log 12$$

so that

$$p = \frac{\log 12}{\log 2} \approx 3.584963.$$

The desired power function is therefore $y = k \cdot x^{3.584963}$. To determine the value of k, we use the first point $(2, 5)$ and obtain

$$5 = k \cdot 2^{3.584963}.$$

Therefore

$$k = \frac{5}{2^{3.584963}} \approx 0.41667,$$

and the power function, which is shown in Figure 2.59, is

$$y = 0.41667x^{3.584963}.$$

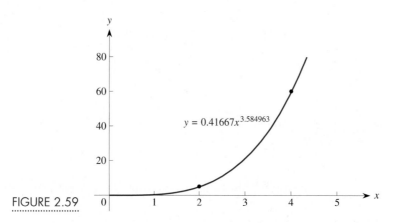

FIGURE 2.59

EXAMPLE 5

Biologists have long observed that the larger the area of a region, the more species inhabit it. The relationship is best modeled by a power function. The island of Puerto Rico contains 40 species of reptiles and amphibians on its 3459 square miles. The nearby island of Hispaniola (comprising Haiti and the Dominican Republic) contains 84 species on 29,418 square miles.

a. Determine a power function that relates the number of species of reptiles and amphibians on a Caribbean island to its area.

b. Use the relationship from part (a) to predict the number of species of reptiles and amphibians on Cuba, which measures 44,218 square miles.

Solution

a. We want a power function of the form $S = kA^p$, where S is the number of species, A is the area in square miles, and k and p are constants that must be determined. Using the information on Puerto Rico, where $S = 40$ and $A = 3459$, we have

$$k \cdot (3459)^p = 40, \qquad (1)$$

which involves both p and k. The data on Hispaniola, $A = 29,418$ and $S = 84$, give

$$k \cdot (29,418)^p = 84, \qquad (2)$$

so we now have two *nonlinear* equations in the two unknowns k and p. We can eliminate the unknown k by dividing Equation (2) by Equation (1). We then get

$$\frac{k \cdot (29418)^p}{k \cdot (3459)^p} = \frac{84}{40} = 2.1.$$

We cancel the common factor k to get

$$\frac{29418^p}{3459^p} = 2.1.$$

Using one of the properties of exponents, we find that this equation reduces to

$$\left(\frac{29418}{3459}\right)^p = (8.505)^p = 2.1.$$

To solve for p, we take logs of both sides of this exponential equation to get

$$\log(8.505)^p = p \log(8.505) = \log(2.1),$$

from which we find that

$$p = \frac{\log(2.1)}{\log(8.505)} \approx 0.3466.$$

Substituting this value into Equation (1) gives

$$k \cdot (3459)^{0.3466} = 40,$$

so that

$$k = \frac{40}{(3459)^{0.3466}} \approx 2.3739.$$

Thus the power function that models the number of species of reptiles and amphibians on a Caribbean island having area A is

$$S = 2.3739 A^{0.3466}.$$

Note from the graph of this function shown in Figure 2.60 that it is an increasing, concave down function, which is what we would expect from a power function with $p = 0.3466$.

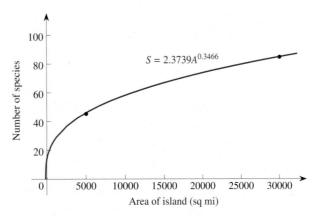

FIGURE 2.60

b. For the area of Cuba, 44,218 square miles, we use this formula to estimate that there are

$$S = 2.3739(44218)^{0.3466} \approx 96.745,$$

or about 97 reptile and amphibian species on Cuba.

Power Functions with Integer Powers

So far, we have restricted our attention to power functions with $x \geq 0$ because most power functions of the form $y = x^{m/n}$ with rational exponents are not defined when $x < 0$. However, if the power is an integer, either positive or negative, the power function $y = x^n$ (n an integer) is well defined for all values of x, both positive and negative. So, when the power is an integer, we extend the domain to include all real x.

Let's look at the case where the power n is a positive integer so that these power functions include $y = x$, $y = x^2$, $y = x^3$, $y = x^4$, $y = x^5, \ldots$, for all real x. Note from Figure 2.61 that the graphs of these power functions fall into two groups: functions with odd powers and functions with even powers.

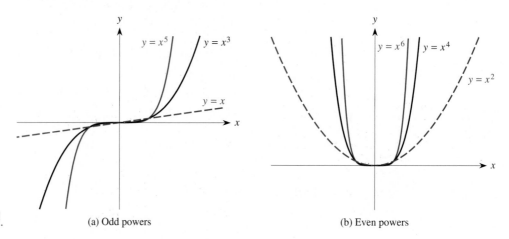

FIGURE 2.61

(a) Odd powers   (b) Even powers

Even Positive Integer Powers

We look at the even power functions: $y = x^2$, $y = x^4$, $y = x^6, \ldots$, as shown in Figure 2.61(b). Their common characteristics are the following.

♦ All even power functions decrease at first (until $x = 0$) and then increase as x increases from left to right, so all are U-shaped with a turning point at the origin.

♦ The higher the power n, the flatter the curve is as it passes through the origin.

♦ All the even powers are concave up everywhere, so they do not have a point of inflection.

♦ Not only does each curve pass through the origin and the point $(1, 1)$, but each one also passes through the point $(-1, 1)$ [because $(-1)^n = 1$ if n is an even integer].

♦ All even power functions are symmetric about the y-axis (the left and right halves of the curves are mirror images.) (See Appendix D.)

Examine some of the even power functions on your own, using your function grapher, to convince yourself that these properties are valid.

Odd Positive Integer Powers

Next, let's examine the power functions with odd powers: $y = x$, $y = x^3$, $y = x^5, \ldots$, as shown in Figure 2.61(a). They have the following characteristics in common.

◆ They all are increasing everywhere as x increases from left to right.

◆ If $n > 1$, all the odd power functions are concave down when $x < 0$ and concave up when $x > 0$. This change in concavity at $x = 0$ means that every odd power function except the line $y = x^1 = x$ has a point of inflection at the origin, where it is growing most slowly.

◆ The higher the power n, the flatter the curve is as it passes through the origin.

◆ Not only does each curve pass through the origin and the point $(1, 1)$, but each one also passes through the point $(-1, -1)$ [because $(-1)^n = -1$ if n is an odd integer].

◆ All odd power functions are symmetric about the origin; that is, the portion of each curve in the third quadrant is the mirror image of the corresponding portion in the first quadrant.

Examine some of the odd power functions on your own, using your function grapher, to convince yourself that these properties are valid.

Problems

1. Which of the pairs of points shown can determine a power function of the form $y = kx^p$ and which cannot. For those that do, sketch the graph of the power function, indicate the sign of the coefficient k, and tell whether the power p is less than 0, between 0 and 1, or greater than 1.

(a)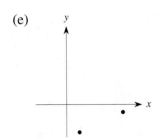

(b)

(c)

(d)

(e)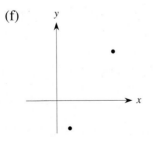

(f)

2. Match each formula for a power function with one of the graphs. Explain the reasons for your decisions.

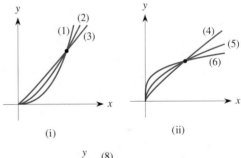

(i)

(ii)

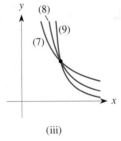

(iii)

a. $y = x^{-0.8}$ b. $y = x^{-0.2}$ c. $y = x^{-1.4}$
d. $y = x^{1.8}$ e. $y = x^{1.6}$ f. $y = x^{1.4}$
g. $y = x^{0.8}$ h. $y = x^{0.6}$ i. $y = x^{0.4}$

3. Identify which of the functions in parts (a)–(n) are exponential functions, which are power functions, and which are neither. Give the reasons for your decisions.

a. $f(x) = 40x^{1.05}$ b. $f(x) = 40(1.05)^x$

c. $f(x) = \dfrac{1}{(1.4)^x}$ d. $f(x) = -\dfrac{3}{x^{2.4}}$

e. $f(t) = 5t^{-3.7}$ f. $f(q) = 1.09q - 4.37$

g. $f(t) = 12(0.35)^{-t}$ **h.** $f(t) = 5\sqrt{t}$
i. $f(s) = \sqrt{s^2 + 3}$ **j.** $f(r) = \frac{4}{3}\pi r^3$
k. $f(z) = z \cdot z^{3/5}$ **l.** $f(x) = x^x$
m. $f(w) = w^2 \cdot 3^w$ **n.** $f(u) = 7(1.62)^{-u}$

4. Match each formula with its corresponding table of values.

   a. $y = 4x^{1.2}$ **b.** $y = 5x^{0.8}$ **c.** $y = 4(1.2)^x$.

   i.

x	2	3	4	5	6
$f(x)$	8.71	12.04	15.16	18.12	20.96

   ii.

x	1	2	3	4	5
$g(x)$	4.80	5.76	6.91	8.29	9.95

   iii.

x	2	4	6	8	10
$h(x)$	9.19	21.11	34.34	48.50	63.40

5. Data from three different functions are shown in the tables of values. One function is exponential, one has the form $y = ax^2$, and one has the form $y = bx^3$. Which function is which?

   i.

x	3	3.5	4	4.5	5
$f(x)$	28.8	39.2	51.2	64.8	80.0

   ii.

x	3	3.5	4	4.5	5
$g(x)$	4.39	5.01	5.71	6.51	7.42

   iii.

x	3	3.5	4	4.5	5
$h(x)$	10.80	17.15	25.60	36.45	50.00

6. For each relationship, **(i)** identify which of the quantities should be considered the independent variable and which the dependent variable; **(ii)** write an equation expressing the dependent variable in terms of the independent variable to create a power function that represents the relationship; and **(iii)** sketch a rough graph of the function based on the value of the power p.

   a. If a car is traveling at a constant rate, the distance d that it travels is proportional to the time t that it travels.

   b. The distance d that an object falls under the influence of gravity is proportional to the square of the time t that it is falling.

   c. According to the ideal gas law, when a gas is kept at a constant temperature, the pressure P is inversely proportional to the volume V.

   d. The force F of gravity between two objects is inversely proportional to the square of the distance d between them.

   e. The square of the diameter d of the long bone in the leg of many animals is proportional to the cube of the length L of the bone.

   f. The cube of the surface area S of many vertebrate mammals is proportional to the square of their body mass m.

   g. The fourth power of the rate R at which air flows into and out of the lungs of many vertebrate mammals is proportional to the cube of their body mass m.

   h. The fourth power of the speed s at which many mammals can trot is proportional to their body mass m.

   i. The cube of the speed s at which most birds fly is proportional to the square of their body mass m.

   j. The square of the swimming speed s of most species of fish is proportional to the length L of their bodies.

   k. The fifth power of the radius r of the shock wave after the explosion of a nuclear bomb is proportional to the square of the time t since the bomb exploded.

7. Use the formula relating the weight of a large flying bird to its wingspan to explain why a 15 pound turkey with a wingspan of about 2.5 feet can't soar like an eagle.

8. A full grown African vulture has a 9-foot wingspan. Based on the model relating weight to wingspan, how much does a vulture weigh?

9. The largest known flying creature, with a wingspan of 40 feet, was the pterosaur that lived 65 million years ago. Assuming that the formula for birds also applies to flying dinosaurs, estimate the weight of an adult pterosaur. What can you conclude from your answer?

10. Find a formula expressing the volume V of a sphere as a function of its surface area S.

11. Find the power function that passes through the following pairs of points.

    a. $(1, 3)$ and $(4, 6)$ **b.** $(1, 3)$ and $(4, 8)$
    c. $(1, 3)$ and $(4, 10)$ **d.** $(5, 20)$ and $(6, 30)$
    e. $(1, 10)$ and $(4, 5)$ **f.** $(2, 20)$ and $(5, 8)$

12. Police sometimes use the formula $s = \sqrt{30kd}$ to estimate the speed s in miles per hour that a car was

going if it left a set of skid marks *d* feet long. The coefficient *k* depends on the road conditions (dry or wet) and the type of pavement. For instance, $k = 0.8$ for dry concrete, $k = 0.4$ for wet concrete, $k = 1.0$ for dry tar, and $k = 0.5$ for wet tar.

a. A car left a set of skid marks 120 feet long on dry concrete. How fast was it going?

b. Suppose that the concrete pavement in part (a) was wet. How fast was the car going?

c. If the car in part (a) left skid marks 240 feet long, how fast was it going?

d. Suppose that a car is going 50 mph on a dry tar surface when the driver slams on the brakes. How far will it skid?

e. Suppose that the tar pavement in part (d) was wet. How far will the car skid?

13. Scientists are actively investigating the potential of using windmills to generate electricity. They have found that, for moderate wind speeds, the power *P* in watts generated by a windmill is related to the wind speed *v* in miles per hour according to the equation

$$P = 0.015v^3.$$

a. How much power is generated by a steady wind at 10 mph?

b. How much power is generated by a steady wind at 20 mph?

c. Based on your results in parts (a) and (b), by what factor does doubling the wind speed increase the power generated?

d. Compare the power generated by a steady wind at 5 mph to that of a steady wind at 10 mph. Does doubling of the wind speed increase the power generated by the same factor found in part (c)?

e. Suppose that a certain community has power needs for an additional 250 kilowatts of electricity and can anticipate winds on the average of 12 mph. How many windmills would be needed to meet the added electric demand?

f. What wind speed would be needed to light a 100-watt light bulb?

14. **a.** Use your function grapher to plot on the same screen the graphs of the power functions $y = x^2$, x^5, and x^8 for the interval $-0.2 \le x \le 0.2$. Determine an appropriate range for *y* so that all powers will be distinguishable in the viewing rectangle.

b. Plot the same graphs for $-2 \le x \le 2$ and determine an appropriate range for *y*.

c. Plot the same graphs for $-20 \le x \le 20$ and determine an appropriate range for *y*.

15. **a.** Use your function grapher to plot on the same screen the graphs of the power functions $y = x^{1/2}$, $x^{1/3}$, and $x^{1/4}$ for the interval $0 \le x \le 0.2$. Determine an appropriate range for *y* so that all powers will be distinguishable in the viewing rectangle.

b. Plot the same graphs for $0 \le x \le 2$ and determine an appropriate range for *y*.

c. Plot the same graphs for $0 \le x \le 20$ and determine an appropriate range for *y*.

d. What happens if you use the interval $-2 \le x \le 2$?

16. What happens to

a. x^3 as $x \to \infty$? as $x \to -\infty$?

b. $-x^3$ as $x \to \infty$? as $x \to -\infty$?

c. $x^{1/3}$ as $x \to \infty$? as $x \to -\infty$?

d. $-x^{1/3}$ as $x \to \infty$? as $x \to -\infty$?

e. x^{-3} as $x \to \infty$? as $x \to -\infty$?

f. x^{-3} as $x \to 0$?

17. In 1990, 442.2 million prerecorded audio cassette tapes were sold, and 865.7 million CDs were sold in the United States. In 1998, 158.5 million cassette tapes were sold, and 1,124.3 million CDs were sold. Assume for now that the patterns of sales for both items are power functions.

a. Find the equation for the number of cassette tapes sold as a power function of time.

b. Find the equation for the number of CDs sold as a power function of time.

c. If the trends in sales of both items were indeed power functions, find when the number of CDs sold overtook the number of cassette tapes sold.

18. In the accompanying figure let *R* be the radius of the Earth (about 3960 miles). Find an expression for the distance *D* to the horizon from a point at a height of *H* miles above the Earth's surface. (*Hint:*

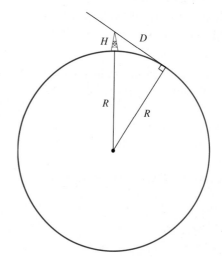

Recall that, for a circle, any tangent line is perpendicular to a radius.)

19. The observation deck of the Empire State Building in New York is 1250 feet high. If you're standing there, complete the phrase: "On a clear day, you can see. . . ." (*Hint:* Use the formula you created in Problem 18.)

20. Ultra high frequency (UHF) TV transmissions travel along a line of sight from a transmitter as far as the horizon. In the Chicago area, the UHF stations broadcast from a transmitter atop the 1454-foot (= 0.275 mile ≈ $\frac{1}{4}$ mile) high Sears tower. What is the greatest distance that someone could receive a UHF signal from the tower?

21. Suppose that a mast 250 feet (about $\frac{1}{20}$ mile) high is being planned for the Sears tower to extend the broadcast range of UHF stations. How much farther would the signal extend? How much larger a receiving area would be covered?

22. NASA's space shuttles orbit the Earth at altitudes of about 200 miles. Find the maximum line-of-sight transmission distance from the shuttle to the surface of the Earth. Approximately how large a receiving area on the Earth is in range of this shuttle?

23. Communications satellites orbit the Earth in geosynchronous orbits (carefully chosen heights and velocities so that they appear to be permanently above a fixed point on the surface of the Earth as the Earth rotates). Suppose that such a satellite is in orbit at a height of 23,000 miles above a point on the equator. The radius of the Earth is about $R = 3960$ miles, so the distance around the equator is approximately $2\pi R = 24{,}880$ miles. Consequently, a point on the equator is rotating at a velocity of about 1037 mph. Find the orbital velocity of such a communication satellite in a geosynchronous orbit.

24. Explain why it isn't possible to have a communications satellite whose signals cover a full half the Earth's surface.

25. Using $R = 3960$ miles for the radius of the Earth, the formula you found in Problem 18 for the line-of-sight distance to the horizon from a height of H miles is $D(H) = \sqrt{H^2 + 2RH} = \sqrt{H^2 + 7920H}$. When H is small, the term H^2 seemingly has little effect on

the value of D, so you might be tempted to approximate the distance D using the simpler formula $D \approx \sqrt{7920H} \approx 89\sqrt{H}$. Determine whether using this approximation is reasonable by completing the following table comparing the estimated value for this distance with the actual value.

H	$D \approx 89\sqrt{H}$	$D = \sqrt{H^2 + 7920H}$
0.1 mile		
1 mile		
10 miles		
100 miles		

26. **a.** Find, correct to three decimal places, *all* values of x for which $x^4 = 3^x$ by graphing the two functions $y = x^4$ and $y = 3^x$. (*Hint:* Use different windows to convince yourself that you have located all points of intersection of the two curves.)
 b. Repeat part (a) by creating the function $y = x^4 - 3^x$ and looking for all the points where $y = 0$.

27. Consider the function $f(x) = x^2$ and let P be the point on the curve where $x = 0$, R be the point where $x = 2$, and Q be the midpoint where $x = 1$. Find the slopes of the three line segments PQ, QR, and PR. How does the slope of PR compare to the slopes of the other two segments?

28. Repeat Problem 27, using the function $g(x) = x^3$. Does the relationship among the three slopes you found in Problem 27 also hold for g?

29. Consider the function $f(x) = x^2$ and let P be the point where $x = a$, Q be the point where $x = a + h$, and R be the point where $x = a + 2h$, for any quantity $h > 0$. Find the slopes of the three line segments PQ, QR, and PR. Show that the slope of PR is the average of the other two slopes.

30. Consider the sequence of values 10^5, 10^4, 10^3, 10^2, and 10^1 and use it to provide a reason for defining $10^0 = 1$. What about 10^{-1}?

31. By trial and error, determine the largest power of 10 that your calculator can handle. What is the smallest positive number?

Exercising Your Algebra Skills

Use the properties of exponents to evaluate each term (do not use a calculator).

1. $9^{1/2}$

2. $9^{-1/2}$

3. $8^{4/3}$

4. $8^{-4/3}$

Simplify.

5. $x^4 \cdot x^3$

6. $x^6 \cdot x^{-8}$

7. $\dfrac{r^8}{r^4}$ 8. $\dfrac{z^{12}}{z^{-9}}$ 11. $\dfrac{x^{3/4}}{x^{5/4}}$ 12. $\dfrac{a^{-2/3}}{a^{5/3}}$

9. $x^{3/4} \cdot x^{5/4}$ 10. $a^{-2/3} \cdot a^{5/3}$ 13. $\dfrac{a^{5/3}}{a^{-2/3}}$

2.8   Comparing Rates of Growth and Decay

Most of the families of functions that we've discussed in this chapter—linear, exponential, and logarithmic—are either strictly increasing or strictly decreasing. If we restrict our attention to nonnegative values of x, power functions also are either strictly increasing or strictly decreasing. Let's summarize what we know so far.

Function	Equation	Behavior	Graph
Linear	$y = mx + b$	Strictly increasing when $m > 0$. The more positive the slope, the faster the rate of increase. Strictly decreasing when $m < 0$. The more negative the slope, the greater the rate of decrease.	
Exponential	$y = c^x$ (for $c > 0$)	Strictly increasing when the growth factor $c > 1$. The larger c is, the faster the function grows. Strictly decreasing when the decay factor $0 < c < 1$. The smaller c is, the faster the function decays toward zero. Exponential graphs are always concave up.	
Power	$y = x^p$	Strictly increasing when $p > 0$. The larger p is, the faster the function grows beyond $x = 1$. If $p > 1$, the graph is concave up—it grows more and more rapidly. If $0 < p < 1$, the graph is concave down—it grows more and more slowly.	

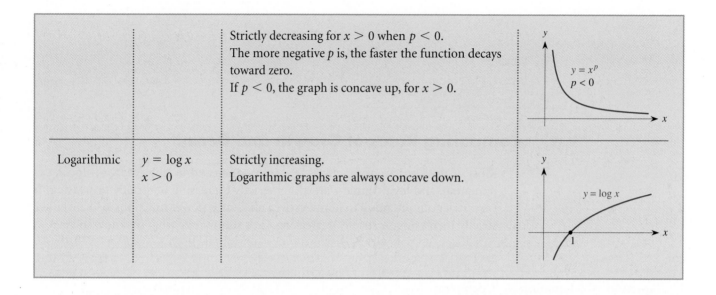

		Strictly decreasing for $x > 0$ when $p < 0$. The more negative p is, the faster the function decays toward zero. If $p < 0$, the graph is concave up, for $x > 0$.	$y = x^p$ $p < 0$
Logarithmic	$y = \log x$ $x > 0$	Strictly increasing. Logarithmic graphs are always concave down.	$y = \log x$

This summary of information compares the growth or decay rate of one function in a family to that of other functions in the same family. In this section, we look at two other ways to compare rates of growth. At a local level, we look at how fast a single function is growing or decaying at different points. At a global level, we look at how quickly functions in one family grow or decay compared to how quickly functions in a different family grow or decay. In particular, we want to answer two questions: (1) which family of functions grows fastest? and (2) which family of functions decays to zero fastest?

Exponential Versus Linear Functions: Which Grow Faster?

We saw in Section 2.4 that any exponential growth function $y = kc^x$ with $c > 1$ will eventually grow faster than any linear function with positive slope. The reason is that the multiplicative effect of the growth factor c in an exponential function is greater than the additive effect of the slope in a linear function. Similarly, any exponential decay function will eventually decrease more slowly than any linear function with a negative slope.

Exponential Versus Power Functions: Which Grows Faster?

Power functions of the form $y = x^p$, $p > 1$ and exponential growth functions of the form $y = c^x$, $c > 1$ both grow rapidly as x increases. But, do they grow at roughly the same rate, or does one grow much faster than the other?

Consider $y = 2^x$ and $y = x^4$ for $x \geq 0$. We know that every power function of the form $y = x^p$ with $p > 0$ passes through the origin and that every exponential curve of the form $y = c^x$ crosses the vertical axis at $y = 1$. So let's begin by comparing these two functions for small values of x. The local, or close-up, view, in Figure 2.62 shows that, between $x = 0$ and $x = 1$, the graph of $y = 2^x$ is above the graph of $y = x^4$ but that the power function seems to be growing more rapidly. If we extend the interval somewhat, we find that by $x = 2$ the power function has surpassed the exponential function. (Where does that happen?) On a somewhat larger scale, Figure 2.63 reveals that the power function continues to pull away from the exponential function.

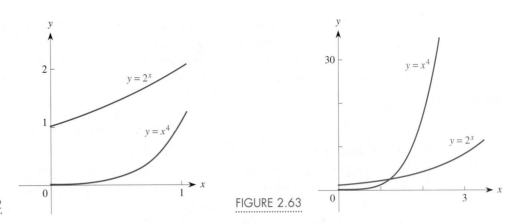

FIGURE 2.62

FIGURE 2.63

However, in Figure 2.64, which shows the interval from $x = 0$ to $x = 20$, the exponential curve has again overtaken the power curve. (Where does that happen?) The still larger view from $x = 0$ to $x = 25$ in Figure 2.65 shows that, for large x-values, $y = x^4$ is insignificant compared to $y = 2^x$. In fact, $y = 2^x$ is growing so much faster than $y = x^4$ that its graph appears almost vertical in comparison to the relatively slow growth of $y = x^4$. Verify this comparison numerically by trying several different values of x—say, $x = 1$, $x = 2$, $x = 10$, and $x = 50$ (but don't go too far because you may exceed your calculator's capacity).

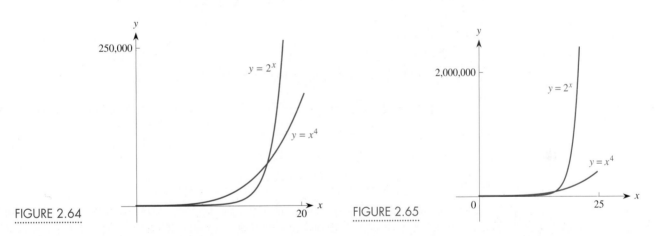

FIGURE 2.64

FIGURE 2.65

Think About This

Use your function grapher to find, correct to two decimal places, all the points where $y = 2^x$ and $y = x^4$ intersect. ▢

This behavior pattern is typical of any power function $y = x^p$, with $p > 1$ compared to any exponential function $y = c^x$, for $c > 1$. Although the exponential function starts more slowly than the power function for small values of x, the exponential function eventually dominates $y = x^p$ for any value of $p > 1$, and so it always wins the race toward infinity.

Think About This

Plot $y = 3^x$ and $y = x^5$ for $0 \leq x \leq 20$ with $0 \leq y \leq 300,000$ to see where the exponential function overtakes the power function. ▢

You have already seen that a positive constant multiple doesn't change the overall shape or behavior of a function. For instance, the power function $y = 5000x^4$ has the same shape as the power function $y = x^4$, but it grows more rapidly because the first is 5000 times larger for any x value. We've already shown that the exponential function $y = 2^x$ eventually overtakes the power function $y = x^4$. It also eventually overtakes the power function $y = 5000x^4$; it just takes longer. The only question is: Where does that happen? In the long run, the exponential function invariably wins the race to infinity.

EXAMPLE 1

Estimate the point x where $f(x) = 1.05^x$ finally overtakes $g(x) = x^{10}$.

Solution   We know that the power function $g(x) = x^{10}$ grows very rapidly and that the exponential function $f(x) = 1.05^x$ has a fairly small growth factor of 1.05. Let's look at their respective function values numerically for different values of x.

x	$f(x) = (1.05)^x$	$g(x) = x^{10}$
10	1.62889	10^{10}
100	131.501	$100^{10} = (10^2)^{10} = 10^{20}$
1000	1.5463×10^{21}	$1000^{10} = (10^3)^{10} = 10^{30}$
10,000	7.816×10^{211}	$10,000^{10} = (10^4)^{10} = 10^{40}$

From this comparison, it is evident that the exponential function has overtaken the power function sometime after $x = 1000$ but long before $x = 10,000$, where $f(x) = (1.05)^x$ has far exceeded the value of $g(x) = x^{10}$. Suppose we narrow our search by trying a few additional values of x.

x	$f(x) = (1.05)^x$	$g(x) = x^{10}$
2000	2.3911×10^{42}	1.024×10^{33}
1500	6.0806×10^{31}	5.7665×10^{31}
1499	5.7911×10^{31}	5.7282×10^{31}
1498	5.5153×10^{31}	5.6901×10^{31}

We conclude that the exponential function $f(x) = (1.05)^x$ finally overtakes the power function $g(x) = x^{10}$ just before $x = 1499$, as illustrated in Figure 2.66. If we zoom in, either numerically on the table or geometrically on the graph, we find that the point of intersection of the two functions occurs near $x = 1498.718$ and $y = 5.71169 \times 10^{31}$.

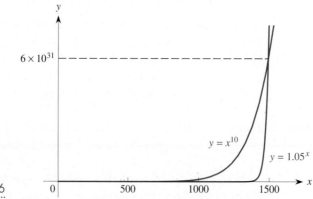

FIGURE 2.66

Now let's compare a decaying exponential function and a power function with a negative exponent. Consider $y = \left(\frac{1}{2}\right)^x = (0.5)^x$ and $y = x^{-2} = 1/x^2$. Both graphs eventually approach the x-axis as a horizontal asymptote, but which one approaches the x-axis faster? We could compare them graphically or numerically by trying several large values of x, such as $x = 100$ or $x = 1000$. Alternatively, we can reason that, because 2^x is eventually larger than x^2, we know that $(0.5)^x = \left(\frac{1}{2}\right)^x = 1/2^x$ is eventually smaller than $x^{-2} = 1/x^2$. So the graph of $y = \left(\frac{1}{2}\right)^x$ eventually falls below the graph of $y = x^{-2}$, as shown in Figure 2.67.

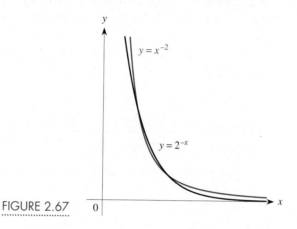

FIGURE 2.67

This behavior pattern is typical. All decaying exponential functions invariably approach 0 faster than any power function with a negative exponent as $x \to \infty$. A power function could begin dropping at a faster rate when compared to an exponential function; for instance, compare $y = x^{-100}$ with $y = 0.9^x$. However, as $x \to \infty$, the exponential function eventually decays faster than the power function to win the race toward 0. The only question is: When does the decaying exponential function overtake the decaying power function on the way to zero? As we have demonstrated, this point of intersection can be approximated with any desired degree of accuracy by using either numerical or graphical methods.

Logarithmic Functions Versus Power Functions: Which Grows Faster?

We know that a logarithmic function and a power function with power $0 < p < 1$ both increase and are concave down. A table of values for $f(x) = \log x$ reveals that the logarithm grows very slowly as x increases beyond $x = 1$. Figure 2.68 shows the

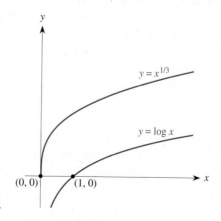

FIGURE 2.68

graphs of $y = \log x$ and $y = x^{1/3}$. In fact, $\log x$ grows more slowly than any positive power of x. In the long run, any power function with $0 < p < 1$ beats the logarithmic function in the race toward infinity.

Think About This

Compare the behavior of different power functions (for example, $y = x^{1/2}$ or $y = x^{1/10}$) to $y = \log x$ on your function grapher to convince yourself of how slowly the log function grows. ⌐

Furthermore, both logarithmic functions and power functions with power $0 < p < 1$ grow more slowly than linear functions with slope $m > 0$. The reason is that the linear function $y = x$ is a power function with $p = 1$, which grows faster than any power function with a smaller power p.

In summary, we have the following facts:

Comparisons When x is Large

Concave Up Growth Functions

> Power functions with power $p > 1$ grow faster than linear functions with slope $m > 0$.

> Exponential functions with growth factor $c > 1$ grow faster than power functions with power $p > 1$.

Concave Down Growth Functions

> Power functions with power $0 < p < 1$ grow more slowly than linear functions with slope $m > 0$.

> Logarithmic functions grow more slowly than power functions with power $0 < p < 1$.

Decay Functions

> Exponential functions with decay factor $0 < c < 1$ decay more rapidly than power functions with power $p < 0$.

Average Rate of Growth

We have often described an increasing, concave up function as "increasing faster and faster" or "increasing at an increasing rate," although we never precisely defined what this means. We now formalize this concept by building on what we know about lines. One of the main characteristics of a linear function is that it grows (or decays) at a constant rate. That is, for each fixed increase (say, Δx) in the independent variable x, the line rises (or falls) the same amount Δy no matter what point on the line we use, as shown in Figure 2.69. The constant ratio $\Delta y / \Delta x$ is the slope of the line.

Now let's try this with an exponential growth function $y = f(x) = kc^x$. At different points on the curve, move the same horizontal distance Δx to the right and determine the corresponding vertical change Δy, as shown in Figure 2.70. At point P, there is a relatively small change in y; at point Q, the corresponding change in y is somewhat larger; and by the time we get to point S, the corresponding change in y is considerably larger.

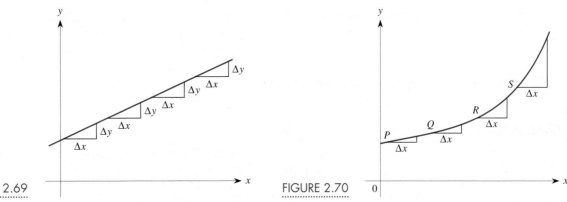

FIGURE 2.69

FIGURE 2.70

For instance, suppose that $y = f(x) = 1.2^x$ and that we take steps of size $\Delta x = 0.5$. We start at $x = 0$, where $y = 1.2^0 = 1$. When we move $\Delta x = 0.5$ to the right, we get to $x = 0.5$ at a height of $1.2^{0.5} = 1.095$. The change in height is $\Delta y = 1.095 - 1 = 0.095$, as shown in Figure 2.71.

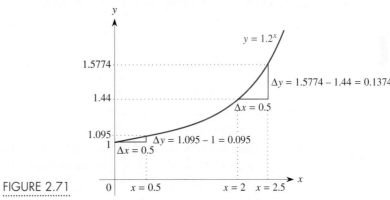

FIGURE 2.71

We now repeat this process, starting at $x = 2$ where $y = f(2) = 1.2^2 = 1.44$. We again move $\Delta x = 0.5$ to the right, getting to the point $x = 2.5$ and $y = f(2.5) = 1.2^{2.5} = 1.5774$. The corresponding change in height is now $\Delta y = 1.5774 - 1.44 = 0.1374$, as shown in Figure 2.71. Thus the same size step to the right has resulted in a considerably larger increase in the change in height of the exponential growth function.

Think About This

Repeat the preceding argument by starting from the point $x = 3$ and moving to the right by $\Delta x = 0.5$. How does the change in y compare to the values of Δy that we found? ⬜

To measure how rapidly this, or any other function $y = f(x)$ is increasing (or decreasing), we consider the ratio $\Delta y / \Delta x$, called the *average rate of change* of the function over an interval, which we discussed briefly in Section 1.1. We have

$$\text{Average rate of change of } f \text{ from } x_1 \text{ to } x_2 = \frac{\Delta y}{\Delta x} = \frac{f(x_2) - f(x_1)}{x_2 - x_1}.$$

In the special case where the function is linear, this ratio is simply the slope of the line.

EXAMPLE 1

Find the average rate of change of the linear function $f(x) = 3x + 5$ between $x = 1$ and $x = 9$.

Solution The average rate of change is

$$\frac{f(9) - f(1)}{\Delta x} = \frac{(3 \cdot 9 + 5) - (3 \cdot 1 + 5)}{9 - 1} = \frac{32 - 8}{8} = \frac{24}{8} = 3,$$

which is the slope of the line.

EXAMPLE 2

Find the average rate of change of the exponential function $f(x) = 1.2^x$ **(a)** between $x = 0$ and $x = 0.5$ and **(b)** between $x = 2$ and $x = 2.5$.

Solution

a. The average rate of change of $f(x) = 1.2^x$ between $x = 0$ and $x = 0.5$ is

$$\frac{f(0.5) - f(0)}{\Delta x} = \frac{1.095 - 1}{0.5} = 0.19.$$

b. The average rate of change of $f(x) = 1.2^x$ between $x = 2$ and $x = 2.5$ is

$$\frac{f(2.5) - f(2)}{\Delta x} = \frac{1.5774 - 1.44}{0.5} = \frac{0.13744}{0.5} = 0.27488.$$

Note how the average rate of change between $x = 2$ and 2.5 is considerably larger than that between $x = 0$ and 0.5. The function is growing ever faster as we move to the right.

Instead of thinking of the average rate of change of a function $y = f(x)$ from x_1 to x_2, we can also think of it as the average rate of change from any point $x = x_0$ to $x = x_0 + \Delta x$. We then have

$$\text{Average rate of change} = \frac{f(x_0 + \Delta x) - f(x_0)}{\Delta x},$$

as illustrated in Figure 2.72. Geometrically, the average rate of change is the slope of the line segment through the two points $(x_0, f(x_0))$ and $(x_0 + \Delta x, f(x_0 + \Delta x))$. The more positive this slope is, the faster the curve is increasing. Thus, for an increasing, concave up function, the average rate of change increases as we move from left to right.

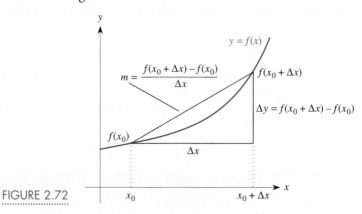

FIGURE 2.72

If $f(x)$ is a decreasing function, the average rate of change will be negative, just as the slope is negative for a decreasing linear function.

Think About This   What is the average rate of change of $f(x) = x^{-1} = 1/x$ from $x = 1$ to $x = 1.02$? ▭

EXAMPLE 3

Find the average rate of change of $f(x) = \log x$ between $x = 2$ and $x = 2.4$.

Solution   In going from $x = 2$ to $x = 2.4$, we have a step of $\Delta x = 0.4$, as depicted in Figure 2.73. Therefore the average rate of change between $x = 2$ and $x = 2.4$ is

$$\frac{f(2.4) - f(2)}{\Delta x} = \frac{\log 2.4 - \log 2}{0.4} \approx 0.19795.$$

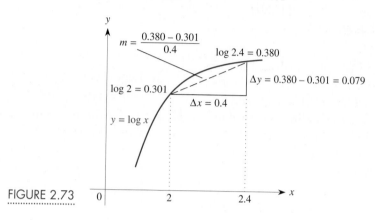

FIGURE 2.73

Problems

1. Use your function grapher to graph $y = x^3$ and $y = 2^x$. Determine appropriate x- and y-scales to obtain the diagrams shown below.

2. **a.** For what values of x is $2^x > x^2$?
   b. For what values of x is $3^x > x^3$?
   c. For what values of x is $4^x > x^4$?

3. Use your function grapher to estimate when $y = (0.6)^x$ overtakes $y = x^{-6}$ as they both decay to zero.

4. Estimate where $f(x) = x^{0.15}$ finally overtakes $g(x) = \log x$ on their "turtle versus snail" race toward infinity.

5. For each linear function, find the average rate of change on the indicated intervals.

   a. $f(x) = 4x - 9$ between $x = 2$ and $x = 5$
   b. $f(x) = 4x - 9$ between $x = -2$ and $x = 3$
   c. $f(x) = -3x + 4$ between $x = 2$ and $x = 5$
   d. $f(x) = -3x + 4$ between $x = -2$ and $x = 3$
   e. $4x - 3y = 12$ between $x = 2$ and $x = 5$

6. Prove that the average rate of change for any linear function $f(x) = mx + b$ on any interval from x_1 to x_2 is equal to the slope m.

7. Find the average rate of change of $f(x) = x^2$ **(a)** between $x = 0$ and $x = 1$, **(b)** between $x = 0$ and $x = 2$, and **(c)** between $x = 1$ and $x = 2$. Put them in ascending numerical order.

(a)
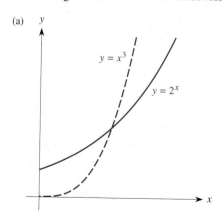

(b)
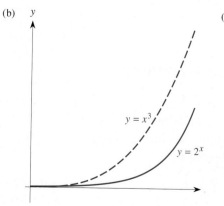

(c)
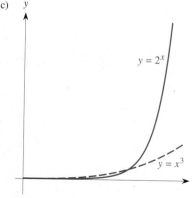

8. Consider the function $f(x) = x^2$.

   a. Find the average rate of change of f between $x = 0$ and 0.01.

   b. Find the average rate of change of f between $x = 1$ and 1.01.

   c. Find the average rate of change of f between $x = 2$ and 2.01.

   d. Based on your results from parts (a)–(c), can you predict the average rate of change of f between $x = 3$ and 3.01? between $x = 4$ and 4.01?

   e. What happens to your answers in parts (a)–(c) if $\Delta x = 0.001$ instead of 0.01?

9. a. In parts (b), (c), and (d), you are asked to find the average rate of change of $f(x) = \sqrt{x}$ between $x = 0$ and 1, between 1 and 2, and between 0 and 2. Before calculating these values, predict the numerical order, from smallest to largest, of these three quantities.

   b. Find the average rate of change of f between $x = 0$ and 1.

   c. Find the average rate of change of f between $x = 1$ and 2.

   d. Find the average rate of change of f between $x = 0$ and 2.

10. The functions $y = x^2$ and $y = 2^x$ intersect at $x = 2$ and at $x = 4$; the functions $y = x^3$ and $y = 3^x$ intersect at $x = 3$ and at $x \approx 2.478$. In general, for $x \geq 0$ the graphs of $f(x) = x^p$ and $g(x) = p^x$ intersect at two points except for one specific value of p (with $p > 1$) for which the curves intersect at only one point. Use your function grapher and trial and error to locate the one special value of p (accurate to two decimal places) for which the curves $y = x^p$ and $y = p^x$ intersect at only one point.

11. a. Use the three points P, Q, and R shown on the accompanying graph of $y = f(x)$ to determine the three line segments PQ, QR, and PR. List these line segments in the order of *increasing* slope (smallest to largest).

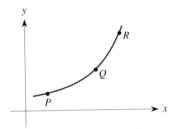

   b. Repeat part (a) if the function is increasing and concave down instead.

12. Consider the function f shown and the point A at $x = a$ on the curve. Determine the point:

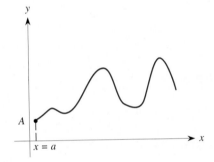

   a. B at $x = b$ giving the interval from a to b over which the change in f is least.

   b. C at $x = c$ giving the interval from a to c over which the change in f is greatest.

   c. D at $x = d$ giving the interval from a to d over which the average rate of change in f is least.

   d. E at $x = e$ giving the interval from a to d over which the average rate of change in f is greatest.

2.9 ········Inverse Functions

Countries using the metric system report temperatures in degrees Celsius. Thus an American visiting Canada who wants to know the temperature in degrees Fahrenheit must be able to convert the Celsius readings to Fahrenheit readings by using the formula

$$F = \frac{9}{5}C + 32.$$

In this relationship the Fahrenheit measurement is a function of the Celsius measurement. Canadian visitors to the United States face the reverse problem: They must convert Fahrenheit readings to Celsius readings. They can do so by solving the preceding formula algebraically for C as a function of F, getting the related function

$$C = \frac{5}{9}(F - 32).$$

These two functions have the effect of undoing each other. For that reason, they are called **inverse functions.**

In general, we write the inverse of a function f as f^{-1} and read it as "f inverse." The inverse of a function f is a function that reverses or undoes f. The two functions relating F and C, which give temperature conversions between the two systems of measurements, represent a pair of inverse functions.

Suppose that we have a function $y = f(x)$ that represents some quantity or process of interest to us. Typically, we can ask two types of predictive questions. The first question is: Determine the value of y corresponding to a particular value of x. All we need do is substitute the value of x in the expression for the function. The second question is: Determine when the quantity achieves a particular level—that is, find the value of the independent variable x that produces a given value for the dependent variable y. Here we must undo the given function, which is what the inverse function is all about. To be able to answer this question requires the existence of an inverse function and the ability either to find its equation algebraically or to estimate its values graphically or numerically.

When a function is given in a table, finding the inverse function is trivial, as we demonstrate in Example 1.

EXAMPLE 1

Table 2.3 gives the average distance D from the sun (in millions of miles) for each of the planets as a function of its average speed S (in miles per hour). Find the inverse function.

Solution For this function, we think of the average speed S of each planet as the independent variable and the average distance D from the sun as the dependent variable, so

TABLE 2.3

Planet	Speed	Distance $D = f(S)$
Mercury	4,110	36.0
Venus	7,671	67.2
Earth	10,605	92.9
Mars	16,153	141.5
Jupiter	55,171	483.3
Saturn	101,164	886.2
Uranus	203,459	1782.3
Neptune	318,790	2792.6
Pluto	418,744	3668.2

$D = f(S)$. The corresponding inverse function f^{-1} simply reverses the role of the two variables. Thus the average distance D from the sun of each planet is now the independent variable and the average speed S of each planet is the dependent variable. We write $S = f^{-1}(D)$. Table 2.4 simply interchanges the columns for S and D from Table 2.3, as shown.

TABLE 2.4

Planet	Distance	Speed $S = f^{-1}(D)$
Mercury	36.0	4,110
Venus	67.2	7,671
Earth	92.9	10,605
Mars	141.5	16,153
Jupiter	483.3	55,171
Saturn	886.2	101,164
Uranus	1782.3	203,459
Neptune	2792.6	318,790
Pluto	3668.2	418,744

In general, if (a, b) is a point on the graph of a function $y = f(x)$, then (b, a) must be a point on the inverse function $x = f^{-1}(y)$. Thus, finding the inverse for any function given in a table is trivial, assuming that the inverse function exists.

For relatively simple functions given by formulas, determining the inverse function f^{-1} for a given function f is straightforward. We just solve the original expression algebraically for the independent variable in terms of the dependent variable, as illustrated in Example 2.

EXAMPLE 2

Find the inverse function to the Celsius to Fahrenheit conversion function $F = \frac{9}{5}C + 32$.

Solution To find the inverse function, we first subtract 32 from both sides of the equation and get

$$\frac{9}{5}C = F - 32.$$

Multiplying both sides of the equation by $\frac{5}{9}$ gives

$$C = \frac{5}{9}(F - 32).$$

So $C = \frac{5}{9}(F - 32)$ is the inverse function.

Many functions, however, do not have an inverse, as we show later. Even when a function f does have an inverse, it is not always possible to find a formula for the inverse f^{-1} algebraically. Fortunately, as Examples 3 and 4 indicate, most of the common functions that we have discussed do have inverses and we can find expressions for them algebraically.

EXAMPLE 3

a. Find the inverse function for the exponential function $P(t) = 12.94(1.029)^t$ that models the population of Florida, where t is the number of years since 1990 and P is the population in millions.

b. What does this inverse function tell us?

c. What are reasonable values for the domain and range of this inverse function?

Solution

a. Since $P = 12.94(1.029)^t$, we have to solve for t as a function of P. First, we divide both sides by 12.94:

$$\frac{P}{12.94} = (1.029)^t.$$

To solve for t, we take logs of both sides of this equation:

$$\log\left(\frac{P}{12.94}\right) = \log(1.029)^t = t\log(1.029).$$

Therefore

$$t = \frac{\log(P/12.94)}{\log(1.029)} = f^{-1}(P).$$

We can stop here or we can use properties of logarithms to simplify this expression. Using the property that the logarithm of a quotient is the difference of the logs, we get

$$t = f^{-1}(P) = \frac{\log(P) - \log(12.94)}{\log(1.029)}$$

or, using the approximate values of log 12.94 and log 1.029, we have

$$t \approx \frac{\log(P) - 1.1119}{0.0124} \approx 80.545\log(P) - 89.669.$$

This logarithmic function is the inverse to the function modeling Florida's population.

b. The inverse function gives the number of years since 1990 (the value of t) that it takes for the population of Florida to reach any given level P.

c. For the inverse function, the independent variable is the value of the population P and the dependent variable is the number of years t since 1990. Therefore the domain of the inverse function consists of all reasonable values for P—say, from 5 million to a maximum of 50 million people. The range consists of all corresponding values of t. To find these values, we use the equation for the inverse function that we obtained in part a. If we substitute $P = 5$ into the preceding equation, we get

$$t = f^{-1}(5) \approx 80.545\log(5) - 89.669 \approx -33.4.$$

According to this model, the population of Florida was 5 million about 33 years before 1990, or in 1957. Similarly, substituting $P = 50$ yields

$$t = f^{-1}(50) \approx 80.545\log(50) - 89.669 \approx 47.2,$$

so the population of Florida will reach 50 million about 47 years after 1990, or in early 2037. Consequently, a reasonable range for the inverse function is t from -33.4 to 47.2 years, which corresponds to about 1957 to about 2037.

◆

In general, the inverse of any exponential function will be a logarithmic function (and vice versa) because the logarithm undoes the exponential function.

EXAMPLE 4

We have shown that the power function $W = f(S) = 0.15S^{9/4}$ can be used to model the weight W of large birds as a function of their wingspan S.

a. Find the inverse function for f.

b. What does the inverse function tell us?

c. What wingspan would allow a 15 pound turkey to fly?

d. What is a realistic domain and range for the inverse function?

Solution

a. We have

$$W = f(S) = 0.15S^{9/4}.$$

To solve for S, we first divide both sides of the equation by 0.15:

$$S^{9/4} = W/0.15 \approx 6.667\ W.$$

To find S, we must undo the $\frac{9}{4}$ power, so we raise both sides of this equation to the $\frac{4}{9}$ power:

$$(S^{9/4})^{4/9} = S = (6.667W)^{4/9},$$

using properties of exponents. Therefore the inverse function is

$$S = (6.667)^{4/9}\ W^{4/9} \approx 2.324\ W^{4/9}.$$

b. The inverse function gives the wingspan S in feet needed to support in flight a bird that weighs W pounds.

c. If a turkey weighs 15 pounds, this formula predicts that, in order for the turkey to fly, it would need a wingspan of

$$S = 2.324(15)^{4/9} \approx 7.7436 \text{ feet.}$$

Since this is about three times the actual wingspan of a turkey, it isn't able to fly.

d. For the original function f, the independent variable is the wingspan S and the dependent variable is the weight W of a bird. For the inverse function f^{-1}, the independent variable is the weight W and the dependent variable is the wingspan S. If we consider reasonably large birds that weigh between 2 pounds and 20 pounds, say, the domain for the inverse function f^{-1} will be between $W = 2$ and $W = 20$. To find the corresponding range, we use the equation for S we obtained in part (a):

$$f^{-1}(2) = 2.324(2)^{4/9} \approx 3.16;$$
$$f^{-1}(20) = 2.324(20)^{4/9} \approx 8.80.$$

Thus the range of the inverse function is from about 3 feet to almost 9 feet.

◆

Note that the inverse function to the power function $W = f(S) = 0.15S^{9/4}$ turned out to be another power function, $S = f^{-1}(W) = 2.324W^{4/9}$. In general, the inverse of any power function, if it exists, is a power function.

Further, note how the powers of the two functions $W = f(S) = 0.15S^{9/4}$ and $S = f^{-1}(W) = 2.324W^{4/9}$ compare algebraically: Each power is the reciprocal of the other. This result is analogous to what happens with $y = x^3$: We solve for x by extracting the cube root to get $x = y^{1/3} = \sqrt[3]{y}$. Thus the inverse function for $y = f(x) = x^3$ is $x = y^{1/3} = f^{-1}(y)$. Similarly, if we need to find the value of x for which

$$x^{175} = 200, \quad \text{then} \quad x = \sqrt[175]{200} = 200^{1/175} \approx 1.03074.$$

If we need to determine the value of the base b for which

$$b^{2004} = \frac{1}{2}, \quad \text{then} \quad b = \sqrt[2004]{\frac{1}{2}} = \left(\frac{1}{2}\right)^{1/2004} \approx 0.99965.$$

To verify these results, just calculate $(1.03074)^{175}$ and $(0.99965)^{2004}$. Although these numbers may seem bizarre to you, we perform such operations routinely in later chapters because they allow us to answer some interesting questions.

Examples 3 and 4 illustrate two important, though different, situations. To extract an unknown variable that appears as the base in a power function $y = x^p$, take the corresponding pth root of both sides of the equation. To extract an unknown variable that appears in the exponent of an exponential function $y = c^x$, take the logarithm of both sides of the equation. We summarize this information as follows. Be sure that you understand the difference between these two situations.

> To solve for the variable x from a power function $y = x^p$, we extract the pth root. That is,
>
> if $y = f(x) = x^p$,   then $x = f^{-1}(y) = y^{1/p}$.
>
> If p is a fraction whose denominator is even, we must have $x \geq 0$ as the domain for f.
>
> The inverse of a power function is another power function.
>
> To solve for the variable x from an exponential function $y = c^x$, we take logarithms. That is, for $c = 10$,
>
> if $y = f(x) = 10^x$,   then $x = f^{-1}(y) = \log y$.
>
> The inverse of an exponential function is a logarithmic function and vice versa.

Extracting the appropriate root from the preceding functions was quite simple. Unfortunately, complications can arise, depending on the behavior of the function, as we illustrate in Example 5.

◆ **EXAMPLE 5**

The function $y = f(t) = -16t^2 + 48t + 6$ models the height y of a ball thrown straight up as a function of time t. Find how long it takes the ball to reach a height of 35 feet.

Solution For the desired height of $y = 35$ feet, we want to find the corresponding value of t, so we have to solve the equation

$$-16t^2 + 48t + 6 = 35.$$

Figure 2.74 shows the graph of the function $f(t)$. Note that the horizontal line corresponding to the height $y = 35$ crosses the curve twice, once when $t \approx 0.84$ seconds (on the way up) and again when $t \approx 2.16$ seconds (on the way down).

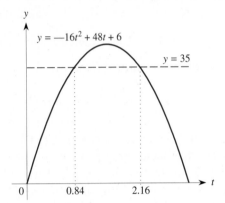

$y = -16t^2 + 48t + 6$

$y = 35$

FIGURE 2.74 0 0.84 2.16 t

Example 5 illustrates the fact that not every function has an inverse. In this case, two different values of t correspond to a height of 35 feet. Consequently, the function $y = f(t) = -16t^2 + 48t + 6$ does not have an inverse because t is *not* a function of y—at least one value of y ($y = 35$) leads to two different values of t. That is, we can't undo the effects of the original function f uniquely. We discuss this situation in more detail later.

We have said that a function f and its inverse f^{-1} undo each other. To show what this means, suppose that we start with a number x in the domain of f. The function f carries x into the corresponding value of $y = f(x)$ in the range of f, as illustrated in Figure 2.75. Similarly, if we start with any value of y in the range of f, then f^{-1} maps y into the value of x associated with it so that $f^{-1}(y) = x$. That is,

$$x = f^{-1}(y) \quad \text{and} \quad y = f(x).$$

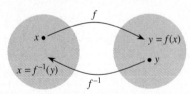

f

$x \bullet$ $y = f(x)$

$\bullet y$

$x = f^{-1}(y)$

FIGURE 2.75 f^{-1}

Again, consider Figure 2.75. For the original function f, the circle on the left represents the domain of f (the allowable values of x) and the circle on the right represents the range of f (the corresponding values of y). For the inverse function f^{-1}, the circle on the right represents the domain of f^{-1} (the allowable values of y) and the circle on the left represents the range of f^{-1} (the corresponding values of x).

In particular, suppose that x_0 is any specific value of x and that $y_0 = f(x_0)$, so that f transforms x_0 into y_0. If we follow this by applying f^{-1} to y_0, we get $f^{-1}(y_0) = x_0$, returning to the original x_0 value. That is, f^{-1} undoes the effect of f on any value x_0. (We consider the idea of applying one function after another in detail in Section 4.6.)

Similarly, if y_0 is any specific value of y and $x_0 = f^{-1}(y_0)$, then $f(x_0) = y_0$. That is, f undoes the effect of f^{-1} on any value y_0.

We can represent these ideas pictorially as follows.

If f and f^{-1} are inverse functions,

for any x: $\quad x \xrightarrow{\ f\ } y = f(x) \xrightarrow{\ f^{-1}\ } x = f^{-1}(y);$

for any y: $\quad y \xrightarrow{\ f^{-1}\ } x = f^{-1}(x) \xrightarrow{\ f\ } y = f(x).$

For instance, the exponential function and the logarithmic function are inverses of each other, which simply restates the relationships

$$\log(10^x) = x \quad \text{and} \quad 10^{\log y} = y.$$

Think About This

Consider the model for the population of Florida $P = f(t) = 12.94(1.029)^t$ and its inverse

$$t = f^{-1}(P) = \frac{\log(P) - \log(12.94)}{\log(1.029)}$$

(from Example 3.) Select any year—say, 1996 when $t = 6$—and verify that if $P = f(t)$, then $t = f^{-1}(P)$. ◻

Determining the Existence of an Inverse Function

Based on the results of Example 5 on the height of a ball thrown vertically upward, we know that not every function f has an inverse f^{-1}. In particular, no power function with an even power can have an inverse. To see why, consider the function $f(x) = x^2$, whose graph is the parabola shown in Figure 2.76. The inverse function, if it exists, should give the value of x that produces any given height y. Suppose that we start at a height of $y = 25$. Clearly, there are two different points on the parabola at a height of 25, one corresponding to $x = -5$ and the other corresponding to $x = +5$. We cannot reverse the process uniquely to get only one value of x for a given y, so y is *not* a function of x and $f(x) = x^2$ does not have an inverse.

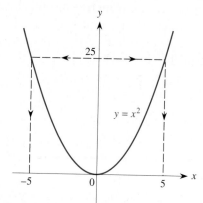

FIGURE 2.76

However, we can restrict the domain of this function to produce a partial inverse. Suppose that we limit our attention to nonnegative values of x and consider the function $g(x) = x^2$, for $x \geq 0$. In this case, if we take any positive value for y

(say, $y = 25$,) we can undo the function g by taking the square root and accepting only the nonnegative value, $x = 5$. Thus the function $g(x) = x^2$ with domain restricted to $x \geq 0$ has an inverse, $g^{-1}(y) = \sqrt{y}$. Alternatively, if we restrict our domain to values of $x \leq 0$, we could also uniquely undo the results of squaring and get the inverse $h^{-1}(y) = -\sqrt{y}$.

Is there a simple criterion to determine whether a function f has an inverse? Definitely! Again, we know that the function $f(x) = x^2$ has an inverse if we restrict its domain to either $x \geq 0$ or $x \leq 0$. It does not have an inverse if we allow the domain to include both positive and negative values for x. When we restrict the domain to $x \geq 0$, we consider only the right-hand side of the parabola where the function is strictly increasing. When we restrict the domain to $x \leq 0$, we consider only the left-hand side of the parabola where the function is strictly decreasing. In both instances, the restricted function has an inverse. On the one hand, when we allow both positive and negative values for x, the function first decreases and then increases and thus does not have an inverse. On the other hand, the function $h(x) = x^3$ has an inverse $x = h^{-1}(y) = \sqrt[3]{y}$ without any restrictions on x. We also know that this function is strictly increasing for all values of x, as shown in Figure 2.77.

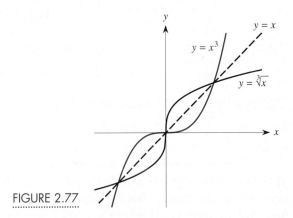

FIGURE 2.77

These observations suggest a simple criterion for functions to have an inverse: The function must be either strictly increasing or strictly decreasing. We call such a function *monotonic*. Compare the two functions shown in Figure 2.78. The one on the left is strictly increasing and, for any desired height y, we can undo the effect of the function and come back to a unique x that produced that particular y. In contrast, the one on the right increases and decreases over different intervals. There are some heights, such as y_3 and y_4, that occur several times, so different x-values correspond to the same y-value. Thus it is not possible to find the unique x that produces a particular y-value.

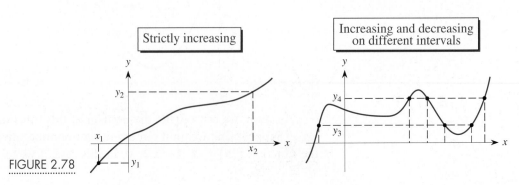

FIGURE 2.78

In summary, all linear, exponential, and logarithmic functions have inverses because they are either strictly increasing or strictly decreasing. Any power function that is restricted to $x > 0$ also has an inverse.

There is an alternative criterion analogous to the Vertical Line Test that we discussed in Section 1.4 for determining whether a curve represents a function. Recall that a curve represents a function if every vertical line crosses the curve at most once. In other words, for any value of x, one and only one value of y corresponds to that x. In an analogous way, we can use a Horizontal Line Test to determine whether a function has an inverse: If every horizontal line crosses the curve at most once, a function f has an inverse f^{-1}. In other words, for any height y, one and only one value of x corresponds to that height.

Behavior of the Inverse Function

At times, expressing both f and f^{-1} as functions of the same variable is desirable so that their graphs can be drawn on the same set of axes and their behaviors compared easily. For instance, we previously showed that

$$y = f(x) = x^3 \quad \text{and} \quad x = f^{-1}(y) = \sqrt[3]{y}$$

are inverse functions. Instead, let's write these two related functions so that both are functions of the same independent variable, x:

$$y = f(x) = x^3 \quad \text{and} \quad y = f^{-1}(x) = \sqrt[3]{x}.$$

Figure 2.77 displayed the graphs of both f and f^{-1}. Notice that the graphs of the function $f(x) = x^3$ and its inverse $f^{-1}(x) = \sqrt[3]{x} = x^{1/3}$ are mirror images of each other about the line $y = x$. (Note that, if we didn't interchange x and y for the inverse function, the two formulas $y = x^3$ and $x = \sqrt[3]{y} = y^{1/3}$ would represent identical curves, so that we would see only one curve.)

Similarly, the exponential function $y = 10^x$ and the logarithmic function $y = \log x$ are inverse functions. If $y = f(x) = 10^x$, we can solve for x by taking the logarithm of both sides to get

$$\log y = \log 10^x = x = f^{-1}(y).$$

We now interchange the variables so that x is also the independent variable for the inverse function and write $y = f^{-1}(x) = \log x$. Again, note that the graphs of these two functions are mirror images of each other about the line $y = x$, as shown in Figure 2.79.

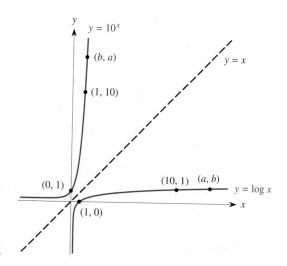

FIGURE 2.79

In general, the graphs of a function f and its inverse f^{-1} are always mirror images of each other about the line $y = x$. For the inverse to exist, the function f must be either strictly increasing or strictly decreasing. Consequently, the graph of the inverse function f^{-1} also is monotonic, and both functions increase or both decrease.

However, there are no clear patterns for the concavity of f and f^{-1}. Both f and f^{-1} can be concave up, both can be concave down, each can have opposite concavity, or both can have no concavity (if their graphs are lines).

Note that $f^{-1}(x)$ is not the same as $1/f(x)$. For instance,

$$\text{if } f(x) = x^3, \quad \text{then} \quad f^{-1}(x) = x^{1/3},$$

whereas $1/f(x) = 1/x^3 = x^{-3}$, which is not the same as $x^{1/3}$. (Check their graphs to convince yourself.) Similarly,

$$\text{if } g(x) = 10^x, \quad \text{then} \quad g^{-1}(x) = \log x,$$

but $1/g(x) = 1/10^x = 10^{-x}$, which is not the same as $\log x$. (Check their graphs to convince yourself.) Figure 2.80 shows the graphs of $g(x) = 10^x$ and $1/g(x) = 10^{-x}$. The graphs are not mirror images of each other about the line $y = x$. Using the symmetry condition, describe where the graph of the inverse function of g would be in Figure 2.80.

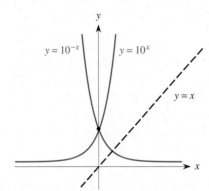

FIGURE 2.80

Finding the Inverse Function

Suppose that we know that a function f has an inverse because it is either strictly increasing or strictly decreasing. Can we always find the inverse function f^{-1}? If the formula for the function is quite simple, we might be able to undo the equation algebraically to obtain a formula for f^{-1}. For instance, we demonstrated earlier that we can undo the conversion formulas between the Fahrenheit and Celsius temperature scales, getting

$$f(x) = \frac{9}{5}x + 32 \quad \text{and} \quad f^{-1}(x) = \frac{5}{9}(x - 32);$$

the algebra is simple because the relationship is linear. Similarly, we can undo the equation of an exponential function to get the logarithmic function and vice versa so that

$$g(x) = 10^x \quad \text{and} \quad g^{-1}(x) = \log x, \qquad \text{for } x > 0.$$

Also, we can undo the relationship between the squaring and square root functions algebraically, obtaining

$$h(x) = x^2 \quad \text{and} \quad h^{-1}(x) = \sqrt{x}, \qquad \text{for } x > 0.$$

In Example 6, we illustrate these ideas with a somewhat more complicated function.

EXAMPLE 6

For the function $y = g(x) = 1 + 1/x$ with domain $x > 0$, find g^{-1}. Analyze the behavior of g and g^{-1}.

Solution   As shown in Figure 2.81, the graph of the function g is strictly decreasing, so its inverse exists. And because the domain of g is $x > 0$, we know that $1/x > 0$. Thus $1 + 1/x$ must be larger than 1, and therefore the range of g must be $y > 1$.

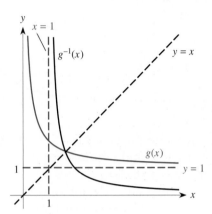

FIGURE 2.81

Next, let's find a formula for the inverse function. If $y = 1 + 1/x$,

$$\frac{1}{x} = y - 1.$$

Taking the reciprocal of both sides, we get

$$x = \frac{1}{y - 1} = g^{-1}(y), \qquad y > 1.$$

Interchanging the roles of x and y to use the same independent variable yields

$$y = \frac{1}{x - 1} = g^{-1}(x), \qquad x > 1.$$

The graphs of the function g and its inverse g^{-1} are also shown in Figure 2.81. As expected, they are mirror images of the other about the line $y = x$.

Further, the original function g is a decreasing function; it decays from a vertical asymptote at $x = 0$ toward a horizontal asymptote of $y = 1$. The graph of g^{-1} also decreases from a vertical asymptote at $x = 1$ toward a horizontal asymptote of $y = 0$. Note that the vertical and horizontal asymptotes are interchanged and that both curves are concave up.

Unfortunately, solving for an inverse function algebraically, as we did in Example 6, usually is not possible. Hence, we usually have to resort to numerical or graphical methods to estimate values for the inverse function; that is, given a particular value for y, we can determine the corresponding value of x by examining either the graph of the original function or successive numerical estimates.

EXAMPLE 7

For the function $f(x) = 2^x + 3^x$, **(a)** explain why f^{-1} exists and **(b)** estimate the value of $f^{-1}(10)$.

Solution

a. As shown in the graph of f in Figure 2.82, the function appears to be strictly increasing, which means that the inverse function exists.

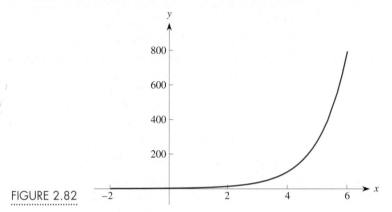

FIGURE 2.82

b. Unfortunately, for the expression $y = 2^x + 3^x$, it is not possible to solve for x in terms of y, so we are unable to find a formula for f^{-1}, even though we do know that f^{-1} exists. Graphically, the curve clearly passes the level $y = 10$ at some point x, so we must estimate, either numerically or graphically, the value of x for which $f(x) = 10$. We know that $f(1) = 2 + 3 = 5$ and that $f(2) = 2^2 + 3^2 = 4 + 9 = 13$, so the desired value of x must be between 1 and 2. We can zoom in either by checking further numerical values or by examining the graph of the function between 1 and 2, as shown in Figure 2.83. Either way, we find that $x \approx 1.73$, so $f(1.73) = 2^{1.73} + 3^{1.73} \approx 10.007$.

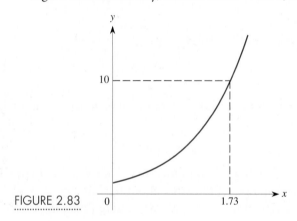

FIGURE 2.83

Problems

1. Which of the following functions have inverses? Explain why or why not. For any function having an inverse, describe what the inverse function tells you.

   a. The height of water after t minutes in a child's pool that you are filling at a steady rate, using a garden hose.

   b. Your distance from New York on an airplane flight from New York to San Francisco as a function of the time t since takeoff.

   c. The height of the student who is numbered n on your instructor's class roster.

   d. The amount that the nth customer in line at Burger Heaven pays for lunch.

e. The length of the fingernail on your right index finger if t is the number of hours since you last clipped your nails.

f. The amount spent by a family to heat their home if T is the temperature at which they set the thermostat.

g. The depth of the snow on a person's front lawn in Buffalo as a function of the time t elapsed from October 1 to the following March 1.

h. The total amount of snow that falls on the person's lawn in part (g) as a function of the time t elapsed from October 1 to the following March 1.

2. For the function f shown in the accompanying figure, estimate the value for x that corresponds to

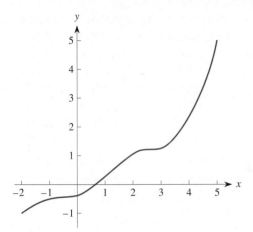

a. $y = 0$;   **b.** $y = 2$;

c. $y = 5$;   **d.** $y = -1$.

Then plot the resulting points and use them to sketch the graph of the inverse function f^{-1}.

3. Consider the function f with values given in the following table.

x	0	1	2	3	4	5
$f(x)$	2.94	2.48	2.05	1.84	1.44	1.12

a. What is the domain of f? What is the range?

b. Create a table of values for f^{-1}. What are its domain and range?

4. Use your function grapher to decide which functions have inverses. For those functions that do, estimate the value for $f^{-1}(10)$.

a. $f(x) = x^3 - 9x^2 + 5x - 5$

b. $f(x) = x^3 - 2x^2 + 5x - 5$

c. $f(x) = 2x + x^2$ **d.** $f(x) = 2x - x^2$

e. $f(x) = 2x + x^3$ **f.** $f(x) = 2x - x^3$

5. The table of values gives the time T needed for a Trans Am to accelerate from zero to the indicated final speed v.

Final speed, v (mph)	30	40	50	60	70
Time, T (sec)	3.00	4.29	5.52	7.38	9.81

a. Explain why this set of data represents a function and why it has an inverse.

b. Explain what the inverse function tells you. What is $f^{-1}(5.52)$? Estimate the value of $f^{-1}(7)$.

6. We know that 1 inch is equivalent to about 2.54 centimeters.

a. Write a formula for the function f that gives an object's length C in centimeters as a function of its length I in inches.

b. Find a formula for the inverse function f^{-1} and explain what f^{-1} tells you, in practical terms.

7. Find the inverse function of $p(t) = (1.04)^t$.

8. Find the inverse function of $f(t) = 50(10)^{0.1t}$.

9. Suppose that the temperature of an object is being measured to the nearest degree on both the Fahrenheit and Celsius scales. In general, which reading would you expect to be more accurate? Why?

10. For each function f shown, sketch the graph of the inverse function f^{-1} on the same set of axes.

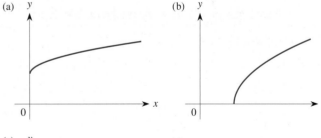

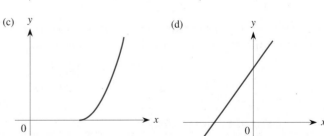

11. Suppose that a function f is increasing and concave up. By thinking of its inverse as the reflection about the line $y = x$, explain why f^{-1} is also increasing. Is it concave up or concave down? What happens if f is increasing and concave down?

12. Repeat Problem 11 if the function f is decreasing and concave up; the function f is decreasing and concave down.

13. On the same set of axes, sketch the graph of f^{-1} that corresponds to the function f shown in the accompanying figure.

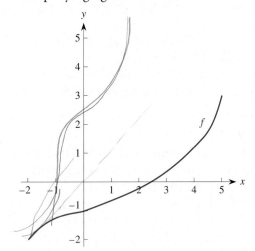

14. Use the quadratic formula

$$t = \frac{-b \pm \sqrt{b^2 - 4ac}}{2a}$$

to solve the quadratic equation $-16t^2 + 48t + 6 = 35$ from Example 5 modeling the height of a ball. What is the significance of the two roots of this equation?

15. The level of Prozac in the blood can be modeled by the function $P(t) = 80(0.75)^t$.
    a. Find a formula for the inverse function.
    b. Use the inverse function to determine how long it will take until the level of Prozac drops to 25 mg.

16. The temperature of a chicken cooking in an oven can be modeled by the function $T(t) = 350 - 310(0.99)^t$.
    a. Find a formula for the inverse function.
    b. Use the inverse function to determine how long it will take until the temperature of the chicken reaches 175°.

17. In Problem 20 of Section 1.3 we introduced a function f that represents a simple replacement code in which each letter of the alphabet is replaced by a different letter according to $f(A) = M$, $f(B) = D$, $f(C) = K$, $f(D) = V$, $f(E) = X$, $f(F) = B$, $f(G) = P$, $f(H) = T$, $f(I) = J$, $f(J) = S$, $f(K) = Z$, $f(L) = Q$, $f(M) = H$, $f(N) = O$, $f(O) = A$, $f(P) = L$, $f(Q) = W$, $f(R) = C$, $f(S) = F$, $f(T) = Y$, $f(U) = R$, $f(V) = G$, $f(W) = I$, $f(X) = U$, $f(Y) = N$, and $f(Z) = E$.
    a. Explain why this function has an inverse f^{-1}.
    b. Use the inverse to decode the message

    JF YTJF HMYT?

Exercising Your Algebra Skills

Solve for the unknown in each equation.

1. $c^{25} = 14$
2. $0.07^t = 3$
3. $0.84^k = 0.20$
4. $m^{1995} = 4$
5. $17b^8 = 32$
6. $25c^9 = 8$
7. $4(1.02)^x = 7$
8. $2(0.75)^t = 1$

For Problems 9–12, solve for the independent variable in terms of the dependent variable for each function.

9. $y = f(x) = 12x^{7/2}$
10. $y = f(t) = 12(1.06)^t$
11. $Q = g(w) = 27w^{-3/4}$
12. $L = g(t) = 125(0.92)^t$

For Problems 13–16, solve for the indicated variable in each formula to find the inverse function.

13. $F = ma$, for a
14. $E = mc^2$, for m
15. $P = kVT$, for V
16. $K = \frac{1}{2}mv^2$, for v

Chapter Summary

In this chapter, we covered the following ideas and approaches relating to families of functions.

◆ Important behavior characteristics of four important families of functions— linear functions, exponential functions, logarithmic functions, and power functions.

◆ How to find the slope and the equation of a line.

◆ What the slope of a line means.

◆ A criterion for knowing when a set of data follows a linear pattern.

◆ How to estimate the equation of a line that captures the linear pattern in a set of data.

◆ How to set up and solve problems involving linear processes.

◆ The behavior of exponential growth and exponential decay functions.

◆ What the growth and decay rates and the growth and decay factors mean.

◆ A criterion for knowing when a set of data follows an exponential pattern.

◆ How to find the doubling time for an exponential growth process and the half-life for an exponential decay process.

◆ How exponential behavior compares to linear behavior.

◆ How to find the exponential function that passes through two points.

◆ How to set up and solve problems involving exponential processes.

◆ The behavior of logarithmic functions.

◆ How to set up and solve problems involving logarithmic functions.

◆ How to use logarithms with bases other than 10.

◆ The behavior of power functions when the power is greater than 1, between 0 and 1, and negative.

◆ How to find the power function that passes through two points.

◆ How to set up and solve problems involving power functions.

◆ How power function behavior compares to exponential behavior.

◆ How logarithmic function behavior compares to power function behavior.

◆ How to determine whether a function has an inverse.

◆ What the inverse function tells you.

◆ How to set up and solve problems involving inverse functions.

Review Problems

1. For each of the curves shown, suggest any types of functions that might have the indicated behavior pattern. If you suggest an exponential function, indicate whether the base c is greater than 1 or less than 1. If you suggest a power function, indicate whether the power p is positive or negative and whether p is greater than 1 or less than 1.

(a)
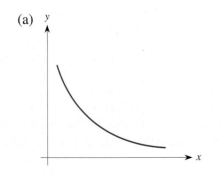

(b)
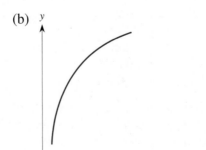

(c)
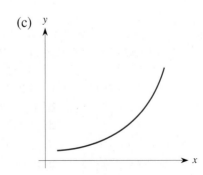

2. Identify each function as linear, exponential, logarithmic, or power. In each case, explain your reasoning.

a.
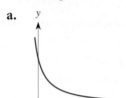

b.
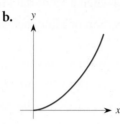

c.
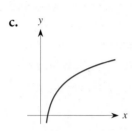

d.
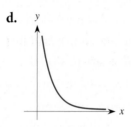

e.
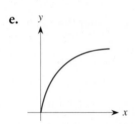

f.
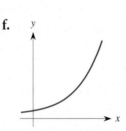

g. $y = 1.05^x$

h. $y = x^{1.05}$

i. $y = (0.7)^x$

j. $y = x^{0.7}$

k. $y = 1/\sqrt{x}$

l. $5x - 3y = 15$

m.

x	y
0	3
1	5.1
2	7.2
3	9.3

n.

x	y
0	5
1	7
2	9.8
3	13.72

i. $y = 5(1.08)^x$ **j.** $y = x^{2.5}$

k. $y = x^{-2.5}$ **l.** $y = x^{0.25}$

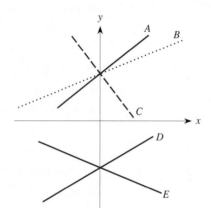

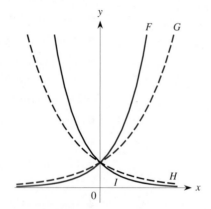

3. Match each formula for a function with one of the graphs (A)–(L). Because more than one function from the same family appears, match each member of that family to the most appropriate graph.

a. $y = 3x + 3$ **b.** $y = 2x - 3$

c. $y = 3 - 2x$ **d.** $y = 2x + 3$

e. $y = -x - 3$ **f.** $y = 5(0.92)^x$

g. $y = 5(0.97)^x$ **h.** $y = 5(1.03)^x$

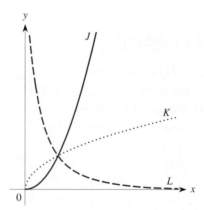

4. The accompanying figure shows the graphs of the values of shares of 7 stocks as functions of time. Match each scenario with one of the graphs and write a brief scenario for each of the remaining graphs.

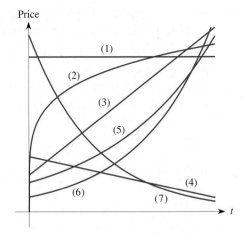

a. The value of the stock increased 12% per year.
b. The value of the stock increased 8% per year.
c. The value of the stock dropped by $4 each year.
d. The value of the stock increased by $6 each year.
e. The value of the stock remained steady.

5. The points $(1, 1)$, $(2, 5)$, $(3, 20)$, $(4, 50)$, and $(5, 100)$ are plotted in the accompanying figure and a line is drawn through them.

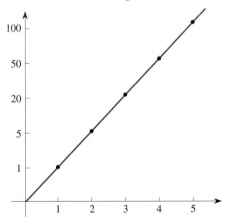

a. What is wrong with the graph as it is drawn?
b. Draw a correct graph of these points and sketch a smooth curve that passes through them.
c. Describe the behavior pattern in the function based on these points. Identify a possible type of function that might be an appropriate model for these values. Explain your reasoning.

6. A function $F(t)$ is exponential with known values $F(0) = 5$ and $F(3) = 8.5$. Determine the function and give the growth factor.

7. **a.** The profits of Alamo Paper Company are growing by $100,000 each year. In 1990, its profits were $1.5 million. Determine the profit function $P(t)$, where t represents the number of years since 1990. Draw a graph of $y = P(t)$ and determine the year in which profits first exceed $2 million.
   b. Ord Paper Company had profits of $950,000 in 1990, and its profits are growing at the rate of 10% each year. Determine when its profits first exceed $2 million.
   c. Use your function grapher to determine when the profits of Ord first exceed the profits of Alamo.

8. When the bald eagle was formally put on the endangered species list in 1967, there were about 800 of the eagles in the United States. As a result of eagle protection and restoration efforts, the bald eagle was removed from the list in 1994 when its population was just over 8000.

   a. Determine the doubling time of the eagle population assuming the growth pattern is exponential.
   b. Estimate the number of bald eagles in the United States in 2005, assuming that the growth trend continues.
   c. When can you expect the eagle population to reach 20,000?

9. In preparing a holiday cranberry mold, a cook added boiling water at 212°F to the fruit and gelatin mixture, which was then poured into the mold and put into a 40° refrigerator. After 30 minutes, the temperature of the mixture was 148°. The temperature $F(t)$ at time t (in minutes) is given by $F(t) = 172(1 + a)^t + 40$, where a is a constant. What is the temperature of the mixture 3 hours after the mold was put in the refrigerator?

10. From 1980 to 1998, the number of workers, in millions, covered by Social Security can be approximated by a linear function of time t. Use the data in the table below and the black thread method to find the equation of a line that fits the data.
    Use this linear function to estimate the number of workers covered by Social Security in 2005.

Year	1980	1985	1990	1992	1993	1994	1995	1996	1997	1998
Workers (millions)	140.4	150.9	164.0	167.5	169.1	170.7	172.9	174.8	177.0	179.1

Source: *2000 Statistical Abstract of the United States*

11. The following table shows the number of deaths in the United States resulting from accidental poisoning by drugs and medications in various years. Use the black thread method to find the equation of a linear function that fits the data. What is the slope of the line? What does this slope represent? Use the equation to predict the number of deaths from accidental poisoning in 2003.

Year	1980	1990	1994	1995	1996
Deaths	2492	4506	7828	8000	8431

Source: *2000 Statistical Abstract of the United States*

12. Find possible equations for the function represented by each graph.

    a.
    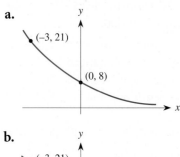

    b.
    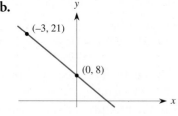

    c.
    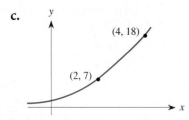

13. In 1990 (when $t = 0$), the IRS collected $1055 billion in taxes. In 1995, the IRS collected $1573 billion.

    a. Construct the linear function giving the amount of taxes collected by the IRS as a function of time t.
    b. Use the linear function to estimate the amount of taxes collected in 2003.
    c. Construct the exponential function giving the amount of taxes collected as a function of time t.
    d. Use the exponential function to estimate the amount of taxes collected in 2003.

    e. Use both models to predict the amount of taxes collected in 2010. Which model seems more accurate? Explain.

14. The median family income I in the United States was about $17,700 in 1980 and rose to about $37,000 in 1997. Let t be the number of years since 1980.

t	Linear Model	Exponential Model
0	$17,700	$17,700
5	?	?
10	?	?
15	?	?
17	$37,000	$37,000
25	?	?
30	?	?

    a. If you assume that the increase in family income has been linear, find an equation for the line representing income I in terms of t. Use this equation to complete the second column of the table.
    b. If you assume that the increase in family income has been exponential, find an equation of the form $I = I_0 c^t$ to represent family income levels since 1980 and complete the third column of the table.
    c. On the same set of axes, sketch the graphs of the functions you obtained in parts (a) and (b).
    d. Use the equations from parts (a) and (b) to predict the median family income in 2003 for both types of growth.
    e. Suppose that both predictions from part (d) seem unreasonable. Can you suggest any other types of functions that might be a better fit?

15. (Continuation of Problem 14) Suppose that you learn that the median family income in 1990 was about $28,900.

    a. Which of the two models in Problem 14 now seems more accurate?
    b. If you plot the three data points corresponding to 1980, 1990, and 1997, how would you describe the shape of the graph of median family income as a function of time? What is the significance of this shape?

16. For each function, draw the inverse function, if one exists, on the same axes. If the function has no inverse, explain why.

    a.

    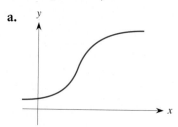

    b.

    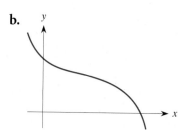

    c.

    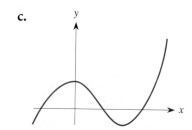

17. For each function, give its domain and find the inverse function.

    a. $f(t) = 0.5\log(2t - 4)$ **b.** $g(x) = x^3 + 6$

18. The function $F(x)$ is either linear or exponential. From the values in the table, decide which is the correct type and find a formula for F.

x	1	2	3	4	5
$F(x)$	6	9	13.5	20.25	30.375

19. Match each of the functions f, g, and h in the table to the behavior described as

    a. increasing, straight line.
    b. increasing, concave down.
    c. increasing, concave up.

x	1	2	3	4	5
$f(x)$	2.70	3.64	4.92	6.34	8.96
$g(x)$	1.4	4.6	7.8	11.0	14.2
$h(x)$	5.10	5.19	5.27	5.34	5.40

20. **a.** Since 1960, the price of an ice cream cone in one southern city has been growing approximately exponentially according to the function $f(t) = Ac^t$. If the price of a one-scoop cone was 20¢ in 1960 and $1.80 in 2000, (i) determine the function f and (ii) predict what the price of such a cone will be in 2005.

    b. The average price of a ticket to a first-run movie was $2.00 in 1960. This price has been growing exponentially and in 2000 was $9.00. Which of the prices, for ice cream or for movies, is growing faster?

    c. When can you expect the ice cream and movie prices to be the same if they each continue to grow in the same way?

    d. How much would a ticket to the movies cost at the time you found in part (c)?

    e. Would you use your model to predict the answer you got in parts (c) and (d)? Explain.

21. The aim of a college administration is to reduce the number of students who need remedial work in English by 10% each year. At the time the policy was put into place, 1600 students were enrolled in remedial English classes. If this program is successful, how many students will be enrolled in remedial English in 3 years? How long will it take for the number of students enrolled in such classes to be reduced to one section of 15 students?

22. The level of a drug in the bloodstream decreases at a rate of 30% of the drug per hour. Assume that the initial dose is 150 mg. How long does it take to bring the drug level down to under 20 mg? How long does it take to bring the drug level down to 5% of the original level?

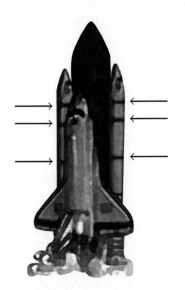

Fitting Functions to Data

FIGURE 3.1

3.1 · Introduction to Data Analysis

When the space shuttle *Challenger* exploded some 90 seconds after launch on January 28, 1986, it left the entire nation in a state of shock. The shuttle accident was especially traumatic to the many school children who were watching the launch live on television because Christa McAuliffe, a teacher, was a member of the crew.

The *Challenger* explosion also left behind many unanswered questions. Among the most important were:

- ◆ What went wrong?
- ◆ Could the problem have been anticipated?
- ◆ If it could have been anticipated, why wasn't it?

The first question is technical, the second is mathematical, and the third is political. In this section, we address the first two questions.

The *Challenger* disaster involved two factors: a component known as an O-ring and the air temperature at launch. The O-ring is a very thin ring (37.5 feet in diameter but only 0.28 inches thick) that seals the connections, or joints, between different sections of the shuttle engines. The locations of the six O-rings are indicated by the arrows shown in Figure 3.1. On the morning of the *Challenger* launch, the air temperature was 31°F, which was considerably colder than the temperature at any previous launch. In fact, the coldest temperature recorded at any previous launch was 53°F.

The Rogers Commission, which studied the *Challenger* disaster, focused on the O-rings as a possible cause of the explosion because there had been problems with O-rings on previous flights. In fact, the night before the launch, some of the project engineers, as part of the standard prelaunch routine, had thought about the six O-rings and questioned whether the *Challenger* should lift off because of the predicted overnight temperatures. The reasoning that went into the flight decision is worth considering because it demonstrates the important role that mathematical analysis can play in making informed decisions.

Twenty-four shuttle flights preceded the *Challenger* flight. On seven of them, relatively minor problems had occurred with the O-rings. In reviewing these previous incidents, the engineers examined the data shown in Figure 3.2. Note that the

161

horizontal axis indicates the air temperature at launch, and the vertical axis gives the number of O-rings affected. Each black dot represents a particular shuttle launch. Thus the shuttle launched at 53°F experienced problems with three different O-rings. Even though the graph shows one dot for 70°, two shuttles actually were launched at this temperature and both had problems with one O-ring. Nevertheless, examination of this graph (as the shuttle engineers probably did) reveals no consistent pattern indicating that the lower the air temperature at launch, the more likely there will be O-ring problems. In fact, note that the shuttle launch at 75°F had problems with two different O-rings. (Interestingly, it was the previous launch of the *Challenger.*) Consequently, the data didn't give the engineers any solid reason for canceling the *Challenger* launch the following morning.

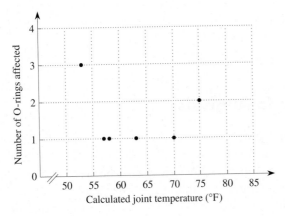

FIGURE 3.2

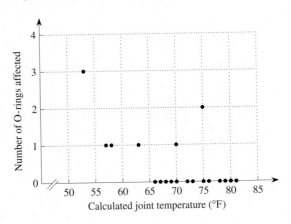

FIGURE 3.3

Unfortunately, the engineers didn't realize that the data in Figure 3.2 is just part of the story. It reflects only those launches during which there were O-ring problems. What is missing are those launches that were trouble-free. From the full set of data for all 24 previous shuttle flights shown in Figure 3.3, a striking pattern emerges: Almost all the problem launches occurred at low temperatures, and all the problem-free launches occurred at temperatures above 65°F. These results suggest that problems with O-rings are unlikely on warm days, but there may likely be a problem on a cool day. And, the predicted temperature of 31°F when the *Challenger* was due to lift off was far colder than the temperature for any previous launch. Had the engineers looked at all the data, there is no way that they could have allowed the *Challenger* to lift off.

Moreover, we can go beyond just an eyeball examination of the data in looking for trends. Recall that in Chapter 2 we introduced the "black thread method" for estimating the line that is the best fit to a set of data points. The graph of the data points is known as a **scatterplot.** In Figure 3.3 the scatterplot represents the number of O-ring problems as a function of air temperature. A line is not a good fit to this set of data because the pattern is clearly concave up. Instead, a curve such as the one shown superimposed over the data points in Figure 3.4 suggests a decaying exponential function or a power function with $p < 0$. Although extrapolating beyond the region of the data points (in this case from 53°F to 80°F) usually is risky, there is little doubt that launching the *Challenger* at 31°F was even riskier!

Most scientific, engineering, and technical work involves collecting and working with data and, more important, basing decisions on what can be inferred from

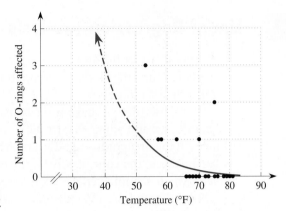

FIGURE 3.4

the data. To make intelligent decisions, you must understand all the information that a set of data imparts. Thus you need techniques for displaying the data in a form that makes it easy to extract the relevant information. Unfortunately, these techniques haven't been a major focus in past mathematical and technical training.

The *Challenger* situation illustrates the changing role of mathematics and how people use it. Applying mathematics is not a matter of "Here's an expression. Factor it." or "Here's an equation. Solve it." Rather, in the real world you will often face a situation about which a decision must be made. You need to be able to view that situation mathematically (i. e., create an appropriate mathematical model), identify the appropriate question to ask, obtain the solution (often with some electronic tool), interpret the solution in terms of the original problem, and communicate that solution effectively to others. The emphasis is much more on reasoning and judgment, not just on mechanical operations.

In Chapter 2 we discussed families of functions that have various behavior patterns. In this chapter, we consider the problem of finding the function that best fits a set of data. If the data fall into a roughly linear pattern, either increasing or decreasing, we want to find the line that is the best possible fit to the data. Similarly, if the data fall into certain nonlinear patterns, we want to find the function that best fits the data. Figure 3.5 shows three different data sets that clearly are not linear patterns. Figure 3.5(a) shows an increasing, concave up pattern that could be modeled by either an exponential growth function with base $c > 1$ or a power function with power $p > 1$. Figure 3.5(b) shows a decreasing, concave up pattern that could be modeled by either an exponential decay function with base $c > 1$ or a power function with power $p < 0$. Figure 3.5(c) shows an increasing, concave down pattern that could be modeled by either a power function with power $0 < p < 1$ or a logarithmic function. Once we have written a formula for the function that fits a set of data, we can then use it to answer questions of a predictive nature.

The techniques that we discuss here are the methods by which we construct functions. With this approach, preventable disasters such as the *Challenger* explosion aren't likely to recur.

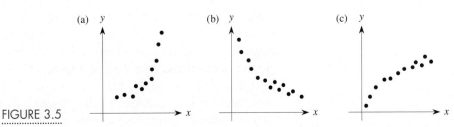

FIGURE 3.5

3.2 ·······Linear Regression Analysis

We now consider fitting a line to a set of data from a somewhat more sophisticated viewpoint than the black thread method. We now want to determine the single *best* line to fit a set of points.

Suppose that we have a set of n measurements on two presumably related quantities x and y: (x_1, y_1), (x_2, y_2), ..., (x_n, y_n), where x_1 and y_1 are the coordinates of the first point, x_2 and y_2 are the coordinates of the second point, and so on. For instance, the coordinates might represent people's heights and weights; students' high school averages and college GPAs; or the gross receipts (box office, cable and broadcast TV, videotapes, etc.) of the movie industry, in billions of dollars, in different years, as given in the following table.

Year	1990	1993	1994	1995	1996	1997
Gross receipts	39.98	49.80	53.50	57.18	60.28	63.01

Source: *2000 Statistical Abstract of the United States.*

We begin by drawing a scatterplot of the data, as shown in Figure 3.6(a). The data appear to fall in a roughly linear pattern with a positive slope. We can draw many different lines that all seem to capture the overall pattern of the points in the scatterplot, as shown in Figure 3.6(b). However, none of them can possibly pass through all the points. In fact, a *good fit* line may not necessarily pass through any of the points. Our objective is to determine the *best* linear fit to this set of data: the one line that comes closest to all the data points.

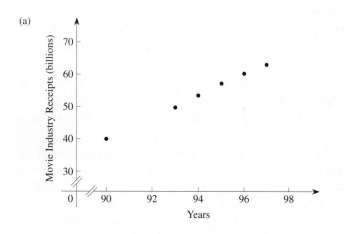

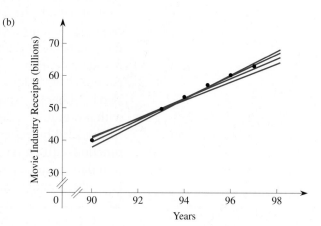

FIGURE 3.6

We now write the equation of a line in the form $y = ax + b$ instead of the more usual form $y = mx + b$ to give a better match to the display on most calculators. As always, the coefficient of the independent variable represents the slope, and the constant gives the vertical intercept regardless of the letters used.

The key to finding this line is the phrase "comes closest to all the points." The most common interpretation of this phrase is that the sum of the squares of all the vertical distances from the points to the line should be a minimum, as shown in

Figure 3.7. (If we used only the actual vertical distances, rather than their squares, some would be positive, others would be negative, and they would tend to cancel each other when added.) We discuss these vertical distances, called *residuals,* in more detail in Section 3.6.

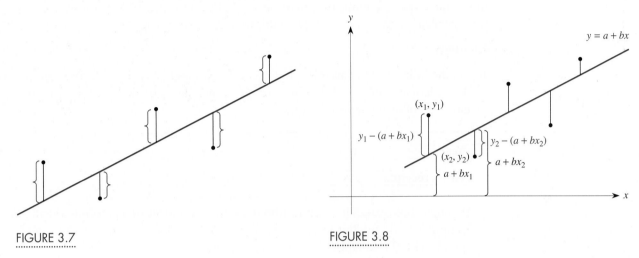

FIGURE 3.7   FIGURE 3.8

Suppose that the equation of the best-fit line is $y = ax + b$, where a and b are, for now, unknown constants. For the first data point (x_1, y_1), the point on the line with the same x-coordinate $x = x_1$ has height $y = ax_1 + b$. Hence the vertical distance from the data point to the line is $y_1 - (ax_1 + b)$, as illustrated in Figure 3.8. For the second data point (x_2, y_2), the vertical distance to the line is similarly $y_2 - (ax_2 + b)$, and so on for the rest of the data points. The corresponding squares of these vertical distances, for each of the n points, are

$$[y_1 - (ax_1 + b)]^2, \qquad [y_2 - (ax_2 + b)]^2, \quad \ldots, \quad [y_n - (ax_n + b)]^2.$$

To measure how close the line comes to all n data points, statisticians use the *sum of the squares* of these differences:

$$[y_1 - (ax_1 + b)]^2 + [y_2 - (ax_2 + b)]^2 + \cdots + [y_n - (ax_n + b)]^2.$$

Exactly one line corresponds to a *minimum* value for the sum of these squares. It is known as the *least squares line,* or more commonly, the **regression** line. The formulas[1] for the coefficients a and b in the equation of the regression line are built into most sophisticated calculators, usually under the STAT (statistics) menu and may be marked as LinReg or LinR. Routines for these calculations also are widely available in many computer packages, including spreadsheets such as Excel™.

To find the equation of this line with your calculator, you typically have to access the STAT menu, select the data option, clear any old data, enter the new data, and then go back to the main STAT menu and select calculate linear regression. Most calculators will give you values for the coefficients a and b in the regression equation $y = ax + b$, and the value of a third quantity that we discuss later in this section. See the detailed instructions in your calculator's manual.

---

[1]The regression coefficients are

$$a = \frac{n\sum(xy) - (\sum x)(\sum y)}{n(\sum x^2) - (\sum x)^2} \quad \text{and} \quad b = \frac{(\sum x^2)(\sum y) - (\sum xy)(\sum x)}{n(\sum x^2) - (\sum x)^2},$$

where n = number of data pairs (x, y), $\sum x^2$ = the sum of the squares of the x's, $(\sum x)^2$ = the square of the sum of the x's, and $\sum(xy)$ = the sum of the products of x and y in each pair.

We illustrate how to apply these ideas in Example 1.

EXAMPLE 1

(a) Find the equation of the regression line that best fits the data on gross receipts of the movie industry, in billions of dollars, as a function of time. (b) What does the slope of this line mean in this context? (c) Use this function to estimate the gross receipts in 2003.

Solution

a. Let the independent variable x represent the number of years since 1990 and let the dependent variable y be the gross receipts, in billions of dollars, for each year. We have the following table of values.

Year	0	3	4	5	6	7
Gross receipts	39.98	49.80	53.50	57.18	60.28	63.01

We enter these values for x and y into the statistics routine of a graphing calculator or computer and run the linear regression routine. The calculator or computer responds with the equation of the regression line

$$y = 3.3435x + 40.027.$$

The graph of this line superimposed over the scatterplot of the data is shown in Figure 3.9. Note that the line seems to be an excellent fit to all the data points.

b. The slope, 3.3435, of this regression line tells us that the gross receipts of movies are increasing, on average, by about $3\frac{1}{3}$ billion each year.

c. Having this linear model for the gross receipts as a function of time, we use it to predict the receipts in 2003, when $x = 13$ years since 1990. We then get

$$y = 3.3435(13) + 40.027 = \$83.49 \text{ billion.}$$

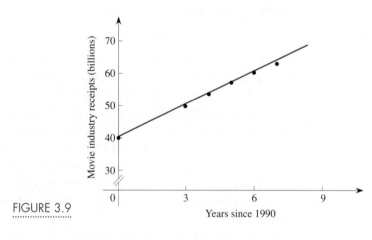

FIGURE 3.9

In general, when working with years or other "large" numbers, scaling them down as we did in Example 1 usually is helpful. For instance, we let x be the number of years since 1990.

We next look at a somewhat more complicated example based on the data presented in Table 3.1 giving world record times (in minutes: seconds) for the mile run and the year the record was set.

◆ **EXAMPLE 2**

(a) Find the equation of the regression line that best fits the data in Table 3.1 on the world records for the mile run as a function of time. (b) What does the slope of this line mean in this context? (c) Use this function to estimate the world record in 2005.

TABLE 3.1 *World Records in the Mile Run*

Year	Record time	Winner
1911	4:15.4	John Paul Jones, United States
1913	4:14.6	John Paul Jones, United States
1915	4:12.6	Norman Taber, United States
1923	4:10.4	Paavo Nurmi, Finland
1931	4:09.2	Jules Ladoumegue, France
1933	4:07.6	Jack Lovelock, New Zealand
1934	4:06.8	Glen Cunningham, United States
1937	4:06.4	Sidney Wooderson, Great Britain
1942	4:06.2	Gunder Haegg, Sweden
1942	4:06.2	Arne Andersson, Sweden
1942	4:04.6	Gunder Haegg, Sweden
1943	4:02.6	Arne Andersson, Sweden
1944	4:01.6	Arne Andersson, Sweden
1945	4:01.4	Gunder Haegg, Sweden
1954	3:59.4	Roger Bannister, Great Britain
1954	3:58.0	John Landy, Australia
1957	3:57.2	Derek Ibbotson, Great Britain
1958	3:54.5	Herb Elliott, Australia
1962	3:54.4	Peter Snell, New Zealand
1964	3:54.1	Peter Snell, New Zealand
1965	3:53.6	Michel Jazy, France
1966	3:51.3	Jim Ryun, United States
1967	3:51.1	Jim Ryun, United States
1975	3:51.0	Filbert Bayi, Tanzania
1975	3:49.4	John Walker, New Zealand
1979	3:49.0	Sebastian Coe, Great Britain
1980	3:48.9	Steve Ovett, Great Britain
1981	3:48.8	Sebastian Coe, Great Britain
1981	3:48.7	Steve Ovett, Great Britain
1981	3:47.6	Sebastian Coe, Great Britain
1985	3:46.5	Steve Cram, Great Britain
1993	3:44.4	Noureddine Morceli, Algeria

Solution

a. Let the independent variable x represent the number of years since 1900 and let the dependent variable y be the time, in minutes, of each record-breaking mile run. The times given in Table 3.1 are in the form minutes : seconds, so it is necessary to recalculate each value in terms of minutes. For instance, the first entry of 4:15.4, or 4 minutes and 15.4 seconds, is equivalent to $4 + 15.4/60 \approx 4.2567$ minutes. Making this conversion for all values in the table, we obtain

x	11	13	15	...	93
y	4.2567	4.2433	4.2100	...	3.7400

Using either a calculator or computer program, we get the equation of the regression line as

$$y = -0.00663x + 4.335.$$

Figure 3.10 shows the graph of this line superimposed over the scatterplot of the data; the line seems to fit all the data points well.

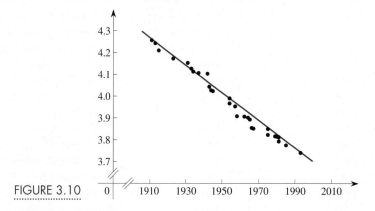

FIGURE 3.10

b. The slope, -0.00663, of this regression line tells us that the world record is dropping, on average, by 0.00663 minutes, or 0.498 seconds, each year.

c. Assuming that this linear model for the world-record times in the mile run continues to hold, we can use it to predict the world record in 2005, when $x = 105$ years since 1900. We then get

$$y = -0.00663(105) + 4.335 = 3.639 \text{ minutes}$$

or about 3 minutes and 38.3 seconds.

Think About This   Does the world record time of 3 minutes and 38.3 seconds in 2005 calculated in Example 2 seem reasonable? What is the predicted value for the world record in the mile run in 2500 based on the linear model? Is that value reasonable? Similarly, what happens if you predict back to 1492? Do you get a reasonable value? Why or why not?

Recall that, if you use the equation of the regression line to predict values of y corresponding to values of x within the interval of data values (called *interpolating*), the results are usually quite reasonable. Recall also that, if you try to use the regression equation well beyond the interval of data values (called *extrapolating*), the results become extremely questionable. Thus you should not extrapolate too far into either the future or the past. The domain of the mathematical model should not be extended too far beyond the given data.

The Correlation Coefficient

There is another major concern regarding the use of the regression equation. If you take any set of measurements relating two variables, you can always construct the regression equation based on the data. However, the results are completely meaningless if the two variables are not linearly related or even are totally unrelated. For example, you could collect data on people's telephone numbers and their social security numbers, construct a scatterplot, and calculate the corresponding regression equation. But the two variables are unrelated, so the results of predicting people's social security numbers from their phone numbers would be of no value.

Therefore we need a way to determine whether, in fact, a linear relationship exists between two quantities. The most common way of detecting such a relationship is by using a quantity known as the *linear correlation coefficient,* or simply the **correlation coefficient.** This quantity is denoted by r and is always a number between -1 and $+1$.

◆ Values of r close to $+1$ indicate a high degree of positive correlation between x and y. That is, they are likely related via a linear relationship, and the regression line will have positive slope. For example, we would expect a high positive correlation between a company's profits and its sales: As sales go up, profits usually go up also.

◆ Values of r close to -1 indicate a high degree of negative correlation between x and y. They are likely related by a linear relationship, and the regression line will have negative slope. For instance, there is a high negative correlation between a car's gas mileage and its weight—as weight goes up, gas mileage goes down, and vice versa. Similarly, there is a high negative correlation between the literacy rate and the infant mortality rate in any nation.

◆ Values of r close to 0 indicate little or no correlation, and we would conclude that there is no linear relationship between the variables. For instance, there is no correlation between students' social security numbers and their telephone numbers.

You can visualize the different cases by looking at the scatterplots shown in Figure 3.11. In Figure 3.11(a), the data points lie more or less along a rising line and the value of the correlation coefficient will be positive and relatively close to 1. In Figure 3.11(b), the points are scattered about a downward-sloping line, which means that the correlation coefficient is negative and relatively close to -1. In Figure 3.11(c), there is no clear pattern for the points, and so the correlation coefficient is relatively close to 0, indicating that there is no linear relationship between the variables.

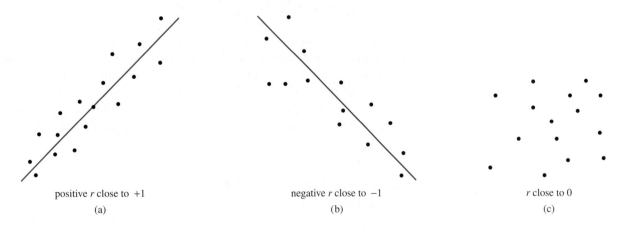

| positive r close to $+1$ | negative r close to -1 | r close to 0 |
| (a) | (b) | (c) |

FIGURE 3.11

As with the calculation of the regression coefficients, the correlation coefficient[2] usually is obtained either with a computer program or a calculator, not by hand. The important thing is knowing how to interpret the results, not how to calculate the regression equation and the correlation coefficient.

◆ EXAMPLE 3

Find the correlation coefficient for the data on the mile run records and interpret it.

Solution When we enter the recalculated data values from Example 2 in a calculator or program, we get a correlation coefficient of $r = -0.9899$. The fact that r is negative simply reflects the downward trend in the data and the negative slope in the equation of the regression line $y = -0.00663x + 4.335$. The value of $r = -0.9899$ is extremely close to -1, which suggests that there is negative correlation between the world record in the mile run and the year.

◆

Incidentally, if the data points all lie exactly on a line, the regression line found by the calculator or program actually passes through the points and the correlation coefficient is either 1 or -1, depending on whether the slope is positive or negative.

Determining When the Level of Correlation Is Significant

To use the correlation coefficient properly, you must be able to distinguish between significant correlation and no correlation. Typically a set of data represents just a sample from a much larger population of possible data values. For instance, if we were studying the relationship between gas mileage and vehicle weight, we certainly couldn't collect data on every car on the road—only on a random sample of all cars. Presumably the sample would be representative of all cars, but there is no guarantee of that. It is conceivable that we chose an extremely unrepresentative sample. What's worse, there usually is no way of telling simply by looking at the values that a sample is unrepresentative.

No statistical conclusion is ever 100% certain because that would require encompassing *every* conceivable sample from a population and that is not possible. Statisticians have developed a set of techniques that allow us to draw conclusions that are correct with 95% certainty or, equivalently, that are correct 95% of the time, to account for the possibility that a sample may contain several very unrepresentative observations. (Statisticians also work with 90% and 99% levels of certainty, but we consider only 95% certainty here.)

Suppose that you have a random sample of n data points and use either a computer program or a graphing calculator to graph the scatterplot and calculate both the regression line and the correlation coefficient r. If the points on the scatterplot clearly do not appear to fall in a linear pattern, you should expect that there is little or no linear correlation between the two variables. Even if the pattern seems to be roughly linear, you must still use the information provided by the correlation coefficient to decide whether it indicates a significant level of correlation between the two variables.

---

[2]The correlation coefficient is:

$$r = \frac{n(\Sigma xy) - (\Sigma x)(\Sigma y)}{\sqrt{n(\Sigma x^2) - (\Sigma x)^2}\ \sqrt{n(\Sigma y^2) - (\Sigma y)^2}}$$

The size of the sample also comes into play here. If we take a relatively small sample—say, $n = 5$ values—and this sample contains a nonrepresentative data point, that point will have a significant impact on the results. If the sample contains $n = 50$ random values, the effect of a single nonrepresentative data point will likely be diluted. If the sample consists of $n = 500$ random values, the effect of any single nonrepresentative data point is almost certain to be negligible.

Table 3.2. contains a partial set of so-called *critical values* for the correlation coefficient r. These critical values separate what we interpret as correlation from what we interpret as no correlation. These critical values change, depending on the size of the sample, n—the larger the sample size, the smaller the critical value. We write r_n to indicate the critical value of r based on n data points.

TABLE 3.2 *Critical Values of r*

n	r_n	n	r_n	n	r_n
3	0.997	13	0.553	27	0.381
4	0.950	14	0.532	32	0.349
5	0.878	15	0.514	37	0.325
6	0.811	16	0.497	42	0.304
7	0.754	17	0.482	47	0.288
8	0.707	18	0.468	52	0.273
9	0.666	19	0.456	62	0.250
10	0.632	20	0.444	72	0.232
11	0.602	21	0.433	82	0.217
12	0.576	22	0.423	92	0.205

To use the table, compare the value of r from the sample data to the corresponding critical value r_n shown in the table. For instance, a sample of size $n = 10$ has a critical value of $r_{10} = 0.632$. If the correlation coefficient for the data is *greater than* 0.632—say, $r = 0.758$—we can conclude with 95% certainty that there is positive correlation between the two variables. If the correlation coefficient for the data is *less than* -0.632—say, $r = -0.685$—we can conclude with 95% certainty that there is negative correlation between the two quantities being studied. But, if the value for r is between -0.632 and $+0.632$—say, $r = 0.446$ or $r = -0.583$—we cannot conclude that there is any linear correlation between the two variables. In other words there does not appear to be a *linear* relationship between x and y. Figure 3.12 illustrates the process. Of course, there still may be a nonlinear functional relationship between x and y; we examine such cases later in this chapter.

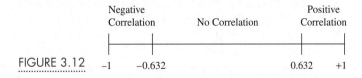

FIGURE 3.12

Note that the critical value for a sample of size $n = 20$ is $r_{20} = 0.444$, whereas the critical value for a sample of size $n = 10$ is $r_{10} = 0.632$. As we pointed out, a larger sample is more likely to be representative of the population, so the evidence for correlation does not have to be quite as great. With a small sample, the value for r must be very close to 1 or -1 for us to conclude with any certainty that there is correlation. For instance, the critical value for a sample of size $n = 3$ is $r_3 = 0.997$.

A more extensive table of critical values for r is reproduced on the inside back cover for easy reference.

EXAMPLE 4

Decide whether there is a significant level of correlation between the world record times for the mile run and the year in which the record was set in Examples 2 and 3.

Solution   In Example 3 we found that the value for the correlation coefficient was $r = -0.9899$, based on the $n = 32$ entries in Table 3.1. The corresponding critical value is then $r_{32} = -0.349$. Because the value for the data is more negative than this critical value (i.e., closer to -1), we conclude that a significant level of correlation exists between the two variables. ◆

In our examples, we consistently use X as the independent variable and Y as the dependent variable for the output from either the computer or calculator. We then have to interpret what the X- and Y-variables represent in the context of each individual situation.

EXAMPLE 5

A study is conducted to determine the relationship between a person's height in inches and shoe size, based on the following set of data pairs:

$(66, 9), (63, 7), (67, 8\frac{1}{2}), (71, 10), (62, 6), (65, 8\frac{1}{2}), (72, 12), (68, 10\frac{1}{2}), (60, 5\frac{1}{2}), (66, 8)$.

a. Determine the value of the correlation coefficient and decide whether it is significant.
b. Find the equation of the regression line relating height to shoe size, based on this sample.
c. Use the equation to predict the most likely shoe size for a person who is 70 inches tall and for someone who is 61 inches tall.
d. Use the equation to predict the most likely height of a person whose shoe size is 9.

Solution

a. Using a calculator or software package, we find that the correlation coefficient for this set of data is $r = 0.951$, which suggests a high degree of positive correlation between a person's height and shoe size. Because there are $n = 10$ data points, the critical value for r is $r_{10} = 0.632$ from Table 3.2. Because $0.951 > 0.632$, we can conclude with 95% certainty that a significant level of positive correlation exists between a person's height and shoe size.
b. Figure 3.13 shows the scatterplot of the data and the graph of the associated regression line. The equation of the regression line, as given by either a calculator or computer, is

$$Y = 0.51X - 25.016,$$

which is equivalent to

$$S = 0.51H - 25.016$$

for a person's shoe size S as a function of the person's height H in inches.
c. Using this regression equation for a person who is 70 inches tall, we estimate that the shoe size corresponding to $H = 70$ is

$$S = 0.51(70) - 25.016 = 10.684 \approx 10\frac{1}{2}.$$

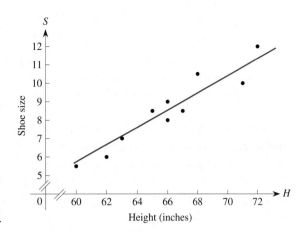

FIGURE 3.13

Similarly, for $H = 61$, we estimate that the corresponding shoe size is

$$S = 0.51(61) - 25.016 = 6.094 \approx 6.$$

d. For a shoe size of $S = 9$, we need to solve for H in

$$S = 0.51H - 25.016 = 9.$$

We add 25.016 to both sides of the equation to get

$$0.51H = 9 + 25.016 = 34.016,$$

so that

$$H = 34.016/0.51 \approx 66.7.$$

We conclude that the person is about $66 \frac{3}{4}$ inches tall.

In the context of Example 5, it seems sensible to think of shoe size as a function of a person's height. Thus we chose height H as the independent variable. However, it is also appropriate to interchange the roles of S and H and think of height H as a function of shoe size S.

Suppose that you have determined, with 95% certainty, that there is a high degree of linear correlation between two variables. What does that tell you? In Example 5, a high correlation coefficient means that there is a linear relationship between H and S. That is, the two variables can be related with a linear equation—namely, the regression equation. However, the fact that two quantities are correlated and that a linear relationship exists does not mean that there is a cause-and-effect relationship between them. For example, several studies have found a high positive correlation between teacher salaries in a school district and the amount of alcohol consumed by the students. That is, if teacher salaries are low, the level of student drinking also is low; if teacher salaries are high, considerable drinking is going on. Is it reasonable to conclude that one causes the other? Could the level of student drinking be reduced by lowering teacher salaries in a district?

A positive correlation between two quantities says nothing more than that a linear relationship exists between them. Other factors could contribute to both. For instance, high teacher salaries typically reflect a school district in a relatively affluent community where the students are likely to have relatively large amounts of their own money to spend on alcohol.

EXAMPLE 6

One of the most famous moments in the history of science was Galileo's reported experiment of dropping various objects from the top of the 180 foot high Leaning Tower of Pisa and discovering that they fell at the same rate, regardless of their weight. The following table gives the speed v, in feet per second, of an object dropped from the top of the tower measured at half second intervals until it hits the ground. The negative values for the speed simply reflect the convention that velocity is considered positive when an object is moving upward and negative when an object is moving downward. Note how the object starts to fall slowly and then accelerates. (Incidentally, these values are considerably more accurate than anything Galileo could have measured at the end of the fourteenth century.)

Time, sec	0	0.5	1.0	1.5	2.0	2.5	3.0
Speed, ft/sec	0	-16.0	-32.1	-47.9	-64.1	-80.2	-96.1

a. Does the correlation coefficient indicate a significant level of correlation between the two variables?

b. Find the line that best fits the experimental data, giving the speed v as a function of time t.

c. What is the significance of the slope in the equation of the line that fits this data?

Solution

a. We start with the scatterplot of the data as shown in Figure 3.14 and observe that the pattern looks extremely linear. The value for the correlation coefficient is $r = -0.999997$. The correlation coefficient is negative because the trend in the data is downward, so the slope of the regression line should be negative. The fact that the value for r is so close to -1 means that there is a significant level of correlation between the two variables and that the regression line almost perfectly fits the data points. The speed of the falling object at any time is a linear function of t.

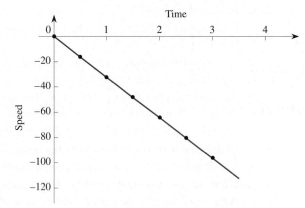

FIGURE 3.14

b. Using a calculator or computer package, we obtain the equation of the line that best fits this data as

$$Y = -32.05X + 0.01786,$$

or, in terms of the variables in this situation,

$$v = -32.05t + 0.01786.$$

Note that the constant term 0.01786 is very close to zero, and we know that the velocity of any falling object at time zero is zero. The slight discrepancy occurs because of possible inaccuracies in measurements or rounding errors. We conclude that a better model for the speed of a falling object might be

$$v = -32.05t.$$

c. The slope of the regression line is -32.05. Because v is measured in feet per second and t is measured in seconds, the units for the slope are feet per second per second or feet per second squared, written ft/sec^2. The slope is negative because the object is falling downward ever faster. The specific value for the slope, approximately $-32 \ ft/sec^2$, is the acceleration due to gravity. ◆

EXAMPLE 7

Psychologists claim that the human mind has a preference for rectangles having a certain shape with the longer side approximately 1.6 times the shorter side. (The exact value used is known as the *golden ratio*, which is equal to $(1 + \sqrt{5})/2 \approx 1.618$.) If this claim is correct, we might expect that artists would have naturally used this ratio in their paintings because it is visually pleasing. Table 3.3 shows a selection of art masterpieces and their actual dimensions, in centimeters. Draw the scatterplot and find the regression line that best fits the data. What evidence does this result provide that artists instinctively use the golden ratio?

TABLE 3.3

Artist	Title	Dimensions (cm)
da Vinci	Mona Lisa	77 × 53.5
Rubens	Landscape with the Chateau of Steen	134.5 × 236.7
Goya	The Third of May, 1808	270 × 410
Rembrandt	Self Portrait	133.6 × 103.8
Vermeer	Young Woman with a Water Jug	45.7 × 40.6
T. Rousseau	A Meadow Bordered by Trees	41.6 × 61.9
Millet	The Gleaners	83.8 × 111.8
Constable	The Haywain	130.1 × 185.4
Manet	The Fifer	160 × 97.5
Monet	Regatta at Argenteuil	47.5 × 74.3
Renoir	Le Moulin de la Galette	130.7 × 175.3
Degas	Prima Ballerina	58.3 × 42
Cassatt	The Bath	100.3 × 66
Cezanne	Still Life with Apples	43.5 × 54
Seurat	The Bathers	202 × 300.3
Van Gogh	The Starry Night	73.7 × 92.1
Picasso	The Old Guitarist	121.3 × 82.7
H. Rousseau	The Dream	200 × 300
Matisse	The Joy of Life	171.3 × 238
Chagall	I and the Village	191.5 × 151
Dali	The Persistence of Memory	24 × 33

Solution   To simplify things, we ignore the fact that some paintings are in portrait style (taller than they are wide) while others are in landscape style (wider than they are tall). We use the shorter dimension W (for width) as the independent variable and the longer dimension H (for height) as the dependent variable. The scatterplot for the data is shown in Figure 3.15. The corresponding correlation coefficient is $r = 0.9894$, which indicates a high degree of linear correlation between the two variables. The associated regression line, which is superimposed over the data in the scatterplot, is

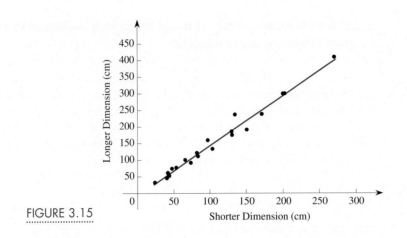

FIGURE 3.15

$$H = 1.5104W - 7.0017.$$

The slope of the regression line is 1.5104, which is fairly close to the golden ratio 1.618. Also, the line passes relatively close to the origin, so the slope is approximately equal to the ratio of height to width. Note that the points are all clustered fairly tightly about the regression line, so we can conclude that most of these artists used a proportion close to the golden ratio for these canvases.

So far we have investigated only the possibility that a linear relationship exists between two variables. We then used the correlation coefficient to detect and measure only the strength of the linear relationship. In the following sections, we consider ways of detecting, measuring, and calculating a nonlinear (exponential, power, or logarithmic) relationship between two quantities.

Problems

1. Match each value for the correlation coefficient r with its scatterplot (i)–(v).

   a. $r = -0.962$ **b.** $r = -0.781$
   c. $r = 0.434$ **d.** $r = 0.837$
   e. $r = 0.998$

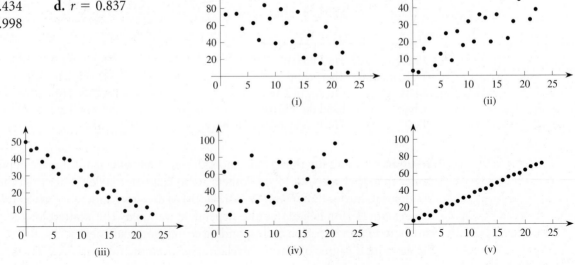

2. Suppose that you have data on each of the following pairs of variables. For which pairs would you expect the correlation coefficient to be close to +1, for which would you expect r to be close to −1, and for which would you expect r to be close to 0?

a. A young child's height and weight.
b. The age of a car and its book value.
c. A child's shoe size and the number of words in the child's vocabulary.
d. The number of hours per week that a student studies and the amount of money spent on food each week.
e. The number of hours per week that a student studies and the resulting GPA.
f. The number of cartons of cigarettes sold in the United States and the tax revenue on cigarettes.
g. The number of hours a person sleeps nightly, on average, and the number of push-ups the person can do.

3. What is wrong with each statement?

a. The correlation coefficient for the number of push-ups an athlete can do and the time it takes the athlete to run a mile is 1.25.
b. For a set of (x, y) data, the regression equation is $Y = 6.592 - 2.158X$ and the correlation coefficient is $r = 0.87$.
c. If the correlation coefficient for a set of (x, y) data is negative, it follows that the smaller x is, the smaller y will be.

4. The following table gives some measurements on the rate of chirping (per second) of the striped ground cricket as a function of the temperature.

T (°F)	89	72	93	84	81	75	70
Chirps	20	16	20	18	17	16	15
T (°F)	82	69	83	80	83	81	84	76
Chirps	17	15	16	15	17	16	17	14

Source: Adapted from George W. Pierce, *The Songs of Insects.* Boston: Harvard University Press, 1948.

Determine the equation of the line that best fits this set of data. How does it compare to the equation we estimated by eye in Example 4 of Section 2.3? Does the value of the correlation coefficient indicate a high degree of correlation between chirp rate and air temperature?

5. According to a leading road and track magazine, the following information gives the time in seconds for a Mercedes to accelerate from zero to the indicated speed in miles per hour:

Speed (mph)	30	40	50	60	70	80
Time (sec)	3.6	5.0	7.0	9.1	11.9	15.2

Does the corresponding correlation coefficient indicate a significant level of linear correlation between the two variables? If so, determine the equation of the regression line that best fits the data. Estimate how long it would take a Mercedes to accelerate to 45 mph and to 90 mph. Which is more likely to be accurate? Why?

6. The following table gives the percentage of the U. S. gross domestic product (GDP) spent on health care over the years.

Year	1960	1965	1970	1975
Percentage	4.4	5.7	7.7	8.3
Year	1980	1985	1990	1995
Percentage	10	10.2	12.2	13.7

Source: *2000 Statistical Abstract of the United States.*

Is there a significant level of linear correlation between these two variables? If so, what is the regression line? Estimate the percentage of GDP spent on health care in 2000. If current trends continue, when will 20% of GDP be spent on health care?

7. Repeat Problem 6, using 60, 65, 70, . . . , 90 for the years instead of 1960, 1965, . . . How do the results for the equation of the line of best fit and for the correlation coefficient change? What would happen if you used 0, 5, 10, . . . , 30 for the years instead?

8. The electrical resistance R of a piece of metal depends on the temperature T of the metal. An experiment was conducted by measuring the resistance in ohms in a piece of wire at different temperatures in degrees Celsius. The results were as follows:

T	33.2	40.6	45.3	51.8	58.4
R	4.71	4.80	4.93	5.02	5.17
T	63.8	71.0	76.9	80.6	90.1
R	5.34	5.39	5.52	5.55	5.75

Determine the correlation coefficient and the equation of the regression line that best fits the data. Does the value for the correlation coefficient suggest a linear relationship between the two quantities?

9. Consider the data in Example 5 on shoe size versus height.

   a. Interchange the roles of H and S to consider S as the independent variable and then find the corresponding linear regression equation and correlation coefficient.

   b. How is this correlation coefficient related to the one we obtained in Example 5?

   c. In Example 5, we found $S = 0.51H - 25.016$. Solve this equation algebraically for H as a function of S.

   d. Explain why the results in (a) and (c) differ. (*Hint:* The linear regression equation is based on minimizing the sum of squares of the *vertical* distances from the points to the line.)

10. Many baseball fans are concerned that the team owners with the most money are able to "buy" the best players and so dominate their leagues. The table below shows the total payroll, in millions of dollars, for each of the major league teams during the 2000 baseball season and the number of games each team won out of the maximum of 162.

   a. Find the equation of the line that best fits the number of wins during the 2000 baseball season as a function of a team's payroll.

   b. Is there a significant level of correlation between the two variables?

   c. Based on your linear model in part (a), how many wins would $100 million have bought during this season?

   d. Do the regression equation and the correlation coefficient support or contradict the claim that the teams with larger bankrolls are buying more wins?

   e. Write a short essay with your views on this issue. Be sure to include appropriate mathematical arguments, based on your findings in parts (a) and (b).

11. The great chemist Mendeleev once conducted an experiment relating the solubility of sodium nitrate in water to the temperature of the water, in degrees Celsius. He obtained the data on the next page.

Team	Payroll	Wins	Team	Payroll	Wins
New York Yankees	$113.37	87	Toronto Blue Jays	59.22	83
Atlanta Braves	95.01	95	Tampa Bay Devil Rays	55.16	69
Los Angeles Dodgers	94.22	86	San Francisco Giants	54.24	97
Boston Red Sox	93.87	85	Houston Astros	52.00	72
New York Mets	89.75	94	Chicago Cubs	51.08	65
Arizona Diamondbacks	80.76	85	Chicago White Sox	36.94	95
Cleveland Indians	78.72	90	Philadelphia Phillies	36.68	65
St. Louis Cardinals	72.38	95	Cincinnati Reds	35.13	85
Seattle Mariners	62.55	91	Milwaukee Brewers	33.77	73
Texas Rangers	61.36	71	Oakland A's	32.69	92
Detroit Tigers	60.60	79	Pittsburgh Pirates	31.94	69
Baltimore Orioles	59.22	74	Montreal Expos	27.97	67
Anaheim Angels	58.74	82	Florida Marlins	25.86	79
Colorado Rockies	56.05	82	Kansas City Royals	24.47	77
San Diego Padres	54.68	76	Minnesota Twins	15.82	69

Temperature	0	4	10	15	
Solubility	66.7	71.0	76.3	80.6	
Temperature	21	29	36	51	68
Solubility	85.7	92.9	99.4	113.6	125.1

a. Based on the scatterplot of the data and the value of the correlation coefficient, is it reasonable to use a linear function to model this relationship?

b. Find the equation of the linear function that best fits these data.

c. Use your model to estimate the solubility at a water temperature of 40°C.

d. Explain why using the linear model to predict a value for the solubility S when $T = -5$ would not be appropriate.

e. What might be appropriate values for the domain and range of this linear function?

12. The table at the bottom of the following page gives Olympic gold medal times, in seconds, for the 100-meter freestyle for men and women. For each data set, find the best-fit line. What is the practical significance of each line's slope? Because the slopes are different, determine the point where the two lines intersect and tell what this point means. Is it reasonable?

13. In a standard experiment conducted in introductory physics class, an object was dropped and its speed v measured every tenth of a second. The following table gives the speed, in centimeters per second, at different times. Find the equation of the line that best fits the data. What are the units for the slope?

t	0	0.1	0.2	0.3	0.4	0.5	0.6	0.7	0.8	0.9	1.0
v	0	−98	−196	−295.6	−391.8	−490.2	−587.8	−606.1	−783.8	−881.7	−979.9

14. Use the fact that 1 inch equals 2.54 cm to verify that the value you obtained for the slope in Problem 13 (the acceleration due to gravity in the metric system) is equivalent to the value we obtained in Example 4 (U. S. customary units of measurement).

15. A spring is mounted from the ceiling and hangs straight down. When a mass is attached to the end of the spring, as shown in the accompanying figure, the spring lengthens. The following data values were obtained in an experiment where different masses m, in grams, were attached and the associated lengths L, in centimeters, of the spring were recorded.

m	0	100	200	300
L	0	3.9	7.9	12.0
m	400	500	600	700
L	16.0	20.1	24.1	28.2

a. Decide which variable is independent and which is dependent.

b. Does the correlation coefficient indicate a significant level of linear correlation between the two variables?

c. Find the equation of the line that best fits the experimental data.

d. What are the units for the slope of the line? What is the significance of the slope?

e. Adjust the equation of the line so that it passes through the origin. (*Note:* The resulting relationship is known as Hooke's law after the British scientist Benjamin Hooke and the slope, which depends on the particular spring used, is called the spring constant.)

16. The following table gives estimates for the average temperature, in degrees Celsius, at the Earth's surface, worldwide, in different years.

Year	1880	1900	1920	
Temperature	13.8	13.95	13.9	
Year	1940	1960	1980	1999
Temperature	14.15	14.0	14.2	14.4

a. Find the equation of the line that best fits these data.

b. Assuming that the trend continues, predict the average surface temperature in 2020.

17. The following table shows the trend in worldwide grain production (wheat, rice, and corn, primarily), in millions of tons.

Year	1965	1970	1975	1980
Amount	905	1079	1237	1430
Year	1985	1990	1995	1999
Amount	1647	1769	1713	1855

Source: Lester R. Brown et al., *Vital Signs 2000: The Environmental Trends That Are Shaping Our Future.*

a. Find the equation of the linear function that best fits these data.
b. What does the model predict for the amount of grain produced in 2010?
c. When does the model predict that the total amount of grain produced will reach 2000 million tons?
d. Write a paragraph describing the implications of the fact that grain production is growing roughly linearly while the population is growing roughly exponentially.

Olympic Gold Medal Times in Swimming

MEN'S TIMES			WOMEN'S TIMES		
Year	Swimmer	Time	Year	Swimmer	Time
1908	Charles Daniels (US)	65.60	1912	Fanny Durack (AUST)	82.20
1912	Duke Kahanamoku (US)	63.40	1920	Ethelda Bleibtrey (US)	73.60
1920	Duke Kahanamoku (US)	61.40	1924	Ethel Lackie (US)	72.40
1924	Johnny Weissmuller (US)	59.00	1928	Albina Osipowich (US)	71.00
1928	Johnny Weissmuller (US)	58.60	1932	Helene Madison (US)	66.80
1932	Yasuji Miyazaki (JAP)	58.20	1936	Hendrika Mastenbroek (NETH)	65.90
1936	Ferenc Csik (HUN)	57.60	1948	Greta Andersen (DEN)	66.30
1948	Walter Ris (US)	57.30	1952	Katalin Szoke (HUN)	66.80
1952	Clarke Scholes (US)	57.40	1956	Dawn Fraser (AUST)	62.00
1956	Jon Hendricks (AUS)	55.40	1960	Dawn Fraser (AUST)	61.20
1960	John Devitt (AUS)	55.20	1964	Dawn Fraser (AUST)	59.50
1964	Don Schollander (US)	53.40	1968	Margo Jan Henne (US)	60.00
1968	Michael Wenden (AUS)	52.20	1972	Sandra Neilson (US)	58.59
1972	Mark Spitz (US)	51.22	1976	Kornelia Ender (E GER)	55.65
1976	Jim Montgomery (US)	49.99	1980	Barbara Krause (E GER)	54.79
1980	Jorg Woithe (E GER)	50.40	1984	Nancy Hogshead (US)	55.92
1984	Rowdy Gaines (US)	49.80	1988	Kristin Otto (E GER)	54.93
1988	Matt Biondi (US)	48.63	1992	Zhuang Yong (CHI)	54.65
1992	Aleksandr Popov (USSR)	49.02	1996	Le Jingyi (CHI)	54.50
1996	Aleksandr Popov (USSR)	49.02	2000	Inge de Bruijn (NETH)	53.83
2000	Pieter van den Hoogenband (NETH)	48.30			

18. Use of the automobile has sparked immense changes in human culture, both for good and for bad. The following table shows the growth in the number of cars in use throughout the world, in millions, since 1950.

Year	1950	1960	1970	1975	1980
Cars	8	13	23	25	29

Year	1985	1990	1993	1995	1997	1999
Cars	32	36	34	36	38	39

Source: Lester R. Brown et al., *Vital Signs 2000: The Environmental Trends That Are Shaping Our Future.*

a. Find the equation of the linear function that best fits these data.

b. What is the significance of the slope of this line?

c. What does the model predict for the number of cars in use worldwide in 2025?

d. When does the model predict that the number of cars in use will reach 50 million?

19. An ad for wireless phone service listed the following prices for monthly service:

$19.99 for 325 free minutes

$29.99 for 750 free minutes

$39.99 for 1100 free minutes

$79.99 for 3000 free minutes

a. Find the equation of the linear function that best fits these data.

b. What is the practical significance of the slope and the vertical intercept?

c. According to this model, predict the cost of a plan that provides 2000 free minutes.

20. Consider the four points $(0, 0)$, $(0, 1)$, $(1, 0)$ and $(1, 1)$. Two reasonable guesses for the best-fit line are (a) the diagonal line passing through $(0, 0)$ and $(1, 1)$ and (b) the horizontal line $y = \frac{1}{2}$. Using the criterion that the best-fit line is the one with the minimal value for the sum of the squares of the vertical distances from each of the points to the line, decide which of these two lines is a better fit. Find the actual best-fit line by using your graphing calculator. How high is the value for the correlation coefficient?

21. The accompanying graph shows the scatterplot for the points $(1, 48)$, $(2, 68)$, $(3, 93)$ and $(4, 114)$ along with the line $y = 20x + 30$.

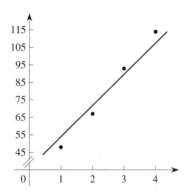

a. Find the sum of the squares associated with this line.

b. What changes would you make (increase or decrease) to the slope and/or the vertical intercept to get a better fit?

3.3 Fitting Nonlinear Functions to Data

Although linear regression and correlation are extremely powerful methods, not all relationships between two quantities are linear. If a scatterplot suggests a known nonlinear pattern, you need the ability to fit the appropriate function, such as an exponential function, a power function, or a logarithmic function, to the data.

Just as graphing calculators and software packages can calculate the equation of the line that best fits a set of data, they can also find the equation of: (1) an exponential function of the form $y = ac^x$ (or the form $y = ab^x$, which is the format used by most calculators); (2) a power function of the form $y = kx^p$ (or the form $y = ax^b$, which is what most calculators show); or (3) a logarithmic function of the form $y = a + b \ln x$ that fits the data. We show how to find such functions in the examples in this section.

EXAMPLE 1 The Growth of the U.S. Population from 1780 to 1900

The population of the United States, in millions, from 1780 to 1900, is shown in Table 3.4. The corresponding scatterplot of population versus time is shown in Figure 3.16.

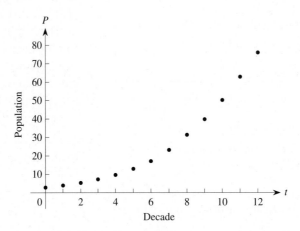

FIGURE 3.16

a. Find an appropriate function that best fits these population data and discuss its characteristics.

b. Find the correlation coefficient and discuss its significance.

Solution

a. To simplify the calculations, we let the independent variable t represent the number of decades since the year 1780. That is, 1780 corresponds to $t = 0$, 1790 corresponds to $t = 1$, and so on until 1900, which corresponds to $t = 12$.

The growth pattern for the U.S. population shown in Figure 3.16 clearly is not linear; it is concave up. Based on our discussions of population growth in Chapter 2, we expect the pattern is likely exponential. To check, we calculate the ratio of successive terms, as shown in Table 3.4. The ratios are roughly constant, which suggests that the pattern is roughly exponential. The discrepancies may be due to other factors (political or economic, say), which could give the population a spurt in one decade while slowing its growth during another time period, such as during the Civil War.

TABLE 3.4 *U.S. Population (1780–1900)*

Decade	Year	Population	Ratio
0	1780	2.8	1.39 = 3.9/2.8
1	1790	3.9	1.36
2	1800	5.3	1.36
3	1810	7.2	1.33
4	1820	9.6	1.34
5	1830	12.9	1.33
6	1840	17.1	1.36
7	1850	23.2	1.35
8	1860	31.4	1.27
9	1870	39.8	1.26
10	1880	50.2	1.25
11	1890	62.9	1.21
12	1900	76.0	

To determine an exponential function that fits these population values, we enter the data in a calculator, say, and select exponential regression. The calculator responds with the values of the parameters $a = 3.069$ and $b = 1.321$ for an exponential function of the form $y = ab^x$, so the function is

$$Y = 3.069(1.321)^X$$

or equivalently, in terms of our variables,

$$P(t) = 3.069(1.321)^t,$$

where t is the number of decades since 1780.

The growth factor, 1.321, is very close to most of the ratios of successive population values we calculated in Table 3.4. This value indicates that the U.S. population was growing at a rate of about 32.1% per decade from 1780 to 1900, which is about 3% per year.

b. Figure 3.17 shows the original population data with this exponential function superimposed. The curve fits the data well. Moreover, the corresponding correlation coefficient $r = 0.998$ is very close to 1. For the $n = 13$ data points, the table on the inside back cover gives the critical value of $r_{13} = 0.553$, so the level of correlation is definitely significant. Therefore the exponential function is a good fit to the population data from 1780 through 1900.

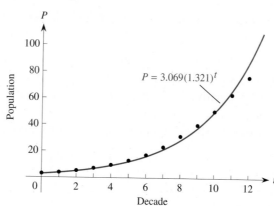

FIGURE 3.17

Think About This

For each of the decades $t = 0, 1, \ldots, 12$ since 1780, use the exponential function $P(t) = 3.069(1.321)^t$ to create a table of predictions for the U.S. population based on this model. Use these values to then calculate the *error* between the actual value of the population and this predicted value for each decade. ◻

We could also analyze this set of population data by letting t represent the year itself, ($t = 1780, 1790, \ldots$) or letting t represent the number of years since 1780 ($t = 0, 10, 20, \ldots$). Each produces equivalent exponential functions with different bases, but each function is an appropriate model for the situation. As a result, you must be careful to keep track of what the variables used represent.

You may wonder why we have considered only the U.S. population up to 1900 and not beyond. The reason is that the population does not follow an exponential pattern quite as closely thereafter. Various factors, such as limitations on immigration, changes in lifestyle reflected in smaller families, and the end of westward expansion came into play during the twentieth century to slow the rate of population growth. We introduce a more sophisticated mathematical model for population growth that takes such factors into account in Chapter 5 and consider the question of fitting a better mathematical function to the data on the U.S. population up to the present day.

The methods we used in Example 1 also apply to situations that can be modeled by an exponential decay function, as illustrated in Example 2.

◆EXAMPLE 2 Level of a Drug in the Body

L-Dopa is a drug used to control the symptoms of Parkinson's disease. The following data give the amount L of L-Dopa in the bloodstream, in nanograms per milliliter, t minutes after the drug was absorbed into the blood.

t	0	20	40	60	80	100	120	140	160	180	240	300
L	2950	2600	1550	1100	900	725	600	510	440	300	250	225

Source: Douglas Brown & Tom Timchek; unpublished manuscript.

Find an exponential function that models the level of L-Dopa in the blood as a function of time and check its correlation coefficient for significance.

Solution The scatterplot shown in Figure 3.18 suggests that a decaying exponential function is a reasonable model. An alternative might be to use a decaying power function with a negative power p. But a power function has a vertical asymptote at $t = 0$ and we are told that a fixed amount of the drug is in the blood at time $t = 0$. Therefore a power function would not be an appropriate choice.

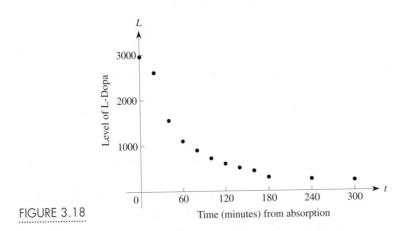

FIGURE 3.18

Using a calculator, we find that an exponential decay function that fits these data is

$$L(t) = 2147.8(0.9909)^t,$$

which indicates that the level of L-dopa in the blood decreases by almost 1% every minute. This function is superimposed over the data in Figure 3.19 demonstrating a reasonably good fit. The corresponding correlation coefficient is $r = -0.9545$; it is negative because the trend in the data is decreasing. Also, the critical value of r with $n = 12$ data points is $r_{12} = 0.576$, indicating that the fit is good.

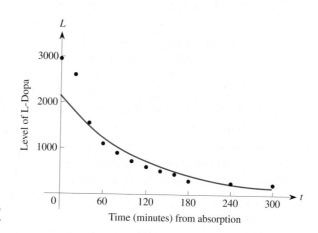

FIGURE 3.19

We next consider fitting a power function to a set of data.

EXAMPLE 3   Number of Species on an Island

Biologists have long observed that the larger the area of a region, the more species that inhabit it. The following table gives some data on the area (in square miles) of various Caribbean islands in the Greater and Lesser Antilles and estimates on the number of reptile and amphibian species living on each. (Note that Trinidad has been omitted because the island is exceptionally rich in species and tends to distort the data. Trinidad is only 7 miles off the coast of Venezuela, and many species have been able to emigrate easily from the mainland to the island.)

a. Determine a function that models the relationship between the number of species N living on one of these islands and the area A of the island and find the correlation coefficient.

b. The area of Barbados is 166 square miles. Estimate the number of species of reptiles and amphibians living there.

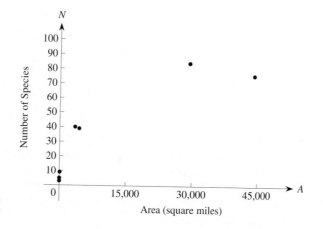

FIGURE 3.20

Island	Area	N
Redonda	1	3
Saba	4	5
Montserrat	40	9
Puerto Rico	3459	40
Jamaica	4411	39
Hispaniola	29,418	84
Cuba	44,218	76

Solution

a. It is reasonable to think that the number of species depends on the area of the island, so area A is the independent variable and the number of species N is the dependent variable. Figure 3.20 shows the scatterplot of the data. Don't be misled by the fact that the last point is somewhat lower than the preceding one; it is not reasonable to expect that a larger area will necessarily be home to fewer species, so the trend should not turn back down. Note that the points corresponding to the first few entries in the table are very close together because of the large horizontal scale needed. The points appear to lie on top of one another, which can very easily distort our perception of the actual pattern in the data.

The overall data pattern suggests either a power function with a positive power $p < 1$ or a logarithmic function, both of which are increasing and concave down. However, a theoretical island of zero area would be home to no species, so the function should pass through the origin. Consequently, the model of choice is a power function.

Using a calculator to construct a power function to fit these data, we obtain

$$N = 3.055\, A^{0.310}.$$

In Figure 3.21 this function is superimposed on the original data and captures the trend in the data reasonably well. The corresponding correlation coefficient $r = 0.998$ indicates a very high level of correlation between the variables.

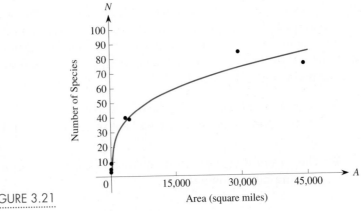

FIGURE 3.21

b. The area of Barbados is $A = 166$ square miles, so the model predicts that

$$N = 3.055\,(166)^{0.310} \approx 14.902,$$

or about 15 species of reptiles and amphibians live there.

Biologists have found that comparable results apply for virtually any other environment, large or small, and for any other species. Typically the power p is relatively close to 0.3. Because $10^{0.3} = 1.995$, biologists use the rule of thumb that a tenfold increase in the size of an environment leads to roughly double the number of species inhabiting it.

EXAMPLE 4   The Spread of AIDS

The following table shows the accumulated total number of reported cases of AIDS in the United States since 1983. Find an exponential function that can be used to model the spread of AIDS and check its correlation coefficient for significance.

Year	1983	1984	1985	1986	1987	1988	1989	1990
Number of AIDS Cases	4,589	10,750	22,399	41,256	69,592	104,644	146,574	193,878
Year	1991	1992	1993	1994	1995	1996	1997	1998
Number of AIDS Cases	251,638	326,648	399,613	457,280	528,144	594,641	653,084	701,353

Source: U.S. Centers for Disease Control and Prevention.

Solution   The scatterplot of the data points is shown in Figure 3.22. The pattern is increasing and concave up for the most part. However, note that the rate of growth seems to be diminishing in the last couple of years shown. It is reasonable to model the growth in the total number of reported cases of AIDS with an exponential function where the independent variable is the time t since 1980 and the dependent variable is the number of cases of AIDS A.

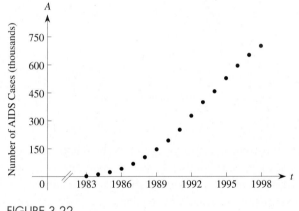

FIGURE 3.22

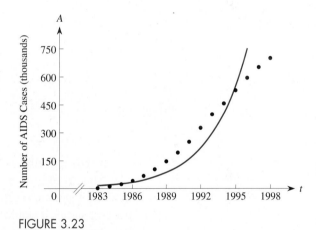

FIGURE 3.23

The exponential function to fit this data, obtained by using a calculator, is

$$A(t) = 5413.5(1.3626)^t,$$

where t is measured in years since 1980. The growth factor of 1.3626 indicates that the overall growth rate is over 36% per year. The corresponding correlation coefficient is $r = 0.9483$. Based on $n = 16$ data points, the critical value for r is $r_{16} = 0.497$, so there is a significant level of correlation, which suggests that this function fits the data well. However, when we superimpose this exponential function over the data, as shown in Figure 3.23, the curve doesn't match the pattern of the data. In particular, the exponential function grows more slowly than the number of cases of AIDS from 1987 to 1994 and thereafter grows much more rapidly.   ◆

Because the spread of AIDS has been slower than exponential growth, it makes sense to try a different model—say, a power function—which we illustrate in Example 5.

◆ **EXAMPLE 5** **The Spread of AIDS (Continued)**

Find a power function that can be used to model the spread of AIDS and compare the results to those obtained with the exponential function in Example 4.

Solution We use the data in Example 4, with t representing the number of years since 1980 and A the total number of cases of AIDS in the United States. Using a calculator, we obtain the power function that fits these data:

$$A(t) = 233.287t^{2.865}.$$

The corresponding correlation coefficient is $r = 0.9961$, which is considerably higher than the value of $r = 0.9483$ for the exponential function. This result suggests that the power function fits the data better than the exponential function. Superimposing this power function over the data, as shown in Figure 3.24, confirms that the power function fits the data better than the exponential function. However, like the exponential function, the power function also is growing more rapidly than the number of cases of AIDS over the last few years shown.

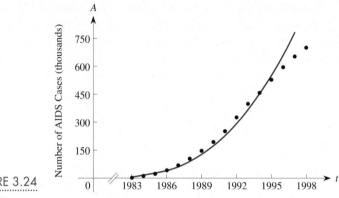

FIGURE 3.24

Together, Examples 4 and 5 illustrate an important point: When faced with a set of data, you should not be content to consider only one type of function as a potential model if there are several different families of functions that have the same behavior pattern. Rather, it would make sense for you to examine what happens when you fit various types of functions to the data and then decide which one is the best fit. We return to this example later in the book to show the effects of using other functions as models.

In Section 1.1, we discussed the growth in life expectancy over the years since the beginning of the twentieth century. The graph showing this trend is presented in Figure 3.25. Let's investigate this situation.

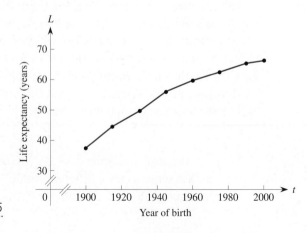

FIGURE 3.25

EXAMPLE 6   Increase in Life Expectancy

The following table gives the life expectancy for children born in the United States in the years shown since the beginning of the twentieth century.

a. Find an appropriate function to fit these data and check its correlation coefficient for significance.

b. Predict the life expectancy of a child born in 2008.

Year, t	1900	1915	1930	1945	1960	1975	1990	2000
Life Expectancy, L	47.3	54.5	59.7	65.9	69.7	72.6	75.4	76.4

Source: *2000 Statistical Abstract of the United States.*

Solution

a. The increasing, concave down pattern shown in Figure 3.26 suggests that either a power function with $0 < p < 1$ or a logarithmic function would be an appropriate choice for a model. However, the function modeling life expectancy clearly cannot pass through the origin (life expectancy was not 0 in the year 0), so a power function is not an appropriate choice. Note that the data values are growing more and more slowly over time, so we might expect that a logarithmic function would be a reasonable model for the data. Using a calculator, we find that the logarithmic function that fits these data is

$$L = 561.93 \ln t - 4192.2,$$

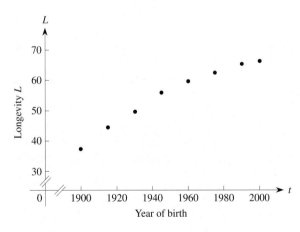

FIGURE 3.26

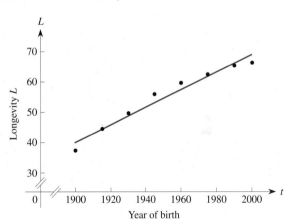

FIGURE 3.27

where t is the actual year. The result is given in terms of the base $e \approx 2.71828$, which we discussed in Section 2.6. Superimposed over the data in Figure 3.27, this function is a reasonably good fit. The corresponding correlation coefficient is $r = 0.9820$. The critical value for $n = 8$ data points is $r_8 = 0.707$, so we have a high level of positive correlation.

b. When $t = 2008$, the logarithmic function predicts a life expectancy for someone born in 2008 of

$$L(2008) = 561.93 \ln(2008) - 4192.2 = 81.2 \text{ years.}$$

EXAMPLE 7

An object is dropped from the top of the 1450-foot-high Sears Tower in Chicago. The following set of measurements show how far the object has fallen after each second.

Time (sec)	1	2	3	4	5	6	7	8
Distance Fallen (ft)	16	64	144	256	401	574	786	1022

Find a function that gives the distance the object has fallen after any time t.

Solution   From the scatterplot shown in Figure 3.28, it is clear that our best choices are either an exponential function or a power function because the pattern is increasing and concave up. However, at time $t = 0$ the object has fallen a distance of 0 feet, so the curve should pass through the origin. Thus an exponential function is not appropriate. Using a calculator, we find that the power function fit is

$$D = 16.004t^{1.99977}$$

with a correlation coefficient $r = 0.999999$, which suggests a virtually perfect fit.

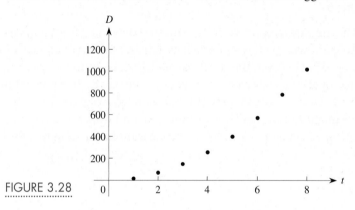

FIGURE 3.28

In fact, the actual formula for the distance fallen, based on an application of Newton's laws of motion, is precisely

$$D = 16t^2.$$

The coefficient 16 is measured in feet per second per second. Because this coefficient actually is one half the quantity known as the *acceleration due to gravity* (denoted by $g = 32$ ft/sec^2), we can write the formula as

$$D = \frac{1}{2}gt^2.$$

Usually, knowing how high the falling object in Example 7 is above the ground is far more important than how far it has fallen. If the object is dropped from a height of 1450 feet and the distance it falls in t seconds is $16t^2$, then subtracting the distance fallen from the initial height of 1450 feet gives

$$H = 1450 - 16t^2$$

as the object's height above the ground at any time t. We now can easily answer the question of how long it takes for the object to hit the ground. We set the height $H = 0$, and solve the resulting equation for t:

$$16t^2 = 1450$$

$$t^2 = \frac{1450}{16} = 90.625$$

$$t = \sqrt{90.625} \approx 9.5 \text{ sec.}$$

Thus it takes about 9.5 seconds until impact.

In general, if an object is dropped from any initial height y_0, its height y above ground level at any time t is given by

$$y = y_0 - 16t^2.$$

In the preceding examples, we illustrated the power of regression methods to create a function to model a set of data. However, a number of other important issues remain. First, we didn't discuss *how* the different functions are calculated, based on the data. Second, a number of problems can arise if you enter certain kinds of numbers into either a calculator or computer program and then ask the machine to calculate the best-fit function. In order to know what kinds of numbers to avoid, you have to understand how the different functions are calculated. Third, if certain kinds of numbers have to be avoided, you need a way to circumvent the problems that can arise and create functions to fit the actual data values. Fourth, most calculators use one method to find the equations of the functions that fit a set of data, whereas some mathematical software packages, such as Derive™ and Maple™, use a very different method. As a result, different functions from the same family of functions will be obtained, depending on the technology employed. Again, which function is the best to use requires an understanding of the methods employed. Fifth, in each of the preceding examples, we were able to decide on a single function to use. However, recall how when we weren't satisfied with the results of fitting an exponential function to the data on the spread of AIDS, we then tried a power function. When you try different functions, how do you decide which function is the best choice if all are appropriate mathematical models?

We discuss all these issues in the following sections. In this section, we focused on creating a single function to fit a set of data and using that function to answer predictive questions in that context.

Problems

1. For each scatterplot, select the type of function— exponential, power, or logarithmic—that is the most reasonable candidate to fit the data or decide that none are appropriate.

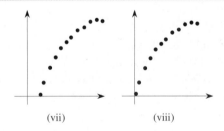

(vii)   (viii)

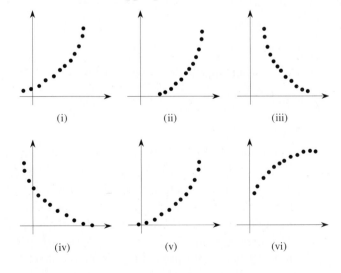

(i)   (ii)   (iii)

(iv)   (v)   (vi)

2. For each table of values, select the type of function—linear, exponential, power, or logarithmic— that is most likely to fit the data.

a.

x	y
1	14.8
2	7.4
3	3.6
4	1.9
5	0.9

b.

x	y
1	14.5
2	11.3
3	8.2
4	4.9
5	1.6

c.

x	y
1	2
2	9.8
3	25.1
4	48.2
5	161.5

d.

x	y
1	2.8
2	4.8
3	8.1
4	14.3
5	24.5

e.

x	y
1	0
2	5.5
3	8.7
4	11.2
5	12.8

f.

x	y
1	5
2	6.0
3	6.8
4	7.4
5	7.7

3. In Example 1, we created an exponential model for the growth of the U.S. population from 1780 to 1900.

   a. Use the model to predict the population in 1820 and 1850.

   b. Calculate the error of using this model to predict the population in 1820 and 1850 by comparing the values of the function to the actual population values in the table.

   c. Use the model to predict the U.S. population in 1920 and 1950.

   d. How large are the errors in these estimates? (The actual U.S. population was 105.7 million in 1920 and 150.7 million in 1950.)

   e. Based on these values, has the U.S. population been growing faster or slower than the exponential model?

4. The Caribbean island of Guadeloupe covers 687 square miles. Estimate the number of species of reptiles and amphibians that live on the island according to the model in Example 3.

5. How large a Caribbean island would be necessary, according to the model created in Example 3, to support 25 species of reptiles and amphibians?

6. Use the model created in Example 4 to estimate (**a**) the total number of cases of AIDS in the United States through 2001 and (**b**) the year in which the total number of cases reaches 1 million.

7. Use the model created in Example 2 to estimate (**a**) the amount of L-dopa in the blood after $t =$

200 minutes and (**b**) the length of time required for the level of L-dopa in the blood to drop to 100 nanograms per milliliter?

8. Use the model created in Example 6 to estimate (**a**) the life expectancy of a child born in 2004 and (**b**) the year in which the life expectancy of a newborn will be 85 years?

9. The table shows the number of species of nonflying mammals living on various islands in the English Channel, as well as the areas of the islands, in square kilometers.

Island	Herm	Sark	Alderney	Guernsey	Jersey
Area	1.3	5.2	7.9	63.5	116.3
Species	2	2	3	5	9

Source: Kevin Mitchell et al., *Mathematical Models of Biological Systems.* 1998.

   a. Find the power function that models the number of species on these islands as a function of area.

   b. How large an island would be needed to support 15 species of nonflying mammals?

10. The table shows estimates for the number of species of plants in different coastal regions of California and the area, in square miles, of each region.

Region	Area	Species
Tiburon Peninsula	5.9	370
San Francisco	45	640
Santa Barbara area	110	680
Santa Monica Mountains	320	640
Marin County	529	1060
Santa Cruz Mountains	1386	1200
Monterey County	3324	1400
San Diego County	4260	1450

Source: Kevin Mitchell et al., *Mathematical Models of Biological Systems.* 1998.

   a. Find the power function that models the number of plant species in coastal regions of California as a function of the area of the region.

   b. The area of Baja California is 24,210 square miles. What is your prediction for the number of species of plants in Baja?

c. The actual number of plant species in Baja California is about 1450. Because Baja California is considerably south of the other regions listed in the table, what does that suggest to you about the importance of latitude (north–south distance) to the validity of the model?

11. The accompanying table shows the number of households, in millions, with cable television in various years.

Year	1977	1980	1983	1986	1988	
Number	12.1	17.7	34.1	42.2	48.6	

Year	1990	1992	1994	1996	1998	2000
Number	54.9	57.2	60.5	64.7	67.0	68.5

Source: *World Almanac and Book of Facts.*

a. Find a linear function, an exponential function, and a power function that fit these data using t as the number of years since 1975.

b. Using each function, predict the total number of households with cable TV in the year 2005.

c. Which prediction do you think is most reasonable? Explain.

12. The table gives the total number, in thousands, of high school graduates in the indicated years since 1900.

Year	High School Grads
1900	95
1910	156
1920	311
1930	667
1940	1221
1950	1200
1960	1858
1970	2889
1980	3043
1990	2586
2000	2839

Source: *Digest of Education Statistics 2000,*
U.S. Department of Education.

a. Let t be the number of years since 1890. Determine the best linear, exponential, and power functions to model the number of high school graduates as a function of t.

b. Use each function to predict the number of high school graduates in 2010. Which prediction seems the most reasonable? the least reasonable?

c. Use each function to predict the year in which there will be 5 million high school graduates.

d. Which function seems to give the most reasonable prediction? the least reasonable?

13. The table gives the number of violent crimes per 100,000 people in the United States since 1960.

Year	1960	1965	1970	1975
Number of Crimes	175	200	360	490
Year	1980	1985	1990	1995
Number of Crimes	580	550	750	685

Source: FBI, *Crime in the United States.*

a. Find the best linear and exponential fits to this set of data, where t represents the number of years since 1950.

b. Use both models to predict the number of violent crimes, per 100,000 people, that occurred in 2000.

c. With both models, predict when the number of violent crimes will reach 1000 per 100,000 people.

d. What is the doubling time for the exponential model?

14. According to *Motor Trend* magazine, the following data are the times (in seconds) it takes a Trans Am to accelerate from zero to the indicated speed.

Speed (mph)	30	40	50	60	70
Time (sec)	3.00	4.29	5.52	7.38	9.81

a. Determine the best linear, exponential, and power functions to model the acceleration time as a function of the final speed.

b. Estimate how long it will take a Trans Am to accelerate to 45 mph; to 80 mph; to 90 mph.

c. Which estimated time is most likely to be accurate? (You might want to compare these data to the corresponding information about the Mercedes in Problem 5 of Section 3.2.)

15. According to the U.S. Department of Education, the following data are the numbers, in thousands, of college degrees awarded during the indicated year from 1900 to 1995.

Year	1900	1910	1920	
College Graduates	30	54	73	
Year	1930	1940	1950	1960
College Graduates	123	223	432	530
Year	1970	1980	1990	1995
College Graduates	878	935	1017	1165

Source: U.S. National Center for Educational Statistics.

a. Determine the best linear, exponential, power, and logarithmic functions to model the number of college graduates as a function of t, the number of years since 1890.
b. Use each of the four functions to predict the number of college graduates in 2005.
c. Which of the four predictions seems the most reasonable? the least reasonable?
d. Use each of the four models to predict the year in which there will be 2 million college graduates.
e. What is the doubling time for the exponential model?

16. a. Compare the results in Problem 15 to the results in Problem 12 concerning the number of high school graduates as a function of time. In particular, for the two exponential growth models, which model is growing faster?
b. If you project forward indefinitely with both models, when will the number of college diplomas awarded surpass the number of high school diplomas awarded? Discuss the reasonableness of this scenario.

17. a. Use the set of data from Problem 12 on high school diplomas awarded and the data from Problem 15 on college degrees awarded to create a table consisting of the number of college degrees and the number of high school diplomas awarded each year from 1900 to 1990.
b. Determine the linear, exponential, power, and logarithmic functions that fit these data to construct four models relating the number of college degrees awarded as a function of the number of high school diplomas awarded the same year.

c. What is the significance of the positive slope for the linear fit?
d. What is the significance of the growth factor in the exponential fit being greater than 1?

18. The Dow-Jones average of 30 major industrial stocks is probably the most widely watched measure of the stock market. The following table shows the Dow-Jones average at the beginning of each year since 1980.

Year	1980	1981	1982	1983	1984	1985	1986
Dow	839	964	875	1047	1259	1212	1547
Year	1987	1988	1989	1990	1991	1992	1993
Dow	1896	1939	2169	2753	2634	3169	3301
Year	1994	1995	1996	1997	1998	1999	2000
Dow	3758	3834	5177	6447	7965	9184	11,358

Source: *Wall Street Journal.*

a. Find the best linear, exponential, and power functions to fit these values for the Dow, using t to represent the number of years since 1979.
b. Which of the three functions seems to best fit these data? Explain.
c. What is your prediction, based on this function, for the current value of the Dow? Check a newspaper or listen to the business news on the radio or television to find out how close your prediction is to the actual value.
d. The Dow closed above 4000 for the first time on February 23, 1995. Which of the three models comes closest to predicting that date?
e. The Dow closed above 5000 for the first time on November 21, 1995. Which of the three models comes closest to predicting that date?
f. Based on your best-fitting function, when do you predict the Dow will first reach 14,000?

19. The following table shows the cumulative number of HIV infections, worldwide, in millions, since 1980.

a. Assuming that the number of people infected with HIV is growing exponentially, find the exponential function that fits these data, where the variable t represents the number of years since 1979.
b. What is the growth factor in your function and what does it mean?
c. Predict the number of people who will have been infected with HIV by 2005.

Year	1980	1981	1982	1983	1984
Number	0.1	0.3	0.7	1.2	1.7
Year	1985	1986	1987	1988	1989
Number	2.4	3.4	4.5	5.9	7.8
Year	1990	1991	1992	1993	1994
Number	10.0	12.8	16.1	19.7	23.8
Year	1995	1996	1997	1998	1999
Number	28.3	33.5	38.9	44.1	49.9

Source: Lester R. Brown et al., *Vital Signs 2000: The Environmental Trends That Are Shaping Our Future.*

d. How does the growth in the number of people infected with HIV compare to the growth of this exponential function? That is, does the function grow faster or slower than the actual data?

e. Find the power function that fits these data, where the variable t represents the number of years since 1979.

f. Predict the number of people who will have been infected with HIV by 2005.

g. How does the growth in the number of people infected with HIV compare to the growth of this power function? That is, does the power function grow faster or slower than the actual data?

h. Does the exponential function or the power function seem to fit the data better? Explain your reasoning.

20. The following table shows the cumulative number of deaths from AIDS worldwide, in millions, since 1984.

Year	1984	1985	1986	1987		
Number	0.1	0.2	0.3	0.5		
Year	1988	1989	1990	1991	1992	1993
Number	0.8	1.2	1.7	2.4	3.3	4.4
Year	1994	1995	1996	1997	1998	1999
Number	5.7	7.3	9.2	11.3	13.7	16.3

Source: Lester R. Brown et al., *Vital Signs 2000: The Environmental Trends That Are Shaping Our Future.*

a. Assuming that the number of deaths from AIDS is growing exponentially, find the exponential

function that fits the data, where the variable t represents the number of years since 1979.

b. What is the growth factor in your function and what does it mean?

c. Predict the total number of people who will have died from AIDS by 2005.

d. Find the power function that fits the data, where the variable t represents the number of years since 1979.

e. Predict the number of people who will have died from AIDS by 2005.

f. Does the exponential function or the power function seem to fit the data better? Explain.

21. A growing percentage of the world's population is now living in urban areas, as shown in the following table.

Year	1950	1955	1960	1965	1970	
Percentage	29.7	31.6	33.6	35.5	36.7	
Year	1975	1980	1985	1990	1995	1999
Percentage	37.8	39.4	41.2	43.2	45.3	47.0

Source: Lester R. Brown et al., *Vital Signs 2000: The Environmental Trends That Are Shaping Our Future.*

a. Find the equation of the linear function that best fits the data, where t is the number of years since 1945.

b. Predict the percentage of the world's population that will live in urban areas in 2020, based on the linear function.

c. Find the equation of the exponential function that fits the data.

d. Predict the percentage of the world's population that will live in urban areas in 2020, based on the exponential function.

e. Find the equation of the power function that fits the data.

f. Predict the percentage of the world's population that will live in urban areas in 2020, based on the power function.

g. Which of the three functions appears to fit the data best? Which of the three functions appears to give the most reasonable prediction for the percentage of the world's population living in urban areas in 2020? Which gives what appears to be the least reasonable prediction? Explain.

22. The following table shows worldwide consumption of natural gas, measured in the equivalent of millions of tons of oil, over time.

Year	1950	1960	1970	1975	
Natural Gas Used	187	444	1022	1199	
Year		1980	1985	1990	1995	1999
Natural Gas Used		1406	1640	1942	2116	2301

Source: Lester R. Brown et al., *Vital Signs 2000: The Environmental Trends That Are Shaping Our Future.*

a. Find the equation of the linear function that best fits the data, where *t* is the number of years since 1940.
b. Predict the worldwide consumption of natural gas in 2020, based on the linear function.
c. Find the equation of the exponential function that fits the data.
d. Predict the worldwide consumption of natural gas in 2020, based on the exponential function.
e. Find the equation of the power function that fits the data.

f. Predict the worldwide consumption of natural gas in 2020, based on the power function.
g. Which of the three functions appears to fit the data best? Which of the three functions appears to give the most reasonable prediction for the worldwide consumption of natural gas in 2020? Which gives what appears to be the least reasonable prediction? Explain.

23. As the pressure of a liquid goes up in a confined space (say, in a pressure cooker), the boiling point also goes up. The following table gives the temperature *T* of the boiling point of water, in degrees Celsius, at various vapor pressures, *P*, in kilo-pascals.

a. Find the logarithmic function that fits the data. How good is the fit?
b. Use your function from part (a) to find the boiling point of water when the vapor pressure is 6.2 kilo-pascals.
c. What vapor pressure is needed if the boiling point of the water is 120°C?

Pressure	0.61	1.22	2.34	4.25	7.38	12.34	19.93	31.18	47.37	70.12	101.32
Temp.	0°	10°	20°	30°	40°	50°	60°	70°	80°	90°	100°

Source: CRC *Handbook of Chemistry and Physics,* 1996.

3.4 How to Fit Exponential and Logarithmic Functions to Data

In Section 3.2, we developed the ideas of correlation and linear regression that allow us to determine whether a linear relationship exists between two variables, and if so, to find the equation of the regression line that best fits the data. In Section 3.3, we considered the natural extension of these ideas to fit nonlinear functions such as exponential, power, and logarithmic functions to a set of data. However, the approach we used there was just to quote the results obtained from a calculator or a spreadsheet. Let's now look at how the calculator finds these equations. In the process, you will come to understand some of the cautions we mentioned at the end of Section 3.3.

One of the key approaches used in practice when a set of data does not fall in a linear pattern is the following three-step process.

1. *Transform* the data in some way so that the resulting transformed values fall in a roughly linear pattern. We call this *linearizing the data.*

2. Once the data have been linearized, find the regression line and correlation coefficient for the transformed data.

3. Undo the transformation by using the appropriate inverse function to produce the equation of the nonlinear function that fits the original set of data.

Here, we consider modeling a data pattern with an exponential or a logarithmic function; in Section 3.5, we consider modeling a data pattern with a power function.

Fitting Exponential Functions to Data

Let's begin by considering the population of the United States, in millions, from 1780 to 1900, as shown in Table 3.4 (Example 1 of Section 3.3) with the corresponding scatterplot in Figure 3.29. Again, the independent variable t is the number of decades since 1780. As we showed there, the successive ratios are fairly constant, which suggests that the pattern is roughly exponential. The calculator gives the exponential function as $P(t) = 3.069\,(1.321)^t$ with a correlation coefficient of $r = 0.998$. Now let's see how this function is actually obtained, using transformations.

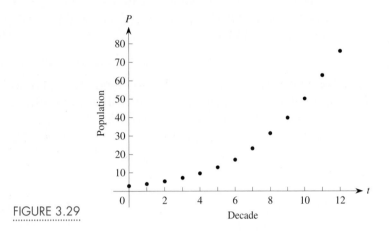

FIGURE 3.29

EXAMPLE 1

Construct the exponential function that fits the data on the U.S. population from 1780–1900, using the transformation approach.

Solution To determine the exponential function that fits the population values shown in Table 3.4, we first transform the population data by taking the logarithm of each population value, as shown in Table 3.5. (We discuss the reason for doing so later.)

The resulting scatterplot of log(population) versus time in decades, shown in Figure 3.30, now indicates a roughly linear pattern. Checking the differences in successive values under log(population) in Table 3.5 reveals that they are roughly equal. We therefore find that the best linear fit to this set of transformed data, using the ideas from Section 3.2, is

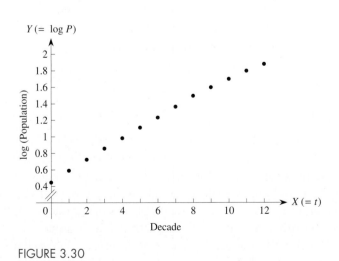

FIGURE 3.30

TABLE 3.5 *Logarithms of Population Values*

Decade	Population	log(population)
0	2.8	0.447
1	3.9	0.591
2	5.3	0.724
3	7.2	0.857
4	9.6	0.982
5	12.9	1.111
6	17.1	1.233
7	23.2	1.365
8	31.4	1.497
9	39.8	1.600
10	50.2	1.701
11	62.9	1.799
12	76.0	1.881

$$Y = 0.121X + 0.487,$$

where $Y = \log(\text{population})$ and X is the number of decades since 1780.

The corresponding correlation coefficient is $r = 0.998$. Comparing this value to the critical value of $r_{13} = 0.553$ from the table on the inside back cover shows that, with 95% certainty, there is positive correlation between the decade and the logarithm of the population from 1780 through 1900. Note that the value for the correlation coefficient, $r = 0.998$, for the linear fit to the transformed data values is exactly the same value that a calculator gives as part of its exponential regression calculations.

However, in the regression equation $Y = 0.121X + 0.487$, the independent variable X represents t and the dependent variable Y represents $\log P$. That is, the regression equation actually represents

$$\log P = 0.121t + 0.487.$$

We now undo the original transformation (the logarithm) of the data by applying the inverse function, which is an exponential function with base 10, and by using the appropriate properties of exponents. We obtain

$$
\begin{aligned}
P = 10^{\log P} &= 10^{0.121t + 0.487} \\
&= 10^{0.121t} \cdot 10^{0.487} \\
&= 3.069(1.321)^t,
\end{aligned}
\qquad
\begin{aligned}
&10^{\log u} = u \\
&10^{u+v} = 10^u \cdot 10^v \\
&10^{u \cdot p} = (10^u)^p
\end{aligned}
$$

where t is the number of decades since 1780. This new equation has the form $P_0 c^t$ for an exponential function. This expression is identical to the exponential function that the calculator gives automatically.

◆

Why the Transformation Approach Works

Let's explore why we took the logarithm of the population values. Suppose that the scatterplot for a set of data appears to follow an exponential pattern so that we hope to fit an exponential function of the form $y = f(x) = kc^x$ to the data for some constants c and k. If we take logarithms of both sides of this equation and use properties of logs, we get

$$
\begin{aligned}
\log y = \log(kc^x) & \\
= \log k + \log(c^x) &\qquad \log AB = \log A + \log B \\
= \log k + x \log c &\qquad \log A^p = p \log A \\
= (\log c)x + \log k. &
\end{aligned}
$$

Because $\log c$ and $\log k$ are constants, this expression has the form

$$Y = aX + b,$$

where

$$Y = \log y, \qquad a = \log c, \quad \text{and} \quad b = \log k.$$

Thus, if y is an exponential function of x, then $\log y$ is a linear function of x, and this transformation (taking the logarithm of the y values) linearizes the data, as illustrated in Figure 3.31. We then find the coefficients a and b using the linear regression technique. Finally, we undo the transformation by taking powers of 10 of both sides of the equation and apply the appropriate properties of exponents and logarithms to get the desired exponential function, as illustrated in Example 1.

Example 2 demonstrates that this approach—based on transforming the data values from (x, y) to $(x, \log y)$—actually works.

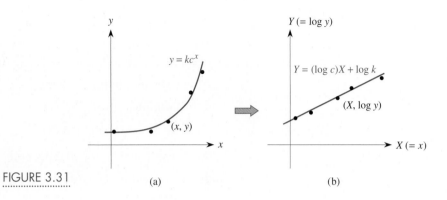

FIGURE 3.31   (a)   (b)

◆ **EXAMPLE 2**

Show that the linearization technique works for points on the exponential function $y = 5 \cdot 2^x$.

Solution   We select the points on the curve $y = 5 \cdot 2^x$ corresponding to $x = 0$ (so that $y = 5 \cdot 2^0 = 5$), $x = 1$, $x = 2$, $x = 3$, and $x = 4$, as shown in Figure 3.32(a). These points give us the first two columns in the following table. When we transform this set of data by taking the logarithms of the y-values, we get the third column.

The scatterplot of the transformed data points $(x, \log y)$ shown in Figure 3.32(b) suggests that the points appear to lie exactly on a straight line. If we take any two of the

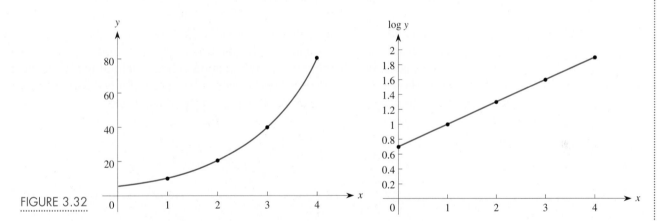

FIGURE 3.32

x	y	$\log y$
0	5	0.699
1	10	1
2	20	1.301
3	40	1.602
4	80	1.903

transformed points, we find that the slope of the line joining them is 0.301. Furthermore, the y-intercept is $b = 0.699$. Thus the equation of the line through the transformed data points is $Y = 0.301X + 0.699$, which is equivalent to

$$\log y = 0.301x + 0.699.$$

We now undo the logarithm by taking powers of 10 on both sides of this equation:

$$
\begin{aligned}
10^{\log y} = y &= 10^{0.301x + 0.699} \\
&= (10^{0.301x})(10^{0.699}) \\
&= (10^{0.301})^x (5.0003) \\
&= 5.0003(1.9999)^x,
\end{aligned}
\qquad
\begin{aligned}
10^{\log u} &= u \\
10^{u+v} &= 10^u \cdot 10^v \\
10^{u \cdot p} &= (10^u)^p
\end{aligned}
$$

which is effectively the exponential function $y = 5 \cdot 2^x$ that we started with when rounding errors are eliminated.   ◆

We summarize this approach as follows.

> If the (x, y) data appear to follow an exponential pattern $y = kc^x$,
>   1. plot $\log y$ versus x to linearize the data;
>   2. find the best linear fit,
>
> $$\log y = ax + b,$$
>
>   to the transformed $(x, \log y)$ data; and
>   3. undo the logarithmic transformation by taking powers of 10.

Virtually every graphing calculator has a built-in routine as part of its statistics capabilities to find the exponential function that fits a set of data based on this procedure.

Caution: In the process of transforming to $\log y$ versus x, all the y-values *must* be positive because the logarithm is not defined for negative or zero values.

Many scientific software packages, as well as most spreadsheets, contain routines that will fit an exponential (or other) function to a set of data. However, some mathematical software packages, such as Derive™, Maple™ and Mathematica™, calculate a best fitting exponential function by applying the least squares criterion directly. They find the values for k and c in $y = kc^x$ that minimize the sum of the squares of the vertical deviations between the data points and the curve, as illustrated in Figure 3.33. The results obtained by using this method can be somewhat different from the results obtained with the transformation approach to linearize the data values. We use the transformation approach throughout the book and all answers shown are based on it. Also, whenever we speak of a *best* fitting (nonlinear) function, it is in the context of the transformation approach.

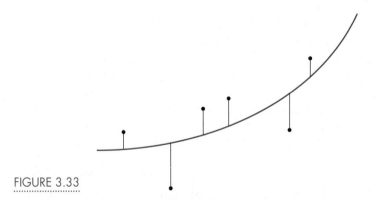

FIGURE 3.33

The transformation method also applies to situations that can be modeled by an exponential decay function, as demonstrated in Example 3.

EXAMPLE 3

The following table and the graph shown in Figure 3.34 give the amount L of L-Dopa in the bloodstream, in nanograms per milliliter, t minutes after the drug was absorbed into the blood. Find an exponential function that models the level of L-Dopa in the blood as a function of time, using the transformation approach.

t	0	20	40	60	80	100	120	140	160	180	240	300
L	2950	2600	1550	1100	900	725	600	510	440	300	250	225

Source: Douglas Brown and Tom Timchek; unpublished manuscript.

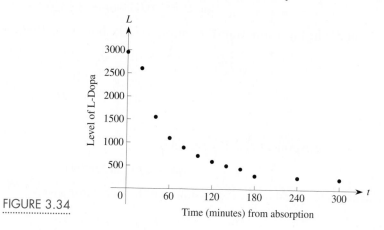

FIGURE 3.34

Solution   In Example 2 of Section 3.3, we found that the calculator gave

$$L(t) = 2147.8(0.9909)^t$$

as the exponential decay function that best fits these data.

To find this exponential function by using the transformation approach, we first look at log L versus t.

t	0	20	40	60	80	100	120	140	160	180	240	300
log L	3.470	3.415	3.190	3.041	2.954	2.860	2.778	2.708	2.643	2.477	2.398	2.352

The corresponding scatterplot, shown in Figure 3.35, is reasonably linear. Applying a calculator's linear regression routine to this transformed data, we find that the resulting regression equation that is the best linear fit to this transformed data is

$$Y = -0.00396X + 3.3320,$$

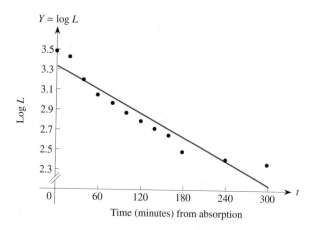

FIGURE 3.35

where $Y = \log L$. The corresponding correlation coefficient is $r = -0.9545$. Note that both the slope and the correlation coefficient are negative because the trend in the data is decreasing. Also, the critical value is $r_{12} = 0.576$, indicating a significant level of correlation.

Because we transformed the data by plotting $\log L$ versus t, the linear function we got actually represents

$$\log L = -0.00396t + 3.3320.$$

To undo the logarithm transformation, we take powers of 10 on both sides:

$$
\begin{aligned}
L = 10^{\log L} &= 10^{-0.00396t + 3.3320} & 10^{\log u} &= u \\
&= 10^{-0.00396t} \cdot 10^{3.3320} & 10^{u+v} &= 10^u 10^v \\
&\approx (10^{-0.00396})^t \cdot 2147.8 & 10^{u \cdot p} &= (10^u)^p \\
&\approx 2147.8 \cdot (0.9909)^t
\end{aligned}
$$

so that the exponential decay model is

$$L(t) = 2147.8(0.9909)^t,$$

which is identical to the expression that the calculator gave.

Fitting Logarithmic Functions to Data

Suppose now that a set of data falls in the increasing, concave down pattern shown in Figure 3.36. Such a shape might suggest a logarithmic function of the form

$$y = f(x) = a \log x + b.$$

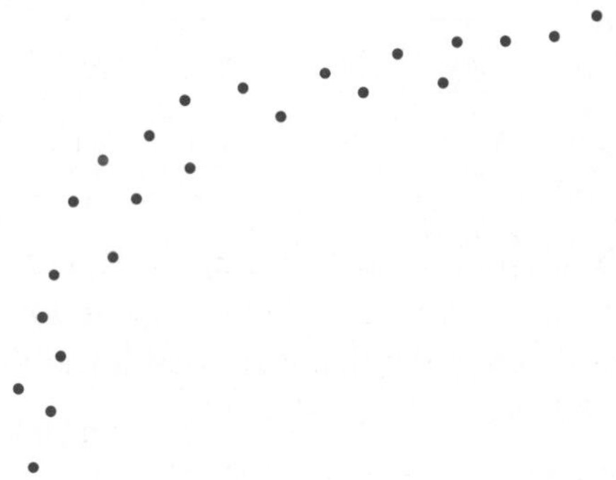

FIGURE 3.36

(This scatterplot may also suggest a power function with $0 < p < 1$; we consider such cases in Section 3.5.) This expression for the logarithmic function suggests that we should transform the original data set by comparing y to $\log x$ rather than comparing y to x.

However, most calculators and computer packages that perform this calculation use natural logarithms with base e rather than logarithms with base 10. Thus we actually fit logarithmic functions of the form

$$y = f(x) = a \ln x + b$$

to the data. So, to linearize data that follows a logarithmic pattern, we plot y versus $\ln x$.

In Example 6 of Section 3.3, we created the logarithmic function $L = 561.93 \ln t - 4192.2$ to model the growth in life expectancy over the years since the beginning of the twentieth century. The graph of this trend is shown in Figure 3.37.

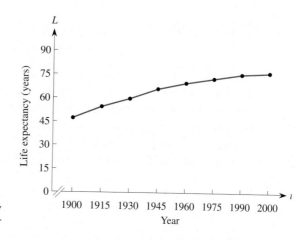

FIGURE 3.37

EXAMPLE 4

The following table shows the trend in life expectancy for children born in the years shown since the beginning of the twentieth century. Find the best logarithmic function to fit these data.

Year, t	1900	1915	1930	1945	1960	1975	1990	2000
Life Expectancy, L	47.3	54.5	59.7	65.9	69.7	72.6	75.4	76.4

Source: *2000 Statistical Abstract of the United States.*

Solution   To fit a logarithmic function to the data, we need to compare life expectancy L to the natural logarithm $\ln t$ of the year, as follows.

ln t	7.5496	7.5575	7.5653	7.5730	7.5807	7.5883	7.5959	7.6009
L	47.3	54.5	59.7	65.9	69.7	72.6	75.4	76.4

The graph of the transformed data $(\ln t, L)$ shown in Figure 3.38 appears to be reasonably linear and the best linear fit to this transformed data is

$$Y = 561.93 X - 4192.2,$$

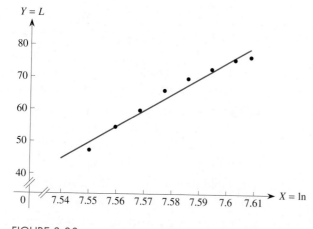

FIGURE 3.38

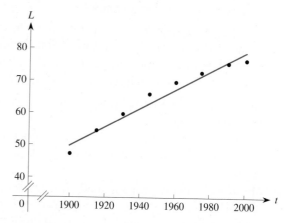

FIGURE 3.39

which really represents

$$L = 561.93 \ln t - 4192.2.$$

Because this expression already gives L as a function of t, we need not detransform the equation in order to solve for L. Figure 3.39 shows this function superimposed over the data, and it is a reasonably good fit, though it doesn't bend as much as the data set does. The corresponding correlation coefficient is $r = 0.9820$, which indicates a high level of positive correlation compared to the critical value $r_8 = 0.707$. Incidentally, note that this function is identical to the one found by calculator in Example 6 of Section 3.3. ◆

We summarize this approach as follows.

> If the data appear to follow a logarithmic pattern,
> 1. plot y versus $\ln x$ to linearize the data; and
> 2. find the best linear fit,
> $$y = a \ln x + b,$$
> to the transformed data.

Incidentally, a plot of either $\log y$ versus x (for an exponential fit) or y versus $\log x$ (for a logarithmic fit) is known as a *semi-log plot*. You may have used semi-log paper in some of your laboratory courses to plot experimental data that follow either an exponential or a logarithmic pattern.

Problems

In each problem use the transformation approach to find the exponential or logarithmic function that best fits the data.

1. (a) Repeat the analysis of the growth of the U.S. population presented in Example 1, but concentrate on the period from 1780 to 1890 instead of 1780 to 1900. (b) Does this exponential function give a better fit? How do you know?

2. The accompanying graph shows the number of deaths per 100,000 women in the United States from both stomach cancer and lung cancer since 1930. Both sets of data appear to be exponential functions.

   a. Find the best exponential function that fits each set of data.

   b. What are your estimates for the number of deaths per 100,000 women from each type of cancer in 2005?

   c. About how many women died in 1990 from stomach and lung cancer according to these models if there were approximately 100 million adult women in the United States at that time?

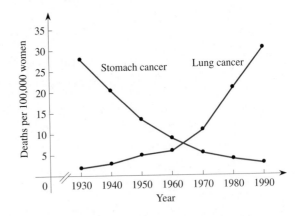

d. Can you give any reasons why the trend for stomach cancer (like that for most other cancers) is decreasing exponentially, while the trend for lung cancer is rising exponentially?

3. The table shows the growth of the federal debt in billions of dollars from 1940 to 2000.

   a. Determine the exponential function that best fits these data.

b. Use this model to predict the federal debt in 2005.

Year	1940	1950	1960	1970	1980	1990	2000
Debt	651	257	291	381	909	3207	5686

Source: *2000 Statistical Abstract of the United States.*

4. The table shows the growth of the U.S. population in millions from 1940 to 1990.

   a. Determine the exponential function that best fits these data.

   b. Use this model to predict the U.S. population in 2000.

   c. According to the 2000 census, the actual U.S. population in 2000 was 281.4 million. Did the model give a reasonably accurate prediction? Explain.

Year	1940	1950	1960
Population	131.7	150.7	179.3
Year	1970	1980	1990
Population	203.3	226.5	248.7

5. Use the data in Problems 3 and 4 including the fact that the U.S. population was 281.4 million in 2000 to construct a set of values representing the average amount of the national debt per person in the United States every 10 years from 1940 to 2000. Determine an appropriate function that best fits these data. If these trends continue, estimate your share of the debt in 2005.

6. The table at the bottom of the page shows the median family income, in current dollars, in the United States from 1970 to 1992.

   a. Determine the best exponential function to fit these data on median family income by year.

   b. Use your result in part (a) to predict the median family income in 2000 and in 2005.

   c. Using your model, determine when median family income will reach $50,000.

   d. What is the doubling time for the exponential model?

e. Consult the current edition of the *Statistical Abstract of the United States* to find the correct value for the median family income in 2000. How close is your prediction in part (a) to the actual value?

7. Suppose that the pattern you found in Problem 6 for the growth in median family income continues without change. In addition, suppose that inflation "remains under control" for the foreseeable future and is limited to about 3% per year. Write a short interpretation of what these two trends, if they continue without change, mean in terms of the standard of living in 20 years.

8. The table shows the Dow-Jones average for 30 industrial stocks at the beginning of each year since 1990.

Year	1990	1991	1992	1993	1994	
Dow	2753	2634	3169	3301	3758	
Year	1995	1996	1997	1998	1999	2000
Dow	3834	5177	6447	7965	9184	11,358

Source: *Wall Street Journal.*

   a. Find the exponential function that best fits these data.

   b. What is your prediction, based on this function, for the current value of the Dow? Check a newspaper or listen to the business news on the radio or television to find out how close the prediction is to the actual value.

   c. The Dow closed above 4000 for the first time on February 23, 1995. How close did the model come to predicting that date?

   d. The Dow closed above 5000 for the first time on November 21, 1995. How close did the model come to predicting that date?

   e. Based on your best fitting function, when do you predict the Dow will first reach 14,000?

   f. How do the results of this problem compare to those of Problem 18 in Section 3.3, which was based on the value of the Dow-Jones average from 1980 to 2000?

Year	1970	1975	1980	1985	1990	1993	1995	1997	1999
Median family income	9,867	13,719	21,023	27,735	35,353	36,959	40,611	44,568	48,950

Sources: *2000 Statistical Abstract of the United States* and U.S. Bureau of the Census.

9. In Problem 18 of Section 3.3, you were asked to find the exponential function that best fits the Dow-Jones average from 1980–2000.

   a. Suppose that your grandmother wants to invest heavily in the stock market. Write a paragraph based on the results of Problem 8 above or Problem 18 of Section 3.3 and your interpretations that might convince her to be more conservative.

   b. Suppose that you are the aggressive stockbroker who is trying to convince her to invest heavily. Write another paragraph based on the results from either Problem 8 above or Problem 18 of Section 3.3 that might convince her to let you invest her life savings for her.

10. The table shows worldwide car production, in millions.

Year	1950	1960	1970	1980	1990	1999
Cars	9	14	22	30	38	39

Source: Lester R. Brown et al., *Vital Signs 2000: The Environmental Trends That Are Shaping Our Future.*

   a. Find the exponential function that best fits this data.

   b. According to your model, what do you predict for the number of cars produced in 2010?

11. The table shows worldwide wind energy generating capacity, in megawatts, over time.

Year	1980	1985	1988	1990
Wind Energy	10	1020	1580	1930
Year	1992	1995	1997	1999
Wind Energy	2510	4820	7640	13840

Source: Lester R. Brown et al., *Vital Signs 2000: The Environmental Trends That Are Shaping Our Future.*

   a. Find the exponential function that best fits the data.

   b. What is the doubling time for this exponential function? Explain what it means.

   c. According to your model, what do you predict for the total wind energy generating capacity in 2010?

12. The table shows the worldwide production of photovoltaic cells used for collecting solar energy over time. The units are the energy equivalent of the cells in megawatts.

Year	1975	1980	1985	1990
Solar energy	1.8	6.5	22.8	46.5
Year	1993	1995	1997	1999
Solar energy	60.1	78.6	125.8	201.3

Source: Lester R. Brown et al., *Vital Signs 2000: The Environmental Trends That Are Shaping Our Future.*

   a. Find the exponential function that best fits these data.

   b. What is the doubling time for this exponential function? Explain what it means.

   c. Write a paragraph comparing the growth rate for the total energy equivalent of the photovoltaic cells to the growth rate for the total wind energy generating capacity in Problem 11. What is the likely long-term significance of this difference?

   d. According to your model, what do you predict for the total energy equivalent of the photovoltaic cells produced in 2010?

13. There is considerable discussion worldwide about the growing levels of carbon dioxide (CO_2) in the atmosphere because of its effects on global warming. The table below shows the atmospheric concentrations of CO_2, in parts per million, over time.

   a. Find the equation of the exponential function that best fits these data.

   b. What is the doubling time for this exponential function? Explain what it means.

   c. What does the model predict for the atmospheric concentration of CO_2 in 2010?

   d. When does the model predict that the carbon dioxide concentration will reach 400 parts per million?

Year	1960	1970	1975	1980	1985	1990	1993	1995	1997	1999
CO_2 Concentration	316.7	325.5	332.0	338.5	345.7	354.0	357.0	358.8	363.9	368.4

Source: Lester R. Brown et al., *Vital Signs 2000: The Environmental Trends That Are Shaping Our Future.*

14. Tourism is a booming global pastime and has become an important economic base for many nations. The following table shows the growth in the number of international tourist arrivals, in millions, since 1950.

Year	1950	1960	1970	1975	1980
Tourists	25	69	166	223	286

Year	1985	1990	1993	1995	1997	1999
Tourists	328	459	519	569	620	657

Source: Lester R. Brown et al., *Vital Signs 2000: The Environmental Trends That Are Shaping Our Future.*

   a. Find the equation of the exponential function that best fits these data.
   b. What is the doubling time for this exponential function? Explain what it means.
   c. What does the model predict for the number of international tourists in 2010?
   d. When does the model predict that the number of international tourists will reach 1 billion?

15. The table shows the growth in the number of telephones, in millions, in use throughout the world since 1960.

Year	1960	1970	1975	1980	1985
Telephones	89	156	229	311	407

Year	1990	1993	1995	1997	1998
Telephones	520	606	691	788	844

Source: Lester R. Brown et al., *Vital Signs 2000: The Environmental Trends That Are Shaping Our Future.*

   a. Find the equation of the exponential function that best fits these data.
   b. What is the doubling time for this exponential function? Explain what it means.
   c. What does the model predict for the number of phones in use in 2020?

   d. When does the model predict that the number of phones in use will reach 1 billion?
   e. The exponential function $P(t) = 3.6(1.013)^t$ can be used to model the growth of the world's population. Clearly, the growth factor for the population is considerably smaller than the growth factor for the number of phones. If both trends continue, when will there be one phone for every person living? (Hint: Think about units.)

16. The table shows the growth in the number of cell phones, in millions, in use throughout the world since 1985.

Year	1985	1988	1990	1991	1992
Cell Phones	1	4	11	16	23

Year	1993	1994	1995	1996	1997	1998
Cell Phones	34	55	91	142	215	319

Source: Lester R. Brown et al., *Vital Signs 2000: The Environmental Trends That Are Shaping Our Future.*

   a. Find the equation of the exponential function that best fits these data.
   b. What is the doubling time for this exponential function? Explain what it means.
   c. What does the model predict for the number of cell phones in use in 2020?
   d. When does the model predict that the number of cell phones in use will reach 1 billion?

17. The table below shows the growth in the number of computers connected to the Internet, in thousands, throughout the world since 1985.

   a. Find the equation of the exponential function that best fits these data.
   b. What is the doubling time for this exponential function? Explain what it means.
   c. What does the model predict for the number of computers that can access the Internet in 2020?
   d. When does the model predict that the number of computers that can access the Internet will reach 250 million?

Year	1985	1988	1990	1991	1992	1993	1994	1995	1996	1997	1998	1999
Internet Connections	2.3	80	376	727	1313	2217	5846	14,352	21,819	29,670	43,230	72,398

Source: Lester R. Brown et al., *Vital Signs 2000: The Environmental Trends That Are Shaping Our Future.*

Exercising Your Algebra Skills

Several sets of data have been linearized to find the best exponential fit. The calculator gives the following equations for the lines that best fit the transformed data. Undo the transformations to get the corresponding exponential functions.

1. $Y = 0.7782 + 0.0219X$

2. $Y = 1.3010 + 0.0128X$

3. $Y = 1.0729 - 0.0223X$

4. $Y = -0.3010 - 0.0706X$

5. $Y = 0.3522 + 1.0843X$

6. $Y = -1.3015 + 0.7840X$

7. $Y = 0.8525 - 1.2733X$

8. $Y = -1.581 - 0.903X$

3.5 How to Fit Power Functions to Data

Suppose that we have a growth pattern in a set of data that seems to be less extreme than exponential growth or exponential decay or one that is concave down. We might then suspect that a power function of the form $y = f(x) = kx^p$, where k and p are two constants to be determined, is the most appropriate model. As with the exponential function in Section 3.4, we need a way to linearize the data with an appropriate transformation so that we can apply linear regression techniques. We then undo the transformation to find the power function that fits the original data.

To begin, we take logarithms of both sides of the equation $y = k \cdot x^p$ and use properties of logs:

$$\log y = \log(kx^p)$$
$$= \log k + \log(x^p) \qquad \log(u \cdot v) = \log u + \log v$$
$$= \log k + p \log x \qquad \log(u^p) = p \log u$$

Because both $\log k$ and p are constants, we can interpret the equation

$$\log y = \log k + p \log x$$

as saying that $\log y$ is a linear function of $\log x$. That is,

$$\log y = a \log x + b$$

is of the form

$$Y = aX + b,$$

where

$$a = p \quad \text{and} \quad b = \log k.$$

Therefore, if y is a power function of x, we can linearize the original data set by taking the logarithms of *both* the x- and y-values. Thus, if y versus x follows a power function pattern, $\log y$ versus $\log x$ will follow a linear pattern. We then apply the linear regression technique to determine a and b and finally undo the transformation to get the desired values for $p = a$ and $k = 10^b$.

◆ **EXAMPLE 1**

Show that the linearization technique works for points on the power function curve $y = 5x^2$.

Solution   We select the points on the curve $y = 5x^2$ corresponding to $x = 1$ (so $y = 5 \cdot 1^2 = 5$), $x = 2$, $x = 3$, and $x = 4$. These points give the first two columns in the following table. Figure 3.40 shows the curve through these points. When we transform this set of data by taking the logarithms of both the x- and y-values, we get the other two columns in the table.

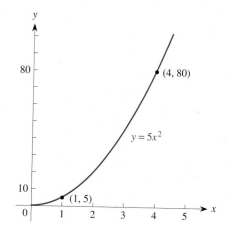

FIGURE 3.40

x	y	$\log x$	$\log y$
1	5	0	0.699
2	20	0.301	1.301
3	45	0.477	1.653
4	80	0.602	1.903

Figure 3.41 shows the transformed data points $(\log x, \log y)$, and they appear to lie on a line. Taking any two of the transformed points gives the slope of the line as 2. Furthermore, the y-intercept is $b = 0.699$, so

$$\log y = 2 \log x + 0.699$$
$$= \log x^2 + 0.699. \qquad p \log x = \log x^p$$

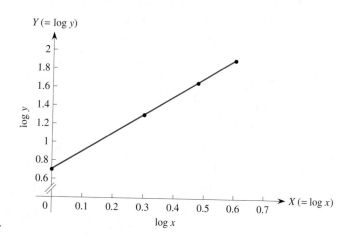

FIGURE 3.41

We undo the logs by taking powers of 10 on both sides of the equation:

$$10^{\log y} = y = 10^{(\log x^2 + 0.699)} \qquad 10^{\log u} = u$$
$$= 10^{0.699} \cdot 10^{\log x^2} \qquad 10^{u+v} = 10^u \cdot 10^v$$
$$= 5 \cdot x^2. \qquad 10^{\log u} = u$$

So the original data points do indeed lie on the curve $y = 5x^2$.

We now apply these ideas to a variety of real world examples.

EXAMPLE 2

In Example 3 of Section 3.3, we used data on the number of species of reptiles and amphibians living on different Caribbean islands in the Greater and Lesser Antilles to find a relationship between the number of species N and the area A of the habitat. We found the power function $N = 3.055A^{0.310}$. Find this function, using the transformation approach for the data given in the following table.

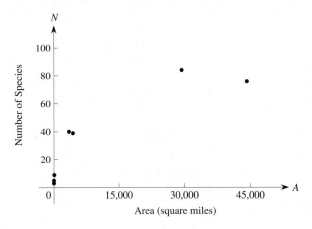

FIGURE 3.42

Island	Area	N
Redonda	1	3
Saba	4	5
Montserrat	40	9
Puerto Rico	3459	40
Jamaica	4411	39
Hispaniola	29,418	84
Cuba	44,218	76

Solution Figure 3.42 shows the scatterplot for these data. The increasing and concave down pattern suggests a power function with $0 < p < 1$ as the model of choice, as discussed in Example 3 of Section 3.3. This choice is reinforced because such a function passes through the origin and, if there were an island with $A = 0$ area, it would be home to $N = 0$ species.

To fit a power function $N = kA^p$ to the original data using the transformation approach, we extend the preceding table to include two additional columns, one for $\log N$ and the other for $\log A$. We calculate these quantities and then plot $\log N$ versus $\log A$, as shown in Figure 3.43.

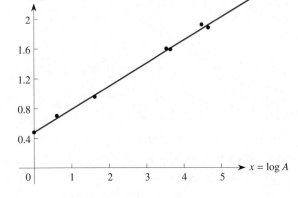

FIGURE 3.43

Island	Area	N	log A	log N
Redonda	1	3	0	0.48
Saba	4	5	0.602	0.70
Montserrat	40	9	1.602	0.95
Puerto Rico	3459	40	3.539	1.60
Jamaica	4411	39	3.645	1.59
Hispaniola	29,418	84	4.469	1.92
Cuba	44,218	76	4.646	1.88

These points fall in a roughly linear pattern. Using the linear regression routine on a calculator, we find that the equation of the regression line for the transformed data points $(\log A, \log N)$ is

$$Y = 0.310X + 0.485.$$

In the present context, this equation is equivalent to

$$\log N = 0.310 \log A + 0.485$$
$$= \log(A^{0.310}) + 0.485. \qquad p \log u = \log u^p$$

To solve for N, we use properties of exponents and logarithms:

$$N = 10^{\log N} = 10^{[\log(A^{0.310}) + 0.485]} \qquad\qquad 10^{\log u} = u$$
$$= 10^{\log(A^{0.310})} \cdot 10^{0.485} \qquad\qquad 10^{u+v} = 10^u \cdot 10^v$$
$$= A^{0.310}(3.055). \qquad\qquad 10^{\log u} = u$$

Thus we have the power function $N = 3.055A^{0.310}$, which is the same function we found in Section 3.3 using a calculator regression routine for the power function. When we superimpose this power function over the original data values as shown in Figure 3.44, it fits the data well. In fact, the correlation coefficient for the data is $r = 0.9982$, which also indicates that the fit is extremely good.

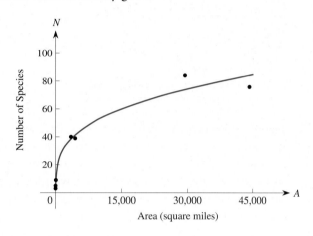

FIGURE 3.44

We summarize this approach as follows.

> If the data appear to follow a power function pattern $y = kx^p$,
> 1. plot $\log y$ versus $\log x$ to linearize the data;
> 2. find the best linear fit,
>
> $$\log y = a \log x + b,$$
>
> to the transformed data; and
> 3. undo the logarithmic transformation by taking powers of 10.

EXAMPLE 3

The following table gives the takeoff weights, in thousands of pounds, of various jet liners (weight of the plane plus fuel plus passengers) and their wingspans, in feet.

a. Decide which variable is the independent variable and which is the dependent variable.

b. Find a power function that fits these data, using the transformation approach.

Solution

a. An aeronautical engineer designing a new plane would likely start with the desired load, obtain the total weight, and then calculate the wingspan needed to support that load. Thus the weight W is the independent variable and the wingspan S is the dependent variable.

Airplane	Weight	Wingspan
Boeing 707	330	145.7
Boeing 727	209.5	108
Boeing 737	117	93
Boeing 747	805	195.7
Boeing 757	300	156.1
DC8	350	148.5
DC9	121	93.5
DC10	572	165.4

Source: Michael Taylor (ed.), *Jane's Encyclopedia of Aviation.* Crescent Books, 1993.

b. Figure 3.45 shows the scatterplot of the data with S as a function of W. The pattern is increasing and concave down, so a power function with a power p between 0 and 1 is reasonable. Also, it makes sense to have a function that passes through the origin—a plane that weighs 0 pounds will have a wingspan of 0 feet. To find the power function, we first transform the data by plotting log S versus log W, as shown in Figure 3.46. The pattern of the transformed data is quite linear, and the line that best fits it is

$$Y = 0.3942X + 1.157$$

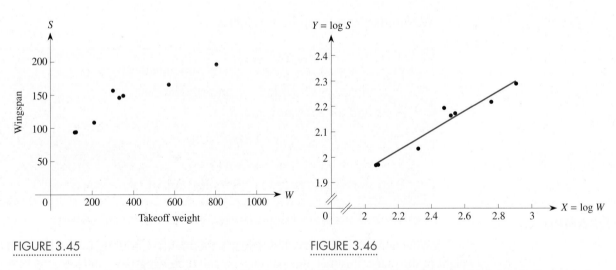

FIGURE 3.45

FIGURE 3.46

The correlation coefficient is $r = 0.9688$, which indicates a high level of correlation between the variables. In this context, this linear function is equivalent to

$$\log S = 0.3942 \log W + 1.157$$
$$= \log(W^{0.3942}) + 1.157. \qquad p \log u = \log u^p$$

We find S by taking powers of 10 on both sides of this equation:

$$10^{\log S} = S = 10^{[\log(W^{0.3942}) + 1.157]} \qquad 10^{\log u} = u$$
$$= 10^{\log(W^{0.3942})} \cdot 10^{1.157} \qquad 10^{u+v} = 10^u \cdot 10^v$$
$$= W^{0.3942}(14.35), \qquad 10^{\log u} = u$$

which is the desired power function. Figure 3.47 shows this function superimposed over the original data points, and it fits the measurements well.

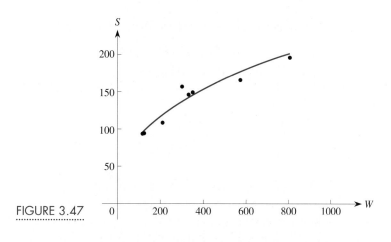

FIGURE 3.47

Summary of Curve Fitting Procedures

If the data appear to follow an exponential pattern, $y = kc^x$,
1. plot log y versus x to linearize the data;
2. find the best linear fit, log $y = ax + b$, to the transformed data; and
3. undo the transformation, using the inverse function.

If the data appear to follow a power function pattern, $y = kx^p$,
1. plot log y versus log x to linearize the data;
2. find the best linear fit, log $y = a \log x + b$, to the transformed data;
3. undo the transformation, using the inverse function.

If the data appear to follow a logarithmic pattern, $y = k \ln x + d$,
1. plot y versus ln x to linearize the data; and
2. find the best linear fit, $y = a \ln x + b$, to the transformed data.

Cautions When Transforming Data with Logarithms

You should be aware of one major problem when using any of the curve fitting procedures involving logarithmic transformations of the data. Recall that the logarithmic function $y = \log x$ is defined only for values of x greater than 0. Thus, if any of your data values are 0 or negative, you cannot take their logarithms. Often you can circumvent this difficulty by redefining the independent variable. For example, suppose that the data represent values of a quantity versus time starting in 1950. You could count the years since 1950, but then 1950 corresponds to $t = 0$, which causes a problem when you take logs. Alternatively, you could count the years since 1900 because 1950 then corresponds to $t = 50$, which circumvents the problem of log 0 being undefined. For that matter, you simply could count the years from 1949 so that 1950 corresponds to $t = 1$. You could even use the full year 1950 itself, although having $t = 1950$ creates potential rounding errors owing to the size of the numbers, as we previously discussed. Whatever you do, be sure to keep track of what your independent variable represents.

Cautions when Fitting Power Functions to Data

Some serious problems can arise when you are fitting power functions to data. Consider the data values shown.

x	1	2	3	4	5
y	400	410	430	460	500

The scatterplot shown in Figure 3.48 indicates clearly that the pattern in the data is increasing and concave up. Using a calculator, we find that the power function that best fits these data is $y = 387.43\, x^{0.129}$, with correlation coefficient $r = 0.907$. The correlation coefficient indicates a fairly good fit (the critical value for r with $n = 5$ data points is $r_5 = 0.878$). However, superimposing the graph of this power function on the scatterplot, as shown in Figure 3.49, indicates that this function is concave down. This result is also evident because the power $p = 0.129 < 1$. Clearly, this power function completely misses the trend in the data. Let's see why.

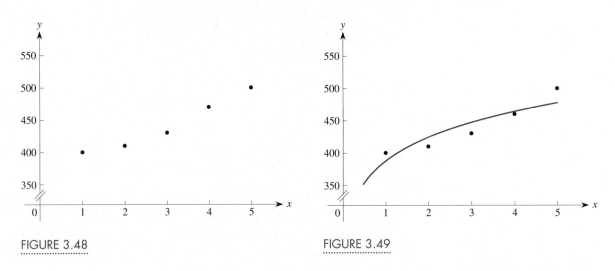

FIGURE 3.48

FIGURE 3.49

We know that every power function with a positive power p passes through the origin. Although five points are listed in the preceding table, the act of fitting a power function to the data automatically introduces the origin as an additional point, as shown in the following table.

x	0	1	2	3	4	5
y	0	400	410	430	460	500

Worse, considering the large jump from the origin to the point $(1, 400)$, we see that any power function that attempts to follow the trend in *all* six points must be concave down. As a result, a power function is not an appropriate model to use for this set of data.

In general, whenever you attempt to fit a power function to a set of data, you must look at the numbers and the scatterplot to decide whether a power function is an appropriate model. You must also keep in mind that the point $(0, 0)$ is automatically added to the data if the pattern is increasing.

You can circumvent this problem if you adjust the original data values *before* fitting a power function and then undo the adjustment *after* the fact. To avoid the large jump from the origin to the given data values, first adjust or *shift* the y-values down (we discuss such shifts formally in Section 4.4) by reducing each y-value in the original table by 399. We then get the values in the following table.

x	1	2	3	4	5
y	400	410	430	460	500
$Y = y - 399$	1	11	31	61	101

Using a calculator, we find that the power function that best fits these adjusted values is $Y = y - 399 = 1.1877x^{2.86}$. The corresponding correlation coefficient $r = 0.995$ is considerably larger than the value $r = 0.907$ obtained for the unadjusted values. More important, the power is now $p = 2.86$, which is greater than 1 and this function is concave up and so better captures the trend in the data. Finally, having shifted all the y-values down by 399, we now adjust the function up by adding the same amount, 399 to the expression and so obtain

$$y = 1.1877 x^{2.86} + 399.$$

The result is not a pure power function because of the additional term 399. Figure 3.50 shows this function superimposed over the original data points; it is a much better fit to the original data, especially the first few data points.

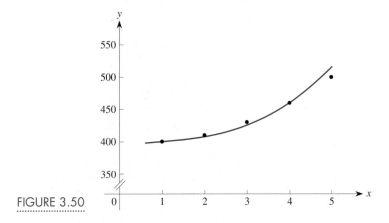

FIGURE 3.50

In general, we recommend the following approach when fitting a power function to a set of data. Unless there is a natural origin for the data (as there was in Example 2 on the number of species of reptiles and amphibians on an island—a theoretical island with zero area would have zero species), you should think about adjusting your data values by shifting the y-values down (or up) by some appropriate amount. Unfortunately, there is no simple rule to know by how much to adjust the y-values—it depends on the data you have. In practice, it may be a good idea to try different adjustments in the y-values to see which accomplishes the best fit.

After you have found the best power function for the adjusted data values, undo the shift by adding (or subtracting) the same amount back in the expression.

There is another complication to keep in mind when you fit a power function to data. In Examples 2 and 3 of Section 2.3, we considered data on U.S. imports in 1990 ($495 billion) and in 1998 ($912 billion) and constructed the line through these two points, using first $t = 0$ in 1990, then $t = 0$ in 1900, and finally $t = 0$ in year 0. In each case, the slope of the resulting line was the same, although the vertical intercept changed to reflect the starting point. Further, no matter which equation we use, we get the same prediction for a particular year. This principle holds in general for the line through any points.

Similarly, in Problem 5 of Section 2.4, you were asked to examine the same issue with exponential functions. You should have found that in all three cases the growth factor remained the same, although the vertical intercept for the three exponential functions did change to reflect the starting point. Also, you will get the same predictions for a given year by using any of these equations.

Unfortunately, this principle does not carry over to power functions. If you change the meaning of the independent variable (say, t) the function changes totally. Recall that every power function with power $p > 0$ passes through the origin. Thus, when you find the equation of an increasing power function passing through two points or the power function that fits three or more points, that function automatically passes through an additional point at the origin.

Imagine what happens when you have data points between the years 1991 and 2000, say. Your choice of what t represents means that you are repositioning the "origin" and the resulting power function is forced through a different extra point (the new "origin"). For instance, if you let t represent the number of years since 1990, the origin is very close to the data points. If you let t represent the number of years since 1900, the origin is quite far from the data points. And if you let t represent the year itself, the origin is extremely far from the data points. In each case, you will get a power function with very different values for the parameters p and k in $y = kx^p$. Moreover, any predictions based on one expression will be different from those based on a different expression. We illustrate this by fitting a power function to the two data points on U.S. imports.

◀ **EXAMPLE 4** ....

In 1990, the United States imported $495 billion worth of goods from abroad. In 1998, the United States imported $912 billion worth. Assuming that the growth in imports follows a power function pattern, find an equation of the power function that models U.S. imports by using the independent variable t to represent (**a**) the number of years since 1989, (**b**) the number of years since 1900, and (**c**) the number of years since year 0. (**d**) Compare the results obtained from the three approaches. (**e**) Use each model to predict the amount of imports in 2005.

Solution

a. With $t = 0$ in 1989, we have the points $(1, 495)$ and $(9, 912)$. Using a calculator, we find that the corresponding power function is $G_1(t) = 495t^{0.278}$ with correlation coefficient $r = 1$.

b. With $t = 0$ in 1900, the points are $(90, 495)$ and $(98, 912)$. Using a calculator, we find that the corresponding power function is $G_2(t) = 4.69 \cdot 10^{-12}t^{7.176}$ with correlation coefficient $r = 1$.

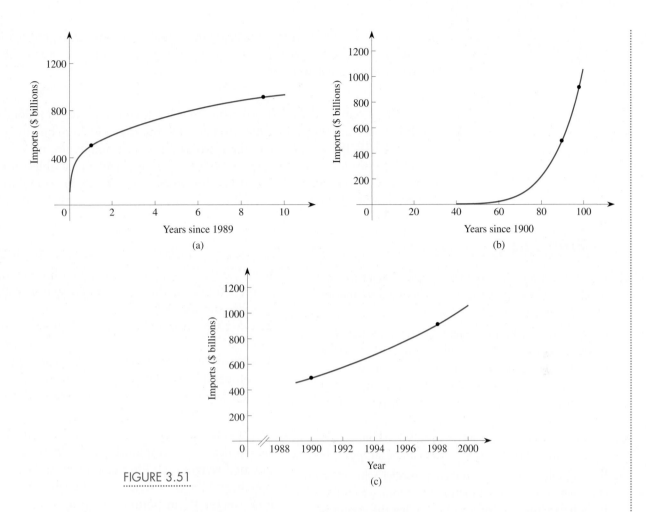

FIGURE 3.51

c. With $t = 0$ in the year 0, the points are $(1990, 495)$ and $(1998, 912)$. Some calculators give the corresponding power function as $G_3(t) = 1.736 \cdot 10^{-500} t^{152.312}$ with correlation coefficient $r = 1$. Other calculators give an overflow error because the numbers are too large.

d. Note that all three choices led to a correlation coefficient of $r = 1$, which suggests that each of the three functions perfectly fit the data. However the three coefficients and the three powers bear no relationship to one another. In fact, the power function in part (a) has $p = 0.278$, which is between 0 and 1, so that power function is increasing and concave down. The power functions in parts (b) and (c), in comparison, both have powers greater than 1, so they are increasing and concave up. Figures 3.51(a)–(c) illustrate these results.

e. With $t = 0$ in 1989, the year 2005 corresponds to $t = 16$ and the function from part (a) yields

$$G_1(16) = 495(16^{0.278}) = 1069.918.$$

With $t = 0$ in 1900, 2005 corresponds to $t = 105$ and the function from part (b) yields

$$G_2(105) = 4.69(10^{-12}) \cdot 105^{7.176} = 1497.019.$$

With $t = 0$ in the year 0, we have $t = 2005$ and the function in part (c) gives

$$G_3(2005) = 1.736(10^{-500})(2005^{152.312}) = 1553.083.$$

The results clearly are very different and, given the actual data values, the prediction based on the formula from part (a) with t representing the number of years since 1989 appears to be the most accurate.

Example 4 emphasizes that, whenever you work with a power function, you must think very carefully about the variable you are using and consequently your choice of origin. If you have more than two data points, it might also make sense for you to try some different interpretations for the independent variable—for instance, years since 1989, years since 1988, and years since 1987 to determine which function best fits the data.

Problems

1. Use the model created in Example 3 relating the wingspan of a jet aircraft to its total load to answer the following questions.
   a. If a new jet is being designed to have a total take-off load of 500 thousand pounds, what wingspan will be necessary to support the load?
   b. A super-jumbo jet is being designed to carry a total weight of 1.2 million pounds. What wingspan will be needed to support it?
   c. The wingspan of a jet is 175 feet. What is the maximum takeoff weight that can be supported?

2. In Example 3, we constructed a model where the wingspan S of a jet is a function of its total weight W. Interchange the roles of S and W so that the weight is a function of the wingspan. Use the data in Example 3 to find the corresponding power function. How do the parameters in this new power function compare to those in the power function in Example 3?

In Problems 3–13, use the transformation approach to find a power function that fits the data.

3. Marc notices that the radio frequency numbers on the AM dial of his stereo don't seem to lie in a linear pattern. He measures the distances, in centimeters, from the extreme left end of the dial to each of the numbers printed and gets the readings shown in the accompanying figure.

   a. Determine the power function that relates the distance D to the station numbers n shown.

   b. Using this function, estimate the distance from the left end of the dial to radio station 880; to radio station 1270.
   c. What station would be 6 cm from the left end of the dial?
   d. Determine whether the same function fits the comparable set of readings on your radio.
   e. Determine the function that fits the readings on the FM band on your radio.

4. Weight-lifting competitions are divided into body weight classes. Championships are often based on the total weight each lifter can lift in the press, the snatch, and the clean-and-jerk. The table shows the total weight W, in pounds, lifted in several body-weight classes as a function of the body weight B, in pounds, of the lifter.

Body Weight	123	130	148	166	182	197
Weight Lifted	750	776	851	912	966	1023

Source: Thomas A. McMahon and John Tyler Bonner, *On Size and Life.* Scientific American Library: 1983.

   a. Explain why a power function would be an appropriate fit for these data.
   b. Find the power function that best fits these data.
   c. What total weight could be lifted by a 225 pound super-heavyweight lifter according to this model?
   d. What total weight could be lifted by the proverbial 98-pound weakling?

5. The running speed of animals appears to be related to their overall body lengths. The table gives the lengths L of various organisms, in centimeters, and their top running speed S in centimeters per second.

   a. Explain why a power function would be the function of choice to model this data.

Animal	Length (cm)	Speed (cm/sec)
Clover mite	0.08	0.85
Ant	0.42	6.5
Deer mouse	9.0	250
Zebra-tail lizard	15.0	720
Eastern chipmunk	16.0	480
Iguana	24.0	730
Gray squirrel	25.0	760
Red fox	60.0	2000
Cheetah	120.0	2900

Source: Thomas A. McMahon and John Tyler Bonner, *On Size and Life*. Scientific American Library: 1983.

b. Find the power function that best fits these data according to this model.
c. Estimate the length of an animal that could run at a speed of 1500 cm/sec.
d. If this model applies to humans, estimate the best running speed for a man who is 6 feet tall. (*Hint:* 1 inch = 2.54 cm.) Based on your answer, does this model apply to humans? Explain.

6. The flying speed of animals also appears to be related to their overall body length. The accompanying table gives the lengths L of various organisms, in

Animal	Length (cm)	Speed (cm/sec)
Fruit fly	0.2	190
Horse fly	1.3	660
Hummingbird	8.1	1120
Dragonfly	8.5	1000
Bat	11.0	690
Common swift	17.0	2550
Flying fish	34.0	1560
Pintail duck	56.0	2280
Swan	120.0	1880
Pelican	160.0	2280

Source: Thomas A. McMahon and John Tyler Bonner, *On Size and Life*. Scientific American Library: 1983.

centimeters, and their top flying speed, S, in centimeters per second.

a. Find the power function that best fits these data.
b. Estimate the top flying speed for a bird that is 80 cm long according to this model.
c. Estimate the length of a bird that could fly at a speed of 2000 cm/sec.
d. Estimate the flying speed of a pterodactyl that was approximately 240 cm long.

7. The swimming speed of animals also appears to be related to their overall body length. The table gives the lengths L of various organisms, in centimeters, and their top swimming speed S in centimeters per second.

Animal	Length (cm)	Speed (cm/sec)
Paramecium	0.02	0.1
Water mite	0.13	0.4
Goldfish	0.7	75
European dace	10.0	130
Herring	30.0	440
Penguin	75.0	380
Tuna	98.0	2080
Dolphin	220.0	1030
Blue whale	2600	1030

Source: Thomas A. McMahon and John Tyler Bonner, *On Size and Life*, Scientific American Library: 1983.

a. Find the power function that best fits these data.
b. Estimate the best swimming speed for a fish that is 50 cm long according to this model.
c. Estimate the length of a fish that could swim at a speed of 1000 cm/sec.

8. The table gives the weights, in pounds, of a variety of full-grown flying birds and their wingspans in feet.

a. Decide which variable is the independent variable and which is the dependent variable.
b. Find a power function that fits these data.
c. If a falcon has a wingspan of 3.3 feet, predict its weight based on your model.
d. If a turkey weighs 15 pounds, predict the wingspan it would need in order to fly.

Bird	Weight	Wingspan
Vulture	18.7	9.3
Eagle	12	7.5
Horned Owl	5	5.0
Golden Eagle	13	7.3
Red tailed hawk	4	4.0
Harris hawk	2.6	3.2
Turkey vulture	6.5	6.0
Whooping Crane	16.1	7.5
Bald Eagle	16	7.5
Condor	22	9.9
Blue-footed booby	4	3.0
Crow	1	2.9

Source: U.S. Fish and Wildlife Services.

9. The table gives the number of species of birds living on various islands in the East Indies versus the area, in thousands of square miles, of each island.

Island	Area	Number of Species
New Guinea	312,000	540
Borneo	290,000	420
Philippines	144,000	368
Celebes	70,000	220
Java	48,000	337
Ceylon	25,000	232
Flores	8870	143
Timor	18,000	137

Source: Kevin Mitchell et al., *Mathematical Models of Biological Systems.* 1998.

a. Decide which variable is the independent variable and which is the dependent variable.
b. Find a power function that fits these data.
c. The island of Sumba, which has an area of 4600 square miles, is in this region. Estimate how many species of birds live on Sumba.

d. A habitat for 200 species of birds is to be established in this region of the world. How large must the habitat be to support that many species?

10. Consider a bicyclist riding around a circular track with a constant radius. To keep making the turn, the bicyclist and the bike must lean sideways toward the center of the circle. The angle A of the lean from the vertical is related to the velocity of the bike. The table shows the velocity v in meters per second and the angle A to the vertical in degrees.

Angle	3	5.7	13.7	22.9	28.6	40.1
Velocity	1.3	2.0	3.1	4.0	4.7	5.7

Source: Thomas A. McMahon and John Tyler Bonner, *On Size and Life*, Scientific American Library: 1983.

a. What is the independent variable and what is the dependent variable? Explain.
b. What is the range for the function relating these two variables?
c. Find the power function that best fits these data.
d. What velocity is a bicyclist going if the angle of lean is 50° from the vertical?
e. If a bicyclist is going 3.5 meters per second, what angle from the vertical is needed to stay on the circle?

11. Consider a bicyclist riding around a circular track with a constant velocity. To keep making the turn, the bicyclist and the bike must lean sideways toward the center of the circle. The angle of lean to the vertical is related to the radius of the circle. The table shows the radius r in meters, and the angle A to the vertical in degrees.

a. What is the independent variable and what is the dependent variable? Explain.
b. What is the range for the function relating these two variables?
c. Find the power function that best fits these data.

Angle	38.4	34.4	29.8	24.6	20.6
Radius	0.8	0.9	1.05	1.3	1.74

Source: Thomas A. McMahon and John Tyler Bonner, *On Size and Life*, Scientific American Library: 1983.

d. What radius circle is needed for a bicyclist leaning 15° from the vertical?

e. If a bicyclist is going around a circle that has a radius of 1.5 meters, what angle from the vertical is needed for the byciclist to stay on the circle?

12. The following table shows the total worldwide carbon emissions, in millions of tons of carbon, from fossil fuel burning over the years.

a. Find the equation of the power function that best fits these data.

b. What does the model predict for the amount of carbon emitted by burning fossil fuels in 2010?

c. When does the model predict that the total amount of carbon emitted from burning fossil fuels will reach 7000 million tons?

Year	1950	1960	1970	1975	1980	
Carbon Emissions	1612	2535	3998	4518	5156	
Year	1985	1990	1993	1995	1997	1999
Carbon Emissions	5271	5946	5896	6212	6349	6307

Source: Lester R. Brown et al., *Vital Signs 2000: The Environmental Trends That Are Shaping Our Future.*

13. Dennis is a gunnery officer for M1A1 Abrams tanks in the National Guard. He has determined the distance, in meters, to a potential target and the time, in seconds, needed to locate, identify, and fire at that target.

Range	250	500	1000	1500	2000	2500	3000
Time	3	4	5	6.5	9	11	16

Source: Student project.

a. Determine the power function that fits these data.

b. An enemy tank is at a distance of 1750 meters. How long will it take for an M1A1 tank crew to locate, identify, and then fire at the target?

c. If it takes 10 seconds for an M1A1 tank crew to fire at a target, what is the range to the target?

d. An allied tank is 3000 meters away. How long does the crew of an M1A1 tank have to realize that the target is actually a friend and avoid firing at it?

Exercising Your Algebra Skills

Several sets of data have been linearized to find the best power fit. The following equations are for the lines that best fit the transformed data. What are the corresponding power functions?

1. $Y = 0.9031 + 1.5X$
2. $Y = 0.6990 + 0.7X$
3. $Y = 1.0792 - 1.5X$
4. $Y = -0.2218 - 0.4X$
5. $Y = 0.3522 + 1.0843X$
6. $Y = -1.3015 + 0.7840X$
7. $Y = 0.8525 - 1.2733X$
8. $Y = -0.817 - 2.015X$

3.6   How Good Is the Fit?

Throughout this chapter, we have looked for the function within each appropriate family of functions that best fits a set of data values in the least squares sense. If the data points fall in a linear pattern, we use the regression line. If the data points fall in some other recognizable pattern and if appropriate software is available, we can apply the least squares criterion directly to construct the best fit function in any particular family. Alternatively, we can use the built-in routines of graphing calculators or spreadsheets to find the function that best fits the data, based on the transformation approaches described in Sections 3.3 and 3.4.

In doing any of these things, we face two problems:

1. being able to assess how well a particular curve fits the set of data; and
2. deciding which function is a better fit when there are several reasonable candidates.

Unfortunately, there is no single, clear-cut way of resolving these problems. In this section, we look at several different ways to make such decisions.

Interpreting the Correlation Coefficient

Comparing the quality of the best fit in one family of functions to the best fit in another involves some deep and subtle issues and we can only touch the surface here. First, recall that the correlation coefficient measures only how well the best *linear* function fits any set of data, be it the original or the transformed data. If r is not close enough to either $+1$ or -1, there is no point in using that function as a model to describe the pattern in the data.

Suppose that we have sufficiently high values for r for several different families of functions—say, linear, exponential, and power. It is tempting to think that the function having the largest value for r is the best fit among them, but unfortunately that is not necessarily the case, especially if the values for r are relatively close—say, 0.93, 0.95, and 0.90. Thus you should *not* make a decision about which function is the best fit by using r as the only criterion.

Interpreting the Residuals

The residuals associated with a particular fit provide considerably more information about how good the fit is and how to compare different fits for the same set of data. Recall that, for a linear fit to the original data, the residuals are the differences between the actual y-values and the predicted values based on the regression line, as shown in Figure 3.52(a). When we transform the data and obtain the regression line for the transformed data, then the residuals are the differences between the *transformed* data values (Y) and the predicted values based on the regression line for the transformed data $(y = mX + b)$. The differences between the original y-values and the predicted values based on the *detransformed* function—say, an exponential function—are not residuals; they are called *errors,* as shown in Figure 3.52(b).

For a linear fit, if the actual height is y and the predicted value given by the regression line is $y_p = ax + b$, the corresponding residual is $y - y_p = y - (ax + b)$. If a data point lies above the regression line, the residual is positive; if a data point lies below the regression line, the residual is negative, as shown in Figure 3.53. If the fit is good, roughly half the residuals should be positive and roughly half should be negative because the regression line should pass, more or less, midway between the data points.

We plot the residuals corresponding to the data in the scatterplot shown in Figure 3.53 on a different set of axes as shown in Figure 3.54. This plot is called a **residual plot,** Points on the horizontal axis of the residual plot represent data points with zero residual—that is, data points that fall exactly *on* the regression line. Whenever a data point lies above the regression line, the residual is positive and the associated point in the residual plot is above the horizontal axis by that amount. Similarly, whenever a data

Data Set (Linear Pattern)

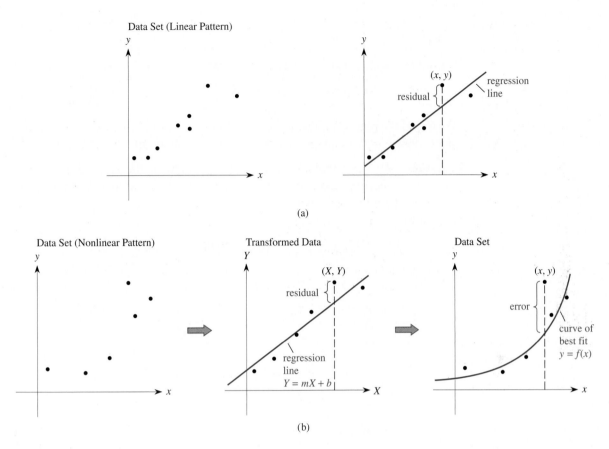

(a)

(b)

FIGURE 3.52

point lies below the regression line, the residual is negative and the associated point in the residual plot is below the horizontal axis by that amount. (Note that the horizontal axis shown in the residual plot has no direct relationship to the regression line.)

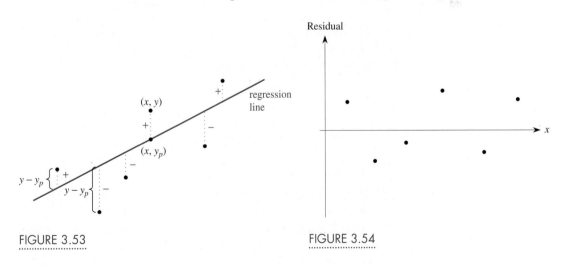

FIGURE 3.53

FIGURE 3.54

EXAMPLE 1

In Example 1 of Section 3.2, we looked at the gross receipts of the movie industry, in billions of dollars, in different years since 1990 based on the following table.

Year	0	3	4	5	6	7
Total Receipts	39.98	49.80	53.50	57.18	60.28	63.01

Source: *2000 Statistical Abstract of the United States.*

We found the regression line to be $y = 3.3435t + 40.027$, where t is the number of years since 1990. Use this equation to calculate the residuals and create the residual plot.

Solution We show the scatterplot of the data with the regression line $y = 3.3435t + 40.027$ superimposed in Figure 3.55. Now consider the first data point, where $t = 0$ and $y = 39.98$ billion dollars. The predicted height of the corresponding point on the regression line is

$$y_p = 3.3435(0) + 40.027 = 40.027 \approx 40.03.$$

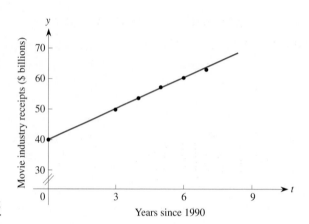

FIGURE 3.55

This predicted value (the height to the line when $t = 0$) is larger than the actual value 39.98, so the data point lies below the regression line. The residual for this point is then

$$y - y_p = 39.98 - 40.03 = -0.05,$$

so the corresponding point in the residual plot that we draw later will lie below the horizontal axis there. We perform comparable calculations for each of the remaining data points and list the results in the following table.

Year	0	3	4	5	6	7
Total Receipts y	39.98	49.80	53.50	57.18	60.28	63.01
Predicted y_p	40.03	50.06	53.40	56.74	60.09	63.43
Residual	−0.05	−0.26	0.10	0.44	0.19	−0.42

We now plot the residuals in the corresponding residual plot shown in Figure 3.56. As we indicated the first residual point lies slightly below the horizontal axis because the actual data point lies slightly below the regression line. The same is true of the second point. The third, fourth, and fifth points all lie above the horizontal axis because each corresponds to a point that lies above the regression line in Figure 3.55. Finally, the last point lies below the horizontal axis because the actual data point lies below the height predicted by the regression line.

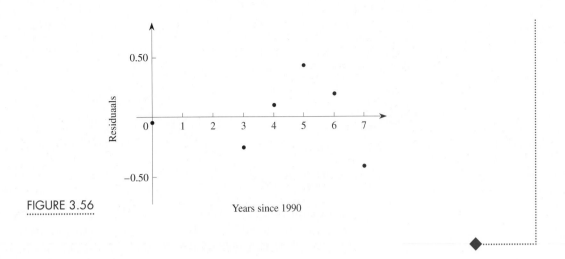

FIGURE 3.56

Interpreting a Scatterplot

Figure 3.57 shows both a scatterplot and the associated residual plot for a set of data in which a linear function fits the data well. Note that the regression line essentially passes through the middle of the cluster of data points so that roughly half the points lie above the regression line and roughly half lie below it. Also, note that in the residual plot roughly half the residuals are positive (corresponding to data points above the regression line) and roughly half are negative (corresponding to data points below the regression line). Moreover, the residuals seem to be *scattered randomly* both above and below the horizontal axis; there does not appear to be any pattern to their locations.

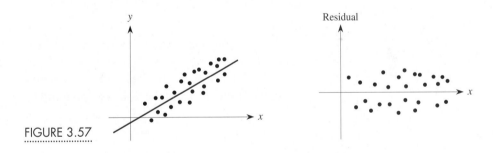

FIGURE 3.57

In contrast, consider the scatterplot and the associated residual plot shown in Figure 3.58. The data fall in a linear pattern, but the regression line does not appear to be a particularly good fit; it is distorted by the presence of two points that are far from the line. How does this poor fit show up in the residual plot? Again, note that roughly half the residuals are above the horizontal axis and roughly half are below it. However, this time the fact that most of the residuals on the left are negative (because most of the data points on the left fall below the regression line) is significant. Similarly, most of the residuals on the right are positive. Thus rather than being scattered randomly above and below the horizontal axis, the residuals have a pattern. This pattern in the residuals indicates that the fit is not good.

Consider now the scatterplot of a different set of data and its associated residual plot shown in Figure 3.59. The data values can likely be modeled by either an exponential growth function or a power function with $p > 1$. For comparison, the regression line is superimposed on the scatterplot, even though the line does not

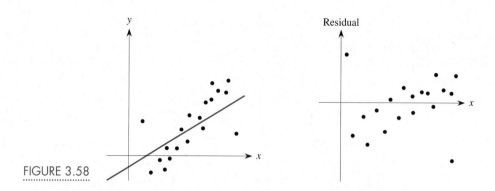

FIGURE 3.58

capture the trend in the data. Note that the data points on the left lie above the line, the points in the middle lie below the line, and the points on the right lie above the line. This behavior is reinforced by the residual plot in which the points fall into a U-shaped pattern with the middle residual points falling below the horizontal axis. As was the case in Figure 3.58, the existence of this pattern indicates that the linear fit is not a good one.

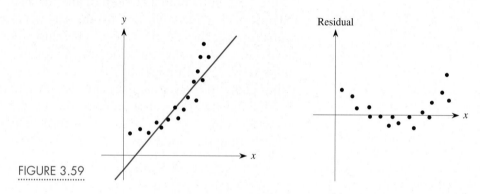

FIGURE 3.59

Figure 3.60 shows the best exponential fit to the data in Figure 3.59 and the residual plot associated with the transformed data. The exponential curve is a good fit to the data. Also, the residuals are small with no obvious pattern; they are scattered about the horizontal axis in an apparently random pattern.

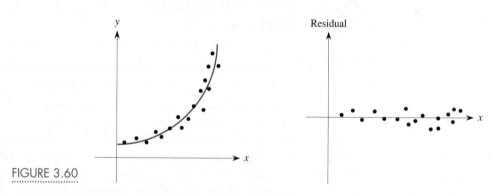

FIGURE 3.60

In general, just by looking at the scatterplot, you may not be able to see that a particular curve fits a set of data well. We have demonstrated previously that the scales of the graph can distort an image when either or both variables extend over a very large interval of values, preventing us from recognizing either a good or a bad fit. We can overcome this difficulty by examining and interpreting the residual plot associated with the regression line for the transformed data.

- ◆ If a particular fit is good, the residuals should display no pattern.

- ◆ If the residuals do display a pattern (e.g., rising and falling, U-shaped, or periodic), the fit is likely not a good one and you should look further for a better fit among other families of functions.

- ◆ If the residuals are extremely large numerically, you should be suspicious of the fit, even if the fit on the scatterplot looks good.

Furthermore, if only one or two residuals are extremely large compared to all the others, that might indicate an error in keying in data values, errors in the actual measurement, or highly unrepresentative values, called **outliers.** Check your values to be sure that there is no input error. If there is none, you might want to experiment by removing the outliers from the data set and recomputing the regression equation to determine whether there is a dramatic improvement in the fit. However, any final report you write on your results should mention the apparently unrepresentative outliers, even if you don't use them in the calculations.

Once you have convinced yourself that you have found an appropriate fit for the data, the linear correlation coefficient r provides a measure of the degree of agreement. However, keep in mind that you can calculate a correlation coefficient for *any* set of data, either a set of original values or a transformed set of values. Several different fits for the same set of data may have a significant level of correlation. (Think about the graphs of functions: Over a relatively small interval, a line, an exponential function, and a power function may appear almost identical, but when you extend them sufficiently, they clearly move away from one another.)

Thus a good strategy is first to find the *best* fit and only then consider the associated correlation coefficient to verify that it is significant. You can, and probably should, try different fits for a set of data, one after the other, until you come up with the best curve among the families of linear, exponential, power, and logarithmic functions. Of course, a set of data may not fall in any of these patterns, but a great many reasonable data sets do. In later chapters, we extend these ideas to cases in which the possible patterns include quadratic and higher degree polynomials, periodic functions, or other common patterns. Two or more different patterns may apply to different portions of a set of data. For instance, a pattern could start with linear growth and later appear to have exponential growth.

Interpreting the Sum of the Squares

Probably the best way to compare two or more functions to decide which is the better fit to a set of data is based on the least squares criterion. The objective in fitting any function, whether linear or nonlinear, to a set of data is to minimize the sum of the squares of the differences between the actual heights y of the data points and the predicted heights $y_p = f(x)$ based on the function. That is, for each of the data points $(x_1, y_1), (x_2, y_2), \ldots, (x_n, y_n)$, look at the vertical distances, $y_i - f(x_i)$, for $i = 1, 2, 3, \ldots, n$, from the point to the curve and then calculate the sum of the squares of these vertical differences:

$$\sum_{i=1}^{n} [y_i - f(x_i)]^2$$

(where $\sum_{i=1}^{n}$ indicates the sum of all the terms for each value of i from 1 through n as discussed in Appendix A3.) This quantity is often calculated automatically in many mathematical software packages. It is also fairly easy to calculate using a graphing calculator in the statistics or table mode. If functions from several different families fit a set of data well, you could then compare the sum of the squares for each function and select the function having the smallest sum of squares value as the best fit among the choices.

However, the sum of the squares should be only one of several criteria used. The fact that the sum of the squares is smaller for one function than for another function doesn't necessarily mean that it is the best fit to the data. Other factors about the situation may not be reflected in the data. For instance, you may have measurements on the growth of bacteria in a test tube that might suggest exponential growth, but you know that such growth cannot continue indefinitely, so an exponential function will model the population only for a short while. You should examine the scatterplot for the presence of outliers that will distort any predicted model. You should look at the residual plots to see if there is any pattern to the residuals that will provide additional insight into which function is the best fit.

Another quantity is often used in the sciences to measure how well a function fits a set of data. Instead of looking at just the sum of the squares of the vertical distances between the data points and the approximating function, finding the average of this sum over all the data points and then taking the square root of the result may be convenient. Doing so gives the *root-mean-square* (RMS) value associated with the function. It is calculated from the formula

$$\sqrt{\frac{\sum_{i=1}^{n}[y_i - f(x_i)]^2}{n}}.$$

In some sense, this measures the "average" distance of the data points from the curve. The smaller the average distance is, the better the fit.

Statisticians use a slightly different measure of the average distance of the points from the curve, called the *standard error of the estimate*. It is given by

$$\sqrt{\frac{\sum_{i=1}^{n}[y_i - f(x_i)]^2}{n - 2}}.$$

This discussion about which fit is the best—and the lack of consensus about how to determine it—reinforces the point we made at the beginning of this section that the issues are subtle and clearly depend on the set of data you are studying. Each set of data has to be approached and examined on its own merits.

Kepler's Third Law of Planetary Motion: An Application

In 1619, Johannes Kepler published his third law concerning the motion of the planets around the sun. This law expresses the relationship between the average distance D of a planet from the sun and the period t (the length of a year) for that planet to complete a full orbit about the sun. Kepler's work was based on the best experimental data available at the time. In fact, astronomers then weren't aware of the existence of the three outer planets: Uranus, Neptune, and Pluto. We use the following current data to determine the relationship between D and t, where D is the average distance of each planet from the sun, in millions of miles, and t is the length of its year, in days. To give you some perspective, the Earth takes 365 days (its year) to complete a full orbit about the sun and the average distance to the sun is 92.9 million miles.

First, we must decide which is the independent variable and which is the dependent variable. Although time is usually the independent variable, in this case it makes more sense to think of the length of the year t for a planet as a function of

Planet	Period t	Distance D
Mercury	88	36.0
Venus	225	67.2
Earth	365	92.9
Mars	687	141.5
Jupiter	4,329	483.3
Saturn	10,753	886.2
Uranus	30,660	1782.3
Neptune	60,150	2792.6
Pluto	90,670	3668.2

the planet's distance D from the sun—the farther a planet is from the sun, the longer it takes to complete a full orbit.

We next look at the scatterplot of the data, as shown in Figure 3.61 where the average distance D from the sun is plotted along the horizontal axis and the period t is plotted along the vertical axis. By eye, it appears that the points fall in a roughly linear pattern, so we first investigate the results of fitting a linear function to this data.

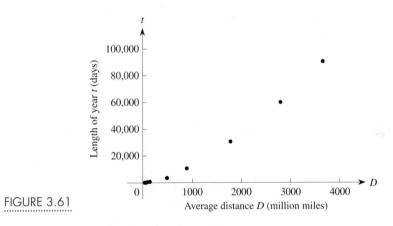

FIGURE 3.61

EXAMPLE 2

Find the linear function that best fits the data on planetary motion and discuss how good the fit is.

Solution   The linear regression equation for this data set is

$$Y = 24.038X - 4584.52,$$

or equivalently in terms of our variables here,

$$t = 24.038D - 4584.52.$$

The correlation coefficient is $r = 0.9887$, which indicates a high level of correlation because the critical value for $n = 9$ data points is $r_9 = 0.666$.

Figure 3.62 shows this line superimposed over the data points. Although at first glance it appears that the data points fall in a linear pattern, actually they do not. Consider the scale for the scatterplot. The horizontal scale extends from 36 million miles out to 3668 million (or 3.668 billion) miles. Furthermore, all the data points fall rather far from the regression line. These discrepancies may seem small to the eye, but they actually are enormous considering the size of the quantities involved. Hence the residuals—the differences between the actual values and the predictions based on the linear function—may be quite large.

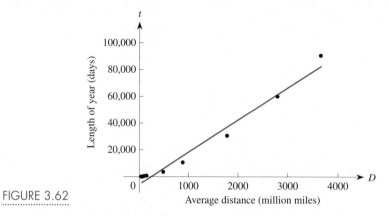

FIGURE 3.62

Using the linear regression equation, we predict values for the length of the year for each planet based on its distance from the sun and so find the corresponding residuals. They are shown in the following table.

Planet	Distance D	Actual period t	Predicted period $t = 24.038D - 4584.5$	Residual = actual − predicted
Mercury	36.0	88	−3719.1	3807.1
Venus	67.2	225	−2969.1	3194.1
Earth	92.9	365	−2351.4	2716.4
Mars	141.5	687	−1183.1	1870.1
Jupiter	483.3	4,329	7033.1	−2704.1
Saturn	886.2	10,753	16,718.0	−5965.0
Uranus	1782.3	30,660	38,258.4	−7598.4
Neptune	2792.6	60,150	62,544.0	−2394.0
Pluto	3668.2	90,670	83,591.7	7078.3

Note that the predictions for the length of a year for the four innermost planets are negative, which means that the year has a negative number of days! In fact, the most accurate prediction based on the linear regression model (i.e., the smallest residual) is off by 1870 days! As we said previously, the apparently small discrepancies between the data points on the scatterplot and the regression line actually are immense when we take the scale into account. Consequently, we conclude that the linear fit is definitely not appropriate for this set of data, despite the high level of correlation.

When we examine the scatterplot of the planetary data with the regression line superimposed, as shown in Figure 3.62, there clearly is a concave up pattern to the data. This pattern suggests that we should use either an exponential growth function or a power function with $p > 1$. In Example 3, we show what happens with the exponential function.

EXAMPLE 3

Find the exponential function that best fits the data on planetary motion and discuss how good the fit is.

Solution   The exponential regression equation for this data set is

$$Y = 484.176(1.00172)^x$$

or equivalently in terms of our variables here,

$$t = 484.176(1.00172)^D.$$

The correlation coefficient is $r = 0.8937$, which indicates a relatively high level of correlation. Comparing it to the critical value $r_9 = 0.666$ based on $n = 9$ data points indicates a significant level of correlation, although the correlation is considerably lower than that for the linear fit. We show the exponential function superimposed over the scatterplot in Figure 3.63, and observe that it is not a particularly good fit.

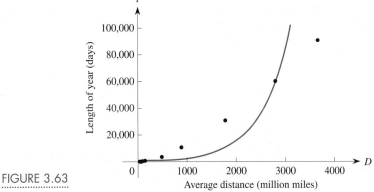

FIGURE 3.63

To check the accuracy of this function, we look at the errors—the differences between the actual values and the predicted values—we get when using this exponential formula. Recall that, when we are working with a linear function, the errors are the same as the residuals but the term *residual* applies only to a linear fit. We show these errors in the following table.

Planet	Distance D	Actual Period t	Predicted Period $t = 484.176\,(1.00172)^D$	Error = Actual − Predicted
Mercury	36.0	88	515.1	−427.1
Venus	67.2	225	543.4	−318.4
Earth	92.9	365	568.0	−203.0
Mars	141.5	687	617.5	69.5
Jupiter	483.3	4,329	1,111.0	3218.0
Saturn	886.2	10,753	2,220.3	8532.7
Uranus	1782.3	30,660	10,356.5	20,303.5
Neptune	2792.6	60,150	58,782.2	1367.8
Pluto	3668.2	90,670	264,694.8	−174,024.8

The best prediction, for the length of the year on Mars, is off by almost 70 days. The worst prediction, for Pluto, is off by more than 174,000 days and this error is almost twice the actual length of the year there. Consequently, we conclude that the exponential fit is very poor for this set of data, despite the fact that the correlation coefficient is significant.

◆

In Example 4 we show what happens with the power function.

◆ **EXAMPLE 4**

Find the power function that best fits the data on planetary motion and discuss how good the fit is.

Solution The power regression equation for this data set is

$$Y = 0.4079 \ X^{1.4999},$$

or equivalently in terms of our variables here,

$$t = 0.4079 \ D^{1.4999}.$$

The correlation coefficient is $r = 0.9999999$, which indicates an almost perfect fit. The power function superimposed over the scatterplot, as shown in Figure 3.64, certainly appears to be an excellent fit.

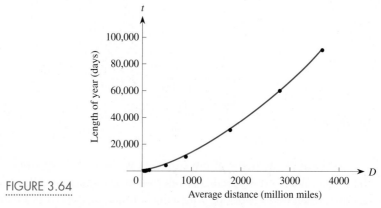

FIGURE 3.64

To verify the accuracy of this fit, we again look at the errors obtained when we use the power function to predict the length of each planet's year.

Planet	Distance D	Actual Year t	Predicted year $t = 0.4079 \ D^{1.4999}$	Error = actual − predicted
Mercury	36.0	88	88.1	−0.1
Venus	67.2	225	224.6	0.4
Earth	92.9	365	365.1	−0.1
Mars	141.5	687	686.2	0.8
Jupiter	483.3	4,329	4,331.2	−2.2
Saturn	886.2	10,753	10,753.7	−0.7
Uranus	1782.3	30,660	30,669.0	−9.0
Neptune	2792.6	60,150	60,148.1	1.9
Pluto	3668.2	90,670	90,547.6	122.4

The predictions for five of the planets are accurate to within a fraction of a day. Even the worst prediction—that for Pluto—is off by only 122.4 days; although that may seem like a fairly large amount, consider it in terms of the magnitude of the correct value, 90,670 days. As a percentage, this result is off by only $0.00135 = 0.135\%$ or about one tenth of 1%.

Finally, we compare how well each of the three functions created in Examples 2, 3, and 4 fit the data on planetary motion by considering the sum of the squares of the deviations. For the linear function, we need the vertical distance from each of the points to the line (the residuals shown in the last column of the table in Example 2). We square these residual values and then add them. Therefore, for the linear function, we have

$$\text{Sum of squares} = 3807.1^2 + 3194.1^2 + \cdots + 7078.3^2 = 192{,}035{,}629.$$

For the exponential function, we similarly find the vertical distance from each point to the curve (the error value shown in the last column of the table in Example 3), square each of these error values, and then add them. The result is

$$\text{Sum of squares} = (-427.1)^2 + (-318.4)^2$$
$$+ \cdots + (-174{,}024.8)^2 = 3.0782 \times 10^{10}.$$

Finally, for the power function, we likewise sum the squares of the error values from the last column of the table in Example 4 to obtain

$$\text{Sum of squares} = (-0.1)^2 + (0.4)^2 + \cdots + (122.4)^2 = 15{,}081.48.$$

Summarizing these results, we have the following.

For the linear function:	sum of the squares $= 192{,}035{,}629$
For the exponential function:	sum of the squares $= 3.0782 \times 10^{10}$
For the power function:	sum of the squares $= 15{,}081.48$.

The value of the sum of the squares for the power function is quite small, especially considering the size of the numbers involved. (In fact, almost the entire value of 15,081.48 comes from the contribution of Pluto, for which the error is 122.4, so that $122.4^2 = 14{,}981.8$.) In contrast the sum of the squares for the linear function is extremely large and that for the exponential function is astronomical. Therefore, based on the sum of squares criterion, the power function is also by far the best fit to these data.

Let's look a little closer at the equation for the power function obtained in Example 4, $t = 0.4079D^{1.4999}$. The power 1.4999 is certainly greater than 1 as we expected. However, numerically it is virtually equal to 1.5, so we write the equation as

$$t = 0.4079D^{1.5}.$$

This equation is known as Kepler's third law of planetary motion.

Think About This

In practice, measuring the length of the year for a planet is relatively easy, especially for the inner planets whose years are reasonably short. You do so by determining how long it takes to make a complete revolution about the sun. However, measuring the distance from the sun is much harder. Rewrite the equation for Kepler's third law to express the distance D as a function of the length of the year t. ◻

Actually, there is a more interesting form for the relationship between the length of the year and the average distance of the planets from the sun. The power 1.5 in the function can be written as the fraction $\frac{3}{2}$ so that

$$t = 0.4079D^{3/2}.$$

Squaring both sides of this equation, we get

$$t^2 = (0.4079D^{3/2})^2 = 0.1664D^3,$$

which is the form in which Kepler's third law usually is expressed.

Astronomers discovered the planet Pluto in 1930 after observing some minor perturbations in the orbit of Neptune. They hypothesized that these discrepancies could be accounted for by the existence of a previously unknown outer planet. Knowing the timing of the perturbation, the astronomers knew approximately where to look for this unknown planet by using predictions based on Kepler's third law. Similarly, they have used the law to estimate how far newly discovered planets are in their orbits about other stars.

Problems

1. The best-fit line is constructed for each of four sets of nonlinear data, whose patterns can be roughly described as

   a. increasing and concave up;
   b. increasing and concave down;
   c. decreasing and concave up;
   d. decreasing and concave down.

Match each description with one of the two residual plots (i) and (ii) shown in the column at the left. Explain your answer in each case.

2. Three different types of functions are fitted to a set of data based on the three residual plots (a)–(c) shown below and at the top of the next column. Decide which function best fits the data.

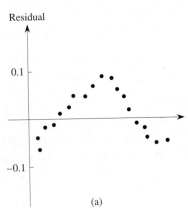

(a)

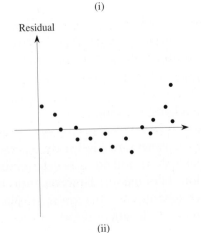

(i)

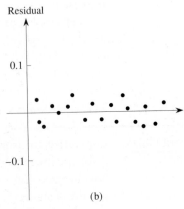

(b)

(ii)

Residual

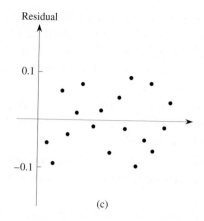

0.1

−0.1

(c)

3. Consider the points $(0, 1)$, $(1, 3)$, $(2, 6)$, $(3, 8)$ and $(4, 10)$.

   a. The line $y = 2x + 1$ fits these points reasonably well. Calculate by hand the residuals and the sum of their squares.

   b. The line $y = 2.5x + 1$ fits the data slightly better. Calculate the sum of the squares and compare it to the value you found in part (a).

   c. Find the equation of the regression line for the data. Then calculate the sum of the squares for it. How does it compare to your results in parts (a) and (b)?

4. A line is fit to different sets of data, and the following residual plots are obtained. Assume that in each case the residuals are all relatively small, so you are to consider only the patterns.

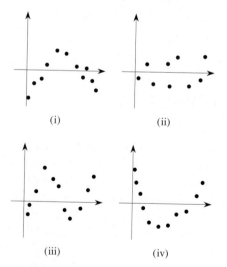

(i)   (ii)

(iii)   (iv)

   a. Decide which, if any, of the residual plots indicate that a linear fit is appropriate for the actual data.

   b. For any residual plot that indicates that a linear fit is likely not appropriate, decide whether the pattern in the original data is increasing and con-

cave up, increasing and concave down, decreasing and concave up, decreasing and concave down, or none of the above.

   c. From among the families of functions that you have studied so far, decide which types of functions would be appropriate for each of the actual data sets that you listed in part (b). State what you would expect of the values of the parameters for exponential functions and for power functions.

5. The *Statistical Abstract of the United States* contains data on the median starting salaries of people based on their level of education in six categories: did not complete high school, high school graduate, associate's degree, bachelor's degree, master's degree, and Ph.D. Those who didn't complete high school had approximately $E = 10$ years of education; those who have only a high school diploma have $E = 12$ years of education. Similarly, $E = 14$ for an associate's degree, $E = 16$ for a bachelor's degree, $E = 18$ for a master's degree, and $E = 20$ for a Ph.D. The following table relates the median starting salary S in 1999 to the number of years E of education:

E	10	12	14
S	$16,053	$23,594	$32,468
E	16	18	20
S	$43,782	$52,794	$74,712

   a. Find the equation of the regression line for the data. What is the meaning of the slope in this context?

   b. Use the equation from part (a) to calculate the residuals associated with each data value and construct the residual plot. Does the residual plot indicate that the regression line fits the data well? Explain.

   c. Calculate the sum of the squares associated with the regression line.

6. For the data in Problem 5

   a. find the equation of the exponential function that fits the data and state the meaning of the growth factor in this context;

   b. calculate the sum of the squares associated with the exponential function you found in part (a); and

   c. the residuals for the exponential function $S = Ac^E$ in part (a) are based on the linear equation,

$\log S = \log A + (\log c)E$. Calculate the residuals and draw the associated residual plot. Does the residual plot indicate that the exponential function fits the data well? Explain.

7. For the data in Problem 5:

a. Find the equation of the power function that fits the data.

b. Calculate the sum of the squares associated with the power function you found in part (a); and

c. The residuals for the power function fit $S = AE^p$ in part (a) are based on the equation, $\log S = \log A + p \cdot (\log E)$. Calculate the residuals and draw the associated residual plot. Does the residual plot indicate that the power function fits the data well? Explain.

8. Use the data in Problem 5 and the results from Problems 5–7:

a. By looking at the way the linear function from Problem 5, the exponential function from Problem 6, and the power function from Problem 7 fit the data, which of the three functions appears to be the best fit?

b. By comparing the correlation coefficients for the linear function from Problem 5, the exponential function from Problem 6, and the power function from Problem 7, which of the three functions appears to be the best fit?

c. By comparing the sum of the squares for the linear function from Problem 5, the exponential function from Problem 6, and the power function from Problem 7, which of the three functions appears to be the best fit?

d. By comparing the residual plot for the linear function from Problem 5, the exponential function from Problem 6, and the power function from Problem 7, which of the three functions appears to be the best fit?

e. Based on the criteria in parts (a)–(d), which of the three functions do you conclude is the best fit? Explain your decision.

9. Assume that each of the planets from Mercury to Neptune revolves about the sun in a roughly circular orbit.

a. Extend the first table in Example 2 to include the average speed of each planet in its orbit. (*Hint:* If you know the average distance from a planet to the sun, what is the approximate distance that it travels in its orbit around the sun?)

b. Find the best fit to this set of data on the speed of a planet as a function of the length of the planet's year from among linear, exponential, and power functions.

c. Explain how the formula you found in part (b) can be directly determined algebraically from Kepler's third law.

10. Using the transformation approach of Section 3.5, the power function in Example 4 can be found from

$$\log D = \log(t^{0.667}) + 0.2596.$$

Show that you get the same result by using properties of logarithms when you write this function as

$$\log D - \log(t^{0.667}) = 0.2596.$$

3.7 ····· Linear Models with Several Variables

So far, we have studied situations in which one variable (y) depends on another (x). In the real world, things aren't always this simple and we often encounter situations in which one quantity depends on two or more other independent variables. For example, when the weather report gives the windchill factor during the winter, that value depends on both the air temperature and the wind speed, so it is a function of two independent variables. Similarly, when you take out a car loan, the monthly payment depends on the amount borrowed, the interest rate, and the length of the loan, so there are three independent variables. A study conducted at a college found that student performance in math classes could be accurately predicted from a combination of the student's score on a placement test, the student's age, gender, SAT score, high school GPA, number of years since the previous math course, grade in that course, and several other variables. Actually, each of a dozen independent variables contributed valuable information, and a model based on all of them produced the most accurate predictions.

In general, suppose that we have a single variable y that depends on any number of independent variables $x_1, x_2, \ldots, x_n$. Statisticians often refer to independent variables as *explanatory variables*. To keep things simple, however, we focus on the case of a single variable y that depends on two independent variables x_1 and x_2, so that $y = f(x_1, x_2)$.

Suppose that we want to determine whether a relationship exists between a person's serum cholesterol level and the person's weight and systolic blood pressure. (The systolic reading is the first, or higher, number in a blood pressure measurement, such as 120 over 80; the smaller number is the diastolic reading.) We collect a set of data on a random sample of individuals from some population group—say, young males. The data in the accompanying table, for a sample of 11 apparently normal males between the ages of 13 and 16, shows the weight of each of these individuals in kilograms (independent variable x_1) and the systolic blood pressure of each (independent variable x_2), as well as their serum cholesterol level in mg/100 cc (the dependent variable y). If there is a relationship between serum cholesterol level and the other two variables, we can use it to predict serum cholesterol levels for other members of this population group, based on weight and blood pressure.

Serum Cholesterol y	Weight x_1	Systolic Blood Pressure x_2
152.2	59.0	108
158.0	52.3	111
157.0	56.0	115
155.0	53.5	116
156.0	58.7	117
159.4	60.1	120
169.1	59.0	124
181.0	62.4	127
174.9	65.7	122
180.2	63.2	131
174.0	64.2	125

Source: Wayne W. Daniel, *Biostatistics: A Foundation for Analysis in the Health Sciences*, 4th ed., New York: John Wiley & Sons, 1987.

Recall that we fit a line to a set of (x, y) data points by using the least squares criterion and constructing the corresponding linear function $y = ax + b$. To do so we minimize the sum of the squares of the vertical distances to the desired line to obtain the line of best fit, as shown in Figure 3.65.

Here, we have a set of (x_1, x_2, y) data points in three dimensions and want to construct the linear function that best fits the data. The equation of a linear function of two independent variables x_1 and x_2 can be written as

$$y = ax_1 + bx_2 + c,$$

where a, b, and c are three constants. The graph of any such equation is a *plane* in three-dimensional space. Now, instead of fitting a line to a set of (x, y) data

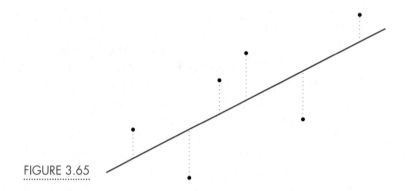

FIGURE 3.65

points in two dimensions, we need to fit a plane to a set of (x_1, x_2, y) data points in three dimensions. We want to find the best-fit plane—the specific plane for which the sum of the squares of the vertical distances to the desired plane is a minimum, as shown in Figure 3.66.

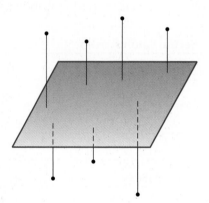

FIGURE 3.66

It turns out that there is a single plane that satisfies this criterion of coming closest to all the data points in three-dimensions; it is known as the *regression plane.* The process of finding the equation of this plane is called *multivariate regression* or *multiple regression.* (If there are more than two independent variables, a natural extension is used, although it isn't possible to visualize the "hyperplane" in four or more dimensions that best fits these data points.)

The arithmetic involved in any multivariate regression tends to be extremely tedious, but the method is so widely used that a routine is available in almost any statistical software package and in most spreadsheets. (It is not yet available on the common hand-held calculators.) We aren't concerned with the mechanics of calculating these quantities but simply cite the results obtained by using appropriate software. In the problems at the end of this section, we similarly assume that you have access to software that will perform the calculations for you.

◆ **EXAMPLE 1**

(a) Find the multivariate regression equation expressing serum cholesterol level as a function of both body weight and systolic blood pressure. (b) Interpret the coefficients in the multivariate regression equation. (c) Use the regression equation to predict the serum cholesterol level of a male in the 13- to 16-year-old age group who weighs 60 kg and whose systolic blood pressure is 123.

Solution

a. Using an appropriate software package, such as the Excel™ spreadsheet (we present details on how to use it at the end of this section), we find that the multivariate regression equation is

$$y = -7.6419 + 0.6297x_1 + 1.1315x_2,$$

or equivalently

Cholesterol level $= -7.6419 + 0.6297$ (weight) $+ 1.1315$ (systolic blood pressure).

b. To interpret this equation, let's see what happens to the values for the cholesterol level when we change each of the independent variables, weight or systolic blood pressure. In particular, suppose that a person's weight x_1 increases by 1 kg. According to the formula, the cholesterol level y will increase by 0.6297 units. Alternatively, suppose that a person's systolic blood pressure x_2 increases by 1 unit. In that case, the cholesterol level y will increase by 1.1315 units. Thus an increase of 1 unit in blood pressure has a considerably larger impact on cholesterol level than an increase of 1 unit in weight has on cholesterol level. That is, the systolic blood pressure apparently has a greater effect on cholesterol level than weight does for a 1-unit increase in each quantity.

c. According to this model, we predict that the serum cholesterol level for an individual who weighs $x_1 = 60$ kg and whose systolic blood pressure $x_2 = 123$ is

$$y = -7.6419 + 0.6297\,(60) + 1.1315\,(123) = 169.315,$$

or about 169 mg/100 cc.

◆

Note how the coefficients $a = 0.6297$ and $b = 1.1315$, play the same role in this multivariate regression equation $y = ax_1 + bx_2 + c$ that the slope m plays in the equation of a line $y = mx + b$. We can think of the regression plane as having two different slopes, one in the x_1 direction and another in the x_2 direction, and they indicate how quickly the dependent variable increases for a given change in either x_1 or x_2, respectively. Imagine balancing a book with one corner on a table by holding it at the opposite corner. Suppose that the point on the table is the origin and that x_1- and x_2-axes are drawn on the table. The cover of the book forms a plane in three-dimensional space and you draw two lines on the cover of the book, one parallel to the x_1-axis and the other parallel to the x_2-axis, as shown in Figure 3.67. The lines are inclined at two different angles and so have different slopes, one with respect to the x_1-axis and the other with respect to the x_2-axis. For that reason, you may want to think of the equation of the regression plane in the form

$$y = m_1x_1 + m_2x_2 + c,$$

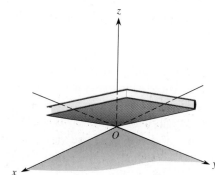

FIGURE 3.67

where m_1 and m_2 are the slopes in the x_1 and x_2 directions, respectively.

Also note that, in making the prediction in part (c) of Example 1, we took values for the independent variables x_1 and x_2 that were within the ranges of the data set. Had we used values that were well outside the ranges of the data—say, $x_1 = 70$ kg and $x_2 = 140$—we would have considerably less confidence in the accuracy of the prediction.

The Multiple Correlation Coefficient

In addition to finding the equation of the plane that best fits the set of data points in three-dimensional space, we also need a way of measuring how good the fit is. With (x, y) data points (where we think of y as a function of x), we used the linear correlation coefficient r as such a measure. With two or more independent variables, we instead use a quantity known as the *multiple correlation coefficient*, denoted by R. Like r, R also takes on values between -1 and 1, and the closer it is to either $+1$ or -1, the stronger the linear relationship between y and the linear combination of the independent variables.

Furthermore, the square of the multiple correlation coefficient, R^2, is known as the *coefficient of determination*. The value of R^2 provides some extremely useful information about the extent to which the multivariate regression equation explains the relationship between the dependent variable y and the independent variables x_1 and x_2 (if there are two) or $x_1, x_2, \ldots, x_n$ (if there are more than two). In particular, the value of R^2 indicates the percentage of the variation in the data that is explained by the linear relationship. For instance, if $R^2 = 0.85$, say, then 85% of the variation is explained by the linear function. (A similar interpretation applies to r^2, the square of the usual correlation coefficient with (x, y) data points.)

As with the multivariate regression equation, both the multiple correlation coefficient and the coefficient of determination are typically calculated as part of the output of both spreadsheets and statistical software.

EXAMPLE 2

For the data in the preceding table relating serum cholesterol level to an individual's weight and systolic blood pressure, find and interpret both the multiple correlation coefficient and the coefficient of determination.

Solution Using an appropriate software package, we find that the coefficient of determination is

$$R^2 = 0.8508$$

and so the multiple correlation coefficient is

$$R = 0.9224.$$

The multiple correlation coefficient is reasonably close to 1, so we conclude that a high degree of linear correlation exists between the dependent variable y and the two independent variables x_1 and x_2. (Statisticians have developed a table of critical values for R comparable to the critical values for r, but we won't go into that here.) Moreover, the value $R^2 = 0.8508$ indicates that about 85% of the variation in serum cholesterol levels in this population group can be "explained" by these two variables.

In Example 2 the value for the coefficient of determination $R^2 = 0.8508$ indicates that about 85% of the variation observed is explained by the two variables, weight and systolic blood pressure. Thus, about 15% of the variation is *not* explained by these two variables. We can explain a greater amount of the variation by introducing an additional variable, perhaps the number of hours that each individual in the study spends exercising each week, as we might expect that cholesterol levels also depend on the level of physical activity. In that case, we would be working with three independent variables x_1, x_2, and x_3 and would obtain a linear regression equation that relates serum cholesterol level to all three. Incidentally, when we do so, the coefficients of the original two independent variables will almost certainly change.

Each time we introduce an additional variable, we get a larger value for the coefficient of determination, although the increase might be minimal. At the same time, we also increase the complexity of the model, and there are often drawbacks to doing so.

Performing Multivariate Regression in Excel

Because the Excel spreadsheet is so widely available, we briefly introduce its use for performing a multivariate regression analysis. In Excel the dependent variable is always denoted by y, and the different independent variables are always denoted by x. Think of the two independent variables as x_1 and x_2. You begin by entering the given data values in the first three columns labeled A, B, and C, as shown in Figure 3.68.

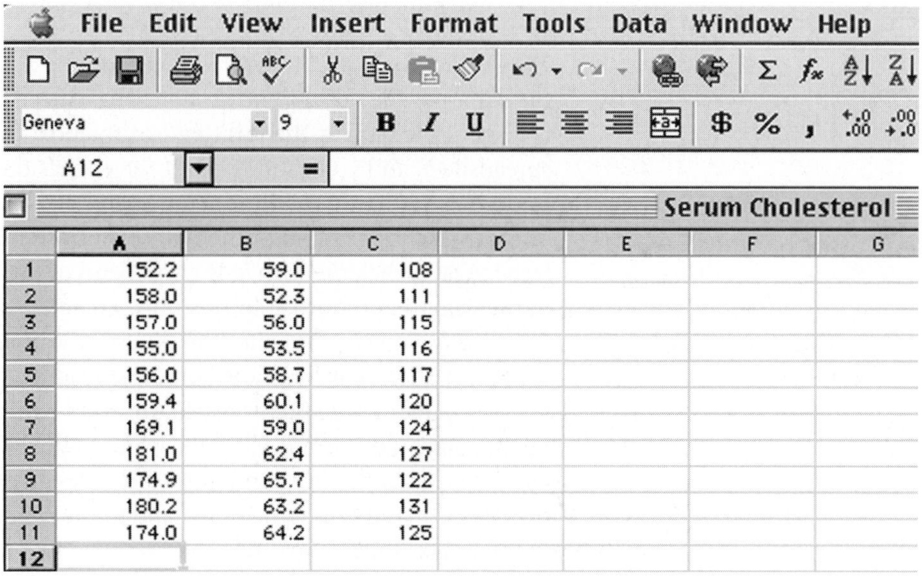

	A	B	C	D	E	F	G
1	152.2	59.0	108				
2	158.0	52.3	111				
3	157.0	56.0	115				
4	155.0	53.5	116				
5	156.0	58.7	117				
6	159.4	60.1	120				
7	169.1	59.0	124				
8	181.0	62.4	127				
9	174.9	65.7	122				
10	180.2	63.2	131				
11	174.0	64.2	125				
12							

FIGURE 3.68

Once you have entered all the data values, click `Tools` on the top line and scroll down to the last entry, `Data Analysis. . . .` (We indicate the computer displays in a different font, for emphasis.) If `Data Analysis . . .` doesn't appear, you will have to install the Excel Analysis ToolPak™ before proceeding. To install it, go to the `Tools` menu and click `Add-ins`. If `Analysis ToolPak` is listed, just click it to permanently install it. If `Analysis ToolPak` isn't listed in the `Add-Ins` dialog box, click `Browse` and locate the drive and folder names and the file name `Analys32.xll` for the `Analysis ToolPak`—it usually is located in the `Library\Analysis` folder.

When you click `Data Analysis . . .`, you will see a long list of available statistical procedures in alphabetical order. Scroll down until you reach

Regression and then either double click it or single click and then click OK. Doing so brings up the window shown in Figure 3.69.

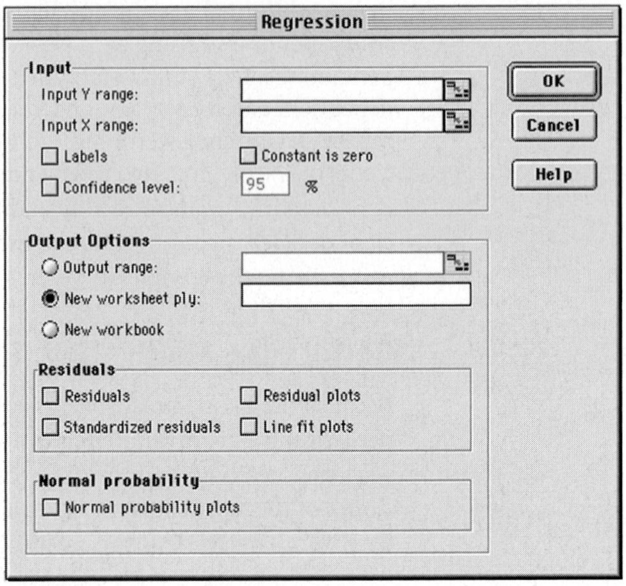

FIGURE 3.69

In this window, you first have to enter the Input Y Range—the cells in which the values of the dependent variable y have been entered. The simplest way to do so is to click the icon at the right end of the box; that brings you back to the original spreadsheet, and you can highlight the entries down the first column (the y values) under A and then press Enter. You then have to enter the Input X Range—the cells in which the values of the two (or more) independent variables have been entered. Again, click the icon at the right end of the box and then highlight the entries in the second and third columns under B and C and press Enter.

Finally, you have to enter the Output Range—where the results of the regression calculations will be printed on the spreadsheet. You don't want them printed over the data values in the first three columns, so you probably would want them printed starting, say, in column E. Click the white circle to the left of Output Range and then click the icon at the right end of the box. Designate a block of cells starting at the top of column E and extending down and to the right by highlighting the first cell under E; then press Enter. Finally, in the Regression window, press OK.

Excel will then print a large amount of information, as shown in Figure 3.70. Only a few of these entries are of interest to us; the rest are used for more sophisticated statistical analysis. In particular, note that the first block of output is called Regression Statistics and that the first two numbers under it are labeled Multiple R and R Square—the values for the multiple correlation coefficient R and the coefficient of determination, R^2.

The third block of output starts with three lines labeled Intercept, X Variable 1, and X Variable 2. The entries to their right give the coefficients. In particular, if the regression equation is $y = ax_1 + bx_2 + c$, the intercept is the constant coefficient c, the value corresponding to the first X Variable is the coefficient a for x_1, and the value corresponding to the second X Variable is the coefficient b for x_2. Once you have these values, you can write the multivariate regression equation.

	A	B	C	D	E	F	G
							Serum Cholesterol.xls
1	SUMMARY OUTPUT						
2							
3	*Regression Statistics*						
4	Multiple R	0.92237661					
5	R Square	0.85077861					
6	Adjusted R Square	0.81347326					
7	Standard Error	4.67509989					
8	Observations	11					
9							
10	ANOVA						
11		*df*	*SS*	*MS*	*F*	*Significance F*	
12	Regression	2	996.9129824	498.456491	22.8058081	0.00049582	
13	Residual	8	174.8524722	21.856559			
14	Total	10	1171.765455				
15							
16		*Coefficients*	*Standard Error*	*t Stat*	*P-value*	*Lower 95%*	*Upper 95%*
17	Intercept	-7.6418777	25.64715303	-0.297962	0.77332377	-66.784357	51.5006014
18	X Variable 1	0.62965272	0.481162227	1.30860795	0.22700791	-0.4799101	1.73921552
19	X Variable 2	1.13146262	0.295454277	3.82956929	0.00502099	0.4501434	1.81278185

FIGURE 3.70

Problems

1. A study was conducted relating the heights of teenage boys y to the length of their radius bones x_1 and the length of their femur bone x_2, as shown in the accompanying table. (All measurements are in centimeters.)

y	x_1	x_2
149.0	21.00	42.50
152.0	21.79	43.70
155.7	22.40	44.75
159.0	23.00	46.00
163.3	23.70	47.00
166.0	24.30	47.90
169.0	24.92	48.95
174.5	25.80	50.30
176.1	26.01	50.90
176.5	26.15	50.85
179.0	26.30	51.10

Source: Wayne W. Daniel, *Biostatistics: A Foundation for Analysis in the Health Sciences,* 4th ed., New York: John Wiley & Sons, 1987.

a. Use an appropriate software package to calculate the coefficient of determination and the multiple correlation coefficient. How much of the variation in height is explained by the length of the two bones?

b. Find the equation of the plane that best fits the data.

c. Use your equation from part (b) to predict the height of a teenage boy whose radius measures 25.50 cm and whose femur measures 49.90 cm. How close does your prediction come to the boy's actual height of 172 cm?

2. The windchill factor is an adjustment made to temperature readings to take into account the effects of the wind and so indicate how cold it feels. The following table gives the windchill factors associated with different combinations of air temperature in degrees Fahrenheit and wind speeds in miles per hour.

a. Use an appropriate software package to calculate the coefficient of determination and the multiple correlation coefficient. How much of the variation in the windchill factor is explained by the two independent variables?

b. Find the equation of the plane that best fits the data. (*Note:* The actual formula used to calculate the windchill factors isn't a linear function.)

c. Use your equation from part (b) to predict the windchill factor corresponding to a temperature of 18°F and a 22 mph wind.

Wind Speed	Temperature							
	35	30	25	20	15	10	5	0
5	33	27	21	16	12	7	0	−5
10	22	16	10	3	−3	−9	−15	−22
15	16	9	2	−5	−11	−18	−25	−31
20	12	4	−3	−10	−17	−24	−31	−39
25	8	1	−7	−15	−22	−29	−36	−44

d. Suppose that the air temperature is 8°F and the windchill factor is −30. Use the equation of the plane to predict the wind speed.

e. Which variable, temperature or wind speed, has a greater effect on the windchill factor? Explain.

3. A study was conducted on a group of people with pulmonary function problems to relate their forced expiratory volume, in liters per second, to their vital lung capacity x_1, in liters, and to their total lung capacity x_2, in liters.

y	x_1	x_2
1.6	2.2	2.5
1.0	1.5	3.2
1.4	1.6	5.0
2.6	3.4	4.4
1.2	2.0	4.4
1.5	1.9	3.3
1.6	2.2	3.2
2.3	3.3	3.3
2.1	2.4	3.7
0.7	0.9	3.6

Source: Wayne W. Daniel, *Biostatistics: A Foundation for Analysis in the Health Sciences*, 4th ed., New York: John Wiley & Sons, 1987.

a. Use an appropriate software package to calculate the coefficient of determination and the multiple correlation coefficient. How much of the variation in forced expiratory volume is ex-plained by the two independent variables?

b. Find the equation of the plane that best fits the data.

c. Use your equation from part (b) to predict the forced expiratory volume of an individual whose vital capacity is 2.6 liters and whose total lung capacity is 3.9 liters?

d. Which variable, vital lung capacity or total lung capacity, has a greater effect on forced expiratory volume? Explain.

4. A study was conducted on the weight w, in kilograms, of 14 patients with primary type II hyperlipoproteinemia (a genetic disorder that leads to massively high cholesterol levels) just prior to the start of medical treatment. For each patient, the total cholesterol level x_1 and the triglyceride level x_2 were recorded, in milligrams per 100 cubic centimeters, as shown in the table at the top of the next page.

a. Use an appropriate software package to calculate the coefficient of determination and the multiple correlation coefficient. How much of the variation in weight is explained by the levels of cholesterol and triglyceride?

b. Find the equation of the plane that best fits the data.

c. Use your equation from part (b) to predict the weight of a patient with primary type II hyperlipoproteinemia who has a cholesterol level of 332 and a triglyceride level of 186. How close does your prediction come to the actual reading for this patient, who weighed 78 kg?

5. As part of a study to investigate the relationship between stress y and several other variables, the accompanying data were collected on a random sample of 14 corporate executives. In this table, x_1 measures the size of the firm (the number of

x_1	302	336	220	300	382	379	331	332	426	399	279	410	389	302
x_2	139	101	57	56	113	42	84	186	164	205	230	160	153	139
w	76	97	83	52	70	67	75	78	70	99	75	70	77	76

Source: Wayne W. Daniel, *Biostatistics: A Foundation for Analysis in the Health Sciences*, 4th ed., New York: John Wiley & Sons, 1987.

employees), x_2 measures the number of years in the present position, x_3 is the annual salary in thousands, and x_4 is the person's age.

a. Use an appropriate software package to calculate the coefficient of determination and the multiple correlation coefficient when y is considered as a function of x_1 and x_2 only. How much of the variation in the measure of stress is explained by these two variables only?

b. Find the equation of the plane that best fits the data on y as a function of x_1 and x_2 only.

c. Use your equation from part (b) to predict the stress level of a corporate executive whose company has 484 employees and who has been in the present position for 8 years.

d. Repeat parts (a)–(c) if you now take into account the three independent variables, x_1, x_2, and x_3. In particular, how much more of the variation in stress level is explained by including the third variable? Predict the stress level for the individual in part (c) if $x_3 = 81$.

e. Repeat parts (a)–(c) if you now take into account all four independent variables. In particular, how much more of the variation in stress level is explained by including the fourth variable? Predict the stress level of the individual in parts (c) and (d) if $x_4 = 40$.

y	x_1	x_2	x_3	x_4
101	812	15	70	38
60	334	8	60	52
10	377	5	60	27
27	303	10	94	36
89	505	13	92	34
60	401	4	67	45
16	177	6	66	50
184	598	9	92	60
34	412	16	74	44
17	127	2	68	39
78	601	8	82	41
141	297	11	124	58
11	205	4	71	51
104	603	5	78	63

Source: Wayne W. Daniel, *Biostatistics: A Foundation for Analysis in the Health Sciences*, 4th ed., New York: John Wiley & Sons, 1987.

Chapter Summary

In this chapter, we introduced you to regression analysis. Specifically, we showed you how to find the best fit line or curve for the various families of functions and interpret the results in the following ways.

◆ How to find the regression, or least squares, line that is the best linear fit to a set of data.

◆ How to interpret the correlation coefficient as a measure of how good the linear fit is.

◆ How to use the linear regression equation for making predictions.

◆ How to find the exponential function, power function, or logarithmic function that best fits a set of data.

◆ How to transform a set of data to linearize it if the underlying pattern is an exponential function, a power function, or a logarithmic function.

◆ How to undo the transformation to produce the best-fit exponential or power function.

◆ How to interpret the residuals to assess how well a linear function fits the original data or the transformed data.

◆ How to interpret the sum of squares to assess how well a function fits a set of data.

◆ How to use the best fitting nonlinear function for making predictions.

◆ How to fit a linear function of several variables to a set of data.

Review Problems

1. The table shows the budget and attendance (both in millions) at 15 zoological parks in the United States. Find the best-fit function from among linear, exponential, power, and logarithmic functions for the attendance as a function of the budget. How good is the fit?

Budget	10.0	3.4	27.0	6.2	9.7
Attendance	1.0	0.5	2.0	0.6	1.3
Budget	7.0	4.8	18.0	6.5	13.0
Attendance	1.0	1.1	4.0	0.6	3.0
Budget	9.0	15.7	7.0	3.2	14.7
Attendance	0.5	1.3	1.0	0.5	2.7

Problems 2–6 are based on the data below showing total expenditures, in billions of dollars, for both health and public education in the United States for the years shown.

2. (a) Find the best linear, exponential, power, and logarithmic functions that can be used to model the total health expenditures in the United States as a function of the number of years since 1979. (b) Use each model to predict the amount spent on health expenditures in 2004. (c) Of the four predictions, which seems the most accurate and which is the least accurate. Explain your reasoning.

3. (a) Find the best linear, exponential, power, and logarithmic functions that can be used to model the total expenditures on public education in the United States as a function of time since 1979. (b) Use each model to predict the amount spent on public education in 2004. (c) Of the four predictions, which seems the most accurate and which the least accurate. Explain your reasoning.

4. Consider the linear functions you created in Problems 2 and 3.
   a. Interpret the slope in each case.
   b. Use the two functions to estimate when expenditures for health care first exceeded those for public education.

5. Consider the exponential functions you created in Problems 2 and 3.
   a. Interpret the base in each case.
   b. Use the two functions to estimate when expenditures for health care first exceeded those for public education.

6. Find the linear function that best fits health care expenditures as a function of education expenditures. What is the meaning of the slope of this line?

7. The table on the next page gives the relationship between the average longevity (in years) and the gestation period (in days) for a sample of animals. The data indicate that the animals' average longevity can be predicted reasonably well as a function of the gestation period.

Year	1980	1985	1990	1992	1993	1994	1995	1996	1997	1998
Health Expenditures	247.3	428.7	699.4	836.5	898.5	947.7	993.7	1042.5	1092.4	1146.1
Public Education	345.1	378.7	486.0	506.7	517.6	527.9	541.9	554.4	569.5	583.8

Source: *2000 Statistical Abstract of the United States.*

a. Find the linear, exponential, power, or logarithmic function that best predicts longevity as a function of gestation period.

b. Use your function grapher to graph the best-fit function.

Animal	Gestation	Longevity
baboon	187	20
black bear	219	18
house cat	63	12
dog	61	12
cow	284	15
elephant	660	35
giraffe	425	10
gorilla	258	20
horse	330	20
kangaroo	36	7
lion	100	15
monkey	166	15
mouse	19	3
opossum	13	1
rabbit	31	5
sea lion	350	12
squirrel	44	10
human	270	76

Source: *The Universal Almanac.*

8. The U.S. Postal Service charges 37¢ for first-class postage for the first ounce of mail and 23¢ for every additional ounce. What linear function, based on weights 1, 2, ..., 10 oz, best models this situation? Explain why this function does not give the exact charge for an 8.5-oz letter.

9. Consumer credit data from 1970 through 1995 show the following amounts of outstanding consumer credit (in billions of dollars) in the United States at the end of each year.

a. Find the linear function that best fits the data.

b. Use the model from part (a) to predict the amount of outstanding consumer credit at the end of 1992.

c. Determine the year in which the outstanding consumer credit first will exceed $1.5 trillion.

Year	1970	1975	1980	1985	1987	1988
Credit	133.8	207.5	355.4	601.6	692.0	742.1
Year	1989	1990	1991	1993	1994	1995
Credit	791.8	809.3	796.7	863.9	988.8	1131.9

Source: *World Almanac and Book of Facts.*

10. The number of airline flights generally has risen during the past 20 years. Flights (in millions) are given for the years 1983–2000 in the table below:

a. Find the exponential function that fits the data.

b. How well does the exponential function fit the data?

c. Predict when the number of airline flights will first exceed 10 million per year.

11. The table on the next page shows the height H in feet and the number of stories n of some notable buildings.

a. Find the linear, exponential, power, or logarithmic function that best relates a building's height to its number of stories.

b. What is the significance of the slope of the linear function?

Year	1983	1984	1985	1986	1987	1988	1989	1990	1991
Number of flights	5.0	5.4	5.8	6.4	6.6	6.7	6.6	6.9	6.8
Year	1992	1993	1994	1995	1996	1997	1998	1999	2000
Number of flights	7.1	7.2	7.5	8.1	8.2	8.2	8.3	8.6	9.0

Source: *2000 Statistical Abstract of the United States.*

	H	n
Empire State Building (New York)	1250	102
John Hancock Tower (Boston)	788	61
Sears Tower (Chicago)	1450	110
NationsBank Plaza (Dallas)	921	72
TCBY Tower (Little Rock)	546	40
Peachtree Center (Atlanta)	374	31
Place Ville Marie (Montreal)	620	45

Source: *World Almanac and Book of Facts.*

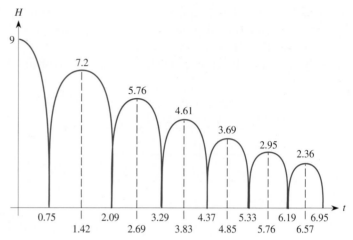

c. On average, how many feet are allocated to each story?

12. Draw a scatterplot for each of the functions f, g, and h in the table. For each set of data, decide whether the pattern of data is linear, exponential, or logarithmic. Explain your choices.

x	1	2	3	4	5	6
$f(x)$	6	4	2	0	-2	-4
$g(x)$	5.4	4.86	4.374	3.937	3.543	3.189
$h(x)$	-2	-0.194	0.863	1.612	2.194	2.669

13. An experiment is conducted in which a ball is dropped from an initial height of 9 feet and its subsequent height above floor level as a function of time is recorded and displayed, as illustrated in the accompanying figure. When the curve is traced, the measurements shown on the graph indicate the times when the ball hits the floor, the times when the ball reaches its maximum heights, and the values of these maximum heights.

a. Note that the times when the ball hits the floor appear to follow a linear pattern. Find the best linear fit to these times as a function of the number of the bounce; that is, bounce number $n = 1$ occurs at time $t = 0.75$, and so on. What is the meaning of the slope?

b. Note that the times when the ball reaches its maximum heights also appear to follow a linear pattern. Find the best linear fit to these times as a function of n, where now $H = 9$ when $n = 0$, etc. What is the meaning of the slope?

c. Note that the maximum heights don't follow a linear pattern. Find the best nonlinear function that fits these data values, as a function of n, from among the families of functions you have studied in this chapter. For the exponential fit, what is the significance of the base you obtain?

d. Find the best fit for the maximum heights H as a function of time t.

e. Use the results you obtained in parts (a)–(d) to predict the corresponding values for the time and height on the next bounce of the ball.

14. In October 2002, astronomers reported the discovery of a new body half the size of Pluto in an orbit that takes 105,120 days to complete a full revolution about the sun at a distance of about 4 billion miles from the sun. Verify whether this object, named Quaoar after a Native American god, satisfies Kepler's Third Law from Example 4 in Section 3.6.

15. In October 2002, a pilot flying a small plane in southwest Alaska reported spotting a bird with a wingspan of 14 feet. Based on Problem 8 in Section 3.5, how much would this newly discovered bird weigh?

Extended Families of Functions

4.1 ## Introduction to Polynomial Functions

Samantha has been keeping track of the price of the stock of HyperTech Corporation since her grandmother gave her several shares as a gift. She has plotted the stock values, as shown in Figure 4.1, and wants to construct a mathematical model that represents the price of the stock. Clearly, a linear function, an exponential function, a power function, or a logarithmic function is not a reasonable candidate because none have this kind of behavior pattern. To better capture the trend in the stock prices, Samantha needs a function that changes both its direction and its concavity, as illustrated in Figure 4.2.

Note that the graph increases, then decreases, and finally increases again. Thus, the graph has two turning points, one at the local maximum point and the other at the local minimum point. Also, the curve initially is concave down and then is concave up, so the graph has one point of inflection, where the concavity changes.

In this section, we introduce a new family of functions, the *polynomial functions,* that possess this type of more complicated behavior. A **polynomial function,** or **polynomial,** is any finite sum of power functions with nonnegative integer powers. For instance,

$$y = 3x - 5; \tag{1}$$

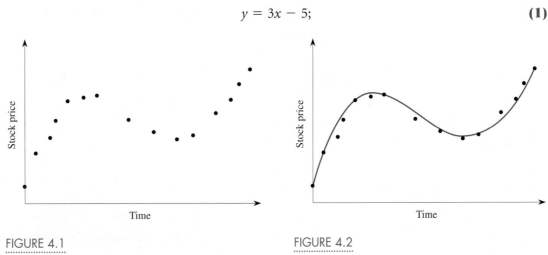

FIGURE 4.1

FIGURE 4.2

$$y = 6x^2 + x - 7; \tag{2}$$

$$y = 6 + 8x - 3x^2; \tag{3}$$

$$y = 4x^3 + 5x^2 - 7x + 12; \tag{4}$$

$$y = 10x^8 - 7x^5 + 3 \tag{5}$$

are all polynomials.

The **degree of a polynomial** is the highest power of the variable present. Hence, the degree of Polynomial (1) is 1, the degree of Polynomials (2) and (3) is 2, the degree of Polynomial (4) is 3, and the degree of Polynomial (5) is 8.

The constant multiples in each term in any polynomial are called its **coefficients.** In particular, the coefficient of the highest power term is the **leading coefficient.** Thus, in Polynomial (1), the coefficients are 3 and -5 and the leading coefficient is 3; in Polynomial (2), the coefficients are 6, 1, and -7 and the leading coefficient is 6. Note that, in Polynomial (3), the leading coefficient is -3 (it is *not* necessarily the first coefficient). As we show in Section 4.2, the sign of the leading coefficient determines the overall behavior of the polynomial.

Another way to describe a polynomial is to say that it is a *linear combination* of power functions because, as we noted, it is made up of a sum of power functions. In this sense, power functions are the basic building blocks we use to construct any polynomial.

If a polynomial has degree 1, it is a linear function of the form, $y = ax + b$, where a and b are constants and $a \neq 0$. Its graph is a line with slope $m = a$ and vertical intercept b.

If a polynomial has degree 2, it is called a **quadratic function** and it has the form

$$y = ax^2 + bx + c,$$

where a, b, and c are constants and $a \neq 0$. With three coefficients in the equation, the set of all quadratic functions is a three parameter family of functions. The graph of any quadratic function is a curve called a **parabola.** Such curves abound in the real world—in the path of a fly ball in baseball, in the shape of the main support cable in a suspension bridge such as the Golden Gate Bridge or the George Washington Bridge, or in the cross sections of a TV satellite dish, as depicted in Figure 4.3.

If a polynomial has degree 3, it is called a **cubic function** and its graph is called a **cubic.** In general, a cubic function has the form

$$y = ax^3 + bx^2 + cx + d,$$

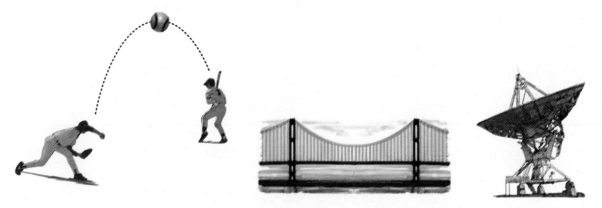

FIGURE 4.3

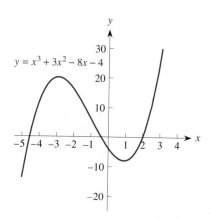

FIGURE 4.4

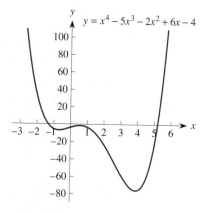

FIGURE 4.5

where a, b, c, and d are constants and $a \neq 0$. For example, the graph of the cubic $y = x^3 + 3x^2 - 8x - 4$ is shown in Figure 4.4. This graph is typical of a cubic function, having two turning points and one inflection point.

If a polynomial has degree 4, it is called a **quartic function** and its graph is called a **quartic.** In general, a quartic polynomial has the form

$$y = ax^4 + bx^3 + cx^2 + dx + e,$$

where a, b, c, d, and e are constants and $a \neq 0$. The graph of the quartic $y = x^4 - 5x^3 - 2x^2 + 6x - 4$ is shown in Figure 4.5. This graph is typical of a quartic polynomial. Notice that it has three turning points and two inflection points.

Think About This

How many parameters are there in the family of cubic polynomials? In the family of quartic polynomials? In the family of polynomials of degree n, for any n? ▭

The Zeros of a Polynomial

A key piece of information about any polynomial function is the value or values of the variable x that make the function zero. These values are known as the **zeros** of the polynomial. For instance, the zeros of the quadratic polynomial $P(x) = x^2 - 6x + 8$ are $x = 2$ and $x = 4$ because

$$P(2) = 2^2 - 6(2) + 8 = 0 \quad \text{and} \quad P(4) = 4^2 - 6(4) + 8 = 0.$$

From a different point of view, if we set the expression for the polynomial function equal to zero, we have an equation and the solutions to this equation are called the **roots.** So, corresponding to the quadratic *polynomial* $P(x) = x^2 - 6x + 8$, we have the quadratic *equation*

$$x^2 - 6x + 8 = 0.$$

Factoring this expression gives

$$(x - 2)(x - 4) = 0.$$

The two solutions of this equation, $x = 2$ and $x = 4$, are the roots of the quadratic function.

Note that a *function has zeros,* that an *equation has roots,* and that there is a direct correspondence between them. The zeros of a function f occur at precisely the same points as the roots of the equation $f(x) = 0$.

EXAMPLE 1

Find the zeros of the quadratic function $P(x) = x^2 - 5x + 6$ and the roots of the corresponding quadratic equation $x^2 - 5x + 6 = 0$, both graphically and algebraically.

Solution The graph of the quadratic function $y = x^2 - 5x + 6$ is shown in Figure 4.6. Note that the graph crosses the x-axis twice: once when $x = 2$ and again when $x = 3$. So, graphically, we conclude that these are the zeros of the function. If we consider the associated quadratic equation

$$x^2 - 5x + 6 = 0,$$

its roots are $x = 2$ and $x = 3$.

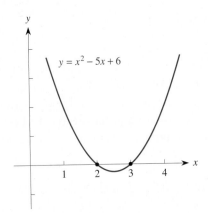

FIGURE 4.6

Alternatively, we can solve this equation algebraically. We start with the associated quadratic equation

$$x^2 - 5x + 6 = 0.$$

We can factor the quadratic expression on the left as

$$(x - 2)(x - 3) = 0.$$

Recall that, when the product of two factors is zero, one or the other or both must be zero, so we have either $x - 2 = 0$ or $x - 3 = 0$, leading to the roots $x = 2$ and $x = 3$. Because they are the roots of the quadratic equation, they are also the zeros of the quadratic polynomial $P(x) = x^2 - 5x + 6$. ◆

If the coefficients in a quadratic are appropriately chosen, we *may* be able to find the roots of a quadratic equation by algebraic factoring, as we did in Example 1. If the coefficients are not just right—say, $4.35709x^2 + 15.46031x - 11.02013 = 0$, or even $5x^2 + 3x - 17 = 0$—the factoring approach won't work. The same principle applies to polynomials of higher degree, but the algebra typically becomes much more complicated as the degree of the polynomial increases. Consequently, factoring is far less likely to work when the degree of a polynomial is 3 or higher.

The two roots for any quadratic equation

$$ax^2 + bx + c = 0, \qquad a \neq 0,$$

always can be found from the *quadratic formula*.

The Quadratic Formula

$$x = \frac{-b \pm \sqrt{b^2 - 4ac}}{2a}$$

This formula is derived in any algebra textbook.

EXAMPLE 2

Find the zeros of the quadratic polynomial $P(x) = x^2 - 3x - 5$.

Solution   With $a = 1$, $b = -3$, and $c = -5$, the quadratic formula gives the roots of the associated quadratic equation $x^2 - 3x - 5 = 0$ as

$$x = \frac{-(-3) \pm \sqrt{(-3)^2 - 4(1)(-5)}}{2(1)}$$

$$= \frac{3 \pm \sqrt{9 - (-20)}}{2}$$

$$= \frac{3 \pm \sqrt{29}}{2}.$$

The result is a pair of irrational numbers. Thus, the roots are

$$x = \frac{3 + \sqrt{29}}{2} \approx 4.19258 \quad \text{and} \quad x = \frac{3 - \sqrt{29}}{2} \approx -1.19258.$$

The quadratic formula was essentially known to the ancient Babylonians, some 4000 years ago. However, not until about 1540 did Italian mathematicians Tartaglia and Cardano discover a comparable, although considerably more complicated, formula for the three roots of any cubic equation. Not long after that, another Italian mathematician, Ferrari, discovered an even more complicated formula that gives the four roots of any quartic equation. (These formulas are programmed into some calculators and software packages.) Finally, in 1824, Danish mathematician Abel proved that no general formula could exist that would give the roots of any polynomial equation of fifth or higher degree. When we encounter polynomials of higher degree, we usually have to resort to numerical methods to find the roots. We illustrate this approach in Example 3 for a polynomial of degree 3.

EXAMPLE 3

Find, correct to four decimal places, all the zeros of the cubic polynomial $y = Q(x) = x^3 + 3x^2 - 8x - 4$.

Solution   The graph of this polynomial is shown in Figure 4.7. Note that it crosses the x-axis three times and that each of these points is a zero of the polynomial. By zooming in on each point in turn, using a function grapher, we estimate that the zeros are located at approximately $x = -4.56155$, $x = -0.43845$, and $x = 2.00000$. This last value for x suggests that the third zero may be $x = 2$ precisely. To determine whether that is indeed the case, we substitute $x = 2$ into the formula for the cubic and find that

$$Q(2) = (2)^3 + 3(2)^2 - 8(2) - 4 = 8 + 12 - 16 - 4 = 0,$$

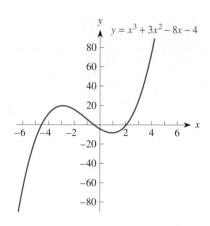

FIGURE 4.7

so $x = 2$ is precisely the zero. Because we were asked to give the three zeros correct to four decimal places, we conclude that $x \approx -4.5616$, $x \approx -0.4385$, and $x = 2$.

We found three zeros for the cubic polynomial in Example 3 based on its graph in Figure 4.7. But, how do we really know that there are no additional zeros? We could expand the viewing window and examine the graph from $x = -100$ to $x = 100$, say, or perhaps from $x = -1000$ to $x = 1000$, and maybe other zeros will come into view. Unfortunately, this kind of exploratory approach never completely closes the door on the possibility that other zeros might exist if only we look further. Instead, we need to know something about the behavior of polynomials in general, which will give us information on how many zeros a particular polynomial has and some indication of where to look for them. We discuss this in the next section.

Problems

1. What is the degree of each polynomial?
   a. $P(x) = 6x^3 - 5x^2 + 8$
   b. $P(x) = 5x^4 + 6x^3 + 7x - 11$
   c. $P(x) = 6x - 5x^2$
   d. $P(x) = x^5 - x^8$
   e. $P(x) = -4x^3 - 9x^2 + 12$
   f. $P(x) = 10 - 4x + 5x^3 + 3x^6$

2. What is the leading coefficient of each polynomial in Problem 1?

3. Which values of x are zeros of the polynomial $P(x) = x^3 + 2x^2 - 3x$ and which are not?
   a. $x = 3$ b. $x = 2$ c. $x = 1$
   d. $x = 0$ e. $x = -1$ f. $x = -2$
   g. $x = -3$

4. Consider these polynomials.
   a. $P(x) = 6x^3 - 5x^2 + 8$
   b. $P(x) = -6 - 5x^2 + 8x^3$
   c. $P(x) = 8 - 5x^2 - 6x^3$

   d. $P(x) = -8x^3 - 5x^2 + 6$
   e. $P(x) = 6x^3 - 5x^2 + 8$
   f. $P(x) = 3x^2 - 5x + 4$
   g. $P(x) = -4x^2 - 5x - 8$
   h. $P(x) = 4 - 3x^2$
   i. $P(x) = 8 - 5x + 4x^2$
   j. $P(x) = 6x^4 - 5x^3 + 8x - 3$
   k. $P(x) = 3x^4 - 5x^3 + 8x - 6$
   l. $P(x) = 3 - 6x^4$

   For each polynomial in (a)–(l), indicate whether it is a
   i.   quadratic polynomial,
   ii. cubic polynomial,
   iii. quartic polynomial,
   iv. quadratic polynomial whose leading coefficient is 4,
   v.   cubic polynomial whose leading coefficient is -6, or
   vi. quartic polynomial whose leading coefficient is -6.

5. The table gives some values for a polynomial P. Identify possible roots of the corresponding polynomial equation.

x	-5	-4	-3	-2	-1		
$P(x)$	227	21	0	-8	0		
x	0	1	2	3	4	5
$P(x)$	16	23	0	-32	0	166

6. The figure to the right shows the graph of a polynomial. How many zeros does it have?

7. Estimate, correct to three decimal places, all the zeros of the polynomial $P(x) = 2x^3 - 6x^2 + 5x - 3$.

8. Estimate, correct to three decimal places, all the zeros of the polynomial $P(x) = x^4 - 4x^3 + 5x - 1$.

9. For the cubic polynomial in Problem 7, how many turning points are there? how many inflection points? Estimate, correct to two decimal places, all the turning points.

10. For the quartic polynomial in Problem 8, how many turning points are there? how many inflection points? Estimate, correct to two decimal places, all the turning points.

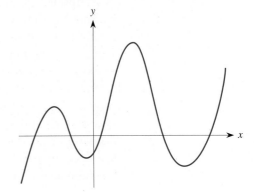

Exercising Your Algebra Skills

Add or subtract each pair of polynomials by combining like terms.

1. $(6x^3 - 5x^2 + 8) + (6x - 5x^2)$
2. $(6x^3 - 5x^2 + 8) - (6x - 5x^2)$
3. $(5x^4 + 6x^3 + 7x - 11) + (-4x^3 - 9x^2 + 12)$
4. $(5x^4 + 6x^3 + 7x - 11) - (-4x^3 - 9x^2 + 12)$
5. $(10 - 4x + 5x^3 + 3x^4) + (5x^4 + 6x^3 + 7x - 11)$
6. $(10 - 4x + 5x^3 + 3x^4) - (5x^4 + 6x^3 + 7x - 11)$

Multiply each pair of polynomials.

7. $x(3x - 5)$
8. $x(4x + 2)$
9. $x(7 + 3x)$
10. $x(6 - 5x)$
11. $(x - 1)(x - 3)$
12. $(x - 2)(x - 5)$
13. $(x - 2)(x + 3)$
14. $(x + 4)(x + 3)$
15. $(x + 5)(x - 5)$
16. $(x - 3)(x + 3)$
17. $(x + 2)(x - 2)$
18. $(x - 21)(x + 21)$

Raise each polynomial to the indicated power.

19. $(x - 1)^2$
20. $(x - 3)^2$
21. $(x + 2)^2$
22. $(2x + 5)^2$
23. $(2x - 6)^2$
24. $(x + 10)^2$

4.2 The Behavior of Polynomial Functions

The behavior of a polynomial depends on the ideas we introduced in Section 4.1: the degree, the zeros, and the sign of the leading coefficient of the polynomial. Let's see how.

Quadratic Polynomials

We begin by analyzing the behavior of quadratic functions. The graph of any quadratic function $y = ax^2 + bx + c$ is a parabola that opens either upward or downward. The sign of the leading coefficient a in

$$y = ax^2 + bx + c$$

determines whether the parabola opens upward or downward and so determines the overall behavior of the parabola. When the leading coefficient is positive, the parabola opens upward and is concave up. When the leading coefficient is negative, the parabola opens downward and is concave down. To understand why, think about what happens when x gets very large—say, $x = 100$ or $x = 1000$. Then x^2 is much larger, on the order of 10,000 or 1,000,000. Therefore the term ax^2 eventually overwhelms any contribution from the linear term bx or the constant term c. Thus, when a is positive, the quadratic term is extremely positive and the parabola opens upward. Similarly, when a is negative, the quadratic term is extremely negative and the parabola opens downward.

For instance, the parabola $y = 5x^2 - 20x - 300$ has a positive leading coefficient and so opens upward—when x becomes large, either positively or negatively, the overall effect is positive. In contrast the parabola $y = 20 - 4x^2$ has a negative leading coefficient and so opens downward—when x becomes large, either positively or negatively, the overall effect is negative. Check the graphs of both functions on your function grapher to convince yourself of the behavior in each case. Moreover, whichever way the parabola opens, as x increases indefinitely in either direction, the parabola either increases toward infinity or decreases toward negative infinity, as illustrated in Figure 4.8.

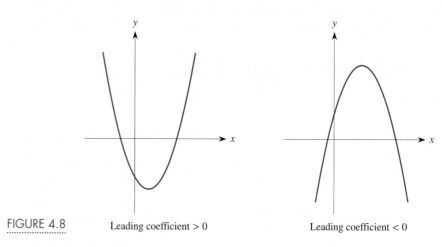

FIGURE 4.8   Leading coefficient > 0   Leading coefficient < 0

Every parabola has one turning point—also called its **vertex.** For instance, the parabola $y = x^2$ has its vertex at the origin, because that is the location of the turning point. If a parabola opens upward, the turning point corresponds to the minimum value of the function. If a parabola opens downward, the turning point corresponds to the maximum value of the function. In addition, the parabola is always symmetric about the vertical line through its turning point, so the left and right halves of the parabola are mirror images of one another. (See Appendix D for a discussion of symmetry.)

Next, let's examine the effects of the other two terms (the linear term and the constant term) in the formula for a quadratic function. Figure 4.9 shows the graphs associated with the quadratic functions $y = x^2$, $y = x^2 + 6$, $y = x^2 - 5x + 6$, and $y = x^2 + 5x + 6$. The leading term determines the basic behavior of the quadratic function, so all four open upward. However, the other terms affect the location of the graph. The constant term 6 in $y = x^2 + 6$ raises the parabola $y = x^2$ by 6 units (if the constant term is negative, the parabola would be lowered instead). Use your function grapher to experiment with this effect on the graph of the parabola by changing the constant term. For instance, how do the graphs of $y = x^2 + 5x + 7$ and $y = x^2 + 5x - 2$ compare to the graph of $y = x^2 + 5x + 6$? Be sure to look at

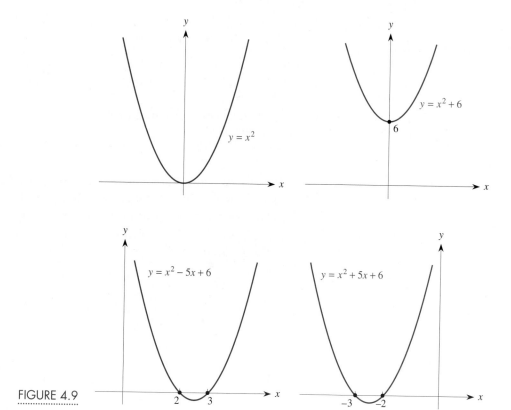

FIGURE 4.9

enough graphs to convince yourself of the effect of the constant term. We investigate this effect in detail in Section 4.7.

The effect of the linear term is more complicated because it involves both vertical and horizontal shifting of the parabola. We don't go into that here but do so in Section 4.7.

Furthermore, as we showed in Section 4.1, the two roots of any quadratic equation

$$ax^2 + bx + c = 0$$

can always be found from the quadratic formula

$$x = \frac{-b \pm \sqrt{b^2 - 4ac}}{2a}.$$

These two roots could be real numbers (as in Examples 1 and 2 in Section 4.1) or the roots could be a pair of complex numbers of the form $x = \alpha + \beta i$ and $x = \alpha - \beta i$ where $i = \sqrt{-1}$ (α and β are the Greek letters alpha and beta, respectively). A pair of complex numbers such as these is called a pair of *complex conjugates*. Complex numbers are discussed in Appendix E.

◣**EXAMPLE 1** ....

Find the roots of the quadratic equation $x^2 - 2x + 2 = 0$.

Solution   Using $a = 1$, $b = -2$, and $c = 2$ in the quadratic formula, we get

$$x = \frac{2 \pm \sqrt{(-2)^2 - 4(1)(2)}}{2(1)}$$

$$= \frac{2 \pm \sqrt{4 - 8}}{2}$$

$$= \frac{2 \pm \sqrt{-4}}{2} = \frac{2 \pm 2i}{2}.$$

We now divide through by 2 and find that the two complex roots of the quadratic are $x = 1 + i$ and $x = 1 - i$.

Because every quadratic equation has two roots, every quadratic function has exactly two zeros. Let's see what this means in terms of the graph of the quadratic function. Consider again the quadratic polynomial $P(x) = x^2 - 5x + 6$, whose graph is shown in Figure 4.10. This polynomial has zeros at $x = 2$ and $x = 3$ because we can factor the quadratic as

$$x^2 - 5x + 6 = (x - 2)(x - 3).$$

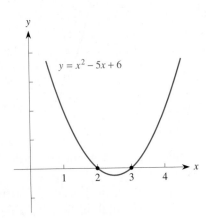

FIGURE 4.10

But the graph shows that the parabola crosses the x-axis at the points $x = 2$ and $x = 3$. Thus, just as the point at which a line crosses the x-axis gives the root of a linear equation, the points at which a parabola crosses the x-axis give the *real roots* of a quadratic equation, as illustrated in Figure 4.11 (a) and (b), respectively.

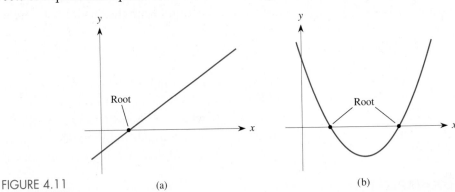

FIGURE 4.11   (a)   (b)

If we know that a parabola crosses the x-axis at a point $x = r$, then $x = r$ is a zero of the associated quadratic function and $x - r$ is a factor of the quadratic expression. You can locate the real roots of any quadratic to any desired level of accuracy with your graphing calculator or with the quadratic formula, so you can always find the linear factors.

In general, for any quadratic function $f(x) = ax^2 + bx + c$,

♦ the real roots of the quadratic equation $ax^2 + bx + c = 0$ correspond graphically to the points where the associated parabola crosses the x-axis, and

♦ the real roots of the quadratic equation $ax^2 + bx + c = 0$ correspond algebraically to the linear factors of the quadratic polynomial.

Depending on the orientation of the parabola (opening up or down) and the position of the turning point, a parabola may not touch the x-axis at all. This is the case with the graph of $y = x^2 + 6$, as shown in Figure 4.12. For such a parabola, the corresponding quadratic equation still has two roots, but they are complex roots. If a quadratic equation has complex roots, they must occur in conjugate pairs of the form $\alpha \pm \beta i$. This property follows directly from the quadratic formula for the case where the term inside the radical, $b^2 - 4ac$, is negative. The expression $b^2 - 4ac$ is called the **discriminant** of the quadratic. When the discriminant is positive, the two roots are real numbers. When the discriminant is negative, the two roots are complex numbers. Finally, when the discriminant is zero, there is a *double real root*.

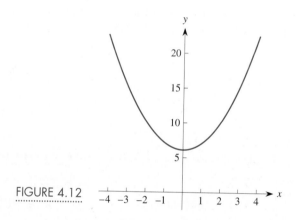

FIGURE 4.12

For instance, the discriminant for $y = x^2 + 6$ is $0^2 - 4(1)(6) = -24 < 0$, so the two roots are complex and they occur in pairs. From the quadratic formula, the roots are

$$x = \frac{-0 \pm \sqrt{0^2 - 4(1)(6)}}{2(1)}$$

$$= \frac{\pm \sqrt{-24}}{2}$$

$$= \pm \frac{2\sqrt{-6}}{2} = \pm \sqrt{6}i.$$

We have already demonstrated that a parabola can cross the x-axis at two points (corresponding to two real roots) or that it may not ever cross the x-axis (corresponding to a pair of complex conjugate roots). A third possibility is that the parabola could be tangent to the x-axis; that is, it can touch the axis and bounce back without ever crossing the axis. For instance, consider the quadratic

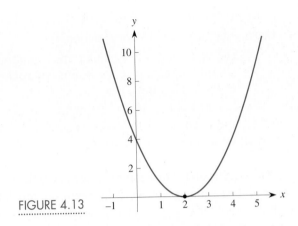

FIGURE 4.13

function $y = x^2 - 4x + 4$. If you apply the quadratic formula, you will find that the discriminant $b^2 - 4ac = (-4)^2 - 4(1)(4) = 0$ and the two roots are $x = 2$ and $x = 2$, as shown in Figure 4.13. Use your function grapher to zoom in on this point and note how the parabola just touches the x-axis at $x = 2$. Thus, the quadratic function has two roots, but because they are equal, $x = 2$ is a double root.

Think About This

Try changing the value of the constant term in $y = x^2 - 4x + 4$ slightly from 4— say, to 4.01 and then to 3.99. What happens to the graph in each case? What is the value of the discriminant in each case? ⊐

We summarize these ideas about quadratic polynomials as follows.

Characteristics of Quadratic Polynomials $y = ax^2 + bx + c$

- ◆ A quadratic polynomial has degree 2 and has precisely 2 zeros.
- ◆ A parabola has precisely one turning point, its vertex.
- ◆ A parabola opens upward if the leading coefficient a is positive; it opens downward if the leading coefficient a is negative, as shown in Figures 4.14 (a)–(f).
- ◆ The corresponding quadratic equation of degree 2 has precisely 2 roots. They may be real or complex.
- ◆ The complex roots occur in pairs of complex conjugates, $x = \alpha \pm \beta i$, where $i = \sqrt{-1}$, as illustrated in Figures 4.14 (a) and (d).
- ◆ The real roots correspond to the points where the parabola crosses the x-axis, as illustrated in Figures 4.14 (b) and (e), or where the parabola touches the x-axis, as illustrated in Figures 4.14 (c) and (f).
- ◆ You can always find the real roots graphically by using your function grapher to zoom in on the points where the parabola crosses or touches the x-axis.
- ◆ The real roots correspond to the linear factors of the quadratic expression.
- ◆ You can always find the roots, real or complex, by using the quadratic formula.

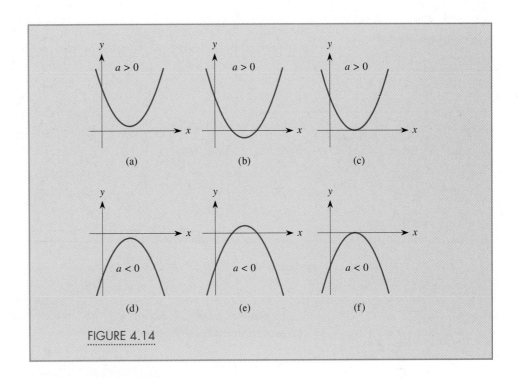

FIGURE 4.14

Cubic Polynomials

Next, we consider the characteristics of cubic polynomials having the form

$$y = ax^3 + bx^2 + cx + d,$$

where a, b, c, and d are constants and $a \neq 0$. The graph of the cubic polynomial $y = x^3 + 3x^2 - 8x - 4$ is shown in Figure 4.15, which is typical of a cubic function. The cubic rises toward positive infinity in one direction and drops toward negative infinity in the other. Also, this particular cubic has two turning points and crosses the x-axis at three points, so it has three real zeros. Moreover, the curve has one point of inflection, is concave down on one side of the point of inflection, and is concave up on the other.

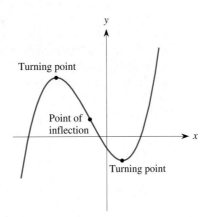

FIGURE 4.15

As with a quadratic function, the sign of the leading coefficient in a cubic always determines the overall behavior pattern of the function. If the leading coefficient is positive, the cubic increases as x increases (except possibly for a relatively small dip between the two turning points), as shown on the left in Figure 4.16. If the leading coefficient is negative, the cubic decreases as x increases (except for a possible rise between the two turning points), as shown on the right in Figure 4.16.

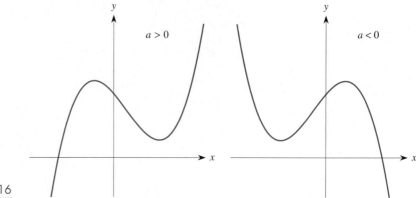

FIGURE 4.16

In general, a cubic function $y = ax^3 + bx^2 + cx + d$ has three zeros, and the corresponding cubic equation

$$ax^3 + bx^2 + cx + d = 0$$

has three roots. The roots can all be real numbers or can consist of a single real number and a pair of complex conjugate numbers. Each of the real roots corresponds to a linear factor of the corresponding cubic expression. Any complex conjugate roots must occur in pairs and correspond to a quadratic factor of the cubic polynomial.

> The real roots of a cubic equation correspond graphically to the points at which the associated cubic curve crosses the x-axis.
>
> The real roots of a cubic equation correspond algebraically to the linear factors of the cubic polynomial.

If a cubic has three real roots, its curve crosses the x-axis at the corresponding three points. If it has only one real root, the curve crosses the x-axis only once, as shown in Figure 4.17.

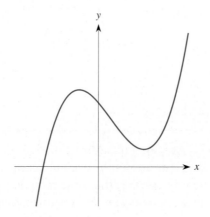

FIGURE 4.17

EXAMPLE 2

Analyze the behavior of the cubic function

$$f(x) = (x - 1)(x + 2)(x + 5) = x^3 + 6x^2 + 3x - 10.$$

Solution   The cubic has the three linear factors—$(x - 1)$, $(x + 2)$, and $(x + 5)$—so it has three real zeros: at $x = 1$, $x = -2$, and $x = -5$, corresponding to each of the three factors. Consequently, its graph crosses the x-axis at $x = 1$, -2, and -5, as shown in Figure 4.18. Further, the leading term $x^3 = 1 \cdot x^3$, being the highest power present, eventually dominates the other terms as x increases. Because the leading coefficient 1 is positive, the cubic must increase toward $+\infty$ as $x \to \infty$ and decrease toward $-\infty$ as $x \to -\infty$. Verify this result graphically by using your function grapher and numerically by substituting some large positive and negative values for x.

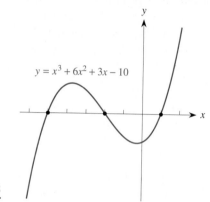

$y = x^3 + 6x^2 + 3x - 10$

FIGURE 4.18

Although there is a formula for calculating the roots of a cubic equation, it is considerably more complicated than the quadratic formula and is seldom used. If the cubic polynomial happens to factor simply, you can find the zeros directly because each factor corresponds to a zero. However, that is not likely to happen. Usually, the simplest way to find the real roots of a cubic equation is to approximate them by using your function grapher—just keep zooming in on the points where the curve crosses the x-axis until you find the roots to whatever degree of accuracy you desire.

We summarize these ideas about cubic polynomials as follows.

Characteristics of Cubic Polynomials
$y = ax^3 + bx^2 + cx + d$

♦ A cubic polynomial of degree 3 has precisely 3 zeros.
♦ A cubic has at most two turning points.
♦ A cubic typically has one inflection point.
♦ A cubic increases (rises upward) to the right as x increases if the leading coefficient a is positive; it decreases (falls downward) to the right as x increases if the leading coefficient a is negative.
♦ The corresponding cubic equation of degree 3 has precisely 3 roots. The roots may all be real or one real and two complex.

- The complex roots occur in pairs of complex conjugates, $\alpha \pm \beta i$, where $i = \sqrt{-1}$.
- The real roots correspond to the points where the cubic crosses the x-axis.
- You can always find the real roots graphically by using your function grapher to zoom in on the points where the cubic crosses the x-axis.
- The real roots correspond to linear factors of the cubic expression.

Figure 4.19 illustrates most of the possible cases for a cubic polynomial. In Figures 4.19 (a) and (b) there are three distinct real roots when the leading coefficient a is either positive or negative. In Figures 4.19 (c) and (d) there are three real roots, but one of them is repeated, so the x-axis is tangent to the cubic at the corresponding point. These two graphs correspond to when the leading coefficient $a > 0$; similar graphs can be drawn when $a < 0$. Figure 4.19 (e) shows a cubic with a triple real root and $a > 0$; note how the curve flattens as it crosses the x-axis. Think about the graph of $y = x^3$ as it passes through the origin. Finally, in Figures 4.19 (f)–(h) there is one real root and a pair of complex roots, again when $a > 0$. You can draw similar graphs when $a < 0$.

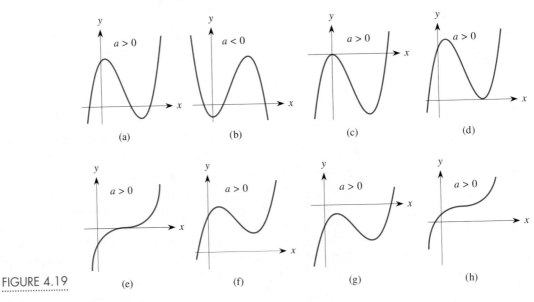

FIGURE 4.19

Moreover, it turns out that every cubic is symmetric about its inflection point. Imagine a cubic with a hinge at its inflection point—if you take either half of the curve and rotate it about that hinge, it will eventually be a perfect fit to the other half of the curve.

Think About This Prove that any cubic polynomial of the form $f(x) = ax^3$ is symmetric about its inflection point at the origin by showing that for any value of x—say, $x = h > 0$—then $f(-h) = -f(h)$. ▫

Polynomials of Degree n

The ideas discussed for polynomials of degree 2 (quadratics) and degree 3 (cubics) can be extended to polynomials of any degree n, where n is a positive integer. In particular, they have the following characteristics.

> ## Characteristics of Polynomials of Degree *n*
>
> ◆ A polynomial of degree *n* has precisely *n* zeros.
> ◆ A polynomial of degree *n* has at most $n - 1$ turning points.
> ◆ A polynomial of degree *n* has at most $n - 2$ points of inflection.
> ◆ If the leading coefficient is positive, the polynomial rises toward $+\infty$ to the right as $x \to \infty$, and if the leading coefficient is negative, the polynomial falls toward $-\infty$ to the right as $x \to \infty$.
> ◆ The corresponding polynomial equation of degree *n* has precisely *n* roots, which may be real or complex.
> ◆ The complex roots occur in pairs of complex conjugates, $\alpha \pm \beta i$, where $i = \sqrt{-1}$.
> ◆ The real roots correspond to the points where the curve crosses or touches the *x*-axis.
> ◆ You can always find the real roots graphically to any desired level of accuracy by using your function grapher to zoom in on the points where the curve crosses the *x*-axis.
> ◆ The real roots correspond to linear factors of the polynomial expression.

Think About This   Sketch the graph of a fifth degree polynomial with five real roots and a positive leading coefficient. Sketch the graph of a fifth degree polynomial with three real roots and a negative leading coefficient. ◻

◆ EXAMPLE 3

Suppose that a polynomial has roots at $x = -4, -1, 1, 3$, and 6. Find a possible formula for it and describe its behavior.

Solution   The polynomial has five real roots, so its degree must be at least 5; it might be higher if there are complex roots or repeated roots. The five corresponding linear factors are $(x - (-4)) = (x + 4)$, $(x + 1)$, $(x - 1)$, $(x - 3)$, and $(x - 6)$. If these are the only roots, one possible formula for this polynomial is

$$P(x) = (x + 4)(x + 1)(x - 1)(x - 3)(x - 6),$$

although any constant multiple *A* of this expression would be an alternative formula.

We can determine the value of the multiple *A* if we know the vertical intercept of the polynomial—or any other point on the curve. If the multiple *A* is positive, the graph of the polynomial has the behavior shown in Figure 4.20. Note that $P(x)$ rises toward ∞

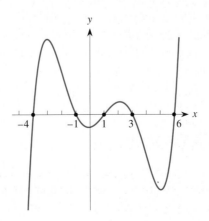

FIGURE 4.20

as $x \to \infty$ and that $P(x)$ falls toward $-\infty$ as $x \to -\infty$. Alternatively, if the constant multiple A is negative, this behavior is reversed; the graph drops toward $-\infty$ as $x \to \infty$ and rises toward $+\infty$ as $x \to -\infty$. Can you explain why this is the case?

What if a polynomial has a double or repeated factor? For instance,

$$P(x) = (x + 1)(x-2)(x-4)^2 = 0$$

has roots at $x = -1$, 2, and 4, but $x = 4$ is a double root because $(x - 4)^2 = (x - 4)(x - 4)$ is a repeated factor. Note that its graph, as shown in Figure 4.21, falls to touch the x-axis at $x = 4$ where it flattens and then rises again. Zooming in on the curve about this point reveals that the x-axis is tangent to the graph at $x = 4$, just as the x-axis is tangent to the parabola $y = x^2$ at the origin.

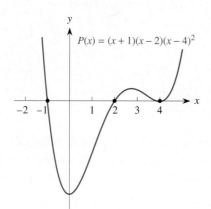

FIGURE 4.21

Think About This   The polynomial $y = (x + 1)(x - 2)^3$ has a triple factor. Examine its graph to see what happens near that triple root. First, try to predict what will happen, based on your knowledge of the behavior of $y = x^3$ near the origin.

◆ **EXAMPLE 4**

Use the fifth degree polynomial from Example 3 to demonstrate why it must have four turning points and three inflection points.

Solution   Let's trace the polynomial's curve in Figure 4.20 slowly from left to right. The function starts rising as we move to the right and crosses the x-axis at the first root at $x = -4$. It must cross the x-axis again at $x = -1$, so there must be a turning point between these two roots. Similarly, there must be a turning point between the roots at $x = -1$ and $x = 1$, and in fact, there is a turning point between each successive pair of roots. Because there are five real roots, there must be four turning points.

Now let's consider inflection points. We begin with the first two turning points, one near $x = -3$ where the curve is concave down and the next near $x = -0.2$ where the curve is concave up. The change in concavity means that there must be an inflection point between the successive turning points. In fact, between each successive pair of turning points, there is a point of inflection. Because there are four turning points, there must be three inflection points.

Things may not be quite so simple if there are complex roots or multiple roots.

The End Behavior of a Polynomial

The end behavior of any polynomial depends on the sign of the leading term because, as x approaches $+\infty$ or $-\infty$, the leading term eventually dominates all other terms. For instance, consider the polynomial $P(x) = 2x^4 - 6x^3 + 7x - 10$. When x is very large, the term $2x^4$ will overwhelm all other terms. The graphs of the functions $P(x) = 2x^4 - 6x^3 + 7x - 10$ and $Q(x) = 2x^4$ for x between -3 and 4 and y between -25 and 100 are shown in Figure 4.22; the two curves look quite different. Figure 4.23 presents a slightly larger view, where x is between -6 and 6 and y is between -50 and 950. Here the two curves look more similar than in the preceding view. In the much larger view presented in Figure 4.24, where x is between -25 and 25 and y is between 0 and $500,000$, there no longer is much difference between the two curves. The term $2x^4$ dominates the behavior of the polynomial, and the effect of the lower power terms is negligible.

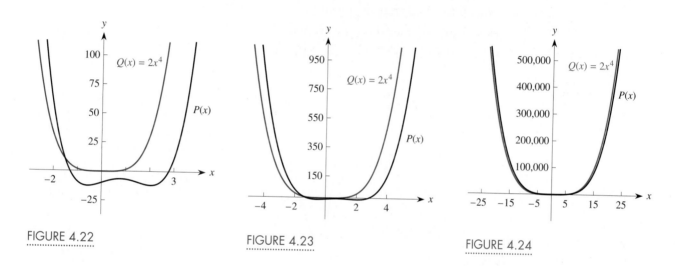

FIGURE 4.22

FIGURE 4.23

FIGURE 4.24

In general, for any polynomial, when x is large enough, the curve is indistinguishable from the curve corresponding to just the leading term. In other words, in the large, the behavior of any polynomial is virtually identical to that of the power function consisting of the leading term. You can see the *end behavior* easily on your function grapher if you use a reasonably large viewing window.

On the one hand, if the viewing window is too large, the location of the turning points and the zeros of a polynomial is a *local* aspect of the graph and can be easily missed. On the other hand, if the viewing window is too small, the overall growth pattern of the polynomial is lost. For instance, by focusing too closely on one particular turning point or root, you may lose sight of all the others. Rarely does a single view suffice to show all the important details of a function. Therefore, as a matter of routine, you should use the information given in several different views on your calculator or computer to sketch a rough hand-drawn graph of the function, called the *complete graph*, which highlights the key information, even if you intentionally do *not* draw it to scale.

We expect you to use your function grapher to produce the graph of a polynomial, but you should interpret with care what the calculator or computer shows. Usually, the important characteristics of any function—and a polynomial in particular—are

◆ the end behavior (is it increasing or decreasing as $x \to \infty$ or $x \to -\infty$?),
◆ the intervals over which the function is increasing or decreasing,
◆ the locations of the turning points,
◆ the intervals over which the function is concave up or concave down,
◆ the locations of the points of inflection, and
◆ the locations of the real zeros.

EXAMPLE 5

For the polynomial shown in Figure 4.25, answer the following questions.

a. How many real roots are there?
b. How many turning points are there?
c. How many inflection points are there?
d. What is the minimum degree of the polynomial?
e. How many complex roots does it have?
f. What is the sign of the leading coefficient?

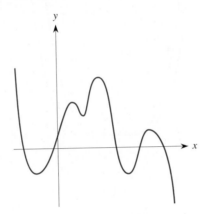

FIGURE 4.25

Solution

a.–c. The graph shown in Figure 4.25 reveals five real roots that correspond to the five points where the curve crosses the x-axis. It also shows six turning points and five inflection points.

d. Because the number of turning points is typically 1 less than the degree of the polynomial and the number of inflection points is 2 less than the degree, we conclude that the polynomial shown is at least a seventh degree polynomial.

e. Because there are five real roots and the degree of the polynomial is at least seven, there must be at least two complex roots.

f. The graph eventually falls toward $-\infty$ as $x \to \infty$, so we conclude that the leading coefficient must be negative.

EXAMPLE 6

Factor the polynomial $P(x) = x^4 - 5x^3 - 7x^2 + 8x + 3$.

Solution This polynomial is a quartic, so it has precisely four roots. We know that the linear factors of the polynomial correspond to its real roots, and the graph shown in Figure 4.26 reveals that there are four real roots. As a result, there cannot be any complex roots. We can locate each of these real roots to any desired level of accuracy, using either numerical or graphical methods. Correct to four decimal places, the roots are $x \approx -1.6272$, $x \approx -0.3105$, $x \approx 1.0000$, and $x \approx 5.9377$. The third of these results, $x \approx 1.0000$, suggests that the root might be $x = 1$ precisely. To verify that this is true, we substitute into the formula for the polynomial and find that

$$P(1) = (1)^4 - 5(1)^3 - 7(1)^2 + 8(1) + 3 = 1 - 5 - 7 + 8 + 3 = 0.$$

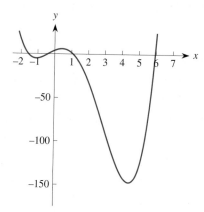

FIGURE 4.26

(If you do the same with the other three values, which are just approximations to the roots, the value of the polynomial will only be close to, but not quite equal to, zero.)

The corresponding linear factors are therefore roughly $(x + 1.6272)$, $(x + 0.3105)$, $(x - 1)$, and $(x - 5.9377)$, so the polynomial can be factored, approximately, as

$$P(x) = x^4 - 5x^3 - 7x^2 + 8x + 3 \approx (x + 1.6272)(x + 0.3105)(x - 1)(x - 5.9377).$$

Problems

1. The overall trend in the growth of the gross domestic product (GDP) has been upward except for a small dip. Sketch a graph representing the value of the GDP as a function of time. What type of function might model it? What can you conclude about any of the coefficients?

2. The overall pattern in the growth of the Dow-Jones Industrial Average over the past 10 years has been one of increase except for three sharp, but relatively short-term, drops. Sketch a graph representing the value of the Dow as a function of time. What type of function might model it? What can you conclude about any of the coefficients?

3. Each graph represents a polynomial. For each one:

   a. What is the minimum possible degree of the polynomial? Why?

   b. Is the leading coefficient of the polynomial positive or negative? Why?

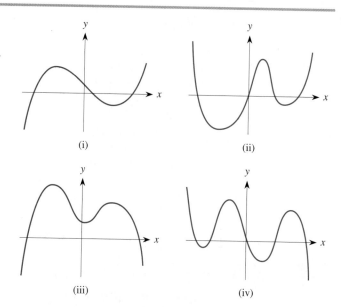

4. Each table gives some values for a polynomial. What is the minimum degree of each polynomial? Based

on the values given, what can you conclude about the sign of the leading coefficient in each case?

a.

x	-3	-2	-1	0	1	2	3	4	5
y	145	16	-5	-9	12	-3	21	2	-48

b.

x	-3	-2	-1	0	1	2	3	4	5
y	145	16	27	-11	24	16	41	-9	78

5. Match each polynomial expression a–f with its graph (i)–(vi). Use your knowledge about roots; do not use your function grapher.

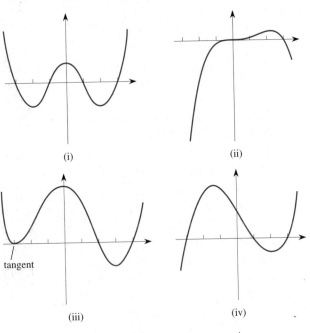

(i)   (ii)

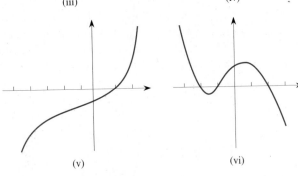

tangent

(iii)   (iv)

(v)   (vi)

  a. $f(x) = (x - 1)(x - 3)(x + 3)$
  b. $f(x) = (x + 1)(x + 2)(2 - x)$
  c. $f(x) = (x - 1)(x^2 + 4)$
  d. $f(x) = (x - 1)(x + 1)(x - 3)(x + 3)$
  e. $f(x) = 3x^3 - x^4$
  f. $f(x) = (x - 2)(x - 4)(x + 3)^2$

Based on your knowledge about roots and factors, sketch the graph of each polynomial function in Problems 6–9. Do not use your function grapher.

  6. $f(x) = (x + 2)(x - 1)(x - 3)$
  7. $f(x) = 5(x^2 - 4)(x^2 - 25)$
  8. $f(x) = -5(x^2 - 4)(x^2 - 25)$
  9. $f(x) = 5(x - 4)^2(x^2 - 25)$
  10. The polynomial $P(x) = 2x^6 + 5x^5 - 8x^4 - 21x^3 - 12x^2 + 22x + 12$ can be factored as $P(x) = (x - 2)(x - 1)(2x + 1)(x + 3)(x^2 + 2x + 2)$.
     a. What is the degree of the polynomial?
     b. What are the real roots? The complex roots?
     c. What happens as $x \to \infty$? As $x \to -\infty$?
     d. What is the maximum number of turning points you expect? Explain.
     e. What is the maximum number of points of inflection? Explain.
  11. Determine cubic polynomials that represent the accompanying graphs.

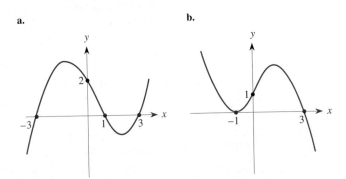

a.   b.

12. Each graph represents a function. For each one, **(i)** read off approximate intervals over which the function is increasing and over which it is decreasing; **(ii)** estimate intervals over which the function is concave up and concave down; and **(iii)** find a possible formula for the function.

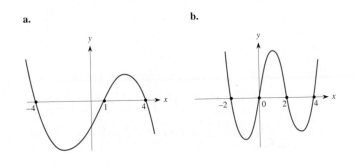

a.   b.

13. For each polynomial, (a) determine the number of real roots and the number of complex roots; and (b) find all real roots correct to three decimal places.

   i. $P(x) = x^4 - 8x^2 - 9$

   ii. $P(x) = x^5 - 4x^4 - 6x^3 + 6x^2 - 27x + 27$

   iii. $P(x) = x^6 - 4x^5 + 6x^4 - 16x^3 + 11x^2 - 12x + 6$

   iv. $P(x) = x^6 - 9x^5 + 26x^4 - 41x^3 + 71x^2 - 42x + 6$

14. Determine which of the graphs suggest the end behavior for each polynomial.

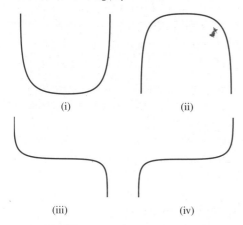

    (i)   (ii)

    (iii)   (iv)

   a. $y = 5x^5 - 8x^4 + 2x^2 + 3x - 4$

   b. $y = -4x^6 + 3x^4 + 7x^3 - 8x^2 - 4x$

   c. $y = 3x^8 + 4x^5 + 6x^3 - 5x^2 + 6$

   d. $y = -x^7 - 4x^6 + 3x^4 - 6x^3 + 7x - 9$

   e. $y = -4x^9 + 6x^6 - 5x^3 + 35$

   f. $y = 100 - x^4$ g. $y = (9 - 6x^2)^2$

   h. $y = (9 - 6x^3)^2$ i. $y = (9 - 6x^2)^3$

   j. $y = (9 - 6x^3)^3$

15. a. The graph of the polynomial $P(x) = 2x^4 - 6x^3 + 7x - 10$ in Figure 4.22 suggests that there are three turning points. Use your function grapher to locate them to 3 decimal place accuracy by zooming in on the graph.

   b. Estimate all intervals over which $P(x)$ is increasing or decreasing.

   c. Estimate the locations of all points of inflection.

   d. Estimate all intervals over which $P(x)$ is concave up or concave down.

   e. Estimate all real roots.

16. Describe the end behavior of each function. Specifically for the graph of each function f, (i) as $x \to \infty$, does $f(x) \to \infty$ or $-\infty$? why? and (ii) as $x \to -\infty$, does $f(x) \to \infty$ or $-\infty$? why?

   a. $f(x) = -3x^3 + 70x^2 - 20$.

   b. $f(x) = 20x^4 + 3x^3 + x^2 + 1000$.

   c. $f(x) = -3x^4 + 20x^3 - 5x^2 + x - 20$.

   d. $f(x) = x^4 + x^5$.

   e. $f(x) = 4x^4 + 5x^5 - 6x^6$.

17. Find the equation of a quadratic polynomial that has a real root at $x = 2$ and a turning point at $(1, 5)$.

18. A cubic polynomial P has turning points at $(1, 4)$ and $(5, 12)$.

   a. What is the behavior of $P(x)$ as $x \to \infty$?

   b. Where is the point of inflection? (*Hint*: Recall that a cubic is symmetric about its point of inflection.)

19. Suppose that a quadratic polynomial has roots at $x = 6$ and $x = -2$.

   a. Write a possible formula for the quadratic function.

   b. Use the fact that a quadratic is symmetric about the vertical line through its turning point to determine the x-coordinate of the turning point of this quadratic function.

   c. Suppose that the quadratic has a maximum value of 20. What must be its equation?

   d. Suppose that the quadratic has a minimum value of -20 instead. What must be its equation?

20. An apple is tossed from ground level straight up at time $t = 0$ with velocity 64 ft/sec. Its height at time t is $f(t) = -16t^2 + 64t$. Find the time when it hits the ground and the instant when it reaches its highest point. What is the maximum height?

21. The height s (in cm) of an object above the ground at time t (in seconds) is given by

$$s = v_0 t - \frac{1}{2}gt^2,$$

where v_0 represents the initial velocity and g is a constant, the acceleration due to gravity.

   a. At what height does the object start?

   b. How long is the object in the air before it hits the ground?

   c. When will the object reach its maximum height?

   d. What is that maximum height?

22. a. Sketch a smooth graph of today's air temperature from midnight to midnight.

   b. When is it a minimum? A maximum?

   c. When does it have a point of inflection?

   d. What type of polynomial might be a good match to the curve you drew?

e. What function would be a better choice if you expand the domain to include the temperatures for yesterday and tomorrow?

23. Factor the polynomial $P(x) = x^3 - 5x^2 + 3x + 7$, using zeros that are correct to two decimal places.

24. **a.** Prove that any cubic polynomial of the form $f(x) = ax^3 + cx$ is symmetric about its inflection point at the origin by showing algebraically that, if you take any value of x—say, $x = h > 0$—then $f(-h) = -f(h)$.

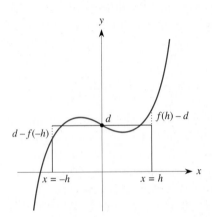

b. Prove that any cubic polynomial of the form $f(x) = ax^3 + cx + d$ is symmetric about its inflection point at $(0, d)$ by showing algebraically that, if you take any value of x—say, $x = h > 0$—then $f(h) - d = d - f(-h)$, as illustrated in the accompanying figure. (*Note:* The same ideas apply to an arbitrary cubic polynomial when the bx^2 term is present, but the proof is considerably more complicated.)

25. Recall that the *average rate of change* of a function f over an interval $x = a$ to $x = b$ (see Section 2.8) is defined as the slope of the line segment connecting the endpoints of the curve on that interval, or

$$\frac{\Delta y}{\Delta x} = \frac{f(b) - f(a)}{b - a},$$

as illustrated in the accompanying figure. The table gives some values for the function $f(x) = x^2 - 4x$.

a. Find the average rate of change of f from $x = -2$ to $x = 3$.

b. Calculate the average rate of change of f between each successive pair of points in the table; that is,

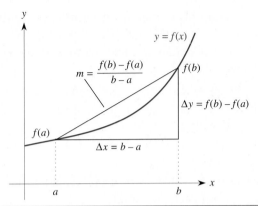

x				-2	-1	0	1	2	3
$y = f(x) = x^3 - 4x$				0	3	0	-3	0	15

between $x = -2$ and $x = -1$, between $x = -1$ and $x = 0$, and so on. What is the average value of all these slopes?

c. Extend the table to include the point $x = 4$ and repeat parts (a) and (b). Does the same result hold?

d. Extend the table farther to include $x = -3$. Show that the same result holds.

e. Does the same result hold for any function and any set of points? State this result as a potential theorem.

26. Prove the result you conjectured in part (e) of Problem 25. Let f be defined on an interval from a to b. The average rate of change of f is

$$\frac{f(b) - f(a)}{b - a}.$$

Let

$$x_0 = a, x_1, x_2, \ldots, x_n = b$$

be any set of uniformly spaced points so that

$$\Delta x = \frac{b - a}{n}.$$

27. Find all polynomials p of degree ≤ 2 that satisfy each set of conditions.

a. $p(0) = p(1) = p(2) = 1$
b. $p(0) = p(1) = 1$ and $p(2) = 2$
c. $p(0) = p(1) = 1$
d. $p(0) = p(1)$

(*Hint:* Think about the graphs.)

Exercising Your Algebra Skills

Factor each of the following polynomial expressions as completely as possible.

1. $x^2 + 7x + 12$
2. $x^2 - 4x - 5$
3. $x^2 - 7x + 12$
4. $x^2 - x - 12$
5. $x^2 + x - 12$
6. $x^2 - 4x + 4$
7. $x^2 - 6x + 9$
8. $x^2 - 25$
9. $x^2 - 100$
10. $x^2 + 36$
11. $x^3 + x^2 - 20x$
12. $x^3 - 4x^2 + 3x$
13. $x^3 + 10x^2 + 25x$
14. $x^3 - 36x$

4.3 Modeling with Polynomial Functions

As we mentioned in Example 4 of Section 3.2, one of the most famous moments in the history of science was Galileo's reported experiment of dropping various objects from the top of the 180-foot high Leaning Tower of Pisa and discovering that they fell at the same rate, regardless of their weight. Instead of looking at the speed of a falling object, we now look at the height H, in feet, of an object falling from the top of the tower at various times t, as given in the table.

Time	0	0.5	1.0	1.5	2.0	2.5	3.0
Height	180	176	164	144	116	80	36

Note how the object starts falling slowly and then accelerates. (Incidentally, these values are considerably more accurate than anything Galileo could have measured at the end of the fourteenth century.)

The ideas we introduced in Chapter 3 on fitting linear, exponential, power, and logarithmic functions to a set of data can be extended to fitting polynomial functions to data. All graphing calculators have the capability to fit quadratic, cubic, and quartic polynomials to any set of data; spreadsheets such as Excel™ can fit polynomials up to degree 6, and specialized software packages allow polynomials of any finite degree. However, the approach used to determine a best-fit polynomial is different from the types of transformations we used in Sections 3.4 and 3.5. In fact, it is based on the idea of fitting a linear function of several variables to a set of data, as we discussed in Section 3.7. As we also discussed there, the correlation coefficient does not apply directly. Instead, statisticians have developed a comparable measure of the goodness of fit, known as the *coefficient of determination*, which is denoted by R^2. Its value is provided by most calculators and software. It always lies between 0 and 1, and the closer R^2 is to 1, the better is the fit; a value of 1 indicates a perfect fit.

EXAMPLE 1

(a) Find an equation for the height of an object falling from the top of the 180-foot high Leaning Tower of Pisa as a function of time. (b) Then use the formula to calculate how long it takes for the object to hit the ground.

Solution

a. We show the scatterplot of the data for height H as a function of time t in Figure 4.27 and observe that the pattern in the data resembles the right half of a parabola with

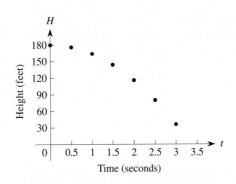

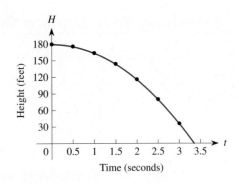

FIGURE 4.27

FIGURE 4.28

negative leading coefficient. Using the quadratic function regression routine on a calculator, we obtain the equation

$$H = -16t^2 + 180.$$

The corresponding value for the coefficient of determination is $R^2 = 1.00$, suggesting that the parabola apparently is a perfect fit to the data, as shown in Figure 4.28.

b. The object hits the ground when $H = 0$. To find how long it takes, we must find the value of t for which

$$H = -16t^2 + 180 = 0.$$

We can solve this quadratic equation graphically, with the quadratic formula, or by direct algebraic means. Algebraically, we add $16t^2$ to both sides of this equation to obtain

$$16t^2 = 180$$

so that

$$t^2 = \frac{180}{16} = 11.25.$$

When we take the square root of both sides, we get $t \approx \pm 3.35$. Because $t = -3.35$ seconds makes no real-world sense, we conclude that it takes about 3.35 seconds for the object to hit the ground.

◆

Let's look at the equation $H = -16t^2 + 180$ for the height at any time when the object is falling from the top of the 180-foot high tower. Note that the constant term 180 equals the height of the tower. We rewrite the function as

$$H(t) = 180 - 16t^2,$$

which indicates that the height starts at 180 feet, when $t = 0$, and decreases thereafter. In general, if an object is dropped from any initial height H_0 and is affected only by the force of gravity, its height at any time t is given by

$$H(t) = H_0 - 16t^2.$$

Now suppose that an object is not simply dropped but instead is tossed upward with some initial velocity—say, 40 ft/sec. What do we expect? Obviously, the object starts off rising until it reaches a maximum height and then falls back until it hits the ground. The larger the initial velocity, the higher the object goes. In Example 2, we construct a function to model such a situation.

EXAMPLE 2

When an object is thrown vertically upward with an initial velocity of 40 ft/sec from the top of the 180-foot high Tower of Pisa, the following set of measurements of its height as a function of time are obtained.

t	0	0.5	1.0	1.5	2.0	2.5	3.0	3.5	4.0	4.5
H	180	196	204	204	196	180	156	124	84	36

a. Find an equation of a function that can be used to model the height of the object as a function of time.

b. Estimate how long it takes for the object to reach its maximum height and what that maximum height is.

c. How long does it take for the object to fall back to the ground?

Solution

a. The scatterplot of the data shown in Figure 4.29 indicates that the pattern for the height H as a function of time t looks like a portion of a parabola with a negative leading coefficient. Using a calculator to fit a quadratic function, we find that the quadratic function that best fits the data is

$$H(t) = 180 + 40t - 16t^2.$$

Note that the coefficients of the constant and linear terms are essentially the same as the initial height 180 feet and the initial velocity 40 feet per second, respectively. Moreover, the coefficient of the quadratic term is the same, -16, as in Example 1. This function superimposed over the scatterplot shown in Figure 4.30 reveals that it is an excellent fit to the data. The associated coefficient of determination is $R^2 = 1.0$, providing additional evidence that the fit is virtually perfect.

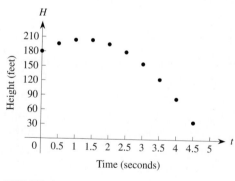

FIGURE 4.29

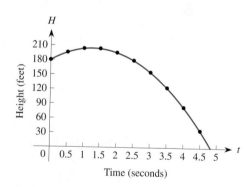

FIGURE 4.30

b. To estimate the time it takes for the object to reach its maximum height and the value for that maximum height, we need merely trace along the curve to find the coordinates of the turning point; or we can use the routine for locating the maximum for a function that is on many calculators. Either way, the coordinates are $t \approx 1.25$ seconds and $H \approx 205$ feet.

c. To find the time it takes for the object to return to the ground, we solve the equation

$$H(t) = 180 + 40t - 16t^2 = 0.$$

We can do this either graphically or by the quadratic formula. Using the quadratic formula with $a = -16$, $b = 40$, and $c = 180$ gives

$$t = \frac{-40 \pm \sqrt{40^2 - 4(-16)(180)}}{2(-16)}$$

$$= \frac{-40 \pm \sqrt{1600 + 11520}}{-32}$$

$$= \frac{-40 \pm \sqrt{13120}}{-32}.$$

Consequently, we get two possible values for t: $t \approx 4.83$ seconds and $t \approx -2.33$ seconds. The second value makes no sense physically, so the realistic solution is $t \approx 4.83$ seconds.

In general, we can say the following.

> The height of an object thrown vertically upward with initial velocity v_0 from an initial height H_0 at any time t is
>
> $$H(t) = -16t^2 + v_0 t + H_0.$$

If there is no initial velocity, so that $v_0 = 0$, this formula reduces to the expression we had previously for the height of any object falling under the influence of gravity.

The questions that we would want to answer about any object thrown upward into the air are:

1. How high does it go?
2. How long does it take to reach its maximum height?
3. How long does it take to return to the ground?

The Path of a Projectile

Picture the path of a long home run in baseball or the path of a perfect pass in football or the arch of the high-pressure stream of water from a supershooter water gun. In each case, the path looks something like the curve shown in Figure 4.31, whose shape suggests a parabola or possibly some higher degree polynomial curve with a negative leading coefficient. (If a strong wind is blowing, the path may not be quite so symmetric and the analysis of the shape of the path is considerably more complicated than that described here.)

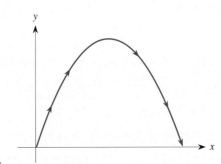

FIGURE 4.31

Using various kinds of technology, such as time-lapse photography or a video camera, we can capture a set of data on the path of such a projectile. For instance, the following set of data consists of measurements for the path of a long fly ball in baseball, where the height y of the ball depends on the distance x from home plate. Both sets of measurements are in feet.

x	0	30	60	90	120	150	180	210	240	270	300	330	360	390
y	4	37	65	88	105	117	123	124	119	109	94	73	47	16

The ball rises to a maximum height of about 124 feet. More important, the ball travels a horizontal distance of about 400 feet until it comes back down into the outfielder's glove, hits the ground or fence, or lands in the stands. To determine what happens, we need an equation for the path of the ball, which we find in Example 3.

EXAMPLE 3

(a) Determine the equation of a function that models the path of the baseball based on the preceding data. (b) If the fence 400 feet from home plate is 8 feet high, will the ball clear the fence to be a home run?

Solution

a. Because the shape of the data, as shown in the scatterplot in Figure 4.32, suggests a parabola, we begin by fitting a quadratic function to the data. The result is the quadratic function

$$y = -0.003x^2 + 1.202x + 3.936,$$

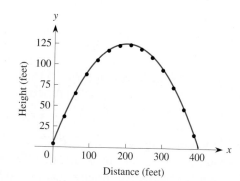

FIGURE 4.32

which is shown superimposed over the scatterplot and is an outstanding fit to the data. As expected, the leading coefficient is negative. Moreover, the corresponding value for the coefficient of determination is $R^2 = 0.9999$, which provides additional evidence that the quadratic function is an excellent model to use.

b. In order for the ball to be a home run, it must clear the 8-foot high fence when it is 400 feet from home plate. Therefore we substitute $x = 400$ into the equation of the parabola and find that

$$y = -0.003(400)^2 + 1.202(400) + 3.936 = 4.736.$$

That is, when it reaches the fence, the ball's height is somewhat less than 5 feet, so it wouldn't be a home run, as shown in the smaller view in Figure 4.33.

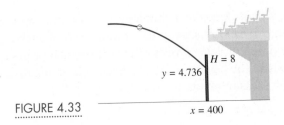

FIGURE 4.33

Fitting Polynomials to Data

The concept of fitting a polynomial function to data is one that applies in all walks of life, not just in the physical situations we encountered in Examples 1–3. We illustrate two other cases in Examples 4 and 5.

EXAMPLE 4

The table shows the accumulated total number of reported cases of AIDS in the United States since 1983.

Year	1983	1984	1985	1986	1987	1988	1989	1990
Number of AIDS Cases	4589	10,750	22,399	41,256	69,592	104,644	146,574	193,878
Year	1991	1992	1993	1994	1995	1996	1997	1998
Number of AIDS Cases	251,638	326,648	399,613	457,280	528,144	594,641	653,084	701,353

Source: U.S. Centers for Disease Control and Prevention.

Determine a function that fits the data well and interpret the behavior of the function.

Solution In Example 4 of Section 3.3, we explored the possibility that the growth in the total number of reported cases of AIDS in the United States follows an exponential pattern. The resulting best-fit exponential function, found with a calculator, was

$$A = 5413.5(1.3626)^t,$$

where t is measured in years since 1980. The corresponding correlation coefficient $r = 0.9483$ is quite close to 1, suggesting that this function is a very good fit. But, when we superimpose this exponential function over the data points, as shown in Figure 4.34, the curve doesn't fit the data well.

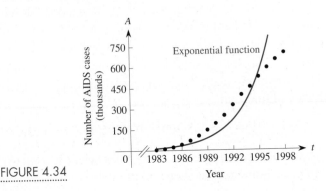

FIGURE 4.34

Alternatively, suppose that we use the capability of the calculator to fit a polynomial to this data. Most calculators allow us to fit polynomials of degree 2, 3, or 4 to a set of data, and we can easily experiment with different degrees. When we do so, we find that a cubic polynomial is an excellent fit to this set of data. The calculator gives the best cubic function, rounded to one decimal place, as

$$A = -221.9t^3 + 9261.8t^2 - 62275.9t + 122988.9,$$

where t is again the number of years since 1980. When we superimpose this polynomial over the AIDS data points shown in Figure 4.35, we get an exceptionally good fit, which certainly is a far better fit than the exponential function shown in Figure 4.34.

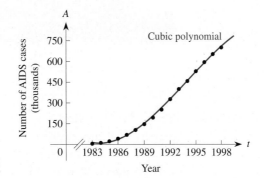

FIGURE 4.35

This graph strongly suggests that the number of cases in the spread of AIDS follows a cubic pattern. (When scientists discovered this several years ago, they were excited because polynomial growth is much slower than exponential growth, which is the trend that they too had expected.) The corresponding coefficient of determination, $R^2 = 0.99996$, provides further evidence of how well the cubic function fits the data.

We know from the formula for the cubic that the leading coefficient is negative, so the cubic will eventually approach $-\infty$. The larger view in Figure 4.36 suggests that the cubic passed its inflection point in about 1995 or 1996 and that the growth in AIDS has begun to slow somewhat since then. The graph also shows that the function will reach a turning point in about 2003.

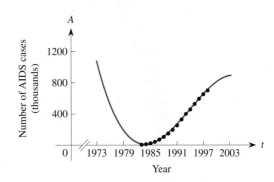

FIGURE 4.36

However, recall that the data represent the total number of AIDS cases reported in the United States, so the cubic can't actually turn and begin to decline; it can only slow and, at best, eventually level off. Thus we demonstrate again how dangerous extrapolation with a mathematical model can be. The model only describes the situation based on the data points; it is not a guarantee of the actual process, especially for extrapolating into the future or the past.

Let's look at another example of fitting polynomials to data. Figure 4.37 shows a picture of the famous Gateway Arch in St. Louis. Its shape suggests a portion of a downward opening parabola. Let's see if we can determine a specific function that best models the arch.

FIGURE 4.37

EXAMPLE 5

Determine a polynomial function that fits the Gateway Arch well.

Solution To find an appropriate function, we need some measurements for the arch. Overall, the arch stands 630 feet tall, and the distance between its two legs also is 630 feet. We superimpose a grid on the arch, as shown in Figure 4.38, and choose the coordinate system so that the vertical axis passes through the center of the arch. We then construct the following table of estimates of the height H corresponding to various horizontal distances x. We make our estimates from the middle of the arch; slightly different results might occur if we use values from the inner edge or the outer edge. We ask you to investigate these possibilities in the problems at the end of this section.

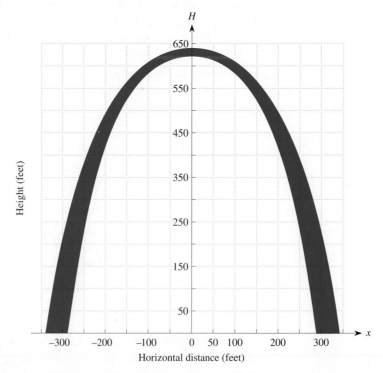

FIGURE 4.38

x	-325	-300	-250	-200	-150	-100	0	100	150	200	250	300	325
H	0	100	330	500	570	610	630	610	570	500	330	100	0

The first thing we notice from both the figure and the table is that the measurements are symmetric about the vertical axis $x = 0$. As a result, we would expect that the best-fit parabola has no x term. When we enter the data into the quadratic regression routine of a calculator, we find the quadratic function that best fits the data is

$$H = -0.0064x^2 + 0x + 699.01.$$

We plot this function over the data points, as illustrated in Figure 4.39, and conclude that it is a reasonably good fit, though certainly not a great one. Among other things, the curve rises much too high above the central data point and the pattern of data points flattens out far more than the parabola does near the center.

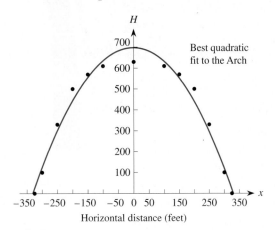

FIGURE 4.39

In our discussion of power functions with integer powers in Section 2.7, we pointed out that the higher the power, the flatter the curve as it passes through the origin. This result suggests that we should use a higher degree polynomial than a quadratic. From the basic shape of the arch, we know that a cubic would not be appropriate—it doesn't have the correct behavior. How about a quartic polynomial? When we try it, the calculator responds with the equation

$$H = (-3.27 \times 10^{-8})x^4 + 0x^3 - 0.00282x^2 + 0x + 644.25.$$

When we superimpose this function over the data points, as shown in Figure 4.40, it appears visually to be an exceptionally good fit to the shape of the arch. The coefficient of determination for this fit is $R^2 = 0.9953$, which also indicates that it is a very good fit. (Actually, the true shape of the arch is a curve known as a hyperbolic cosine, which you may encounter in calculus.)

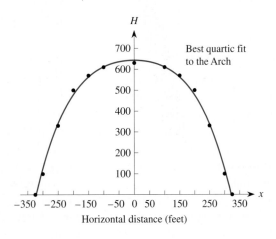

FIGURE 4.40

Problems

1. We showed in the text that the cubic function

   $$y = -221.9t^3 + 9261.8t^2 - 62275.9t + 122988.9$$

   is an excellent fit to the total number of reported cases of AIDS in the United States from 1983 to 1998, where t is the number of years since 1980.

   a. Based on this model, what is the prediction for the total number of cases through 2000?

   b. Check a recent copy of the *Statistical Abstract of the United States* or an almanac to see how accurate the prediction in part (a) is.

   c. If this cubic pattern continues, how many total cases would you expect by 2004?

   d. When would you expect a total of 850,000 cases of AIDS, based on this model?

2. Find the equations of the best quadratic and quartic functions to fit measurements taken at the outer edge of the Gateway Arch instead of at the middle.

3. Repeat Problem 2 with measurements taken at the inner edge of the arch instead of at the middle.

4. The table shows the percentage of the U.S. population that is foreign born in various years.

   a. What is the minimum degree polynomial that you would use to model this data?

   b. Find that polynomial and use it to estimate the time when the percentage of foreign-born people in the United States was a minimum. What was that minimum percentage?

Year	1950	1960	1970	1980	1990	2000
Percentage	6.9	5.4	4.8	6.2	7.9	10.4

Source: *2000 Statistical Abstract of the United States.*

5. The accompanying figure shows a grid superimposed on the image of the McDonald's arches.

   a. Decide on a scale that you can use to estimate measurements on the arches. (*Hint*: Think about where you want to set up your coordinate axes.)

   b. Use your estimated measurements to determine the equation of a polynomial that best fits one of the arches. (*Hint*: Think again about where you want to set up your coordinate axes.)

   c. Can you use the formula you obtained for one of the arches to construct a formula for the other arch? Explain.

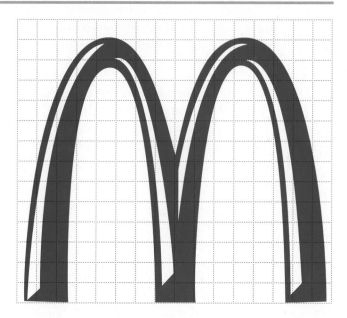

6. The table gives the horsepower generated on a Chevy 383 car engine at different rpm.

Horsepower	138	172	203	216	
Revolutions per Minute	2000	2500	3000	3500	
Horsepower	209	182	144	98	42
Revolutions per Minute	4000	4500	5000	5500	6000

Source: Student project.

   a. Which variable is the independent variable and which is the dependent variable?

   b. What is the equation of the quadratic function that relates these two quantities?

   c. What does your model predict for the horsepower generated by this engine at 4800 rpm?

   d. If the engine puts out 165 horsepower, what is the possible value for the rpm according to this model?

7. Car enthusiasts know that it's not horsepower that is significant, but rather the amount of torque that an engine puts out that really matters in how quickly a car moves forward. The table gives the torque, in foot-pounds, generated at different rpm values for a Chevy 383 engine. From among the usual families of functions (linear, exponential, power,

quadratic, and cubic), find the one that seems to be the best fit to these data.

Torque	363	361	355	324	
Revolutions per Minute	2000	2500	3000	3500	
Torque	275	213	151	93	36
Revolutions per Minute	4000	4500	5000	5500	6000

Source: Student project.

8. **a.** Create a single table based on the information given in Problems 6 and 7 relating the amount of torque generated to the horsepower for the Chevy 383 engine.
    b. From among the usual families of functions (linear, exponential, power, quadratic, and cubic), find the one that seems to be the best fit to this data.

9. The table shows the number of 18- to 24-year-olds in the United States in recent years. Find the quadratic function that best fits this data set and use it to predict the number of people in this age range in **(a)** 2000 and **(b)** 2005. Which prediction would you have more confidence in?

Year	1970	1975	1980
Population (millions)	24.71	28.76	30.35
Year	1985	1990	1995
Population (millions)	29.48	26.14	24.85

Source: *2000 Statistical Abstract of the United States.*

10. According to the theory of relativity, the mass M of an object increases as its velocity v increases so that $M = f(v)$. Suppose that the mass of an object is 1 unit when it is at rest ($v = 0$). The table gives the mass of the object at different speeds that are expressed as fractions of c, the speed of light (about 186,280 miles per second). Find the best quadratic fit to this set of data.

Velocity (fraction of c)	0	0.1	0.2
Mass	1	1.0050	1.0206
Velocity (fraction of c)	0.3	0.4	0.5
Mass	1.0483	1.0911	1.1547

11. While approaching the Verrazano Bridge in New York City, Ken noticed that the main cable looks like a parabola, as illustrated in the accompanying figure. As his car crawled across the bridge in heavy traffic, he estimated the following heights, in feet, of the cable above the road and the distance, in feet, starting from one of the vertical support columns.

Distance from Support Column	0	1000	2150
Estimated Height	500	150	20
Distance from Support Column	3000	4000	4300
Estimated Height	100	400	500

Find an equation of the parabola that best fits Ken's estimates. (Think how to set up the coordinate axes.)

12. The table shows the price of a barrel of oil, in dollars, in different years.

Year	1960	1970	1975	1980
Price	11	9	37	64
Year	1985	1990	1995	2000
Price	40	28	20	32

Source: Lester R. Brown et al., *Vital Signs 2000: The Environmental Trends That Are Shaping Our Future.*

   a. What type of function is reasonable to use as a model for the price of oil as a function of time?
   b. Find the equation of the polynomial function of appropriate degree to fit the data.
   c. What does your model predict for the price of a barrel of oil in 2005?
   d. Use the graph of your function to estimate the location of the turning points for the function. According to this model, what was the maximum price of a barrel of oil between 1960 and 2000 and when did it occur? What was the minimum price and when did it occur?

13. The table shows the trend in worldwide grain production (wheat, rice, and corn, primarily), in kilograms per person. The pattern in the data suggests that a quadratic function is an appropriate model for grain production per person as a function of the year.

Year	1965	1970	1975	1980
Amount per Person	270	291	303	321
Year	1985	1990	1995	1999
Amount per Person	339	335	301	309

Source: Lester R. Brown et al., *Vital Signs 2000: The Environmental Trends That Are Shaping Our Future.*

a. Find the equation of the quadratic that best fits these data.
b. Based on the model, what was the maximum level of grain production per person worldwide?
c. What does the model predict for the amount of grain produced per person in 2010?
d. Write a paragraph describing the possible reasons for this trend and the implications if the trend continues.

14. The table gives the total number, in thousands, of high school graduates in the indicated years since 1900. In Problem 12 of Section 3.3, we asked you to find the best linear, exponential, and power functions to fit these data. If you examine the data carefully, you should expect that a polynomial function would be a better fit.

Year	1900	1910	1920	
High School Grads	95	156	311	
Year	1930	1940	1950	1960
High School Grads	667	1221	1200	1858
Year	1970	1980	1990	2000
High School Grads	2889	3043	2586	2839

Source: *Digest of Education Statistics 2000*, U.S. Department of Education.

a. What degree polynomial function is a good candidate to fit these values? Explain.

b. Let t be the number of years since 1890. Determine the best polynomial function of the degree that you decided was appropriate in part (a) to model the number of high school graduates as a function of time t.
c. Use this function to predict the number of high school graduates in 2010.
d. Use this function to predict the year in which there will be 5 million high school graduates.

15. The table, collected from a chemistry lab experiment, gives the density D of water, in grams per milliliter, at various temperatures T, in °C.

Temperature, T	0°	4°	10°
Density, D	0.99987	1.00000	0.99973
Temperature, T	20°	30°	40°
Density, D	0.99823	0.99567	0.99224
Temperature, T	60°	80°	100°
Density, D	0.98324	0.97183	0.95838

Source: John R. Holum, *Elements of General and Biological Chemistry*, 8th ed. New York: John Wiley & Sons, 1991.

a. Find a quadratic function that fits these data.
b. Use your function from part (a) to find the density of water at 70°C.
c. Find the temperature at which the density of water is 0.99100 grams per milliliter.

16. The height of an object falling from an initial height of y_0 is given by the formula

$$y = y_0 - 16t^2,$$

with units of feet and seconds. What is the equivalent formula based on the metric system of units with meters and seconds? (*Hint*: 1 foot = 0.3048 meters.)

17. Galileo conducted his famous experiment in which he dropped objects from the top of the 180-foot high Leaning Tower of Pisa in about 1590. His goal was to obtain experimental data to show that all bodies fall with equal velocities. How long did it take for the objects that he dropped from the tower to hit the ground?

18. The Eiffel Tower is 300 meters tall. How long would it take an object dropped from its top to hit the ground?

4.4 The Roots of Polynomial Equations: Real or Complex?

The Roots of Quadratics

In Section 4.1, we stated that, for any quadratic equation,

$$ax^2 + bx + c = 0, \qquad a \neq 0,$$

we can always find its roots by using the quadratic formula

$$x = \frac{-b \pm \sqrt{b^2 - 4ac}}{2a}.$$

Further, the roots may be two distinct real numbers, a repeated real root, or a pair of complex conjugate numbers of the form $\alpha \pm \beta i$, where $i = \sqrt{-1}$, as illustrated in Figure 4.41.

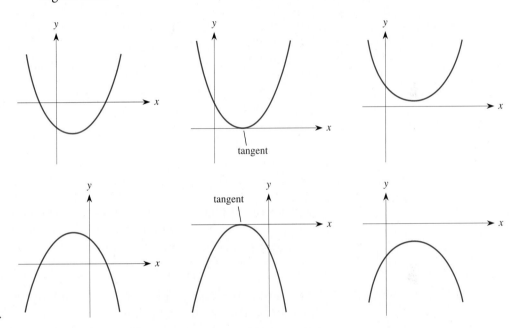

FIGURE 4.41

Most students think that complex roots occur very rarely. In this section we investigate how frequently they do arise. To do so, we consider many different quadratic equations and find the percentage of them that do have complex roots. A quadratic equation $ax^2 + bx + c = 0$ has complex roots when its discriminant, $b^2 - 4ac$, is negative. The quadratic formula then requires taking the square root of that negative discriminant to produce two complex numbers. For instance, for the quadratic equation $x^2 - 2x + 2 = 0$, the discriminant is $(-2)^2 - 4(1)(2) = -4$, so the roots will be complex. The quadratic formula gives the roots as

$$x = \frac{-(-2) \pm \sqrt{4 - 8}}{2} = \frac{2 \pm \sqrt{-4}}{2} = \frac{2 \pm 2i}{2} = 1 \pm i,$$

or $x = 1 + i$ and $x = 1 - i$. Thus we can use the sign of the discriminant as the criterion to decide whether any particular quadratic has complex roots.

To come to any meaningful conclusions about the percentage of quadratics that have complex roots, we must examine a very large number of quadratics. Doing so requires using a computer or calculator program rather than hand computation.

Even the simplest case—when the quadratic has integer coefficients—has infinitely many possible quadratics, so the best we can do is examine a finite selection of them. Let's examine all possible quadratics $y = ax^2 + bx + c$ where the coefficients a, b, and c are integers from 0 to 5, say, but $a \neq 0$. We write this in *interval notation* as $[0, 5]$. We then use a computer program that considers all possible integer values for a, b, and c in this interval and keeps track of how many of the quadratics have complex roots, using the discriminant criterion. Similarly, we can investigate all possible integer coefficients in various other intervals, the results of which are shown in Table 4.1.

TABLE 4.1

Interval for a, b, and c, $a \neq 0$	Percentage with Complex Roots
All in $[0, 5]$	70
All in $[0, 10]$	73
All in $[0, 20]$	74
All in $[0, 50]$	74.5
All in $[-3, 3]$	37.4
All in $[-5, 5]$	37.5
All in $[-10, 10]$	37.8
All in $[-20, 20]$	37.7
All in $[-50, 50]$	37.5
$[0, 5], [0, 5], [-5, 0]$	0
$[0, 5], [-5, 0], [0, 5]$	70

Therefore, rather than being a rarity, complex roots actually occur with surprising frequency. In fact, almost three-fourths of quadratics whose coefficients are all nonnegative integers have complex roots. Even allowing for negative values almost 40% have complex roots.

Think About This

There is one exception in Table 4.1. If the constant coefficient c is negative while a and b are both positive, the quadratic apparently always has two real roots. Can you explain why? Can you give another example where the quadratic always has two real roots? Look at the discriminant. (Note that we have checked only specific integer values for a and b between 0 and 5 and c between -5 and 0, so we can't generalize to what may happen over all similar intervals of values.) ▱

We suggest that you conduct your own investigations of these ideas if an appropriate program is available or if you want to write a fairly short program for your calculator. Think about the following questions.

♦ With integer coefficients, what happens as the size of the interval increases? Does the frequency of complex roots stay roughly the same or does it increase or decrease significantly?

♦ What happens if you use different ranges of values for each coefficient?

Don't be too generous in your choices when you begin; such systematic processes tend to take a long time. For example, if you want to check all quadratics where a, b, and c are integers between 0 and 10, say, you are actually having the computer or calculator investigate 1210 different equations. (There are 10 possible values for *a* since the equation would not be quadratic if *a* were zero. There are 11 possible values for *b* and 11 for *c*, which leads to $10 \times 11 \times 11 = 1210$ different cases.) If you ask for all integers from 0 to 100 on each of the coefficients, the computer or calculator will investigate 100×101^2 different quadratics. It may take all night to complete this study of more than one million cases.

We should also find out what happens when the quadratic has noninteger coefficients, either rational numbers or irrational numbers. In such cases, we can't simply check all possible equations because there are infinitely many possibilities, even for any finite interval. Instead, we use a random selection process to generate large numbers of quadratics with randomly selected (noninteger) coefficients in desired intervals, test each for the nature of its roots, and keep track of how many of the roots are complex. (We perform just such an analysis in Supplementary Section 11.3 as part of our study of probability.)

The Roots of a Cubic Function

We next consider an arbitrary cubic equation

$$ax^3 + bx^2 + cx + d = 0,$$

where *a*, *b*, *c*, and *d* are any four real numbers and $a \neq 0$. Recall that, just as a quadratic equation has two roots, a cubic equation has three. They can be either real or complex roots. Recall also that any complex roots must occur as a pair of complex conjugates, $\alpha + \beta i$ and $\alpha - \beta i$. Thus, for any cubic equation, the three roots may be either three real numbers, or a single real number and a pair of complex conjugate numbers.

Moreover, we know that the real roots correspond geometrically to points where the cubic crosses the *x*-axis. If there are three distinct real roots, the cubic crosses the *x*-axis in three places, as illustrated in Figures 4.42(a) and 4.42(b). If there is a double real root and a separate real root, the *x*-axis is tangent to the cubic at the point corresponding to the double root and the curve crosses the *x*-axis at

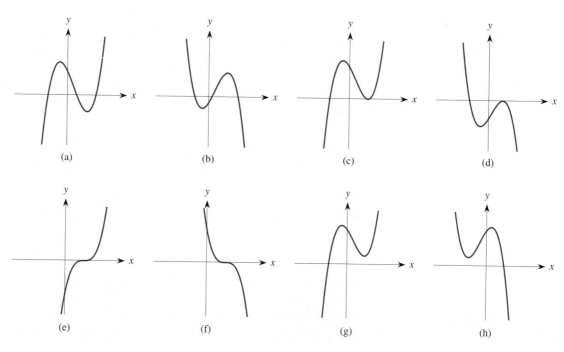

FIGURE 4.42

the point corresponding to the other real root, as depicted in Figures 4.42(c) and 4.42(d). If there is a triple real root (as with $y = x^3$), the cubic flattens as it crosses the x-axis at the single point, as shown in Figures 4.42(e) and 4.42(f). Finally, if there is a single real root and a pair of complex conjugate roots, the cubic crosses the x-axis once, as illustrated in Figures 4.42(g) and 4.42(h). Thus a cubic can have either three real roots or one real root.

We have demonstrated that quadratic equations are likely to have complex roots. How likely is it for a cubic equation to have complex roots? To answer this question, we again use a computer program to investigate many different cubics. First, though, we must devise a test comparable to using the sign of the discriminant in the quadratic formula to decide whether a particular cubic has complex roots.

Suppose that a cubic has three real roots. In that case, the curve crosses the x-axis at three points if the three roots are distinct, it crosses the axis at two points if there is a double real root, and it crosses the axis at one point if there is a triple real root. The cubics shown in Figure 4.43 all have the same shape; the only difference is the height of the turning points. The cubic on the left has its first turning point above the x-axis and its second below; therefore it has three real roots. The second cubic has both turning points above the x-axis and so must have one real root and a pair of complex roots. The third cubic has both turning points below the x-axis, so it also must have one real root and a pair of complex roots.

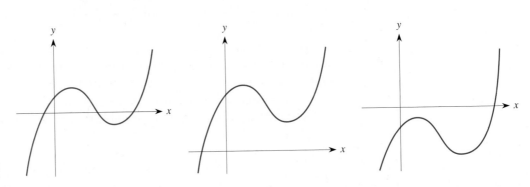

FIGURE 4.43

A further case occurs when the x-axis is tangent to the curve at one of the turning points; such a cubic has a double real root, so it cannot have a pair of complex roots and its third root must be real. The final case is when the two turning points coincide along the x-axis; this case corresponds to a triple real root. Therefore, in order to have two complex roots, a cubic must have both turning points above the x-axis or both below it.

When you study calculus, you will be able to determine that the two turning points of the cubic $y = ax^3 + bx^2 + cx + d = 0$ are located at

$$x = \frac{-b \pm \sqrt{b^2 - 3ac}}{3a},$$

provided that $b^2 - 3ac \geq 0$. (This formula clearly resembles the quadratic formula.)

Think About This   Verify graphically that this formula gives the approximate location of the turning points of the cubic $y = x^3 - 4x^2 + 4x + 5$. ▭

Call these two x-values x_1 and x_2. Because we know the equation of the cubic curve,

$$y = f(x) = ax^3 + bx^2 + cx + d,$$

we can determine the heights of the two turning points:

$$y_1 = f(x_1) \quad \text{and} \quad y_2 = f(x_2).$$

Once we have calculated these values, we need only check whether both are positive or both are negative to conclude that the cubic has complex roots, as illustrated in Figure 4.44. If the two y-values have opposite signs or if either is zero, the cubic has three real roots. We use this criterion in our investigation.

We apply this criterion to cubics with integer coefficients a, b, c, and d within various intervals of values. In Supplementary Section 11.3 we investigate cases with randomly generated noninteger values for a, b, c, and d within any desired intervals of values, provided that $a \neq 0$.

In Table 4.2 we list the results of performing this investigation with all possible *integer* coefficients in the indicated intervals of values. This table indicates that a cubic with integer coefficients seems even more likely to have complex roots than a quadratic does.

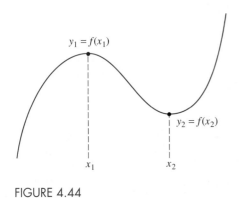

FIGURE 4.44

TABLE 4.2

Intervals for a, b, c, and d, $a \neq 0$	Percentage of Complex Roots
All in $[0, 5]$	94.54
All in $[-3, 3]$	78.43
All in $[-4, 4]$	78.74
All in $[-5, 5]$	78.93
$[0, 4], [0, 4], [0, 4], [-4, 0]$	88.4
$[0, 4], [0, 4], [-4, 0], [0, 4]$	74.8
$[0, 4], [-4, 0], [0, 4], [0, 4]$	88.4
$[0, 4], [0, 4], [-4, 0], [-4, 0],$	44
$[0, 4], [-4, 0], [-4, 0], [0, 4]$	44
$[0, 4], [-4, 0], [-4, 0], [-4, 0]$	74.8

Think About This   In intervals of the form $[-k, k]$ for all four coefficients, the proportion of complex roots seems to be essentially the same regardless of the value of k. Does that make sense? Imagine what would happen if you have a particular cubic and multiply each coefficient by 10, say. Wouldn't you expect the same type of roots? In fact, wouldn't you expect the identical roots? ▭

Think About This   When we studied the nature of the roots of quadratics, we saw that the two roots are always real whenever $c < 0$ and $a > 0$. Are there any simple combinations of values for the coefficients a, b, c, and d in a cubic that likewise guarantee real roots? (What about $d = 0$, $c < 0$, and $a > 0$?) ▭

It turns out that for polynomials of higher degree, the likelihood of complex roots is even greater than for quadratics or cubics, but we won't investigate these cases.

Using Information on the Nature of the Roots

We next turn to an application for which knowing the nature of the roots of a polynomial is crucial. Home thermostats and automobile cruise controls are examples of *control systems* that engineers use to control a process. In such devices, when the system deviates slightly from the specified level, it should return to that level automatically—the temperature shuts off or the car stops accelerating. Such a system is called *stable*. Often, control systems are described mathematically by a polynomial. A control system is stable if

1. all the real roots are negative, and
2. all the complex roots have negative real parts.

A control system having any positive real roots or having complex roots whose real parts are positive is *unstable*. That is, the system does not return to the specified level when small changes are introduced.

EXAMPLE

A control system is described by the cubic polynomial $P(s) = s^3 + 3s^2 + 4s + 2$. Determine whether the system is stable or unstable.

Solution   The graph of this cubic polynomial is shown in Figure 4.45. Its associated cubic equation $s^3 + 3s^2 + 4s + 2 = 0$ has only one real root, so it must therefore have a pair of complex conjugate roots. Moreover, it is evident that the real root is negative. If we zoom in on the point where the curve crosses the s-axis, we find that the root appears to be located near $s = -1$. We can determine whether the root is $s = -1$ exactly by evaluating

$$P(-1) = (-1)^3 + 3(-1)^2 + 4(-1) + 2$$
$$= -1 + 3 - 4 + 2 = 0,$$

which shows that the root is precisely $s = -1$.

The problem we now face is to determine the complex roots. We know the real root $s = -1$, so the corresponding linear factor is $(s + 1)$. We can therefore factor the polynomial by dividing it by $(s + 1)$, using the technique of long division for polynomials from algebra:

$$
\begin{array}{r}
s^2 + 2s + 2 \\
(s+1)\overline{\smash{\big)}\,s^3 + 3s^2 + 4s + 2} \\
\underline{s^3 + s^2} \\
2s^2 + 4s \\
\underline{2s^2 + 2s} \\
2s + 2 \\
\underline{2s + 2} \\
0
\end{array}
$$

Thus, $(s^2 + 2s + 2)$ is the quadratic factor, so that the original cubic polynomial is

$$P(s) = s^3 + 3s^2 + 4s + 2 = (s + 1)(s^2 + 2s + 2).$$

We now apply the quadratic formula to find the complex roots of the quadratic factor:

$$s = \frac{-2 \pm \sqrt{2^2 - 4 \cdot 1 \cdot 2}}{2 \cdot 1}$$

$$= \frac{-2 \pm \sqrt{-4}}{2} = \frac{-2 \pm 2i}{2}.$$

The two complex roots are therefore $s = -1 + i$ and $s = -1 - i$. Because the real parts of both complex roots are negative, and the real root is negative also, the control system is stable.

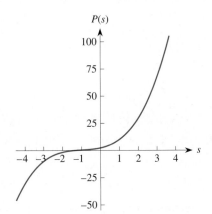

FIGURE 4.45

Problems

1. Consider all quadratics with $a = 1$ and the coefficient of the linear term $b = 0$ so that they take the form $y = x^2 + c$. What percentage of these quadratics should have two real roots?

2. Consider all quadratics of the form $y = ax^2 + bx$ with $c = 0$. What percentage of them should have real roots?

3. Show that, if each coefficient in the quadratic $y = ax^2 + bx + c$ is multiplied by 10, the resulting discriminant is multiplied by 100. What would you expect to happen to the discriminant if each coefficient were multiplied by the same number k? How do the roots of the two quadratics compare?

4. Consider all fourth degree polynomials of the form
$$y = ax^4 + bx^3 + cx^2 + dx + e,$$
where, for simplicity, you may consider $a > 0$.

   a. Based on the general graph shown without axes, how likely do you think it is (roughly 10%, 25%, 50%, 75%, or 90%) for such a polynomial to

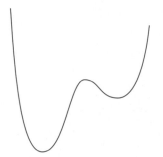

have four real roots? four complex roots? Explain your answers.

   b. How would your answers change if $e = 0$?

5. For each of the following cubic equations, use your function grapher to produce the graph and zoom in to estimate where the two turning points are located. Then apply the formula
$$x = \frac{-b \pm \sqrt{b^2 - 3ac}}{3a}$$
from the text (based on $y = ax^3 + bx^2 + cx + d$) to verify that the values given by the formula match the points you found graphically.

   a. $y = x^3 + 4x^2 - 8x + 3$
   b. $y = x^3 - 7x^2 - 2x + 6$
   c. $y = 5x^3 - 3x^2 - 6x + 8$
   d. $y = -4x^3 + 3x^2 + 5x - 4$
   e. $y = -4x^3 + 3x^2 - 5x - 4$

6. a. Determine the location of the turning points for the cubic $y = x^3 - 3x^2 + 2x + 10$. What are the maximum and minimum values for this function?

   b. Use the fact that a cubic is symmetric about its point of inflection to determine the location of the point of inflection of the cubic in part (a).

7. If a different control system is described mathematically by each polynomial, determine whether it is stable or unstable.

a. $P(s) = s^2 + 6s + 8$
b. $P(s) = s^2 + 5s - 12$
c. $P(s) = s^2 + 5s + 3$

d. $P(s) = s^3 - 4s^2 - 12s$
e. $P(s) = s^3 + 3s^2 + 7s + 5$

4.5 ⋯⋯Finding Polynomial Patterns

In Section 2.2, we developed a criterion for determining whether a set of m points $(x_1, y_1), (x_2, y_2), \ldots, (x_m, y_m)$ follows a linear pattern when the x-values are uniformly spaced.

> A set of points lies on a line if the differences between successive y-values are all equal when the x-values are uniformly spaced. The slope of that line
>
> $$m = \frac{\Delta y}{\Delta x}$$
>
> is the constant difference between successive y-values divided by the uniform spacing between successive x-values.

x	y	Δy
0	1	
		1
1	2	
		3
2	5	
		5
3	10	
		7
4	17	
		9
5	26	

We now consider the related problem of determining whether a set of points follows a quadratic, a cubic, or a higher degree polynomial pattern. Suppose that we have the points $(0, 1), (1, 2), (2, 5), (3, 10), (4, 17)$, and $(5, 26)$, which actually lie on the parabola $y = x^2 + 1$. We construct the table at the left of differences of the y-values. Obviously, the Δy values are not constant. In fact, they clearly follow a linear pattern because the differences between successive Δy values (the differences of the differences) are all constant. The differences of the differences, $\Delta(\Delta y)$, are called the *second differences* and are written $\Delta^2 y$. If we extend the previous table to include the second differences of the y-values, as shown in the table below, we get a constant value for all the second differences.

In general, we have the following criterion based on uniformly spaced x-values.

> A set of points $(x_1, y_1), (x_2, y_2), \ldots, (x_m, y_m)$ lies on a quadratic $y = ax^2 + bx + c$ if the second differences of the y-values are all constant when the x-values are uniformly spaced.

x	y	Δy	$\Delta^2 y$
0	1		
		1	
1	2		$2 = 3 - 1$
		3	
2	5		2
		5	
3	10		2
		7	
4	17		2
		9	
5	26		

In the problems at the end of this section we ask you to explore the significance of this constant second difference.

EXAMPLE 1

Show that the points $(0, 2)$, $(1, 0)$, $(2, 4)$, $(3, 14)$, $(4, 30)$, and $(5, 52)$ lie on a parabola. Then find the equation of the parabola by using regressions methods.

Solution   We construct a table of second differences.

x	y	Δy	$\Delta^2 y$
0	2		
		$-2 = 0 - 2$	
			$6 = 4 - (-2)$
1	0		
		4	
			6
2	4		
		10	
			6
3	14		
		16	
			6
4	30		
		22	
5	52		

Because the differences of the differences are constant, the points follow a quadratic pattern of the form

$$y = ax^2 + bx + c,$$

where the coefficients a, b, and c must be determined.

Thinking of the points as data values and using the curve fitting routines of a calculator, we find that the quadratic function that best fits the data is

$$y = 3x^2 - 5x + 2.$$

The corresponding coefficient of determination is $R^2 = 1$, which suggests a perfect fit to the data. Figure 4.46 shows that the graph of this parabola apparently passes through all six points. Test this result by substituting each value in the formula that we created.

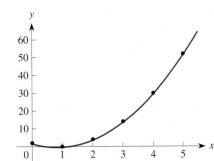

FIGURE 4.46

Alternatively, we could find the equation of the quadratic function that fits these points by using algebraic methods, as we demonstrate in Example 2.

EXAMPLE 2

Find the equation of the parabola that passes through the points $(0, 2)$, $(1, 0)$, $(2, 4)$, $(3, 14)$, $(4, 30)$, and $(5, 52)$, using algebraic methods.

Solution As in Example 1, we have to find the three coefficients a, b, and c in the equation of the quadratic function $y = ax^2 + bx + c$. Substituting the coordinates from the first point $x = 0$ and $y = 2$ gives

$$2 = a \cdot (0) + b \cdot (0) + c,$$

so $c = 2$ and therefore the equation of the parabola becomes $y = ax^2 + bx + 2$. Using the second point $(1, 0)$, we get

$$0 = a \cdot (1^2) + b \cdot (1) + 2 = a + b + 2,$$

and so

$$a + b = -2 \tag{1}$$

Using the third point $(2, 4)$, we get

$$4 = a \cdot (2^2) + b \cdot (2) + 2$$
$$= 4a + 2b + 2,$$

or

$$4a + 2b = 2.$$

Dividing both sides of this equation by 2 yields

$$2a + b = 1. \tag{2}$$

Equations (1) and (2) are a system of two linear equations in two unknowns.

We can solve for a and b by using the usual algebraic methods. We subtract Equation (1) from Equation (2) to get

$$a = 3.$$

Substituting this value into Equation (1) gives

$$3 + b = -2 \quad \text{or} \quad b = -5.$$

So, as before, the desired quadratic is

$$y = 3x^2 - 5x + 2.$$

You can easily verify that the last three points satisfy this function.

Alternatively, we can solve this system of two equations in two unknowns by using the matrix methods described briefly in Appendix C and also find that $a = 3$ and $b = -5$. Thus the equation of the parabola again is $y = 3x^2 - 5x + 2$.   ◆

We can extend these ideas to develop similar criteria for deciding when a set of m points $(x_1, y_1), (x_2, y_2), \ldots, (x_m, y_m)$ follow a polynomial pattern of degree n for any n. For instance, we have the following criterion for $n = 3$.

> A set of m points (x_1, y_1), (x_2, y_2), ..., (x_m, y_m) lies on a cubic $y = ax^3 + bx^2 + cx + d$ if the third differences (differences of the differences of the differences) of the y-values are all constant when the x-values are uniformly spaced.

Think About This   Show that the points $(-3, -17)$, $(-2, 0)$, $(-1, 5)$, $(0, 4)$, $(1, 3)$, $(2, 8)$, and $(3, 25)$ lie on a cubic polynomial by creating a difference table that extends to the third differences. ◻

Sums of Integers

We use the preceding ideas on differences and polynomial patterns to develop a number of formulas involving sums of numbers that arise frequently in mathematics. Among them are the sum of the first n integers

$$1 + 2 + 3 + \cdots + n$$

and the sum of the squares of the first n integers

$$1^2 + 2^2 + 3^2 + \cdots + n^2.$$

Let's begin with the expression for the sum of the integers. We let S_n denote the sum of the first n integers:

$$S_n = 1 + 2 + 3 + \cdots + n.$$

For instance, $S_4 = 1 + 2 + 3 + 4 = 10$. We want a formula for S_n for any value of n. We derive it in two ways.

The first is a particularly simple way that involves a nice trick. If

$$S_n = 1 + 2 + 3 + \cdots + (n - 2) + (n - 1) + n,$$

we can also write this sum in the reverse order as

$$S_n = n + (n - 1) + (n - 2) + \cdots + 3 + 2 + 1.$$

We now add these two equations together term by term in the following way:

$$S_n + S_n = [1 + n] + [2 + (n - 1)] + [3 + (n - 2)] + \cdots + [(n - 1) + 2] + [n + 1]$$
$$= \underbrace{(n + 1) + (n + 1) \qquad + (n + 1) + \cdots \qquad + (n + 1) + (n + 1).}_{n \text{ times}}$$

Because there are n of these terms on the right side, we have

$$2S_n = n(n + 1).$$

Dividing both sides by 2, we obtain

$$S_n = \frac{n(n + 1)}{2},$$

which gives the following general result.

> The sum of the first n integers is
>
> $$1 + 2 + 3 + \cdots + n = \frac{n(n + 1)}{2}. \qquad \textbf{(3)}$$

EXAMPLE 3

Find the sum of the first 100 integers: $1 + 2 + 3 + \cdots + 100$.

Solution Using Formula (3) with $n = 100$, we get

$$1 + 2 + 3 + \cdots + 100 = \frac{100(101)}{2} = 5050.$$

We can also write Formula (3) in *summation notation* (see Appendix A3):

$$\sum_{k=1}^{n} k = 1 + 2 + 3 + \cdots + n = \frac{n(n+1)}{2}.$$

Alternatively, we can derive this result by using either the ideas on fitting functions to data or algebraic methods, as shown in Example 4. The advantage of deriving this formula in other ways is that it demonstrates techniques that can be applied to more complicated cases for which the trick we used previously doesn't work.

EXAMPLE 4

Derive the formula for the sum of the first n integers by using (**a**) curve fitting methods and (**b**) algebraic methods.

Solution We again write the sum of the first n integers as $S_n = 1 + 2 + 3 + \cdots + n$. Thus the sum of the first integer is $S_1 = 1$; the sum of the first two integers is $S_2 = 1 + 2 = 3$; the sum of the first three integers is $S_3 = 1 + 2 + 3 = 6$, and when we continue $S_4 = 10$, $S_5 = 15$, $S_6 = 21$, and so on. If we form a table of second differences with these entries, we get the following.

The second differences $\Delta^2 S_n$ are all constant, so the desired pattern is a quadratic function of n. Thus $S_n = an^2 + bn + c$, where a, b, and c are constants that we must now determine.

a. Using the regression features of a calculator, we find that the quadratic function that best fits these points is

$$S_n = 0.5n^2 + 0.5n + 0$$
$$= \frac{1}{2}n^2 + \frac{1}{2}n$$
$$= \frac{1}{2}n(n+1),$$

as before. All the points lie on this curve, as we can verify by substituting the coordinates of the points into the equation.

b. Using the point $n = 1$, $S_1 = 1$ in the quadratic function $S_n = an^2 + bn + c$, we get

$$S_1 = 1 = a \cdot (1^2) + b \cdot (1) + c,$$

and so

$$a + b + c = 1.$$

When $n = 2$, we have $S_2 = 3$ so that

$$S_2 = 3 = a \cdot (2^2) + b \cdot (2) + c,$$

and hence

$$4a + 2b + c = 3.$$

Similarly, when $n = 3$ and $S_3 = 6$, we have

$$S_3 = 6 = a \cdot (3^2) + b \cdot (3) + c,$$

so that

$$9a + 3b + c = 6.$$

We therefore have a system of three linear equations in three unknowns:

$$a + b + c = 1;$$
$$4a + 2b + c = 3;$$
$$9a + 3b + c = 6.$$

Using the matrix techniques from Appendix C, we find that

$$a = \frac{1}{2}, \qquad b = \frac{1}{2}, \quad \text{and} \quad c = 0,$$

which is the same set of coefficients we found in part (a).

     There is one difficulty with both derivations in Example 3. We used both methods to derive a formula for the sum of the first n integers, which is supposed to be true for *any* n. But, in fact, we based both derivations on just the first six values that we calculated for $S_1, S_2, \ldots, S_6$, which we showed followed a quadratic pattern by looking at a table of second differences. The catch is that we can't know for sure, just by looking at examples, that *all* subsequent values for S_n continue to follow a quadratic pattern. So the "proof" really isn't legitimate unless we can demonstrate that it applies to every value of n, not just the first six. We do so in Example 5.

EXAMPLE 5

Show that all the values for $S_n = 1 + 2 + 3 + \cdots + n$, for all values of n, fall in a quadratic pattern.

Solution To show that all values of S_n fall in a quadratic pattern, we must demonstrate that the second differences are always constant for any value of n. Let's consider any value of n, so that the sum of the first n integers is

$$S_n = 1 + 2 + \cdots + n.$$

If we take the next integer, $n + 1$, and form the sum of the first $n + 1$ integers, we get

$$S_{n+1} = (1 + 2 + \cdots + n) + (n + 1).$$

The difference between S_n and S_{n+1} is

$$\Delta S_n = S_{n+1} - S_n = n + 1,$$

because all other terms cancel.

    Similarly, the sum of the first $n + 2$ integers is

$$S_{n+2} = (1 + 2 + \cdots + n) + (n + 1) + (n + 2).$$

The difference between this total and S_{n+1} is

$$S_{n+2} - S_{n+1} = \Delta S_{n+1} = n + 2,$$

because, again, all other terms cancel. As a result, the second difference, or difference of the differences, is just

$$\Delta S_{n+1} - \Delta S_n = (n + 2) - (n + 1) = 1,$$

for any value of n. Therefore *all* values of S_n have a constant second difference and consequently, no matter what value of n we select, the sum of all the differences must follow a quadratic pattern.

Sums of Squares of Integers

We now find a formula for the sum of the first n squares,

$$S_n = 1^2 + 2^2 + 3^2 + 4^2 + \cdots + n^2.$$

We have

$$S_1 = 1, \qquad S_2 = 1^2 + 2^2 = 5, \qquad S_3 = 1^2 + 2^2 + 3^2 = 14,$$

$$S_4 = 30, \qquad S_5 = 55, \quad \text{and} \quad S_6 = 91,$$

and so on. For simplicity, we also use the sum of the squares of the first zero terms, $S_0 = 0^2 = 0$. Arranging these values in a table, we obtain the following.

n	S_n	ΔS_n	$\Delta^2 S_n$	$\Delta^3 S_n$
0	0			
		$1 = 1 - 0$		
1	1		$3 = 4 - 1$	
		4		$2 = 5 - 3$
2	5		5	
		9		2
3	14		7	
		16		2
4	30		9	
		25		2
5	55		11	
		36		
6	91			

The third differences $\Delta^3 S_n$ are all constant, so these data values follow a cubic pattern; that is, the formula for the sum of the squares of the first n integers is a cubic function

$$S_n = an^3 + bn^2 + cn + d.$$

Using polynomial regression, we find the cubic polynomial that fits the points $(0, 0)$, $(1, 1)$, $(2, 5)$, $(3, 14)$, $(4, 30)$, $(5, 55)$, and $(6, 91)$ has coefficients $a = 0.33333333$ (or $\frac{1}{3}$), $b = 0.5 = \frac{1}{2}$, $c = 0.16666667$ (or $\frac{1}{6}$) and $d = -3.5E(-12) = -3.5 \times 10^{-12} = -0.0000000000035$, which essentially is 0. Therefore the cubic function that fits the data is

$$S_n = \left(\frac{1}{3}\right)n^3 + \left(\frac{1}{2}\right)n^2 + \left(\frac{1}{6}\right)n + 0.$$

We factor out the common factors n and $\frac{1}{6}$ and then factor the resulting quadratic to get

$$S_n = n\left[\left(\frac{1}{3}\right)n^2 + \left(\frac{1}{2}\right)n + \left(\frac{1}{6}\right)\right]$$

$$= \left(\frac{1}{6}\right)n[2n^2 + 3n + 1]$$

$$= \left(\frac{1}{6}\right)n(2n + 1)(n + 1),$$

which is more commonly written as

$$S_n = \frac{n(n + 1)(2n + 1)}{6}.$$

Alternatively, we could solve for the coefficients of the cubic polynomial $S_n = an^3 + bn^2 + cn + d$ algebraically.

Using $n = 0$ and $S_0 = 0$, we get $0 = d$.

Therefore we have

$$S_n = an^3 + bn^2 + cn.$$

Further

when $n = 1$ and $S_1 = 1$: $a + b + c = 1$

when $n = 2$ and $S_2 = 5$: $8a + 4b + 2c = 5$

when $n = 3$ and $S_3 = 14$: $27a + 9b + 3c = 14$

These results give a system of three equations in the three unknowns a, b, and c; we have already determined that $d = 0$. Using matrix methods to solve this system of equations, we again get $a = \frac{1}{3}$, $b = \frac{1}{2}$, and $c = \frac{1}{6}$.

In general, we have the following formula.

The sum of the squares of the first n integers is

$$\sum_{k=1}^{n} k^2 = 1^2 + 2^2 + 3^2 + \cdots + n^2 = \frac{n(n + 1)(2n + 1)}{6}. \qquad \text{(4)}$$

EXAMPLE 6

Find the sum of the squares of the first 100 integers: $1^2 + 2^2 + \cdots + 100^2$.

Solution Using Formula (4) with $n = 100$, we get

$$1^2 + 2^2 + \cdots + 100^2 = \frac{100(101)(201)}{6} = 338{,}350.$$

Note that, although Formula (4) is true for *all* values of n, we have only established it for $n = 0, 1, \ldots, 6$ by using both of these approaches. As with the sum of the first n integers, we must prove that the sum of the squares of the first n integers follows a cubic pattern for every possible value of n. We do so in Example 7.

EXAMPLE 7

Prove that the sum of the squares of the first n integers, for any n, follows a cubic pattern.

Solution To do so, we have to show that the third differences of S_n are all constant, for any value of n. We write

$$S_n = 1^2 + 2^2 + \cdots + n^2$$

so that

$$S_{n+1} = 1^2 + 2^2 + \cdots + n^2 + (n+1)^2;$$
$$S_{n+2} = 1^2 + 2^2 + \cdots + n^2 + (n+1)^2 + (n+2)^2;$$
$$S_{n+3} = 1^2 + 2^2 + \cdots + n^2 + (n+1)^2 + (n+2)^2 + (n+3)^2.$$

We begin by forming the first differences of each successive pair. In each case, all terms but one cancel, leaving us with

$$S_{n+1} - S_n = \Delta S_n = (n+1)^2 = n^2 + 2n + 1;$$
$$S_{n+2} - S_{n+1} = \Delta S_{n+1} = (n+2)^2 = n^2 + 4n + 4;$$
$$S_{n+3} - S_{n+2} = \Delta S_{n+2} = (n+3)^2 = n^2 + 6n + 9.$$

Each of these first differences is a quadratic function of n. We now form the second differences by taking the difference of each successive pair of first differences:

$$\Delta S_{n+1} - \Delta S_n = \Delta^2 S_n = (n^2 + 4n + 4) - (n^2 + 2n + 1) = 2n + 3;$$
$$\Delta S_{n+2} - \Delta S_{n+1} = \Delta^2 S_{n+1} = (n^2 + 6n + 9) - (n^2 + 4n + 4) = 2n + 5.$$

Finally, we find the third differences by forming the difference between these last two expressions and get

$$\Delta^2 S_{n+1} - \Delta^2 S_n = \Delta^3 S_n = (2n + 5) - (2n + 3) = 2,$$

which is constant for *all* values of n. That is, the sum of the squares of the first n integers follows a cubic pattern for every value of n. ◆

EXAMPLE 8

When cannonballs are stacked in a pyramidal pile, as shown in the accompanying figure, they are organized from the top layer down as follows: A single ball is at the top of the pile; four balls are in the second layer, arranged in a square to support the single ball on top; nine balls are in the third layer, arranged in a square of size 3 by 3 that supports the second layer; and so on. How many cannonballs are in a pile that is 10 layers high?

Solution The number of cannonballs is

$$1^2 + 2^2 + 3^2 + \cdots + 10^2.$$

We can evaluate this total using Formula (4) for the sum of the squares of the first n integers with $n = 10$. Thus

$$1^2 + 2^2 + 3^2 + \cdots + 10^2 = \frac{10(10+1)[2(10)+1]}{6}$$
$$= \frac{10(11)(21)}{6}$$
$$= 385 \text{ cannonballs.}$$

◆

Example 9 illustrates some additional applications of these ideas to find the total for a quantity when the individual amounts are known. In it we use the following basic properties of sums of numbers:

$$\sum_{k=1}^{n} (a_k + b_k) = \sum_{k=1}^{n} a_k + \sum_{k=1}^{n} b_k \qquad \textbf{(5)}$$

and

$$\sum_{k=1}^{n} (m \cdot a_k) = m \cdot \sum_{k=1}^{n} a_k, \qquad \text{for any constant } m. \quad \textbf{(6)}$$

We ask you to prove these two results in the problems at the end of this section.

EXAMPLE 9

A study of the financial records of a company finds that its monthly revenues, in thousands of dollars, are modeled by the function $R(x) = 0.001x^2 + 0.02x + 32$, where x is the number of months since the start of the study and $x \geq 1$. Find the total revenue for this company over its first 10 years of operation.

Solution   The 10-year period is equivalent to 120 months. We need to add the revenues $R(1)$ in month 1, $R(2)$ in month 2, $R(3)$ in month 3, ..., $R(120)$ in month 120. Doing so, we get

$$R = R(1) + R(2) + \cdots + R(120) = \sum_{k=1}^{120} (0.001k^2 + 0.02k + 32),$$

where the variable k takes on all values between 1 and 120. Using Property (5) of sums, we simplify the preceding equation and get

$$R = \sum_{k=1}^{120} 0.001k^2 + \sum_{k=1}^{120} 0.02k + \sum_{k=1}^{120} 32,$$

Using Property (6), we get

$$R = 0.001 \sum_{k=1}^{120} k^2 + 0.02 \sum_{k=1}^{120} k + 32 \sum_{k=1}^{120} 1.$$

The first term involves the sum of the squares of the first 120 integers, so

$$\sum_{k=1}^{120} k^2 = \frac{(120)(120 + 1)(2 \cdot 120 + 1)}{6} = 583{,}220.$$

The second term involves the sum of the first 120 integers, so

$$\sum_{k=1}^{120} k = \frac{(120)(121)}{2} = 7260.$$

The third term involves the sum of 120 ones, so

$$\sum_{k=1}^{120} 1 = 120 \cdot (1) = 120.$$

Therefore the total revenue for this company over the 10-year period is

$$R = 0.001(583{,}220) + 0.02(7260) + 32(120) = 4568.42$$

thousand dollars, or about $4.568 million.

Problems

1. In Examples 1 and 2, we found the parabola that passes through the points $(0, 2)$, $(1, 0)$, $(2, 4)$, $(3, 14)$, $(4, 30)$, and $(5, 52)$. Suppose now that the points are $(0, 2)$, $(1, 0)$, $(2, 4)$, $(3, 15)$, $(4, 30)$, and $(5, 52)$ instead.

   a. Show that these points do not lie on a parabola.

   b. Attempt to repeat the procedure used in Example 2 to see what goes wrong.

2. Determine which sets of values come from a quadratic function and which come from a cubic function. For those that come from a quadratic function, determine the equation of the quadratic.

x	$f(x)$	$g(x)$	$h(x)$	$k(x)$
0	0	1	1	3
1	-2	6	0	1
2	2	13	5	3
3	12	22	22	9
4	28	33	57	19
5	50	46	116	33

3. The following measurements were taken on a quantity that follows a cubic pattern. However, one of the values was recorded in error. Find the incorrect entry and correct it. (*Hint*: It isn't necessary to actually determine the formula for the cubic.)

x	0	1	2	3	4	5	6	7
y	40	34	24	22	40	90	184	344

4. Consider the array of numbers known as Pascal's triangle in which each row begins and ends with 1 and each intermediate entry is simply the sum of the two numbers diagonally above it in the previous row.

$$
\begin{array}{ccccccccccc}
 & & & & 1 & & 1 & & & & \\
 & & & 1 & & 2 & & 1 & & & \\
 & & 1 & & 3 & & 3 & & 1 & & \\
 & 1 & & 4 & & 6 & & 4 & & 1 & \\
1 & & 5 & & 10 & & 10 & & 5 & & 1 \\
\end{array}
$$
$$1 \quad 6 \quad 15 \quad 20 \quad 15 \quad 6 \quad 1$$

   The rows are numbered $n = 1, 2, \ldots$ The second diagonal consists of the entries 1, 2, 3, 4, 5, and 6, $\ldots$

   a. Find a formula for the terms in the third diagonal: 1, 3, 6, 10, 15, $\ldots$, in terms of the row number n.

   b. Find a formula for the terms in the fourth diagonal 1, 4, 10, 20, $\ldots$, in terms of the row number n.

5. Construct the quadratic polynomial that passes through the points $(0, 1)$, $(1, 4)$, and $(2, 9)$. Use it to estimate the value of the underlying function when $x = 0.5$ and when $x = 3$.

6. The main support cable of a suspension bridge is a parabola. For the Golden Gate bridge, suppose that the cable's lowest point is 15 feet above the roadway. Use the dimensions shown in the accompanying figure to find an equation of the cable for the Golden Gate bridge.

7. Find (a) the sum of the first 25 integers, (b) the sum of the first 100 integers, and (c) the sum of the first 1000 integers.

8. Find (a) the sum of the squares of the first 25 integers and (b) the sum of the squares of the first 50 integers.

9. Suppose that the produce manager in a supermarket receives a delivery of 1000 large grapefruit, which he wants to display in a pyramid with a square base. How many layers are needed?

10. a. Find the sum of the integers from 83 through 225, inclusive.

    b. Find the sum of the squares of these integers.

11. The annual rainfall R, in inches, in a particular region in year t since the start of the last century can be modeled by the formula $R(t) = -0.02t^2 + 1.8t + 42$. Find the total rainfall from 1900 (when $t = 0$) through 2000 in that region.

12. Cannonballs are sometimes stacked in rectangular piles. The accompanying figure shows the fourth layer of a stack of n rectangular layers.

    a. Suppose that such a stack ends with a single row of two balls at the top. Devise a formula in sum-

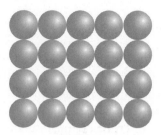

Fourth layer

mation notation for the number of balls in a stack n layers high.

b. Use the properties of summations to expand the formula you found in part (a).

c. Suppose that a stack of cannonballs ends with a single row of three balls as the top layer. Devise a formula for the number of balls in a stack n layers high.

d. Use the summation formulas from parts (a) and (c) to predict the result if the top layer consists of a single row of four balls.

13. a. Consider the function $y = ax^2$. Construct a table of values for the function if $x = -2, -1, 0, 1, 2, 3$ and extend it to a table of differences until you can construct a formula for $\Delta^2 y$ for this function.

b. Repeat part (a) for the function $y = ax^3$ to devise a formula for $\Delta^3 y$.

c. Repeat part (a) for the function $y = ax^4$ to devise a formula for $\Delta^4 y$.

d. Based on your results in parts (a)–(c), predict a formula for $\Delta^5 y$ when $y = ax^5$.

14. For the sequence of numbers $\{y_0, y_1, y_2, \ldots, y_n, y_{n+1}, y_{n+2}, \ldots\}$, show that

a. $\Delta^2 y_0 = y_2 - 2y_1 + y_0$.

b. $\Delta^2 y_n = y_{n+2} - 2y_{n+1} + y_n$, for any n.

15. Suppose that a set of data values (x_0, y_0), (x_1, y_1), (x_2, y_2), ... has uniformly spaced x-values (Δx) and constant second differences $\Delta^2 y = k$ so that the points follow a quadratic pattern $y = ax^2 + bx + c$. Use the result of Problem 14 to show that the leading coefficient is

$$a = \frac{1}{2} \frac{\Delta^2 y}{(\Delta x)^2}.$$

(*Hint*: Write $x_1 = x_0 + \Delta x$ and $x_2 = x_0 + 2\Delta x$ and use the first three points to construct a system of linear equations in a, b, and c.)

16. Because the sum of the first n integers follows a quadratic pattern and the sum of the squares of the first n integers follows a cubic pattern, you might

conjecture that the sum of the cubes of the first n integers

$$S_n = \sum_{k=1}^{n} k^3 = 1^3 + 2^3 + 3^3 + \cdots + n^3$$

follows a quartic polynomial pattern.

a. Calculate values for S_n for $n = 0, 1, 2, \ldots, 7$.

b. Use a table of differences to show that these values follow a quartic pattern.

c. Find a formula for the sum of the cubes of the first n integers.

17. Find the sum of the cubes of the first 25 integers.

18. By writing out

$$\sum_{k=1}^{n} a_k = a_1 + a_2 + \cdots + a_n,$$

$$\sum_{k=1}^{n} b_k = b_1 + b_2 + \cdots + b_n, \quad \text{and} \quad \sum_{k=1}^{n} (m \cdot a_k),$$

show that

$$\sum_{k=1}^{n} (a_k + b_k) = \sum_{k=1}^{n} a_k + \sum_{k=1}^{n} b_k$$

and

$$\sum_{k=1}^{n} (m \cdot a_k) = m \sum_{k=1}^{n} a_k, \qquad \text{for any constant } m.$$

19. A Pythagorean triple is a set of three integers a, b, and c that satisfy the Pythagorean theorem $a^2 + b^2 = c^2$ and hence represent the sides of a right triangle. The following is a list of the first five Pythagorean triples (a_n, b_n, c_n).

n	a_n	b_n	c_n
1	3	4	5
2	5	12	13
3	7	24	25
4	9	40	41
5	11	60	61

(There are infinitely many Pythagorean triples.) Notice that, for any n, $a_n = 2n + 1$ and $c_n = b_n + 1$.

a. Construct a table of differences to determine the pattern in the b_n terms.

b. Find a formula for b_n for each value of n, based on the pattern from part (a).

c. Show that the resulting triple (a_n, b_n, c_n), forms a Pythagorean triple for any integer n.

d. What is the next Pythagorean triple following the ones shown in the table?

4.6 Building New Functions from Old: Operations on Functions

The functions that we've considered so far, such as $y = \sqrt{x}$, $y = x^5$, $y = 10^x$, and $y = \log x$, can be thought of as building blocks from which we can construct other, more complicated functions. In a simple case, we can take the power functions $y = x$ and $y = x^2$ and the constants 3, 4, and -5 to create the quadratic function $f(x) = 3x^2 + 4x - 5$ as a linear combination of power functions. In fact, we can think of any polynomial as a linear combination of power functions. In this section, we investigate how to generate larger classes of functions by applying simple operations (e.g., addition, subtraction, multiplication, and division) to the basic families of functions that we already have discussed.

Sums and Differences

Let's begin with the sum of two functions. The function

$$f(x) = x^2 + 2^{-x}$$

is the sum of the two functions $y = x^2$ and $y = 2^{-x} = 1/2^x = \left(\frac{1}{2}\right)^x$. Their individual graphs are shown in Figure 4.47. If we "pile" one set of y-values on top of the other and add, we get the graph of the sum as shown in Figure 4.48. You can verify that this result is indeed the case by plotting the sum function on your function grapher.

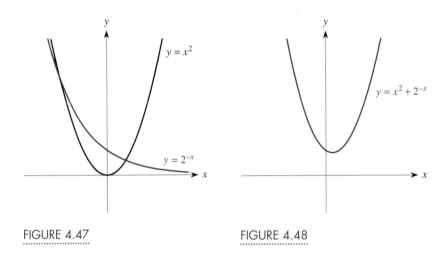

FIGURE 4.47   FIGURE 4.48

Note the behavior of the sum $y = x^2 + 2^{-x}$. Because $y = 2^{-x} = \left(\frac{1}{2}\right)^x$ decays rapidly as x increases, its contribution becomes less and less significant, and the quadratic term eventually dominates in the sum when x is large. As a result, toward the right, the graph quickly becomes indistinguishable from a parabola. For negative values of x, both functions become large, but 2^{-x} grows much faster than x^2 does and so the exponential term dominates on the left.

In general, we write

$$S(x) = f(x) + g(x)$$

for the sum of two functions $f(x)$ and $g(x)$. That is, for each value of x, we add the values of $f(x)$ and $g(x)$ to produce the value of $S(x)$. For instance, if $f(3) = 15$ and $g(3) = 4$, then $S(3) = f(3) + g(3) = 15 + 4 = 19$.

Similarly, we construct the difference of two functions by taking the difference between their values for each possible value of x. In general, we write

$$D(x) = f(x) - g(x)$$

for the difference of two functions. Thus, if $f(3) = 15$ and $g(3) = 4$, then $D(3) = f(3) - g(3) = 15 - 4 = 11$. Graphically, if we subtract $g(x)$ from $f(x)$ and $f(x) > g(x)$, the difference $D(x)$ is just the difference in height between the two curves for each value of x.

Products of Functions

For the product of the two functions $f(x) = x^2$ and $g(x) = 2^{-x}$, we use the same interpretation as with sums and differences of functions. Thus the product of the two functions

$$P(x) = f(x) \cdot g(x)$$

means that, for each permissible value of x, we multiply the corresponding function values. So, if $f(3) = 15$ and $g(3) = 4$, then $P(3) = f(3) \cdot g(3) = 15 \cdot 4 = 60$.

What does the graph of the product function look like? Unlike the sum and difference of two functions, there is rarely a direct graphical interpretation of the product of two functions. However, you can produce the graph of the product of two functions on your function grapher and then analyze the behavior of that graph. For instance, consider

$$P(x) = f(x) \cdot g(x) = x^2 \cdot 2^{-x}.$$

We know that, for large positive x, $y = x^2$ grows ever larger and $y = 2^{-x}$ approaches zero. We also know that an exponential function with a positive exponent grows much faster than a power function does. Similarly, an exponential function with a negative exponent decays much faster than a power function with a negative power. Together, these facts indicate that, in the product $x^2 \cdot 2^{-x}$, the exponential term drives the product toward zero as x increases. For values of $x < 0$, both functions grow without bound, so their product becomes infinitely large. By using your function grapher, you can obtain the result shown in Figure 4.49.

Let's look at a real-life example of a product of two functions. Lyme disease is caused by a bacterial infection transmitted by blood-sucking ticks. When a person is infected, the body produces antibodies to fight the bacteria. Figure 4.50 shows the level of concentration of the antibody in the bloodstream as a function of the number of weeks since the first infection. Note that the pattern is remarkably similar to

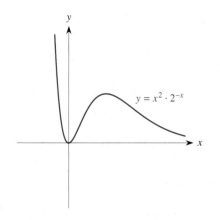

$y = x^2 \cdot 2^{-x}$

FIGURE 4.49

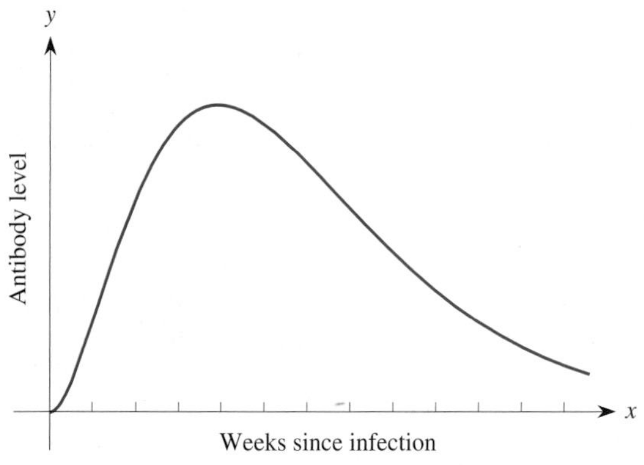

FIGURE 4.50

Antibody level

Weeks since infection

the behavior of the function $f(x) = x^2 \cdot 2^{-x}$ for $x > 0$. This pattern suggests that such functions, known as *surge functions,* might be appropriate as mathematical models to describe the antibody level, both for Lyme disease and possibly other infections. We explore some applications of surge functions in Section 4.9.

Quotient of Functions

When we consider the quotient of two functions

$$Q(x) = \frac{f(x)}{g(x)},$$

there is a complication that we must take into account. The quotient is undefined at any point where the denominator $g(x)$ is zero, and typically a vertical asymptote occurs there. We illustrate this behavior in Example 1.

EXAMPLE 1

Sketch the graph of the function

$$Q(x) = \frac{x^2 + 1}{x^2 - 1}.$$

Solution   We begin analyzing the behavior of this function by looking at what happens when the denominator $x^2 - 1 = (x - 1)(x + 1)$ is zero. That occurs when $x = 1$ and $x = -1$, so the quotient $Q(x)$ is not defined there. When you take values of x very close to either of these two points, the corresponding values for the quotient $Q(x)$ become extremely large, positively or negatively. To see this result, first consider points near $x = 1$. Suppose that x is slightly larger than 1. So

if $x = 1.001$, then $y = Q(1.001) \approx 1000.5$;

if $x = 1.0001$, then $y = Q(1.0001) \approx 10000.5$;

if $x = 1.000001$, then $y = Q(1.000001) \approx 1,000,000.5$.

Hence, as x approaches 1 from the right (or from above) through values of x that are slightly larger than 1, y becomes ever larger and approaches ∞.

Now suppose that x is slightly smaller than 1. So

if $x = 0.999$, then $y = Q(0.999) \approx -999.5$;

if $x = 0.9999$, then $y = Q(0.9999) \approx -9999.5$;

if $x = 0.999999$, then $y = Q(0.999999) \approx -999,999.5$.

Hence, as x approaches 1 from the left (or from below) through values of x that are slightly smaller than 1, y approaches $-\infty$.

By a similar analysis around the point $x = -1$, you can verify that, as x approaches -1 from the right, the function approaches $-\infty$, whereas, if x approaches -1 from the left, the function approaches $+\infty$. Therefore it is not surprising that this quotient function has vertical asymptotes at $x = 1$ and $x = -1$.

We next analyze the *end behavior*—what happens to this function as x becomes large, both positively and negatively. Suppose, for instance, that $x = 1000$. The value of the function then is

$$Q(1000) = \frac{1,000,001}{999,999} \approx 1.000002,$$

which is extremely close to 1. Actually, adding 1 to x^2 in the numerator and subtracting 1 from x^2 in the denominator really has little effect on the value of the function when x is 1000. If x were even larger—say, 1,000,000—adding or subtracting 1 from x^2 would have a negligible effect. Thus for large values of x, the numerator is dominated by the x^2 term and the denominator is dominated by the x^2 term, so the quotient behaves like

$$Q(x) \approx \frac{x^2}{x^2} = 1$$

when x is large. As a result, this quotient function gets closer and closer to a height of 1, so that it has a horizontal asymptote of $y = 1$ as x approaches ∞.

What happens as x approaches $-\infty$? Again, for large negative values of x, the 1 in the numerator and the -1 in the denominator are negligible and x^2 again dominates both the numerator and the denominator. Thus there is also a horizontal asymptote of $y = 1$ as x approaches $-\infty$.

We next look for the points where the curve crosses the two axes. It crosses the y-axis when $x = 0$, so $Q(0) = 1/(-1) = -1$. Where does the curve cross the x-axis? For that to happen, y must equal 0, so the numerator has to be 0. Because the numerator for $Q(x)$, $x^2 + 1$, is never 0 for real values of x, the quotient cannot be 0 anywhere. Hence the curve never crosses the x-axis. The complete graph of this function is shown in Figure 4.51.

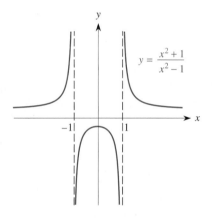

$$y = \frac{x^2 + 1}{x^2 - 1}$$

FIGURE 4.51

Rational Functions

Example 1 illustrates most of the ideas involving the behavior of quotients of functions in general and quotients of polynomials in particular. The quotient (or ratio) of two polynomials is called a **rational function.** We assume that any common factors in the numerator and the denominator have been canceled and therefore that the rational function is expressed in simplest form.

The following are some of the important facts about rational functions.

> ## Behavior of Rational Functions $R(x) = P(x)/Q(x)$
>
> ◆ The zeros of the numerator $P(x)$ correspond to zeros of the rational function $R(x)$; its graph crosses the x-axis at these points.

> ◆ The zeros of the denominator $Q(x)$ correspond to the points where the rational function $R(x)$ is not defined; its graph usually has a vertical asymptote at these points.
> ◆ The highest power term in the numerator $P(x)$ dominates the numerator for *large values of x*, either positive or negative.
> ◆ The highest power term in the denominator $Q(x)$ dominates the denominator for *large values of x*, either positive or negative.
> ◆ For *large values of x*, either positive or negative, the rational function $R(x)$ behaves like the highest power term of the numerator divided by the highest power term of the denominator. The result may be a horizontal asymptote or the values may approach ∞ or $-\infty$ as x increases either positively or negatively.

We illustrate these ideas in Examples 2 and 3.

EXAMPLE 2

Analyze the behavior of the rational function

$$R(x) = \frac{x^2 - 1}{x - 2}.$$

Solution   Here, $R(x)$ has zeros when its numerator $x^2 - 1 = 0$, so that $x = \pm 1$, and the graph crosses the x-axis at these two points. Also, the denominator is zero when $x = 2$, which creates a vertical asymptote there. Suppose that x approaches 2 from the right (with values slightly larger than 2); for instance,

if $x = 2.01$, then $y = R(2.01) = 304.01$;

if $x = 2.001$, then $y = R(2.001) = 3004.001$;

if $x = 2.00001$, then $y = R(2.00001) = 300004.00001$.

Thus, when x approaches 2 from the right, $R(x)$ approaches $+\infty$. Similarly, when x approaches 2 from the left, $R(x)$ approaches $-\infty$ (try some values of x slightly less than 2—say, $x = 1.99$ or $x = 1.9999$).

You can locate the vertical asymptotes of a rational function by finding the roots of the denominator, but you must check what happens on either side (in this case at $x = 2.001$ and $x = 1.999$, for example) to determine the sign of the function on each side of the vertical asymptote. Doing so lets you decide whether the curve rises toward $+\infty$ or drops toward $-\infty$ on each side of the vertical asymptote.

Next, consider the end behavior of $R(x)$. For large values of x, the numerator is dominated by the leading x^2 term and the denominator is dominated by the leading x term. As a result, for large values of x, the quotient behaves like $y = R(x) \approx x^2/x = x$. For instance,

if $x = 10$, then $R(10) = 12.375$;

if $x = 100$, then $R(100) = 102.0306$;

if $x = 1000$, then $R(1000) = 1002.003006$.

The larger x is, the closer $R(x)$ is to x and, for large positive values of x, the graph increases toward $+\infty$.

Similarly, for large negative values, the quotient $R(x)$ behaves like $y = x^2/x = x$ and the graph tends toward $-\infty$.

Figure 4.52 displays all this behavior in the complete graph of R.

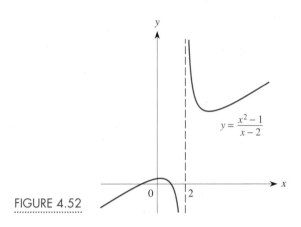

$$y = \frac{x^2 - 1}{x - 2}$$

FIGURE 4.52

Think About This
Graph both the original quotient function $R(x) = (x^2 - 1)/(x - 2)$ and the *limiting function* $y = x$ in the same fairly large viewing window—say, from -1000 to 1000 for both x and y—on your function grapher. What do you observe?

Getting all the important details on the behavior of a rational function from a single view in your function grapher is often almost impossible. Try it for the function $R(x)$ in Example 2 and see what types of information may be lost because of the scale you use for the domain and range.

EXAMPLE 3

Analyze the behavior of the rational function

$$S(x) = \frac{x - 2}{x^2 - 1}.$$

Solution Here, $S(x)$ was formed by interchanging the numerator and denominator of the rational function $R(x)$ in Example 2, but the behavior of the two functions is quite different.

Note that $S(x)$ has only one zero at $x = 2$ when the numerator is zero. It has two vertical asymptotes, one at $x = 1$ and the other at $x = -1$ when the denominator is zero. Let's see what happens on either side of the asymptotes. When $x = 1.001$, say, we have $S(1.001) = -499.25$, so we conclude that the curve drops toward $-\infty$ as x approaches 1 from the right. Similarly, when $x = 0.999$, we have $S(0.999) = 500.75$ and the curve rises toward $+\infty$ as x approaches 1 from the left. Similarly, when $x = -1.001$, we have $S(-1.001) = -1499.75$ and the curve drops toward $-\infty$ as x approaches -1 from the left. Also, when $x = -0.999$, $S(-0.999) = 1500.25$, and the curve rises toward $+\infty$ as x approaches -1 from the right. Use your calculator to check these conclusions numerically with other values of x on either side of $x = 1$ and on either side of $x = -1$.

Further, the numerator is dominated by x and the denominator is dominated by x^2, so for large values of x, the rational function behaves like $y = x/x^2 = 1/x$. Therefore, for large positive values of x, the function is positive and decays toward the x-axis as a horizontal asymptote. Similarly, for large negative values of x, the function is negative and rises toward the x-axis as a horizontal asymptote.

The complete graph of $S(x)$ is displayed in Figure 4.53.

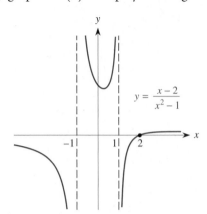

$$y = \frac{x-2}{x^2-1}$$

FIGURE 4.53

As before, though, we urge you to examine the behavior carefully with your function grapher to see how viewing the overall characteristics depends on the window you use.

Think About This   Examine the graphs of the quotient function $S(x)$ and the limiting function $y = 1/x$ in the same large viewing window. What do you observe?

We next consider a real-world application that involves rational functions.

EXAMPLE 4

According to the law of universal gravitation, the gravitational force between any two objects of mass m_1 and m_2 is

$$F = \frac{Gm_1 m_2}{r^2},$$

where r is the distance between the objects and G is the gravitational constant. Envision a spacecraft traveling from the Earth to the moon, a distance of about 240,000 miles. Because the mass of the Earth is roughly 81 times that of the moon, the Earth's gravitational effect on the spacecraft will be greater than that of the moon's until the spacecraft is quite close to the moon, when it's gravity becomes dominant. Determine the distance from the Earth when the two gravitational forces exactly balance each other.

Solution   We begin with a sketch of the situation, as shown in Figure 4.54, where r represents the distance, in thousands of miles, from the Earth to the spacecraft. Hence

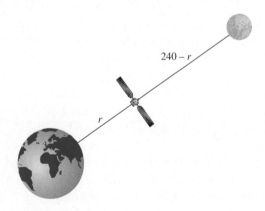

$240 - r$

r

FIGURE 4.54

$240 - r$ is the distance from the moon to the spacecraft. Let m_0 be the mass of the spacecraft, m_1 be the mass of the Earth, and m_2 be the mass of the moon. The Earth's gravitational force on the spacecraft F_e is

$$F_e = \frac{Gm_0m_1}{r^2},$$

and the moon's gravitational force on the spacecraft F_m is

$$F_m = \frac{Gm_0m_2}{(240 - r)^2}.$$

Both F_e and F_m are rational functions of r. Because the Earth is 81 times as massive as the moon, $m_1 = 81m_2$. We rewrite F_e as

$$F_e = \frac{Gm_0(81m_2)}{r^2}.$$

The two gravitational forces are equal when

$$\frac{81Gm_0m_2}{r^2} = \frac{Gm_0m_2}{(240 - r)^2}.$$

Dividing both sides of this equation by Gm_0m_2 (because none of these quantities are zero) gives

$$\frac{81}{r^2} = \frac{1}{(240 - r)^2}.$$

Cross-multiplying yields

$$r^2 = 81(240 - r)^2.$$

We expand the expression on the right by squaring the binomial term and obtain

$$r^2 = 81(240^2 - 480r + r^2) = 81(240)^2 - 81(480)r + 81r^2.$$

Collecting like terms and simplifying, we have the quadratic equation

$$80r^2 - 38{,}880r + 4{,}665{,}600 = 0.$$

Dividing through by the common factor 80, we get

$$r^2 - 486r + 58{,}320 = 0.$$

Using the quadratic formula, we find that the roots of this quadratic equation are $r = 216$ and $r = 270$. These answers are distances in thousands of miles from the Earth. Because the moon is about 240 thousand miles from the Earth, the only reasonable answer is the first. Therefore the two forces balance at a point about 216 thousand miles from the Earth and about 24,000 miles this side of the moon. The second solution, 270,000 miles from the Earth, corresponds to a point beyond the moon where the effects of the moon's gravity and the Earth's gravity are numerically the same, though both forces are in the same direction. ◆

A Function of a Function

There is yet another way in which we can construct new functions from simpler functions. In Example 5 of Section 2.2, we showed that the rate R at which a snow tree cricket chirps is a function of the temperature T, and we found a mathematical model for this relationship as the function $R = f(T) = 4T - 160$. However, the air temperature doesn't remain constant, but actually varies with the time of the day, so the temperature T is really a function of time t: $T = g(t)$. As a result, the chirp rate, though a function of the temperature T, is actually a function of time t. That is, we have two functions:

$$R = f(T) = 4T - 160 \quad \text{and} \quad T = g(t).$$

If we substitute $T = g(t)$ into the expression $R = f(T)$, we get

$$R = f(T) = f(g(t)).$$

We call this type of situation a **function of a function** or a **composite function.**

Let's look at this notion from a different perspective. Consider the function $f(x) = \sqrt{x^3 + 1}$. To see what it means, suppose that $x = 1$. Then

$$f(1) = \sqrt{1^3 + 1} = \sqrt{2}.$$

For $x = 2$,

$$f(2) = \sqrt{2^3 + 1} = \sqrt{9} = 3.$$

To evaluate this function in each case, we actually performed two *successive* steps: (1) for each value of x, we evaluated the expression $x^3 + 1$; and (2), we took the square root of the result. The reason is that we are really working with two functions successively: first the "inner" function $x^3 + 1$ and then the "outer" function $\sqrt{u}$, where $u = x^3 + 1$. The final function f is therefore a function of a function.

Let's set up the mathematical framework for this concept. Suppose that we let $y = F(u)$, where $u = G(x)$. Here, $y = F(u) = \sqrt{u}$, where in turn $u = G(x) = x^3 + 1$. Consequently,

$$y = F(u) = F(G(x)) = F(x^3 + 1) = \sqrt{x^3 + 1}.$$

Our original function f is the result of applying the functions G and F successively. This composite function $y = F(G(x))$ is sometimes written as $F \circ G$ and read "F of G".

In general, for two functions F and G, the composite function $F(G(x))$ is the result of evaluating the two functions successively, as depicted in Figure 4.55. We start with a value of x, which is carried into a value u by the first, or inner, function G, which in turn is carried to a value y by the second, or outer, function F. For this method to make sense mathematically, the domain of the outer function F must include the range of the inner function G.

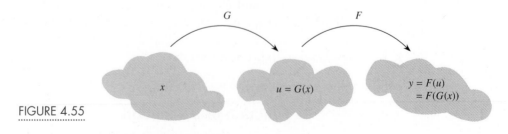

FIGURE 4.55

Using composite functions, we can construct many other types of functions by using the basic functions as building blocks.

EXAMPLE 5

Find two functions f and g so that $y = f(g(x)) = 10^{3x}$.

Solution   Think about how you would evaluate this function for any value of x—first, triple the x value because of the $3x$ term and then take that power of 10. That is, the linear function $3x$ is used as the exponent for the exponential function with base 10. So the first, or inner, function is $g(x) = 3x$ followed by the second, or outer, function $y = f(x) = 10^x$. The result gives $f(g(x)) = f(3x) = 10^{3x}$, as required. ◆

EXAMPLE 6

Find two functions f and g so that $y = f(g(x)) = \log(x^2 - 5x + 2)$.

Solution   Here, the quadratic function $x^2 - 5x + 2$ is used as the argument of the log function. So the first, or inner, function is the quadratic $g(x) = x^2 - 5x + 2$ and the second, or outer, function is the log function $y = f(x) = \log x$. Using the same approach as in Example 5, we get $f(g(x)) = f(x^2 - 5x + 2) = \log(x^2 - 5x + 2)$, as required. ◆

Are F ∘ G and G ∘ F the same?

Is the order important in forming the composition of two functions? That is, is $F \circ G$ the same as $G \circ F$? Again consider

$$f(x) = \sqrt{x^3 + 1} = F(G(x)) = F \circ G(x),$$

where

$$u = G(x) = x^3 + 1 \quad \text{and} \quad y = F(u) = \sqrt{u}.$$

If we interchange the order to form $G(F(x))$, we get

$$G(F(x)) = G(\sqrt{x}) = (\sqrt{x})^3 + 1 = x^{3/2} + 1,$$

which clearly is not the same as $F(G(x)) = \sqrt{x^3 + 1}$. By substituting a couple of values for x—say, $x = 1$ or $x = 2$, you can see that the results are numerically different. In general, except in rare cases,

$$G(F(x)) \neq F(G(x)).$$

However, if F and G are inverse functions, the equality does hold.

EXAMPLE 7

In Example 6 we chose $f(x) = \log x$ and $g(x) = x^2 - 5x + 2$. Find $f(g(x))$ and $g(f(x))$.

Solution   We have

$$f(g(x)) = f(x^2 - 5x + 2) = \log(x^2 - 5x + 2),$$

whereas

$$g(f(x)) = g(\log x) = (\log x)^2 - 5 \log x + 2.$$

Clearly, they are very different functions. ◆

Applications of Composite Functions

We next consider a real-world application of composite functions in Example 8.

EXAMPLE 8

When a kicker punts a football, it's path can be modeled by the quadratic function $y = f(x) = -x^2/27 + 1.92x + 1$, where the height y and the horizontal distance downfield x from the point where the ball is kicked are measured in yards. Furthermore, the horizontal distance x from the kicker is given by $x = g(t) = 12t$, where t is measured in seconds.

a. Find an equation giving the height of the football as a composite function of time t.

b. Determine the hang-time for the football—how long it remains in the air after being punted.

Solution

a. The path of the ball is the parabola shown in Figure 4.56, where $y = f(x) = -x^2/27 + 1.92x + 1$. The graph shows that the ball carries somewhat more than 50 yards from the point where it is kicked, which is usually about 10 yards behind the line of scrimmage. Using the formula $x = g(t) = 12t$ for x as a function of t, we can form the composite function giving the height y as a function of t:

$$y = f(x) = f(g(t)) = -\frac{(12t)^2}{27} + 1.92(12t) + 1$$

$$\approx \frac{-144t^2}{27} + 23t + 1$$

$$\approx -5.33t^2 + 23t + 1.$$

Note that this is also a quadratic function of t with a negative leading coefficient.

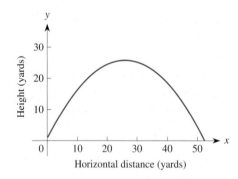

FIGURE 4.56

b. The hang-time for the football is the value of t when the ball comes back to the ground. It is the zero of the composite function, so we must solve the quadratic equation

$$-5.33t^2 + 23t + 1 = 0.$$

Equivalently, if we multiply both sides by -1, we get

$$5.33t^2 - 23t - 1 = 0.$$

Using either graphical methods or the quadratic formula, we find that $t \approx 4.36$ seconds.(A second solution to the quadratic equation gives a negative value for t, which makes no sense in this context.)

EXAMPLE 9

Two functions f and g are defined in the following table. Use the values given in the table to complete it. (If any operations are not defined, write "UNDEF.")

x	$f(x)$	$g(x)$	$f(x) - g(x)$	$f(x) \cdot g(x)$	$f(x)/g(x)$	$f(g(x))$	$g(f(x))$
0	1	3					
1	0	1					
2	3	0					
3	2	2					

Solution The values of the functions for four specific values of x—namely, $x = 0, 1, 2$, and 3—are defined in the table. The first open column asks for the difference between the two functions for each value of x. For instance, when $x = 0$, the first entry for this column is $f(0) - g(0) = 1 - 3 = -2$, and so on down that column. The second open column asks for the product of the two functions for each value of x. When $x = 0$, we get $f(0) \cdot g(0) = 1 \cdot 3 = 3$, and so on down the column.

The third open column asks for the quotient of the two functions. When $x = 0$, we have $f(0)/g(0) = 1/3$, and so on. However, because $g(2) = 0$, the quotient is not defined when $x = 2$, so we enter UNDEF in the corresponding position in the table.

The fourth and fifth open columns ask for values for the composite functions $f(g(x))$ and $g(f(x))$. In the fourth column, the function g is applied first and then the function f is applied. When $x = 0$, we need to form $f(g(0))$. To do so we evaluate $g(0) = 3$ first and then take $f(g(0)) = f(3) = 2$, so the first entry in the fourth column is 2. For the next entry, we start with $x = 1$ and form $f(g(1))$. Because $g(1) = 1$, we get $f(g(1)) = f(1) = 0$. Similarly, we get the remaining two entries in this column.

To fill in the entries in the last column, we reverse the order of operations of the two functions and apply first f, followed by g. Starting with $x = 0$, we now need $g(f(0))$. Because $f(0) = 1$, we have $g(f(0)) = g(1) = 1$. Similarly, when $x = 1$, we need $g(f(1))$. Because $f(1) = 0$, we have $g(f(1)) = g(0) = 3$. Incidentally, for each of the four values of x, $f(g(x)) \neq (g(f(x)))$.

We now have the completed table.

x	$f(x)$	$g(x)$	$f(x) - g(x)$	$f(x) \cdot g(x)$	$f(x)/g(x)$	$f(g(x))$	$g(f(x))$
0	1	3	-2	3	$\frac{1}{3}$	2	1
1	0	1	-1	0	0	0	3
2	3	0	3	0	UNDEF	1	2
3	2	2	0	4	1	3	0

Problems

1. For $f(x) = 3x - 4$ and $g(x) = \frac{1}{x}$, find

   a. $f(5) + g(5)$.
   b. $f(5) - g(5)$.

   c. $f(5) \cdot g(5)$.
   d. $\dfrac{f(5)}{g(5)}$.

   e. $f(g(5))$.
   f. $g(f(5))$.
   g. $f(f(5))$.
   h. $g(g(5))$.
   i. $f(x) + g(x)$.
   j. $f(x) - g(x)$.

   k. $f(x) \cdot g(x)$.
   l. $\dfrac{f(x)}{g(x)}$.

   m. $f(g(x))$.
   n. $g(f(x))$.
   o. $f(f(x))$.
   p. $g(g(x))$.

2. Repeat Problem 1 for $f(x) = x^2 + 4$ and $g(x) = \sqrt{x}$.

3. Repeat Problem 1 for $f(x) = 10^x$ and $g(x) = \log x$.

4. Two functions f and g are defined in the table at the bottom. Use the values given to complete the table. If any of the entries are not defined, write "UNDEF."

5. The functions f and g have the values $f(2) = 10$, $f(4) = 20$, $f(6) = 35$, $g(2) = 8$, $g(4) = 4$, and $g(6) = 2$. Which expressions, (a)–(g), are correct, which are incorrect, and which, if any, are not defined?

   a. $f(6) - f(4) = 2$
   b. $f(g(6)) = 35$
   c. $g(g(6)) = 8$
   d. $f(2) - g(6) = 8$
   e. $f(4) - g(4) = 0$
   f. $f(4) \cdot g(4) = 16$
   g. $f(4)/g(4) = 5$

6. Two functions f and g are given in the accompanying figure. The six graphs (a)–(f) represent $f + g$,

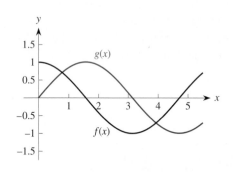

$f - g$, $g - f$, $f \cdot g$, f/g, and g/f. Decide which is which and give reasons for your answers.

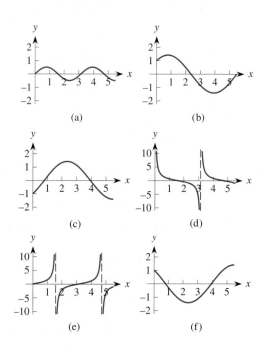

7. For the pairs of functions f and g shown, sketch the graph of the function $y = f(x) + g(x)$.

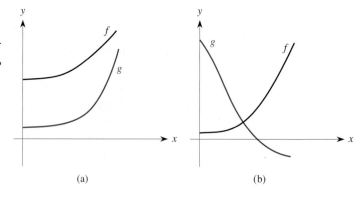

8. For the pairs of functions f and g shown, sketch the graph of the function $y = f(x) - g(x)$.

x	$f(x)$	$g(x)$	$f(x) - g(x)$	$f(x)/g(x)$	$g(x)/f(x)$	$f(x) \cdot g(x)$	$f(g(x))$	$g(f(x))$	$f^{-1}(x)$
0	1	0							
1	2	3							
2	3	1							
3	0	2							

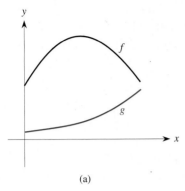

(a)

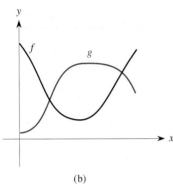

(b)

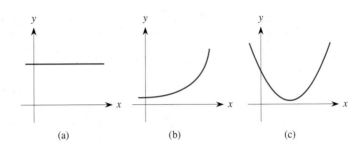

(a)   (b)   (c)

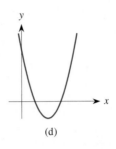

(d)

11. Often in technical books and articles, graphs are shown for log y as a function of x, as in the following graphs. In each case, given the graph of log y as a function of x, sketch the graph of y as a function of x.

9. The graphs of three functions, **(a)**–**(c)**, are shown in the accompanying figure.

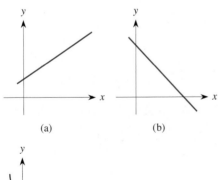

(a)   (b)

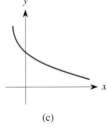

(c)

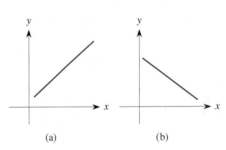

(a)   (b)

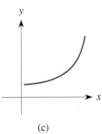

(c)

Sketch a rough graph of **(i)** 2^f, **(ii)** log f, and **(iii)** f^2. If any portion of a graph is not defined, mark it on the x-axis.

10. For each function **(a)**–**(d)**, sketch the graph of log f. If any portion of a graph is not defined, mark it along the x-axis.

12. Match each function with its graph.

a. $y = \dfrac{x^2 - 1}{x^2 - x - 6}$ **b.** $y = \dfrac{x^2 + 1}{x^2 - x - 6}$

c. $y = \dfrac{9 - x^2}{x^2 - 4}$ **d.** $y = \dfrac{x^2 - x - 6}{x^2 - 1}$

e. $y = \dfrac{x^3 - x}{x^2 - 4}$ **f.** $y = \dfrac{(x - 1)(x - 4)}{x^2 - 4}$

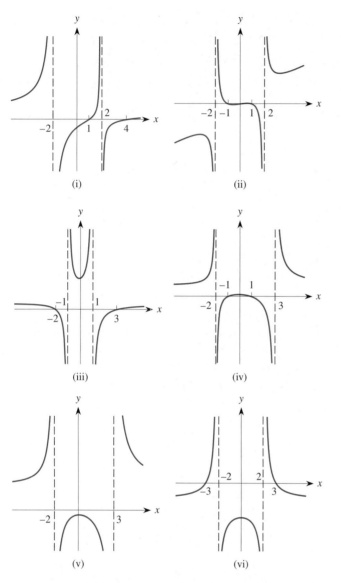

(i)
(ii)
(iii)
(iv)
(v)
(vi)

13. For the two functions f and g that are defined by the graphs shown, find **(a)** $f(g(1))$; **(b)** $g(f(1))$; **(c)** $f(g(-1))$; and **(d)** $g(f(-1))$.

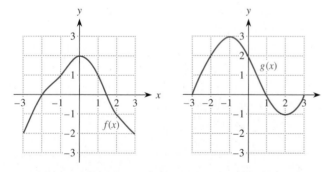

14. For the functions f and g that are defined by the graphs in Problem 13, sketch the graph of **(a)** $f(g(x))$ and **(b)** $g(f(x))$.

For Problems 15–18, determine functions F and G such that $h(x) = F(G(x))$. There are different correct answers to this question; however, do not use $F(x) = x$ or $G(x) = x$.

15. $h(x) = x^4 + 5$ **16.** $h(x) = (x + 5)^4$

17. $h(x) = \log(x + 3)$ **18.** $h(x) = 3 + \log x$

19. The time t that a traffic light should remain yellow depends on the speed limit s on the road. The function $t = 1 + s/20 + 70/s$, where t is measured in seconds and s is the speed in feet per second, is used to determine the length of the yellow cycle. Note that 30 mph = 44 ft/sec.

a. How long is the yellow cycle if the posted speed limit is 30 mph?

b. How long is the yellow cycle if the posted speed limit is 50 mph?

c. What are reasonable values for the domain and range of this function?

d. Suppose that the traffic department using this formula wants to increase the length of the yellow cycle somewhat. Should it increase or decrease the values of each of the two parameters 20 and 70 to do so?

e. Rewrite the formula for t as a rational function by combining all the terms over a common denominator.

20. According to Einstein's theory of relativity, the mass M of an object increases as its speed increases according to the formula

$$M = f(v) = \frac{M_0}{\sqrt{1 - (v^2/c^2)}} = M_0\left(1 - \frac{v^2}{c^2}\right)^{-1/2},$$

where M_0 is the mass of the object when it is at rest ($v = 0$) and c is the speed of light (about 186,282 miles per second). Suppose that an object has a rest mass of $M_0 = 1$ unit.

a. Construct a table of values for the mass of the object for each of the following speeds expressed as a fraction of the speed of light: $v = 0, 0.5c, 0.9c, 0.95c, 0.99c,$ and $0.999c$.

b. Sketch a graph showing the behavior of the mass of an object as its speed approaches the speed of light.

c. What is the mathematical significance of the speed of light? What is the physical significance of the speed of light in the context of the speeds of moving objects?

21. Some physicists hypothesize the existence of particles called *tachyons* that exist only at speeds greater than that of light. The slower that a tachyon moves, the greater is its mass; the speed of light is a lower limit on the possible speed of a tachyon. Sketch a graph of the mass as a function of speed for all possible values of $v \geq 0$. Indicate which region corresponds to normal particles and which to tachyons.

22. According to Newton's laws of motion, the speed of an object can be changed only by applying a force. Also, the greater the mass of an object, the more force is needed to accelerate it to a given velocity in a fixed amount of time. Suppose that an object is to be accelerated from speed 0 to almost the speed of light.

    a. Sketch the graph of the force needed to accelerate it as a function of the velocity v. Pay careful attention to concavity.
    b. Sketch the graph of the velocity as a function of the force needed, paying careful attention to concavity.

23. a. Graph the two functions $y = \sqrt{x^2 + 25}$ and $x + 5$. Are they the same?
    b. Repeat part (a) with $y = \sqrt{x^2 + 4}$ and $y = x + 2$. Are they the same?
    c. Can you find any value for a for which
    $$\sqrt{x^2 + a^2} = x + a?$$

24. a. Graph the two functions $y = \dfrac{1}{x + 4}$ and $y = \dfrac{1}{x} + \dfrac{1}{4}$. Are they the same?

    b. Repeat part (a) with $y = \dfrac{1}{x - 5}$ and $y = \dfrac{1}{x} - \dfrac{1}{5}$. Are they the same?

    c. Can you find any value for a so that
    $$\frac{1}{x + a} = \frac{1}{x} + \frac{1}{a}?$$

25. For the function $f(x) = \frac{x + 1}{x}$, (a) what is $f(1)$? (b) $f(f(1))$? (c) $f(f(f(1)))$? (d) Continue to apply the function f repeatedly to the previous result, expressing all your answers as fractions. Do you observe any pattern in the values for the numerators and denominators of the fractions that you're generating? (e) Now look at the decimal representations of the fractions that you generated in parts (a)–(d). Do they appear to be approaching a fixed value?

26. For any two linear functions $f(x) = ax + b$ and $g(x) = cx + d$, is $f \circ g$ the same as $g \circ f$?

27. The volume of a sphere is given by $V = \frac{4}{3}\pi r^3$ and its surface area is given by $S = 4\pi r^2$.

    a. Find a formula for the volume as a function of the surface area. Interpret the result in terms of a composite function.
    b. Find a formula for the inverse function of the function you found in part (a). What does it tell you?

28. The degree of polynomial P is m and the degree of polynomial Q is n, where $m < n$.

    a. What is the degree of $P + Q$?
    b. What is the degree of $P - Q$?
    c. What is the degree of $P \cdot Q$?
    d. What is the end behavior of P/Q?
    e. What is the end behavior of Q/P?

29. In Problem 20 of Section 1.3, we introduced a function f that represents a simple replacement code in which each letter of the alphabet is replaced by a different letter according to $f(A) = M$, $f(B) = D$, $f(C) = K$, $f(D) = V$, $f(E) = X$, $f(F) = B$, $f(G) = P$, $f(H) = T$, $f(I) = J$, $f(J) = S$, $f(K) = Z$, $f(L) = Q$, $f(M) = H$, $f(N) = O$, $f(O) = A$, $f(P) = L$, $f(Q) = W$, $f(R) = C$, $f(S) = F$, $f(T) = Y$, $f(U) = R$, $f(V) = G$, $f(W) = I$, $f(X) = U$, $f(Y) = N$, and $f(Z) = E$.

    Suppose that we now have a second such code defined by the function g:

    $g(A) = P$, $g(B) = K$, $g(C) = T$, $g(D) = E$, $g(E) = L$, $g(F) = U$, $g(G) = H$, $g(H) = N$, $g(I) = Y$, $g(J) = C$, $g(K) = R$, $g(L) = W$, $g(M) = G$, $g(N) = Z$, $g(O) = B$, $g(P) = J$, $g(Q) = A$, $g(R) = X$, $g(S) = Q$, $g(T) = D$, $g(U) = S$, $g(V) = M$, $g(W) = V$, $g(X) = I$, $g(Y) = O$, and $g(Z) = F$.

    a. Find $g(f(A))$.   b. Find $f(g(A))$.
    c. Find $f(f(P))$.   d. Find $g(g(K))$.
    e. Find $f^{-1}(g^{-1}(A))$.

30. The algebraic method of elimination for solving a system of linear equations involves adding a multiple of one equation to another equation to eliminate one of the variables. Consider the system of two equations in two unknowns:

    $$y = 4x - 3 \qquad \textbf{(1)}$$
    $$y = 7 - x. \qquad \textbf{(2)}$$

    a. Plot the two lines carefully on a sheet of graph paper and determine the point of intersection.
    b. Solve the two equations algebraically.

c. Add two times Equation (2) to Equation (1) to get a new linear function. Plot its graph on the same graph you created in part (a). What do you observe about the three lines?

d. Add three times Equation (2) to Equation (1) and plot that function on the same graph. What do you observe about the four lines?

e. Add four times Equation (2) to Equation (1) and plot that function on the same graph. What can you conclude from this result?

f. Find an appropriate multiple of Equation (2) that, when added to Equation (1), will eliminate the x term. What will the graph of the resulting line look like when x is eliminated?

4.7 Building New Functions from Old: Shifting, Stretching, and Shrinking

In Section 4.6, we created new functions from known functions by extending the standard arithmetic operations of addition, subtraction, multiplication, and division to functions. We also created new functions by using composition of functions. In this section we introduce several other ways in which we can build new functions from a single function. Suppose that we have the function $y = f(x)$. We can form a related function by changing either the independent variable x or the dependent variable y by multiplying it by a constant or by adding or subtracting a constant from it.

Shifting Functions

We can shift functions up and down or left and right. The former involves transforming the y-variable, and the latter involves transforming the x-variable.

Shifting Up and Down   We first investigate the effect on any function $y = f(x)$ of adding a constant to y or subtracting a constant from y.

◆ **EXAMPLE 1**

Consider $y = f(x) = x^2$ and the related functions $y = x^2 + 1$, $y = x^2 + 3$, $y = x^2 - 2$, and $y = x^2 - 5$. What is the effect of the constant in each case?

Solution   All these functions are shown in Figure 4.57. Clearly, each constant shifts the basic parabola $y = x^2$ up or down by the corresponding amount that is added or subtracted. For instance, the curve $y = x^2 + 1$ lies 1 unit above $y = x^2$ for each value of x, whereas $y = x^2 - 2$ lies 2 units below it.

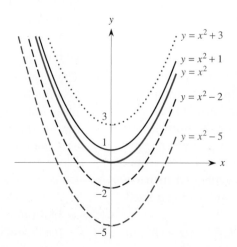

FIGURE 4.57

In general, the following principle holds for any function of x, assuming $b > 0$.

Vertical Shift

Replacing $f(x)$ with $f(x) + b$ shifts the graph of $f(x)$ *up* by the amount b.

Replacing $f(x)$ with $f(x) - b$ shifts the graph of $f(x)$ *down* by the amount b.

We can get a different feel for what is happening if we rewrite each of these expressions by moving the constant term to the left side. For instance, $y = x^2 + 1$ is equivalent to $y - 1 = x^2$, which emphasizes the fact that it is the variable y, or the height, which is being affected by the constant.

We can therefore rephrase the vertical shift principle for any function of x, assuming $b > 0$, as follows.

Vertical Shift

Replacing y with $y - b$ shifts the graph of $f(x)$ *up* by the amount b.

Replacing y with $y + b$ shifts the graph of $f(x)$ *down* by the amount b.

Shifting Left and Right   Next we investigate the effect on $y = f(x)$ of adding a constant to x or subtracting a constant from x.

EXAMPLE 2

Consider $y = f(x) = x^2$ and the related functions $y = (x - 1)^2$, $y = (x - 3)^2$, and $y = (x + 2)^2$, where we replace x by $(x - 1)$, $(x - 3)$, or $(x + 2)$, respectively. What is the effect of the constant in each case?

Solution   The resulting graphs are shown in Figure 4.58. Each of these changes causes a horizontal shift. For instance, $y = (x - 1)^2$ has a double zero at $x = 1$, so the graph of $y = x^2$ is shifted to the right by 1 unit. Similarly, $y = (x + 2)^2$ has a double zero at $x = -2$, so the graph of $y = x^2$ is shifted to the left by 2 units.

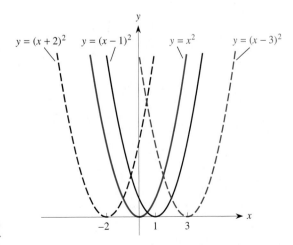

FIGURE 4.58

In general, the following principle holds for any function of x.

Horizontal Shift

Replacing x with $x - a$ shifts the graph of $f(x)$ to the *right* by the amount $a > 0$.
Replacing x with $x + a$ shifts the graph of $f(x)$ to the *left* by the amount $a > 0$.

Thus, for instance, the graph of $y = 10^{x-2}$ has the identical shape as the graph of $y = 10^x$, but is shifted to the right by 2 units. Similarly, the graph of $y = \sqrt{x + 3}$ has the same shape as the graph of $y = \sqrt{x}$, but is shifted to the *left* by 3 units. Check these and other graphs on your function grapher.

In summary, when we replace x by $x - a$ or $x + a$, we are changing x and so produce a horizontal effect. When we replace y with $y - b$ or $y + b$, we produce a vertical effect.

When we combine a horizontal shift (replace x by $x - a$) and a vertical shift (replace y by $y - b$), we effectively have a diagonal shift. For example, consider the graph of $y = (x - 4)^2 + 7$, or equivalently $y - 7 = (x - 4)^2$. It involves a change in x (x is replaced by $x - 4$) and a change in y (y is replaced by $y - 7$). So, $y = (x - 4)^2 + 7$ corresponds to shifting the parabola $y = x^2$ four units to the right and seven units up. This produces a parabola whose vertex is at $(4, 7)$, as shown in Figure 4.59.

Similarly,

$$x^2 + y^2 = r^2$$

is the equation of a circle with radius r centered at the origin (see Appendix A6). We should then expect that

$$(x - 5)^2 + (y - 3)^2 = r^2$$

is the graph of a circle with radius r that has been shifted 5 units to the right and 3 units up. It is therefore the equation of a circle with radius r centered at the point $(5, 3)$. The new circle is produced from the original circle by a combination of a horizontal shift (5 units to the right) and a vertical shift (3 units up), as shown in Figure 4.60.

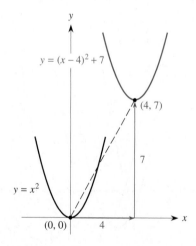

FIGURE 4.59

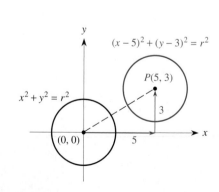

FIGURE 4.60

Stretching and Shrinking Functions

We can stretch and shrink functions vertically and horizontally. Vertical stretching and shrinking involves multiplying a function by a constant, whereas horizontal stretching and shrinking involves multiplying the independent variable by a constant.

A Constant Multiple of a Function   We can also create a new function from a given function by multiplying the function, or equivalently the *y*-value, by a constant. For example, consider the two functions $y = 2^{-x}$ and $y = 5 \cdot 2^{-x}$. Both are exponential decay functions, as shown in Figure 4.61. The function $y = 2^{-x}$ passes through the point $(0, 1)$, whereas the transformed function $y = 5 \cdot 2^{-x}$ passes through the point $(0, 5)$, so you might be tempted to think of the second function as resulting from a vertical shift of the first. However, think about what each looks like for large values of *x*; both curves have the *x*-axis as a horizontal asymptote. Therefore the relationship between them cannot be a vertical shift. In particular, the height for every point on the curve $y = 5 \cdot 2^{-x}$ is five times the height of the corresponding point (with the same value for *x*) on the curve $y = 2^{-x}$. The effect of the constant multiple 5 is to increase the height all along the curve by a factor of 5. If we multiply the original function by 20, the curve will be *stretched* to a new curve that is everywhere 20 times as tall.

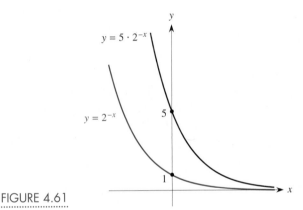

FIGURE 4.61

If instead we multiply the original function by $\frac{1}{4}$, the curve will *shrink* to a new curve that is one fourth the original height for each value of *x*. Finally, if we multiply the function by a negative constant, such as -3, the curve is stretched by a factor of 3, but it is also flipped upside down across the horizontal axis. Figure 4.62 shows the graphs of $y = \sqrt{x}$ and $y = -3\sqrt{x}$. Not only is the graph of the second function flipped upside down across the *x*-axis, but it also moves downward much faster (three times as fast) than the first function rises. Verify this result on your function grapher with some other functions.

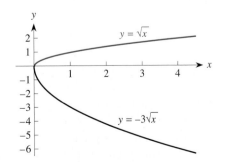

FIGURE 4.62

In general, we have the following principle.

Vertical Stretching and Shrinking

Multiplying a function by a constant changes the height of its graph by that multiple, but it does not change the general shape.

If the multiple is greater than 1, the height is increased.

If the multiple is a number between 0 and 1, the height is decreased.

If the multiple is negative, the curve is flipped over across the horizontal axis.

We illustrate an application of some of these ideas in Example 3.

EXAMPLE 3

Suppose that a chicken is taken from the freezer at 0°F and put directly into an oven kept at a constant temperature of 350°F. After 30 minutes, the temperature of the chicken is 110°F. Construct a function to model the temperature of the chicken as it cooks in the oven.

Solution   The temperature of the chicken rises rapidly at first and then increases ever more slowly the closer the chicken's temperature comes to the oven temperature of 350°. Eventually, the temperature of the chicken levels off at the temperature setting for the oven. The temperature T, in °F, plotted against time t, in hours, looks like the graph shown in Figure 4.63. (This description is actually an oversimplification because the temperature rise will temporarily stop at the freezing point of 32° while the ice melts. Also, the chicken should be removed from the oven when its temperature reaches about 180°, or it will begin to burn.)

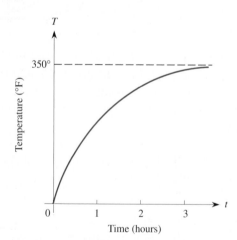

FIGURE 4.63

The horizontal line representing the temperature of 350° is a horizontal asymptote because the curve gets even closer to this line as time goes by, but never quite reaches it. The rate at which the temperature of the chicken increases slows as it approaches 350° (if we left the chicken in the oven that long), so the curve is concave down.

We want to model this process by creating a formula giving the temperature T as a function of the time t. Simplistically we will find a mathematical model by inspecting the graph of the process and deciding which type of function has the right shape. In Section 5.4 we demonstrate how to construct such a function directly.

The graph in Figure 4.63 appears to be an exponential decay function turned upside down so that it rises toward the oven temperature 350° instead of dropping asymptotically toward the horizontal axis. We can form such a function from a pure exponential function $y = Ac^x$ by using a negative coefficient (to turn the curve upside down) and a vertical shift so that the curve approaches 350 instead of 0. Thus a formula for T might look like

$$T = 350 - Ac^t,$$

where t is in hours and $0 < c < 1$. As t increases, the term c^t approaches 0, and the entire expression $350 - Ac^t$ approaches 350.

What might be possible values for A and c? We know that at time $t = 0$, the chicken's temperature is $T = 0$ when it comes out of the freezer, so

$$T(0) = 350 - Ac^0 = 350 - A = 0.$$

Thus $A = 350$ and the formula becomes

$$T = 350 - 350(c^t).$$

Furthermore, the temperature of the chicken after half an hour is $T(\frac{1}{2}) = 110°$. This value yields

$$T(\tfrac{1}{2}) = 350 - 350(c^{1/2}) = 110.$$

So we have

$$350(c^{1/2}) = 350 - 110 = 240;$$
$$c^{1/2} = 240/350 = 0.686.$$

Squaring both sides of this equation gives

$$c = 0.47.$$

Consequently, our formula for the temperature becomes

$$T = 350 - 350(0.47)^t = 350[1 - (0.47)^t],$$

where t is measured in hours.

This function is an upside down exponential: As t increases, $(0.47)^t$ gets ever smaller, so $1 - (0.47)^t$ increases and gets ever closer to 1. That is, $1 - (0.47)^t \to 1$ as $t \to \infty$. Consequently,

$$T = 350[1 - (0.47)^t] \to 350, \qquad \text{as } t \to \infty,$$

confirming that the graph has a horizontal asymptote at $T = 350$.

Think About This   Verify the behavior of the preceding function on your function grapher. Look at the overall shape and then zoom in to verify the height of the asymptote. Estimate by eye from the graph when T reaches 180°, when it reaches 250°, and when it reaches 300°, 340°, and 349°. ⌐

In general, consider the function $y = f(t) = L + Ac^t$, where $0 < c < 1$. We know that as t increases, c^t decays toward zero so that the function approaches a limiting value of L. The question is: How does it approach L—from above or from below? First, whenever $A < 0$, the values of the function are less than L. As the term Ac^t decreases, the amount subtracted from L decreases, and the values of the function increase toward L in a concave down manner, as illustrated by the upper curve in Figure 4.64. This was the case with the temperature of the chicken in the oven. Second, whenever $A > 0$, the values of the function are greater than L and so decrease toward it in a concave up manner, as illustrated by the lower curve in Figure 4.64.

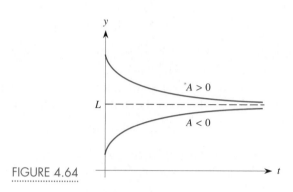

FIGURE 4.64

This type of function is used as the mathematical model for many different real-world processes.

A Constant Multiple of the Independent Variable   Finally, we investigate the effects of multiplying the independent variable x by a constant.

EXAMPLE 4

Consider the cubic function $y = f(x) = x^3 - 12x$ and the related functions $y = f(2x)$, $y = f(4x)$, and $y = f(\frac{1}{2}x)$. What is the effect of the constant multiple in each case?

Solution   Figure 4.65 shows the graphs of $y = f(x) = x^3 - 12x$ and $y = f(2x) = (2x)^3 - 12(2x)$. The cubic $y = f(x) = x^3 - 12x$ passes through the origin and has two turning points. If you trace along the curve, you will find that one turning point is at $x = 2$ and the other at $x = -2$. (We could also locate the turning points by using the formula presented in Section 4.4.) The corresponding local maximum (at $x = -2$) is at a height of $y = 16$ and the local minimum (at $x = 2$) is at a height of $y = -16$.

The cubic $y = f(2x) = (2x)^3 - 12(2x) = 8x^3 - 24x$ also passes through the origin and has two turning points, one at $x = 1$ and the other at $x = -1$. The corresponding local maximum is at $y = 16$, and the local minimum is at $y = -16$. Hence the heights are the same; they just occur sooner. In fact, the curve for $y = f(2x)$ traces out the identical vertical values as $f(x)$, but does so twice as fast.

Figure 4.66 shows the graphs of $y = f(x) = x^3 - 12x$ and $y = f(4x) = (4x)^3 - 12(4x) = 64x^3 - 48x$. The local maximum for $y = f(4x)$ now occurs at $x = -\frac{1}{2}$ and the local minimum occurs at $x = \frac{1}{2}$. Again, the same heights are achieved, but the curve $y = f(4x)$ is traced out four times as fast as $y = f(x)$.

Figure 4.67, shows the graphs of $y = f(x) = x^3 - 12x$ and $y = f(\frac{1}{2}x) = (\frac{1}{2}x)^3 - 12(\frac{1}{2}x) = \frac{1}{8}x^3 - 6x$, but we had to extend the window to show the details. The function $y = f(\frac{1}{2}x)$ achieves its local maximum at $x = -4$ and its local minimum at $x = 4$. The curve $y = f(\frac{1}{2}x)$ traces out the identical heights, but does so half as fast as $y = f(x)$.

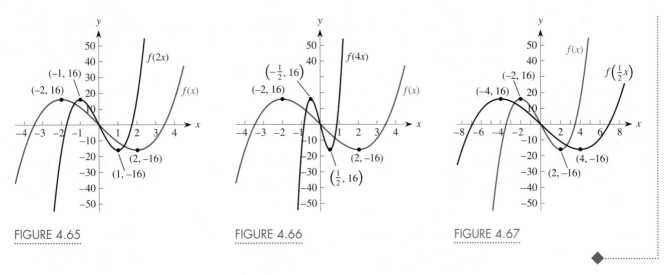

FIGURE 4.65   FIGURE 4.66   FIGURE 4.67

We summarize the ideas on stretching and shifting functions horizontally as follows.

> ### Horizontal Stretching
>
> Multiplying the independent variable x by a constant k changes the speed at which the graph is traced out, but it does not change the general shape.
>
> If the multiple k is greater than 1, the graph of $y = f(kx)$ is traced out k times faster than $y = f(x)$.
>
> If the multiple k is between 0 and 1, the graph of $y = f(kx)$ is traced out more slowly than $y = f(x)$.
>
> If the multiple k is negative, then the curve $y = f(kx)$ is reflected across the y-axis.

EXAMPLE 5

For the function $f(x) = x^3 - 12x$, draw the graph of $f(-3x)$ and locate its turning points.

Solution   Figure 4.68 shows the graphs of the two functions. The graph of $y = f(-3x)$ has the same basic shape as the graph of $y = f(x)$, but is flipped upside down across the

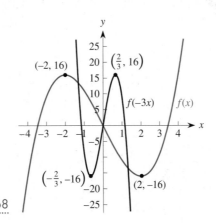

FIGURE 4.68

x-axis and is traced out 3 times as fast. As before, the turning points for $y = f(x)$ are at $(-2, 16)$ and $(2, -16)$. The turning points for $y = f(-3x)$ are at $(-\frac{2}{3}, -16)$, a local minimum, and at $(\frac{2}{3}, 16)$, a local maximum.

EXAMPLE 6

A function *f* is defined in the following table. Use the values given in the table to complete it. If any entries are not defined, mark them "UNDEF."

x	$f(x)$	$f(x) - 1$	$f(x - 1)$	$f(2x)$	$3f(x)$
0	1				
1	0				
2	3				
3	2				

Solution   The values of the function for $x = 0, 1, 2,$ and 3 are defined in the table. The first open column asks for a vertical shift when the function's values are reduced by 1 for each value of *x*. For instance, when $x = 0$, the first entry is asking for $f(0) - 1 = 1 - 1 = 0$, and so on down that column.

The second open column asks for a horizontal shift of 1 unit to the right, because *x* is replaced by $x - 1$. Thus, when $x = 0$, we want $f(0 - 1) = f(-1)$, but there is no way to determine this value from the information given in the table; that is, the function is not defined for $x = -1$, so we record it in the table as "UNDEF." However, when $x = 1$, we want $f(1 - 1) = f(0) = 1$, and so on down the column.

The third open column asks for values when the independent variable is doubled. So, when $x = 0$, we need $f(2 \cdot 0) = f(0) = 1$; similarly, when $x = 1$, we need $f(2 \cdot 1) = f(2) = 3$. However, when $x = 2$, $f(2 \cdot 2) = f(4)$, which is not defined. When $x = 3$, $f(2 \cdot 3) = f(6)$ is also not defined.

Finally, the last open column asks for 3 times the value of the function. When $x = 0$, we need $3 \cdot f(0) = 3 \cdot 1 = 3$, and so on down the column. The completed table follows.

x	$f(x)$	$f(x) - 1$	$f(x - 1)$	$f(2x)$	$3f(x)$
0	1	0	UNDEF	1	3
1	0	-1	1	3	0
2	3	2	0	UNDEF	9
3	2	1	3	UNDEF	6

Problems

x	f(x)	5f(x)	f(x) + 3	f(x − 1)	[f(x)]²
3	5				
4	2				
5	−1				
6	3				
7	8				

1. A function f is defined in the table above. Use the values given to complete the table. If any of the entries are not defined, write "UNDEF."

2. A function $y = f(x)$ is defined by the accompanying graph. Match each transformation of f with one of the graphs (i)–(vi).

   a. $y = f(2x)$ **b.** $y = 2f(x)$
   c. $y = f(x) + 2$ **d.** $y = f(x + 2)$
   e. $y = f(x) - 2$ **f.** $y = f(x - 2)$

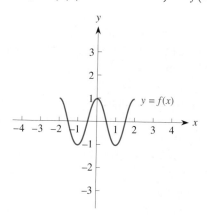

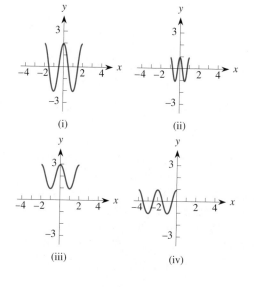

(i)   (ii)

(iii)   (iv)

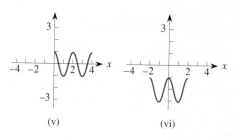

(v)   (vi)

3. For the functions f and g that are defined by the graphs shown, sketch the graph of

   a. $2g(x)$.   **b.** $g(2x)$.
   c. $f(x + 1)$.   **d.** $f(x - 1)$.
   e. $f(x)+1$.   **f.** $g\left(\dfrac{1}{2}x\right)$.

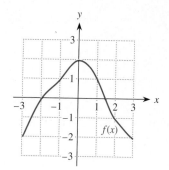

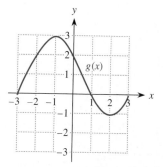

4. Consider the function $y = f(x) = x^2$.

   a. Write an equation for the function that you get when you stretch the graph of f by a factor of 2 and then shift it up 3 units. Call this new function F and sketch its graph.

   b. What is the equation you get if you reverse the order of the two operations in part (a)? Call this new function G and sketch it.

   c. What is $F - G$?

5. a. Translate the line $y = mx$ to a line with slope m that passes through the point $(5, 12)$.

   b. Repeat part (a) if the new line passes through the point (x_0, y_0). What do you call this new equation?

6. a. Translate the parabola $y = x^2$ to a parabola with vertex at $(5, 12)$.

   b. Repeat part (a) if the new parabola has its vertex at the point (x_0, y_0).

7. For the function f shown, sketch the graph of

   a. $y = -f(x)$ **b.** $y = 2f(x)$
   c. $y = f(x) - 1$ **d.** $y = f(x - 1)$
   e. $y = f(x + 1)$ **f.** $y = f(x) + 1$

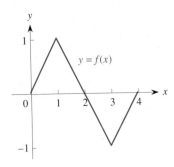

8. The graphs of three functions (a)–(c) are shown in the accompanying figures. Sketch the graph of

   (i) $y = -f(x)$ **(ii)** $y = 2f(x)$
   (iii) $y = -2f(x)$ **(iv)** $y = f(x + 2)$
   (v) $y = f(x) + 2$ **(vi)** $y = f(x) - 2$
   (vii) $y = f(x - 2)$

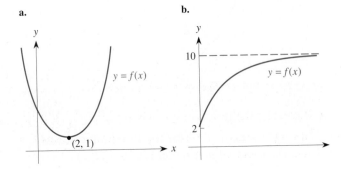

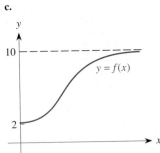

9. Consider the function f in the table. If this function is shifted 4 units to the right and 7 units upward, construct the corresponding table for the transformed function.

x	-1	0	1	2	3
y	24	16	11	8	9
x	4	5	6	7	
y	15	27	39	35	

10. a. Use the graphs of $f(x) = x$ and $g(x) = \log x$ to sketch a rough graph of the product $P(x) = x \log x$.

   b. Estimate the values of x for which $\log x < x$ and the values for which $\log x > x$.

   c. Because $\log x$ grows exceedingly slowly, the product $x \log x$ grows only slightly faster than x does. Use your function grapher to decide whether $x \log x$ ever grows faster than $x^{1.5}$, than $x^{1.1}$, than $x^{1.05}$. What does this investigation suggest to you about the rate of growth of $x \log x$ compared to power functions x^p?

11. If $f(x) = x^2 - 3x + 4$ and h is a constant, find

   a. $f(x) + h$ **b.** $f(x + h)$

   c. $f(x + h) - f(x)$ **d.** $\dfrac{f(x + h) - f(x)}{h}$

   e. What is the value of the expression in part (d) if $x = 5$ and if $h = 0.1$? if $h = 0.01$? if $h = 0.0001$?

12. a. An unbaked apple pie is taken from the counter in a kitchen where the temperature is 70°F and placed in an oven. Suppose that, after 60 minutes, the temperature of the pie is 180°F. Sketch a graph of the temperature of the pie as a function of time.

   b. The pie is removed from the oven and placed back on the counter. Suppose that it takes another 60 minutes for its temperature to come back down to 70°F. Sketch a graph of the temperature of the pie as a function of time.

c. When the first pie is removed from the oven, a second, unbaked pie is put in the oven to bake. Sketch a graph of the sum of the temperatures of the two pies as a function of time over the 60-minute period.

d. Find a formula that models the temperature of the pie, while it cools, as a function of time.

13. A Thanksgiving turkey is taken from the refrigerator at a temperature of 40°F and placed in a hot oven at 350°F to cook. After 1 hour, the internal temperature of the bird is 124°F. Write a possible formula for the temperature of the turkey as a function of time, in minutes.

14. In an attempt to claim responsibility for winning the war against the growing national balance of trade deficit, the president presented a graph similar to the one shown to illustrate the trend in the *annual* deficit.

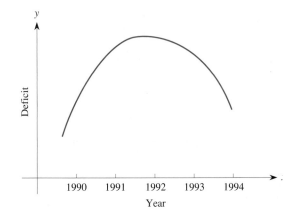

a. Based on this graph, sketch the graph of the *total* national debt as a function of time.

b. Does your graph have any points of inflection? If so, what do they represent?

c. Do you agree or disagree with the president's assertion that the war has been won? Explain.

15. Use your function grapher to graph the functions $f(x) = x^n(0.5)^x$, for $n = 1, 2, 3, 4, 5$, and estimate the location of the turning point for each curve for $x > 0$. Then perform a linear regression analysis on the x-values of these turning points, as functions of n. Is the linear fit appropriate? What does it predict for $n = 1.5$? Is it accurate compared to the actual graph?

16. Use your function grapher to graph the functions $f(x) = x^2 a^x$, for $a = 0.3, 0.4, 0.5, 0.6$, and 0.7. Estimate the location of the turning point for each curve by zooming in on it. Then determine the function from among the usual families of functions—linear, exponential, and power—that best fits these data as a function of the base a.

17. Describe how you might use the results of Problems 15 and 16 to find a function of the form $f(x) = x^p a^x$ that matches the function for the level of Lyme disease antibody in the bloodstream discussed in Section 4.6 (see Figure 4.50).

18. Find conditions on the coefficients a, b, and c in $P(x) = ax^2 + bx + c$ if P is to satisfy each equation for all values of x.

a. $P(x) = P(-x)$ b. $P(x) = -P(x)$
c. $P(2x) = 2P(x)$

Exercising Your Algebra Skills

For the function $f(x) = x^2 - 5x + 3$, find a simplified expression for

1. $f(2x)$.

2. $f(3x)$.

3. $f(4x)$.

4. $f\left(\dfrac{1}{2}x\right)$.

5. $f(x + 1)$.

6. $f(x - 2)$.

7. $f(2x - 1)$.

8. $f(x^2)$.

4.8 ⸱⸱⸱⸱⸱ Using Shifting and Stretching to Analyze Data

The ideas on shifting and stretching functions in Section 4.7 can be applied to create functions that fit sets of data that do not quite fall into the standard behavior patterns, such as exponential growth or decay, that we have discussed.

Analyzing a Cooling Experiment

Suppose that an experiment is conducted to study the rate at which temperature changes. A temperature probe (a thermometer connected to a calculator) is first heated in a cup of hot water and then removed and placed in a cup of cold water, as illustrated in Figure 4.69. The temperature of the probe, in °C, is measured every second for 36 seconds and recorded in Table 4.3; the data are also displayed in the scatterplot in Figure 4.70.

FIGURE 4.69

TABLE 4.3 *Experimental Data: Temperature (°C) versus Time*

Time	Temperature	Time	Temperature	Time	Temperature
1	42.3	13	12.51	25	9.29
2	36.03	14	11.91	26	9.16
3	30.85	15	11.54	27	9.16
4	26.77	16	11.17	28	9.04
5	23.58	17	10.67	29	8.91
6	20.93	18	10.42	30	8.83
7	18.79	19	10.17	31	8.78
8	17.08	20	9.92	32	8.78
9	15.82	21	9.8	33	8.78
10	14.77	22	9.67	34	8.78
11	13.82	23	9.54	35	8.66
12	13.11	24	9.42	36	8.66

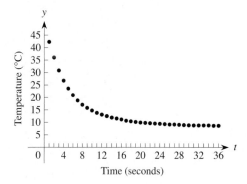

FIGURE 4.70

EXAMPLE 1

Find a function that fits the data from the temperature cooling experiment.

Solution The pattern depicted in Figure 4.70 is that of a decreasing, concave up function, so we might consider either a decaying exponential function or a power function with $p < 0$. However, a power function is not a good model for the process because it has a vertical asymptote at time $t = 0$, whereas the function we want must have a finite value when $t = 0$. So the more appropriate model would be an exponential decay function.

But there is a catch. Any exponential decay function decreases to zero, but the temperature readings decay to the temperature of the cold water (which cannot be 0°C, for then the water would be frozen). From the experimental data, the temperature of the cold water is about 8.6°C. How do we construct a function that decays to about 8.6 rather than to 0? Probably the most reasonable approach is to subtract 8.6 from each of the temperature readings to obtain a new set of data that decays to zero. This approach is equivalent to performing a vertical shift downward of 8.6 (i.e., replacing the temperature T with $T - 8.6$) to produce the transformed data shown in Table 4.4.

TABLE 4.4   *Transformed Data:* $(T - 8.6)$ *versus Time*

Time	$T - 8.6$	Time	$T - 8.6$	Time	$T - 8.6$
1	33.7	13	3.91	25	0.69
2	27.43	14	3.31	26	0.56
3	22.25	15	2.94	27	0.56
4	18.17	16	2.57	28	0.44
5	14.98	17	2.07	29	0.31
6	12.33	18	1.82	30	0.23
7	10.19	19	1.57	31	0.18
8	8.48	20	1.32	32	0.18
9	7.22	21	1.20	33	0.18
10	6.17	22	1.07	34	0.18
11	5.22	23	0.94	35	0.06
12	4.51	24	0.82	36	0.06

The scatterplot of the transformed data, shown in Figure 4.71, looks like an exponential decay pattern that tends toward 0. Using a calculator, we find that the exponential function that best fits the transformed data is

$$y = T - 8.6 = 35.4394(0.848)^t.$$

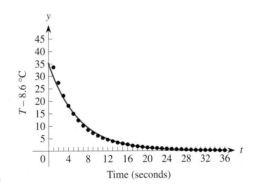

FIGURE 4.71

The graph of this function is shown superimposed over the transformed data in Figure 4.71, and there appears to be extremely close agreement. The corresponding correlation coefficient is $r = -0.9948$, which is very close to -1.

Having found the exponential function that best fits the transformed data, we now have to undo the transformation. We simply add the same amount, 8.6, to the function $y = T - 8.6$ to create the final expression

$$T(t) = 8.6 + 35.4394(0.848)^t.$$

This function is shown superimposed over the original temperature data in Figure 4.72, and it is an exceptionally good fit to the temperature readings. In particular, note how this function approaches the limiting value of about 8.6 for the temperature readings as t increases.

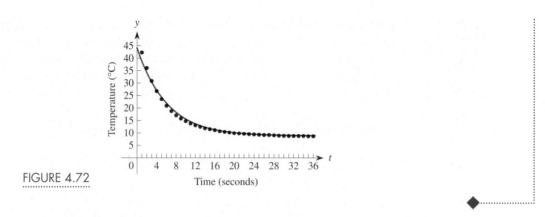

FIGURE 4.72

Analyzing the Challenger Data

At the beginning of Chapter 3, we considered data relating the number of incidents involving O-ring problems on space shuttle launches to the air temperature at launch. These data eventually were used to identify the O-rings as the likely cause of the *Challenger* disaster. We now use this set of data as a case study to illustrate the process of data analysis when it is necessary to shift the data values.

Recall that the dependent variable was the number N of O-ring problems or "incidents" as a function of launch temperature T. The data are shown in the following table.

T	53	57	58	63	66	67	67	67	68	69	70	70
N	3	1	1	1	0	0	0	0	0	0	1	1
T	70	70	72	73	75	75	76	76	78	79	80	81
N	0	0	0	0	2	0	0	0	0	0	0	0

Figure 4.73 shows the scatterplot for this data along with a curve superimposed over the data points to indicate the nature of the relationship, which appears to be a decaying exponential. However, this curve is only an artist's rendering of the apparent relationship. We want to obtain a formula for such a function.

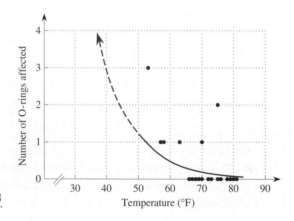

FIGURE 4.73

EXAMPLE 2

Find a function that can model the data on the number of O-ring incidents as a function of the air temperature.

Solution   The decreasing, concave up pattern in the scatterplot in Figure 4.73 suggests either a decaying exponential function or a power function with $p < 0$. The power function does not make sense, however, because there is no vertical asymptote at $T = 0$. Fitting an exponential function to the set of data by using the transformation approach used by calculators and spreadsheets involves plotting the logarithm of the number of incidents log N versus the temperature T. But because the values for N include $N = 0$, we cannot take the logarithm of 0—it is not defined!

One way to circumvent this problem is to shift the data values up to avoid the zeros. The simplest approach is to increase each value of N by 1, replacing N by $N + 1$ and then comparing $N + 1$ to T. We first construct the exponential function that best fits the resulting set of data to obtain the exponential regression equation relating $N + 1$ to T. We then shift back down to obtain an expression for N in terms of T. The data values that we work with are given in the following table, and the associated scatterplot of $N + 1$ versus T is shown in Figure 4.74.

T	N	$N + 1$	T	N	$N + 1$
53	3	4	70	0	1
57	1	2	70	0	1
58	1	2	72	0	1
63	1	2	73	0	1
66	0	1	75	2	3
67	0	1	75	0	1
67	0	1	76	0	1
67	0	1	76	0	1
68	0	1	78	0	1
69	0	1	79	0	1
70	1	2	80	0	1
70	1	2	81	0	1

The resulting exponential regression equation giving $N + 1$ as a function of T is

$$N + 1 = 13.41(0.967)^T,$$

which is shown superimposed on the scatterplot, in Figure 4.74. Finally, we solve for N by subtracting 1 from both sides to get the exponential function that can be used to model N as a function of T:

$$N = 13.41(0.967)^T - 1,$$

which is shown superimposed over the original scatterplot in Figure 4.75. It appears to capture the trend in the data reasonably well.

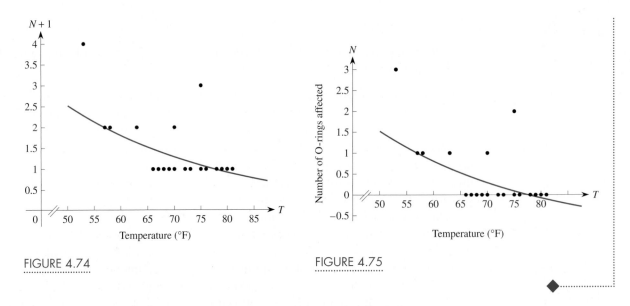

FIGURE 4.74

FIGURE 4.75

The graph certainly suggests that the likelihood of trouble with the O-rings will increase dramatically with falling temperature. However, we know that there is a danger in extrapolating far beyond the range of data values. But the overall trend is so dramatic and the potential loss in terms of both human life and hardware is so extreme that there shouldn't have been a launch if the data had been analyzed in this way.

Terminal Velocity in Skydiving

Matthew is a skydiving enthusiast. He knows, from his reading and from first-hand experience, that the faster he is falling, especially in a spread-eagled position, the greater the air resistance, so that eventually his speed reaches a maximum, known as the *terminal velocity*. He also found the following set of data on the downward velocity v, in feet per second, of a skydiver at different times t, in seconds, after jumping out of a plane.

t	0	1	2	3	4	5	6	7	8	9	10	11	12
v	0	16	46	76	104	124	138	148	156	163	167	171	174

Source: Student project.

EXAMPLE 3

(a) Find a function that fits these data on velocity as a function of time from among the usual candidates. (b) Improve on the fit by using an appropriate shift.

Solution

a. The data falls in an increasing, concave down pattern, as shown in Figure 4.76. The potential candidates for a function having such a pattern are either a power function with $0 < p < 1$ or a logarithmic function. However, a log function is not defined at $t = 0$. Also, both functions grow indefinitely, while the values for the skydiver's velocity approach terminal velocity, which is a horizontal asymptote. What's worse, we don't know what this limiting value for the terminal velocity is. Thus neither function can be a good fit.

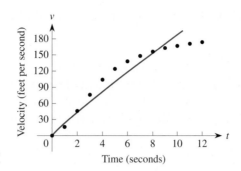

FIGURE 4.76

Moreover, we cannot use a calculator or spreadsheet program to fit a power function to the data because of the first point $(0, 0)$—their regression routines all use the transformation approach, which involves having to take the log of 0. However, if we delete the point $(0, 0)$, we can fit a power function to the remaining data. Figure 4.76 shows the graph of the best-fit power function $v = 23.2t^{0.908}$ (obtained using a calculator) superimposed over the scatterplot of the data. The corresponding correlation coefficient is $r = 0.962$, which is fairly close to 1. The power function is a reasonable fit, but it clearly becomes less good when extended to the right.

b. The pattern in the data suggests an upside down exponential decay function that rises toward a horizontal asymptote. Suppose that we conjecture a value for the terminal velocity by mentally extending the preceding table. We might extrapolate that the limiting value is about 180 ft/sec. We then subtract each velocity value from this supposed limit (replacing v with $180 - v$) to obtain the transformed data shown in the following table. Effectively, this transformation is equivalent to a vertical shift with a flip across the horizontal axis due to the multiple of -1.

t	0	1	2	3	4	5	6	7	8	9	10	11	12
$180 - v$	180	164	134	104	76	56	42	32	24	17	13	9	6

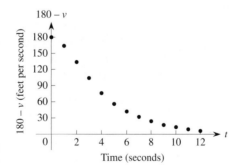

FIGURE 4.77

The scatterplot of these transformed data is shown in Figure 4.77. The decreasing, concave up pattern in this transformed data suggests either a decaying exponential function or a power function with $p < 0$; however, the latter has a vertical asymptote at zero, so it is not an appropriate candidate. A calculator gives the exponential function that best fits this transformed data as

$$y = 180 - v = 226.25(0.7492)^t$$

with a correlation coefficient of $r = -0.9963$. Figure 4.78 shows this function superimposed over the transformed data, and it is a very good fit.

We undo the transformation algebraically by solving for the velocity v to get

$$v = 180 - 226.25(0.7492)^t.$$

The graph of this function is shown superimposed over the original data in Figure 4.79, demonstrating a much better fit than the power function in Figure 4.76. This conclusion is further borne out by the correlation coefficient $r = -0.9963$, which is considerably closer to -1 than the correlation coefficient $r = 0.962$ for the power function was to 1.

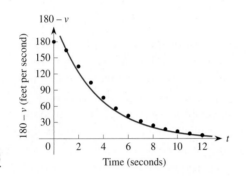

FIGURE 4.78

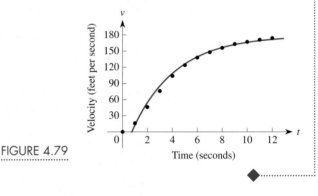

FIGURE 4.79

The value of 180 ft/sec that we chose for the terminal velocity was reasonable, but it was just an intelligent guess. Had we chosen a somewhat different value, we would have obtained a somewhat different function. With a little experimentation, you should be able to get a still better fit.

In Example 3, an exponential function was a very good fit to the transformed data, although the values for $180 - v$ did not fall precisely in an exponential decay pattern. Sometimes, a set of values fall precisely in an exponential pattern as they grow or decay toward a horizontal asymptote. The problem we face in such cases is not knowing exactly what that horizontal asymptote is, as was the case in Example 3. If the transformed data do fall in an exponential pattern, we can determine the limiting value precisely.

Suppose that a set of values $x_0, x_1, x_2, x_3, x_4, \ldots$ is such that the values either fall toward an unknown limiting value L or rise toward L in a purely exponential manner, as shown in Figure 4.80. In particular, suppose that each of the values is below the unknown horizontal asymptote L, so that the set of transformed values

$$L - x_0, \qquad L - x_1, \qquad L - x_2, \qquad L - x_3, \qquad L - x_4, \ldots$$

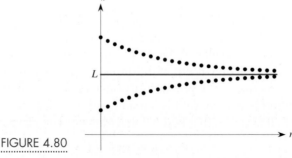

FIGURE 4.80

decays toward zero in an exponential decay pattern. As a result, we know that the successive ratios should be a constant, say k. That is,

$$\frac{L - x_1}{L - x_0} = \frac{L - x_2}{L - x_1} = \frac{L - x_3}{L - x_2} = \cdots = k,$$

where k is the constant, although unknown, ratio. Consider the first of these equalities:

$$\frac{L - x_1}{L - x_0} = \frac{L - x_2}{L - x_1}.$$

We can solve this equation algebraically for the unknown limiting value L by first cross-multiplying to get

$$(L - x_1)^2 = (L - x_0)(L - x_2).$$

Expanding these terms gives

$$L^2 - 2x_1L + x_1^2 = L^2 - x_0L - x_2L + x_0 x_2.$$

Subtracting L^2 from both sides of this equation and then collecting like terms yields

$$(x_0 - 2x_1 + x_2)L = x_0x_2 - x_1^2,$$

so that

$$L = \frac{x_0x_2 - x_1^2}{x_0 - 2x_1 + x_2}, \tag{1}$$

provided that the denominator $x_0 - 2x_1 + x_2 \neq 0$. In fact, if the numbers x_0, x_1, x_2, x_3, x_4, ... approach L in an exponentially decaying manner precisely, the comparable expression—using any three successive values of the x's, not just the first three—gives the same value for L. If the values are not exact, however—even if the discrepancies are due to rounding—quite different values could arise with every group of three successive values for the x's.

EXAMPLE 4

Prozac is prescribed for individuals suffering from depression. Typically, a patient takes a dose of Prozac once a day and, for extreme depression, the dosage is 80 mg. The levels of Prozac in the blood on successive days following the start of treatment are given in the following table. (Note that the last two values are rounded to four decimal places.) It turns out (we investigate this result in detail in Section 5.1) that the level of Prozac P rises toward a horizontal asymptote in a precisely upside down exponential decay manner as a function of the number of days n. Find the value of this horizontal asymptote, assuming that the course of treatment continues.

n	0	1	2	3	4	5	6	...
P	80	140	185	218.75	244.0625	263.0469	277.2852	...

Solution We start with a scatterplot of the data, as shown in Figure 4.81, where the points appear to be approaching a horizontal asymptote at a level somewhat above 300 mg. We call this level L.

These values fall in an upside down decaying exponential pattern as they rise toward the horizontal asymptote, so we can use Equation (1) with the first three values $x_0 = 80$, $x_1 = 140$, and $x_2 = 185$ to find that

$$L = \frac{x_0x_2 - x_1^2}{x_0 - 2x_1 + x_2} = \frac{80(185) - 140^2}{80 - 2(140) + 185} = 320.$$

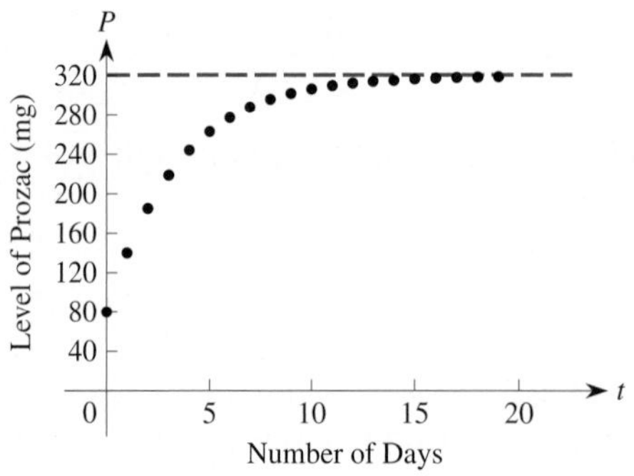

FIGURE 4.81

If instead we use the second, third, and fourth values, so that $x_0 = 140$, $x_1 = 185$, and $x_2 = 218.75$, we obtain

$$L = \frac{x_0 x_2 - x_1^2}{x_0 - 2x_1 + x_2} = \frac{140(218.75) - 185^2}{140 - 2(185) + 218.75} = 320.$$

If we use the last three values shown, so that $x_0 = 244.0625$, $x_1 = 263.0469$, and $x_2 = 277.2852$, we obtain, in the same way, $L = 320.0001$. As you will see in Section 5.1 when we develop a complete mathematical model for the level of Prozac in the blood, the limiting value is 320 mg.

◆

Horizontal Shifts

We next consider some applications involving horizontal shifts. As Example 5 demonstrates, that's just what we've been doing when we changed the scale in the independent variable.

EXAMPLE 5

The following data fall in a linear pattern. Determine the line that passes through the points **(a)** when t represents the number of years since 1980; **(b)** when t represents the number of years since 1900; **(c)** when t represents the number of years since year 0. **(d)** Explain how the three expressions compare by using ideas on shifting functions.

t	1980	1985	1990	1995	2000
y	30	40	50	60	70

Solution

a. We rescale the values of the independent variable so that t represents the number of years since 1980.

t	0	5	10	15	20
y	30	40	50	60	70

Note that each 5 years, the value of y increases by 10, so we have a line with slope

$$m = \frac{\Delta y}{\Delta t} = \frac{10}{5} = 2.$$

The equation of the line then is

$$y - 30 = 2(t - 0), \qquad t = \text{number of years since 1980.}$$

b. We now rescale the values in the table so that t represents the number of years since 1900.

t	80	85	90	95	100
y	30	40	50	60	70

These data values also lie on a line whose slope is $m = 2$, so the equation of the line is

$$y - 30 = 2(t - 80), \qquad t = \text{number of years since 1900.}$$

c. Finally, we use the original values given in the table where t represents the number of years since the year 0. The slope is still $m = 2$, so the corresponding equation of the line is

$$y - 30 = 2(t - 1980), \qquad t = \text{number of years since year 0.}$$

d. We now compare the three equations. In each case, the slope is $m = 2$ because all three lines increase at the same rate. If we expand all the equations to put them in slope–intercept form, we get

$$y = 2t + 30, \qquad y = 2t - 130, \quad \text{and} \quad y = 2t - 3930,$$

respectively. Note the great differences in the vertical intercepts for the three lines.

Let's focus on the equation in part (a), $y - 30 = 2t$, as a baseline, where t represents the number of years since 1980. We first compare it to the equation in part (b), $y - 30 = 2(t - 80)$. The second equation is the result of a horizontal shift to the right of 80 years—moving from a "starting point" of $t = 0$ in 1980 to a "starting point" of $t = 0$ in 1900. Similarly, compare the first equation to the third equation in part (c), $y - 30 = 2(t - 1980)$, which involves a horizontal shift of 1980 years to the right. So scaling the values of the independent variable is equivalent to a horizontal shift by the amount of the scaling.

Let's look at a more realistic example to see how these ideas on horizontal shifts apply when we fit an exponential function to data.

EXAMPLE 6

The following table shows the growth, in millions, of cellular phone users since 1985. Find the exponential function that best fits these values (a) when t represents the number of years since 1985; (b) when t represents the number of years since 1900; (c) when t represents the number of years since year 0. (d) Explain how the three expressions compare, using ideas on shifting functions.

t	1985	1988	1990	1991	1992	1993	1994	1995	1996	1997	1998
C	1	4	11	16	23	34	55	91	142	215	319

Source: Lester R. Brown et al., *Vital Signs 2000: The Environmental Trends That Are Shaping Our Future.*

Solution

a. We first scale the years so that t represents the number of years since 1985, giving the transformed set of data.

t	0	3	5	6	7	8	9	10	11	12	13
C	1	4	11	16	23	34	55	91	142	215	319

A calculator gives the exponential function that best fits the data as

$$C = 1.063(1.555218)^t, \quad t = \text{number of years since 1985.}$$

The corresponding correlation coefficient is $r = 0.99925$.

b. We next scale the years in the original data so that t represents the number of years since 1900.

t	85	88	90	91	92	93	94	95	96	97	98
C	1	4	11	16	23	34	55	91	142	215	319

Again, a calculator gives the exponential function that best fits the modified data as

$$C = 5.2999 \times 10^{-17}(1.555218)^t, \quad t = \text{number of years since 1900.}$$

The corresponding correlation coefficient again is $r = 0.99925$.

c. Finally, the exponential function that best fits the original data where t represents the number of years since the year 0 is

$$C = 2.092 \times 10^{-381}(1.555218)^t, \quad t = \text{number of years since year 0.}$$

The corresponding correlation coefficient once more is $r = 0.99925$.

d. The growth factor, 1.555218, is the same in all three expressions. It indicates that the use of cellular phones is growing, on average, by 55.5% per year, whichever model we construct. The correlation coefficient $r = 0.99925$ is also the same in all three models. It indicates that the fit in all three cases is equally excellent. Only the constant coefficient changes from one expression to the next, and it reflects the vertical intercept of each curve.

We now look at the equation for the exponential function $C = 1.063(1.555218)^t$ in part (a), where t represents the number of years since 1985. If we perform a horizontal shift of 85 years to the right so that t represents the number of years since 1900, the formula for the function becomes

$$C = 1.063(1.555218)^{t-85} = 1.063(1.555218)^t \cdot (1.555218)^{-85} \quad b^{u+v} = b^u b^v$$
$$= [1.063 \times (1.555218)^{-85}](1.555218)^t$$
$$= 5.299988 \times 10^{-17}(1.555218)^t,$$

which is virtually identical to the expression in part (b).

Similarly, if we perform a horizontal shift of 1985 to the right in the equation in part (a), so that t represents the number of years since the year 0, the formula for the exponential function becomes

$$C = 1.063(1.555218)^{t-1985} = 1.063(1.555218)^t \cdot (1.555218)^{-1985} \quad b^{u+v} = b^u b^v$$
$$= [1.063 \times (1.555218)^{-1985}](1.555218)^t$$
$$= 2.0934 \times 10^{-381}(1.555218)^t,$$

which again is virtually identical to the expression $C = 2.092 \times 10^{-381}(1.555218)^t$ in part (c). (The differences are due to rounding.)

◆

This principle—scaling the values of the independent variable is equivalent to a horizontal shift in the function—applies to the linear, exponential, and polynomial families of functions, but it does *not* apply to power functions. Let's see why.

Recall that an increasing power function always passes through the origin, and that a decreasing power function always has a vertical asymptote at 0. When we fit a power function to a set of increasing data, the origin is automatically added as an extra point. Suppose that we now scale the values of the independent variable x to form a new independent variable X. When we then attempt to fit a power function to the transformed values, a different "origin" is added automatically. This new "origin" for X is much closer to the shifted data set than the origin for the original data was, as illustrated in Figure 4.82.

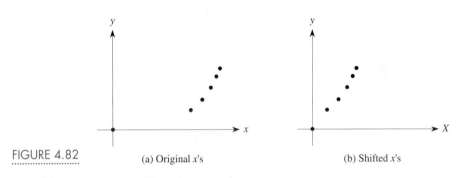

FIGURE 4.82
(a) Original x's   (b) Shifted x's

Consider, for instance, the following data.

x	1	2	3	4
y	1	4	9	16

Clearly, these are points on the curve $y = x^2$ and, if we applied a power function regression routine, that is precisely the equation we would get. This curve certainly passes through the origin $(0, 0)$ for the original variable x. It also passes through each of the data points, as shown in Figure 4.83.

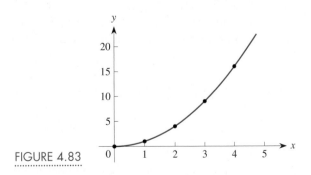

FIGURE 4.83

Let's now shift the data horizontally to the right by 10 units to get the corresponding table of values for the new variable $X = x + 10$.

X	11	12	13	14
y	1	4	9	16

If we apply a power function regression routine, we get the function $y = 1.496 \times 10^{-12} X^{11.43}$. This function passes through the new origin for X, as shown in Figure 4.84. But it misses all the data points. The resulting curve has been flattened enormously to force the new origin to become a point on the graph. As a result, the power for this transformed function is 11.43 rather than 2 for the original function $y = x^2$. Recall that, for power functions $y = x^p$, the higher the power p, the flatter the curve as it passes through the origin. Thus the resulting new power function is distorted compared to the original power function, reflecting the different origin.

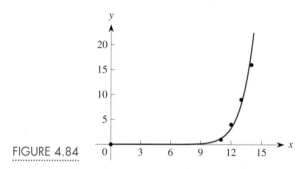

FIGURE 4.84

As we mentioned in Chapter 3, if we use different scales for the independent variable with a power function, not only does the appearance of the function change, but also, and far more important, the predictions based on using the different forms differ. Let's look at what happens when we use the two preceding functions to predict the next value in each table. In the first case, $y = x^2$. When $x = 5$, we get $y = 25$. In the second case, $y = 1.496 \times 10^{-12} X^{11.43}$. The corresponding value of X is $X = x + 10 = 15$, and we get $y = 41.462$, a dramatically different prediction. If we shifted x by other amounts, we would get still different predictions each time. The farther we shift from the original data, the worse this difference gets.

Thus, although power functions are useful, you must use them with extreme caution and careful thought.

Using Stretches

The ideas on stretching functions from Section 4.7 also have direct application when we fit functions to data. In Example 1 in Section 3.3, we created the function

$$P(t) = 3.069(1.321)^t$$

to model the growth of the U.S. population from 1780 to 1900, where P is measured in millions and t is measured in decades since 1780. Actually, we measured P in millions for convenience. If we count the number of people, the corresponding function would then be

$$P(t) = 3{,}069{,}000(1.321)^t.$$

Clearly, these two expressions differ by a factor of 1,000,000, and one function is therefore stretched into the other by this constant multiple of the function.

Moreover, when we created the function for the U.S. population, it was convenient to use t to represent the number of decades since 1780. However, it might be more meaningful to have a function in which the independent variable represents the number of years since 1780 instead. If we use the data values for P from Section 3.3 but count the years $t = 0, 10, 20, \ldots$ rather than decades, we get the function

$$P_1(t) = 3.069(1.02823)^t, \quad t = \text{number of years since 1780,}$$

compared to

$$P_2(t) = 3.069(1.321)^t, \quad t = \text{number of decades since 1780.}$$

How do these two expressions compare? We know that each decade consists of 10 years, which suggests a constant multiple of 10 for the number of years. So, if we start with the first expression $P_1(t)$ for the population where t is measured in years and multiply the independent variable t by 10, we get

$$P_1(10t) = 3.069(1.02823)^{10t}$$
$$= 3.069[(1.02823)^{10}]^t \qquad a^{pu} = (a^p)^u$$
$$= 3.069(1.321)^t,$$

which is the same expression as $P_2(t) = 3.069(1.321)^t$. Whenever we convert the units for the independent variable, from years to centuries, from hours to days, from inches to centimeters, and so on, we actually are stretching or shrinking the function horizontally.

Problems

1. In the analysis of the data on the cooling experiment, we assumed that the water temperature was 8.6°C and so subtracted 8.6 from each of the data values. Assume that the water temperature is 8.65°C instead. Find the corresponding function to fit the original data. Does it appear to be a better or worse fit to the data?

2. Instead of adding 1 to each value of N, as we did with the *Challenger* data in this section, suppose that you add some other quantity (say, 2) to each value. How do the results compare to those obtained earlier?

3. A cup of hot coffee at 200°F is left on the table in a 70°F room to cool. The temperature readings on the coffee at different times as it cools to 70°F are as follows.

Time, t	0	5	10	15	20
Temperature, T	200	163	139	118	108

Find the exponential function that best fits the data.

4. While watching his VCR, Ken noticed that the counter seems to move much faster near the beginning of the tape than toward the end of the tape, so he knows that the readings are not linear. To find the actual pattern, he records the counter reading every 15 minutes and obtains the following set of data relating the counter reading to the elapsed time, in hours.

Time	0	0.25	0.50	0.75	1.0	
Reading	0	445	817	1162	1448	
Time	1.25	1.5	1.75	2.0	2.25	2.5
Reading	1732	2005	2260	2503	2721	2942

a. From among exponential, power, and logarithmic functions, find the function that best fits the data giving the VCR counter reading in terms of the elapsed time.

b. Using the function from part (a), what would you predict the reading to be after 3 hours?

c. Suppose that the label on a VCR tape indicates that a certain program Ken recorded runs from 1600 through 3400 on the counter. How long will that program run?

d. Suppose that the VCR tape is a 6-hour tape. Programs already recorded end at a counter reading of 4200. How much time is left on the tape for the next recording?

5. In Problem 23 of Section 3.3 we looked at how the boiling point (the temperature T) of water in a confined space (say, in a pressure cooker) depends on the pressure of the vapor water. The table gives the boiling point of water, in degrees Celsius, at various vapor pressures, in kilo-pascals.

a. From a scatterplot of the data of T versus P, it appears that the boiling point of water approaches a horizontal asymptote as the pressure P increases. This behavior might suggest an upside down exponential function of the form $y = A + Bc^t$, with $c < 1$. Assume that the horizontal asymptote is at $T = 110°$. Use this value to transform the data and find the corresponding exponential function.

b. Use your function from part (a) to find the boiling point of water when the vapor pressure is 6.2 kilo-pascals.

c. What vapor pressure is needed if the boiling point of the water is 105°C?

Pressure, P	0.61	1.22	2.34	4.25	7.38	12.34	19.93	31.18	47.37	70.12	101.32
Temperature, T	0°	10°	20°	30°	40°	50°	60°	70°	80°	90°	100°

Source: *CRC Handbook of Chemistry and Physics,* 1996.

4.9 The Logistic and Surge Functions

In this section, we consider two other families of functions—the logistic and the surge functions—that frequently arise as mathematical models in a wide variety of applications.

The Logistic Function

A great many processes start out growing exponentially, but eventually other factors come into play to slow the rate of growth, causing a leveling off, as shown in Figure 4.85. Most populations grow in this manner, and many diseases spread in a comparable pattern. The use of new technological innovations, be they new electronic devices such as microwave ovens, cellular phones, or DVD players and new medical products, also spread this way. Such a pattern is called a **logistic curve,** and the associated function is called a **logistic function.**

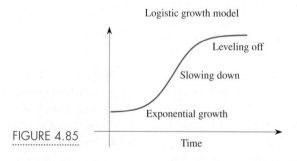

FIGURE 4.85

Logistic processes can be modeled mathematically in several different ways, and we look at one in considerable detail in Section 5.3. For now, we consider the family of functions of the form

$$f(t) = \frac{C}{1 + Ae^{-Bt}},$$

where A, B, and C are positive constants and $e = 2.71828\ldots$ is the base of the natural logarithm system that we introduced in Section 2.6. In most practical situations, the constant A typically is very large, the constant C is fairly large, and the constant B is usually between 0 and 1. Let's first analyze the behavior of this family of functions.

This function actually is the quotient of two functions, so we have to reason in the same way that we analyzed the behavior of rational functions in Section 4.6. In particular, because the numerator is a positive constant, the function has no real roots and thus never crosses the t-axis. Also, in the denominator, both the constant A and the exponential decay function e^{-Bt} are positive, so the denominator is never zero and the function has no vertical asymptotes. Furthermore, when t is negative or when t is positive and relatively small, the term Ae^{-Bt} is extremely large compared to 1. Thus the denominator behaves like $1 + Ae^{-Bt} \approx Ae^{-Bt}$, and therefore the function $f(t)$ behaves like

$$f(t) = \frac{C}{1 + Ae^{-Bt}} \approx \frac{C}{Ae^{-Bt}} = \frac{Ce^{Bt}}{A}.$$

At first (when t is small) this function grows like an exponential function: To the left, it approaches 0 as $t \to -\infty$, and to the right, as t increases, it is increasing and concave up. As t gets larger, however, the term e^{-Bt} decays toward 0, so that the function behaves as if

$$f(t) = \frac{C}{1 + Ae^{-Bt}} \approx \frac{C}{1 + A \cdot 0} = C,$$

which is a constant. That is, the function eventually (when t is larger) grows more slowly, so there is an inflection point. Beyond that point, the curve levels off and approaches a limiting value at the height of C. Thus this type of function has the shape shown in Figure 4.85 and so is called a *logistic function*. In Figure 4.86 we show the graph of the function

$$f(t) = \frac{500}{1 + 200e^{-0.5t}}.$$

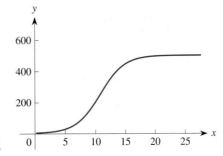

FIGURE 4.86

It has the shape of a logistic curve, eventually leveling off at a height of about 500.

In Example 1 of Section 3.3, we found that the growth in the U.S. population from 1780 to 1900 closely followed an exponential growth pattern with a growth rate of 32.1% per decade. Corresponding to the best fit exponential curve, we had a correlation coefficient of $r = 0.998$. However, we pointed out that this exponential pattern doesn't apply during the twentieth century because the growth rate has slowed dramatically for various reasons. This behavior suggests that a logistic function may be a better choice than an exponential function for modeling the U.S. population over the entire time period since 1780.

EXAMPLE 1

(a) Find a logistic function to fit the following data on the growth of the U.S. population, in millions, since 1780. Let t represent the number of decades since 1780. (b) What does the function predict about the eventual maximum population of the United States? (c) Use the function to predict the U.S. population in 2020.

Year	Population	Year	Population
1780	2.8	1900	76.0
1790	3.9	1910	92.0
1800	5.3	1920	105.7
1810	7.2	1930	122.8
1820	9.6	1940	131.7
1830	12.9	1950	150.7
1840	17.1	1960	179.3
1850	23.2	1970	203.3
1860	31.4	1980	226.5
1870	39.8	1990	248.7
1880	50.2	2000	281.4
1890	62.9		

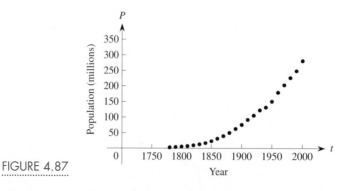

FIGURE 4.87

Solution

a. We begin with the scatterplot of the data shown in Figure 4.87, which indicates that population growth appeared to slow during the latter part of the twentieth century. The successive ratios of the population values also indicate that the rate of population growth slowed from over 20% per decade at the beginning of the twentieth century to about 10% per decade at the end.

We now want to fit a logistic curve to these data. Some calculators have the capability of fitting the best logistic function of the form discussed here to a set of data in the least squares sense. When we use this routine on the U.S. population values, we get the function

$$y = P(t) = \frac{659.45}{1 + 92.05e^{-0.198t}}.$$

This function, superimposed over the population data in Figure 4.88, appears to be an excellent fit to the data.

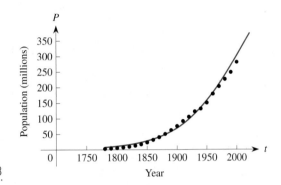

FIGURE 4.88

b. To find the limiting population predicted by this logistic function, we have to determine what happens as $t \to \infty$. As t increases, the term $e^{-0.198t}$ approaches 0, so that the quotient approaches 659.45 million people.

c. Based on this model, the population in 2020, when $t = 24$ decades, will be

$$P(24) = \frac{659.45}{1 + 92.05e^{-0.198(24)}} \approx 367.42 \text{ million people.}$$

The Surge Function

Picture what happens when a medication is first administered to a patient. The effective level of the drug in the bloodstream initially is zero. The drug level then rises rapidly toward a maximum as it is absorbed into the blood. Finally, the drug level decays slowly as it is washed out of the body by the kidneys that filter impurities from the blood. The overall pattern has the shape shown in Figure 4.89. Similarly, a new advertising campaign produces an immediate jump in sales, but

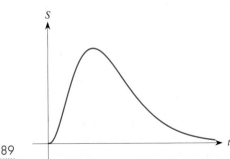

FIGURE 4.89

then the effects of the ad campaign tend to die out slowly over time. The resulting pattern can also be represented by a curve like that shown in Figure 4.89.

Both of these processes are examples of a **surge function,** which can be written as

$$S(t) = At^p b^t,$$

where A, p, and $b < 1$ are three parameters. For realistic situations, we consider only $t \geq 0$. This formula for a surge function actually is the product of a power function t^p and an exponential decay function b^t because $b < 1$. For instance, Figure 4.90 shows the graph of the surge function $S(t) = 100t^{2.5}(0.75)^t$.

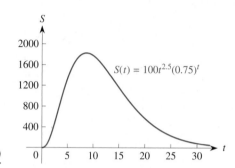

FIGURE 4.90

The coefficient A determines the maximum height of the curve. For the surge function shown in Figure 4.90, this maximum is slightly more than 1800. The power function term t^p reflects the initial impetus and, in fact, the power p determines the location of the maximum value of the function. For this surge function, the maximum occurs at about $t = 8.5$. The decaying exponential term b^t is responsible for the eventual slow decay. Also, remember that an exponential function dominates any power function for large t so that, in the product of the two functions, the exponential decay eventually overwhelms the growth in the power function term.

EXAMPLE 2

The drug L-dopa is administered to people suffering from Parkinson's disease to relieve symptoms such as extreme tremors and rigidity. To be effective, fairly high doses are required because only a small portion of a dose actually lasts in the body long enough to be effective. The side effects of the large doses can be reduced by administering another drug in conjunction with L-dopa. The following table shows the level of L-dopa L in the blood, in nanograms per milliliter, as a function of time t, in minutes.

t	0	20	40	60	80	100	120	140	160	180	200	220	240	300	360
L	0	300	2700	2950	2600	1550	1100	900	725	600	510	440	300	250	225

A plot of these points is shown in Figure 4.91, which suggests the pattern for a surge function. Find the equation of a surge function that models the data.

Solution   The plot of the data indicates that the surge function reaches its maximum at about $t = 60$, where the maximum value is approximately 3000. The function also has two points of inflection, one on either side of the peak. From the table of data, the greatest increase in L occurs between $t = 20$ and $t = 40$, so we estimate that one inflection

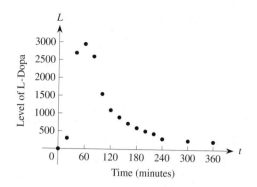

FIGURE 4.91

point occurs at $t = 30$, say. The greatest decrease in L occurs between $t = 80$ and $t = 100$, so we estimate that the other inflection point occurs at about $t = 90$.

We write $L(t) = At^p b^t$ as the general equation of a surge function, where A, p, and b are the three parameters whose values we have to determine. Unfortunately, routines to find these parameters are not built into any calculators or directly into any software packages, so we have to find an indirect way of estimating their values. To do so we apply a transformation approach similar to that used in Sections 3.4 and 3.5. Thus, if $L(t) = At^p b^t$, when we take logs of both sides, we get

$$\log L = \log(At^p b^t) = \log A + \log t^p + \log b^t \qquad \log(u \cdot v \cdot w) = \log u + \log v + \log w$$
$$= \log A + p \log t + t \log b. \qquad \log(u^p) = p \log u$$

Therefore, if L is a surge function of t, $\log L$ is a linear function of t and of $\log t$. Thus, we extend the preceding table to include values for $\log t$ and $\log L$.

t	0	20	40	60	80	100	120	140	160	180	200	220	240	300	360
L	0	300	2700	2950	2600	1550	1100	900	725	600	510	440	300	250	225
$\log t$	UNDEF	1.3	1.60	1.78	1.90	2.00	2.08	2.1	2.2	2.2	2.3	2.3	2.3	2.4	2.56
$\log L$	UNDEF	2.4	3.43	3.47	3.41	3.19	3.04	2.9	2.8	2.7	2.7	2.6	2.4	2.4	2.35

Note that the first entries for $\log t$ and $\log L$ are marked UNDEF because the logarithmic function is not defined.

Because $\log L$ is a linear function of both t and $\log t$, we can use the values from this table in a program that performs multivariate linear regression, as discussed in Section 3.6. The linear function that best fits these data is

$$Y = 1.5004 - 0.00667X_1 + 1.1591X_2,$$

where $X_1 = t$ and $X_2 = \log t$. The regression equation is equivalent to

$$\log L = 1.5004 - 0.00667t + 1.1591 \log t.$$

We undo the transformation, as we did in Sections 3.4 and 3.5, by taking powers of 10 on both sides of this equation:

$$10^{\log L} = L = 10^{1.5004 - 0.00667t + 1.1591 \log t} \qquad 10^{\log u} = u$$
$$= (10^{1.5004})(10^{-0.00667t})(10^{1.1591 \log t}) \qquad 10^{u+v} = 10^u \cdot 10^v$$
$$= (31.65)(10^{-0.00667})^t \cdot (10^{\log t^{1.1591}}) \qquad \log u^p = p \log u$$
$$= 31.65(0.9848)^t t^{1.1591}. \qquad 10^{\log u} = u$$

Therefore our model for the surge function is

$$L(t) = 31.65t^{1.1591}(0.9848)^t.$$

The graph of this function superimposed over the original data set for the level of L-dopa in the blood is shown in Figure 4.92. It is not a particularly good fit to the data, especially for t between 0 and about 100 minutes. It reflects the overall pattern in the data but not very accurately. Moreover, the corresponding coefficient of multiple determination is $R^2 = 0.6876$, so $R = 0.8292$, which is reasonably close to 1. It is close enough for there to be a significant level of correlation.

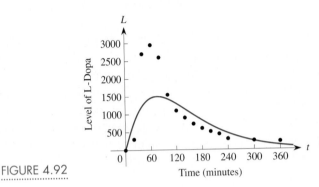

FIGURE 4.92

Problems

1. The growth pattern in human height or weight development from birth through age 18, say, usually follows a logistic growth pattern. The table gives the typical height, in centimeters, of a male and a female in the 50th percentile for height at different ages, in years.

   a. From the table, estimate the typical height of full grown males and females in the 50th percentile (assuming full growth occurs by age 18).

   b. If you have access to a calculator or software package that fits a logistic function to a set of data, find a pair of logistic functions that can be used to model the heights of both males and females as a function of age t for those in this 50th percentile group.

   c. What do the formulas from part (b) predict about the typical heights of full grown males and females in the 50th percentile?

Age	Males	Females
0	50.5	49.9
1	76.1	74.3
2	87.6	86.5
3	96.5	95.6
4	102.9	101.6
5	109.9	108.4
6	116.1	114.6
7	125.0	120.6
8	127.0	126.4
9	132.2	132.2
10	137.5	138.3
11	143.3	144.8
12	149.7	151.5
13	156.5	157.1
14	163.1	160.4
15	169.0	161.8
16	173.5	162.4
17	176.2	163.1
18	176.8	163.7

Source: U.S. Department of Health, Education, and Welfare, *NCHS Growth Curves for Children, Vital and Health Statistics, National Health Survey.* Washington, D.C.: U.S. Government Printing Office.

2. Sweden has one of the longest collections of census records of any country. The table to the right shows the Swedish population, in millions, starting in 1750 when $t = 0$ through 1920 when $t = 170$.

a. From the table, estimate the population of Sweden at the point of inflection.

b. If you have access to a calculator or software package that fits a logistic function to a set of data, find a logistic function that can be used to model the population of Sweden as a function of time t.

c. What does the formula from part (b) predict about the maximum possible population of Sweden?

d. Consult the population table in Appendix G to see how well the logistic function you found in part (b) predicts the actual population in 2002.

3. Consider the surge function $S(t) = 100t^{2.5}(0.75)^t$ (see Figure 4.90). Without using your function grapher, predict how the graph of each surge function (a)–(d) compares to this function in terms of the location of the turning point and the rate at which the function decays to 0.

a. $f(t) = 100t^3(0.75)^t$

b. $f(t) = 100t^2(0.75)^t$

c. $f(t) = 100t^{2.5}(0.70)^t$

d. $f(t) = 100t^{2.5}(0.80)^t$

t	Population
0	1.763
10	1.893
20	2.030
30	2.118
40	2.158
50	2.347
60	2.378
70	2.585
80	2.888
90	3.139
100	3.483
110	3.800
120	4.168
130	4.566
140	4.785
150	5.136
160	5.522
170	5.9004

Source: Raymond Pearl, *The Biology of Population Growth.* New York: Knopf, 1925.

Chapter Summary

In this chapter, we introduced several additional families of functions and ways to build new functions out of old functions. More specifically, we described:

◆ How quadratic, cubic, quartic, and higher degree polynomials behave.

◆ How the real roots of a polynomial equation relate to the linear factors.

◆ How the real roots of a polynomial equation relate to the graph.

◆ How the number of turning points and the number of inflection points relate to the degree of a polynomial.

◆ How the end behavior of a polynomial depends on the sign of the leading coefficient.

◆ How to find the real roots of a polynomial graphically, numerically, and—in the case of quadratic functions—algebraically.

◆ How polynomial functions arise as models in the real world.

◆ How to fit polynomial functions to sets of data.

◆ The relative frequency with which complex roots occur.

◆ How to interpret the higher order differences of a set of numbers to determine polynomial patterns in sets of data.

◆ How to find the sum of the first n integers and the sum of the squares of the first n integers.

◆ What it means to add, subtract, or multiply functions to form new functions.

◆ The behavior of rational functions.

◆ What it means to have a function of a function.

◆ The effects of shifting, stretching, and shrinking on a function.

◆ How to interpret shifting and stretching of functions in terms of fitting functions to data.

◆ How logistic and surge functions behave.

◆ How to use logistic and surge functions as models.

Review Problems

Sketch the graph of each function without using your function grapher.

1. $f(x) = (x + 3)(x - 2)(x - 4)$
2. $g(x) = (2 - x)(x + 3)(x + 1)$
3. $F(x) = (x + 2)(x - 3)(x - 4)(x - 1)$
4. $G(x) = (x + 3)(x - 2)(x - 4)^2$

Factor each polynomial to determine its roots algebraically.

5. $P(x) = x^2 + x - 6$ 6. $Q(x) = 2x^2 + 9x - 5$
7. $R(x) = x^3 - 3x^2 + 2x$

8. Use the quadratic formula to verify your answers to Problems 5–7.

9. A quadratic function f has its vertex at the point $(5, 19)$, and $f(8) = 4$. What is $f(2)$?

10. A cubic function f has its inflection point at $(6, -4)$, and $f(2) = 12$. What is $f(10)$?

11. Estimate the location of the turning points of the graph of the function $y = x^3 + 4x^2 - 5$.

12. Determine the graphs of each pair of functions f and g and use them to draw the graph of $f + g$.
    a. $f(x) = x^2 - 5, g(x) = 3x + 2$
    b. $f(x) = 2x^3 + 4, g(x) = x^2$

13. Analyze the behavior of each rational function including identifying all zeros, vertical asymptotes, and end behavior as x approaches ∞ and $-\infty$. Estimate all turning points graphically.

a. $R(x) = \dfrac{x^2 - 4}{x^2 + 9}$ b. $Q(x) = \dfrac{x^2 - 4}{x^2 - 9}$

c. $S(x) = \dfrac{x^2 + 4}{x^2 + 9}$ d. $T(x) = \dfrac{x^2 + 4}{x^2 - 9}$

14. For each function shown, sketch the graph of
    i. $-f(x)$ ii. $3f(x)$
    iii. $f(x) - 4$ iv. $f(x - 3)$
    v. $f(x + 3)$ vi. $-f(x - 4)$

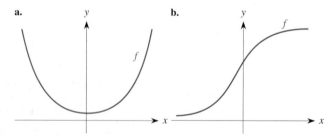

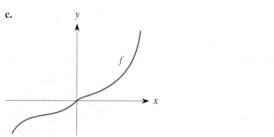

15. Suppose that $f(x) = 2x^2 + 1$ and $g(x) = (x-1)/(x+2)$. Find the following.

   a. $f(3) + g(3)$ **b.** $f(f(3))$
   c. $g(f(3))$ **d.** $g(g(3))$

   e. $g(3)f(3)$ **f.** $\dfrac{f(3)}{g(3)}$

   g. $f(g(x))$ **h.** $f(f(x))$
   i. $g(f(x))$ **j.** $g(g(x))$

   k. $g(x)f(x)$ **l.** $\dfrac{f(x)}{g(x)}$

16. Suppose that $f(0) = 2$, $f(1) = 2$, $f(2) = 3$, $f(3) = 0$ and that $g(0) = 1$, $g(1) = 0$, $g(2) = 2$, $g(3) = 3$. Find the following quantities for $x = 0$, 1, 2, and 3.

   a. $f(g(x))$ **b.** $g(f(x))$

   c. $f(x) + g(x)$ **d.** $\dfrac{f(x)}{g(x)}$

17. Repeat Problem 16(a)–(d) for the functions f and g shown in the graphs below for $x = 1, 2, 3$, and 4.

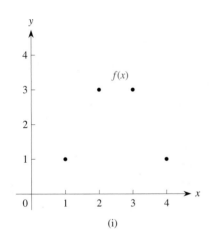

(i)

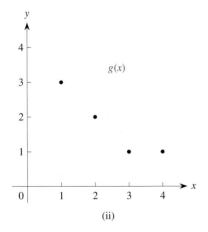

(ii)

18. The return in dollars on an investment seems to be well approximated by the function $F(t) = 2t^2 + t + 4.2$, whereas the return on another investment is modeled by $G(t) = 7.8t + 3.5$. Determine for which values of $t > 0$ the second investment is better than the first.

19. Evaluate the sum
$$3 + 6 + 9 + 12 + 15 + \cdots + 300.$$

20. A polynomial has four turning points.

   a. How many inflection points must it have? Explain.

   b. What is the minimum degree of the polynomial?

   c. What is the minimum number of real roots that the polynomial can have? Explain your answer with a sketch of a polynomial to illustrate what can happen.

   d. What is the maximum number of real roots that the polynomial can have? Explain your answer with a sketch of a polynomial to illustrate what can happen.

   e. Are there any other values for the number of real roots between the minimum number in part (c) and the maximum number in part (d) that the polynomial can have? Explain your answer with a sketch of a polynomial to illustrate what can happen.

21. The accompanying figure shows the graph of a fourth degree polynomial. Use regression methods to find a possible formula for this polynomial.

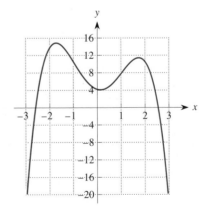

22. The table that follows gives some values, rounded to the nearest integer, for a rational function.

   a. Sketch a possible graph of this rational function $R(x)$.

   b. Find a possible formula for this rational function.

x	-4	-3	-2	-1	0	1
$R(x)$	10	0	UNDEF	0	-3	0
x	2	3	4	5	6	7
$R(x)$	UNDEF	0	18	UNDEF	0	21

23. Each function shown in the accompanying figure can be interpreted as a shift applied to an exponential, a power, or a logarithmic function.

   a. Identify which is which.

   b. Write a possible formula for each function.

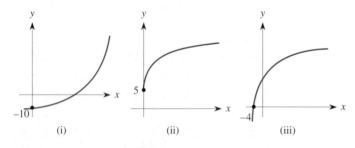

      (i)   (ii)   (iii)

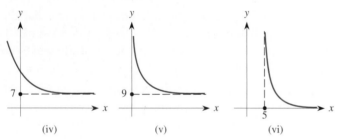

      (iv)   (v)   (vi)

24. The table gives the total number of cell phone subscribers, in millions, in the United States since 1990 and the average local monthly bill, in dollars, for cell phone service.

   a. Find the exponential growth function that best fits the data on the number of subscribers as a function of time since 1990.

   b. Find the exponential decay function that best fits the data on the average monthly bill as a function of time since 1990.

   c. The total industry revenue each year is the product of the number of subscribers and the average monthly bill for service. Use the results of parts (a) and (b) to write a function that models the total cell phone revenue as a function of time since 1990. What is the growth or decay factor for this revenue function?

   d. Extend the table to include a row that gives the total annual revenue in the cell phone industry. Then find the exponential function that best fits the data on the annual revenue as a function of the number of years since 1990. How does this result compare to the one you found in part (c)?

Year	1990	1991	1992	1993	1994	1995	1996	1997	1998	1999
Subscribers	4.37	6.38	8.89	13.07	19.28	28.15	38.20	48.71	60.83	76.28
Average bill	83.94	74.56	68.51	67.31	58.65	52.45	48.84	43.86	39.88	40.24

Source: *2000 Statistical Abstract of the United States.*

Modeling with Difference Equations

Eliminating Drugs from the Body

Prozac is one of the most widely prescribed drugs used to combat extreme depression. Typically, a patient takes a Prozac tablet once a day. As we discussed in Section 2.5, once a medication has been absorbed into the bloodstream, it is washed out of the system by the kidneys, which purify the blood by filtering out foreign chemicals.

For now, let's assume that a person takes a single 80 mg dose of Prozac and that it has been completely absorbed into the blood. During any 24-hour time period, the kidneys eliminate approximately 25% of the Prozac in the bloodstream, so that 75% of the drug remains. As we showed in Section 2.5, the amount of Prozac in the bloodstream based on a single 80 mg dose can be modeled by the exponential decay function

$$D(t) = 80(0.75)^t,$$

where t measures the number of 24 hour periods since the Prozac was taken. The graph of this function is shown in Figure 5.1.

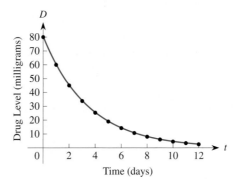

FIGURE 5.1

In particular, $D(0) = 80$. After one 24 hour period, assuming no additional Prozac is taken, $D(1) = 80(0.75) = 60$. After 2 days, $D(2) = 80(0.75)^2 = 45$, and so on. These particular points are highlighted in Figure 5.1.

The collection of values so obtained,

$$\{80, 60, 45, 33.75, 25.3125, \ldots\},$$

is called a **sequence**—it is a set of numbers in a particular order. Thus, a sequence is a function that is defined on a set of nonnegative integers such as $n = 0, 1, 2, \ldots$ The corresponding range is some set of real numbers, such as the levels of Prozac, determined by the rule for the sequence. In other words, for each positive integer, $n = 0, 1, 2, \ldots$, there corresponds a real number $D(n)$ that is the nth term in the sequence. We can write the sequence for the levels of Prozac in the blood as $D(0), D(1), D(2), \ldots$ to emphasize that it is a function.

Although sequences are functions, a special notation is used for them instead of the usual functional notation. For instance, with the Prozac model, we write $D_0 = 80$ instead of $D(0) = 80$, $D_1 = 60$ instead of $D(1) = 60$, and so on. The general, or nth, term in the sequence is then written as D_n instead of $D(n)$, where $n = 0, 1, 2, \ldots$ Unless we're working in a context where some other letters are more appropriate (such as D_n for the level of a drug), we denote the nth term in a sequence by x_n, which represents $\{x_0, x_1, x_2, \ldots\}$.

Keep in mind that, when you work with a sequence such as D_n for the level of Prozac in the system, you are looking only at the value of the function at the start of each 24-hour period. What happens in between isn't considered because the process is basically discrete. Graphs such as the one shown in Figure 5.1 with a smooth curve superimposed over the points in the sequence are convenient for visualizing a trend, but can sometimes be misleading. Such a curve can completely miss anything else that may happen from one value of n to the next.

Repeated Drug Dosages

In practice people often use a drug such as Prozac on a maintenance basis—they take a fixed dose of the medication every time period, rather than a single initial dose. It might be a repeated daily dose of Prozac or some high-blood-pressure medication or a repeated dose of a cold medication every 4 hours. In such cases, the exponential decay model is not realistic. What can we then say about the amount of medication in the body as a function of time?

Suppose that a person takes 80 mg of Prozac each day, starting on some particular day. After the first 24 hours, 25% of the Prozac, or 20 mg, is eliminated, leaving 75%, or 60 mg and then the next day's dose adds 80 mg to that amount. Therefore, after the first 24 hours, the amount of Prozac in the body is

$$D_1 = 60 + 80 = 140 \text{ mg,}$$

or considerably more than the original dose!

During the second 24 hour period, the kidneys eliminate 25% of the 140 mg of Prozac present, or 35 mg, leaving 75% of the 140 mg of Prozac present, or 105 mg. Then the person takes the next dose of 80 mg. Thus, after two 24 hour periods, the amount of Prozac in the body is

$$D_2 = 0.75(140) + 80 = 105 + 80 = 185 \text{ mg.}$$

After 3 days, the level of Prozac is

$$D_3 = 0.75(185) + 80 = 218.75 \text{ mg,}$$

and so on, indefinitely. The corresponding sequence of Prozac levels is

$$\{80, 140, 185, 218.75, 244.0625, 263.0469, \dots \}.$$

Note how the level of the drug keeps rising, but in a concave down manner, as illustrated in Figure 5.2.

Think About This   Does the foregoing explanation suggest that the level of Prozac in the bloodstream keeps rising indefinitely? If so, is that reasonable?

Let's look at a slightly different way to describe this process. The initial dose is $D_0 = 80$ mg. During the first 24 hour time period, 25% of this amount is removed from the blood, leaving 75% of D_0, and the person then takes the next dose of 80 mg. Thus

$$D_1 = 0.75D_0 + 80.$$

Similarly, during the second day, 25% of the Prozac is eliminated, leaving 75% of D_1, and the person then takes another 80 mg dose, so that

$$D_2 = 0.75D_1 + 80.$$

Again, after the third day,

$$D_3 = 0.75D_2 + 80.$$

In general, at the end of $n + 1$ days, for *any* value of n,

$$D_{n+1} = 0.75D_n + 80.$$

This equation shows the relationship between the level of Prozac on any two successive days. In particular, if we know the amount of Prozac in the body after n days, this equation allows us to calculate the amount of Prozac present the following day.

We call an equation such as $D_{n+1} = 0.75D_n + 80$ that relates the successive terms in a sequence a **difference equation.** As you will see, it is an effective way to model drug concentrations in the body. (Note that some authors reserve the term *difference equation* only for equations that explicitly involve the difference $x_{n+1} - x_n$ between successive values in a sequence; we adopt the more common, broader use of the term for any relationship between x_{n+1} and x_n.)

In the drug model, the initial dose of Prozac was $D_0 = 80$ mg. Therefore, if we set $n = 0$ in the difference equation $D_{n+1} = 0.75D_n + 80$, we determine D_1 to be

$$D_1 = 0.75D_0 + 80 = 0.75(80) + 80 = 140 \text{ mg.}$$

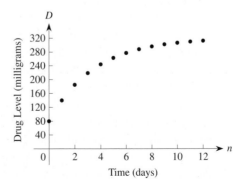

FIGURE 5.2

When $n = 1$ we obtain,

$$D_2 = 0.75D_1 + 80 = 0.75(140) + 80 = 185 \text{ mg}.$$

When $n = 2$ we get,

$$D_3 = 0.75D_2 + 80 = 0.75(185) + 80 = 218.75 \text{ mg}.$$

Continuing in this way, we get

$$D_4 = 0.75D_3 + 80 = 244.0625 \text{ mg};$$
$$D_5 = 0.75D_4 + 80 = 263.0469 \text{ mg};$$
$$D_6 = 0.75D_5 + 80 = 277.2852 \text{ mg};$$

and so on. Over time, the amount of the drug in the body is given by the sequence of numbers

$$\{80, 140, 185, 218.75, 244.0625, 263.0469, 277.2852, \ldots\}.$$

Figure 5.3 shows the graph of these points, connected by a smooth curve. This sequence of numbers is the **solution** to the difference equation; we refer to it as the **solution sequence.**

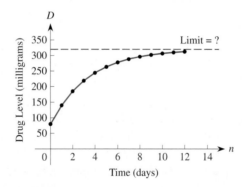

FIGURE 5.3

Note that each successive term, although larger than the preceding value, has grown by somewhat less than the term before. That is, the change from D_0 to D_1 is 60 mg; the change from D_1 to D_2 is 45 mg; the change from D_2 to D_3 is 33.75 mg, and so on. The curve drawn through the points in Figure 5.3 is concave down, and the successive values seem to be leveling off. That is, the drug level continues to rise but at a less steep rate. The overall pattern is characteristic of an upside-down exponential decay process. Rather than gradually dying out toward the horizontal axis as a horizontal asymptote (as an exponential decay function does), this process gradually rises toward the limiting amount of drug in the body as a horizontal asymptote. By continuing this process numerically, we find that the limiting amount L appears to be very close to 320 mg. We say that the terms of the sequence *converge* to this limiting value in the sense that they get closer and closer to L the farther we go in the sequence.

Generating the successive terms of this type of iteration process on either a calculator or a spreadsheet is quite simple. On a calculator, simply enter the starting value for the sequence and press `Enter`. Then enter the iteration formula, using `2nd ANS` as the variable. For instance, using the Prozac model, you would enter the initial dose 80 and then the formula for the difference equation $D_{n+1} = 0.75D_n + 80$ in the form

$$0.75*ANS+80.$$

When you press Enter, you get the next value of the sequence, or 140. Then each time you press Enter, you get the following value. Try this process to see how simple it really is.

In terms of the original problem, the limiting value L represents the maximum level of Prozac that will be reached in the body. This horizontal asymptote is known as the *maintenance level* for the drug. Once that level of Prozac has been reached, the amount in the body will remain constant at that level every 24 hours, so long as the same dose is taken repeatedly every 24 hours.

Think About This   The curve shown in Figure 5.3 is actually incorrect; it was obtained by simply connecting the points to demonstrate the overall pattern. However, it completely ignores what happens during the 24 hours between successive doses. Sketch a more detailed curve that accurately reflects what happens. ▭

In practice, researchers determine the specific level L of a medication that is most effective, considering factors of both safety and effectiveness. An initial dose of, say, 80 mg of the medication means that for some period of time, the amount in the bloodstream is below the optimal level. As a result, doctors often prescribe an initial dose higher than the normal dose so that the drug level approaches the maintenance level L more rapidly. For instance, an initial dose of 240 mg followed by daily doses of 80 mg will achieve the desired level quickly. However, the safety of prescribing such a large dosage, especially as the first dose of the drug, must be considered.

◆ **EXAMPLE 1** ....

Suppose that the initial dose of Prozac is 160 mg (instead of 80 mg) but that all subsequent doses are 80 mg.

a. Find the first six terms of the solution sequence.

b. How does this solution compare to the one we had before?

Solution

a. We still have the same difference equation model,

$$D_{n+1} = 0.75D_n + 80,$$

but now the initial dose is $D_0 = 160$ mg. Therefore

if $n = 0$: $D_1 = 0.75D_0 + 160 = 0.75(160) + 80 = 200$ mg;

if $n = 1$: $D_2 = 0.75D_1 + 80 = 0.75(200) + 80 = 230$ mg;

if $n = 2$: $D_3 = 0.75D_2 + 80 = 0.75(230) + 80 = 252.5$ mg.

Continuing in this manner,

$$D_4 = 0.75D_3 + 80 = 269.375 \text{ mg};$$
$$D_5 = 0.75D_4 + 80 = 282.031 \text{ mg};$$
$$D_6 = 0.75D_5 + 80 = 291.523 \text{ mg};$$

and so on. Thus the levels of Prozac in the bloodstream are now given by the solution sequence

$$\{160, 200, 230, 252.5, 269.375, 282.031, 291.523, 298.643, 303.982, \ldots\}.$$

b. By changing the initial dose D_0, we get a different sequence of values. That is, the difference equation has a different solution sequence that depends on the initial value we choose. However, the behavior of this solution is essentially the same—it is an increasing, concave down function that rises toward the same limiting value of $L = 320$ mg, as shown in Figure 5.4.

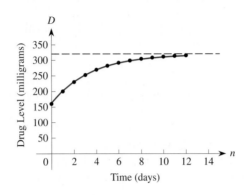

FIGURE 5.4

What happens if the level of drug in the system gets very high? We explore this situation in Example 2.

EXAMPLE 2

Suppose that a person takes an initial dose of 500 mg of Prozac and thereafter takes the usual 80 mg daily.

a. Find the corresponding solution sequence.

b. Discuss the behavior of the solution.

Solution

a. We again use the same difference equation,

$$D_{n+1} = 0.75D_n + 80,$$

but now the initial dose is $D_0 = 500$, so that

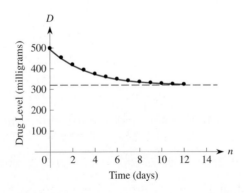

FIGURE 5.5

if $n = 0$: $D_1 = 0.75D_0 + 80 = 455$ mg;

if $n = 1$: $D_2 = 0.75D_1 + 80 = 421.25$ mg;

if $n = 2$: $D_3 = 0.75D_2 + 80 = 395.938$ mg;

and so on. Thus the levels of the drug in the bloodstream are now given by the solution sequence

$$\{500, 455, 421.25, 395.938, 376.953, 362.715, 352.036, 344.027, 338.020, \ldots \}.$$

b. Figure 5.5 shows these points, which fall in a decreasing, concave up pattern that apparently approaches the same limiting value $L = 320$ mg, but from above rather than from below.

◆

Consequently, if the drug level rises too high, some counteracting effects reduce the level. Thus the process that we have described can't lead to an infinite drug level in the body. In fact, our difference equation model predicts that, if the level ever exceeds the maintenance level for the drug, the drug level will decrease as subsequent daily doses are taken.

Again, note that by changing the initial condition D_0 in the difference equation, we obtain a different solution sequence.

Determining the Maintenance Level L

We estimated the maintenance level for Prozac as being about 320 mg by looking at the successive values we calculated in the solution sequence. We now determine the limiting value L for Prozac precisely by using the following argument.

Suppose that, for some value of n, D_n reaches the limit L so that all successive levels of Prozac are the same. Thus for a large enough n, we assume that both $D_{n+1} = L$ and $D_n = L$. Substituting these values into the difference equation for the Prozac drug model,

$$D_{n+1} = 0.75D_n + 80,$$

we obtain

$$L = 0.75L + 80 \quad \text{so that} \quad 0.25L = 80.$$

Hence the limiting value is

$$L = \frac{80}{0.25} = 320 \text{ mg}.$$

We developed all these ideas in the context of a single drug whose level in the bloodstream decreases at a particular rate. In actuality, different drugs are "washed out" of the blood at different rates. For example, aspirin is removed quite rapidly: Its level is reduced by about 50% every 29 minutes. In fact, all drugs are rated by how long it takes for 50% to be eliminated from the body, which is known as the *biological half-life* or simply the **half-life** of the drug. Thus the half-life of aspirin is 29 minutes.

Think About This

In the usual 4-hour time period between successive 325 mg doses of aspirin, how much of the aspirin in the blood would be removed by the kidneys? ▭

Finding a Formula for the Solution

In the preceding situations, we worked with the terms in the solution sequences of the difference equation

$$D_{n+1} = 0.75D_n + 80$$

for different values of the initial dose D_0. Thus, if the initial dose is $D_0 = 80$, the corresponding solution sequence is $D_n = \{80, 140, 185, \ldots\}$. But we did not find a formula for this solution as a function of n. (In Supplementary Chapter 12 we develop a powerful technique for finding formulas for the solution sequences to such difference equations. For now, we construct such a solution by using some of our previous ideas about functions.)

EXAMPLE 3

Find a formula for D_n, the solution sequence to the drug model difference equation for Prozac, based on an initial dose $D_0 = 80$.

Solution   We can think of the values in the solution sequence as a set of data values.

n	0	1	2	3	4	5	6
D_n	80	140	185	218.75	244.0625	263.0469	277.2852

(We could use more values, but these will suffice.) Recall that the shape of the curve shown in Figure 5.2 is like an upside-down exponential decay function. Because the values for D_n approach a limiting value of $L = 320$ mg as a horizontal asymptote, we can shift each D_n value to obtain the corresponding values of $320 - D_n$, as shown in the following table and in Figure 5.6.

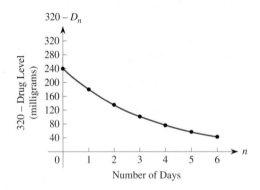

FIGURE 5.6

n	0	1	2	3	4	5	6
D_n	80	140	185	218.75	244.0625	263.0469	277.2852
$320 - D_n$	240	180	135	101.25	75.9375	56.9531	42.7148

Note how the values of $320 - D_n$ in the bottom row decay as n increases. If we continue the process further, these differences approach 0 as n gets larger (because D_n approaches 320 as n increases). It therefore makes sense to fit a decaying exponential function to

$320 - D_n$ as a function of n. (We would not use a decaying power function because the data start with a finite value of 240 when $n = 0$.)

Using a calculator, we find that the exponential function that best fits these data is

$$320 - D_n = 239.99978(0.7500007)^n.$$

The corresponding correlation coefficient is $r = -1$, which suggests a virtually perfect fit. If we solve for the level of Prozac D_n, we obtain

$$D_n = 320 - 239.99978(0.7500007)^n.$$

The numbers in this expression suggest that the "correct" formula for the solution might be

$$D_n = 320 - 240(0.75)^n.$$

We later show that this formula holds for every possible value of n, not just for the few particular values of n we used in constructing the best-fitting exponential function.

Constructing the Solution in General

We now extend the preceding solution formula to solve the comparable difference equation for any medication with any fixed periodic dose. Suppose that the kidneys remove a fixed percentage of a medication every time period, leaving a fraction a, $0 < a < 1$, in the bloodstream. Also, suppose that the repeated dose is an amount B. The corresponding difference equation then is

$$D_{n+1} = aD_n + B.$$

(In our previous development with Prozac, we had $a = 0.75$ and $B = 80$.) For any value of a between 0 and 1 and any positive value of B, the successive terms in the solution sequence for D_n have behavior comparable to that shown in Figure 5.2 assuming D_0 is less than the maintenance level: The solution sequence is an increasing concave down function, and approaches a horizontal asymptote. If your graphing calculator displays solutions of difference equations, select some typical values for a and B and check out the behavior of the solution.

Because the level of the medication in the blood rises toward the maintenance level L, we can solve for L in this general case by realizing that, should this level actually be achieved, then both $D_n = L$ and $D_{n+1} = L$. Therefore

$$L = aL + B, \quad \text{so that} \quad L - aL = L \cdot (1 - a) = B.$$

Hence the maintenance level is

$$L = \frac{B}{1 - a}.$$

The formula for D_n that we constructed previously for Prozac was $D_n = 320 - 240(0.75)^n$, based on $a = 0.75$ and $B = 80$, with an initial value $D_0 = 80$. In this expression, 320 is the limiting value $L = B/(1 - a)$ and 0.75 is the decay factor a. The coefficient 240 is $320 - 80$, the difference between the limiting value L and the initial dose D_0. These results suggest that the general formula for the solution sequence is

$$D_n = L - (L - D_0)a^n$$

for any value of n, with parameters $a, B, L = B/(1 - a)$, and D_0.

EXAMPLE 4

Verify that the preceding expression for D_n is indeed a formula for the solution sequence of the difference equation $D_{n+1} = aD_n + B$ for every value of n.

Solution To verify that the expression for D_n is actually a formula for the solution sequence, we must show that it *satisfies* the difference equation

$$D_{n+1} = aD_n + B$$

for every value of n. Thus we substitute $D_n = L - (L - D_0)a^n$ and the corresponding expression $D_{n+1} = L - (L - D_0)a^{n+1}$ when n is replaced by $n + 1$, into the difference equation. The left-hand side of the difference equation becomes

$$D_{n+1} = L - (L - D_0)a^{n+1}.$$

The right-hand side becomes

$$\begin{aligned} aD_n + B &= a[L - (L - D_0)a^n] + B \\ &= aL - a \cdot (L - D_0)a^n + B \\ &= aL - (L - D_0)a^{n+1} + B. \end{aligned}$$

However,

$$L = \frac{B}{1 - a} \quad \text{so that} \quad B = L \cdot (1 - a).$$

Therefore the right-hand side becomes

$$\begin{aligned} aD_n + B &= aL - (L - D_0)a^{n+1} + B \\ &= aL - (L - D_0)a^{n+1} + L \cdot (1 - a) \\ &= aL - (L - D_0)a^{n+1} + L - aL \\ &= L - (L - D_0)a^{n+1}, \end{aligned}$$

which is identical to the left-hand side. Thus the expression

$$D_n = L - (L - D_0)a^n$$

is a formula for the solution sequence D_n and it holds for all possible values of n.

We summarize the preceding results as follows.

Level of Medication in the Bloodstream

Assumptions
- The kidneys remove a fixed proportion, $1 - a$, of a medication from the bloodstream every time period.
- The repeated dosage of this medication every time period is B.

Mathematical Model
- Difference equation: $D_{n+1} = aD_n + B$
- Maintenance level for the medication: $L = B/(1 - a)$
- Solution: $D_n = L - (L - D_0)a^n$

We developed this mathematical model by simplifying the situation considerably. First, when a person takes a medication, a certain amount of time is needed for it to be completely absorbed into the blood, as well as to reach the intended part of the body. Second, the rate at which a particular drug is washed out of the blood depends on many factors, including a person's weight, metabolism, and the state of the kidneys and liver, all of which are involved in the elimination process. You definitely should not make any medical judgments about the effectiveness of a drug based on the simplified models of this section.

Summary

Let's summarize some of the fundamental ideas about difference equations.

1. A **difference equation** relates the successive terms—say, x_n and x_{n+1}—of a sequence. For instance, it might be

$$x_{n+1} = 1.2x_n - 3.5 \quad \text{or} \quad x_{n+1} = -0.6x_n + 2n.$$

2. The **solution** to a difference equation is a **sequence** (which is a function of n) that satisfies the difference equation. For a given initial value x_0 for the sequence, we can always calculate every successive value of the solution sequence $\{x_0, x_1, x_2, \ldots\}$ by using the difference equation directly.

3. For every possible initial value x_0, there is a different solution sequence to the difference equation. Figure 5.7 shows the graphs of several different solution sequences to the difference equation $x_{n+1} = ax_n + B$ for the drug model, based on different initial doses. Note that, whenever the initial dose x_0 is less than the limiting value L, the pattern is increasing and concave down, approaching L as a horizontal asymptote. Note also that, whenever the initial dose x_0 is greater than L, the pattern is decreasing and concave up and approaches L as a horizontal asymptote.

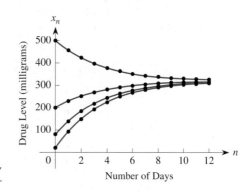

FIGURE 5.7

4. Although the solution sequence can always be expressed in terms of the specific numbers in the sequence, as determined from the difference equation, it is desirable, whenever possible, to write the solution as a formula for x_n in terms of n. Such a formula is called a **closed form expression** for the general term x_n. Thus a solution sequence x_n can be defined

   a. in closed form with a formula for x_n in terms of n,
   b. by the actual sequence of numbers $\{x_0, x_1, x_2, \ldots, x_n, \ldots\}$, or
   c. by a difference equation that relates successive terms of the sequence.

   Think of these ideas in terms of the drug-level model we developed. The same ideas apply to each of the difference equation models we develop throughout the remainder of this chapter.

Finally, the formula $D_n = L - (L - D_0)a^n$ we developed for the solution sequence to the difference equation $D_{n+1} = aD_n + B$ for the drug model can be applied to many other situations. To do so, we rewrite it in a somewhat more general way. Using x_n as the dependent variable gives the difference equation

$$x_{n+1} = ax_n + B.$$

Also, not every difference equation of this form has solutions with a horizontal asymptote. The solutions may grow toward ∞, so we write $B/(1 - a)$ in place of L. With this terminology, the solution to the difference equation becomes

$$x_n = \frac{B}{1 - a} - \left(\frac{B}{1 - a} - x_0\right)a^n$$

$$= \frac{B}{1 - a} + \left(x_0 - \frac{B}{1 - a}\right)a^n.$$

When we multiply out the first equation and collect like terms in a different manner, we get

$$x_n = \frac{B}{1 - a} - \frac{B}{1 - a}a^n + x_0a^n = \frac{B}{1 - a}(1 - a^n) + x_0a^n.$$

In summary we have the following result.

The complete solution to the difference equation

$$x_{n+1} = ax_n + B,$$

where a and B are constants, is

$$x_n = \frac{B}{1 - a}(1 - a^n) + x_0a^n.$$

If $0 < a < 1$, the solution can be written as

$$x_n = L + (x_0 - L)a^n,$$

where

$$L = \frac{B}{1 - a}$$

is the limiting value for the solution as $n \to \infty$.

This formula for the solution applies only to difference equations of the particular form $x_{n+1} = ax_n + B$. Thus, it applies to the difference equation

$$x_{n+1} = 1.05x_n + 3,$$

but it does not apply to the difference equation

$$x_{n+1} = 1.05x_n + 3n$$

because $3n$ is not a constant.

Furthermore, a difference equation of the form

$$x_{n+1} = ax_n + B, \quad n \geq 0,$$

starting at x_0 when $n = 0$, is also known as a *forward difference equation* because it expresses the next value x_{n+1} in a sequence in terms of the current value x_n. You can think of it as "looking forward" to where you are going. We can also write a *backward difference equation* in which the current value x_n in a sequence is written in terms of the previous value x_{n-1}. Think of it as "looking backward" where you came from. The comparable difference equation is

$$x_n = ax_{n-1} + B, \quad n \geq 1.$$

You can convert a forward difference equation to an equivalent backward difference equation by replacing n with $n - 1$ everywhere that n appears, and vice versa. Either way, you get the identical solution sequence, either as a collection of values or as a formula in terms of n.

Many calculators have the capability to calculate successive terms in the solution sequence for a difference equation and to plot the solution as part of their `Sequence` Mode. Check your calculator manual for specific instructions.

Problems

1. Suppose that the kidneys remove 30% of a drug from the bloodstream every 4 hours. If a person takes a single dose of 16 mL, find the amount of the drug in the body after 12 hours and after 24 hours. How long does it take for the level to drop below 1 mL? below 0.01 mL?

2. Suppose that the person in Problem 1 takes repeated doses of 16 mL of the same drug every 4 hours. What is the drug level after 12 hours? after 24 hours? What is the limiting value for the dosage?

3. Suppose that an initial dose of Prozac is $D_0 = 320$ mg followed by daily doses of 80 mg.

   a. Calculate the first six terms of the solution sequence. What do you observe about them?
   b. Can you explain this result?

4. Suppose that the kidneys remove 25% of a drug in the bloodstream every time period and that the initial dose is 48 mL followed by 16 mL doses every time period thereafter. How long will it be until the drug level in the bloodstream exceeds 60 mL?

5. Recall that Figure 5.3 ignores the removal by the kidneys of the Prozac during each 24-hour time period. Sketch a more accurate graph of the level of Prozac in the bloodstream versus time.

6. The accompanying graph shows the level of a medication in the bloodstream just *after* each repeated dose of 16 mL is taken. Sketch the graph of the drug level of the same medication just *before* the next dose

is taken. How do the two graphs compare? Use the two graphs—drawn on the same set of axes—to construct a graph showing the actual level of the medication in the bloodstream at all times, not only at the times just before or just after the medication is taken.

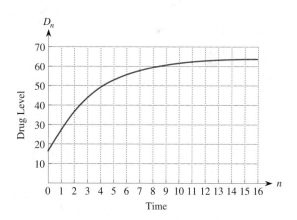

7. The drug dosage for a certain drug is 10 mg per day, and the initial dose is also 10 mg. If the kidneys remove 60% of the drug every 24 hours, find the maintenance level for the medication.

8. Suppose that the daily dosage of the drug in Problem 7 is halved to 5 mg per day. Find the maintenance level. Is the maintenance level also halved?

9. Suppose that the person in Problems 7 and 8 decides to take 10 mg every second day instead of 5 mg each day. Does the patient achieve the same

maintenance level for the medication? Explain why or why not.

10. Two 5-grain aspirin tablets contain 650 mg of the drug. With aspirin's half-life of 29 minutes, how much is left in the bloodstream after 2 hours? How long does it take for the level to be equivalent to 10 mg of aspirin? If an individual takes two tablets every 4 hours, what is the maintenance level of the aspirin?

11. Some studies have shown that taking one aspirin tablet per day significantly reduces a person's risk of heart attack or stroke. If a person follows this regimen, find the maintenance level of the aspirin in the blood.

12. The maintenance level for a certain drug is 600 mg. A patient starts with an initial dose of 100 mg and repeats it daily. Sketch the graph of the level of the drug in the bloodstream as a function of time. Suppose that $D_5 = 400$ and $D_{10} = 520$. Use the graph to determine which values are possible for the drug level and which are impossible.

a. $D_7 = 440$ b. $D_7 = 460$
c. $D_{12} = 540$ d. $D_{12} = 560$

13. The daily dosage for a certain medication is 200 mL and the maintenance level is 500 mL. A person taking this medication reaches a level of 450 mL in 10 days. Let r_1 represent the average daily rate of increase of the drug level over the full 10-day period, let r_2 be the average daily rate of increase over the first 5-day period, and let r_3 be the average daily rate of increase over the last 5 days. Without calculating their values, list these three rates in increasing order. (See Problem 24 of Section 4.1.)

14. Suppose that your car has a 14-gallon gas tank that you fill as soon as the level drops to half-full. Also, every time you fill up, you add one quart ($\frac{1}{4}$ gallon) of an additive that mixes thoroughly with the gas and is then used up along with the gas.

a. Write a difference equation that models the amount of the additive A_n in the tank from one fill-up to the next.
b. Use the difference equation to calculate the amount of additive in the tank over the first 10 fill-ups.
c. Sketch the graph of A_n as a function of n based on the values from part (b). What does the behavior of the function suggest?
d. Find the limiting value for the amount of the additive in the tank as n increases indefinitely.
e. Find the closed form solution of the difference equation.

f. How would the difference equation and the limiting value change if you fill up when the tank is 40% full instead of 50% full?
g. How would the limiting value change if your gas tank holds 16 gallons instead of 14 gallons and you fill up when the tank is half full?

Write the first six terms of each sequence whose general term is given.

15. $x_n = 4n$
16. $x_n = 3n + 5$
17. $x_n = \frac{1}{2}n$
18. $x_n = n^2 + 5$
19. $x_n = n^3 - 10$
20. $x_n = \frac{n^2+1}{n^2+2}$
21. $a_n = \frac{2^n}{3^n}$
22. $a_n = \frac{n^2}{2^n}$
23. $y_n = \frac{1}{n}, \quad n \geq 1$
24. $y_n = \frac{\log n}{n}, \quad n \geq 1$
25. $p_n = 1 - (0.2)^n$
26. $p_n = 1 + (0.2)^n$

27–38. Decide which sequences in Problems 15–26 seem to converge and which clearly do not. Give reasons for your decision. For those that you're not sure about, what could you do to come to a decision?

39–50. Plot the points (n, x_n) for each of the sequences in Problems 15–26. Decide which appear to be strictly increasing or strictly decreasing and which are concave up or concave down.

51. Consider the sequence E whose general term is

$$e_n = \left(1 + \frac{1}{n}\right)^n, \quad \text{for } n = 1, 2, \dots$$

a. Calculate the first 10 terms of this sequence and plot them. Does the graph suggest an eventual limit?
b. Calculate $e_{100}, e_{500}, e_{1000}, e_{10,000}, e_{100,000},$ and $e_{1,000,000}$. What does the limit of the sequence e_n appear to be if you let n increase indefinitely?

52. Consider the sequence $e_n = (1 + 1/n)^n$ again. Use your calculator to evaluate $e_{1000}, e_{1,000,000}, e_{10,000,000},$ and $e_{100,000,000}$. Keep track of all the results obtained. What do you observe about the terms of this sequence? What is your best estimate for the limiting value? Continue the process of taking larger and larger values for n—say, up to $n = 10^{15}$. You will find, depending on your calculator, that the terms eventually jump to 1 instead of continuing as you would expect, owing to calculator round-off. By trial and error, can you find the point where that occurs on your calculator? If so, what is it?

53. Consider the sequence $f_n = (1 - 1/n)^n$. What is the limiting value for the sequence as $n \to \infty$? How is this limiting value related to the one in Problem 51? (*Hint*: There is a simple arithmetic relationship.)

54. Repeat Problem 53, using the sequence $g_n = (1 + 2/n)^n$. How is the limiting value related to the one in Problem 51? Based on this result, conjecture what the limiting value is for $(1 + 5/n)^n$ as $n \to \infty$.

55. What is the limit of the sequence $h_n = (1 + n)^{1/n}$ as $n \to \infty$?

56. Suppose that the successive terms of a sequence are increasing and that the graph drawn through the corresponding points is concave up for all n. Can the sequence converge to a limit? Explain your answer.

5.2   Modeling with Difference Equations

As we have previously discussed, a *mathematical model* for a process is an expression or an equation that represents that process. Sometimes we simply create a formula to fit a set of data. Often, especially in the sciences, we try to determine the underlying principles or assumptions on which a process is based and then attempt to find a relatively simple mathematical relationship that reflects those principles or assumptions. For instance, in football, when a quarterback throws a long pass down the field, the path of the ball can be represented by a simple mathematical formula involving a quadratic function relating the height y to the horizontal distance x—a model based on one of Newton's laws of motion.

Throughout this chapter, the models we develop are difference equations such as $D_{n+1} = 0.75D_n + 80$ that models the level of Prozac in the bloodstream. The solution sequences to such difference equations can be given in two ways: either as a collection of numbers such as $\{80, 140, 185, 218.75, 244.0625, 263.0469, 277.2852, \dots\}$ based on an initial value $D_0 = 80$, say, or as a formula such as

$$D_n = 320 - 240(0.75)^n,$$

a function of n that holds for every value of n.

Exponential Growth and Decay Models

The population of the world passed 6 billion people in 1999 and was growing at a rate of about 1.5% per year, so the growth factor was 1.015. Assuming that this trend continues, we can model the earth's population with the exponential growth function $P(t) = 6(1.015)^t$, where t is the number of years since 1999 and the population is in billions of people. Alternatively, if we use n to represent the number of years since 1999, we can write this exponential function in sequence notation as $P_n = 6(1.015)^n$. We know that a set of numbers follows an exponential pattern if the successive ratios are all constant. That is, for any n,

$$\frac{P_{n+1}}{P_n} = c = 1.015,$$

or when we multiply by P_n,

$$P_{n+1} = cP_n = 1.015P_n,$$

starting with $P_0 = 6$ when $n = 0$ in 1999. This difference equation is a model for the world's population. It tells us that each term in the population sequence $P = \{P_n\}$ is precisely the same multiple, 1.015, of the preceding term.

Let's now look at the reverse of the preceding argument. Suppose that each term in a sequence $X = \{x_n\}$ is a constant multiple c of the preceding term. That is, $x_1 = cx_0, x_2 = cx_1, x_3 = cx_2, \ldots$, or, in general $x_{n+1} = cx_n$, which is a difference equation for the sequence for any value of n. What does this result mean?

EXAMPLE 1

Find a formula for the solution sequence to the difference equation

$$x_{n+1} = cx_n$$

for any constant $c > 0$.

Solution   Starting with the initial term x_0 corresponding to $n = 0$, we have

$$x_1 = cx_0;$$
$$x_2 = cx_1 = c \cdot (cx_0) = c^2x_0;$$
$$x_3 = cx_2 = c \cdot (c^2x_0) = c^3x_0;$$
$$x_4 = cx_3 = c \cdot (c^3x_0) = c^4x_0;$$

and so on. This process leads to an obvious formula for the general term of the solution sequence,

$$x_n = c^nx_0, \quad \text{for any } n = 0, 1, 2, \ldots$$

Thus the solution to any difference equation of the form $x_{n+1} = cx_n$ is an exponential function $x_n = c^nx_0$ or $x_n = x_0c^n$. Such a sequence is known as a **geometric sequence** or an **exponential sequence** in which the ratio of successive terms is a constant, equal to c, because each term is the same multiple of the preceding term.

> The difference equation for exponential growth or decay
>
> $$x_{n+1} = cx_n$$
>
> has as its solution the exponential sequence
>
> $$x_n = x_0c^n,$$
>
> where x_0 is the starting value for the sequence.

Recall that, whenever the constant multiple $c > 1$, the values for x_n get successively larger; whenever c is between 0 and 1, the subsequent values become successively smaller. We presented some illustrations of this property in Chapter 2.

Think About This   What happens if $c < 0$? ▢

From a somewhat different perspective, the equation $P_{n+1} = cP_n$ is a difference equation for exponential behavior and the formula $P_n = P_0c^n$ is the solution sequence. Whenever $c > 1$, we have exponential growth, and whenever $0 < c < 1$, we have exponential decay. Thus in any sequence in which each term is a constant multiple of the preceding term,

$$x_{n+1} = cx_n,$$

x_n is an exponential function of n and $x_n = x_0c^n$.

Alternatively, suppose that $P_{n+1} = cP_n$. If we subtract P_n from both sides of the difference equation, we get

$$P_{n+1} - P_n = cP_n - P_n = (c - 1)P_n = aP_n,$$

where $a = c - 1$. Note that, if $c > 1$, then $a > 0$ and it is the growth rate; if $0 < c < 1$, then $a = c - 1 < 0$ and this gives the decay rate. The expression on the left, $P_{n+1} - P_n$, is simply the difference between successive values, so we can write

$$\Delta P_n = P_{n+1} - P_n = aP_n.$$

Because we can rewrite any such equation as a difference in this way, we call it a *difference* equation.) This difference equation indicates that, for any exponential process, the successive differences ΔP_n are always a fixed multiple aP_n of the quantity itself.

For reference, we recast the previous difference equation $x_{n+1} = cx_n$ for exponential growth or decay in this alternative format.

The difference equation for exponential growth or decay

$$\Delta x_n = ax_n$$

has as its solution the exponential sequence

$$x_n = x_0 \cdot (1 + a)^n = x_0 c^n,$$

where x_0 is the starting value for the sequence.

Writing difference equations in this form is often helpful because the Δx_n formulation lets us think of the *change* in a quantity from one stage to the next instead of how one value depends on the preceding value.

We now consider several examples that demonstrate the use of these ideas relating to difference equations for growth and decay.

EXAMPLE 2

Suppose that you deposit $1000 in a bank account paying 5% interest, compounded annually. Write a difference equation to represent the balance in your account after any number of years and find an expression for the balance at any time from the difference equation.

Solution From the discussion of exponential growth in Chapter 2, we know that the balance in the account after any number of years is given by $b(t) = 1000(1.05)^t$. We now look at this situation from the point of view of difference equations.

Let the original balance be $b_0 = 1000$. After 1 year, the original $1000 balance has earned 5% of $1000, or $0.05(1000) = \$50$ in interest, so the new balance is

$$b_1 = 1000 + 0.05(1000) = (1 + 0.05)1000 = (1.05)1000 = 1050.$$

Symbolically,

$$b_1 = b_0 + 0.05b_0 = 1.05b_0.$$

By the end of the second year, the balance has grown to

$$b_2 = b_1 + 0.05b_1 = 1.05b_1.$$

Similarly, by the end of third year, the balance is

$$b_3 = b_2 + 0.05b_2 = 1.05b_2.$$

In general, by the end of the $(n + 1)$st year, for any n, the balance in the account is

$$b_{n+1} = b_n + 0.05b_n = 1.05b_n,$$

which is the difference equation relating the balance in any year to the balance the next year.

Because this is a difference equation for exponential growth, we immediately know that the solution after n years is given by

$$b_n = b_0 \cdot (1.05)^n.$$

For the initial deposit $b_0 = 1000$, the balance after n years is

$$b_n = 1000(1.05)^n,$$

which is identical to the expression for the exponential growth model. ◆

Before going on, let's recall the formula at the end of Section 5.1.

For the difference equation

$$x_{n+1} = ax_n + B,$$

a formula for the solution sequence is

$$x_n = \frac{B}{1-a} + \left(x_0 - \frac{B}{1-a}\right)a^n,$$

where x_0 is the initial value for the solution sequence.

If the constant B is 0 in the difference equation $x_{n+1} = ax_n + B$, then this equation reduces to $x_{n+1} = ax_n$, which is the difference equation for an exponential process. The formula for the solution sequence similarly reduces to $x_n = x_0a^n$, or the exponential function that we would expect.

Modeling an IRA Account   We next consider how to use a difference equation to model the growth of an IRA account in which a fixed amount of money is invested each year.

EXAMPLE 3

At age 25, Alison sets up an IRA retirement savings account. She invests $3000 annually into an account that earns 5% interest per year.

a. Write a difference equation that models the balance B_n in her account after n years.
b. Find a formula for the solution sequence to the difference equation.
c. Find the balance in her account on her 60th birthday. How much of this balance is attributed to contributions and how much to accrued interest?

Solution

a. We denote the balance in the account after n years by B_n. During the succeeding year, the balance grows by 5% of B_n and is then augmented by the next contribution of $3000. Therefore the balance after $n + 1$ years is

$$B_{n+1} = B_n + 5\% \text{ of } B_n + 3000 = B_n + 0.05B_n + 3000 = 1.05B_n + 3000,$$

which is a difference equation model for the balance in the account.

b. Note that the difference equation $B_{n+1} = 1.05B_n + 3000$ has the same form as the difference equation $x_{n+1} = ax_n + B$, with $a = 1.05$, $B = 3000$, and initial condition $x_0 = 3000 = B_0$. The solution to this difference equation is

$$B_n = \frac{3000}{1 - 1.05} + \left(3000 - \frac{3000}{1 - 1.05}\right)(1.05)^n$$

$$= -\frac{3000}{0.05} + \left(3000 - \frac{3000}{-0.05}\right)(1.05)^n$$

$$= -60,000 + (3000 + 60,000)(1.05)^n$$

$$= 63,000(1.05)^n - 60,000.$$

Figure 5.8 shows the graph of this solution sequence, with the continuous exponential function superimposed.

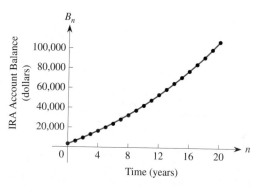

FIGURE 5.8

c. At age 60, Alison's IRA account will have been in existence for $n = 35$ years. Therefore the balance in the account will be

$$B_{35} = 63,000(1.05)^{35} - 60,000 = 287,508.97.$$

Of this total, the amount contributed at $3000 per year is $35(3000) = 105,000$, so the account actually earned about $182,509 in interest.

◆

Modeling a Population with Harvesting   We now introduce a population growth model in which part of the population is removed each time period.

EXAMPLE 4

A poultry farm has 30,000 chickens whose growth rate is 20% per month. Suppose that 5000 chickens are killed (harvested) each month for shipment to stores.

a. Write a difference equation to model this situation.

b. Write a formula for the solution sequence for the difference equation.

c. Discuss the behavior of the solution function and explain the long-term population pattern of the chickens at the farm.

d. What would happen if 8000 chickens are harvested monthly?

e. Determine the number of chickens that should be harvested monthly to maintain a constant population from one month to the next.

Solution

a. Let C_n be the chicken population in the nth month. During that month, the population grows by 20% and 5000 chickens are harvested. Thus the difference equation model is

$$C_{n+1} = 1.20C_n - 5000, \qquad C_0 = 30,000.$$

b. We use the formula for the solution sequence with $a = 1.20$ and $B = -5000$ to get

$$C_n = \frac{-5000}{1 - 1.20} + \left(30,000 - \frac{-5000}{1 - 1.20}\right)(1.20)^n$$

$$= \frac{-5000}{-0.20} + \left(30,000 - \frac{-5000}{-0.20}\right)(1.20^n)$$

$$= 25,000 + (30,000 - 25,000)(1.20)^n$$

$$= 25,000 + 5,000(1.20)^n.$$

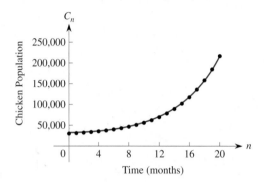

FIGURE 5.9

c. The graph of this function with the continuous function superimposed is shown in Figure 5.9. Note how it grows in a concave up pattern. The solution function is a modified exponential function with a vertical shift of 25,000 and a growth factor of 1.20, so the chicken population at the farm is increasing at an increasing rate. Clearly, this growth pattern can't continue indefinitely.

d. If 8000 chickens are harvested each month instead of 5000, the formula for the solution becomes

$$C_n = \frac{-8000}{1 - 1.20} + \left(30,000 - \frac{-8000}{1 - 1.20}\right)(1.20)^n$$

$$= 40,000 + (30,000 - 40,000)(1.20)^n$$

$$= 40,000 - 10,000(1.20)^n.$$

This solution is an upside-down exponential growth function that has been shifted up by 40,000, as shown in Figure 5.10. Note that it drops ever more quickly, indicating that the chicken population will die out very rapidly.

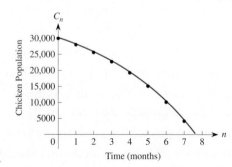

FIGURE 5.10

e. To maintain a constant chicken population from one month to the next requires harvesting the number of chickens that will exactly counterbalance their monthly growth. If the farm starts with 30,000 chickens and they grow at a monthly rate of 20%, 6000 new chickens will be hatched. Therefore, if 6000 older chickens are killed monthly, there is no net increase or decrease.

Modeling the Level of Pollutants   In Examples 5(a)–(e) we consider the level of contaminants in a lake under a variety of circumstances to illustrate how models with difference equations can arise. We also show how the solution to the difference equation model can be used to determine the behavior pattern for the level of contamination over time.

EXAMPLE 5(a)

Initially, 600 lb of a contaminant are dumped into a lake, and 10% of it is washed away each year. Find a formula for the level of contaminant present after any number of years.

Solution   Let C_n represent the level of contaminant present after n years; we know that $C_0 = 600$. Because 10% of the contaminant present is washed out of the lake during any year, 90% of the amount present at the start of any year will be left a year later. Therefore the situation is modeled by the difference equation

$$C_{n+1} = 0.9C_n, \qquad C_0 = 600.$$

This is a difference equation for exponential decay, so we know that the solution is

$$C_n = 600(0.9)^n.$$

This exponential decay function tells us that the level of contaminant will slowly fall over time, as shown in Figure 5.11.

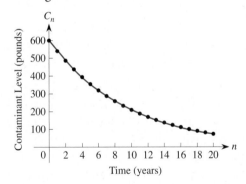

FIGURE 5.11

EXAMPLE 5(b)

Initially, there are 600 lb of the contaminant in the lake, 10% of it is washed out each year, and a manufacturing plant annually dumps 100 lb of the contaminant into a river that feeds into the lake. Find the level of contaminant present after any number of years.

Solution   The amount of contaminant present in the lake is reduced by 10% during a year, so 90% of the amount present each year remains in the following year. However, this amount is then increased by an additional 100 lb each year. This situation is modeled by the difference equation

$$C_{n+1} = 0.9C_n + 100, \qquad C_0 = 600.$$

Note that the form of the difference equation is identical to the equations that we developed previously for the Prozac drug level model and the IRA account balance. Using the formula for the solution to such a difference equation with $a = 0.9$ and $B = 100$—starting with an initial level of $C_0 = 600$—gives a formula for the solution sequence

$$C_n = \frac{100}{1 - 0.9} + \left(600 - \frac{100}{1 - 0.9}\right)(0.9)^n,$$

or

$$C_n = 1000 - 400(0.9)^n.$$

Its graph is shown in Figure 5.12. To understand the graph, note that the exponential term $400(0.9)^n$ dies out as n increases. But, because this term is subtracted from 1000, the solution rises toward 1000 as a horizontal asymptote, which is the limiting value for the contaminant.

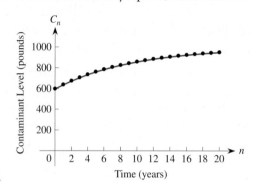

FIGURE 5.12

EXAMPLE 5(c)

The situation is the same as in Example 5(b), but now the plant increases its annual production and thus increases the amount of the contaminant it dumps by 50 lb each year, starting with the initial level of 600 lb.

Solution The 50-pound per year increase in the amount of contaminant dumped into the river means that 100 lb are dumped the initial year, 150 lb the following year, 200 lb the year after that, and so on, following this pattern of linear growth. Thus during the nth year, the company will dump $100 + 50n$ pounds of the contaminant into the river that feeds the lake. As before, 90% of the contaminant in the lake at the start of any year remains at the end of that year and is then augmented by the additional amount dumped into the river. The difference equation that models this situation is

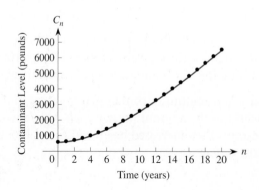

FIGURE 5.13

$$C_{n+1} = 0.9C_n + 100 + 50n, \qquad C_0 = 600.$$

Note that this difference equation has a form different from the preceding two: The term on the right, $100 + 50n$, is no longer a constant but is now a linear function of n, so the formula that we previously developed doesn't apply. We show in Supplementary Section 12.3 that a formula for this solution sequence is

$$C_n = 4600(0.9)^n - 4000 + 500n.$$

Note that this formula consists of a decaying exponential term and a linear term with positive slope. Although the exponential term slowly dies out, the linear term continues to increase as time passes. Over the long term, the level of contaminant eventually increases in a roughly linear pattern with a slope of 500 lb per year, as shown in the graph in Figure 5.13.

◆

EXAMPLE 5(d)

The situation is the same as in Example 5(b), except that the plant now increases the amount of contaminant dumped into the river by 20% per year starting with the initial level of 100 lb.

Solution   The yearly increase in contaminant dumped into the river is now given by the exponential function $100(1.20)^n$, so the difference equation modeling this situation is

$$C_{n+1} = 0.9C_n + 100(1.20)^n, \qquad C_0 = 600.$$

We show in Supplementary Section 12.3 that the solution sequence is

$$C_n = \left(\frac{800}{3}\right)(0.9)^n + \left(\frac{1000}{3}\right)(1.2)^n.$$

The first term is an exponential decay function that eventually dies out, while the second term is an exponential growth function. Early on, the decay term makes a contribution, but because its coefficient (≈ 266.7) is smaller than the growth term's coefficient (≈ 333.3), the contribution is minimal and rather quickly diminishes. The overall behavior pattern is one of roughly exponential growth in the amount of contaminant, as illustrated in Figure 5.14. The eventual exponential growth factor is about 1.2 (verify this result by calculating a pair of successive terms in the solution for moderately large values of n—say, $n = 20$ and $n = 21$), so the annual growth rate is eventually about 20%.

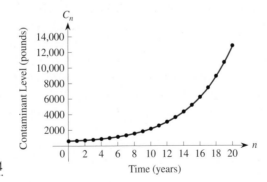

FIGURE 5.14

◆

EXAMPLE 5(e)

The situation is the same as in Example 5(b), with the plant initially dumping 100 lb of the contaminant, but EPA regulations require that it reduce the level of dumping by 25% per year.

Solution The amount dumped into the river is now represented by the decaying exponential function $100(0.75)^n$, so the corresponding difference equation is

$$C_{n+1} = 0.9C_n + 100(0.75)^n, \qquad C_0 = 600.$$

In Supplementary Section 12.3 we show that the solution sequence is

$$C_n = \left(\frac{3800}{3}\right)(0.9)^n - \left(\frac{2000}{3}\right)(0.75)^n.$$

Both terms decay to zero exponentially, so the level of contaminant will eventually die out. The graph of the solution sequence in Figure 5.15 shows a surprising pattern in which the level of contaminant first increases and then dies out. Let's see why. Suppose that we calculate the first few terms of the solution sequence:

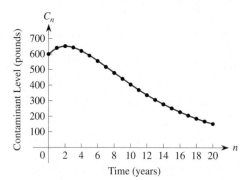

FIGURE 5.15

$$C_1 = \left(\frac{3800}{3}\right)(0.9)^1 - \left(\frac{2000}{3}\right)(0.75)^1 = 1140 - 500 = 640;$$

$$C_2 = \left(\frac{3800}{3}\right)(0.9)^2 - \left(\frac{2000}{3}\right)(0.75)^2 = 1026 - 375 = 651;$$

$$C_3 = \left(\frac{3800}{3}\right)(0.9)^3 - \left(\frac{2000}{3}\right)(0.75)^3 = 923.4 - 281.25 = 642.15;$$

$$C_4 = \left(\frac{3800}{3}\right)(0.9)^4 - \left(\frac{2000}{3}\right)(0.75)^4 = 831.06 - 210.94 = 620.12.$$

Obviously, both $(0.9)^n$ and $(0.75)^n$ decrease as n increases, but the second term decreases more rapidly because its decay factor 0.75 is considerably smaller than the decay factor 0.90 of the first term. As a result, when n is small ($n = 1$ or 2), the second term has a much greater effect on the result than it does when n gets larger—the amount subtracted from the first term is relatively large at first but then quickly diminishes. As less is subtracted away, the solution increases somewhat at first; however, eventually the second term has minimal effect and the first term decays toward zero, as we expect. In particular, $C_{10} = 404.12$, $C_{20} = 151.88$, and $C_{30} = 53.58$.

The Fibonacci Model for Population Growth We next consider one of the earliest mathematical models for a biological process, developed by Italian mathematician Fibonacci, who lived about 1200 A.D. Fibonacci constructed a simple model for predicting the local rabbit population based on the following assumptions.

1. Newborn rabbits mature in 1 month.
2. Once they have matured, rabbits have litters monthly.
3. Each litter consists of one male and one female.

The first two assumptions are fairly accurate, but we can argue about the third assumption on a variety of grounds. First, rabbit litters tend to be considerably larger than 2. However, there is a certain mortality rate for newborn rabbits (they can't run fast enough to escape from the predators in the neighborhood) that lowers the number per litter that survive to maturity. Second, expecting one male and one female to survive from each litter is unreasonable. However, if we consider a large population of rabbits, the numbers of males and females for the entire population average out to about a 50–50 split per litter.

Let's start with one pair of newborn rabbits—one male, the other female—on January 1 of some year, By February 1, the original pair are now mature and ready to do what rabbits do best: produce new rabbits. On March 1, there are two pairs of rabbits, the original pair and their first set of offspring. By April 1, the original pair has produced another litter while their first litter has matured and is ready to enter the family business. Thus there are now three pairs of rabbits, two mature and one newborn. However, by this stage, things are starting to get complicated, so we use the diagram shown in Figure 5.16 to keep track of the rabbits. Let the symbol 🐇 denote an immature pair and 🐇 represent a mature breeding pair.

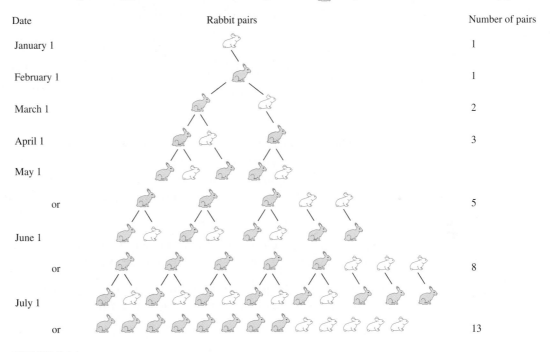

FIGURE 5.16

Consider the rabbit population on July 1, say, which consists of 8 mature pairs and 5 immature pairs. Note that the number of mature pairs, 8, is equal to the total number of pairs in June, the preceding month. Also, the number of immature pairs, 5, is equal to the population 2 months earlier on May 1. Therefore the rabbit population on July 1 is equal to the population on June 1 (mature) plus the population on May 1 (immature). This pattern is not coincidental, and it persists indefinitely—in each month, the rabbit population is equal to the sum of the population values the preceding two months. Let's see why.

The population in the current month P_n consists of breeding pairs and newborn pairs. The number of breeding pairs this month is equal to the total population P_{n-1} last month—all are still alive and all are now mature. Now think about the number of newborn pairs this month. Every rabbit alive two months ago, P_{n-2}, was mature last month and so gave birth to a new litter this month. Therefore the number of newborn pairs this month must be equal to the total population two months ago. So for any $n \geq 2$,

$$P_n = P_{n-1} + P_{n-2}.$$

This is a difference equation relating the successive population values.

Alternatively, instead of considering the current population P_n in terms of the preceding two months' populations (a backward difference equation), we look at the population two months ahead P_{n+2}, which is determined by the current population P_n and next month's population P_{n+1}. As a result, we can rewrite the difference equation in the equivalent forward form

$$P_{n+2} = P_n + P_{n+1}, \qquad n \geq 0,$$

starting with $P_0 = P_1 = 1$. It is known as the **Fibonacci difference equation** or the **Fibonacci model.** Note that with either form we get the population values

$$1, 1, 2, 3, 5, 8, 13, 21, 34, 55, 89, 144, 233, 377, 610, 987, \ldots$$

This particular sequence of numbers is called the **Fibonacci sequence.**

Fibonacci Sequence

$$\{1, 1, 2, 3, 5, 8, 13, 21, 34, 55, 89, 144, \ldots\}$$

These numbers arise in a surprising variety of ways—in nature (e.g., the arrangement of petals on sunflowers and the number of rings in seashells), in economics, in human psychology, and in art (see Example 7 of Section 3.2). However, here we continue to focus on the constantly growing rabbit population of old Italy.

EXAMPLE 6

Construct a table based on Fibonacci's model for the rabbit population over the first 30 months and discuss the growth in this population.

Solution   Starting with the initial values $P_0 = P_1 = 1$, the Fibonacci difference equation gives the values presented in the table and the graph shown in Figure 5.17 for the first 30 months.

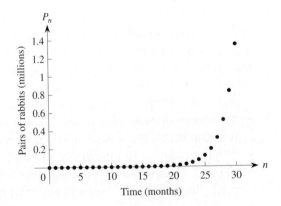

Fibonacci's rabbit model

FIGURE 5.17

Month	Pairs	Month	Pairs	Month	Pairs
0	1	11	144	21	17,711
1	1	12	233	22	28,657
2	2	13	377	23	46,368
3	3	14	610	24	75,025
4	5	15	987	25	121,393
5	8	16	1,597	26	196,418
6	13	17	2,584	27	317,811
7	21	18	4,181	28	514,229
8	34	19	6,765	29	832,040
9	55	20	10,946	30	1,346,269
10	89				

Both the table and Figure 5.17 show a population explosion among the rabbits very quickly. After the first few entries in the table, the ratio of successive terms is approximately 1.618. For example, $233/144 \approx 1.618$, $377/244 \approx 1.618$, and so on. These outcomes suggest that, eventually, the growth pattern is roughly exponential with a growth rate of $0.618 = 61.8\%$ per month.

◆

These numbers are, if anything, conservative because each litter will likely contain more than two rabbits. However, deaths among the rabbits have been ignored. Consequently, the Fibonacci model for the rabbit population may not be a particularly good match to the actual population.

Based on Fibonacci's model, Italy clearly would have had a major overpopulation problem with rabbits in his time, let alone by now. Because that hasn't happened, something is wrong either with the mathematical model or the assumptions on which it is based. Actually, the mathematical model is fairly accurate—at least up to a point. So long as the rabbit population remains relatively small, the model gives numbers that reasonably estimate the population. However, it shouldn't be carried too far because no process can continue to grow exponentially indefinitely. Instead, other factors that act to curb the growing population must be taken into account. For example, as the number of rabbits increases, so too will the number of foxes and other predators that live off them. In turn, the larger numbers of predators eventually reduce the rabbit population. Also, when the rabbit population grows too large, they quickly consume most of the available food supply and there won't be enough food to sustain such a large population. The result is starvation until the population decreases to a more sustainable size. We discuss the mathematical details of this more realistic type of scenario in Section 5.3.

A difference equation that relates one term of a sequence x_{n+1} to the preceding term x_n is called a **first order difference equation.** It is the primary type of difference equation we cover in this chapter and in Supplementary Chapter 12. A difference equation such as Fibonacci's that relates one term x_{n+2} to the preceding two terms x_n and x_{n+1} is called a **second order difference equation.**

Problems

1. In Example 3 about Alison's IRA account, **(a)** what would the account be worth at age 65 instead of age 60? and **(b)** what would it be worth at age 65 if she started the account at age 20 instead of age 25.

2. Repeat the calculations in Example 3 and in Problem 1 if the rate of return for the IRA account is 6% instead of 5%.

3. In Example 4 about the population of chickens, we assumed that there were 30,000 chickens on the farm and a growth rate of 20% per month. We showed that, if the owner kills 8000 chickens a month, the population eventually will die out. Assuming that the owner doesn't notice this developing problem, how long will it be until no chickens are left?

4. Suppose that the population of a certain species of fish in a lake grows exponentially with a growth rate a. Write a difference equation to model each situation.

   a. The fish population grows exponentially.
   b. Fishermen catch and remove 100 fish from the lake each year.
   c. Fishermen catch and remove 40% of the fish from the lake each year.
   d. The number of fish that fishermen remove from the lake each year is proportional to the square root of the number of fish in the lake.
   e. Fishermen remove 40% of the fish each year, and the state's wildlife department restocks the lake with 500 fish each year.

5. Jack and Jill are setting up plans for a retirement fund. Write a difference equation for the balance b_n in this account for each of the scenarios they are contemplating.

   a. They deposit $2000 in an account guaranteed to pay 6% interest per year.
   b. They deposit $2000 initially in an account that pays 6% interest per year and then deposit an additional $1000 each year.
   c. They deposit $2000 initially in the account paying 6% per year and then increase their contribution by $1000 each year.
   d. They deposit $2000 initially into the account paying 6% per year and then increase their contribution by 10% each year.

6. Claire has $80,000 in a retirement fund that pays 6% interest per year.

   a. If she plans to withdraw $10,000 yearly, write a difference equation for the balance in the account.
   b. How long will it take for the balance in the account to be depleted if she withdraws $10,000 every year?
   c. Suppose that she plans to withdraw 20% of the account balance every year. Write a difference equation for the balance in the account.
   d. How long will it take for the balance to be depleted with this withdrawal plan?
   e. Is there a fixed amount she can withdraw from the account every year without diminishing the balance? If so, find it.

7. Marine biologists estimate that there were about 8000 bowhead whales in the waters near Alaska in 1992 and that they were growing at an annual rate of 3%. Alaskan Eskimos are allowed to catch about 50 whales per year.

   a. Write a difference equation giving the population of the whales from one year to the next.
   b. Calculate the projected whale population each year until 2005.
   c. Determine the largest number of whales that the Eskimos could catch each year without the whale population going into decline.
   d. Suppose that the Eskimos are petitioning the government to increase their annual whale harvest and you are acting as their representative before the panel making the decision. What arguments would you use to justify increasing the annual harvest?
   e. Suppose that you were representing a conservation group opposed to increasing the annual whale harvest. What arguments would you use to request a denial of the petition?

8. A company expects the productivity of new employees to increase each day as they gain experience. When a new person starts "cold," the company expects the employee to produce P_0 items per hour. The following day, hourly production should increase by one item per hour to $P_0 + 1$; the day after that, production should increase by 2 items per hour; then by 3 items per hour; and so on.

   a. Write a difference equation to model this situation.
   b. Find an expression for the solution to this difference equation for any number of days n.

c. Write a paragraph discussing whether this expectation for continued improvement seems to be sensible.

9. Psychologists have found that, when a person learns a new body of knowledge, the amount of new knowledge gained in any time period is proportional to the amount that the person does not know. That is, it is easier to improve when you know a little than it is to improve when you know a lot. Suppose that Greg, while preparing for the SAT vocabulary test, is trying to learn 400 new words from a set of flash cards.

   a. Write a difference equation for this learning model based on the number of words W_n that Greg knows out of the 400 total on the nth pass through the deck.

   b. Is the constant of proportionality in the difference equation positive or negative? Is it less than 1 or greater than 1?

   c. Write a paragraph explaining why it is reasonable to expect Greg to learn more words during the first few passes through the deck of cards than through later passes through it.

   d. Sketch a graph of the possible number of new words that Greg learns as a function of the number n of passes through the deck based on this learning model.

10. Repeat the calculations associated with Fibonacci's rabbit model for the first year, assuming an initial population of 10 pairs of newborn rabbits. How do your values relate to those presented in the text?

11. Suppose that a particular breed of rabbits take 2 months to mature instead of 1 month, but that Fibonacci's other assumptions still hold. Calculate the rabbit population each month during the first year.

12. Tribbles are adorable, furry little creatures. The only trouble with tribbles is that they breed like tribbles. Specifically, suppose that a tribble matures in 3 days and then reproduces asexually daily by splitting off a new tribble on the fourth day and every day thereafter.

   a. Construct a difference equation for the tribble population, starting with one newborn tribble, by expressing T_{n+3} in terms of T_n, T_{n+1}, and T_{n+2}. (*Hint*: Let $\triangle$ = newborn tribble, $\square$ = day old immature tribble and $\bigcirc$ = mature tribble and keep track of the number of each over the first 10 days.)

   b. Use the difference equation to calculate the tribble population during the first 15 days, based on an initial newborn tribble the first day.

   c. Examine the ratio of successive terms to determine whether the tribble population appears to be growing exponentially. If it does, what is the exponential growth rate?

   d. Using the result of part (c), estimate the tribble population after a full year.

   e. Assume that the human population of the Earth is currently 6 billion and growing exponentially at an annual rate of 1.7%. Estimate when every human being alive will have a tribble of his or her own.

13. Consider the difference equation $x_{n+1} = x_n + kn$, for any constant multiple k. Show that the solution is always a quadratic function of n. (*Hint*: Use a result from Section 4.5.)

14. Consider the difference equation $x_{n+1} = x_n + kn^2$, for any constant multiple k. Show that the solution is always a cubic function of n.

5.3   **The Logistic or Inhibited Growth Model**

In Section 5.2 we demonstrated that a population P_n growing exponentially satisfies the difference equation

$$P_{n+1} = cP_n$$

for all values of n. A formula for P_n is

$$P_n = c^n P_0,$$

where c is the growth (or decay) factor.

Alternatively, if we subtract P_n from both sides of the difference equation, we get

$$P_{n+1} - P_n = cP_n - P_n = (c - 1)P_n = aP_n,$$

or

$$\Delta P_n = P_{n+1} - P_n = aP_n,$$

where a is the growth (or decay) rate. In this form, the difference equation tells us that, for any exponential process, the successive differences are always a fixed multiple—the growth (or decay) rate—of the quantity itself.

The Logistic Growth Model

The exponential growth model is effective in predicting the growth of most populations over the short run. However, if any growth process were to continue indefinitely, other factors come into play to slow down, or *inhibit*, the rate of growth. We now modify the exponential growth difference equation $\Delta P_n = aP_n$ to model a situation in which a population starts growing exponentially, then slows, and eventually levels off as the population reaches the maximum size that can be sustained by the environment. This type of behavior is illustrated in Figure 5.18.

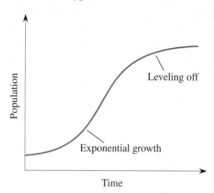

FIGURE 5.18

In the 1830s, the Belgian biologist Verhulst developed a simple extension of the exponential growth model $\Delta P_n = aP_n$ to reflect this situation. He introduced an extra term in the difference equation—one that serves to decrease the rate of growth in the population—to account for this leveling-off effect. Verhulst and other scientists who have since studied such processes have observed that these processes can be modeled by subtracting a term that is proportional to the square of the population. Thus we use the expression for the change in P_n,

$$\Delta P_n = P_{n+1} - P_n = aP_n - bP_n^2,$$

where b is the *inhibiting constant,* which typically is much smaller than the growth rate a. Alternatively, if we add P_n to both sides, we can rewrite this difference equation as

$$P_{n+1} = (1 + a)P_n - bP_n^2.$$

In either form, the resulting process is known as the **logistic growth model** or the **inhibited growth model.** Note that, if $b = 0$, the logistic growth model reduces to the exponential growth model.

> ## The Logistic or Inhibited Growth Model
>
> $$\Delta P_n = P_{n+1} - P_n = aP_n - bP_n^2 \quad \text{or} \quad P_{n+1} = (1 + a)P_n - bP_n^2,$$
>
> where b is much smaller than a.

To work with this logistic growth model, we must get some feel for appropriate values for the **inhibiting constant** b. You may want to experiment, using your graphing calculator (if it has difference equation capabilities) or appropriate computer software. For instance, suppose that the exponential growth rate is $a = 1$ for a species such as rabbits that breeds rapidly and that the inhibiting constant is $b = 0.04$. These values give a graph similar to the one shown in Figure 5.18, and a curve with this shape is called a **logistic curve.** However, if you try $b = 3$ instead, say, the result will be quite different. (We discuss the resulting type of chaotic behavior in Supplementary Section 12.8.) In fact, with a little experimentation, you will find that the model is an effective description of inhibited population growth when b is much smaller than a.

◀ **EXAMPLE 1** ⋯⋯⋯⋯⋯⋯⋯⋯⋯⋯⋯⋯⋯

Consider the logistic model with $a = 1$ and $b = 0.0004$ for a rabbit population, where n represents the number of months. If the initial rabbit population is $P_0 = 1$ pair, use the difference equation to calculate the population over the first 16 months. What does the maximum sustainable population appear to be?

Solution   With $a = 1$ and $b = 0.0004$, the logistic difference equation is

$$P_{n+1} = (1 + a)P_n - bP_n^2 = 2P_n - 0.0004P_n^2.$$

Therefore, using $n = 0$ and $P_0 = 1$, we get

$$P_1 = 2P_0 - 0.0004P_0^2 = 2(1) - 0.0004(1^2) \approx 2.$$

Continuing with $n = 2$ and $n = 3$, we get

$$P_2 = 2P_1 - 0.0004P_1^2 = 2(2) - 0.0004(2^2) \approx 4;$$
$$P_3 = 2P_2 - 0.0004P_2^2 = 2(4) - 0.0004(4^2) \approx 8;$$

and so on. Continuing this process, we obtain the following table of values and the graph shown in Figure 5.19.

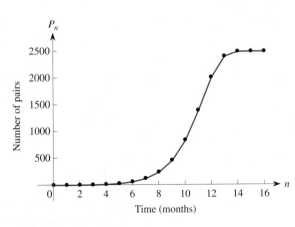

FIGURE 5.19

Month	Number of Pairs	Month	Number of Pairs
0	1	9	463
1	2	10	840
2	4	11	1398
3	8	12	2014
4	16	13	2406
5	32	14	2496
6	63	15	2500
7	125	16	2500
8	243		

Note that the population increases to a maximum of 2500 and seems to remain at that level. Note also that the population grows rapidly to about half this maximum during the first 11 or so months and then grows more slowly thereafter until it reaches the 2500 level. ◆

This behavior pattern is typical of the logistic model: An initial spurt in the population is followed by a slower rate of growth and then an eventual leveling off to a constant fixed population known as the **maximum sustainable population** or the **limit to growth**. (This type of horizontal asymptote is similar to what happens with the maintenance level for a drug discussed in Section 5.1.)

The logistic curve changes from concave up to concave down, so it has one point of inflection. We know that one characteristic of a point of inflection is that the function is growing most rapidly or decreasing most rapidly there. To determine roughly where the point of inflection occurs, we can examine the table to estimate where the largest increase in the population occurs by looking at the differences in successive values of the population. In the preceding table the population has its largest increase during the 11th month when it jumps from 1398 to 2014, an increase of 616.

◆ **EXAMPLE 2**

In illustrating the logistic model, we used $a = 1$ and $b = 0.0004$. How does the behavior of the solution sequence change if we use $a = 1$ and $b = 0.000032$ instead?

Solution   The corresponding results (rounded to the nearest integer) based on the logistic difference equation $P_{n+1} = 2P_n - 0.000032P_n^2$ over the first 20 months are shown in the following table.

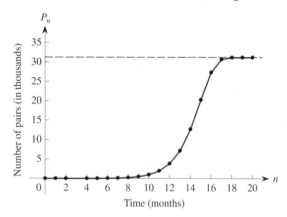

FIGURE 5.20

Month	Number of Pairs	Month	Number of Pairs
0	1	11	1,982
1	2	12	3,839
2	4	13	7,206
3	8	14	12,751
4	16	15	20,299
5	32	16	27,412
6	64	17	30,779
7	128	18	31,243
8	255	19	31,250
9	508	20	31,250
10	1,007		

The resulting pattern, shown in Figure 5.20, has the same logistic shape as that shown in Figure 5.19. In fact, the first few population values are the same as in the previous table because we start with the same initial population $P_0 = 1$, the initial growth rate a is the same, and the inhibiting term has minimal impact while the population is small. Also, because b is now smaller than before, the population grows at a faster rate for a longer time and so the population now has a higher maximum sustainable level—31,250—than in Example 1. Further, the point of inflection now occurs during the 14th month when the population grows from 12,751 to 20,299.

The Maximum Sustainable Population

To understand the behavior of the solution of the logistic difference equation, let's choose a constant b that is much smaller—say, by a factor of $1/1000$—than a. So long as the population P_n remains relatively small, the inhibiting term, $-bP_n^2$, in the logistic difference equation

$$\Delta P_n = aP_n - bP_n^2$$

is negligible compared to the exponential term, aP_n. Therefore, initially, the equation is essentially equivalent to the difference equation

$$\Delta P_n = aP_n$$

for exponential growth. As P_n grows larger, the inhibiting term $-bP_n^2$ becomes larger at a faster rate and so has an ever-greater impact on reducing the value of ΔP_n. Hence the change in P_n begins to decrease. Of course, the fact that the change decreases does not necessarily mean that P_n itself gets smaller; it simply doesn't grow as fast. Moreover, the values for P_n eventually approach a horizontal limit, with no further population growth.

Now let's find a formula for the maximum sustainable population. At this maximum level, there should be no change in P_n, so that the difference between successive terms, $\Delta P_n = P_{n+1} - P_n$, should be zero. Therefore we set the right-hand side of the logistic equation $\Delta P_n = aP_n - bP_n^2$ to 0 and obtain

$$aP_n - bP_n^2 = P_n(a - bP_n) = 0.$$

Because $P_n \neq 0$, we must have $a - bP_n = 0$, so that

$$P_n = \frac{a}{b}.$$

The maximum sustainable population occurs at the level of $P = a/b = L$.

This ratio represents the maximum possible population and is the value, or height, of the horizontal asymptote where the population stabilizes forever. In Example 1 we used $a = 1$ and $b = 0.0004$ so that $L = a/b = 2500$. In Example 2 we used $a = 1$ and $b = 0.000032$ so that $L = a/b = 31{,}250$. These values agree with what we got numerically.

What if the original population P_0 is larger than the maximum sustainable population? For instance, in Example 1, if the initial population were greater than 2500, what would the model predict? In general, for any n, suppose that $P_n > L = a/b$. When we multiply both sides of the inequality $P_n > a/b$ by the positive constant b, we get $bP_n > a$ so that $a - bP_n < 0$. Because $P_n > 0$,

$$\Delta P_n = aP_n - bP_n^2 = (a - bP_n)P_n < 0$$

and the population is decreasing. Thus the phrase *maximum sustainable population* means just what it says: There are too many organisms for the environment to support, and the population will decline due to starvation, predators, and other inhibiting factors. The graph for this scenario is shown in Figure 5.21. You may want to explore this situation further, using a graphing calculator or computer program.

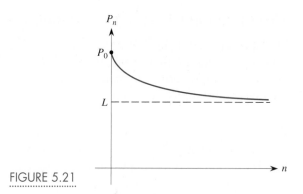

FIGURE 5.21

The Point of Inflection

We next locate the point of inflection, at which the logistic curve changes concavity. It represents the point at which the population is growing most rapidly, so we want to find where the change in the population, $\Delta P_n = P_{n+1} - P_n$, is greatest, as shown in Figure 5.22. To the left of this point, the logistic curve is concave up, so not only is the population increasing, but it is also increasing at an increasing rate $[\Delta(\Delta P_n) > 0]$. To the right of this point, the logistic curve is concave down, so the population is increasing at a decreasing rate $[\Delta(\Delta P_n) < 0]$. At the point of inflection, the population is growing most rapidly, so ΔP_n must be a maximum. Biologically, the most vigorous growth in a population occurs just before decline sets in.

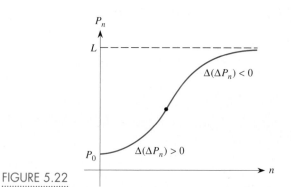

FIGURE 5.22

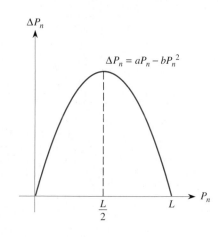

FIGURE 5.23

Let's now find a formula for the location of the point of inflection for the solution sequence of the logistic difference equation,

$$\Delta P_n = aP_n - bP_n^2.$$

Although we typically think of P_n as a function of time n, this equation also expresses the change, ΔP_n, in the population as a quadratic function of the population size P_n, as shown in Figure 5.23. Note that the parabola opens downward because the leading coefficient, $-b$, is negative. Furthermore, the quadratic equation

$$aP_n - bP_n^2 = P_n(a - bP_n) = 0$$

has two real roots, one when $P_n = 0$ and the other when $P_n = L = a/b$, which is the limit to growth. At both of these extremes, ΔP_n is zero and the population isn't growing. Moreover, because a parabola is symmetric about its vertex, the maximum value for ΔP_n occurs at the midpoint of this interval, which is when P_n is half of L, or $\frac{1}{2}(a/b)$. This maximum value for ΔP_n corresponds to the point of inflection.

> The point of inflection occurs at the level of $P = \frac{1}{2}(a/b) = \frac{1}{2}L$.

EXAMPLE 3

Find the point of inflection for the rabbit population in Example 2. How does it match what we observed by comparing values in the table?

Solution   In Example 2, based on $a = 1$ and $b = 0.000032$, the maximum sustainable population was $L = 31{,}250$, so the point of inflection must occur at a height of $\frac{1}{2}L = 15{,}625$. We previously observed that the point of inflection occurred during the 14th month when the population grew most rapidly, jumping from 12,751 to 20,299. Hence the value obtained from the formula agrees well with our previous observation on the data. ◆

The solution sequence of any difference equation consists of a discrete set of numbers, so it isn't reasonable to expect that any one of them will precisely equal $\frac{1}{2}L$. Thus the best we can usually do is to estimate the location of the point of inflection from a table of data values.

We can summarize the logistic growth model as follows.

Summary of the Logistic Growth Model

The logistic difference equation is
$$\Delta P_n = aP_n - bP_n^2 \quad \text{or} \quad P_{n+1} = (1 + a)P_n - bP_n^2,$$
where a is the initial growth rate and b is the inhibiting constant.
The maximum sustainable population is $L = \frac{a}{b}$.
The point of inflection occurs when
$$P_n = \frac{1}{2}L = \frac{1}{2}\left(\frac{a}{b}\right).$$

We've looked at only one specific application of the logistic model concerned with population growth for a single species. The same mathematical model applies to most other species—only the values of the constants a and b change.

EXAMPLE 4

A bacterial culture grows according to the logistic model with $a = 0.4$ and $b = 0.00008$. If there are initially 500 bacteria in the culture, find (**a**) the number present for $n = 1, 2, \ldots, 6$, (**b**) the limiting population for the culture, and (**c**) the location of the point of inflection.

Solution

a. Because the bacterial culture satisfies the logistic model, we know that

$$P_{n+1} = (1 + a)P_n - bP_n^2$$
$$= 1.4P_n - 0.00008P_n^2,$$

starting with $P_0 = 500$. Therefore we find successively that

$$P_1 = 1.4(500) - 0.00008(500)^2 = 700 - 20 = 680;$$
$$P_2 = 1.4(680) - 0.00008(680)^2 = 952 - 37 = 915;$$
$$P_3 = 1.4(915) - 0.00008(915)^2 = 1214;$$
$$P_4 = 1.4(1214) - 0.00008(1214)^2 = 1582;$$
$$P_5 = 1.4(1582) - 0.00008(1582)^2 = 2015;$$
$$P_6 = 1.4(2015) - 0.00008(2015)^2 = 2496.$$

We rounded each successive entry to the nearest whole number because we're dealing with the number of bacteria in the culture.

b. The maximum sustainable population is

$$\frac{a}{b} = \frac{0.4}{0.00008} = 5000.$$

c. The inflection point occurs at

$$\frac{1}{2}\left(\frac{a}{b}\right) = \frac{1}{2}(5000) = 2500,$$

or when the population passes the 2500 level, as depicted in Figure 5.24. Note that the first six terms of the sequence grew to almost one-half the maximum value. However, considerably more than another six terms will be required to get close to the limiting value. For example, $P_{12} = 4680$ and $P_{18} = 4984$.

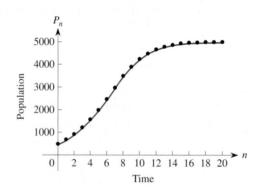

FIGURE 5.24

This type of mathematical modeling is the foundation for most of the projections of limits on world population growth. Moreover, because the growth factor $1 - a$ contains the exponential growth rate a, you might also want to interpret some of these ideas in the context of the values given in the population table in Appendix G. In particular, you might explore the results of using some of the growth rates shown for different countries and assume different values for the inhibiting constant b to see the effects on the limits to growth. Remember, though, that b should be much smaller than a. We discuss how to estimate values for a and b, based on actual population data, later in this section.

Behavior of the Logistic Function

Let's now examine the behavior of the solution to the logistic difference equation in greater detail. Suppose that the maximum sustainable population for an environment is $L = 1000$. The point of inflection then occurs at the level where $P_n = 500$.

Case 1: If the initial population P_0 is less than 500, the population will begin growing in a concave up manner. When it passes a height of 500, the concavity changes and the population continues to grow, but in a concave down manner as it approaches its horizontal asymptote at the level of 1000, as shown in Figure 5.25.

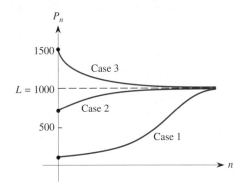

FIGURE 5.25

Case 2: If the initial population P_0 is between 500 and 1000, it starts above the point of inflection. The logistic curve grows toward the limit to growth, 1000, in a concave down manner, as shown in Figure 5.25.

Case 3: Suppose that the initial population P_0 is greater than the maximum sustainable level of 1000. This can occur if there is a sudden influx of immigrants, an unexpected baby boom, or a change in conditions such as a drought or famine that reduces the maximum sustainable level. The resulting logistic curve starts above the limiting value of 1000 and decreases toward 1000 in a concave up manner, as shown in Figure 5.25.

Case 4: Finally, if the initial population value P_0 precisely equals the limiting value L, all subsequent values remain the same. To see this, we consider the logistic difference equation

$$\Delta P_n = aP_n - bP_n^2$$

when $P_n = L = a/b$. Then

$$\Delta P_n = a \cdot \left(\frac{a}{b}\right) - b \cdot \left(\frac{a}{b}\right)^2 = \frac{a^2}{b} - b \cdot \left(\frac{a^2}{b^2}\right) = \frac{a^2}{b} - \frac{a^2}{b} = 0.$$

There is no change, so the population remains constant thereafter. For this reason, the limiting value is also considered the *equilibrium* because the population remains balanced at that level. (The maintenance level for a drug is also an equilibrium.)

Although we typically think of a logistic curve as the S-shaped curve corresponding to initial population values below the inflection point, the other cases are also logistic curves because they are solutions to the logistic difference equation.

More generally, suppose that we have the logistic model

$$\Delta P_n = aP_n - bP_n^2 = (a - bP_n)P_n,$$

with limiting value $L = a/b$ and point of inflection at height $\frac{1}{2}L$. If we factor out the coefficient b from the right-hand side of the difference equation, we get

$$\Delta P_n = b \cdot \left(\frac{a}{b} - P_n\right)P_n = b(L - P_n)P_n.$$

Now suppose that, for some value of n, $P_n > L$; then $L - P_n < 0$. Therefore $\Delta P_n = b(L - P_n)P_n < 0$. Because $\Delta P_n = P_{n+1} - P_n$, we see that P_{n+1} must be smaller than P_n. That is, the solution must be decreasing at such a point. We illustrate this behavior in the top region shown in Figure 5.26 where the arrows are all pointing downward to indicate the direction for the solutions.

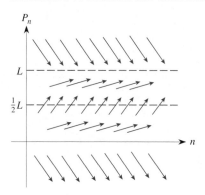

FIGURE 5.26

Next, suppose that, for any value of n, $0 < P_n < L$, so $L - P_n > 0$. Therefore $\Delta P_n = b(L - P_n)P_n > 0$ and so P_{n+1} must be larger than P_n. That is, the solution must be increasing. We show this behavior in the two middle regions of Figure 5.26 where the arrows are all pointing upward to indicate the direction of the solutions. However, the slopes of the arrows crossing the level for the inflection point are steeper than those below or above that level because the logistic curve increases most rapidly at its inflection point.

Finally, although this case is not realistic as a model for population growth, suppose that $P_n < 0$ for some n. Then $L - P_n > 0$ and therefore $\Delta P_n = b(L - P_n)P_n < 0$. Consequently, P_{n+1} must be smaller than P_n and the solution must be decreasing. We show this behavior in the bottom region of Figure 5.26 where all the arrows are pointing downward.

The cases we consider here with the logistic difference equation model are somewhat simplistic. For instance, if the values for a and b are larger than the ones we have used, the solution can overshoot the equilibrium level L for some value of n. However, once there, the succeeding term must decrease toward (or actually even past) the equilibrium. Thus it is possible to obtain a solution that oscillates above and below the equilibrium level while still converging to it. In many ways, this scenario is more realistic than the ideal one we portrayed, in which the solution strictly increases toward L. A real population is likely to grow beyond its limits, then decrease below the maximum sustainable level, and oscillate in this manner thereafter.

Applications of Logistic Growth

The mathematical ideas used in the logistic model for population growth can be applied to many other situations, as Examples 5 and 6 illustrate.

EXAMPLE 5

Suppose that 4000 people are in a university dormitory complex when one student starts a rumor at 9 A.M. Saturday morning. If each person who hears the rumor passes it on to two other people every hour, how long will it be until everyone has heard it?

Solution   At first thought, this situation might seem like an exponential growth model because, seemingly, the number of people who have heard the rumor triples every hour.

In practice, the number of people who hear the rumor does grow roughly exponentially for a while, but eventually the people who are passing the rumor on will find it difficult, if not impossible, to encounter two people who haven't yet heard it.

We must change the underlying exponential formulation to introduce an inhibiting term. The number of new people who hear the rumor after $n + 1$ hours (i.e., the change in the number who have heard it during the past hour) depends not just on how many people P_n have already heard it, but also on the number of people left of the original 4000 people, $(4000 - P_n)$, who haven't heard it. That is, ΔP_n, the change in P_n, is proportional to both P_n and $(4000 - P_n)$, which leads to the difference equation

$$P_{n+1} - P_n = mP_n \cdot (4000 - P_n) = 4000mP_n - mP_n^2,$$

where m is the constant of proportionality. Adding P_n to both sides of the equation, we get

$$P_{n+1} = (1 + 4000m)P_n - mP_n^2 = (1 + a)P_n - bP_n^2,$$

where we have written $a = 4000m$ and $b = m$ to emphasize that this is a logistic model. Initially, $P_0 = 1$ and, because each person tells two new people, $a = 2$ and $1 + a = 3$. The limiting value is $L = a/b = 4000m/m = 4000$, so

$$b = \frac{a}{L} = \frac{2}{4000} = 0.0005.$$

The resulting difference equation is

$$P_{n+1} = 3P_n - 0.0005P_n^2.$$

The following table shows the number of people (rounded to the nearest whole person) who have heard the rumor after each hour, based on this logistic model. The graph of these solution points is shown in Figure 5.27. Thus we conclude that the 4000 students will have heard the rumor in about 9 hours.

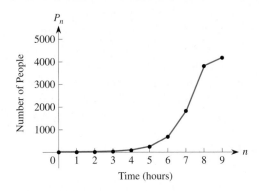

FIGURE 5.27

Time	Number
0	1
1	3
2	9
3	27
4	80
5	238
6	686
7	1823
8	3807
9	4174

The final value displayed, 4174 when $n = 9$, actually overshot the limiting value of 4000. As we said before, once a value of a logistic sequence exceeds L, the following value determined by the difference equation will be smaller because $\Delta P_n < 0$ and, in fact, will usually overshoot downward to produce a value below L.

<u>Think About This</u>   Continue the preceding process repeatedly to verify that the successive values oscillate above and below the limiting value 4000 and eventually converge to it as n increases. ⟞

One problem with the logistic difference equation model is that no one has ever been able to discover a closed form solution for P_n. However, the fact that no explicit solution is known isn't a major handicap. Other than the aesthetic pleasure of having the solution expressed as a formula, a closed form solution wouldn't likely contribute much to this model. The formula would be used primarily to calculate the values for the population at each time. However, the logistic difference equation itself allows us to calculate the values for the solution recursively by using the previous value at each time. There is a slightly different way to create a continuous logistic model using calculus. That model can be solved to give a closed form solution for a logistic curve of the form

$$f(t) = \frac{C}{1 + Ae^{-Bt}},$$

where A, B, and C are positive constants and $e = 2.71828\ldots$ is the base of the natural logarithm system, as we discussed in Section 4.9.

Applications of the logistic growth model include the spread of technological innovations—how fast a new product penetrates the marketplace. For example, when vacuum cleaners were first introduced, they were purchased by relatively few people. Based on word of mouth and advertising, more and more people bought them and the number in use grew rapidly. Eventually, the market for vacuum cleaners became saturated as virtually every household had one, so that the rate of growth diminished, and new sales were basically for replacements. The same logistic pattern applies to items such as televisions and stereos. Newer products such as home computers, VCRs, CD players, DVD players, and cellular phones are now at various stages of the logistic growth process.

Wal-Mart reportedly uses this application of logistic growth to maintain its profit levels. It tracks the sales of lines of merchandise, say, a particular style of jeans. Once sales of that item pass the point of inflection on the logistic curve, Wal-Mart stops ordering that item and orders a different one instead. Thus the company only stocks and sells an item while it is "hot." Hence few, if any, items are sold at clearance prices, which would substantially reduce Wal-Mart's profits.

◆ **EXAMPLE 6** ⋯⋯⋯⋯⋯⋯⋯⋯⋯⋯⋯⋯⋯⋯⋯⋯⋯⋯⋯⋯⋯⋯⋯⋯⋯⋯⋯⋯⋯⋯⋯⋯

The first diesel locomotive was put into service in the United States in 1925. In the early years, the use of diesels among the 25 major railroads then in existence grew at about 25% per year. It took about 30 years for all the railroads to use diesel locomotives. Construct a logistic model for the spread of their use.

Solution   Because the early annual growth rate was 25%, we assume that $a = 0.25$ so that $1 + a = 1.25$. Further, because the limiting value must be $L = 25$ railroads,

$$L = 25 = \frac{a}{b}.$$

We find that

$$b = \frac{a}{L} = \frac{0.25}{25} = 0.01.$$

Thus our logistic model is the difference equation

$$R_{n+1} = 1.25R_n - 0.01R_n^{\,2}.$$

We begin in 1925 when $n = 0$ and $R_0 = 1$ for the one railroad using a diesel locomotive. We then use the difference equation to obtain the values shown in the table and in Figure 5.28.

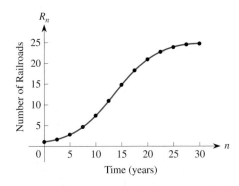

FIGURE 5.28

Year	Number of Diesels	
1	1.24	$=1.25(1) - 0.01(1)^2$
2	1.53	$=1.25(1.24) - 0.01(1.24)^2$
3	1.89	$=1.25(1.53) - 0.01(1.53)^2$
4	2.33	
5	2.86	
10	7.31	
$\vdots$	$\vdots$	
15	14.62	
$\vdots$	$\vdots$	
20	21.03	
$\vdots$	$\vdots$	
25	23.88	
$\vdots$	$\vdots$	
30	24.72	

The model agrees with the historical fact that it took about 30 years after 1925 (or until about 1955) for all 25 of the major railroads to introduce diesels. Incidentally, the other values from the model match the actual spread in the use of diesels quite well. Incidentally, we didn't round the values calculated to the nearest integer because the numbers involved were so small, even though obviously a fraction of a diesel locomotive makes no sense. ◆

Estimating a and b

Suppose that a set of data points fall in a pattern such as the one shown in Figure 5.29 and that we want to fit a logistic curve to it. The hardest part of constructing a logistic model is determining appropriate values for the growth rate a and the inhibiting constant b. To do so, we transform the data so that the resulting points are in a linear pattern, analogous to what we did in Chapter 3.

Suppose that a set of data points $(0, P_0), (1, P_1), (2, P_2), \ldots$ appear to follow a logistic pattern so that P_n satisfies the difference equation

$$\Delta P_n = aP_n - bP_n^{\,2} = (a - bP_n)P_n.$$

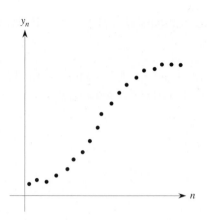

FIGURE 5.29

Dividing both sides by P_n, we obtain

$$\frac{\Delta P_n}{P_n} = a - bP_n.$$

Therefore, if P_n follows a logistic pattern, the quantity $(\Delta P_n)/P_n$ is actually a linear function of P_n. We plot $(\Delta P_n)/P_n$ against P_n (not against n) and find that it falls in a roughly linear pattern, as shown in Figure 5.30. Then we can find the best linear fit to the transformed data. Note that the pattern shown is decreasing, which is the usual case, so the slope of the regression line will be negative. The corresponding regression equation is $Y = a - bX$, which is equivalent to

$$\frac{\Delta P_n}{P_n} = a - bP_n.$$

When we multiply both sides by P_n, we get the logistic difference equation

$$\Delta P_n = aP_n - bP_n^2.$$

Once we have this equation, we have our estimates for the logistic coefficients a and b. We illustrate how to apply these ideas in Examples 7 and 8.

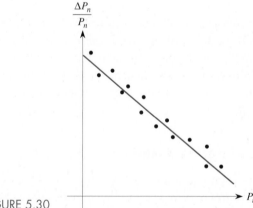

FIGURE 5.30

EXAMPLE 7

German biologist R. Carlson studied the growth of yeast under controlled conditions and obtained the following set of measurements for the weight of yeast, in milligrams, as a function of time, n.

Estimate the parameters a and b to create a logistic model to fit the growth of the yeast population.

n	P_n
1	9.6
2	29.0
3	71.1
4	174.6
5	350.7
6	513.3
7	594.4
8	640.0
9	655.9
10	661.8

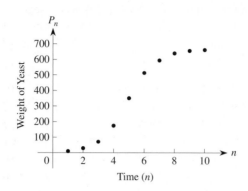

FIGURE 5.31

Solution   By inspecting either the entries in the table or the corresponding scatterplot of the points shown in Figure 5.31, we see that they appear to fall in a logistic pattern. To linearize the data, we add two extra columns to the table, one for the differences ΔP_n of the successive terms and the other for the ratio $(\Delta P_n)/P_n$.

Figure 5.32 shows the plot of the values of $(\Delta P_n)/P_n$ versus P_n (not n). They seem to fall in a roughly linear pattern with negative slope. Using the linear regression routine on a calculator, we find the corresponding regression equation:

$$\frac{\Delta P_n}{P_n} = 1.66 - 0.00271P_n.$$

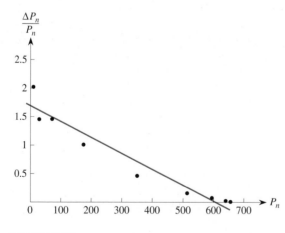

FIGURE 5.32

n	P_n	ΔP_n	$(\Delta P_n)/P_n$	
1	9.6			
		19.4	2.021	(=19.4/9.6)
2	29.0			
		42.1	1.452	
3	71.1			
		103.5	1.456	
4	174.6			
		176.1	1.009	
5	350.7			
		162.6	0.464	
6	513.3			
		81.1	0.158	
7	594.4			
		45.6	0.077	
8	640.0			
		15.9	0.025	
9	655.9			
		5.9	0.009	
10	661.8			

The corresponding correlation coefficient is $r = -0.967$, which suggests a very high degree of negative correlation between $(\Delta P_n)/P_n$ and P_n. Multiplying both sides of the regression equation by P_n yields the logistic difference equation

$$\Delta P_n = 1.66P_n - 0.00271P_n^2,$$

giving $a = 1.66$ and $b = 0.00271$.

The sequence of points based on this difference equation superimposed over the original data is shown in Figure 5.33. Note that the curve connecting the points from the logistic model fits the points of the original data reasonably well, especially the first five points. However, the values of the logistic model level out somewhat below the limiting value for the actual yeast population data.

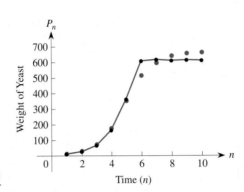

FIGURE 5.33

In Example 1 of Section 3.3, we found that the growth in the U.S. population from 1780 to 1900 closely followed an exponential growth pattern with a growth rate of 32.1% per decade. However, we pointed out that this exponential pattern didn't apply during the twentieth century and, in fact, the rate of growth has slowed considerably since 1900, suggesting a logistic model. In Example 8 we examine the growth in the U.S. population over the entire period since 1780.

EXAMPLE 8

Find a logistic model that fits the data on the U.S. population since 1780 and find the limiting value for the population using this logistic model.

Solution The data on the U.S. population are presented in the table on the next page and in the scatterplot shown in Figure 5.34.

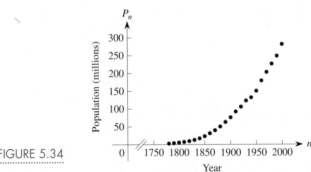

FIGURE 5.34

Both the table and the scatterplot show that the rate of population growth slowed considerably during the twentieth century. Successive ratios also slowly decreased, indicating that the rate of population growth slowed from over 20% per decade at the start of the century to slightly over 10% per decade by the end of the century.

Year	Population	Ratio
1780	2.8	1.39
1790	3.9	1.36
1800	5.3	1.36
1810	7.2	1.33
1820	9.6	1.34
1830	12.9	1.33
1840	17.1	1.36
1850	23.2	1.35
1860	31.4	1.27
1870	39.8	1.26
1880	50.2	1.25
1890	62.9	1.21
1900	76.0	1.21
1910	92.0	1.15
1920	105.7	1.16
1930	122.8	1.07
1940	131.7	1.14
1950	150.7	1.19
1960	179.3	1.13
1970	203.3	1.11
1980	226.5	1.10
1990	248.7	1.13
2000	281.4	

To fit a logistic curve to this data, we must first calculate the differences $\Delta P_n = P_{n+1} - P_n$ and then the ratios

$$\frac{\Delta P_n}{P_n} = \frac{P_{n+1} - P_n}{P_n},$$

the results of which we show in the table on the next page.

When we plot these transformed values $(\Delta P_n)/P_n$ against P_n, as shown in Figure 5.35, we see that they fall in a predominantly decreasing, roughly linear pattern. However, there does seem to be a fair amount of variation about the regression line. Nevertheless, the corresponding plot of the residuals, as shown in Figure 5.36, indicates that the fit is reasonably accurate (the residuals are small and roughly half are below and roughly half are above the baseline), though there does seem to be a pattern to the residuals. This pattern suggests that the logistic model may not fully explain all the variation in the data. The corresponding correlation coefficient, $r = -0.8803$, represents a high degree of negative correlation for the 22 data pairs. Incidentally, we expect a negative correlation between the transformed values and the actual population values because the slope of the regression line is negative. The equation of the regression line for the transformed data is equivalent to

$$\frac{\Delta P_n}{P_n} = 0.3314 - 0.0011 P_n.$$

Multiplying both sides of this equation by P_n gives us the logistic model

$$\Delta P_n = 0.3314 P_n - 0.0011 P_n^2,$$

Year	P_n	ΔP_n	$(\Delta P_n)/P_n$
1780	2.8		
		1.1	0.393 (=1.1/2.8)
1790	3.9		
		1.4	0.359
1800	5.3		
		1.9	0.358
1810	7.2		
		2.4	0.333
1820	9.6		
		3.3	0.344
1830	12.9		
		4.2	0.326
1840	17.1		
		6.1	0.357
1850	23.2		
		8.2	0.353
1860	31.4		
		8.4	0.268
1870	39.8		
		10.4	0.261
1880	50.2		
		12.7	0.253
1890	62.9		
		13.1	0.208
1900	76.0		
		16.0	0.211
1910	92.0		
		13.7	0.149
1920	105.7		
		17.1	0.162
1930	122.8		
		8.9	0.072
1940	131.7		
		19.0	0.144
1950	150.7		
		28.6	0.190
1960	179.3		
		24.0	0.134
1970	203.3		
		23.2	0.114
1980	226.5		
		22.2	0.098
1990	248.7		
		32.7	0.131
2000	281.4		

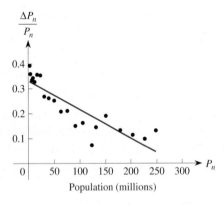

FIGURE 5.35

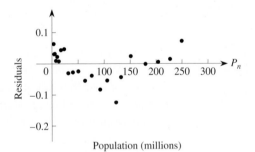

FIGURE 5.36

where $a = 0.3314$ and $b = 0.0011$. Figure 5.37 shows the values predicted by this logistic model, starting from $P_0 = 2.8$; they are the points connected by the smooth curve. For comparison, the original data points for the U.S. population are also shown. The model that we constructed seems to be an excellent fit until about 1930, but the accuracy then seems to break down.

Based on this logistic model, the limiting population for the United States is

$$L = \frac{a}{b} = \frac{0.3314}{0.0011}$$

$$\approx 301.3 \text{ million.}$$

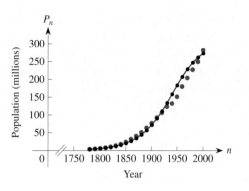

FIGURE 5.37

This prediction is unrealistically low because the U.S. population was fairly close to this level in 2000 and was growing at a rate of about 20 or 30 million people per decade—it isn't reasonable to expect that population growth will slow that much over the next decade. However, recall that, based on the residual plot, we pointed out that the logistic model doesn't seem to account for all the variation in the data. It provides a relatively good fit to the growth of the population, but it isn't an outstanding fit and perhaps a more sophisticated model is needed to give a better match to the population data.

Problems

Calculate the first 10 values predicted for the logistic solution subject to each set of values and plot the points. What is the limiting value for the population? How close do you come to it during the first five time periods?

1. $a = 0.02$, $b = 0.0005$, $P_0 = 10$

2. $a = 0.02$, $b = 0.0005$, $P_0 = 3$

3. $a = 0.02$, $b = 0.0001$, $P_0 = 5$

4. $a = 0.02$, $b = 0.002$, $P_0 = 5$

5. A population grows according to the logistic growth model $\Delta P_n = 0.05P_n - 0.00002P_n^2$. Sketch the behavior of the population if (a) $P_0 = 500$, (b) $P_0 = 1500$, and (c) $P_0 = 3500$.

6. Consider again the rumor spreading through the dorms in Example 2. Suppose that each person repeats it to three new people each hour instead of two people. How long will it be until all 4000 students have heard it?

7. Suppose that a certain population grows according to the logistic model from an initial size of 100 to a final size of 1000.

   a. Sketch the graph of the population as a function of time.

      Use the concavity of your graph from part (a) to answer the following questions.

   b. Suppose that the population is 250 after 10 time periods and 300 after 12 time periods. Use this in-

formation to estimate the size of the population after 11 time periods. Is the actual value higher or lower than your estimate? How do you know?

   c. Suppose that the population is 900 after 30 time periods and 910 after 31 time periods. Use this information to estimate the population after 35 time periods. Is the actual value higher or lower than your estimate? How do you know?

8. Suppose that the deer population in a wildlife refuge follows a logistic growth pattern with an annual growth rate of 40% and an inhibiting rate of 0.02%. Write difference equations to model each situation.

   a. The deer population grows according to the logistic model.

   b. The population grows according to the logistic model, but hunters are allowed to eliminate 120 deer from the region each year.

   c. The logistic model applies, and hunters eliminate 30% of the deer in the region each year.

   d. The logistic model applies, and the number of deer that hunters eliminate in the region each year is proportional to the square root of the number of deer living there.

   e. The logistic model applies, hunters eliminate 40% of the deer, and the state's wildlife department moves 75 new deer into the area each year.

9. The growth of a population follows a logistic trend. The limiting value seems to be about 16,000 and the first few values are 100, 105, 110.2, 115.17, and 121.5. Use this information to estimate both coefficients in the logistic equation and determine how closely the terms you calculate match these observed values.

10. The values for a and b in a logistic model are typically estimated based on a set of observations. As such, they are likely to be somewhat inaccurate. It is important to know how sensitive the results of the logistic model are to slight changes (or errors) in either a or b. Suppose that you estimate $a = 0.05$ and $b = 0.00002$ so that $L = a/b = 2500$. What would be the effect on L if a were actually 10% larger than 0.05 (0.055 instead of 0.05) while b remains fixed? 20% larger? 30% larger? 10% smaller? 20% smaller? How does the value of L depend on the estimate for a if b remains fixed?

11. Repeat Problem 10 by considering the effect on L of changes in b if a is fixed. In particular, what would be the effect on L if b were actually 10% larger than 0.00002? 20% larger? 30% larger? 10% smaller? 20% smaller? How does the value of L depend on the estimate for b if a is fixed?

12. Based on your results in Problems 10 and 11, does the logistic model seem more sensitive to errors or changes in the estimates for a or for b? In estimating values for these two parameters based on a set of data, which do you expect to be more accurate? Explain.

13. The population of a region can be modeled by the logistic growth model with coefficients a and b. A major drought hits the region.

   a. Which coefficient, a or b, is more likely to change because of the drought? Does it become larger or smaller?

   b. Depending on how close the population was to the maximum sustainable population level before the drought, write a short paragraph to describe the effects of the drought on the long-term behavior of the population.

14. Consider the growth model based on the difference equation
$$\Delta P_n = aP_n - bP_n^3,$$
where b is smaller than a.

   a. Sketch the graph of ΔP_n as a function of P_n, for $P_n \geq 0$.

   b. Determine the maximum sustainable population L in terms of a and b for a species modeled by this difference equation.

   c. Use your graph from part (a) to determine the sign of ΔP_n if P_n is between 0 and L. What does it tell you about the behavior of P_n? Explain.

   d. Use your graph from part (a) to determine the sign of ΔP_n if P_n is greater than L. What does it tell you about the behavior of P_n? Explain.

   e. Use your graph from part (a) to show that the solution to this difference equation must have a point of inflection if P_0 is small enough.

   f. The turning points of the general cubic curve $y = Ax^3 + Bx^2 + Cx + D$ occur at
$$x = \frac{-B \pm \sqrt{B^2 - 3AC}}{3A}.$$
Use this fact to determine the location of the point of inflection for the solution to this difference equation.

   g. You know that the point of inflection for the logistic model occurs at half the height of the limiting value. For the model in this problem, does the point of inflection occur at a comparable, a higher, or a lower level? What does that tell you about the kind of behavior for a population that can be well-modeled by this difference equation?

15. Repeat parts (a)–(e) of Problem 14 for the difference equation model
$$\Delta P_n = aP_n - bP_n^4.$$

16. The table gives the average number of words (both spoken and understood) in the vocabulary of the typical child aged 1–6. Find the best fit to these data from among each of the logistic, linear, exponential, and power functions. From among the best fit in each family, which seems to be the overall best fit to the pattern in the data?

Age	1	1.5	2	2.5	3	3.5	4	4.5	5	5.5	6
Vocabulary	50	200	350	600	880	1200	1450	1800	2150	2425	2750

Source: B. A. Moskowitz, Acquisition of Language. *Scientific American*, 1978, vol 279, pp. 92–108.

17. The density of human bones increases through childhood and adolescence. By age 18, 95% of bone density has been achieved. The table shows the percentage of maximum bone density achieved at different ages.

Age	2	4	6	8	10	12	14	16	18
Percentage of Bone Density	43	49	51	56	63	71	82	91	95

Source: Student project.

Find the best fit among each of the logistic, linear, exponential, and power functions for this data. From among the best fit in each family, which seems to be the overall best fit to the pattern in the data?

18. The table shows the accumulated total number of reported cases of AIDS in the United States since 1983.
    a. Determine the logistic fit to this data.
    b. Use the model to predict the number of AIDS cases in 1990.
    c. If the trend is indeed logistic, find the total number of deaths from AIDS that the model predicts in the limit.

Year	1983	1984	1985	1986	1987	1988	1989	1990
Number of AIDS Cases	4589	10,750	22,399	41,256	69,592	104,644	146,574	?
Year	1991	1992	1993	1994	1995	1996	1997	1998
Number of AIDS Cases	251,638	326,648	399,613	457,280	528,144	594,641	653,084	701,353

Source: U.S. Centers for Disease Control and Prevention.

19. Biologists Reed and Holland studied the growth of sunflower plants. They measured the heights H of a number of sunflowers on the same day of successive weeks t and averaged the readings, which are given in centimeters in the table.

    a. Estimate the maximum height to which these sunflowers will grow and the point at which the heights pass their point of inflection.
    b. Estimate the parameters for a logistic difference equation to fit these data.
    c. Use the difference equation to calculate the predicted heights for these sunflowers. How close do the predictions come to the actual values?

Week t	1	2	3	4	5	6
Mean Height, H	17.9	36.4	67.8	98.1	131.0	169.5
Week t	7	8	9	10	11	12
Mean Height, H	205.5	228.3	247.1	250.5	253.8	254.5

Source: H. S. Reed and R. H. Holland, The Growth Rate of an Annual Plant *Helianthus*, Proceedings of the National Academy of Sciences, vol 5, 1919.

20. The table on the next page gives measurements on the weight W in grams of a pumpkin on different days t while it is growing.

    a. Estimate the maximum weight that this pumpkin will reach and the point at which the weight passes its point of inflection.
    b. Estimate the parameters for a logistic difference equation to fit this data.
    c. Use the difference equation to calculate the predicted weight for this pumpkin. How close do the predictions come to the actual values?

t	5	6	7	8	9	10	11	12	13	14	15
W	267	443	658	961	1498	2200	2920	3366	3758	4092	4488

t	16	17	18	19	20	21	22	23	24	25
W	4720	4864	4980	5114	5176	5242	5298	5352	5360	5366

Source: Raymond Pearl, *The Biology of Population Growth.* New York: Alfred A. Knopf, 1925.

21. When *Penicillium chrysogenum,* a strain of penicillin antibiotic, is grown in a batch fermentation process under carefully controlled conditions, the concentration of the cell as measured by its dry weight at various time intervals is given in the table at the bottom left of this page. Determine the coefficients for the logistic fit to the data.

22. From the two tables on the U.S. population in Example 8, compare the entries in the ratio column and the entries in the logistic transformation column $(\Delta P_n)/P_n$ when rounded to two decimal places. What interesting fact do you notice about the two columns of values? Explain how you can account for it mathematically.

23. In Example 7 on the yeast experiment, we constructed the best logistic model to fit the data based on the difference equation $\Delta P_n = 1.66P_n - 0.00271P_n^2$ or $P_{n+1} = 2.66P_n - 0.00271P_n^2$.

    The initial measurement on the yeast was $P_1 = 9.6$. Use this value and the difference equation to construct a table showing the amount of yeast present over the first 10 time periods. Compare these predictions to the actual values measured in the experiment.

24. In Figure 5.32 on the yeast experiment data in Example 7, the scatterplot of $(\Delta P_n)/P_n$ versus P_n may suggest a power function with a negative exponent rather than a linear function.

    a. Using the transformed data, find the power function that best fits the data.

    b. Detransform the results to obtain a first order, nonlinear difference equation relating ΔP_n to P_n.

    c. Use the difference equation from part (b) together with the starting value of $P_0 = 9.6$ grams of yeast to calculate the amount of yeast present for $n = 1, 2, \ldots, 10$. Extend the table you constructed in Problem 23 to include these values. How close do these values come to matching the experimental data shown in Example 7.

25. The scatterplot shown in Figure 5.32 may also suggest a decaying exponential function. Repeat Problem 24, using the exponential function that best fits the transformed data.

26. The growth pattern in human height or weight development from birth through age 18, say, usually follows a logistic growth pattern. The table below gives the typical height, in centimeters, of a male and a female in the 50th percentile for height at different ages, in years. Use the data to construct a pair of logistic functions that model the heights of boys and girls as a function of age t for people in this 50th percentile group.

t	Concentration
0	0.40
22	0.99
46	1.95
70	2.52
94	3.09
118	4.06
142	4.48
166	4.25
190	4.36

Age	0	1	2	3	4	5	6	7	8	9
Boys	50.5	76.1	87.6	96.5	102.9	109.9	116.1	125.0	127.0	132.2
Girls	49.9	74.3	86.5	95.6	101.6	108.4	114.6	120.6	126.4	132.2

Age	10	11	12	13	14	15	16	17	18
Boys	137.5	143.3	149.7	156.5	163.1	169.0	173.5	176.2	176.8
Girls	138.3	144.8	151.5	157.1	160.4	161.8	162.4	163.1	163.7

Source: NCHS Growth Curves for Children. *Vital and Health Statistics, National Health Survey,* U.S. Department of Health, Education and Welfare.

27. The table below shows the worldwide electric generating capacity of nuclear power plants, measured in gigawatts, over time.

    a. Estimate the parameters for the logistic model that fits these data.
    b. What is the limiting value for the worldwide electric generating capacity of nuclear power plants?
    c. From the table, estimate when the worldwide electric generating capacity of nuclear power plants was growing most rapidly.
    d. From your model in part (a), when was the worldwide electric generating capacity of nuclear power plants growing most rapidly?

Year	1960	1965	1970	1975	1980	1985	1990	1995	2000
Electric Capacity	1	5	16	71	135	250	328	340	347

Source: Lester R. Brown et al., *Vital Signs 2000: The Environmental Trends That Are Shaping Our Future.*

28. The table shows the population of the world, in billions, over time. The population appears to be growing more slowly than it was only a few years ago, so it may have passed its point of inflection.

Year	1950	1955	1960	1965	1970	1975
Population	2.556	2.780	3.039	3.345	3.707	4.086
Year	1980	1985	1990	1995	1999	
Population	4.454	4.851	5.277	5.682	6	

Source: Lester R. Brown et al., *Vital Signs 2000: The Environmental Trends That Are Shaping Our Future.*

    a. Estimate the parameters for the logistic model that fits these data.
    b. What is the limiting value for the world's population, based on this model?
    c. From the table, estimate when the world's population was growing most rapidly.
    d. From your model in part (a), when was the world's population growing most rapidly?
    e. Use the difference equation to predict the world's population in 2010.
    f. Use the difference equation to predict when the world's population will reach 7 billion.

29. Recall that the inflection point for the logistic model

$$\Delta P_n = aP_n - bP_n^2$$

occurs at a height of $\frac{1}{2}L$. At this point, the function is growing most rapidly. Show algebraically that the maximum value of ΔP_n is $\frac{1}{4}bL^2$.

5.4 Newton's Laws of Cooling and Heating

In a familiar scene in TV crime shows, the medical examiner studies the homicide victim's body and knowledgeably announces that "Mr. Jones died at approximately 1:30 in the morning." We now develop the mathematics behind this type of conclusion. More generally, we investigate the rate at which any object cools or heats.

Newton's Law of Cooling

Suppose that you heat a pizza in an oven set at 450°F and then remove it to cool on a kitchen counter where the temperature is a constant 70°F. How fast does the temperature of the pizza drop until it reaches room temperature? Clearly, the temperature of the pizza drops most rapidly at first because of the large difference between the pizza temperature and the room temperature. As the pizza cools, the rate at which the temperature decreases slows. That is, the hotter the pizza, the faster the temperature drops; it drops most slowly when the pizza's temperature is close to room temperature. Geometrically, we expect the graph of the temperature as a function of time to be decreasing and concave up, as shown in Figure 5.38. According to a physical principle developed by Isaac Newton, the change in temperature ΔT_n after n time periods is proportional to the difference between the temperature of the pizza (T_n) and the temperature of the surrounding air $(70°F)$, or $T_n - 70$. Thus

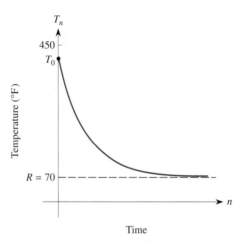

FIGURE 5.38

$$\Delta T_n = -\alpha \cdot (T_n - 70),$$

where the constant of proportionality $\alpha > 0$. Because the temperature change ΔT_n is negative and both $T_n - 70$ and α are positive, there must be a minus sign in front of α. Equivalently, if we add T_n to both sides of the equation, we have the difference equation model

$$T_{n+1} = T_n - \alpha \cdot (T_n - 70)$$
$$= (1 - \alpha)T_n + 70\alpha.$$

More generally, if the room temperature is any constant R, the analogous difference equation is

$$T_{n+1} = (1 - \alpha)T_n + \alpha R. \tag{1}$$

This equation and the principle behind it are both known as **Newton's law of cooling.**

To solve this difference equation, we use the results summarized at the end of Section 5.1. Note that Difference Equation (1) has the same form as

$$x_{n+1} = ax_n + B \tag{2}$$

with

$$a = 1 - \alpha \quad \text{and} \quad B = \alpha R.$$

The solution to Difference Equation (2) is

$$x_n = L + (x_0 - L)a^n,$$

where $L = B/(1 - a)$ is the limiting value.

For Newton's law of cooling, α is a positive fraction, so $a = 1 - \alpha$ is between 0 and 1. Also, we have

$$L = \frac{B}{1 - a} = \frac{\alpha R}{1 - (1 - \alpha)} = \frac{\alpha R}{\alpha} = R.$$

Therefore, if the initial temperature of the pizza (or any other cooling object) is T_0, the solution to Difference Equation (1) is

$$T_n = R + (T_0 - R)(1 - \alpha)^n.$$

The constant α usually is determined from an additional temperature measurement, as we illustrate in Examples 1–3 below.

First, let's examine the behavior of this solution function. The term $(T_0 - R)(1 - \alpha)^n$ is an exponential decay function because $a = 1 - \alpha$ lies between 0 and 1. This term is added to the constant R, so the solution function is decreasing in a concave up manner as it decays to a level of R as n increases. This behavior pattern is precisely what we predicted in Figure 5.38.

We can also see this behavior directly from the formula

$$T_n = R + (T_0 - R)(1 - \alpha)^n.$$

As n increases, the exponential decay term $(1 - \alpha)^n$ approaches 0 because $1 - \alpha$ lies between 0 and 1. So again $T_n \to R$ as $n \to \infty$.

In Example 1 of Section 4.8, we constructed a function to fit a set of data obtained from an experiment on cooling. A temperature probe was warmed to 42.3°C and plunged into cold water at about 8.6°C to cool. The shifted exponential function that best fit the data was $T(t) = 8.6 + 35.4394(0.8480)^t$. Note that this function has the identical form as the solution sequence for the difference equation model.

The mathematical model for Newton's law of cooling may be summarized as follows.

Newton's Law of Cooling

Assumptions
- The temperature R of the medium remains constant.
- The change in temperature is proportional to the difference between the temperature of the object and the temperature of the medium.

Mathematical Model
- Difference equation:
$$\Delta T_n = -\alpha \cdot (T_n - R) \quad \text{or} \quad T_{n+1} = (1 - \alpha)T_n + \alpha R$$
- Solution: $\quad T_n = R + (T_0 - R)(1 - \alpha)^n$

EXAMPLE 1

Suppose that a cake is baking in an oven at 350°F. It is removed when its temperature is 180°F and is left to cool in a kitchen at 70°F. After 10 minutes, the temperature of the cake is 125°F.

a. Find the solution to the difference equation based on Newton's law of cooling.

b. Find the temperature of the cake after 15 minutes.

c. How long does it take the cake to cool to 75°F?

Solution

a. For $T_0 = 180$ and $R = 70$, the difference equation for the temperature at any time n is

$$T_{n+1} = (1 - \alpha)T_n + 70\alpha.$$

The corresponding solution is

$$\begin{aligned} T_n &= (180 - 70)(1 - \alpha)^n + 70 \\ &= 110(1 - \alpha)^n + 70. \end{aligned}$$

Further, when $n = 10$, the temperature $T_{10} = 125$, so

$$T_{10} = 110(1 - \alpha)^{10} + 70 = 125.$$

Subtracting 70 from both sides gives

$$110(1 - \alpha)^{10} = 125 - 70 = 55, \quad \text{so that} \quad (1 - \alpha)^{10} = \frac{55}{110} = 0.5.$$

Taking the tenth root of both sides of this equation yields

$$1 - \alpha = 0.5^{1/10} = 0.933, \quad \text{and so} \quad \alpha = 1 - 0.933 = 0.067,$$

which is between 0 and 1, as we expected. The solution to the difference equation is

$$\begin{aligned} T_n &= 110(1 - \alpha)^n + 70 \\ &= 70 + 110(0.933)^n. \end{aligned}$$

b. After $n = 15$ minutes, the temperature of the cake is

$$\begin{aligned} T_{15} &= 70 + 110(0.933)^{15} \\ &= 70 + 38.87 \approx 108.9°. \end{aligned}$$

c. To find the time needed for the cake to cool to 75°F, we need the value of n when $T_n = 75$. We begin with

$$T_n = 110(0.933)^n + 70 = 75 \quad \text{and so} \quad 110(0.933)^n = 5.$$

We divide both sides by 110 to get

$$(0.933)^n = \frac{5}{110} = 0.045.$$

We take logarithms of both sides of this equation to obtain

$$n \log(0.933) = \log(0.045)$$

and then solve for n:

$$n = \frac{\log(0.045)}{\log(0.933)} \approx 44.7 \text{ minutes}.$$

Thus it takes about three-quarters of an hour for the cake to cool to 75°F.

EXAMPLE 2

Mr. Jones's body was found in his kitchen at 9 A.M. by the police who noted that the body temperature was 77.3°F and that the room temperature was 70°F. According to the medical examiner, the body temperature an hour later was 76.1°F. Assuming that Mr. Jones's body temperature was the normal 98.6°F at the time of death, at what time did he die?

Solution   Using the solution to the difference equation for cooling, the body temperature after n hours is given by

$$T_n = R + (T_0 - R)(1 - \alpha)^n,$$

where $R = 70°$ and $T_0 = 98.6°$. Therefore

$$T_n = 70 + (98.6 - 70)(1 - \alpha)^n$$
$$= 70 + 28.6(1 - \alpha)^n.$$

Because we don't know the time of death (which corresponds to $n = 0$), we don't know the value of n at 9 A.M. However, at 9 A.M., which is n hours after death, the body temperature was 77.3°, so

$$T_n = 70 + 28.6(1 - \alpha)^n = 77.3,$$

or, after subtracting 70 from both sides,

$$28.6(1 - \alpha)^n = 7.3. \tag{3}$$

An hour later at 10 A.M., which is $n + 1$ hours after death, the body temperature was

$$T_{n+1} = 70 + 28.6(1 - \alpha)^{n+1} = 76.1,$$

or, again after subtracting 70 from both sides,

$$28.6(1 - \alpha)^{n+1} = 6.1. \tag{4}$$

Dividing Equation (4) by Equation (3) yields

$$\frac{28.6(1 - \alpha)^{n+1}}{28.6(1 - \alpha)^n} = 1 - \alpha = \frac{6.1}{7.3} \approx 0.8356.$$

Substituting this value into Equation (3) gives

$$28.6(0.8356)^n = 7.3,$$

or, equivalently, when we divide both sides by 28.6,

$$(0.8356)^n = 0.2552.$$

To solve for n, we take logs of both sides of this equation and get

$$n \log(0.8356) = \log(0.2552)$$

so that

$$n = \frac{\log(0.2552)}{\log(0.8356)} \approx 7.6.$$

The body was found at 9 A.M., which is $n = 7.6$ hours after death. Thus the murder occurred 7.6 hours (or 7 hours and 36 minutes) before 9 A.M., so we conclude that Mr. Jones was murdered at approximately 1:24 A.M.

EXAMPLE 3

A cup of hot coffee is left standing on a table in a room where the temperature is 70°F. The temperature T, in °F, of the coffee is measured every minute t, and the results are shown in the following table. Find a model that best fits these temperature readings as a function of time.

t	0	1	2	3	4	5	6	7	8	9	10	11	12	13	14	15
T	186	182	178	175	171	168	165	162	159	156	153	152	148	145	143	141

Solution Based on our knowledge of cooling curves, we would expect the best fit to be an exponential function that decays to 70°F. We can't use our data-fitting techniques from Chapter 3 to fit an exponential curve to the data because the function would decay to 0°F rather than 70°F. As in Example 1 of Section 4.8, we must first subtract 70 from each temperature reading, which is equivalent to introducing a vertical shift for the temperature function, to get the following table.

t	0	1	2	3	4	5	6	7	8	9	10	11	12	13	14	15
$T - 70$	116	112	108	105	101	98	95	92	89	86	83	82	78	75	73	71

The exponential function that best fits the shifted data is

$$T - 70 = 115.6(0.96783)^t,$$

where T is the temperature. The corresponding correlation coefficient is $r = -0.99945$; it is negative because the temperature readings are decreasing. This value for r is extremely close to -1, indicating an extremely good fit. Finally, we undo the vertical shift by adding 70 to both sides of the preceding equation to obtain

$$T = 115.6(0.96783)^t + 70$$

as the best model for the temperature of the coffee, as illustrated in Figure 5.39. Note that this function has the same form as the solution we obtained for the difference equation for Newton's law of cooling.

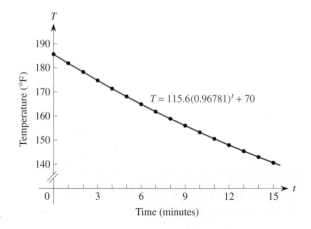

FIGURE 5.39

Newton's Law of Heating

Similar mathematical methods apply if an object is being warmed rather than cooled. The corresponding principle is known as **Newton's law of heating,** which is based on the assumption that the change in temperature ΔT_n (now an increase) is proportional to the difference between the temperature R of the medium (the room, the freezer, the oven, etc.) and the temperature of the object T_n. This assumption leads to the difference equation

$$\Delta T_n = -\alpha \cdot (T_n - R),$$

which is identical to the difference equation for Newton's law of cooling. Let's see why it is the same. Because the object is being heated, the change in temperature ΔT_n of the object must be positive. Also, the temperature T_n of the object is always less than R, so $T_n - R$ is negative. Finally, because the constant of proportionality α is positive, the product of the terms on the right-hand side will be positive.

We solve this difference equation the same way we solved the one for cooling to get

$$T_n = R + (T_0 - R)(1 - \alpha)^n.$$

EXAMPLE 4

A chicken is removed from the refrigerator at a temperature of 40°F and placed into an oven kept at a constant temperature of 350°F. After 10 minutes, the temperature of the chicken is 70°F. The chicken is considered cooked when its temperature reaches 180°F. How long must it remain in the oven until it is cooked?

Solution The solution of the difference equation $\Delta T_n = -\alpha \cdot (T_n - 350)$, with $R = 350$ and $T_0 = 40$, is

$$T_n = 350 + (40 - 350)(1 - \alpha)^n = 350 - 310(1 - \alpha)^n.$$

After 10 minutes,

$$T_{10} = 350 - 310(1 - \alpha)^{10} = 70.$$

Therefore

$$310(1 - \alpha)^{10} = 350 - 70 = 280 \quad \text{so that} \quad (1 - \alpha)^{10} = \frac{280}{310} \approx 0.903.$$

Taking the tenth root of both sides, we obtain

$$1 - \alpha \approx 0.9898.$$

Consequently, the solution to the difference equation for the temperature of the chicken is

$$T_n = 350 - 310(0.9898)^n.$$

We must now find how long it takes the temperature to reach 180°F. Doing so requires finding the value of n for which

$$T_n = 350 - 310(0.9898)^n = 180 \quad \text{so that} \quad 310(0.9898)^n = 350 - 180 = 170.$$

Dividing by 310 gives

$$(0.9898)^n = \frac{170}{310} \approx 0.5484.$$

Taking logs of both sides of this equation yields

$$n \log(0.9898) = \log(0.5484),$$

and so

$$n = \frac{\log(0.5484)}{\log(0.9898)} \approx 58.6 \text{ minutes.}$$

Hence the chicken will be ready in just under an hour.

Problems

1. In Example 2 we assumed that the initial body temperature was a normal 98.6°F. Suppose that Mr. Jones had a slight fever of 100°F when he was murdered. How much difference does that make to the estimated time of death?

2. In Example 2 we also assumed that the room temperature was kept constant at 70°F. Actually, the home heating system probably cycled on and off, so the actual temperature might have oscillated between 67°F and 72°F, say. How might you take this variation into account in predicting the time of death? How much of a difference would it make?

3. In Example 4 we presumed that the temperature of the oven is 350°F for cooking a chicken. Suppose instead that the temperature is set at 325°F. How much difference does that make in the time needed to cook the chicken to 180°F?

4. A pot of bubbling pudding (212°F) is removed from the stove and put immediately into a refrigerator at 40°. After 10 minutes, the temperature of the pudding is 160°. Find the temperature after 1 hour. How long does it take for the temperature to drop to 75°?

5. Sam takes a can of soda at room temperature (70°F) and puts it in a freezer (0°F) to chill quickly. After 10 minutes, the temperature of the soda is 60°. How long does it take the temperature to drop to 40°?

6. A bowl of cold soup is taken out of the refrigerator (36°F) and placed in a heated oven (375°F) to warm. After 10 minutes, the temperature of the soup is 120°. How long does it take for the soup to reach 200°?

7. Professor Smith's body was found in a large walk-in refrigerator in the laboratory at 9 A.M. by the police who noted that the body temperature was 67.3°F and that the refrigerator temperature was 40°. An hour later, with the body still in the refrigerator, the medical examiner found the body temperature to be 63.1°. Assuming that the body temperature at death was 98.6°, at what time did Professor Smith die?

8. A cup of boiling water (100°C) was placed in a refrigerator kept at 7°C at 8 A.M., and the following readings were obtained.

Time (minutes)	1	7	21	45
Temperature (°C)	89	71	53	36
Time (minutes)	73	90	123	152
Temperature (°C)	25	20	14	11

Determine the best exponential fit to these data and use it to predict when the water temperature will be 9°C.

9. The following data are printed on a carton of milk to indicate how many days the milk will last without spoiling at different temperatures.

Temperature (°F)	32	40	45	50	60	70
Time (days)	24	11	5.5	2	1	0.5

a. Determine the linear, exponential, and power functions that best fit these data.

b. Which of the three functions seems to be the best fit to the data?

c. Using the model that you think is the best fit, how long should milk last in a refrigerator kept at 35°F?

10. A cool potato at temperature 60°F is placed in an oven kept at a constant 350°.

a. Sketch the graph of the temperature of the potato as a function of time.

Use the concavity of your graph from part (a) to answer the questions in parts (b)–(d).

b. Suppose that you measure the temperature of the potato after 5 minutes and find that it is 109° and that, after 7 minutes, it is 127°. Use this information to estimate the temperature after 10 minutes.

Is the actual temperature higher or lower than your estimate? How do you know?

c. Suppose that you are told that the temperature of the potato after 12 minutes is 150°. Use this information and the temperature after 5 minutes to estimate the temperature after 10 minutes. Is the actual temperature higher or lower than your estimate? How do you know?

d. How might you use the results from parts (b) and (c) to come up with a better estimate of the temperature after 10 minutes?

11. Your Thanksgiving turkey is taken from a refrigerator at 40°F and is cooked in an oven kept at a constant temperature of 350°. The temperature T of the bird is 70° after 30 minutes and is 96° after 60 minutes. From your knowledge of the pattern of temperature rise, decide which of the following temperature readings are possible and which are impossible.

a. $T(45) = 80°$ b. $T(45) = 85°$
c. $T(75) = 105°$ d. $T(75) = 115°$

12. A cup of hot chocolate (temperature 180°F) is placed on a table where the air temperature is 70°. Suppose that it takes 12 minutes for the drink to cool to 100°. Let r_1 represent the average rate of decrease in temperature per minute over the full 12-minute period, let r_2 be the average rate of decrease over the first 6 minutes, and let r_3 be the average rate of decrease over the last 6 minutes. List these three rates in increasing order.

13. Suppose that it takes t_1 minutes for a raw potato, starting at 70°F, to reach 200° in an oven. At that time, it is removed from the oven and put on the table where it cools. Suppose that the potato takes t_2 minutes to reach 70°. Is $t_1 < t_2, t_1 = t_2,$ or $t_1 > t_2$? Explain.

14. A potato at room temperature (T_0) is placed in an oven at temperature R. Construct the actual solution to the corresponding difference equation for heating in terms of the various parameters. What is the formula for the temperature of the potato if $T = 70°F$ and $R = 350°F$?

15. In Section 4.8, we presented a set of data from a cooling experiment. The temperature readings, in degrees Celsius, were as follows.

Time	Temp	Time	Temp	Time	Temp	Time	Temp
1	42.30	10	14.77	19	10.17	28	9.04
2	36.03	11	13.82	20	9.92	29	8.91
3	30.85	12	13.11	21	9.80	30	8.83
4	26.77	13	12.51	22	9.67	31	8.78
5	23.58	14	11.91	23	9.54	32	8.78
6	20.93	15	11.54	24	9.42	33	8.78
7	18.79	16	11.17	25	9.29	34	8.78
8	17.08	17	10.67	26	9.16	35	8.66
9	15.82	18	10.42	27	9.16	36	8.66

According to Newton's law of cooling, the change in temperature ΔT is proportional to the difference between the temperature of the object and the temperature of the medium (in this case the temperature of the cool water is 8.6°C).

a. Extend the table, with additional columns for $T - 8.6$ and also for the differences ΔT between successive temperature readings.

b. Plot the points $(T - 8.6, \Delta T)$ that you calculated. In what type of pattern do they fall?

c. Find the equation of the line that best fits these points and write a difference equation for ΔT for the temperature data.

d. Solve the difference equation for the temperature T as a function of time.

5.5 Geometric Sequences and Their Sums

Consider the difference equation for exponential growth

$$x_{n+1} = rx_n$$

whose solution is given by

$$x_n = x_0 r^n,$$

where x_0 is any initial value. Using the notation for sequences, we can write this solution as

$$\{x_0, x_0 r, x_0 r^2, x_0 r^3, x_0 r^4, \ldots, x_0 r^n, \ldots\}.$$

Recall that such a sequence is called a *geometric sequence* or *exponential sequence*. For simplicity, suppose that $x_0 = 1$, so that the sequence reduces to

$$\{1, r, r^2, r^3, r^4, \ldots, r^n, \ldots\}.$$

For instance, if $r = 5$, we have the geometric sequence

$$\{1, 5, 5^2, 5^3, 5^4, \ldots, 5^n, \ldots\} = \{1, 5, 25, 125, 625, \ldots, 5^n, \ldots\}.$$

The difference equation $x_{n+1} = rx_n$ reinforces the fact that each term in any geometric sequence is a constant multiple of the preceding term, or, equivalently, there is a common ratio r between successive terms. Alternatively, x_n is an exponential function of n, and we know that the ratio of successive values is a constant—namely, the growth or decay factor r. In the preceding sequence, each term is five times the preceding term, so the common ratio r of each pair of successive terms is 5.

We can deduce a considerable amount of information about the behavior of the terms in a geometric sequence $\{1, r, r^2, r^3, \ldots, r^n, \ldots\}$ from the value of the common ratio r. We summarize this information as follows.

1. If $r > 1$, the terms are successively larger and approach infinity. We say that they *increase monotonically*.
2. If $r = 1$, all the terms are equal to 1.
3. If $0 < r < 1$, the terms are successively smaller and approach zero. We say that they *decrease monotonically*.
4. If $r = 0$, all terms after the initial term, 1, are 0.
5. If $-1 < r < 0$, the terms oscillate between positive and negative values, each is numerically smaller than the preceding term, and they approach 0.
6. If $r = -1$, the terms oscillate between 1 and -1.
7. If $r < -1$, the terms oscillate between positive and negative values, and each term is numerically larger than the preceding term.

If the terms of a sequence approach a single value as n approaches ∞, written $x_n \to \infty$, we say that the sequence *converges* to that value. Otherwise, we say that the sequence *diverges*.

The following seven sequences illustrate these properties.

1. $r = 5$: $\{1, 5, 5^2, 5^3, 5^4, \ldots, 5^n, \ldots\} \to \infty$, so this sequence diverges as $n \to \infty$. (Note that the terms in this sequence increase monotonically.)
2. $r = 1$: $\{1, 1, 1, 1, 1, \ldots\} \to 1$, so this sequence converges to 1 as $n \to \infty$.

3. $r = \frac{1}{2}$: $\left\{ 1, \frac{1}{2}, \left(\frac{1}{2}\right)^2, \left(\frac{1}{2}\right)^3, \left(\frac{1}{2}\right)^4, \ldots, \left(\frac{1}{2}\right)^n, \ldots \right\} = \{1, 0.5, 0.25, 0.125, 0.0625,$
   $\ldots\} \to 0$, so this sequence converges to 0 as $n \to \infty$. (Note that the terms in this sequence decrease monotonically.)

4. $r = 0$: $\{1, 0, 0, 0, 0, \ldots\} \to 0$, so this sequence converges to 0 as $n \to \infty$

5. $r = -\frac{1}{4}$: $\left\{ 1, -\frac{1}{4}, \left(-\frac{1}{4}\right)^2, \left(-\frac{1}{4}\right)^3, \left(-\frac{1}{4}\right)^4, \ldots, \left(-\frac{1}{4}\right)^n, \ldots \right\} = \{1, -0.25,$
   $0.0625, -0.015625, 0.00390625, \ldots\} \to 0$, so this sequence converges to 0 as $n \to \infty$. (Note that the terms in this sequence oscillate between positive and negative values, but that each term is numerically smaller than the previous term; the terms are not monotonic.)

6. $r = -1$: $\{1, -1, (-1)^2, (-1)^3, (-1)^4, \ldots, (-1)^n, \ldots\} = \{1, -1, 1, -1,$
   $1, -1, \ldots\}$, so this sequence does not converge to a single value but diverges as $n \to \infty$.

7. $r = -2$: $\{1, -2, (-2)^2, (-2)^3, (-2)^4, \ldots, (-2)^n, \ldots\} = \{1, -2, 4, -8,$
   $16, -32, 64, -128, \ldots\} \to \pm \infty$, so this sequence diverges as $n \to \infty$. The terms oscillate between positive and negative values, but each term is numerically larger than the previous term; the terms are not monotonic.

The Sum of the Terms in a Geometric Sequence

Geometric sequences arise in a great variety of applications, and usually they are accompanied by the related question: What is the sum of the terms? That is, for any geometric sequence with constant ratio r, what is

$$1 + r + r^2 + r^3 + r^4 + \cdots + r^n$$

for any given value of n?

For instance, there is a very old puzzle in which gold coins are placed on a chessboard according to the pattern: one coin on the first square, two on the second, four on the third, and so on to 2^{63} on the 64th square. The problem then is: Find the total number of coins. That is, find the sum

$$1 + 2 + 2^2 + 2^3 + 2^4 + \cdots + 2^{63}.$$

Before attempting to solve this specific problem, let's look at the more general problem of finding a formula for the sum of the terms in any geometric sequence. We introduce the following notation. Let

$$S_0 = 1, \qquad S_1 = 1 + r, \qquad S_2 = 1 + r + r^2,$$

and, in general, for the sum of the n terms $1, r, r^2, r^3, r^4, \ldots, r^n$,

$$S_n = 1 + r + r^2 + \cdots + r^n.$$

We want to find a formula for the sum S_n for any n. We multiply the expression for S_n by the common ratio r to obtain

$$rS_n = r(1 + r + r^2 + \cdots + r^n) = r + r^2 + r^3 + \cdots + r^n + r^{n+1}.$$

When we subtract the expression for rS_n from the expression for S_n, all the intermediate terms $r, r^2, r^3, \ldots, r^n$ cancel out and we are left with

$$S_n - rS_n = 1 - r^{n+1}.$$

We factor out the term S_n on the left-hand side and get

$$(1 - r)S_n = 1 - r^{n+1}.$$

If $r \neq 1$, we can divide both sides by $1 - r$ to get

$$S_n = \frac{1 - r^{n+1}}{1 - r}.$$

The sum of the terms $1, r, r^2, \ldots, r^n$ of a finite geometric sequence is

$$1 + r + r^2 + r^3 + \cdots + r^n = \frac{1 - r^{n+1}}{1 - r},$$

provided that $r \neq 1$.

EXAMPLE 1

What is the total number of gold coins that would be needed on a chessboard according to the pattern previously described?

Solution The number of coins is

$$1 + 2 + 2^2 + 2^3 + 2^4 + \cdots + 2^{63}.$$

The common ratio in this sum of terms of a geometric sequence is $r = 2$. So, for $n = 63$, we get

$$
\begin{aligned}
S_{63} &= 1 + 2 + 2^2 + \cdots + 2^{63} \\
&= \frac{1 - 2^{64}}{1 - 2} \\
&= 2^{64} - 1 = 1.844674 \times 10^{19},
\end{aligned}
$$

which is more than 18 quintillion.

In two of the seven possible cases for a geometric sequence—namely, $0 < r < 1$ and $-1 < r < 0$—the *terms of the sequence* approach 0. Also, when $r = 0$, all terms after the initial term 1 are 0. In the other 4 cases listed, the terms do not approach 0. Let's see what this means for the *sum of the terms* of the entire geometric sequence, not just the sum of the first n terms.

Using the formula for the sum of the terms from 1 to r^n,

$$S_n = \frac{1 - r^{n+1}}{1 - r},$$

we see that, whenever $-1 < r < 1$, the expression r^{n+1} becomes ever smaller and approaches 0 as n increases. As a result, even though we are adding more and more terms, eventually the additional terms are all so small that together they contribute virtually nothing to the sum. This fact enables us to give meaning to the sum of *all* the terms of an infinite geometric sequence, provided that the terms decrease rapidly enough. In other words,

$$1 + r + r^2 + r^3 + \cdots + r^n + \cdots = \frac{1 - 0}{1 - r} = \frac{1}{1 - r},$$

provided that $-1 < r < 1$, or, equivalently, $|r| < 1$.

However, if $r \geq 1$ or if $r \leq -1$, the values r^{n+1} do *not* approach 0 as n approaches ∞, so no finite value for the sum of the infinite sequence is possible. Thus the sum of an infinite geometric sequence makes sense only when the common ratio r is strictly between -1 and 1.

The sum of the terms of an infinite geometric sequence is

$$1 + r + r^2 + r^3 + \cdots + r^n + \cdots = \frac{1}{1-r},$$

provided that $-1 < r < 1$.

For instance, the sum of the terms in the infinite geometric sequence

$$1 + \frac{1}{2} + \left(\frac{1}{2}\right)^2 + \left(\frac{1}{2}\right)^3 + \cdots = \frac{1}{1 - (1/2)} = \frac{1}{(1/2)} = 2.$$

Think About This   Add enough of the terms from the preceding sequence to convince yourself that this result is reasonable. ▢

The facts and results regarding geometric sequences just described occur frequently in applications of mathematics. We encounter such sequences repeatedly later in this book. For now, let's consider several additional situations in which they arise.

In our discussion of the elimination of a drug from the body in Section 5.1, we saw that the kidneys remove about 25% of any Prozac in the bloodstream every 24 hours and that a person takes the same dose $D_0 = 80$ mg every day. We modeled this situation with the difference equation

$$D_{n+1} = 0.75D_n + 80$$

and created the formula

$$D_n = 320 - 240(0.75)^n$$

for the solution sequence. Let's now find this formula by using our knowledge of the sum of a geometric sequence.

EXAMPLE 2

a. Find a formula for the level of Prozac in the body after any number of days, using the sum of a geometric sequence.

b. Use the result from part (a) to find the level of Prozac in the bloodstream after 5 days and after 10 days.

c. What is the limiting value L for Prozac in the body?

Solution

a. From the difference equation and the initial value $D_0 = 80$, we have

$$D_1 = 0.75D_0 + 80 = 0.75(80) + 80 = (1 + 0.75)(80).$$

Similarly,

$$D_2 = 0.75D_1 + 80$$
$$= 0.75(1 + 0.75)80 + 80 = (1 + 0.75 + 0.75^2)80.$$

Furthermore,

$$D_3 = 0.75D_2 + 80$$
$$= 0.75(1 + 0.75 + 0.75^2)80 + 80 = (1 + 0.75 + 0.75^2 + 0.75^3)80.$$

In general, after n days,

$$D_n = 80(1 + 0.75 + 0.75^2 + 0.75^3 + \cdots + 0.75^n).$$

The expression inside the parentheses is the sum of the terms from 1 to 0.75^n in a geometric sequence with common ratio $r = 0.75$. Therefore the sum of these terms is given by

$$D_n = 80\left[\frac{1 - (0.75)^{n+1}}{1 - 0.75}\right] = 80\left[\frac{1 - (0.75)^{n+1}}{0.25}\right]$$
$$= 320[1 - (0.75)^{n+1}] = 320 - 320(0.75)(0.75)^n$$
$$= 320 - 240(0.75)^n,$$

which is identical to the formula that we created in Section 5.1 using difference equations.

b. After $n = 5$ days, we have

$$D_5 = 320 - 240(0.75)^5 = 263.0469 \text{ mg},$$

which agrees with the value we obtained in Section 5.1. Similarly,

$$D_{10} = 320 - 240(0.75)^{10} = 306.485 \text{ mg}.$$

c. Finally, because $r = 0.75 < 1$, the limiting value for the sum of this geometric sequence is

$$L = D_0 \cdot (1 + r + r^2 + r^3 + \cdots)$$
$$= 80\left(\frac{1}{1 - r}\right) = \frac{80}{1 - 0.75}$$
$$= \frac{80}{0.25} = 320 \text{ mg},$$

which is the same value we found in Section 5.1 for the drug maintenance level.

When Will Our Oil Run Out?

An ongoing political debate having major economic and social implications has to do with energy policy both at home and abroad. One aspect of this debate centers on the use of petroleum, both as a source of energy in vehicles and power plants and as a component of oil-based products such as plastics. Estimated worldwide oil reserves in 2000 were about 2250 billion barrels, and worldwide oil consumption in 2000 was about 26 billion barrels.* How long will the oil last?

*Source: Energy Information Administration, www.eia.doe.gov/ieu/.

If the current rate of consumption remains constant at 26 billion barrels per year, known oil reserves will run out in

$$\frac{2250}{26} \approx 86.5 \text{ years.}$$

But oil consumption hasn't been constant; in fact, some estimates indicate that worldwide oil consumption has been growing at an annual rate of about 2%. How do we then calculate how long the known oil reserves will last?

◆ **EXAMPLE 3**

Calculate how long the estimated 2250 billion barrel oil reserves worldwide will last if oil consumption was 26 billion barrels in 2000 and continues to grow at a 2% annual rate.

Solution   Annual oil consumption C can be modeled by the exponential growth function

$$C(t) = 26(1.02)^t,$$

where t is the number of years since 2000. The total oil consumed from 2000 on is then approximately

$$C(0) + C(1) + C(2) + C(3) + \cdots.$$

Let n be the first year when the total passes the 2250 billion barrel level. We need to solve for the value of n that gives

$$C(0) + C(1) + C(2) + \cdots + C(n) = 2250.$$

This expression is equivalent to

$$26(1.02)^0 + 26(1.02)^1 + 26(1.02)^2 + \cdots + 26(1.02)^n = 2250$$

or, when we factor out the common factor of 26,

$$26[1 + (1.02)^1 + (1.02)^2 + \cdots + (1.02)^n] = 2250.$$

The expression in the brackets is the sum of the first n terms of a geometric sequence with common ratio $r = 1.02$, so

$$26\left(\frac{1 - 1.02^{n+1}}{1 - 1.02}\right) = 26\left(\frac{1 - 1.02^{n+1}}{-0.02}\right)$$

$$= 26\left(\frac{1.02^{n+1} - 1}{0.02}\right) = 1300(1.02^{n+1} - 1).$$

We therefore need to solve the equation

$$1300(1.02^{n+1} - 1) = 2250$$

for n. We divide through by 1300 to get

$$1.02^{n+1} - 1 = 1.7308$$

or, by adding 1 to both sides, we obtain

$$1.02^{n+1} = 2.7308.$$

To extract $n + 1$ from the exponent, we use logarithms to get

$$\log(1.02^{n+1}) = (n + 1)\log 1.02 = \log 2.7308,$$

so that

$$n + 1 = \frac{\log 2.7308}{\log 1.02} \approx 50.73$$

and therefore $n \approx 49.73$. We therefore conclude that all the world's known oil reserves will be depleted in just under 50 years from 2000 if consumption continues to grow at an annual rate of 2%.

◆

The results in Example 3 are startling because all the industrialized nations depend so heavily on oil. Many people therefore advocate major efforts to discover new oil supplies. Let's see what doing so would gain.

EXAMPLE 4

Suppose that, as a result of major efforts to find new sources of oil, the worldwide oil reserves are doubled to 4500 billion barrels. How long will it take to use it all if oil consumption continues to grow at the current 2% annual rate?

Solution   We now have to find the value of n for which

$$C(0) + C(1) + C(2) + \cdots + C(n) = 2(2250) = 4500,$$

which is equivalent to the sum of terms in the geometric sequence

$$26[1 + (1.02)^1 + (1.02)^2 + \cdots + (1.02)^n] = 4500.$$

Following the same algebraic development as in Example 3, we have to solve

$$1300(1.02^{n+1} - 1) = 4500$$

for n. Dividing both sides of the equation by 1300, we get

$$1.02^{n+1} - 1 = 3.4615 \quad \text{or} \quad 1.02^{n+1} = 4.4615.$$

Taking logarithms gives
$$(n + 1)\log 1.02 = \log 4.4615,$$
so that

$$n + 1 = \frac{\log 4.4615}{\log 1.02} \approx 75.52$$

and hence $n \approx 74.52$. Note that doubling the total oil reserve didn't double the 50 years we found in Example 3; it merely added less than 25 more years.

◆

Other people advocate reducing the rate of oil consumption by using alternative energy sources (especially renewable sources), by developing more energy efficient vehicles, by encouraging conservation, and by taking other actions. Let's see what such measures can accomplish.

EXAMPLE 5

Suppose that, as a result of conservation efforts, the annual rate of growth of oil consumption drops to 1.5% from 2%. How long will the current worldwide oil reserve last?

Solution   Our new exponential model for oil consumption is

$$C(t) = 26(1.015)^t,$$

where t is still the number of years since 2000. We now have to solve

$$26[1 + (1.015)^1 + (1.015)^2 + \cdots + (1.015)^n] = 2250$$

for n. When we find the sum of the terms in this geometric sequence and simplify, as we did before, we get

$$1733.33(1.015^{n+1} - 1) = 2250.$$

Dividing by 1733.33, we get

$$1.015^{n+1} - 1 = 1.2981 \quad \text{or} \quad 1.015^{n+1} = 2.2981.$$

Taking logarithms yields

$$(n + 1)\log 1.015 = \log 2.2981,$$

so that

$$n + 1 = \frac{\log 2.2981}{\log 1.015} \approx 55.9,$$

and $n \approx 54.9$ years. Hence decreasing the annual rate at which oil is consumed by one-half percentage point gives the world an additional 5 years of oil reserves. ◆

Think About This   Is it more likely that the rate at which oil consumption is increasing can be reduced through conservation measures or that an amount of undiscovered oil equal to the total known oil reserves can be found? ⌑

A Bouncing Ball

We now consider another application involving the sum of a geometric sequence.

EXAMPLE 6

Suppose that a properly inflated basketball is designed to bounce back to three-quarters of the height from which it is dropped.

a. If such a ball is initially dropped from a height of 10 feet, find the total vertical distance it travels on the first 10 bounces; the first 20 bounces; the first 30 bounces.

b. What total vertical distance does the ball cover if, theoretically, it keeps bouncing indefinitely?

Solution

a. Figure 5.40 shows that the vertical distance the ball travels is 10 feet until the first bounce plus 2 times the distance (up and then down) between the first and second bounces, or

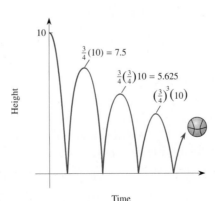

FIGURE 5.40

$$2\left(\frac{3}{4} \cdot 10\right),$$

plus 2 times the distance between the second and third bounces, or

$$2\left[\frac{3}{4}\left(\frac{3}{4} \cdot 10\right)\right] = 2\left(\frac{3}{4}\right)^2 \cdot 10,$$

and so on.

The total vertical distance traveled on the first n bounces is therefore

$$D_n = 10 + 2\left(\frac{3}{4}\right) \cdot 10 + 2\left(\frac{3}{4}\right)^2 \cdot 10 + 2\left(\frac{3}{4}\right)^3 \cdot 10 + \cdots + 2\left(\frac{3}{4}\right)^n \cdot 10$$

$$= 10 + 20\left(\frac{3}{4}\right) + 20\left(\frac{3}{4}\right)^2 + 20\left(\frac{3}{4}\right)^3 + \cdots + 20\left(\frac{3}{4}\right)^n$$

$$= 10 + 20\left[\left(\frac{3}{4}\right) + \left(\frac{3}{4}\right)^2 + \left(\frac{3}{4}\right)^3 + \cdots + \left(\frac{3}{4}\right)^n\right]$$

$$= 10 + 20\left(\frac{3}{4}\right)\left[1 + \left(\frac{3}{4}\right) + \left(\frac{3}{4}\right)^2 + \cdots + \left(\frac{3}{4}\right)^{n-1}\right]$$

$$= 10 + 15\left[1 + \left(\frac{3}{4}\right) + \left(\frac{3}{4}\right)^2 + \cdots + \left(\frac{3}{4}\right)^{n-1}\right].$$

Note that the expression in the brackets on the preceding line is the sum of a finite number of terms of a geometric sequence with common ratio $r = \frac{3}{4}$. Therefore the total vertical distance traveled by the ball during the first n bounces is

$$D_n = 10 + 15\left(\frac{1 - (3/4)^n}{1 - (3/4)}\right) = 10 + 15\left(\frac{1 - (3/4)^n}{1/4}\right)$$

$$= 10 + 60\left[1 - \left(\frac{3}{4}\right)^n\right].$$

Consequently, the vertical distance traveled by the ball during the first 10 bounces is

$$D_{10} = 10 + 60\left[1 - \left(\frac{3}{4}\right)^{10}\right]$$

$$= 10 + 56.621 = 66.621,$$

or about 66.6 feet. During the first 20 bounces, the ball travels

$$D_{20} = 10 + 60\left[1 - \left(\frac{3}{4}\right)^{20}\right] = 69.810,$$

or about 69.8 feet. And during the first 30 bounces, it travels

$$D_{30} = 10 + 60\left[1 - \left(\frac{3}{4}\right)^{30}\right] = 69.989,$$

or about 69.99 feet.

b. Little extra distance is contributed by the 20th through the 30th bounce, and all subsequent bounces will contribute even less. In fact, because the common ratio $r = \frac{3}{4}$ is between -1 and 1, we know that $\left(\frac{3}{4}\right)^n \to 0$ as n increases and so we can sum all of the terms of the infinite geometric sequence to obtain

$$10 + 15\left[1 + \tfrac{3}{4} + \left(\tfrac{3}{4}\right)^2 + \left(\tfrac{3}{4}\right)^3 + \cdots + \left(\tfrac{3}{4}\right)^n + \cdots\right]$$

$$= 10 + 15\left[\tfrac{1}{1 - (3/4)}\right] = 10 + 15\left(\tfrac{1}{1/4}\right)$$

$$= 10 + 15(4) = 70 \text{ feet.}$$

That is, theoretically, if the ball were to continue bouncing forever, the total vertical distance it would travel would be 70 feet.

◆

Problems

1. Find the sum $1 + r + r^2 + \cdots + r^n$, with $r = \frac{1}{2}$, for $n = 10$, $n = 20$, and $n = 30$.

2. Repeat Problem 1 with $r = 0.2$.

3. Repeat Problem 1 with $r = 0.8$.

4. Repeat Problem 1 with $r = -0.8$.

5. Repeat Problem 1 with $r = 1.5$.

6. Repeat Problem 1 with $r = -2.5$.

7. Suppose that 6000 new cases of a certain disease occurred in 1960. If the number of new cases diminished 20% per year since then, what is the *total* number of people who contracted this disease from 1960 through 2000? from 2000 through 2010?

8. Repeat Problem 7 if the number of new cases of the disease increased 20% per year since 1960.

9. In 1980, the United States used approximately 2.5 billion kilowatt-hours of electricity. If electric usage grew by 2% per year, find the total amount of electricity used between 1980 and 2000.

10. The United States produced 195,000 metric tons of wheat in 1984. If production increased by 10% per year find the total amount of wheat produced between 1984 and 2002.

11. The United States produced 70,600 metric tons of rice in 1984. If rice production fell by 9% per year, find the total amount of rice produced between 1984 and 2002.

12. At age 22, Ken gets his first job paying $35,000 per year. If he stays with the same employer and gets an average annual increase of 4% each year, what will be his *total* earnings over his entire career by the time he retires at age 65?

13. In 1986, a total of 70,000 pages of new mathematical research was published. If the amount of research grew at the rate of 8% per year, find the total amount of new mathematics research published between 1986 and 2000.

14. Suppose that several immense new oil fields are discovered that increase the worldwide oil reserves tenfold. If oil consumption continues to grow at an annual rate of 2% and 26 billion barrels were consumed in 2000, determine how long the oil reserves will last.

15. Suppose that, as undeveloped countries become more industrialized, the annual growth rate in oil consumption increases to 2.5%. Based on the current 2250 billion barrel estimate for worldwide oil reserves and the estimated worldwide oil consumption of 26 billion barrels in 2000, how long will it be until all the known oil is used?

16. Repeat Example 6 if the initial height of the ball is 6 feet.

17. Repeat Example 6 if the initial height of the ball is 12 feet.

18. Repeat Example 6 if the ball bounces back to 80% of its height. By how much does the total distance traveled by the ball change compared to a 75% bounce?

19. Repeat Example 6 if the ball bounces back $\frac{2}{3}$ of its height.

20. The repeating decimal $0.222222\ldots$ can be thought of as

$$\frac{2}{10} + \frac{2}{100} + \frac{2}{1000} + \cdots.$$

What is the sum of this geometric sequence?

21. The repeating decimal $0.252525\ldots$ can be thought of as

$$\frac{25}{100} + \frac{25}{10,000} + \frac{25}{1,000,000} + \cdots.$$

What is the sum, as a simple fraction, of this geometric sequence?

22. A geometric sequence is based on the fact that the ratio of successive terms is constant: $x_{n+1}/x_n = r$. Suppose instead that the ratio of successive terms in a sequence is a linear function—say, $x_{n+1}/x_n = rn$, for $n \geq 1$.

   a. How does the growth rate of this sequence compare to that for a geometric sequence?

   b. What is the solution of $x_{n+1}/x_n = n$, for $n \geq 1$?

c. What is the solution of $x_{n+1}/x_n = 5n$, for $n \geq 1$?

d. What is the solution of $x_{n+1}/x_n = rn$ for $n \geq 1$, for any r?

e. What is the solution of $x_{n+1}/x_n = rn + b$, for $n \geq 1$?

23. How would the solution of $x_{n+1}/x_n = n^2$ compare to the solution of $x_{n+1}/x_n = n$, for $n \geq 1$? Construct a formula for this solution.

24. Consider $x_{n+1}/x_n = \frac{1}{n}$. How does its solution behave? How does it compare to the solution of $x_{n+1}/x_n = 1/r$, for any $r > 1$? Construct a formula for this solution.

25. The table below shows the box-office gross, in millions of dollars, of the movie *Star Wars: The Phantom Menace* during its first 10 weeks in the theaters.

a. Find the exponential function that models the box office gross for the movie as a function of the number of weeks in release.

b. Use the exponential function from (a) to estimate the total box office gross for *The Phantom Menace* during its first ten weekends in theaters.

c. How does the estimate in part (b) compare to the actual total gross obtained by adding the entries in the table?

26. The estimated worldwide reserve of natural gas at the end of 1999 was 5200 billion cubic feet. The worldwide consumption of natural gas in 1999 was about 84 billion cubic feet and was growing at an annual rate of about 2.9%. Source: Energy Information Administration, www.eia.doe. gov/ieu/.

a. If this growth rate continues, how long will the known reserves of natural gas last?

b. If new discoveries of natural gas triple the known reserves, how long will it take to deplete the natural gas reserves at the current growth rate in consumption?

c. How long will the current 5200 billion cubic feet of natural gas last if the annual growth rate of consumption rises to 3.5%?

d. How long will the current 5200 billion cubic feet of natural gas last if the annual growth rate of consumption falls to 2%?

Weekend	1	2	3	4	5	6	7	8	9	10
Gross	101.4	48.7	41.2	31.1	23.6	21.5	12.0	8.0	9.6	5.8

Source: Internet Movie Database: http://us.imdb.com.

Chapter Summary

In this chapter we introduced difference equations and their solutions. We also discussed the use of difference equations as mathematical models of population growth and other phenomena. More specifically we showed the following.

◆ How to present the solution sequence to a difference equation either as a closed-form expression for the nth term or as a sequence of numbers generated term-by-term.

◆ The fact that the solution to a difference equation depends on the initial condition with each different initial condition giving rise to a different solution sequence.

◆ How to model the level of a drug in the bloodstream with difference equations.

◆ How to find the maintenance level associated with a drug and what it means.

◆ How to model exponential growth and decay processes with difference equations.

◆ How to model population growth with the Fibonacci difference equation.

◆ How to model inhibited population growth with the logistic model, including finding the maximum sustainable population.

◆ How to interpret the behavior of a logistic curve.

◆ How to estimate the logistic coefficients from a set of data.

◆ How to model temperature decrease with a difference equation based on Newton's law of cooling.

◆ How to model temperature increase with a difference equation based on Newton's law of heating.

◆ How to interpret the behavior of the solutions for the heating and cooling models.

◆ How to sum the first n terms of a finite geometric sequence.

◆ How to sum all the terms of an infinite geometric sequence, provided that the common ratio is between -1 and 1.

◆ How to apply the formulas for the sum of the terms in a geometric sequence.

Review Problems

1. Write the first five terms of each sequence.

   a. $a_n = 6n - 1$

   b. $t_n = \dfrac{3^n}{n}, \qquad n > 0$

   c. $r_n = 1 - (0.3)^n$

2. Determine the first five terms in the solution sequence of each difference equation.

   a. $x_{n+1} = x_n + 8, \qquad x_0 = 2$
   b. $x_{n+1} = x_n - 8, \qquad x_0 = 12$

   c. $x_{n+1} = \dfrac{1}{3}x_n, \qquad x_0 = 5$

   d. $x_{n+1} = x_n + (-3)^n, \qquad x_0 = 10$

3. Determine the first five values in each solution sequence.

   a. $y_{n+2} = y_{n+1} + y_n, \qquad y_0 = 2, \quad y_1 = 7$
   b. $y_{n+2} = y_{n+1} + y_n, \qquad y_0 = 3, \quad y_1 = 7$

4. A drug is administered every 6 hours. The kidneys eliminate 60% of the drug over that period. If the original dose is 100 mg, how much of the drug remains in the body after 8 days?

5. A drug is administered every 4 hours. The kidneys eliminate about 70% of the drug over the 4-hour period. The initial dose of the medicine is 100 mg. How much should the repeated dosage be to ensure a maintenance level of 30 mg?

6. Chlorine is added to a town's water supply reservoir at the rate of 30 lb/day. An estimated 20% of this amount is lost each day through evaporation or filters. What is the maintenance level of chlorine in the reservoir?

7. The difference equation $v_{n+1} = 1.30v_n - 0.00002v_n^2$ models the number of people in a town who have VCRs as a function of the year n. Initially, 40 households had VCRs. Estimate how many people will eventually have VCRs. Determine the year in which half the population has VCRs.

8. A population grows according to the logistic model from an initial size of 1000 to a final size of 12,000. The annual growth rate is 20%. What is the inhibiting constant? Write the difference equation that describes this population for any year n.

9. The size of the fish population in a stream grows in accordance with the logistic model at an annual rate of 30% with an inhibiting constant of 0.04%. Write the difference equation for each situation.

   a. The fish population grows according to the logistic model.

   b. The fish population grows according to the logistic model, but the state game warden allows 2000 fish to be caught and removed from the stream each year.

   c. The fish population grows according to the logistic model, and the state game warden stocks the stream with 400 new fish per year.

   d. The fish population grows according to the logistic model, but about 10% of the population is caught every year.

10. Write the difference equation for each scenario, draw a graph of the behavior of the solution, and answer the question.

    a. Pancake syrup used to be 100% maple syrup, but over the years the amount of maple syrup has been reduced. Suppose that in 1970 a company began reducing the amount of maple syrup in its product by 15% per year. What percentage of maple syrup is in its product in 2005?

    b. John's investment in the stock market has been growing by 10% per year. He adds $2000 a year to his investment. If he had $50,000 invested in 1995, how much does he have invested in 2005?

    c. Advertising in the print media has been less important recently than it was in the past. In 1998, a company decided to decrease its budget for advertising in the print media by $20,000 per year. Also, the company increased its budget for television by 10% per year. If the budgets were each $2 million in 1998, how much is the company budgeting for print and how much for television at the end of 5 years? How is the total advertising budget changing?

11. Chris's old car has a major oil leak. He estimates that it loses about 25% of the oil in the engine every week, so he adds a quart of oil weekly. The capacity of the engine is 6 quarts of oil.

    a. Write a difference equation for the amount of oil in the engine as a function of time, measured in weeks.

    b. Find the solution to the difference equation.

    c. Use the solution to predict when the level of oil in the engine just after the weekly quart is added will be down to 5 quarts.

For each difference equation in Problems 12–19, decide whether the behavior of the solution is an increasing concave up pattern, a decreasing concave up pattern, an increasing concave down pattern, a decreasing concave down pattern, or none of these patterns. In each case, $x_0 = 50$.

12. $\Delta x_n = 5(1.04)^n$

13. $\Delta x_n = 5n^{2.5}$

14. $\Delta x_n = 8(0.85)^n$

15. $\Delta x_n = -4n^{0.3}$

16. $\Delta x_n = 12n^{-1.2}$

17. $\Delta x_n = 5n + 3$

18. $\Delta x_n = 25 - 2n$

19. $\Delta x_n = 12 - 3(0.90)^n$

Introduction to Trigonometry

6.1 ⋯⋯ The Tangent of an Angle

Historically, trigonometry was developed to solve problems involving right triangles. Thus, for the right triangle shown in Figure 6.1, if we know any two of the three sides a, b, and c, we can easily find the third side with the Pythagorean theorem

$$c^2 = a^2 + b^2,$$

where c is the length of the hypotenuse. But, there is no simple way to find the two unknown angles. To find them, we need to use trigonometry.

Similarly, if only one of the three sides and one angle are known, we can easily find the other angle (the two nonright angles must sum to 90° because they are *complementary angles*). However, without trigonometry, there is no simple way to find the lengths of the other two sides.

The basic idea behind trigonometry is a fundamental geometric fact about right triangles. The two right triangles shown in Figure 6.2 share the angle θ (lowercase Greek letter theta). Therefore the remaining angle in both triangles must be the same. We denote it ϕ (the lowercase Greek letter phi). Because all three angles in both triangles are the same, the triangles are *similar* (see Appendix A4). As a consequence, once an angle θ (other than the right angle) has been specified in a right triangle, that triangle is similar to every other right triangle having the same angle θ.

In the smaller triangle shown in Figure 6.2, from the point of view of the angle θ, there is an *adjacent side,* denoted by a_1; an *opposite side,* denoted by b_1; and the *hypotenuse,* denoted by c_1. In the larger triangle, also from the point of view of the angle θ, the *adjacent side* is a_2, the *opposite side* is b_2, and the *hypotenuse* is c_2. The triangles are similar, so that their corresponding sides are proportional, and

$$\frac{a_1}{a_2} = \frac{b_1}{b_2} = \frac{c_1}{c_2}.$$

Equivalently, once an angle θ has been specified, the ratio of corresponding sides of these right triangles will be the same. In particular, among several other comparable ratios,

$$\frac{a_1}{b_1} = \frac{a_2}{b_2}, \qquad \frac{a_1}{c_1} = \frac{a_2}{c_2}, \quad \text{and} \quad \frac{b_1}{c_1} = \frac{b_2}{c_2}.$$

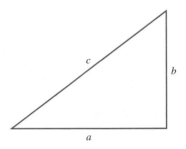

FIGURE 6.1

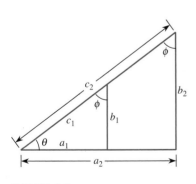

FIGURE 6.2

429

That is, each of these ratios depends solely on the angle θ, not on the dimensions of the triangle. It is these ratios, and their dependence on the angle θ, that form the basis of trigonometry. In this section, we begin by examining one of these three ratios.

The Tangent of an Angle

Suppose that your math instructor has assigned you the task of calculating the height of a tall flagpole in the middle of campus. The direct approach would be to climb to the top, release a string until the bottom reaches the ground, and then measure the length of string. Obviously, this method presents some practical difficulties, and you would likely try to come up with some less physical approach.

Assume that, when you go out to the flagpole, you notice that the pole is casting a 66-foot-long shadow. How can you use this piece of extra information to determine the height of the pole? Suppose that you enlist the aid of a friend Ron, who is exactly six feet tall. Have him stand in the shadow cast by the pole so that the tip of his shadow falls exactly on the same spot A as the tip of the shadow of the flagpole, as illustrated in Figure 6.3. Also, suppose that the length of his shadow is $8\frac{1}{4}$, or 8.25, feet. The two triangles ABC and ADE are similar because the angles are the same, and so the corresponding sides are proportional. Therefore

$$\frac{\text{Ron's height}}{\text{length of his shadow}} = \frac{\text{height of pole}}{\text{length of pole's shadow}}$$

$$\frac{6}{8.25} = \frac{\text{height of pole}}{66}.$$

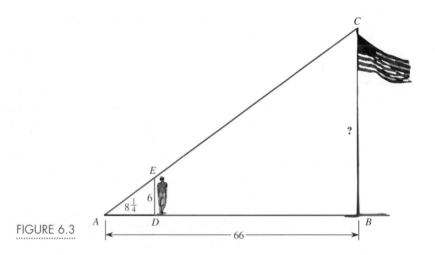

FIGURE 6.3

Multiplying both sides by 66 yields

$$\text{Height of the pole} = \frac{6(66)}{8.25} = 48 \text{ feet.}$$

Is this result correct? You can check it with the help of another friend, Sue, who is five feet tall. Have her stand so that the tip of her shadow matches the end of the pole's shadow. Suppose that the length of her shadow is $6\frac{7}{8}$, or 6.875, feet, which leads to right triangle AFG that is similar to the previous two, as illustrated in Figure 6.4. Because the corresponding sides are proportional, we get

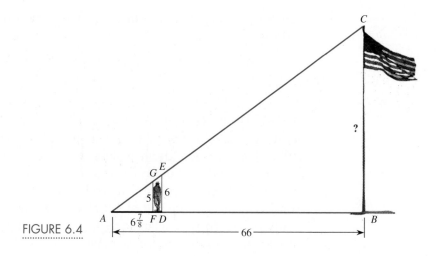

FIGURE 6.4

$$\frac{\text{Sue's height}}{\text{length of her shadow}} = \frac{\text{height of pole}}{\text{length of pole's shadow}}$$

$$\frac{5}{6.875} = \frac{\text{height of pole}}{66}.$$

Again, we find that

$$\text{Height of the pole} = \frac{5(66)}{6.875} = 48 \text{ feet.}$$

Let's look at this situation from a slightly more sophisticated point of view. In each of the three right triangles shown in Figure 6.5, the various lengths are different but the angles in corresponding positions are all the same, so all three triangles are similar. The angle θ (which is the same as angle *CAB*, angle *EAD*, and angle *GAF*) is called the *angle of inclination*. Using a protractor, we measure this angle and find that θ is about 36°. In fact, in *any* right triangle where the angle of inclination is 36°, the ratio of the vertical height (the opposite side) to the horizontal distance or width (the adjacent side) will always be the same; in this case,

$$\frac{\text{Height}}{\text{Width}} = \frac{6}{8.25} = \frac{5}{6\frac{7}{8}} \approx 0.727.$$

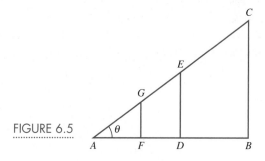

FIGURE 6.5

Of course, if the angle θ has a different value—say, $\theta = 40°$—the configuration of height and width is different and their ratio therefore is different. The ratio of height to width, or opposite side to adjacent side, in a right triangle depends only on the size of the angle θ, so this ratio is a function of the angle. We call this function the **tangent of the angle,** the **tangent ratio,** or the **tangent function** and write it as

$$\tan \theta = \frac{\text{opposite}}{\text{adjacent}} = \frac{\text{height}}{\text{width}}.$$

Use your calculator, in `Degree` mode, to verify that $\tan 36° = 0.7265$. (Note that the values of the tangent function, as well as the other trigonometric functions that we discuss in Section 6.2, typically are irrational numbers, but we usually give the values to three or four decimal places.)

Because we are concerned exclusively with right triangles here, the angle θ must be between 0° and 90°, and so for now the domain of the tangent function consists of all angles $0° < \theta < 90°$. (Later we show how we can extend it to a larger domain.) Also, we can have a right triangle in any possible orientation, as shown in Figure 6.6, so the words *height* and *width* may not be appropriate. Instead, we typically think of the tangent ratio for an angle θ as follows.

$$\tan \theta = \frac{\text{opposite}}{\text{adjacent}}$$

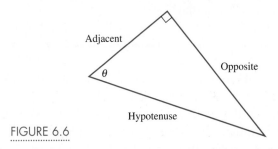

FIGURE 6.6

From the point of view of the other angle ϕ in the triangle, the opposite and adjacent sides are reversed, as depicted in Figure 6.7. Note also that the angles θ and ϕ are complementary angles.

The Tangent of Some "Special" Angles

Recall from geometry that in any 45°–45°–90° right triangle the two sides flanking the hypotenuse are equal and, by the Pythagorean theorem,

$$c^2 = a^2 + a^2 = 2a^2,$$

so $c = \sqrt{2}a$. That is, the hypotenuse must be $\sqrt{2}$ times the length of either side, as illustrated in Figure 6.8. In this triangle with angle $\theta = 45°$ and sides a, a, and $\sqrt{2}a$,

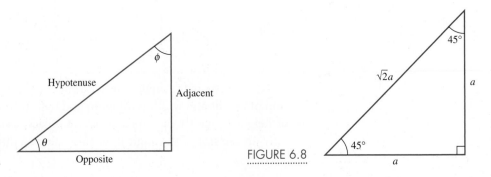

FIGURE 6.7

FIGURE 6.8

$$\tan 45° = \frac{\text{opposite}}{\text{adjacent}} = \frac{a}{a} = 1.$$

You can easily verify that $\tan 45° = 1$ on your calculator. (Be sure that your calculator is set in `Degree` mode.)

Similarly, recall from geometry that in any 30°–60°–90° right triangle, the side opposite the 30° angle is one-half the hypotenuse, or, equivalently, the hypotenuse is twice the side opposite the 30° angle. In such a triangle, suppose that the side opposite the 30° angle has length a so that the hypotenuse has length $2a$, as shown in Figure 6.9. We find the length of the third side from the Pythagorean theorem. Because $a^2 + b^2 = c^2$, we have $b^2 = c^2 - a^2$, so

$$b^2 = (2a)^2 - a^2 = 4a^2 - a^2 = 3a^2 \quad \text{so that} \quad b = \sqrt{3}\,a.$$

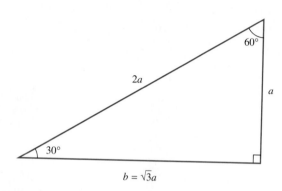

FIGURE 6.9

Consequently, for an angle of 30°, the ratio of the opposite side to the adjacent side is

$$\tan 30° = \frac{a}{\sqrt{3}\,a} = \frac{1}{\sqrt{3}} \approx 0.577.$$

Alternatively, using a calculator, we find $\tan 30° \approx 0.577$.

Similarly, to find the tangent of 60°, we see from Figure 6.9 that the side opposite the 60° angle is $\sqrt{3}\,a$ and the side adjacent to it is a, so that

$$\tan 60° = \frac{\sqrt{3}\,a}{a} = \sqrt{3} \approx 1.732,$$

which you can also check on your calculator.

For any angle θ between 0° and 90°, you can use a calculator to obtain the corresponding value for $\tan \theta$. For instance, to three decimal place accuracy,

$$\tan 10° \approx 0.176,$$
$$\tan 20° \approx 0.364,$$
$$\tan 50° \approx 1.192,$$
$$\tan 80° \approx 5.671.$$

Note that as θ increases toward 90°, the value of $\tan \theta$ also increases; that is, the tangent is an increasing function of θ, at least between 0° and 90°. Does that make sense? Imagine walking toward the 556-foot-high Washington Monument while keeping your eye fixed on the top of the monument, as illustrated in Figure 6.10.

The opposite side (the vertical height) remains the same, 556 feet, while the adjacent side (the horizontal distance) gets smaller and smaller. The closer you get to the monument, the larger the angle of inclination and the larger the ratio of the fixed vertical height to the diminishing horizontal distance. By the time your eye is practically touching the side of the monument, and the angle is virtually 90°, the value for the tangent function has gotten very large indeed. The tangent function is *not defined* for $\theta = 90°$ because the length of the adjacent side would be zero.

What about the tangent of 0°? Suppose that you're standing across the street from a glass elevator that is descending along the outside of a tall building, as illustrated in Figure 6.11. Now the adjacent side (the horizontal distance) is fixed, the opposite side (the vertical height) is decreasing, and the angle θ is decreasing toward 0°. Therefore the value of the tangent function is likewise diminishing because it is the ratio of the decreasing vertical height and the fixed horizontal distance. Clearly, tan 0° is 0. We therefore conclude that the domain of the tangent function can be extended at least to $0° \le \theta < 90°$.

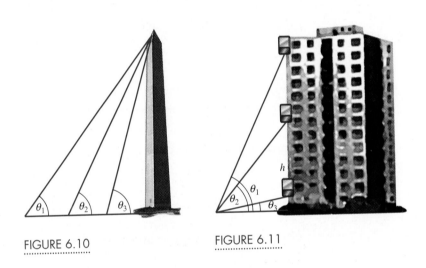

FIGURE 6.10

FIGURE 6.11

Behavior of the Tangent Function

Let's consider the values for the tangent function and investigate their growth pattern. Using a calculator, we obtain the following values.

θ	0°	10°	20°	30°	40°	50°	60°	70°	80°	90°
tan θ	0	0.176	0.364	0.577	0.839	1.192	1.732	2.747	5.671	UNDEF

Note that, as the angle θ increases from 0° to 10° to 20°, and so on, the tangent function is growing ever more quickly, so the function is concave up. The graph of the tangent function $y = \tan \theta$ for angles between 0° and 90° is shown in Figure 6.12. It passes through the origin and grows in a concave up pattern, approaching a vertical asymptote as θ approaches 90°.

How does the growth pattern compare to that of an exponential function? If you examine the successive ratios of the values of tan θ, you will find that they are

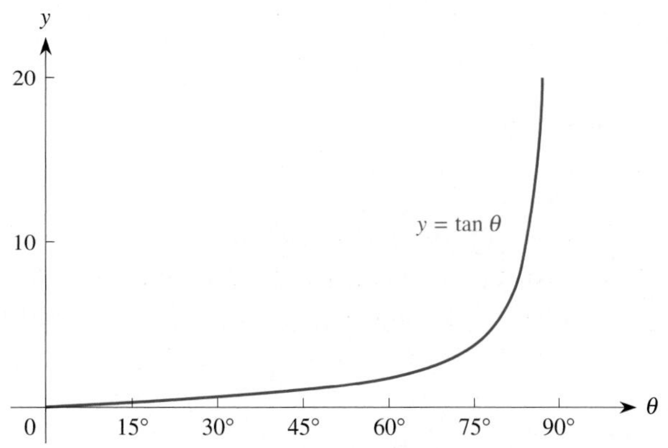

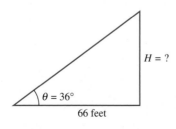

FIGURE 6.12

not constant, but rather are increasing considerably. In fact, the tangent function grows extremely rapidly near $\theta = 90°$ because $\theta = 90°$ is a vertical asymptote for the function. You might want to look at its graph on your function grapher for angles between 0° and somewhat less than 90°. We examine the properties of the tangent function in considerably more detail in Section 7.4.

Think About This   Construct a table of values for the tangent function $y = \tan \theta$ for $\theta = 80°, 81°, 82°, \ldots, 89°$ and plot the points. Repeat for $\theta = 89°, 89.1°, 89.2°, \ldots, 89.9°$. ☐

Using the Tangent Ratio

Suppose that we have a right triangle in which we can measure one of the angles other than the 90° angle with a protractor. We can find the tangent of that angle with a calculator. Then, if we know the length of either the adjacent side or the opposite side, we can easily find the length of the other side without involving Ron, Sue, or anyone else to solve the type of problem we used to begin this discussion.

◄**EXAMPLE 1**

A flagpole casts a shadow of length 66 feet. If the angle of inclination from the tip of the shadow to the top of the flagpole is 36°, find the height of the flagpole.

Solution   Figure 6.13 shows that

$$\tan 36° = \frac{\text{opposite}}{\text{adjacent}} = \frac{H}{66},$$

$H = ?$

$\theta = 36°$

66 feet

FIGURE 6.13

so

$$H = 66 \tan 36° = 47.952,$$

or about 48 feet high.

Note the approach used in Example 1. The first, and key, step was to draw a sketch of the situation, in which we identified all known parts of the right triangle, and marked the unknown parts. We then set up the tangent ratio and used it to find the unknown quantity.

EXAMPLE 2

While hiking through the mountains, you come to the edge of a deep gorge and wonder how far it is to the other side. A vertical tree is rooted on your side at the edge of the gorge. From a point 15 feet up in the tree, you find that the angle of depression (measured down from the horizontal at eye level) to the opposite edge of the gorge is 22°. How far is it across the gorge?

Solution   The height (15 feet) to the point in the tree and the unknown distance D across the gorge form two sides of a right triangle, as depicted in Figure 6.14. Note that the 22° angle of depression is not an angle of the triangle. However, it does determine the measures of the triangle's angles θ and ϕ, based on some simple geometry. First, the angle $\theta = 68°$ because it is the complement of 22°. Second, the angle $\phi = 22°$ because ϕ is the complement of $\theta = 68°$. We therefore have

$$\tan \theta = \tan 68° \approx 2.475.$$

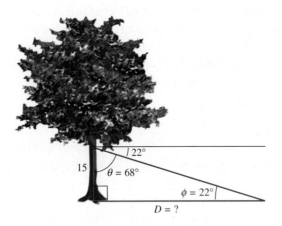

FIGURE 6.14

From the triangle shown in Figure 6.14,

$$\tan \theta = \frac{D}{15} \approx 2.475$$

so that

$$D = 15(2.475) \approx 37.125.$$

Therefore the gorge is about 37 feet across.

Note that if we worked with the angle ϕ instead, we would obtain the same result:

$$\tan \phi = \tan 22° = \frac{15}{D}$$

so that

$$D = \frac{15}{\tan 22°} \approx 37.13.$$

Finding an Angle in a Triangle

We often face the problem of determining an angle θ in a right triangle when we know two of the sides. For example, let the two sides of the right triangle shown in Figure 6.15 be $a = 20$ and $b = 13$, so that

$$\tan \theta = \frac{b}{a} = \frac{13}{20} = 0.65.$$

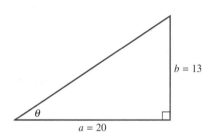

FIGURE 6.15

Suppose that we want to find θ. We know from the table of values we constructed previously for the tangent function that $\tan 30° = 0.577$ and $\tan 40° = 0.839$. Because the values for the tangent are strictly increasing, we expect θ to be between 30° and 40°. We can improve on these rough estimates by trial and error. For instance, using a calculator, we might find that $\tan 35° = 0.7002$ (too high), $\tan 32° = 0.625$ (too low), $\tan 34° = 0.6745$ (slightly too high), and so on.

A far more effective method is to use the inverse of the tangent function, which gives the angle whose tangent has a particular value. (We discuss this inverse function in detail in Section 7.4.) For now, on your calculator simply press either 2nd or INV followed by TAN and then the known tangent value. For this example, INV TAN 0.65 returns 33.024. That is, 33.024° is the angle whose tangent value is 0.65. You can check on your calculator that $\tan 33.024° \approx 0.65$.

The inverse tangent of a number x is usually written as either arctan x or $\text{Tan}^{-1} x$. We will use the first notation, arctan x.

EXAMPLE 3

A ski slope drops 1500 feet vertically in the process of covering 4300 feet horizontally.

a. What is the angle of inclination of the ski slope?

b. What is the actual distance that a skier will ski down the slope?

Solution

a. We start with a sketch of the ski slope, as shown in Figure 6.16. From geometry, the angle θ equals the angle inside the triangle at the end of the ski run (they are alternate angles between parallel lines). Therefore

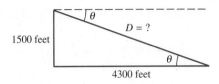

FIGURE 6.16

$$\tan \theta = \frac{\text{opposite}}{\text{adjacent}} = \frac{1500}{4300} \approx 0.3488,$$

so that

$$\theta = \arctan(0.3488) = 19.2288,$$

or about 19.2°.

b. The actual distance skied D is simply the length of the hypotenuse. Therefore, from the Pythagorean theorem,

$$D^2 = 1500^2 + 4300^2,$$

so that

$$\sqrt{1500^2 + 4300^2} = \sqrt{20,740,000} \approx 4554.12,$$

or about 4554 feet.

In general, problems in right angle trigonometry typically involve knowing a small amount of information about a right triangle and using that information intelligently to determine values for the other parts (either the sides or the angles) of the triangle. In fact, there are only a limited number of possibilities. We list these cases (based on the right triangle shown in Figure 6.17) in the following table. We leave the last column for you to complete. Decide on an appropriate strategy for finding each of the missing pieces, based on the information given or previously determined.

Given	Objective	Strategy
a and b	Find c.	
	Find θ.	
a and c	Find b.	
	Find θ.	
b and c	Find a.	
	Find θ.	
a and θ	Find b.	
	Find c.	
b and θ	Find a.	

(continued)

Given	Objective	Strategy
	Find c.	
c and θ	Find a.	Cannot be done simply by using $\tan\theta$.
	Find b.	Cannot be done simply by using $\tan\theta$.

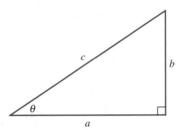

FIGURE 6.17

We examine the last case when c and θ are known—which cannot be solved by using the tangent of an angle—in Section 6.2.

Whenever you face any problem involving a right triangle, your first step should always be to draw a simple picture of the situation to identify the different parts of the triangle and see how they are related. Your drawing will help you determine which strategy, if any, to use to solve for the remaining parts of the triangle.

Problems

1. **a.** Use a ruler to measure, as accurately as possible, the lengths AB, AC, AD, AE, AF, AG, BE, CF, and DG.

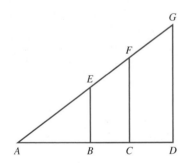

   b. Using your values from part (a), calculate the ratios $\frac{AB}{AE}$, $\frac{AC}{AF}$, $\frac{AD}{AG}$, $\frac{BE}{AB}$, $\frac{BE}{AE}$, $\frac{AC}{AF}$, $\frac{AD}{AG}$, $\frac{CF}{AC}$, and $\frac{DG}{AD}$.

   c. Group the ratios that you calculated in part (b) that appear to be equal. Identify any patterns that help you explain why certain ratios have the same values.

For Problems 2–6, refer to the accompanying figure. Use the information given to find all other parts of the triangle.

2. $\theta = 52°$ and $b = 12$ 3. $\theta = 16°$ and $a = 12$

4. $c = 15$ and $a = 6$ 5. $c = 30$ and $b = 18$

6. $a = 72$ and $b = 47$

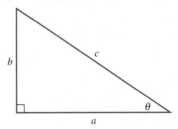

7. The shadow of a flagpole is 50 feet long. A line of sight from the tip of the shadow to the tip of the pole makes an angle of 28° with the ground. How high is the pole?

8. You want to find the distance across a straight, fast-flowing river. You find two vertical trees that are directly across the river from one another at points A and B so that the angle at B is a right angle, as shown in the accompanying diagram. You then measure a distance of 32 feet to another tree at point C on the edge of the river on your side. From the tree at C, you find that the angle ACB is 56°. Find the distance across the river.

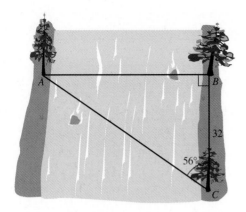

9. The line of sight from the top of a lighthouse to a Jet-Ski out on the water makes an angle of depression of 5° with the horizontal. The lighthouse is 50 feet high.

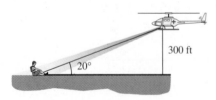

a. How far is the Jet-Ski from the base of the lighthouse?

b. What is the straight-line distance from the top of the lighthouse to the Jet-Ski?

10. A helicopter is hovering over a particular spot with its searchlight trained on an injured hiker on the ground. Because of tricky wind currents, the pilot can't get the copter any closer to the hiker. The angle that the searchlight makes with the ground is 20°. If the copter pilot estimates that she is at a height of 300 feet above the ground, how far away, horizontally, is the injured hiker?

11. A wheelchair ramp is to be built from ground level to a platform 7 feet above the ground. The angle of inclination with the ground is required to be no greater than 15°.

a. What is the shortest length for a ramp that meets this requirement?

b. How far is the start of the ramp from the base of the platform?

12. Jill is standing at the top of a vertical cliff and Jack is standing 25 feet away from the foot of the cliff and estimates that the angle of elevation θ from his position to Jill's is 40°. Approximately how high is the cliff?

13. Suppose that Jack's measurement in Problem 12 of the distance to the cliff is off by 1 foot. How much difference does this error make in the calculated height of the cliff? (*Hint*: Recalculate your answer to Problem 12 two ways, once for a distance of 24 feet and then for a distance of 26 feet.)

14. Suppose that Jack's estimate of the angle of elevation in Problem 12 is off by 1°. How much difference does this error make in the calculated height of the cliff?

15. Suppose that Jill, at the top of the cliff, wants to find the horizontal distance from the foot of the cliff to where Jack is standing without climbing down and measuring it directly. She drops a rock at the end of a long measuring tape down the cliff and finds that the height of the cliff is about 75 feet. Next, she measures the angle of depression from her position to Jack's to be approximately 70°. How far is Jack from the foot of the cliff?

16. The installation instructions for a TV satellite receiver at a particular location call for it to be aimed at an angle of 68° from the horizontal. Unfortunately, your protractor is broken. Devise a strategy that will help you aim the dish in the proper direction.

17. **a.** Find the missing entries in the table.

tan θ	0	0.5	1	1.5	2	2.5	3
θ							

b. Plot the points (tan θ, θ) and connect them with a smooth curve. (This is part of the graph of the function $y = \arctan \theta$.)

6.2 ······ **The Sine and Cosine of an Angle**

Suppose that you're flying a kite at the end of 400 feet of string and are curious about how high the kite is. How can you find its height? Figure 6.18 shows that the length of string is simply the hypotenuse of the right triangle. You can measure the angle of inclination θ that the kite string makes with the horizontal—say, $\theta = 37°$. Recall that this case is the last one presented in the strategy table in Section 6.1, where we pointed out that the tangent function is of no help. We must devise a different strategy to determine the height y of the kite.

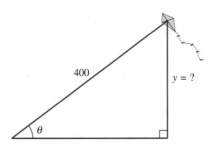

FIGURE 6.18

Using a yardstick, you could measure 5 feet along the kite string and then measure the height from the horizontal to that point on the string; say that you get 3 feet. As shown in Figure 6.19, you have a pair of similar right triangles ABC and ADE, so you know that their corresponding sides are proportional. Consequently,

$$\frac{\text{height of kite}}{\text{length of hypotenuse}} = \frac{\text{height to point on string}}{\text{length of string to that point}}$$

$$\frac{y}{400} = \frac{3}{5}.$$

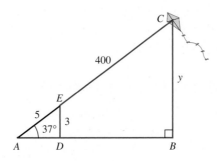

FIGURE 6.19

Therefore the height of the kite is

$$y = 400\left(\frac{3}{5}\right) = 240,$$

or 240 feet above the "horizontal." (If, in fact, you hold the kite string chest-high, say, 4 feet above the ground, the kite is 240 feet above your hand. Hence the kite is actually 244 feet above the ground.)

The Sine of an Angle

The key to solving this problem is to construct the ratio of the height and the hypotenuse of the right triangle. In our discussion of the tangent of an angle, we indicated that, for any angle θ, we can construct infinitely many right triangles that are all similar. Thus the ratio of the height and the hypotenuse will be the same for all these similar triangles, as illustrated in Figure 6.20. Because the ratio changes as the angle changes, this ratio is a function of the angle θ. We define this ratio to be the **sine of the angle,** or the **sine function,** and write it as follows.

$$\sin \theta = \frac{\text{opposite}}{\text{hypotenuse}}$$

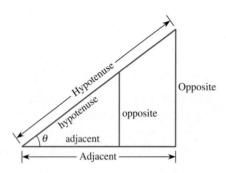

FIGURE 6.20

As with the tangent function, *opposite* refers to the side opposite the angle θ regardless of the orientation of the right triangle. You must think of the opposite side in terms of the angle θ, not as the side of a triangle that is in some particular location, such as the vertical position.

The Behavior of the Sine Function

For now we're concerned only with right triangles, so the domain of the sine function consists of angles between 0° and 90°. The following comments apply only to this situation. In Section 6.3, we consider cases for which this restriction is lifted, leading to more interesting and useful behavior patterns for the sine function.

As with the tangent ratio, you can get the values for the sine of any angle in a right triangle using your calculator in `Degree` mode. For instance,

$$\sin 10° = 0.174,$$
$$\sin 20° = 0.342,$$
$$\sin 30° = 0.5,$$
$$\sin 40° = 0.643,$$
$$\sin 75° = 0.966.$$

Note that the values for the sine function $y = \sin \theta$ are increasing as the angle θ increases from 0° to 90°. Further, the sine function grows more rapidly for small angles and less rapidly as angles get closer to 90°, so that these values follow a concave down pattern. Figure 6.21 shows a graph of the sine function for θ between 0° and 90°. (We discuss the sine's behavior for angles outside this interval in Section 6.3.)

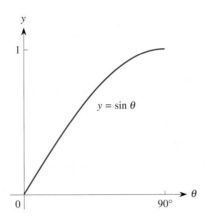

FIGURE 6.21

The Sine of Some "Special" Angles

As we did with the tangent function in Section 6.1, let's consider some of the special angles, notably $\theta = 0°, 30°, 45°, 60°$, and $90°$, from the point of view of the sine function. To begin, think about what happens in a right triangle as the angle shrinks to $0°$ for a fixed hypotenuse c. (Imagine the kite nosediving toward the ground at the end of the taut string as shown in Figure 6.22.) The length of the opposite side also decreases to 0, so

$$\sin 0° = \frac{\text{opposite}}{\text{hypotenuse}} = \frac{0}{c} = 0.$$

What about $\sin 90°$? Again, for a fixed hypotenuse c, think about what happens in a right triangle as the angle increases to $90°$. (Although improbable, imagine the kite moving directly overhead so that the height of the kite becomes equal to the length of the string, as illustrated in Figure 6.23.) The length of the opposite side in the triangle grows until it approaches the length of the hypotenuse, so

$$\sin 90° = \frac{c}{c} = 1.$$

Verify these two facts on your calculator.

Now let's look at the other special angles. As shown in Figure 6.24, when $\theta = 45°$, we have a right triangle with two angles of $45°$ and the two corresponding sides of equal length, say a. Recall that, by the Pythagorean theorem, the length of

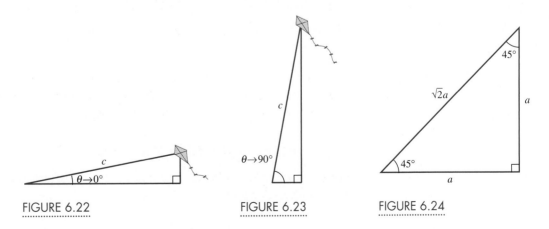

FIGURE 6.22

FIGURE 6.23

FIGURE 6.24

the hypotenuse is $\sqrt{2}\,a$. Hence

$$\sin 45° = \frac{\text{opposite}}{\text{hypotenuse}} = \frac{a}{\sqrt{2}\,a} = \frac{1}{\sqrt{2}} \approx 0.707.$$

Similarly, as shown in Figure 6.25, when $\theta = 30°$, the remaining angle is 60°. Recall that, in any 30°–60°–90° right triangle, the length of the side opposite the 30° angle is half the length of the hypotenuse. If the hypotenuse has length $2a$, the opposite side has length a. Consequently,

$$\sin 30° = \frac{\text{opposite}}{\text{hypotenuse}} = \frac{a}{2a} = 0.5.$$

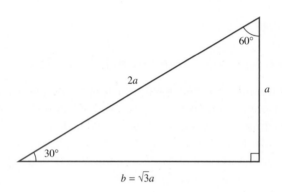

FIGURE 6.25

By the Pythagorean theorem, if the remaining side has length b, then

$$b^2 = (2a)^2 - a^2 = 4a^2 - a^2 = 3a^2,$$

so $b = \sqrt{3}\,a$. We therefore have

$$\sin 60° = \frac{\sqrt{3}\,a}{2a} = \frac{\sqrt{3}}{2} \approx 0.866,$$

as we found previously by using a calculator.

We summarize these findings as follows

θ	0°	30°	45°	60°	90°
$\sin \theta$	0	0.5	$\frac{1}{\sqrt{2}} \approx 0.707$	$\frac{\sqrt{3}}{2} \approx 0.866$	1

Note that the values for the sine of any angle in a right triangle must always lie between 0 and 1. The reason is that the sine is the ratio of the opposite side and the hypotenuse, and in any right triangle the hypotenuse is always the longest side.

Applications of the Sine Function

We apply the sine function in Examples 1–3 to illustrate its use in solving different types of everyday problems.

EXAMPLE 1

A highway through the mountains has a stretch that drops at a grade of 5°. If you drive a distance of 12 miles along this road, how far do you descend vertically?

Solution   To help visualize the situation, we "straighten out" all curves in the road and sketch the situation, as shown in Figure 6.26, which is not to scale. Note that a 5° grade also can be thought of as a 5° angle of descent or a 5° angle of declination. We know that the length of the hypotenuse is 12 miles. We let y be the vertical drop, and get

$$\sin 5° = \frac{y}{12} \quad \text{so that} \quad y = 12(\sin 5°) \approx 1.05.$$

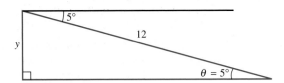

FIGURE 6.26

Consequently, along this stretch of highway, the road drops about 1.05 miles, or about 5544 feet.

EXAMPLE 2

A tall tree has been uprooted during a storm. It is tilted over and supported near its top by a vertical wall, as shown in Figure 6.27. The actual horizontal distance from the tree's roots to the wall is 42 feet and the angle of elevation of the tree is estimated to be 35°.

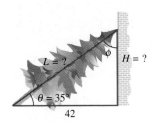

FIGURE 6.27

a. Estimate the length of the tree.
b. Estimate how high on the wall the top of the tree is lodged.

Solution

a. Note that

$$\sin \theta = \sin 35° = \frac{H}{L},$$

which involves two unknowns; thus we cannot solve the equation. Instead, we must work with the remaining angle ϕ in the triangle, which is $\phi = 90° - 35° = 55°$. Therefore we have

$$\sin 55° = \frac{42}{L},$$

so that

$$L = \frac{42}{\sin 55°} \approx 51.27 \text{ feet.}$$

(Note that we could also have used the tangent of 35° to determine the height H of the triangle and then used the Pythagorean theorem to find the length of the hypotenuse.)

b. We now use the Pythagorean theorem to find the height of the triangle:

$$H^2 = 51.27^2 - 42^2 = 864.613,$$

so that

$$H = 29.4 \text{ feet.}$$

Think About This Suppose that the estimate of the angle in Example 2 is off by 5°, either high or low. How much difference would this error make in the answers to parts (a) and (b) of Example 2. ⊐

Often, we face the problem of determining an angle when we know the value of the sine of that angle. For instance, if the hypotenuse of a right triangle is 20 and the side opposite the angle θ is 15, as shown in Figure 6.28, then $\sin \theta = 0.75$. What is the angle θ? We could find it by trial and error (we know that $\sin 45° = 0.707$ and $\sin 60° = 0.866$, so we might try 50°, and so on). A far more effective approach is to use the *inverse sine function*, which gives the angle whose sine has a particular value. We write this inverse function as arcsin x, for any given value x (although $\sin^{-1}x$ is also used). We discuss the inverse sine function in detail in Section 7.3. For now, with your calculator, simply press either 2nd or INV, followed by SIN, and then the known value of the sine function—say, 0.75: the calculator returns 48.590. To verify that it is the correct angle, we check that

$$\sin 48.59 = 0.749996 \approx 0.75.$$

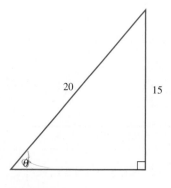

FIGURE 6.28

EXAMPLE 3

FIGURE 6.29

A 12-foot-long ladder is leaning against a wall. If the foot of the ladder is 4 feet from the wall, what is the angle of inclination of the ladder?

Solution   We start with a sketch of the situation, as shown in Figure 6.29. To use the sine function, we have to consider the point of view of the angle ϕ, so

$$\sin \phi = \frac{4}{12} = \frac{1}{3}$$

and therefore

$$\phi = \arcsin \frac{1}{3} = 19.47°.$$

Hence the angle of inclination of the ladder is

$$\theta = 90° - 19.47° = 70.53°.$$

In Section 6.1, we asked you to complete a table outlining strategies for solving for all the parts of a right triangle given various combinations of sides and angles. In all but one of those cases, you could determine all the other parts of the triangle by using the tangent. Now we ask you to complete the table again by deciding on appropriate strategies to determine the parts of a right triangle by using the sine instead of the tangent. Refer to Figure 6.30.

Given	Objective	Strategy
a and b	Find c.	
	Find θ.	
a and c	Find b.	
	Find θ.	
b and c	Find a.	
	Find θ.	
a and θ	Find b.	
	Find c.	
b and θ	Find a.	
	Find c.	
c and θ	Find a.	
	Find b.	

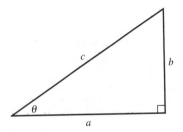

FIGURE 6.30

In practice, which function you apply doesn't matter so long as you use a correct strategy. Thus for most of the cases, a variety of different approaches will give the correct answers. Incidentally, together the tangent function and the sine function allow you to solve for all the parts of any right triangle in all six cases.

The Cosine of an Angle

So far, we've considered two of the six possible ratios among the sides of a right triangle:

$$\tan \theta = \frac{\text{opposite}}{\text{adjacent}} \quad \text{and} \quad \sin \theta = \frac{\text{opposite}}{\text{hypotenuse}}$$

One other ratio is very useful: the ratio of the adjacent side and the hypotenuse, which also depends only on the angle θ. We now define a third trigonometric function, the **cosine of an angle,** or the **cosine function,** as follows.

$$\cos \theta = \frac{\text{adjacent}}{\text{hypotenuse}}$$

The Behavior of the Cosine Function

As with the sine and the tangent, you can get the values for the cosine of any angle in a right triangle using your calculator. For instance,

$$\cos 10° \approx 0.985,$$
$$\cos 20° \approx 0.940,$$
$$\cos 30° \approx 0.866,$$
$$\cos 40° \approx 0.766,$$
$$\cos 50° \approx 0.643.$$

These values are decreasing more slowly for small angles and more rapidly as angles get closer to 90°. Therefore the values for the cosine function decrease in a concave down pattern for angles between 0° and 90°. Figure 6.31 shows a graph of the cosine function $y = \cos \theta$ for θ between 0° and 90°. Note that it starts at a height of 1 and decreases toward 0 in a concave down pattern as θ approaches 90°.

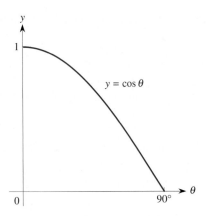

FIGURE 6.31

The Cosine of Some "Special" Angles

Again, let's consider the special angles $\theta = 0°, 30°, 45°, 60°$, and $90°$. To begin, what is cos 0°? Think about a right triangle in which the hypotenuse remains constant and the angle θ shrinks to 0°. (Imagine again the kite as it nosedives toward the ground on a windy day so that the string remains taut, as shown in Figure 6.32.) The hypotenuse gets closer and closer to the adjacent side, so

$$\cos 0° = \frac{\text{adjacent}}{\text{hypotenuse}} = 1.$$

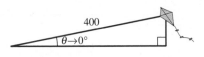

FIGURE 6.32

Similarly, think about a right triangle in which the hypotenuse remains fixed and the angle approaches 90°. (Imagine the kite moving directly overhead, as shown in Figure 6.33.) The adjacent side gets closer to 0, so

$$\cos 90° = 0.$$

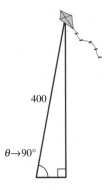

FIGURE 6.33

Next, let's look at the other special angles. As shown in Figure 6.34, when $\theta = 45°$,

$$\cos 45° = \frac{\text{adjacent}}{\text{hypotenuse}} = \frac{a}{\sqrt{2}a} = \frac{1}{\sqrt{2}} \approx 0.707.$$

Think About This

This result is the same value we found for the sine of 45°. Explain why they are the same. ❐

Also, when $\theta = 30°$, Figure 6.35 shows that

$$\cos 30° = \frac{\sqrt{3}a}{2a} = \frac{\sqrt{3}}{2} \approx 0.866,$$

which is the same as sin 60°. Similarly,

$$\cos 60° = \frac{a}{2a} = \frac{1}{2},$$

which is the same as sin 30°.

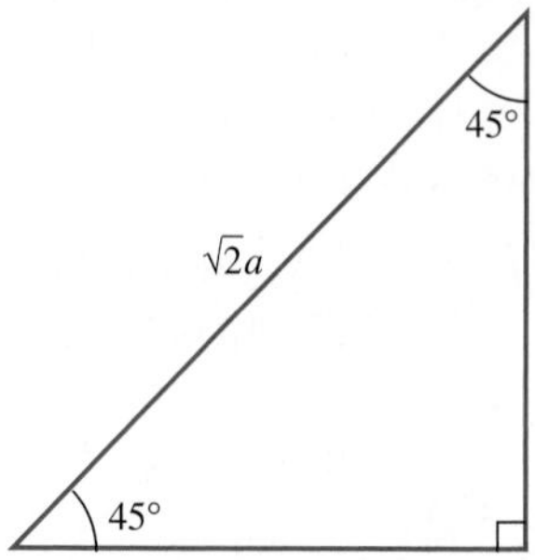

FIGURE 6.34

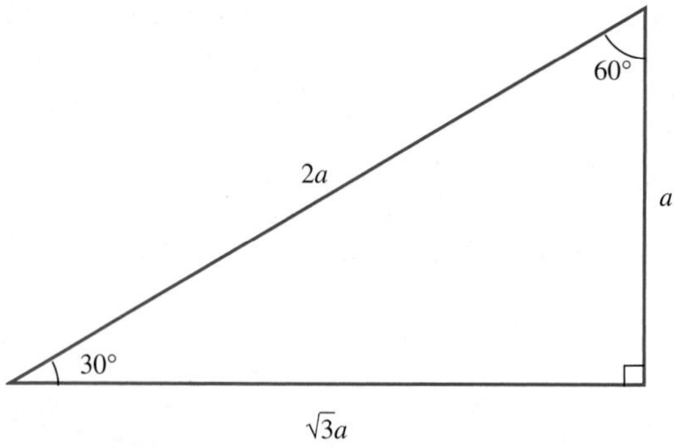

FIGURE 6.35

We summarize these key values for the cosine function as follows.

θ	0°	30°	45°	60°	90°
$\cos\theta$	1	$\frac{\sqrt{3}}{2} \approx 0.866$	$\frac{1}{\sqrt{2}} \approx 0.707$	0.5	1

Think About This

Explain why $\cos 30° = \sin 60°$ and $\cos 60° = \sin 30°$. ❐

Applications of the Cosine Function

We use the cosine function in Examples 4 and 5 to illustrate its value in solving a couple of rather simple problems.

EXAMPLE 4

To get onto a straight water slide at an amusement park requires climbing a flight of steps 60 feet high. The slide itself is inclined downward at a 42° angle. How long is the actual slide?

Solution Figure 6.36 indicates that the angle in the right triangle is 48° and that the adjacent side is 60 feet long. Therefore, to find the length L of the slide, we use

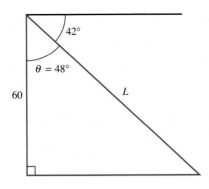

FIGURE 6.36

$$\cos 48° = \frac{\text{adjacent}}{\text{hypotenuse}} = \frac{60}{L}.$$

Thus

$$L = \frac{60}{\cos 48°} \approx 89.67.$$

So the slide is almost 90 feet long.

◆

Think About This   Can you solve Example 4 by using the sine function instead? the tangent function? ▭

◆ **EXAMPLE 5**

A 30 foot ramp extends 24 feet horizontally.

a. What is the angle of elevation of the ramp?

b. How high does the ramp extend?

Solution

a. We start with a sketch of the situation, as shown in Figure 6.37. The hypotenuse has length 30 feet and the base (which is the adjacent side from the point of view of the unknown angle θ) is 24 feet. Therefore

$$\cos \theta = \frac{\text{adjacent}}{\text{hypotenuse}} = \frac{24}{30}.$$

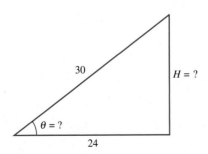

FIGURE 6.37

To find the angle θ, we undo the cosine using the inverse cosine (see Section 7.3), so that

$$\theta = \arccos\left(\frac{24}{30}\right) = 36.87°,$$

or about 37°.

b. We can solve for the height H in a variety of ways. Probably the simplest is to use the Pythagorean theorem, which gives

$$H^2 = 30^2 - 24^2 = 324,$$

so that

$$H = 18 \text{ feet.}$$

Applications from Physics

The trigonometric functions arise frequently in applications of the physical sciences. Many physical quantities, such as force and velocity, involve both a direction and a size. Such quantities are known as *vectors*, and we look at them more formally in Section 10.1. For now, we consider some physical applications informally to illustrate the use of trigonometry.

Imagine pushing against a window that is stuck in order to open it. You exert a certain force, but the effect of that force depends on the angle θ at which you exert it. If the angle is primarily vertical, most of the effect of your effort is applied to push the window upward, as shown in Figure 6.38(a). If the angle is more horizontal, as shown in Figure 6.38(b), only a small portion of your effort is applied to moving the window upward while most of your effort is wasted in the horizontal direction, effectively pushing the window outward. The total force exerted can be broken into two parts, one *horizontal* and the other *vertical*, based on the angle at which the force is applied. We illustrate this principle in Examples 6–8.

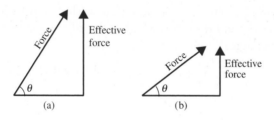

FIGURE 6.38

EXAMPLE 6

A 30 pound force is exerted at an angle of 20° with the vertical to push a stuck window upward. Find the effective value of the force actually exerted to move the window vertically.

Solution   We begin with a sketch of the situation, as shown in Figure 6.39, where the force being exerted is represented by the hypotenuse of the right triangle. We let the lengths of the sides equal the sizes of the forces. Thus the hypotenuse has length 30. The portion F of the force effectively applied to move the window vertically upward is the vertical side of this triangle. From the point of view of the 20° angle, the effective force is exerted along the adjacent side of the triangle, which suggests using the cosine function. In particular,

$$\cos 20° = \frac{F}{30} = 0.9397$$

so that

$$F = 30 \cos 20° = 28.19,$$

or slightly more than 28 pounds of the 30 pounds of the force is applied to moving the window.

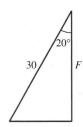

FIGURE 6.39

EXAMPLE 7

A sailboat is out on a still lake where the wind is blowing at a speed of 16 mph from the northeast, as shown in Figure 6.40. How fast is the sailboat moving toward the west? toward the south?

Solution   Because the wind is blowing from the northeast, the angle it makes with the horizontal is 45°. The wind is actually pushing the sailboat toward the southwest at 16 mph, as represented by the hypotenuse of the right triangle shown in Figure 6.40. We want to find the speed w in the westward direction, as indicated by the horizontal side of the triangle, and the speed s in the southward direction, as indicated by the vertical side of the triangle.

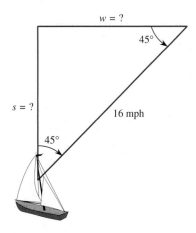

FIGURE 6.40

Let's first determine the sailboat's speed toward the west. Because the side opposite that angle is the unknown and we have the hypotenuse, we use the sine function. Thus

$$\sin 45° = \frac{w}{16}$$

so that

$$w = 16 \sin 45° \approx 11.314.$$

That is, the sailboat is moving at slightly more than 11 mph toward the west.

To find the speed toward the south, we simply observe that, because the angle in this right triangle is 45°, the two sides are equal and so $s \approx 11.314$ also. Thus the sailboat is also moving at slightly more than 11 mph toward the south.

◆

Suppose that the wind is not quite blowing from the northeast but from some angle other than 45°. Example 8 demonstrates how the solution changes accordingly.

EXAMPLE 8

A sailboat is out on a still lake where the wind is blowing at a speed of 16 mph from the northeasterly direction of 32° east of north, as shown in Figure 6.41. How fast is the sailboat moving toward the west? toward the south?

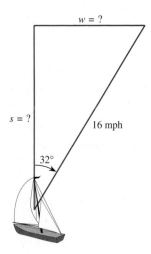

FIGURE 6.41

Solution   As in Example 7, the wind is actually pushing the sailboat in a southwesterly direction at 16 mph, as represented by the hypotenuse in the right triangle in Figure 6.41. But now the angle is 32° instead of 45°. We again indicate the westward speed w along the horizontal side of the triangle and the southward speed s along the vertical side of the triangle.

We first determine the speed toward the west. Using the sine function, we have

$$\sin 32° = \frac{w}{16}$$

so that

$$w = 16 \sin 32° = 8.479.$$

That is, the sailboat is moving at about $8\frac{1}{2}$ mph toward the west.

To find the speed toward the south, we need to determine the remaining side of the right triangle. We do so by using the Pythagorean theorem:

$$s^2 = 16^2 - w^2 = 16^2 - 8.479^2 = 184.107$$

so that

$$s \approx 13.569.$$

Thus the sailboat is moving at about $13\frac{1}{2}$ mph toward the south.

Problems

1. Use a ruler to measure the three sides of the triangle shown. Based on the measurements, what are your best estimates for $\sin\theta$, $\cos\theta$, and $\tan\theta$? What is your estimate for the angle θ?

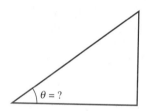

Problems 2–7 refer to the accompanying figure. Use the information given to find all other parts of the triangle.

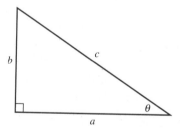

2. $\theta = 52°$ and $b = 15$

3. $\theta = 16°$ and $c = 12$

4. $c = 22$ and $a = 16$

5. $c = 30$ and $b = 8$

6. $a = 12$ and $b = 9$

7. $a = 42$ and $\theta = 72°$

8. A road up a hill is inclined at 11° to the horizontal. A driver starts driving up this hill and, by checking the odometer, discovers that the steep portion of the road extends for three-quarters of a mile. How much has the car gained in altitude?

9. With its radar, an aircraft spots another aircraft 10,000 feet away at an angle of depression of 15°. Find the horizontal distance from one aircraft to the other.

10. As a pendulum of length 21 inches swings back and forth, the maximum angle it makes from the vertical is $\theta = 18°$. What is the greatest height that the end of the pendulum reaches compared to its lowest height when it passes the vertical?

11. From takeoff, an airplane reaches a height of 2 miles (10,560 feet) in the process of covering 20 miles horizontally.

   a. Find the average angle of ascent of the airplane as it climbs.

   b. Is the actual path upward of the airplane a straight line, or is the path curved in a concave up pattern or in a concave down pattern? Explain your reasoning.

   c. If the airplane were to climb along a straight-line path, find the distance it would travel as it goes from the ground to the 2-mile height. Is the distance that the airplane actually travels greater than or less than the distance you calculated? Why?

12. When the space shuttle comes in for a landing at Cape Canaveral, its descent to the ground for the final 10,000 feet of height is at an angle of 19° with the horizontal.

   a. What actual distance does the shuttle traverse along this final glide path?

   b. How far from touchdown, horizontally, should the shuttle be when it passes the 10,000-foot altitude?

13. Jack and Jill are about to climb a 400-foot-high hill. If the angle of ascent is 52° from the horizontal, what is the actual distance they will cover to reach the summit on a straight track?

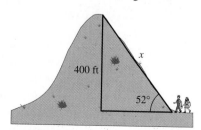

14. When Jack and Jill came tumbling down from the top of the hill in Problem 13, their angle of descent was 61° from the horizontal. What is the actual distance they covered while tumbling down?

15. A javelin is 1 meter long. When it lands after being thrown, its base is 0.6 meters (= 60 cm) vertically above the ground. What angle does the javelin make?

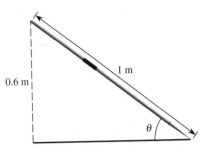

16. Problem 15 is unrealistic because the point of the javelin is going to be embedded in the ground. Suppose that 92 cm of the javelin is visible above the ground, and that its base is still 60 cm vertically above ground level. What angle does the javelin make with the ground?

17. You must hammer a 3-inch-long nail into a piece of wood 2 inches thick. Find the steepest angle at which you can hammer the nail all the way into the wood without it coming out the opposite side.

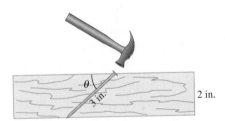

18. When an airplane takes off, it climbs at an angle of 16° at a speed of 180 feet per second. How high is the plane after 1 second? after 2 seconds?

19. The cranberry sauce to go with your holiday turkey comes out of a can and has a diameter of 3 inches. When you slice the roll of cranberry sauce at an angle, most of the slices will be ellipses with a minor axis of 3 inches. Suppose that you slice the roll at an angle of 27° to the vertical. Find the length of the major axis of each elliptical slice. (See Appendix A7; ellipses and their properties are covered in detail in Section 9.3.)

20. An escalator rises at a 26° angle with the horizontal. If it rises 28 feet vertically, what is its length?

21. A safety regulation limits the maximum angle of inclination for the ladder on a fire truck to 72°. If a hook-and-ladder fire truck has a ladder that can extend to a length of 90 feet, what is the maximum height that it can reach?

22. A balloonist is trying to cross the Atlantic Ocean. If the wind is blowing at 40 mph from the northwest, what is the actual airspeed at which the balloon is traveling eastward toward Europe?

23. The wind in Problem 22 now shifts slightly and increases in speed so that it is now blowing at 50 mph from 40° north of west. What is the actual airspeed at which the balloon is moving eastward?

24. **a.** Find the missing entries in the table.

sin θ	0	0.2	0.4	0.6	0.8	1
θ						

b. Plot the points (sin θ, θ) and connect them with a smooth curve.

c. This curve is part of the graph of what function?

25. **a.** Find the missing entries in the table using only your answers to Problem 24.

cos θ	1	0.8	0.6	0.4	0.2	0
θ						

b. Plot the points (cos θ, θ) and connect them with a smooth curve.

c. This curve is part of the graph of what function?

6.3 ······ The Sine, Cosine, and Tangent in General

So far, we've considered the trigonometric functions only for angles in a right triangle—that is, angles between 0° and 90°. However, we often encounter situations in which we need to consider angles larger than 90°. A natural question is: How do we adapt the ideas previously discussed to such cases? Before we address that question, let's choose such an angle—say, θ = 125°—and find out what happens when we use a calculator. We get

$$\sin 125° = 0.819, \qquad \cos 125° = -0.574 \quad \text{and} \quad \tan 125° = -1.428.$$

Let's see how these values are defined and why they have the indicated signs.

Angles Between 90° and 180°

Consider any angle θ between 90° and 180°; imagine it with one side, its **initial side,** along the positive x-axis and the other side, its **terminal side,** in the second quadrant, as depicted in Figure 6.42. By convention, we measure such an angle starting along the positive x-axis and rotating counterclockwise. The terminal side forms the hypotenuse of a right triangle in the second quadrant when we drop a vertical line from the terminal side to the x-axis. Suppose that the hypotenuse of this right triangle is h, the length of the vertical side is y, and the length of the horizontal side is x. By convention, because y extends up from the horizontal axis, we think of it as positive. Because x extends to the left of the vertical axis, we think of it as negative. By convention, the hypotenuse is always considered positive. The angle ϕ in this right triangle is the *supplement* of the angle θ because $\theta + \phi = 180°$; thus $\theta = 180° - \phi$, or $\phi = 180° - \theta$. The angle ϕ is sometimes called the *reference angle.*

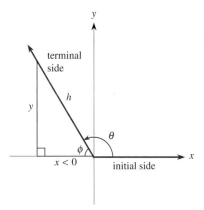

FIGURE 6.42

We define the trigonometric functions when the angle θ is between 90° and 180° in terms of the comparable values for the angle ϕ. Therefore

$$\sin \theta = \frac{\text{opposite}}{\text{hypotenuse}} = \frac{y}{h},$$

$$\cos \theta = \frac{\text{adjacent}}{\text{hypotenuse}} = \frac{x}{h}, \qquad x < 0$$

$$\tan \theta = \frac{\text{opposite}}{\text{adjacent}} = \frac{y}{x}, \qquad x < 0.$$

As previously mentioned, for any angle θ between 90° and 180° with its terminal side in the second quadrant, x is negative and y and h are positive. Thus the cosine and tangent of that angle are negative, whereas the sine is positive, as we saw with $\sin 125° = 0.819$, $\cos 125° = -0.574$, and $\tan 125° = -1.428$.

If $90° < \theta < 180°$,

$$\sin \theta > 0,$$
$$\cos \theta < 0,$$
$$\tan \theta < 0.$$

Angles Between 180° and 270°

What about an angle θ between 180° and 270° whose terminal side is in the third quadrant, as depicted in Figure 6.43. We construct a right triangle in the third quadrant by drawing a vertical line from the terminal side to the x-axis that determines a reference angle ϕ. Now both the x- and y-values are negative, and $\theta = 180° + \phi$. As before, we define the trigonometric functions for θ in terms of this reference angle ϕ by using the appropriate lengths in the right triangle so that

$$\sin \theta = \frac{\text{opposite}}{\text{hypotenuse}} = \frac{y}{h}, \qquad y < 0$$

$$\cos \theta = \frac{\text{adjacent}}{\text{hypotenuse}} = \frac{x}{h}, \qquad x < 0$$

$$\tan \theta = \frac{\text{opposite}}{\text{adjacent}} = \frac{y}{x}, \qquad x < 0 \quad \text{and} \quad y < 0.$$

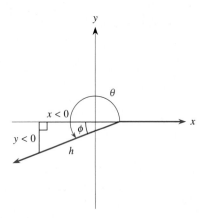

FIGURE 6.43

For any angle between 180° and 270°, the values of x and y are negative, so the sine and cosine are negative. However, for these angles, the tangent is positive because it is the quotient of two negative quantities.

If $180° < \theta < 270°$,
$$\sin \theta < 0,$$
$$\cos \theta < 0,$$
$$\tan \theta > 0.$$

Think About This

Suppose that $\theta = 211°$ so that $\phi = 31°$. Use your calculator to find the values for sin 211°, cos 211°, and tan 211°. How do they compare with sin 31°, cos 31°, and tan 31°?

Angles Between 270° and 360°

Next, consider an angle θ between 270° and 360° whose terminal side is in the fourth quadrant, as shown in Figure 6.44. Once more, we construct a right triangle by drawing a vertical line from the terminal side to the x-axis. We define each of the trigonometric functions in terms of the reference angle ϕ in that triangle. Now the y-value is negative, the x-value is positive, and $\theta + \phi = 360°$, so $\theta = 360° - \phi$. Also,

$$\sin \theta = \frac{\text{opposite}}{\text{hypotenuse}} = \frac{y}{h}, \qquad y < 0$$

$$\cos \theta = \frac{\text{adjacent}}{\text{hypotenuse}} = \frac{x}{h},$$

$$\tan \theta = \frac{\text{opposite}}{\text{adjacent}} = \frac{y}{x}, \qquad y < 0.$$

Therefore, because y is negative, for any angle between 270° and 360°, the cosine is positive and the sine and the tangent are both negative.

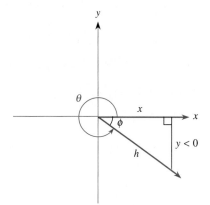

FIGURE 6.44

If $270° < \theta < 360°$,

$$\sin \theta < 0,$$
$$\cos \theta > 0,$$
$$\tan \theta < 0.$$

Angles Greater than 360°

What happens if θ is greater than 360°—say, 410°? As shown in Figure 6.45, we can construct such an angle by looping around a full 360° and then an additional 50°; essentially, this angle is equivalent to an angle of $\phi = 50°$ in the first quadrant. Using a calculator, we find that

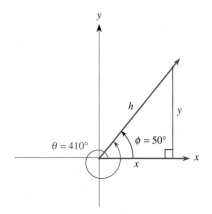

FIGURE 6.45

$$\sin 410° = 0.766 = \sin 50°,$$
$$\cos 410° = 0.643 = \cos 50°,$$
$$\tan 410° = 1.192 = \tan 50°.$$

Similarly, if $\theta = 775°$, we make two full rotations (accounting for $2 \times 360° = 720°$), which leaves an angle $\phi = 55°$, and so

$$\sin 775° = 0.819 = \sin 55°,$$
$$\cos 775° = 0.574 = \cos 55°,$$
$$\tan 775° = 1.428 = \tan 55°.$$

Note that the values for the three trigonometric functions repeat every 360°. Therefore they are **periodic functions** because their behavior repeats. The smallest interval over which the pattern repeats is called the **period.** The periods of the sine and cosine functions are both 360°. In general, for any angle θ, we have the following.

$$\sin(\theta + 360°) = \sin \theta$$
$$\cos(\theta + 360°) = \cos \theta$$

However, the period of the tangent function is 180° because its values repeat every 180°. Thus, for any angle θ, we have the following.

$$\tan(\theta + 180°) = \tan \theta$$

Check these identities on your calculator either numerically with a variety of different values for θ or graphically by comparing the graphs of $y = \tan x$ and $y = \tan(x + 180)$.

Angles Less Than 0°

Finally consider a negative angle—say, $\theta = -30°$—drawn clockwise, as shown in Figure 6.46. This angle is equivalent to a positive angle of 330° because both angles have the same terminal side. Note that y is negative and that x is positive. We therefore have

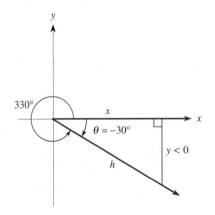

FIGURE 6.46

$$\sin \theta = \frac{y}{h}, \qquad y < 0$$

$$\cos \theta = \frac{x}{h},$$

$$\tan \theta = \frac{y}{x}, \qquad y < 0,$$

as we have already discussed for angles in the fourth quadrant.

Figure 6.47 summarizes the information about the signs of the three trigonometric functions, based on the quadrant containing the terminal side.

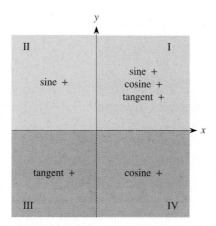

FIGURE 6.47

The Graph of the Sine Function

Let's summarize what we already know about the sine function to determine its overall behavior pattern. For θ between 0° and 90°, $y = \sin \theta$ increases from 0 to 1. For θ between 90° and 180°, $y = \sin \theta$ decreases from 1 to 0. For θ between 180° and 270°, $y = \sin \theta$ continues to decrease from 0 to -1. For θ between 270° and 360°, $y = \sin \theta$ increases from -1 to 0. Thus the sine function has a maximum value of 1 and a minimum value of -1. This oscillatory pattern continues indefinitely in both directions (for $\theta > 360°$ and for $\theta < 0°$). Use your function grapher with θ between $-500°$ and $500°$, say, to observe this pattern.

Think About This   Give a similar summary for the behavior of the cosine function. ⌐

You can visualize the behavior of the trigonometric functions by looking at their graphs. Figure 6.48(a) shows the graph of the function $y = \sin \theta$ for θ between 0° and 360° and how the graph relates to the signs of $\sin \theta$ in the four quadrants shown in Figure 6.47. In Figure 6.48(b) we expand the graph of $y = \sin \theta$ to show its behavior between $-360°$ and 720°. This portion of the curve consists of three full *cycles*, or repetitions, of the *basic sine curve* that occurs between 0° and 360°, which is one full period of the function. Also, the curve oscillates between a minimum height of $y = -1$ and a maximum height of $y = 1$. In particular, the sine curve reaches its maximum when

$$\theta = \ldots, -270°, 90°, 450°, \ldots,$$

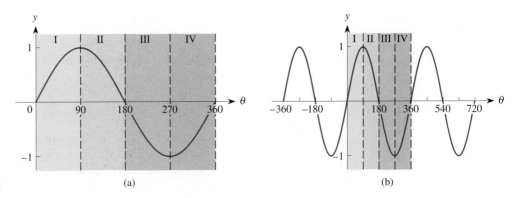

FIGURE 6.48   (a)   (b)

and it reaches its minimum when

$$\theta = \ldots, -90°, 270°, 630°, \ldots$$

In addition, the sine curve is concave down between $\theta = 0°$ and $\theta = 180°$ and is concave up from $\theta = 180°$ to $\theta = 360°$ and again in every other cycle.

You should carefully distinguish between the information shown in Figure 6.47 regarding the *signs* of the sine function in different quadrants and what happens in the *graph* of the sine function shown in Figure 6.48(a). The quadrants referred to in Figure 6.47 are based on a coordinate system with y versus x, and values for the angle θ are measured by rotating the terminal side of the angle. These are not the same quadrants shown in Figure 6.48(a) because that graph shows y as a function of θ, with θ measured horizontally as it takes on values in the different quadrants. In particular, angles in the first quadrant in Figure 6.47 correspond to the portion of the θ-axis in Figure 6.48(a) between $\theta = 0°$ and $\theta = 90°$; the second quadrant in Figure 6.47 corresponds to the portion of the θ-axis between $\theta = 90°$ and $\theta = 180°$ in Figure 6.48(a); and so on. We have marked these differences in Figures 6.48(a) and (b) with Roman numerals and corresponding shadings for the different quadrants to help make the point. Be sure that you understand these subtle differences before going on.

The Graph of the Cosine Function

Figure 6.49 shows the graph of the cosine function $y = \cos \theta$ from $\theta = -360°$ to $\theta = 720°$. Use the graph to answer the following questions: Where is the cosine function increasing? Where is it decreasing? What are its maximum and minimum values? Where do they occur? Where is the cosine function concave up? Where is it concave down? Also, be sure that you understand how the information shown in Figure 6.47 on the sign of the cosine function in the different quadrants relates to the behavior depicted in the graph of the cosine function.

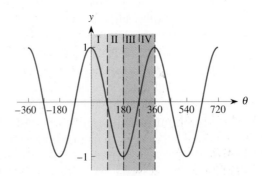

FIGURE 6.49

Problems

In Problems 1–12, find the value of each quantity by using only the information in the table. Do not use the trigonometric function keys on your calculator. (*Hint:* Start by drawing a picture of each angle.)

θ	0°	30°	45°	60°
$\sin \theta$	0	0.5	0.707	0.866
$\cos \theta$	1	0.866	0.707	0.5
$\tan \theta$	0	0.577	1	1.732

1. $\sin 225°$
2. $\cos 210°$
3. $\tan 135°$
4. $\sin 150°$
5. $\sin 330°$
6. $\cos 390°$
7. $\cos 315°$
8. $\tan 240°$
9. $\cos 840°$
10. $\sin (-450°)$
11. $\cos (-240°)$
12. $\tan (-225°)$

Decide whether each quantity is positive, negative, or zero without calculating its value. Give a reason for your answer.

13. $\sin 300°$
14. $\cos 450°$
15. $\cos 240°$
16. $\sin 210°$
17. $\tan 300°$
18. $\tan 450°$
19. $\cos 270°$
20. $\sin 225°$
21. $\sin 215°$
22. $\cos 320°$
23. $\cos 520°$
24. $\sin 885°$
25. $\tan 925°$
26. $\sin 1000°$
27. $\cos 1000°$
28. $\tan 1000°$
29. $\sin (-480°)$
30. $\cos (-500°)$

31. $\tan (-500°)$
32. $\sin (-1000°)$
33. Consider the function $f(x) = \sin x \cos x$.
   a. Determine the sign of $f(x)$ for x between 0° and 90°. between 90° and 180°. between 450° and 540°.
   b. For what values of x between 0° and 540° is $f(x) = 0$?
   c. Use the results of part (a) to sketch a rough graph of $f(x)$.
   d. Does the function appear to be periodic? If so, what is its period?
34. **a.** Find the missing entries in the table below.
   b. Plot the points $(\sin \theta, \theta)$ and connect them with a smooth curve.
   c. This curve is the graph of what function?

$\sin \theta$	−1	−0.75	−0.5	−0.25	0	0.25	0.5	0.75	1
θ									

35. **a.** Find the missing entries in the table below using only your answers to Problem 34.
   b. Plot the points $(\cos \theta, \theta)$ and connect them with a smooth curve.
   c. This curve is the graph of what function?

$\cos \theta$	1	0.75	0.5	0.25	0	−0.25	−0.5	−0.75	−1
θ									

6.4   Relationships among Trigonometric Functions

Consider the following values for the sine and cosine of the special angles between 0° and 90°.

θ	0°	30°	45°	60°	90°
$\sin \theta$	0	0.5	$\dfrac{1}{\sqrt{2}} \approx 0.707$	$\dfrac{\sqrt{3}}{2} \approx 0.866$	0
$\cos \theta$	1	$\dfrac{\sqrt{3}}{2} \approx 0.866$	$\dfrac{1}{\sqrt{2}} \approx 0.707$	0.5	1

Comparing the values of the sine and cosine, you will notice that the values associated with 0° and 90° are reversed, as are the values for 30° and 60°, so that sin 30° = cos 60° and sin 60° = cos 30°. This is no coincidence. In general, for any angle θ,

$$\cos \theta = \sin (90° - \theta) \quad \text{and} \quad \sin \theta = \cos (90° - \theta).$$

Why are these two relationships true? Consider the right triangle shown in Figure 6.50. Side a, opposite angle θ, is the side adjacent to the angle 90° − θ. Similarly, side b, adjacent to angle θ, is the side opposite angle 90° − θ. That is, their roles are reversed, depending on which angle, θ or 90° − θ, you consider.

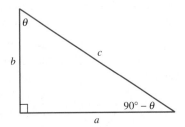

FIGURE 6.50

One reason why the trigonometric functions are so useful is that there are many interrelationships among them, as the two preceding formulas demonstrate. They are called **identities** because they hold for every possible value of the variable θ. But other identities involving relationships between the trigonometric functions are far more important. Let's consider again the right triangle shown in Figure 6.50 and the definitions of the sine and cosine:

$$\sin \theta = \frac{a}{c} \quad \text{and} \quad \cos \theta = \frac{b}{c}.$$

If we multiply both sides of these equations by c, we obtain

$$a = c \sin \theta \quad \text{and} \quad b = c \cos \theta.$$

We substitute these expressions into the formula for the tangent function to get

$$\tan \theta = \frac{a}{b} = \frac{c \sin \theta}{c \cos \theta},$$

provided that $\cos \theta \neq 0$. Therefore, for any such angle θ, we have the following identity.

$$\tan \theta = \frac{\sin \theta}{\cos \theta}, \qquad \text{if } \cos \theta \neq 0$$

For instance, if $\theta = 74°$, sin 74° = 0.9613, cos 74° = 0.2756, and

$$\tan 74° = 3.4874 = \frac{\sin 74°}{\cos 74°} \approx 3.488.$$

If we use more than four digits for sin 74° and cos 74°, the result would be more accurate.

Next let's apply the Pythagorean theorem to the right triangle shown in Figure 6.50:

$$a^2 + b^2 = c^2.$$

When we substitute $a = c \sin \theta$ and $b = c \cos \theta$, we have

$$(c \sin \theta)^2 + (c \cos \theta)^2 = c^2$$

or

$$c^2(\sin \theta)^2 + c^2(\cos \theta)^2 = c^2.$$

Dividing both sides by c^2 (which is not 0) gives

$$(\sin \theta)^2 + (\cos \theta)^2 = 1,$$

which holds for any angle θ. For convenience, it is customary to write

$$(\sin \theta)^2 = \sin^2\theta \quad \text{and} \quad (\cos \theta)^2 = \cos^2\theta$$

and we have the following identity.

The Pythagorean Identity

$$\sin^2\theta + \cos^2\theta = 1$$

Check this result on your calculator by using different values for θ. Be careful to enter the expressions as $(\text{SIN X})\wedge 2$ and $(\text{COS X})\wedge 2$. For instance, if $\theta = 74°$ again, we have

$$\sin^2(74°) + \cos^2(74°) = (0.9613)^2 + (0.2756)^2 = 1.000053.$$

The discrepancy is due to rounding errors. If we use more digits in $\sin 74°$ and $\cos 74°$, the result would be even closer to 1.

Let's now start with the Pythagorean identity and divide both sides by $\cos^2\theta$. We then get

$$\frac{\sin^2\theta}{\cos^2\theta} + \frac{\cos^2\theta}{\cos^2\theta} = \frac{1}{\cos^2\theta},$$

or, equivalently, because $\sin \theta / \cos \theta = \tan \theta$,

$$\tan^2\theta + 1 = \frac{1}{\cos^2\theta}.$$

This relationship is more commonly written in the following form.

$$1 + \tan^2\theta = \frac{1}{\cos^2\theta}$$

We summarize the definitions and special relationships among the three trigonometric functions as follows.

$$\sin \theta = \frac{\text{opposite}}{\text{hypotenuse}}$$

$$\cos \theta = \frac{\text{adjacent}}{\text{hypotenuse}}$$

$$\tan \theta = \frac{\text{opposite}}{\text{adjacent}}$$

$$\tan \theta = \frac{\sin \theta}{\cos \theta}$$

$$\sin^2\theta + \cos^2\theta = 1$$

$$1 + \tan^2\theta = \frac{1}{\cos^2\theta}$$

We investigate many other relationships between these three functions in Section 8.1.

EXAMPLE 1

Suppose that the SIN and TAN keys on your calculator are broken. You can use the COS key to find that cos 20° = 0.940. Determine the values for sin 20° and tan 20°.

Solution   We illustrate three different approaches to solving this problem.

Method 1

Using the Pythagorean relationship,

$$\sin^2\theta + \cos^2\theta = 1,$$

we find that

$$\sin^2\theta = 1 - \cos^2\theta$$

for any angle θ. Therefore, when $\theta = 20°$,

$$\sin^2(20°) = 1 - \cos^2(20°) = 1 - (0.940)^2 = 0.1170.$$

We take the square root of both sides to find

$$\sin 20° = \sqrt{0.1170} = 0.342.$$

Further, we have

$$\tan 20° = \frac{\sin 20°}{\cos 20°} = \frac{0.342}{0.940} = 0.364.$$

Method 2

Figure 6.51 shows that sin 20° = b/c. However, cos 70° = b/c also, and we can use the "broken" calculator to find cos 70° = 0.342. Therefore sin 20° = 0.342 also. Knowing sin 20° and cos 20°, we now can find tan 20° = 0.364, as we did in Method 1.

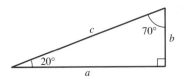

FIGURE 6.51

Method 3

We know that

$$\cos 20° = 0.940 = \frac{\text{adjacent}}{\text{hypotenuse}},$$

so from the triangle shown in Figure 6.52, the ratio a/c must be 0.940. Thus we can assume, for instance, that $a = 94$ and $c = 100$. (There are infinitely many other possibilities; another is $a = 0.940$ and $c = 1$.) Consequently, using the Pythagorean theorem, we can find the third side b:

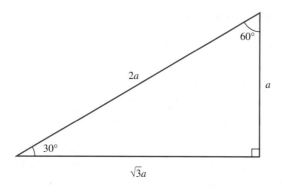

FIGURE 6.52

$$b^2 = c^2 - a^2 = 100^2 - 94^2 = 1164;$$
$$b = \sqrt{1164} = 34.117.$$

As a result, we have

$$\sin 20° = \frac{b}{c} = \frac{34.117}{100} \approx 0.341;$$

$$\tan 20° = \frac{b}{a} = \frac{34.117}{94} \approx 0.363.$$

Both of these values differ slightly from the results in Methods 1 and 2 because of rounding. ◆

EXAMPLE 2

Suppose that the SIN and COS keys on your calculator are broken. Using the TAN key, you find that $\tan 25° = 0.466$. Find the sine and cosine of this angle (**a**) by using trigonometric identities and (**b**) by constructing an appropriate triangle.

Solution

a. We have the relationship

$$1 + \tan^2(25°) = \frac{1}{\cos^2(25°)}$$

so that

$$1 + (0.466)^2 = 1.2172 = \frac{1}{\cos^2(25°)}.$$

Consequently,

$$\cos^2(25°) = \frac{1}{1.2172} = 0.8216.$$

Taking the positive square root, we have

$$\cos 25° = 0.906.$$

Furthermore, because $\cos^2(25°) = 0.8216$, we use the Pythagorean identity to find that

$$\sin^2(25°) = 1 - \cos^2(25°) = 0.178.$$

Taking the positive square root, we have

$$\sin 25° = 0.422.$$

b. Because

$$\tan 25° = \frac{\text{opposite}}{\text{adjacent}} = 0.466$$

we can construct a right triangle in which the length of the side opposite the 25° angle is 466, say, and the adjacent side is 1000, as shown in Figure 6.53. To find the hypotenuse H of this triangle, we have

$$H^2 = 466^2 + 1000^2 = 1,217,156$$

so that, when we take the positive square root,

$$H \approx 1103.25.$$

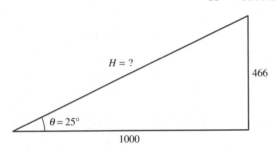

FIGURE 6.53

For this triangle, we now have the desired values

$$\sin 25° = \frac{\text{opposite}}{\text{hypotenuse}} = \frac{466}{1103.25} = 0.422$$

and

$$\cos 25° = \frac{\text{adjacent}}{\text{hypotenuse}} = \frac{1000}{1103.25} = 0.906.$$

EXAMPLE 3

Simplify the expression $\sin^3 x + \sin x \cos^2 x$ by using one of the trigonometric identities.

Solution   We first factor out the common factor of $\sin x$ to get

$$\sin^3 x + \sin x \cos^2 x = \sin x(\sin^2 x + \cos^2 x).$$

Using the Pythagorean identity yields

$$\sin^3 x + \sin x \cos^2 x = \sin x \cdot (1) = \sin x.$$

Think About This   Verify graphically that the given expression $\sin^3 x + \sin x \cos^2 x$ and the final expression $\sin x$ in Example 3 are identically equal for all values of x by graphing both.

Problems

1. Suppose that the COS and TAN keys on your calculator are broken. You can use your SIN key to find that, for some angle θ in the first quadrant, $\sin \theta = 0.3$. Determine the values for $\cos \theta$ and $\tan \theta$. What is the angle θ?

2. Suppose that the SIN and TAN keys on your calculator are broken. You can use your COS key to find that, for some angle θ in the first quadrant, $\cos \theta = 0.4$. Determine the values for $\sin \theta$ and $\tan \theta$ algebraically. What is the angle θ?

3. Suppose that, for a certain angle θ in the first quadrant, $\sin \theta = 0.6$. Using paper and pencil only, find the cosine and tangent of θ.

4. Suppose that, for a certain angle θ in the first quadrant, $\cos \theta = 0.6$. Using paper and pencil only, find the sine and tangent of θ.

5. Suppose that, for a certain angle θ in the first quadrant, $\tan \theta = \frac{3}{4}$. Using paper and pencil only, find the sine and cosine of θ.

6. Suppose that, for a certain angle θ in the first quadrant, $\tan \theta = 1.2$. Find the cosine and sine of θ algebraically.

7. Suppose that, for a certain angle in the second quadrant, $\sin \theta = 0.52$. Find the cosine and tangent of θ algebraically.

8. Suppose that, for a certain angle in the third quadrant, $\tan \theta = 0.75$. Find the cosine and sine of θ algebraically.

9. Suppose that, for a certain angle in the fourth quadrant, $\sin \theta = -0.7$. Find the cosine and tangent of θ algebraically.

10. Simplify the expression $\sin^2 x \cos x + \cos^3 x$ by using one of the trigonometric identities.

11. Consider the two equations:

   i. $\dfrac{\tan x}{\cos x} = \sin x$

   ii. $\dfrac{\sin x}{\tan x} = \cos x$

   a. Determine graphically which of these equations represents an identity that is true for every value of x, except for those points where the denominator is 0, and which is not an identity.

   b. Prove algebraically, using trigonometric identities, that the identity is indeed true.

   c. For the equation that is not an identity, find two different values of x that satisfy the equation.

Exercising Your Algebra Skills

Use appropriate trigonometric identities to simplify each expression.

1. $\cos x \tan x$
2. $(1 - \sin x)(1 + \sin x)$
3. $(1 - \cos x)(1 + \cos x)$
4. $(\sin \theta + \cos \theta)^2$
5. $(\sin \theta - \cos \theta)^2$
6. $\cos^3 x + \sin^2 x \cos x$
7. $\cos x + \tan^2 x \cos x$
8. $\tan^2 \theta - \dfrac{1}{\cos^2 \theta}$
9. $\left(1 - \dfrac{1}{\cos x}\right)\left(1 + \dfrac{1}{\cos x}\right)$

6.5 ⋯⋯ The Law of Sines and the Law of Cosines

When we first introduced trigonometry, our original development was presented in terms of angles in right triangles. We subsequently extended the definitions of the sine, cosine, and tangent to angles larger than 90°. We now consider some additional properties of the sine and cosine in any triangle, not just in a right triangle.

The Law of Sines

Consider the triangle ABC shown in Figure 6.54 where the sides opposite the angles A, B, and C are denoted by a, b, and c, respectively. All three angles are acute; that is, each is less than 90°. Later we consider the case where one angle is greater than 90°.

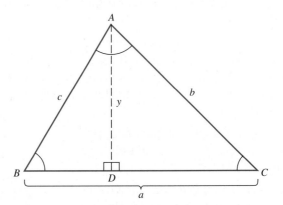

FIGURE 6.54

Suppose that we drop a perpendicular from the vertex at angle A to point D on base a. That perpendicular line AD, whose length we call y, produces two right triangles. In triangle ABD we have

$$\sin B = \frac{y}{c} \quad \text{so that} \quad y = c \sin B.$$

Similarly, in triangle ACD we have

$$\sin C = \frac{y}{b} \quad \text{so that} \quad y = b \sin C.$$

These two expressions for y must be equal, so

$$y = c \sin B = b \sin C,$$

and therefore

$$\frac{\sin B}{b} = \frac{\sin C}{c}.$$

However, we could just as easily have drawn a perpendicular from the vertex at angle B to the opposite side b. In that case, using the same reasoning, we get

$$\frac{\sin A}{a} = \frac{\sin C}{c}.$$

Together, these results yield

$$\frac{\sin A}{a} = \frac{\sin B}{b} = \frac{\sin C}{c}$$

for any triangle with three acute angles.

What about a triangle with an angle greater than 90°? Consider the one shown in Figure 6.55. We can still drop a perpendicular of length y from the vertex at angle A to point D on an extension of side a, as shown. This line forms two right triangles. Clearly, in the large right triangle ACD,

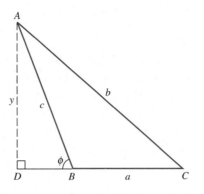

FIGURE 6.55

$$\sin C = \frac{y}{b} \quad \text{so that} \quad y = b \sin C.$$

To determine the sine of angle B, we use the angle ϕ in the smaller right triangle ABD. We thus find that

$$\sin B = \frac{y}{c} \quad \text{so that} \quad y = c \sin B.$$

Consequently,

$$y = c \sin B = b \sin C,$$

so that again we have

$$\frac{\sin B}{b} = \frac{\sin C}{c}.$$

We can similarly drop a perpendicular either from the vertex at angle B onto side b or from the vertex at angle C onto an extension of side c and obtain a similar relationship involving

$$\frac{\sin A}{a}.$$

What we have just proved is called the **law of sines.**

The Law of Sines

In any triangle,

$$\frac{\sin A}{a} = \frac{\sin B}{b} = \frac{\sin C}{c}$$

The law of sines can be used to find all the remaining sides and angles in any triangle if two sides and one angle are known or one side and two angles are known, provided that the known combination of sides and angles includes one angle and the side opposite it.

Think About This

Draw a triangle in which two sides and one angle are known and the law of sines will not apply. Then draw a triangle in which one side and two angles are known and the law of sines will not apply. ▱

We illustrate use of the law of sines in Example 1.

EXAMPLE 1 ⸱⸱⸱

The Federal Communications Commission (FCC) is attempting to locate a pirate radio station by a method called *triangulation.* The FCC set up two monitoring stations 30 miles apart on an east–west line and took simultaneous readings on the direction of the radio signal. The westernmost monitor measured the signal as coming from a direction 42° north of east; the other monitor measured the signal as coming from a point 56° north of west. Where is the pirate station located?

Solution   The information recorded determines the triangle shown in Figure 6.56. The two monitoring stations are located 30 miles apart at the points A and B. The signal directions determine the angles of 42° and 56°. The pirate is located at point C. Hence the angle at C must be $180° - 42° - 56° = 82°$.

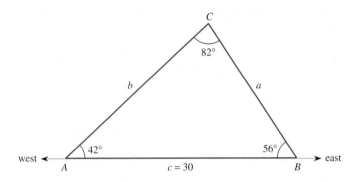

FIGURE 6.56

We now apply the law of sines to find the lengths of sides a and b. Using angles $A = 42°$ and $C = 82°$, we find

$$\frac{\sin A}{a} = \frac{\sin C}{c} \quad \text{or} \quad \frac{\sin 42°}{a} = \frac{\sin 82°}{30}$$

so that

$$a = \frac{30 \sin 42°}{\sin 82°} \approx 20.27.$$

Similarly, to find b we apply the law of sines, using the angles B and C, to get

$$\frac{\sin B}{b} = \frac{\sin C}{c} \quad \text{or} \quad \frac{\sin 56°}{b} = \frac{\sin 82°}{30}$$

so that

$$b = \frac{30 \sin 56°}{\sin 82°} \approx 25.12.$$

Therefore the pirate station is located 25.12 miles from station A in a direction of 42° toward the northeast and 20.27 miles from station B in a direction of 56° toward the northwest. The point C is determined precisely by these two facts. ◆⸱⸱⸱⸱⸱⸱⸱⸱⸱⸱⸱⸱⸱⸱⸱⸱

In Example 1 we used the law of sines when two angles and one side of a triangle are known. The law of sines can also be used when two sides (say, a and b) and the angle opposite one of these sides (either *A* or *B*) are known. However, depending on the sizes of the two known sides, it is possible to obtain either a unique answer or two distinct configurations for the triangle. This *ambiguous case* occurs when we try to find the angle from its sine. Recall that there will be two angles—one less than 90° and the other greater than 90°—that both have the same sine value. We ask you to explore possible ambiguous cases in the Problems at the end of this section.

Another complication may arise when we're using the law of sines if we know two sides and the angle opposite one of them. If in the midst of such a set of calculations, we obtain a sine value greater than 1, it indicates that the values we're working with could not have come from a real triangle. Again, you will encounter such a case in the Problems at the end of this section.

The Law of Cosines

There is one useful relationship involving the cosine function that relates the three sides of any triangle and any one of its three angles. Consider the triangle shown in Figure 6.57 with sides *a*, *b*, and *c*, where the angle opposite side *c* is *C*. The sides and angle are related by the **law of cosines.**

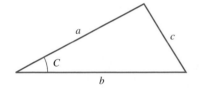

FIGURE 6.57

Law of Cosines

In any triangle,

$$c^2 = a^2 + b^2 - 2ab \cos C$$

Note that the triangle need not be a right triangle; the law of cosines applies to any triangle. The law of cosines allows us to determine (1) the length of the side opposite a known angle if the other two sides are known, or (2) any angle if the three sides of the triangle are known.

We prove this formula for the case where the triangle has three acute angles; a similar argument applies if one of the angles is greater than 90°. Also, to make things easier we assume that one of the vertices is at the origin and that one of the sides of the triangle lies on the *x*-axis, as shown in Figure 6.58. Note that the coordinates of the point *P* are $(a, 0)$ and that the coordinates of the point *Q* are at $x = b \cos C$ and $y = b \sin C$. As a result, the length of side *c* is just the distance from *P* to *Q* and we can find it by using the usual formula for the distance between two points (see Appendix A5):

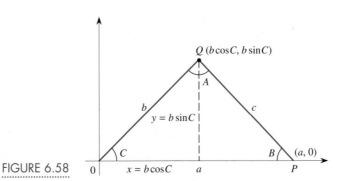

FIGURE 6.58

$$c = \text{distance from } P \text{ to } Q = \sqrt{(b \cos C - a)^2 + (b \sin C)^2}.$$

We square both sides to eliminate the square root and obtain

$$
\begin{aligned}
c^2 &= (b \cos C - a)^2 + (b \sin C)^2 \\
&= (b^2\cos^2 C - 2ab \cos C + a^2) + b^2\sin^2 C \\
&= (b^2\cos^2 C + b^2\sin^2 C) + a^2 - 2ab \cos C \\
&= b^2(\cos^2 C + \sin^2 C) + a^2 - 2ab \cos C \\
&= b^2 + a^2 - 2ab \cos C \qquad \text{Pythagorean identity} \\
&= a^2 + b^2 - 2ab \cos C
\end{aligned}
$$

If we have a right triangle in which angle $C = 90°$, then $\cos C = 0$ and the law of cosines reduces to the Pythagorean theorem $c^2 = a^2 + b^2$.

We use the law of cosines in Examples 2–4.

EXAMPLE 2

Let ABC be a triangle with sides $a = 5$ and $b = 7$, and let the angle C between the two sides a and b be 60°.

a. Find the third side c.

b. Find the other two angles A and B.

Solution

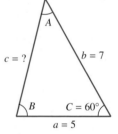

a. We begin with Figure 6.59. Using the law of cosines, we have

$$
\begin{aligned}
c^2 &= a^2 + b^2 - 2ab \cos C \\
&= 5^2 + 7^2 - 2(5)(7) \cos 60° \\
&= 25 + 49 - 70(\tfrac{1}{2}) = 39.
\end{aligned}
$$

FIGURE 6.59

Thus the third side is $c = \sqrt{39} \approx 6.245$.

b. To find the angle A, we use the law of sines:

$$\frac{\sin A}{a} = \frac{\sin C}{c} \quad \text{or} \quad \frac{\sin A}{5} = \frac{\sin 60°}{6.245}.$$

Therefore

$$\sin A = \frac{5 \sin 60°}{6.245} \approx 0.6934,$$

so that

$$A = \arcsin(0.6934) = 43.9°.$$

Consequently,

$$B = 180° - 60° - 43.9° = 76.1°$$

EXAMPLE 3

In a standard baseball infield, the four bases are at the corners of a square whose sides are 90 feet in length. The pitcher's mound is 60 feet, 6 inches, or 60.5 feet, from home plate on a line through second base, as illustrated in Figure 6.60. The distance from the pitcher's mound to second base is about 67 feet.

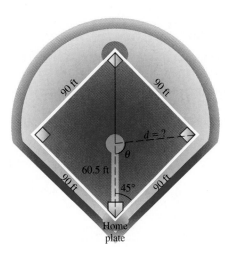

FIGURE 6.60

a. How far is the pitcher's mound from first base?
b. What is the angle at the pitcher's mound between home plate and first base?

Solution

a. Note that the angle at home plate between the mound and first base is 45°, so we know the angle opposite the unknown length d. Therefore, using the law of cosines, we have

$$d^2 = 60.5^2 + 90^2 - 2(60.5)(90)\cos 45°$$
$$= 3660.25 + 8100 - 2(60.5)(90)\cos 45° \approx 4059.86.$$

When we take the square root of both sides, we find that

$$d \approx 63.72 \text{ feet.}$$

b. Note that the angle θ at the pitcher's mound between home plate and first base must be more than 90°. We use the law of sines:

$$\frac{\sin \theta}{90} = \frac{\sin 45°}{d}$$

so that

$$\sin \theta = \frac{90 \sin 45°}{63.72} \approx 0.9987.$$

Therefore, using the inverse sine function, we get $\theta = \arcsin(0.9987) = 87.08°$. Because this is less than 90°, it must be the angle at the pitcher's mound between first base and second base. The desired angle is the supplement, $180° - 87.08° = 92.92°$.

In Section 6.2, we considered some physical examples in which a force or a velocity could be broken into two parts, one horizontal and the other vertical. At the time, we were limited to particularly simple examples where, for instance, a boat was out on a still lake with the wind blowing, but there was no mention of a current. We now look at a more complicated situation.

EXAMPLE 4

The pilot of a small plane is flying due east at its top speed of 200 mph. The wind is blowing out of the northwest at a speed of 40 mph. The wind pushes the plane in a direction south of east and increases its airspeed to more than the 200 mph, as illustrated in Figure 6.61.

a. What is the actual airspeed of the plane due to its own engines and the wind?

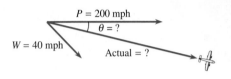

FIGURE 6.61   FIGURE 6.62

b. What is the actual direction that the plane flies?

Solution

a. We first consider the speeds. The plane itself contributes a horizontal airspeed of $P = 200$ mph. The wind comes from the northwest at $W = 40$ mph, so the associated angle inside the triangle in Figure 6.62 is $180° - 45° = 135°$. The actual airspeed s of the plane is the length of the remaining side in the triangle, and we find it by using the law of cosines:

$$s^2 = P^2 + W^2 - 2PW \cos 135°$$
$$= 200^2 + 40^2 - 2(200)(40)(-0.707) = 52{,}912.$$

Taking the positive square root yields

$$s \approx 230.026,$$

or about 230 mph.

b. To find the angle θ in the triangle, we use the law of sines:

$$\frac{\sin \theta}{40} = \frac{\sin 135°}{230},$$

so that

$$\sin \theta = \frac{40 \sin 135°}{230} \approx 0.1230.$$

Therefore

$$\theta = \arcsin(0.1230) = 7.07°,$$

or the plane is actually flying about 7° south of east.

Problems

For Problems 1–6, refer to the notation for the sides and angles in the accompanying figure. Use the information given to find all other parts of the triangle.

1. $A = 26°, B = 63°, b = 12$
2. $A = 47°, C = 72°, c = 60$
3. $A = 35°, B = 65°, c = 24$

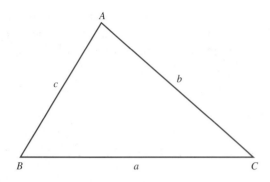

4. $A = 40°$, $a = 10$, $b = 6$ (*Hint*: Is it possible to have two different values for B?)

5. $A = 40°$, $a = 10$, $b = 12$ (*Hint*: Is it possible to have two different values for B?)

6. $A = 40°$, $a = 10$, $b = 18$

7. Two ships at sea are 50 miles apart on a north–south line when they both receive an SOS signal from a third ship in trouble. One ship receives the SOS from a direction of 41° north of east. The other ship receives the signal from a direction of 54° south of east. Where is the third ship?

8. You want to find the distance across a fast-flowing river. You pick two large trees, at points A and B, that are 35 feet apart along the edge of the river on your side. You then spot another tree on the opposite side of the river at point C. The angle CAB at point A is 43°; the angle CBA at point B is 52°. Find the distance across the river.

9. Problems 4–6 involved three cases in which the law of sines works very differently because of the relative sizes of sides a and b. Based on those results, explain the following statements.

  a. Given a value for angle A, there will always be one triangle whenever $b < a$.

  b. There will be two different possible triangles whenever b is somewhat larger than a.

  c. There will be no triangle whenever b is much larger than a.

10. Find the angle C if $a = 7$, $b = 11$, and $c = 5$, as shown in the accompanying figure.

11. Find the angle opposite the side of a triangle whose length is 6 if the lengths of the other two sides are 11 and 8, as shown in the accompanying figure.

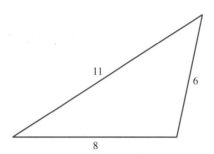

12. A TV camera is positioned 100 feet behind home plate. The center fielder is standing 300 feet from home plate and directly on the line from home plate through second base. The batter hits a long fly ball toward right-center field that the center fielder catches against the wall, 380 feet from home plate. If the TV camera had to pan through an angle of 8° in following the center fielder from the point where he was standing to the point where he made the spectacular catch, how far did he have to run?

13. Using a map of the United States (which ignores the fact that the earth is round), New York City is 1851 miles from Denver and about 10° north of east from Denver. New Orleans is 1282 miles from Denver and about 36° south of east from Denver. Estimate the distance from New York City to New Orleans.

14. Chicago is 695 miles from Atlanta, and Seattle is 2756 miles from Atlanta. The same map of the United States as in Problem 13 shows that Chicago lies at an angle of about 65° north of west from Atlanta and that Seattle lies at an angle of about 29° north of west from Atlanta. Estimate the distance from Chicago to Seattle.

15. A 40 meter vertical tower is to be built and supported by several guy wires anchored to the ground 25 meters from the base of the tower on flat land. Find the length of the guy wires and the angle they make with the ground.

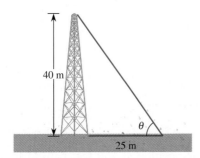

16. The 40 meter tower in Problem 15 is to be built on the side of a hill that slopes upward at an angle of 10° from the horizontal. One guy wire will be positioned directly uphill from the tower and another will be positioned directly downhill from the tower. Each guy wire will be anchored 25 meters from the base of the tower. Find the lengths of the two guy wires and the angles they make with the sloping ground.

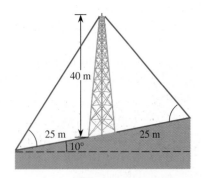

17. A communications satellite passes directly over Phoenix and Los Angeles, which are 340 miles apart. At some point in its orbit when the satellite is between Phoenix and Los Angeles, its angle of elevation from Phoenix is 52° and its angle of elevation from Los Angeles is 72°.

   a. How far is the satellite from Phoenix at that moment?
   b. How high is the satellite above the Earth?

18. Al and Bob are driving toward a moored hot air balloon from opposite sides of the balloon and are in contact via cell phones. When they are 4 miles apart, they both take sightings on the balloon. (Of course, everyone who is about to take a balloon ride carries a protractor.) From Al's position, the angle of elevation to the balloon is 28°; from Bob's position, the angle of elevation is 35°.

   a. How high is the balloon?
   b. How far is each of them from where the balloon is moored?

19. A straight tunnel passes through a mountain. An observer has a clear view of the two ends of the tunnel. The distance from her position to the tunnel entrance toward her left is 520 meters, and the distance to the entrance toward her right is 440 meters. If the angle subtended by the two tunnel entrances is 39°, how long is the tunnel?

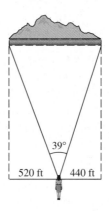

20. Meryl and Bernice are walking along a straight beach when they observe a small island in the distance and wonder if they can swim out to it. To estimate the distance, they separate and walk 400 meters apart. From Meryl's perspective, the angle to the island is 27° and from Bernice's perspective the angle is 39°. How far is the island from the shore?

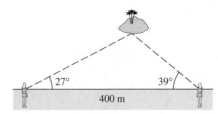

Problems 21 and 22 refer to Figure 6.57 in the text.

21. Find c if $a = 5$, $b = 3$, and $C = 20°$.

22. Find c if $a = 7$, $b = 4$, and $C = 25°$.

Chapter Summary

In this chapter we introduced some of the fundamental ideas and applications of trigonometry as they apply to right triangles. In particular, we discussed:

◆ The definition of the tangent ratio in terms of the sides of a right triangle.

◆ How to use the tangent ratio to solve problems involving right triangles.

◆ The graph of the tangent function between 0° and 90°.

◆ The definition of the sine and cosine ratios in terms of the sides of a right triangle.

◆ How to use the sine and cosine to solve problems involving right triangles.

◆ The graphs of the sine and cosine functions between 0° and 90°.

◆ How to extend the sine, cosine, and tangent functions to angles beyond 0° to 90°.

◆ The fundamental identities that relate the sine, the cosine, and the tangent functions.

◆ How to use the law of sines and the law of cosines to solve various types of problems.

Review Problems

1. A TV cameraman is standing on a platform 75 feet from a straight portion of a race track and is focusing on the runner in the lead as she runs from left to right.

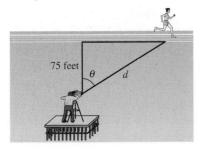

75 feet θ d

   a. Write a formula for the distance d from the camera to the runner as a function of the angle θ, as shown in the accompanying diagram.
   b. Suppose that the maximum distance for which the camera lens can get a good image is 240 feet. Through what interval of angles can the cameraman pan while focusing on the runner?
   c. What might be appropriate values for the domain of this function?

2. The camera in Problem 1 is again focused on the lead runner in the race.

   a. Write a formula for the distance that the runner covers from the instant that she passes closest to the cameraman as a function of the angle θ.

   b. If $\theta = 25°$, how far has the runner gone since she passed the point closest to the cameraman?
   c. When the runner has gone 150 feet past the point closest to the cameraman, through what interval of angles has the cameraman panned while focusing on her?

3. The next assignment for the TV cameraman in Problems 1 and 2 is to videotape the liftoff of the space shuttle. The cameraman is positioned at ground level 500 meters from the launch pad.

   a. Write a formula for the height y of the shuttle as a function of the angle of inclination α.
   b. Find the height of the shuttle when $\alpha = 20°$.
   c. Find the height of the shuttle when $\alpha = 40°$.
   d. Find the angle of inclination when the shuttle is at a height of 2000 meters.

4. A swimming pool is 60 feet long and 25 feet wide. It is 3 feet deep at the shallow end and 12 feet deep at the deep end.

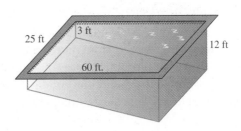

25 ft   3 ft   12 ft   60 ft.

a. Find the angle of depression of the bottom of the pool.

b. Find the equation of the line along the bottom of the pool extending from the shallow end to the deep end along one of the long sides of the pool.

c. Find the equation of the line along the bottom of the pool that extends from one corner to the opposite corner of the pool.

5. A salami is 4 inches in diameter. However, the man in the deli department slices it at an angle of 28° so that each slice comes out oval for a fancier presentation. What is the longest length of each slice of the salami?

6. A piece of metal 96 inches long by 36 inches wide is to be made into a watering trough by bending up 12 inches of the metal along each long side, as shown in the accompanying figure.

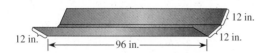

a. If the two metal sides are bent up at angles of 35°, how deep is the trough?

b. If the two metal sides are bent up at angles of 55°, how deep is the trough?

7. The shape of the trough in Problem 6 has a trapezoidal cross section because the top and bottom are parallel. The volume of water that the trough can hold is then its length, 96 inches, times its cross-sectional area, $\frac{1}{2}(b_1 + b_2)h$, where h is the height of the trough and b_1 and b_2 are the horizontal lengths of the top and bottom of each cross section.

a. What volume of water can the trough hold if the two edges are bent up at angles of 35°?

b. What volume of water can the trough hold if the two edges are bent up at angles of 55°?

c. Write a formula for the total volume of water that the trough can hold as a function of the angle θ at which the two sides are bent up.

d. Use your function grapher to estimate the angle that produces a trough that will hold the maximum amount of water.

8. A tall building stands across the street from a hotel—a distance of 220 feet. From one of the hotel windows, a guest in the hotel observes that the angle of inclination to the roof of the building is 36° and that the angle of depression to the base of the building is 23°.

a. How tall is the building?

b. How high is the window in the hotel?

9. The Earth is 93 million miles from the sun, and the moon is 240,000 miles from Earth. When the moon is exactly full, the Earth, the moon, and the sun form a right triangle with the right angle at the moon. Calculate, correct to two decimal places, the angle at the Earth in this triangle.

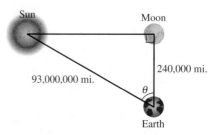

10. To calculate the height of a mountain above a level plain, two measurements are necessary. Suppose that, from a certain point on the plain, the angle of elevation to the top of a mountain is $\alpha = 34°$. The observer then moves 1000 meters closer to the mountain, takes a second reading, and gets an angle of elevation of $\beta = 37°$. How tall is the mountain?

11. As you sit in class waiting for the end of the period, you notice that the length of the minute hand on the wall clock is 10 inches.

a. How far vertically does the point of the minute hand rise from 45 minutes after the hour until 50 minutes after the hour?

b. Explain why the point of the minute hand cannot rise by the same amount from 50 minutes after the hour until 55 minutes after the hour.

c. During what other 5-minute time intervals over the course of an hour does the minute hand either rise or fall vertically by that same amount, as in part (a)?

12. The tangent of some angle in the first quadrant is 1.20.

a. Find the sine and cosine of that angle, using only appropriate trigonometric identities.

b. Find the sine and cosine of that angle by constructing an appropriate triangle.

13. Because of a storm, a tree is inclined at an angle of 10° from the vertical. From a point 70 feet from the base of the tree, the angle of elevation to the top of the tree is 26°, with the tree leaning away from the observer, as shown in the figure on the next page.

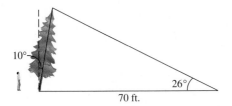

a. Find the height of the tree.
b. Find how high the top of the tree is above ground level.
c. How do your answers to parts (a) and (b) change if the tree is leaning toward instead of away from the observer?

14. Two forest rangers are in observation towers 16 miles apart on an east–west line, and each spots a fire. From the point of view of the ranger in tower *A*, the direction of the fire is 41° north of east. From the point of view of the ranger in tower *B*, the fire is 35° north of west. Find the distance to the fire from each tower.

15. In a football passing play, the wide receiver lines up 12 yards to the right of the quarterback on the line of scrimmage. The quarterback drops straight back 5 yards before throwing the football. The wide receiver runs straight down the field 30 yards before turning to catch the pass.

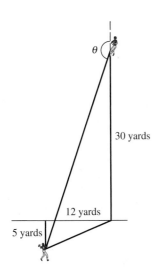

a. At the moment the quarterback throws the ball, what is the distance from the quarterback to the receiver's original position on the line of scrimmage just before the play started?

b. At the moment the quarterback throws the ball, what is the angle between the line of scrimmage and the line from the quarterback to the receiver's original position on the line of scrimmage?
c. Find the distance that the ball travels from the quarterback to the receiver.
d. To catch the ball straight on, the receiver has to turn toward the quarterback. Through what angle θ must the receiver turn in order to face directly toward the quarterback?

16. A motorboat leaves its dock and motors 6 miles due north. It then changes course, heading northeasterly at an angle of 58° east of north for 14 miles. At that point, the pilot decides to turn back and head directly to the dock.

a. How far is it from the turnaround point back to the dock?
b. Through what angle does the motorboat have to turn in order to be pointed directly back to the dock?
c. If the motorboat moves at a roughly constant speed of 18 mph, how long will the return trip take? How long does the entire outing take?
d. The motorboat gets 6 mpg. If the refueling station at the dock charges $2.85 per gallon, how much did the entire outing cost?

17. A steep, snow-covered mountain rises 2700 feet above the surrounding plain and rises at an angle of 68° to the horizontal. A ski lift is to be built from a point 750 feet from the base of the mountain to the summit.

a. What will be the angle of inclination of the cable for the ski lift?
b. What will be the length of the cable?

18. The sides of a triangle have lengths 8 cm, 14 cm, and 19 cm.

a. What is the angle θ?
b. What is the height of the triangle?
c. What is the area of the triangle?
d. Write a formula for the area of any triangle given the lengths *a*, *b*, and *c* of the three sides and one of the angles θ.

Modeling Periodic Behavior

7.1 ·····Introduction to the Sine and Cosine Functions

One of the most common behavior patterns in nature is a *periodic oscillatory effect*—a pattern that repeats over and over. For instance, think about how the ocean level varies at a beach between low tide and high tide approximately every 12 hours. If low tide occurs at midnight, high tide will occur at about 6 A.M., low tide will occur again at about noon, and so on indefinitely. This periodic oscillatory behavior is shown in Figure 7.1. Recall that the word *periodic* refers to the fact that this phenomenon repeats indefinitely and that the *period* is the time needed to complete one full cycle. If it takes 12 hours to complete a full cycle, the period is 12 hours.

Similarly, consider the number of hours of daylight each day in a particular location. The minimum number of hours of daylight occurs on the winter solstice, December 21, the "shortest" day of the year. The number of hours of daylight increases slowly until the maximum daylight occurs on the summer solstice, June 21, the "longest" day, and then decreases to the same minimum the following December 21. This oscillatory behavior repeats year after year. For instance, suppose that, at some location, there are 10 hours of daylight on the shortest day of the year and 14 hours of daylight on the longest day. The number of hours of daylight over the course of several years can be represented by the graph shown in Figure 7.2, which has the same shape as that in Figure 7.1.

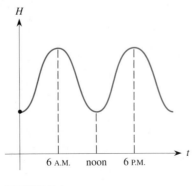

FIGURE 7.1

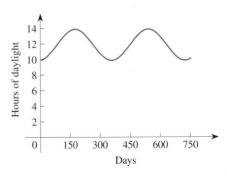

FIGURE 7.2

483

In this chapter, we consider how to model such periodic phenomena by introducing some new functions that have this type of periodic, oscillatory behavior. We can then use the ideas on stretching and shifting functions from Section 4.7 to create two families of functions that are used to model any such periodic phenomenon.

To begin, imagine a clock on a wall that is running backward, so that the hands move counterclockwise, as shown in Figure 7.3. (Unfortunately, the people who originally developed the mathematics relating to periodic, oscillatory behavior chose counterclockwise as their convention and we're stuck with it). In particular, picture the motion of the arrowhead on the minute hand. Let the horizontal axis be the line through the 3 and 9 positions on the clock. Suppose that the minute hand is 5 inches long and that we start the process at the instant the minute hand is pointing straight up. Every 60 minutes, the point of the minute hand moves from a maximum height of 5 inches above the center (when it is pointing straight up) to a minimum height of 5 inches below the center (when it is pointing straight down) and then back up toward to a maximum height of 5 inches again—and it repeats this cycle indefinitely. Thus the height of the arrowhead on the minute hand, as a function of time t, repeatedly traces a path that is periodic and oscillatory. The arrowhead traces the path shown in Figure 7.4 over the first 3 hours, or 180 minutes—it oscillates between -5 and $+5$ every 60 minutes. This type of function is just what we need to model periodic phenomena.

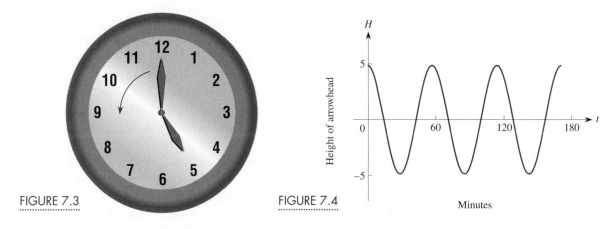

FIGURE 7.3

FIGURE 7.4

To develop this process more formally, we use the **unit circle**—a circle with radius 1 centered at the origin—as shown on the left in Figure 7.5. A point P with coordinates (x, y) lies on this circle if

$$x^2 + y^2 = 1.$$

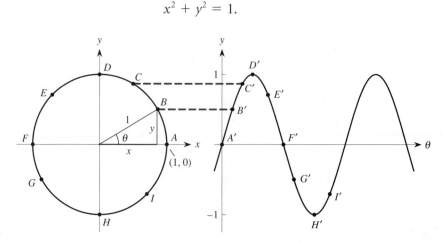

FIGURE 7.5

This point P corresponds to an angle θ measured counterclockwise from the positive x-axis. For instance, when $\theta = 0°$, P is on the positive x-axis; when $\theta = 90°$, P is on the positive y-axis. We consider separately the horizontal distance x to the left or right of the y-axis and the vertical height y above or below the x-axis. Each of these quantities is actually a function of θ, the angle at the center of the circle.

Let's start with the height y. In the triangle at the left in Figure 7.5,

$$\sin \theta = \frac{\text{opposite}}{\text{hypotenuse}} = \frac{y}{1} = y,$$

so the sine of any angle θ precisely equals the y-coordinate of the corresponding point on the unit circle. For instance, consider the point A at $(1, 0)$ at the extreme right of the circle as the starting point; it has height $y = 0$ and an angle of inclination $\theta = 0$, so $y = \sin 0 = 0$. We set up a second coordinate system where y is a function of θ, as shown in the graph on the right in Figure 7.5. When we think of y as a function of θ, the pair $(\theta, y) = (0, 0)$ corresponds to the point A' at the origin in the graph on the right.

Be sure that you distinguish carefully between the coordinate systems in the two diagrams—the circle on the left, where we think of the height y as a function of the horizontal distance x, and the associated graph being created on the right showing the height y as a function of the angle θ. In the circle, x is measured horizontally, whereas in the graph θ is measured horizontally. The position of any point—say, B—on the circle can be specified either (1) in terms of its x- and y-coordinates or (2) in terms of its angle of inclination θ and its vertical distance y above or below the horizontal axis of the circle. It is these (θ, y) pairs that are graphed on the right.

The Sine Function

We know that there are 360° in a circle, so we can trace the complete circle as θ runs from 0° to 360°. Also, we can continue tracing the circle over and over as θ increases beyond 360°; in fact, every 360° we complete another full revolution around the circle.

Let's consider all the points P on the circle. We labeled several specific points as $A, B, C, D, E, \ldots, I$ in Figure 7.5. Each point corresponds to a particular angle θ, and we plot the heights y corresponding to these angles θ in the graph at the right. For instance, the angle θ corresponding to point B is 30° and its height is $\frac{1}{2}$, or 0.5. As point P traces the circle, starting at point A and passing through the points $B, C, D, E, \ldots, I, \ldots$, the height y on the circle rises from 0 (at A) to a maximum of 1 (at D when $\theta = 90°$), then decreases past 0 (at F when $\theta = 180°$) to a minimum height of -1 (at H when $\theta = 270°$), and then back up to 0 as P finally returns to the starting point A, having gone through a full 360°. This motion of P continues as P traces the circle again and again, and the identical pattern of heights recurs repeatedly, every 360°. The oscillatory pattern shown in the right-hand graph is periodic; it repeats forever with a period of 360° and is part of the graph of the **sine function,** $f(\theta) = \sin \theta$, for any angle θ. It is this periodic, oscillatory effect that we need to model periodic phenomena.

Think About This

You can observe the development of the sine function dynamically by using your graphing calculator. Set the mode for radians (we discuss this topic shortly), for parametric graphing (we discuss this topic in Chapter 9), and for simultaneous plotting. The independent variable used is now typically t instead of x. See the instructions for your particular calculator if necessary. Go to the `Y=` menu and enter

$$X1 = \cos t$$
$$Y1 = \sin t$$
$$X2 = t$$
$$Y2 = \sin t$$

For the viewing window, set t between 0 and 7, $\Delta t = 0.1$, set x between -1.5 and 7, and set y between -2 and 2. When the graphs are drawn, the circle (somewhat flattened because of the screen dimensions) and the sine curve are produced simultaneously. Watch how the heights of points on the circle precisely match the heights on the sine curve for any angle t. You may also want to trace this behavior with your fingers on the graphs in Figure 7.5. ◻

We summarize these ideas about the sine function as follows.

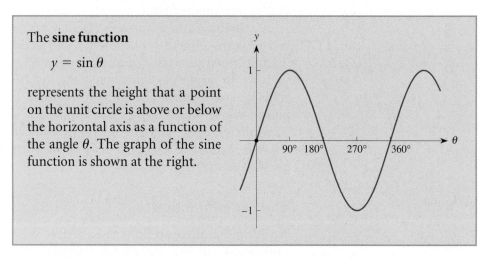

The **sine function**

$$y = \sin \theta$$

represents the height that a point on the unit circle is above or below the horizontal axis as a function of the angle θ. The graph of the sine function is shown at the right.

Again, note that the graph of the sine function oscillates between a maximum height of 1 and a minimum height of -1. Also, the basic shape repeats every 360°, so the behavior pattern you see from 0° to 360° occurs again from $\theta = 360°$ to $\theta = 720°$, again from 720° to 1080°, and so on. Similarly, the same pattern occurs between $\theta = -360°$ and 0°, between $\theta = -720°$ and $-360°$, and so on. Thus the sine function is a periodic function and its period is 360°, as shown in Figure 7.6.

The graph of the sine function passes through the origin and oscillates between -1 and 1 every 360°.

In addition, the sine curve reaches its maximum height of 1 at $\theta = 90°$ and again at $\theta = 450°$ ($=90° + 360°$), 810° ($=90° + 2 \times 360°$), ..., as well as at $\theta = -270°$ ($=90° - 360°$), $-630°$ ($=90° - 2 \times 360°$), ... Similarly, the sine curve reaches its minimum height of -1 at $\theta = 270°$, 630°, ..., and at $\theta = -90°$, $-450°$, $-810°$, ... Note also that the sine curve crosses the horizontal axis at the origin where $\theta = 0°$, again at $\theta = 180°$ (corresponding to the extreme left-hand point on the unit circle), yet again at $\theta = 360°$, and so on indefinitely. In fact, the sine function has zeros at every integer multiple of 180°.

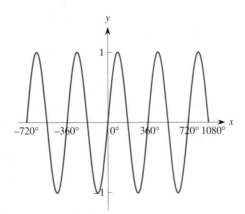

FIGURE 7.6

The Cosine Function

The sine function reflects the *vertical* height y above or below the horizontal axis in the unit circle. Next, we consider the *horizontal* distance x from the vertical axis of the unit circle to points P at (x, y) on the circle, as shown on the left in Figure 7.7. We now treat x as a function of θ. Thus, in the graph of (θ, x) on the right in Figure 7.7, the angle θ is measured along the horizontal axis and the distance x is measured vertically.

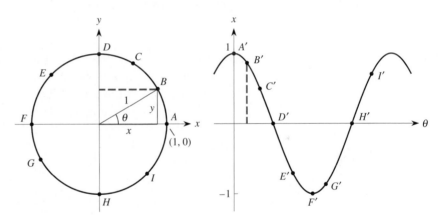

FIGURE 7.7

The initial point A at $(1, 0)$ on the circle lies at a distance of $+1$ to the right of the vertical axis, so the corresponding point A' on the graph at the right has a height of $x = 1$. The succeeding points B and C are closer to the vertical axis of the circle and so, because the x-values are smaller than 1, the corresponding heights on the graph are smaller. The point D is on the vertical axis, so its distance from the vertical axis is 0. The points E and F are to the left of the vertical axis of the circle, so the corresponding heights on the graph are negative. In fact, F is located where $\theta = 180°$ at the extreme left point of the circle at a distance of -1 from the vertical axis, so the corresponding point F' on the graph is a minimum.

As the angle θ continues to increase, the points on the circle approach the vertical axis from the left, and the corresponding points on the graph now rise toward 0. Eventually, the tracing point P on the circle passes the vertical axis at H and approaches the initial point A where $\theta = 360°$. The horizontal distance that P is from the axis changes from negative, to 0, to positive and approaches the distance 1 to the right of the vertical axis, which is where we started. Simultaneously, the graph on the right crosses the θ-axis and rises to its initial starting height of 1. Allowing θ to continue beyond $\theta = 360°$ we see that the previous pattern repeats exactly and indefinitely.

The graph shown on the right in Figure 7.7 is also a periodic, oscillatory curve. This curve, which corresponds to the horizontal distances from the vertical axis of the unit circle to points on the circle is the graph of the **cosine function**, $g(\theta) = \cos\theta$. The cosine function, like the sine function, is periodic and repeats every 360°, so its period is also 360°. The maximum value of the cosine function is 1, which occurs at $\theta = 0°, 360°, 720°, \ldots$, as well as at $\theta = -360°, -720°, \ldots$ The minimum value of the cosine function is -1, which occurs at $\theta = \pm 180°$, $\pm 540°, \ldots$ The cosine function has zeros when $\theta = \pm 90°, \pm 270°, \pm 450°, \ldots$

> The cosine curves passes through the point $(0, 1)$ and oscillates between -1 and 1 every 360°.

We summarize these ideas about the cosine function as follows.

The **cosine function,**

$$y = \cos\theta$$

represents the horizontal distance that a point on the unit circle is to the right or left of the vertical axis as a function of the angle θ. The graph of the cosine function is shown at the right.

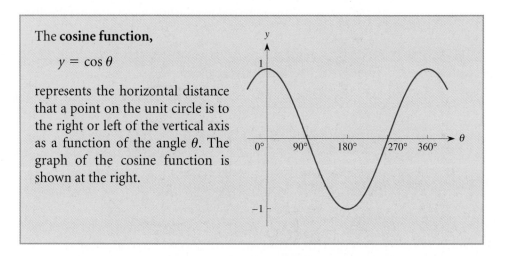

You may find the preceding construction of the cosine curve somewhat easier to visualize by using the following trick. Rotate the circle shown in Figure 7.7 through an angle of 90° counterclockwise, as shown in Figure 7.8. Each horizontal

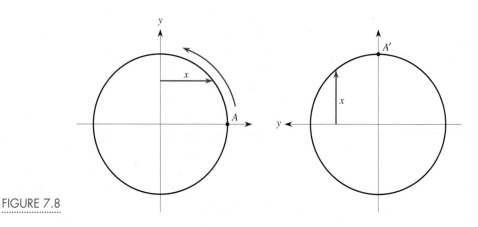

FIGURE 7.8

distance x is then transformed into an equivalent "height" above or below the new horizontal axis. These "heights" produce the heights of the points on the cosine curve shown in the graph on the right in Figure 7.7.

Because of the way that the sine and cosine functions can be defined in terms of the unit circle, they are sometimes called *circular functions.*

Figure 7.9 shows both the sine and cosine graphs from $\theta = 0°$ to $\theta = 540°$. Clearly, these two functions are closely related. Both have the same shape and each can be thought of as arising from the other by an appropriate horizontal shift. If we shift the sine curve to the left by 90°, we get the cosine curve, so

$$\cos \theta = \sin(\theta + 90°).$$

Alternatively, if we shift the cosine curve to the right by 90°, we get the sine curve, so

$$\sin \theta = \cos(\theta - 90°).$$

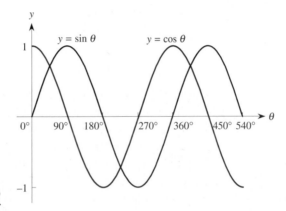

FIGURE 7.9

Moreover, the unit-circle definition suggests what is perhaps the most important relationship between the two functions. Figure 7.10 shows that the vertical height y to a point (x, y) on the unit circle equals $\sin \theta$. Similarly, the horizontal distance x from the vertical axis to the same point equals $\cos \theta$. Because $x = \cos \theta$ and $y = \sin \theta$ must satisfy the equation of the unit circle,

$$x^2 + y^2 = 1,$$

it follows that

$$(\cos \theta)^2 + (\sin \theta)^2 = 1.$$

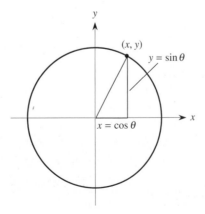

FIGURE 7.10

Recall that we write

$$(\cos \theta)^2 \quad \text{as} \quad \cos^2 \theta \quad \text{and} \quad (\sin \theta)^2 \quad \text{as} \quad \sin^2\theta.$$

Note that this is another way to prove the *Pythagorean identity* that we presented in Section 6.4.

The **Pythagorean identity** is

$$\sin^2\theta + \cos^2 \theta = 1$$

for *any* angle θ.

Think About This Use your function grapher to graph the function $f(x) = \sin^2x + \cos^2x$ for any interval of x values. What does it look like? [You will likely have to enter the function as `(sin x)^2+(cos x)^2.`] ▫

Radian Measure

Because we want to use the sine and cosine functions to model phenomena that are periodic over time, such as the heights of tides or the number of hours of daylight, we need a function of time t rather than a function of an angle θ. Therefore we need a way to avoid angles measured in degrees in our definitions of these functions. To do so, we introduce an alternative unit, called the **radian,** for measuring an angle. In the circle of radius 1 shown in Figure 7.11, we begin on the horizontal axis at the point A at $(1, 0)$. We move counterclockwise around this circle and measure off a distance equal to the radius, or 1. This distance produces an angle α whose size is defined as *one radian.* In degrees, this angle is approximately 57°, as we show shortly.

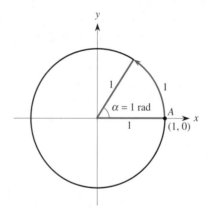

FIGURE 7.11

We next develop a way to convert between radians and degrees. The length of the arc that defines one radian equals the radius of the unit circle. Because $r = 1$, the total circumference of the circle is $2\pi r = 2\pi$. Moreover, the angle α represents a fraction of the full 360° in the circle. As a result, we can set up the proportion

$$\frac{\text{Fraction of the total angle}}{\text{Total angle}} = \frac{\text{Fraction of the total circumference}}{\text{Total circumference}}$$

or

$$\frac{1 \text{ radian}}{360°} = \frac{1}{2\pi}.$$

Cross-multiplying, we get

$$2\pi \text{ radians} = 360°$$

or

$$\boxed{\pi \text{ radians} = 180°}$$

Alternatively, dividing both sides by π gives

$$1 \text{ radian} = \left(\frac{180}{\pi}\right)^{\!\circ} \approx 57.29578°,$$

or about 57.3°. Furthermore, because π radians $= 180°$, we can divide both sides by 180 to get

$$1° = \frac{\pi}{180} \text{ radians}.$$

If we perform the same construction in any circle with radius r—that is, if we measure an arc whose length equals the radius r—the corresponding angle would be the same 1 radian, or about 57.3°. Thus an angle measured in radians is the same no matter what the size of the circle. More important, radians are not tied directly to angles the way degrees are. Using radians, we can consider any variable and apply the sine and cosine functions to it. Thus we can use a variable representing time, height, or any other desired quantity as the independent variable with either the sine or the cosine function.

◆ **EXAMPLE**

Use the fact that $180° = \pi$ radians to obtain the radian measure for the common angles 90°, 60°, 45°, and 30°.

Solution If we divide 180° by 2, we get

$$90° = \frac{180°}{2} = \frac{\pi}{2} \text{ radians} = \frac{\pi}{2}.$$

Similarly,

$$60° = \frac{180°}{3} = \frac{\pi}{3} \text{ radians} = \frac{\pi}{3};$$

$$45° = \frac{180°}{4} = \frac{\pi}{4} \text{ radians} = \frac{\pi}{4};$$

$$30° = \frac{180°}{6} = \frac{\pi}{6} \text{ radians} = \frac{\pi}{6}.$$

To summarize, we have the following relationships.

$$180° = \pi \text{ radians}$$

In particular,

$$30° = \frac{\pi}{6}, \quad 45° = \frac{\pi}{4}, \quad 60° = \frac{\pi}{3}, \quad \text{and} \quad 90° = \frac{\pi}{2}.$$

These results occur often in applications of the trigonometric functions, and you need to know them.

For these standard, or special, angles, we have the following values for the sine and cosine functions (which we derived in Chapter 6).

$$\sin 30° = \sin\left(\frac{\pi}{6}\right) = 0.5 = \frac{1}{2} \qquad \cos 30° = \cos\left(\frac{\pi}{6}\right) = \frac{\sqrt{3}}{2} \approx 0.866$$

$$\sin 45° = \sin\left(\frac{\pi}{4}\right) = \frac{\sqrt{2}}{2} \approx 0.707 \qquad \cos 45° = \cos\left(\frac{\pi}{4}\right) = \frac{\sqrt{2}}{2} \approx 0.707$$

$$\sin 60° = \sin\left(\frac{\pi}{3}\right) = \frac{\sqrt{3}}{2} \approx 0.866 \qquad \cos 60° = \cos\left(\frac{\pi}{3}\right) = \frac{1}{2} = 0.5$$

$$\sin 90° = \sin\left(\frac{\pi}{2}\right) = 1 \qquad \cos 90° = \cos\left(\frac{\pi}{2}\right) = 0$$

Be sure that you know how to use your calculator to obtain the value for the sine or cosine of any argument, both in degrees and radians. We strongly recommend that you permanently set your calculator mode to Radians; we work with radians almost exclusively from this point on.

Note that radians are not always given in terms of π. For instance, we might have $x = 0.5$ radians or $x = 2$ radians or $x = -4.27$ radians.

The Behavior of the Sine and Cosine Functions

Finally, let's consider the important aspects of the behavior of the sine and cosine functions. In general, we want to answer several questions for any function.

1. Where is it increasing?
2. Where is it decreasing?
3. Where is it concave up?
4. Where is it concave down?
5. Where are its points of inflection?
6. Where are its zeros?
7. Where does it achieve its maximum value, and what is that maximum value?
8. Where does it achieve its minimum value, and what is that minimum value?
9. Is it periodic? If so, what is its period?

We can answer all these questions about the sine and cosine functions by examining their graphs and applying ideas developed earlier in this book. However, for other functions that may not be as well known, answering some of these questions requires the use of calculus.

Let's consider the behavior of the sine function $y = \sin x$. It is evident from its graph that the sine curve increases for x between 0 and $\pi/2$, then decreases from $\pi/2$ to $3\pi/2$, and then increases from $3\pi/2$ to 2π—and repeats this cycle thereafter. This behavior is also clear from the unit circle definition of the sine.

Further, the sine curve is concave down for x between 0 and π, concave up for x between π and 2π, and then repeats this cycle thereafter. Consequently, the sine curve has points of inflection at $x = 0, \pm\pi, \pm2\pi, \ldots$, where its concavity changes.

In addition, the sine function has zeros when $x = 0, \pm\pi, \pm2\pi, \ldots$ A special characteristic of the function $f(x) = \sin x$ is that its zeros and its points of inflection are identical, which is not the case for most other common functions.

Finally, the sine function achieves its maximum value of 1 at $x = \pi/2$, at $x = 5\pi/2$, at $x = 9\pi/2, \ldots$ and at $x = -3\pi/2$, at $x = -7\pi/2, \ldots$ The sine function achieves its minimum value of -1 at $x = 3\pi/2$, at $x = 7\pi/2$, at $x = 11\pi/2, \ldots$ and at $x = -\pi/2$, at $x = -5\pi/2, \ldots$

We ask you to describe the behavior of the cosine function in the Problems at the end of this section.

In summary, the key points about the sine and cosine functions are:

The sine function passes through the origin and oscillates between -1 and $+1$ every 2π.

The cosine function passes through the point $(0, 1)$ and oscillates between -1 and $+1$ every 2π.

Problems

1. Janis trims her fingernails every Saturday morning. Sketch the graph of the length of her nails as a function of time. Can this process be modeled by a periodic function? If it is periodic, what is the period?

2. Harry gets a haircut on the first of every month. Sketch the graph of the length of his hair as a function of time. Can this process be modeled by a periodic function? If it is periodic, what is the period?

3. In the accompanying figure, the circle on the left has been subdivided every $15°$ from $\theta = 0°$ to $\theta = 360°$. Use the heights from the horizontal axis to the associated points on the circle to construct the graph of the sine function for $\theta = 0°$ to $\theta = 360°$ on the axes at the right.

4. Convert each angle from degrees to radians.

a. $15°$	**b.** $75°$	**c.** $120°$
d. $150°$	**e.** $225°$	**f.** $315°$
g. $270°$	**h.** $240°$	**i.** $-135°$
j. $-210°$		

5. Convert each angle from radians to degrees.

a. $\dfrac{3\pi}{4}$	**b.** $\dfrac{4\pi}{5}$	**c.** $\dfrac{2\pi}{3}$
d. 1.5	**e.** 2.5	**f.** 3
g. $\dfrac{\pi}{8}$	**h.** $\dfrac{5\pi}{3}$	**i.** $-\dfrac{3\pi}{2}$
j. $-\dfrac{5\pi}{3}$		

6. For $f(\theta) = 5\sin\theta$, evaluate each function.

a. $f(30°)$ **b.** $f(45°)$ **c.** $f(60°)$

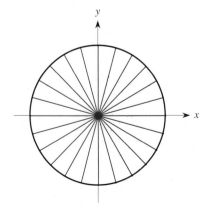

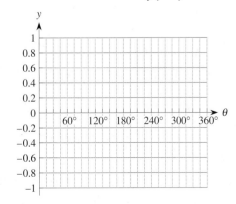

d. $f(120°)$ e. $f(-15°)$ f. $f(873°)$

g. $f\left(\dfrac{\pi}{4}\right)$ h. $f\left(\dfrac{\pi}{3}\right)$ i. $f\left(\dfrac{\pi}{12}\right)$

j. $f\left(-\dfrac{\pi}{6}\right)$ k. $f(5.27)$ l. $f(-25.614)$

7. For $f(\theta) = \sin 2\theta$, evaluate each function.

   a. $f(30°)$ b. $f(45°)$ c. $f(120°)$
   d. $f(225°)$ e. $f(\pi/3)$ f. $f(\pi/12)$
   g. $f(3\pi/8)$ h. $f(2\pi/7)$

8. At the end of this section we posed nine questions about the behavior of any function. Answer these questions for the cosine function.

9. With your calculator set in radians, graph the two functions $y = \sin x$ and $y = \cos(x - \pi/2)$. What do you observe? Explain what you observed.

10. Plot the functions $y = \cos x$ and $y = \sin(x + \pi/2)$. Explain why you see only one graph. (If you see two graphs, check that your calculator MODE is set for Radians.)

11. The population growth patterns of two species are interrelated when one species preys on the other. This situation occurs in northern Canada where lynxes are the predators and hares are the prey. The figure below is based on records kept by the Hudson's Bay Trading Company on the number of each species caught by fur trappers from 1845 through 1935. The graphs indicate that both populations change in roughly periodic cycles.

    a. Estimate the period of the cycle for the lynxes.
    b. Estimate the period of the cycle for the hares.
    c. Estimate the years in which the lynx population reached its maximum and minimum values.
    d. Estimate the years in which the hare population reached its maximum and minimum values.
    e. Can you find any relationship between the lengths of the periods in parts (a) and (b) and the times in parts (c) and (d)? If so, what is it?
    f. Estimate the years when the hare population passed its points of inflection. How do they compare to any of the times you found in parts (c) and (d)?

12. Consider the functions $y = \cos x$ and $y = \sin(\#\$\%\#\$)$. What could #$%#$ represent so that the two graphs are identical? Is there only one correct answer to this question? Explain.

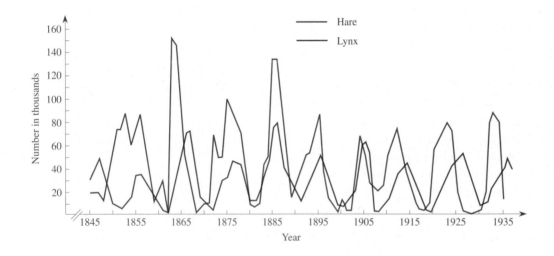

7.2 ⸺ Modeling Periodic Behavior with the Sine and Cosine

Because of their behavior patterns, the sine and cosine functions are used as the mathematical models to represent most periodic phenomena. For example, the number of hours of daylight H any day of the year in San Diego can be modeled by the function

$$H(t) = 12 + 2.4 \sin\left[\dfrac{2\pi}{365}(t - 80)\right],$$

where t is the number of days from the first of the year ($t = 1$ on January 1). We begin by using the formula for some predictions and then see where the formula comes from.

EXAMPLE 1

Based on the model, how many hours of daylight are there in San Diego on (**a**) February 15? (**b**) March 21? (**c**) June 21?

Solution

a. February 15 is the 46th day of the year (31 days in January plus 15 more in February), so $t = 46$. Using a calculator set to radian mode, we find that

$$H(46) = 12 + 2.4 \sin\left[\frac{2\pi}{365}(46 - 80)\right]$$

$$= 12 + 2.4 \sin\left[\frac{2\pi}{365}(-34)\right] = 10.67 \text{ hours.}$$

b. March 21 is the $t = 31 + 28 + 21 = 80$th day of the year, so

$$H(80) = 12 + 2.4 \sin\left[\frac{2\pi}{365}(80 - 80)\right] = 12 \text{ hours.}$$

Incidentally, March 21 is the spring equinox, which means that there are 12 hours of daylight and 12 hours of darkness, so the model gives the right prediction.

c. June 21 is the $t = 31 + 28 + 31 + 30 + 31 + 21 = 172$nd day of the year, so the number of hours of daylight on June 21 (the first day of summer and so the "longest" day of the year) is

$$H(172) = 12 + 2.4 \sin\left[\frac{2\pi}{365}(172 - 80)\right] = 14.40 \text{ hours.}$$

◆

Think About This
Without using a calculator, find the number of hours of daylight in San Diego on December 21, the 355th (and "shortest") day. ▢

Later in this section we show how a similar formula can be developed for any city. For now, let's see what the different numbers 12, 2.4, 365, and 80 in the formula

$$H(t) = 12 + 2.4 \sin\left[\frac{2\pi}{365}(t - 80)\right]$$

for the number of hours of daylight in San Diego actually represent. Obviously, 365 represents the number of days in a year; it tells us how long it takes for a full cycle to be completed. Over a full year, the average number of hours of daylight is 12 hours per day—the days when there are more than 12 hours of daylight exactly counterbalance the days when there are fewer than 12 hours of daylight. Alternatively, averaging the number of hours of daylight for all 365 days gives 12 hours. So the 12 represents the average, or middle, value for the sine function.

Next, as we found in Example 1, the longest day in San Diego has 14.4 hours of daylight and the shortest day has 9.6 hours of daylight. Note that 14.4 is 2.4 hours more than the average level of 12 and that 9.6 is 2.4 hours less than 12. Of course, the maximum and minimum number of hours of daylight at any particular location, as well as the number on any specific date, depend on the location itself and are therefore modeled by a function slightly different from H; think about how long a "day" is during the winter or the summer in the far north, the so-called "land of the midnight sun".

The graph of H, the number of hours of daylight in San Diego over a 3-year interval, is shown in Figure 7.12. It has the same *shape* as the graph of the basic sine or cosine function. However, it does not oscillate about the horizontal axis; rather it oscillates about the horizontal line $H = 12$, which represents the average number of hours of daylight over a full year, so it is shifted up by 12 hours. Also, its maximum and minimum "heights" above the horizontal line $H = 12$ are no longer $+1$ and -1 as with the basic sine and cosine functions. Instead, the graph varies from a minimum of 9.6 hours to a maximum of 14.4 hours, which is 2.4 hours either side of the average 12, so the sine function has been stretched by a factor of 2.4.

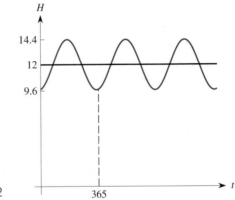

FIGURE 7.12

Note some additional differences: The graph is shifted to the right compared to the sine curve (the curve does not "start" at the vertical axis where $t = 0$ and $H = 12$). Also, the period is 365 days, rather than the usual 2π radians, or 360°. (Incidentally, the ancient Babylonians believed that the length of a year was 360 days. That's why we divide a circle into 360 degrees.)

This particular function $H(t)$ differs from the standard, or base, function $y = \sin x$ that we discussed in Section 7.1 in four ways:

1. a vertical shift,
2. an oscillation other than from -1 to $+1$ (a stretch),
3. the length of a cycle, and
4. the "starting" point of the cycle (a horizontal shift).

Understanding how to incorporate these variations is crucial for applying the sine and cosine functions to describe periodic phenomena. We therefore focus on each in detail.

The equation for the number of hours of daylight in San Diego is

$$H = 12 + 2.4 \sin\left[\frac{2\pi}{365}(t - 80)\right].$$

Consider the more general *sinusoidal function*

$$S(x) = D + A \sin[B(x - C)],$$

where A, B, C, and D are all constants and x is the independent variable. In the San Diego situation, $D = 12$, $A = 2.4$, $B = 2\pi/365$, and $C = 80$. Let's investigate how each of these four *parameters* affects the graph of the basic sine curve. To do so, we consider each parameter separately.

The Vertical Shift or Midline

To show the significance of the D term, we consider the simpler function with $A = 1$, $B = 1$, and $C = 0$:

$$S(x) = D + \sin x.$$

We know that $y = \sin x$ oscillates repeatedly between -1 and $+1$. What is the effect of adding a constant D? From our discussion in Sections 4.6 and 4.7, we know that D raises or lowers the basic sine curve by the amount D. The graph of $S(x) = 2 + \sin x$ has the same shape as the basic sine function but is shifted up 2 units; it oscillates about the horizontal line $y = 2$, between 1 and 3 units above the x-axis as shown in Figure 7.13. Similarly, the graph of $S(x) = -5 + \sin x$ oscillates about the horizontal line $y = -5$, between -6 and -4.

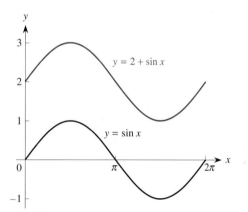

FIGURE 7.13

Thus the effect of the constant D in

$$S(x) = D + \sin x$$

is to produce a sinusoidal curve that oscillates about the horizontal line $y = D$, between $D - 1$ and $D + 1$. If D is positive, the sine curve is shifted upward D units; if D is negative, the curve is shifted downward D units. The number D is the **vertical shift** or the **midline.** In the formula for the number of hours of daylight in San Diego, the vertical shift, or midline, is 12.

The Amplitude

We next investigate the effect of the multiplicative constant A in the general equation of a sinusoidal function $y = D + A \sin[B(x - C)]$. We set $D = 0$, $B = 1$, and $C = 0$ to consider the simpler function

$$S(x) = A \sin x.$$

For example, if $A = 2$, we get $S(x) = 2 \sin x$, whose graph is shown in Figure 7.14, where it is compared to the basic curve for the sine function, $y = \sin x$ (for which

$A = 1$). For comparison, we also show the graph of $T(x) = \frac{1}{2}\sin x$. Although the basic sine function oscillates between -1 and $+1$, the transformed function $S(x) = 2\sin x$ oscillates between -2 and $+2$ and the transformed function $T(x) = \frac{1}{2}\sin x$ oscillates between $-\frac{1}{2}$ and $+\frac{1}{2}$. In general, the effect of multiplying the sine function by a constant A is to increase its vertical height above and below the midline by the factor $|A|$.

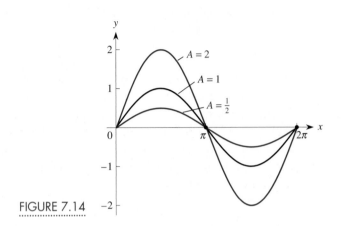

FIGURE 7.14

To show why the absolute value is necessary, we consider the graph of

$$S(x) = -4\sin x,$$

which has the same shape as the basic sine curve, but oscillates between -4 and $+4$. The main difference is that the graph of this function is flipped over the x-axis compared to the graph of $y = \sin x$. We naturally think of it as being four times as high as the base curve, not -4 times as high. (Draw simultaneously the graphs of $y = \sin x$ and $y = -4\sin x$ using your function grapher.) This is the same effect of the negative multiple that we encountered in Section 4.6. However, this curve has the same period ($2\pi = 360°$) and the same zeros ($x = 0, \pm\pi, \pm2\pi, \ldots$) as the basic sine curve.

The quantity $|A|$ is called the **amplitude** of the sine function. In the expression for $H(t)$ modeling the number of hours of daylight in San Diego, the amplitude is 2.4.

In Example 2 we show what happens when we combine the two transformations to construct a new function.

◆**EXAMPLE 2**

Analyze the graph of

$$S(x) = 2 + 3\sin x.$$

Solution   In this formula, 2 is the vertical shift, or midline, and 3 is the amplitude. The effect of multiplying the sine function by 3 is to stretch it vertically by a factor of 3, so that $3\sin x$ oscillates between -3 and 3. Adding the constant 2 to the function $3\sin x$ simply raises the entire curve 2 units vertically. Consequently, the combined effect is to produce a sinusoidal function that oscillates from 3 units below the horizontal line $y = 2$ to 3 units above the line; that is, from -1 to $+5$, as shown in Figure 7.15.

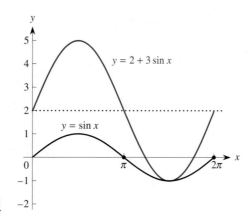

FIGURE 7.15

Incidentally, the function $y = 2 + 3 \sin x$ is not the same as $y = 5 \sin x$; the coefficients cannot be combined because 2 and $3 \sin x$ are not like terms. Graph both functions to see that they produce very different results. Also, $y = 2 + 3 \sin x$ is not the same as $y = 3 + 2 \sin x$—each parameter has its own role to play.

Use your function grapher to examine the graphs of several functions of the form $y = D + A \sin x$ for different values of A and D. Predict and then observe how the different constant values are reflected in the corresponding sinusoidal curve.

The Frequency and the Period

We next consider the effect of the parameter B, which multiplies the term $(x - C)$ in

$$S(x) = D + A \sin[B(x - C)].$$

To concentrate on B only, we take $C = 0$, $D = 0$, and $A = 1$. We also assume that $B > 0$. Consider how the function

$$S(x) = \sin(2x)$$

compares to the basic curve $y = \sin x$, as shown in Figure 7.16. The resulting sinusoidal curve $y = \sin(2x)$ completes two full cycles between $x = 0$ and $x = 2\pi$, compared to one full cycle for the basic sine curve.

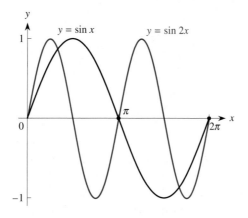

FIGURE 7.16

Similarly, the graph of

$$S(x) = \sin(3x)$$

shown in Figure 7.17 completes three full cycles across the interval from 0 to 2π. Based on these two results, we expect that the graph of

$$S(x) = \sin(nx),$$

for any positive integer n, will complete n full cycles between $x = 0$ and $x = 2\pi$. Try this with your function grapher for values of n such as 5 or 8.

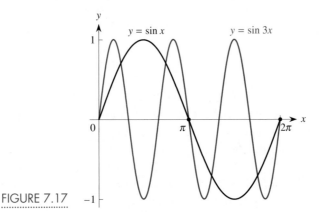

FIGURE 7.17

What happens if the positive multiple B is not an integer? The graph of $y = \sin(\frac{1}{2}x)$ is shown in Figure 7.18. The function $y = \sin(\frac{1}{2}x)$ completes half a complete cycle between 0 and 2π; it actually requires an interval of values for x from 0 to 4π to complete a full cycle.

Similarly, Figure 7.19 shows that the function $y = \sin(2.5x)$ completes 2.5 full cycles between 0 and 2π. (Trace the graph with your finger and count the cycles.) It therefore completes one full cycle in $1/2.5 = 0.4 = 2/5$ of this interval.

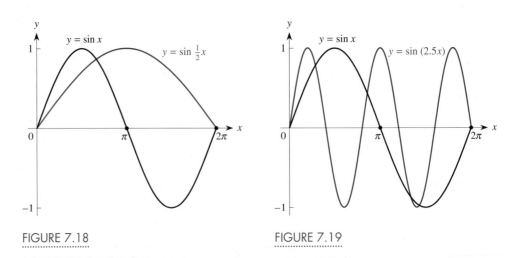

FIGURE 7.18   FIGURE 7.19

Mathematicians call the parameter B in $y = \sin(Bx)$ the **frequency** of the sinusoidal function. It tells us the number of complete cycles that occur between $x = 0$ and $x = 2\pi = 360°$. For instance, the function $y = \sin(6x)$ completes six full cycles across this interval, whereas the function $y = \sin(\frac{3}{8}x)$ completes 3/8 of one cycle.

Note that standard usage is to write $y = \sin 2x$ or $y = \sin 5x$, say, rather than $y = \sin(2x)$ or $y = \sin(5x)$, even though it is less precise a notation. But parentheses are essential on calculators and computers.

As with any periodic function, the *period* of a sinusoidal function $y = \sin Bx$ is the length of the interval needed to complete one full cycle. For $y = \sin 2x$, a full cycle is completed in any interval of x-values of length π (see Figure 7.16), so the period is π radians $= 180°$. For $y = \sin 3x$, the period is

$$\left(\frac{1}{3}\right)2\pi = \frac{2\pi}{3} \quad \text{or} \quad \left(\frac{1}{3}\right)360° = 120°,$$

(see Figure 7.17). For $y = \sin\left(\frac{1}{2}x\right)$, the period is

$$\frac{1}{\frac{1}{2}}(2\pi) = 4\pi \text{ radians},$$

or $720°$ (see Figure 7.18).

In general, the period of $y = \sin Bx$ is

$$\text{Period} = \frac{2\pi}{B} = \frac{2\pi}{\text{frequency}}.$$

Thus, for instance, for $y = \sin(2.5x)$, the period is

$$\text{Period} = \frac{2\pi}{2.5} = \frac{2\pi}{5/2} = \frac{2}{5}(2\pi) = \frac{4}{5}\pi$$

because it is the length of the interval needed for this sinusoidal function to complete one full cycle (see Figure 7.19). This result agrees with our earlier statement that the function $y = \sin(2.5x)$ completes one full cycle in $2/5$ of the interval from 0 to 2π.

We have shown that the period of any periodic function is the length of the interval needed to complete one full cycle. Alternatively, if we start with the period B,

$$\text{Frequency} = \frac{2\pi}{\text{period}}.$$

In the formula for the number of hours of daylight in San Diego

$$H = 12 + 2.4 \sin\left[\frac{2\pi}{365}(t - 80)\right],$$

the frequency of the sinusoidal curve is

$$\text{Frequency} = \frac{2\pi}{365} \approx 0.0172.$$

The period of the sinusoidal curve is

$$\text{Period} = \frac{2\pi}{\text{frequency}} = \frac{2\pi}{(2\pi/365)} = 365 \text{ days}.$$

Thus, as we would expect, the period is 1 year.

In summary we have the following.

$$\text{period} = \frac{2\pi}{\text{frequency}} \qquad \text{frequency} = \frac{2\pi}{\text{period}}$$

Note that in engineering, the frequency is defined somewhat differently. Instead of meaning the number of cycles in 2π radians, engineers consider the number of cycles in a given length of time—say, cycles per second. They then write

$$\text{Frequency} = \frac{1}{\text{period}}$$

and write the sinusoidal function in the form $y = \sin(2\pi Bt)$, where B, not $2\pi B$, is the frequency. Unfortunately, this slight difference in terminology is so deeply embedded in the two fields that it is not possible for either field to change to match the other.

The Phase Shift

Finally, we consider the role of the parameter C in

$$S(x) = D + A \sin[B(x - C)].$$

We simplify the discussion by taking $A = 1$, $B = 1$, and $D = 0$ so that we consider only

$$S(x) = \sin(x - C).$$

From Section 4.7, we know that the term $(x - C)$ should have the effect of a horizontal shift to the right when C is positive and to the left when C is negative.

Figure 7.20 compares the graph of $S(x) = \sin(x + \pi/4)$ to the basic curve $y = \sin x$. The two curves appear similar, but $S(x)$ is shifted to the left (backward) by $\pi/4$, or $\frac{1}{8}$ of 2π (which is one-eighth of a full cycle). Similarly, Figure 7.21 shows the graph of $T(x) = \sin(x - \pi/3)$. It has been shifted to the right (forward) by $\pi/3$, or $\frac{1}{6}$ of 2π (which is one-sixth of a full cycle).

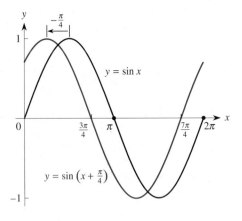

FIGURE 7.20

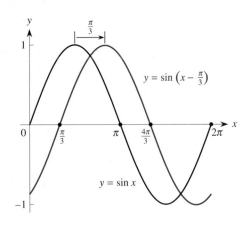

FIGURE 7.21

In general, the parameter C shifts a sinusoidal curve to the left or the right by the amount C. If C is positive in the term $(x - C)$, as in $y = \sin(x - \pi/3)$, the curve is shifted to the right by C; if C is negative in $(x - C)$, as in $y = \sin(x + \pi/4)$, the curve is shifted to the left by C. This parameter is called the **phase shift,** instead of the horizontal shift, in the context of sinusoidal functions. In the expression for the daylight function for San Diego

$$H = 12 + 2.4 \sin\left[\frac{2\pi}{365}(t - 80)\right],$$

the phase shift is 80 days.

EXAMPLE 3

What is the significance of the phase shift in the formula for *H*?

Solution   The phase shift shifts the curve to the right by 80 days. Recall that the 80th day of the year is March 21, which is the spring equinox (the day when there are equal numbers of hours of daylight and darkness). On this day, the graph for the sinusoidal function crosses the midline, or average level, of *D* = 12 hours.

◆

In general, the phase shift for a sine function corresponds to the first point to the right of the origin where the curve crosses the midline while increasing. Equivalently, it occurs midway, horizontally, from a minimum point to a maximum.

We summarize all the results for the San Diego daylight function in Figure 7.22.

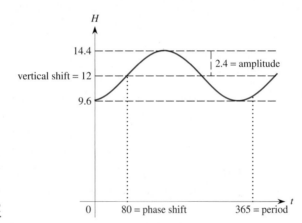

FIGURE 7.22

EXAMPLE 4

The water at a boat dock is 7 feet deep at low tide and 11 feet deep at high tide. On a certain day, low tide occurs at 4 A.M. and high tide at 10 A.M. Find an equation for the height of the tide *y* as a function of time *t*.

Solution   We use the given information to sketch the graph of a sinusoidal curve in Figure 7.23.

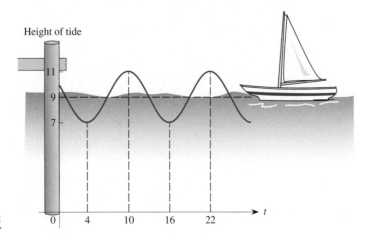

FIGURE 7.23

Because the tide ranges from a minimum height of 7 feet to a maximum height of 11 feet above sea bottom, the curve oscillates about the middle value of 9 feet, which is the vertical shift, or midline. Also, the amplitude of this sinusoidal curve is 2. Further, the time interval between the minimum and maximum heights of the water level is 6 hours; consequently, a complete tide cycle takes 12 hours, so the period is 12 hours. As a result,

$$\text{Frequency} = \frac{2\pi}{\text{period}} = \frac{2\pi}{12} = \frac{\pi}{6}.$$

Finally, because the tide level increases from 4 A.M. to 10 A.M., the curve passes across the middle height of 9 feet halfway between 4 A.M. and 10 A.M. (or at 7 A.M.), which gives the phase shift. (The graph shows that, even though the tide function also crosses the 9-foot level at 1 A.M., the function is decreasing there and so this does not give the phase shift.) Therefore the height y of the water at any time t is modeled by

$$y = 9 + 2 \sin\left[\frac{\pi}{6}(t - 7)\right].$$

EXAMPLE 5

The air conditioning in a home is set to go on when the temperature reaches 74°F and to go off when the temperature drops to 68°. This cycle repeats every 20 minutes. If the temperature in the house at noon is 71° and rising, write a sinusoidal function to model the temperature as a function of the number of minutes t since noon.

Solution A sinusoidal function is of the form $T = D + A \sin[B(t - C)]$, where A, B, C, and D must be determined. We know that the temperature oscillates between 68° and 74°, so it is centered about 71°, which is the vertical shift, or midline, D. Further, because the size of the oscillation above and below this midline is 3, we know that the amplitude $A = 3$. We also know that the length of the cycle is 20 minutes, so the period is 20 and therefore the frequency $B = 2\pi/20 = \pi/10$. Finally, because the house temperature reaches 71°—the level of the vertical shift—at noon when $t = 0$, the phase shift is 0, so $C = 0$. Therefore our model for the temperature of the house as a function of time is

$$T = 71 + 3 \sin\left(\frac{\pi}{10}t\right).$$

Figure 7.24 shows the graph of this sinusoidal function for the first 60 minutes.

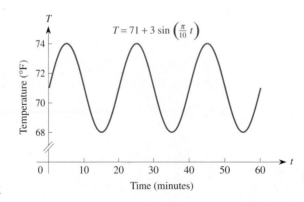

FIGURE 7.24

EXAMPLE 6

For part of the year, the temperature T in the Colorado Rockies can be modeled by the function

$$T(t) = 60 + 10 \sin\left(\frac{\pi}{12}t\right),$$

where t is measured in hours and $t = 0$ is at 9 A.M. In Example 5 of Section 2.2, we used the linear function

$$C(T) = 4T - 160$$

to model the chirp rate C (in chirps per minute) of the snow tree cricket as a function of the air temperature T (in ° F).

a. Express the chirp rate as a function of time.

b. How fast is the cricket chirping at 5 P.M.?

c. What are the domain and range of this function?

Solution

a. The chirp rate in $C = 4T - 160$ is measured in chirps per minute and the time t in the formula for the temperature as a function of time is measured in hours. To make things consistent, we convert the chirp rate to chirps per hour by multiplying by 60 (minutes per hour) to get

$$C = 60(4T - 160) = 240T - 9600.$$

We now have C as a function of T, where T is a function of t, so C is a composite function,

$$\begin{aligned} C = f(t) &= 240T - 9600 \\ &= 240\left[60 + 10 \sin\left(\frac{\pi}{12}t\right)\right] - 9600 \\ &= 14{,}400 + 2400 \sin\left(\frac{\pi}{12}t\right) - 9600 \\ &= 4800 + 2400 \sin\left(\frac{\pi}{12}t\right), \end{aligned}$$

where t is measured in hours since 9 A.M. and C is chirps per hour, as illustrated in Figure 7.25. (Note that we could have gone the other way and converted everything to minutes, but then the numbers get quite large.)

b. At 5 P.M., when $t = 8$ hours after 9 A.M., the chirp rate is

$$C = f(8) = 4800 + 2400 \sin\left(\frac{\pi}{12} \cdot 8\right) \approx 6878 \text{ chirps per hour.}$$

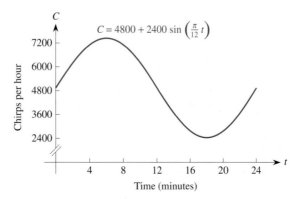

FIGURE 7.25

c. The independent variable is the time t in hours from 9 A.M. on a particular day. The practical domain of the function f depends on how long the functions that are the component models make sense. The model for the temperature is good for only part of the year; let's assume that it applies only for a 30-day period. If the initial time is at the beginning of that time period, the domain would last for the following 30 days; if the initial time is at the middle of that time period, the domain would extend from 15 days before to 15 days after. As for the range, look at the function f. It oscillates above and below 4800 chirps per hour, from a minimum of $4800 - 2400 = 2400$ to a maximum of $4800 + 2400 = 7200$ chirps per hour. So the range is 2400 to 7200.

Identical ideas about vertical shift, amplitude, period, frequency, and phase shift apply to cosine functions of the form

$$y = D + A \cos[B(x - C)],$$

whose behavior also is described as *sinusoidal*. The midline D serves to raise or lower the "center" of the cosine curve; the amplitude A stretches or shrinks the cosine curve vertically about the midline; the frequency B represents the number of cycles over an interval of 2π; and the parameter C is the phase shift, which shifts the cosine curve to the left or the right, depending on the sign of C. The only difference between working with sines and cosines lies in finding the phase shift. For a cosine function, the phase shift corresponds to the first point to the right of the origin where the curve reaches its maximum.

EXAMPLE 7

Describe the graph of the sinusoidal function $y = 5 + 3 \cos\left[2\left(x - \dfrac{\pi}{4}\right)\right]$.

Solution   The basic cosine curve is multiplied by 3, so its amplitude is 3. Because the midline is 5, the function extends from a minimum of $5 - 3 = 2$ to a maximum of $5 + 3 = 8$. Because the frequency is 2, there are two complete cycles between 0 and 2π, so the period is π. Finally, the phase shift is $\pi/4$, so the curve is shifted to the right by $\pi/4$. The graph of this function from $x = 0$ to $x = 2\pi$, shown in Figure 7.26, illustrates all these effects.

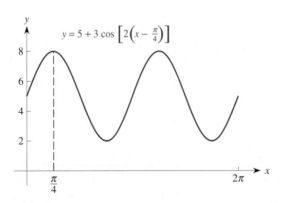

FIGURE 7.26

Combining Sinusoidal and Exponential Functions

Consider a spring hanging vertically from the ceiling with a weight attached at the bottom, as shown in Figure 7.27. The weight is pulled down and then released, so the weight bobs up and down with smaller and smaller oscillations until it settles to a stop in the original rest position, called its equilibrium. Figure 7.28 displays a graph of this vertical displacement y as a function of time t.

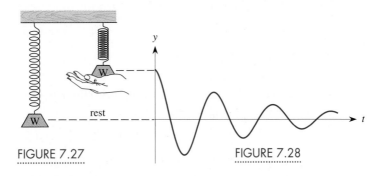

FIGURE 7.27   FIGURE 7.28

EXAMPLE 8

Figure 7.29 shows the results of recording the vertical oscillations of an object attached to a particular spring as a function of time from $t = 0$ to $t = 2\pi$. Construct a function that models this behavior pattern.

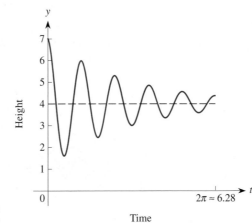

FIGURE 7.29

Solution   What type of function could have this kind of behavior pattern? The oscillatory effect certainly suggests a sinusoidal function, either a sine or a cosine, but the amplitude is not constant. In fact, the oscillations eventually die out. The overall effect of the decreasing amplitude might suggest either a decaying power function or an exponential decay function. Because there is a finite starting value, not a vertical asymptote, at time $t = 0$, a power function is not appropriate. An exponential decay term makes more sense. Moreover, based on our discussion in Section 4.6, we might be tempted to consider the product of such an exponential decay function and a sinusoidal function. Two possible formulas for functions that combine these two behavior patterns are

$$y = Ab^t\sin ct \quad \text{or} \quad y = Ab^t\cos ct,$$

with $b < 1$. You can think of the decaying exponential function as a variable amplitude that decays to 0 over time. In addition, there is a vertical shift, so the possible functions are $y = D + Ab^t \sin ct$ or $y = D + Ab^t \cos ct$. Our task is to determine values for the four parameters A, b, c, and D.

First, we note that the initial height of the object is approximately $y = 7$. Starting from there, the object drops at first. Its height decreases from a maximum, which suggests a cosine function rather than a sine function.

Second, we note that the final, or equilibrium, height for the object is about 4, so the object seems to be oscillating about a height of $y = 4$. The maximum height is 7, or 3 units above this equilibrium level, so the form for the function seems to be

$$y = f(t) = 4 + 3b^t \cos ct,$$

where $f(0) = 4 + 3b^0 \cos 0 = 4 + 3(1)(1) = 7$.

Further, we estimate from the graph that, between $t = 0$ and $t = 2\pi \approx 6.28$, there are about five complete diminishing cycles. Thus the frequency for the cosine function is approximately 5, giving the equation

$$y = f(t) = 4 + 3b^t \cos 5t.$$

Finally, consider the exponential decay curve $g(t) = 3b^t$ that is superimposed over the successive peaks of the decaying sinusoidal function in Figure 7.30. It starts with an initial height of 7 and decays to a final level of 4. Using a ruler, we can estimate that it has dropped halfway (to a height of 5.5) at about $t = 2$, Therefore we use $t = 2$ as an approximation for the half-life of the pure exponential decay function $g(t) = 3b^t$. Thus we must solve the equation

$$g(2) = 3b^2 = \tfrac{1}{2}(3),$$

which gives

$$b^2 = 0.5 \quad \text{so that} \quad b = \sqrt{0.5} \approx 0.707.$$

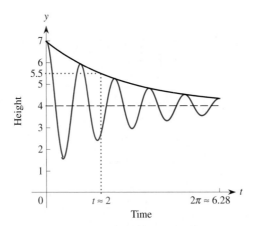

FIGURE 7.30

We can now estimate that the desired function is given by

$$y = f(t) = 4 + 3(0.707)^t \cos 5t.$$

Verify that this function matches the required pattern for the oscillation shown in Figure 7.29 by using your function grapher to graph f between 0 and 2π.

Incidentally, we have implicitly assumed throughout this section that all periodic processes follow a sinusoidal pattern (either a sine or a cosine curve) precisely. In practice, this assumption may be expecting a lot. Think about the length of

Janis's fingernails in Problem 1 from Section 7.1. The nail length is a periodic function, but it is not sinusoidal. Even if you observe that the overall pattern for some periodic process is smooth and appears to be that of a sine curve, you have no guarantee that the behavior is exactly sinusoidal. Nevertheless, sinusoidal functions are your best models for such types of periodic phenomena and consequently are the models that you should use when faced with such behavior.

Finally, just as you can construct linear, exponential, power, and other functions to fit a set of data, you often will be faced with the problem of having a set of data that exhibits a periodic pattern and wanting to find the periodic function that best fits the data. We ask you to explore several such cases in the following Problems.

Problems

1. Decide which of the following functions are periodic. For those that are periodic, what is the period? (Assume that each graph continues in the same pattern indefinitely to the left and right.)

2. Find the number of hours of daylight in San Diego on March 1, on May 12, on July 4.

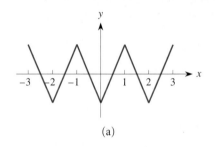

(a)

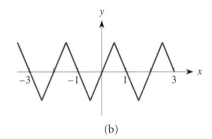

(b)

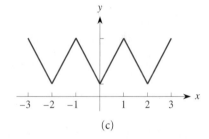

(c)

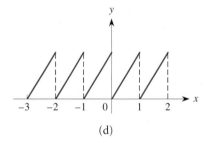

(d)

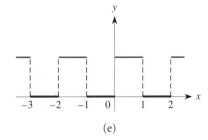

(e)

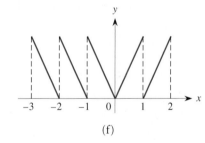

(f)

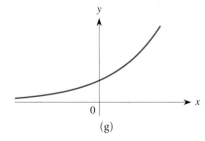

(g)

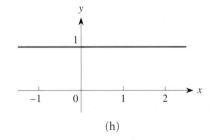

(h)

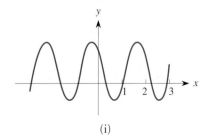

(i)

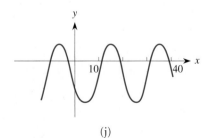

(j)

3. The number of hours of daylight in Montreal is given by

$$H(t) = 12 + 3.6 \sin\left[\frac{2\pi}{365}(t - 80)\right],$$

where t is the number of days from the 1st of the year.

 a. What is the amplitude of this function?
 b. What is the period of this function?
 c. What is the number of hours of daylight on the shortest day of the year?
 d. What is the number of hours of daylight on the longest day of the year?

4. The shortest day of the year in Fairbanks, Alaska, has 3.70 hours of daylight. Find a formula for the number of hours of daylight there on any day of the year.

5. Write a formula giving the number of hours of darkness in San Diego as a function of the day of the year.

6. Consider Example 1 regarding the height of the tide at a dock. Suppose that low tide still occurs at 4 A.M. but that high tide actually occurs at 10:30 A.M. Find an equation for the height of the tide as a function of time t.

7. The Bay of Fundy in eastern Canada is known for the highest tides in the world. The tides there rise and fall by as much as 50 feet. If the tidal cycle takes 11 hours, find a sinusoidal function that models the tides in the bay. For convenience, assume that low tide corresponds to a height of 0.

8. The thermostat in Sylvia's home in Baltimore is set at 66°F. Whenever the temperature drops to 66° (roughly every 30 minutes), the furnace comes on and stays on until the temperature reaches 70°.

 a. Write a sinusoidal function that models this situation.
 b. Gary's thermostat in upper New York State is set the same way. How would the model you created in part (a) change to reflect Gary's climate?
 c. Jodi, who lives in central Florida, likewise has her thermostat set to come on at 66°F. How would you change the models you created for parts (a) and (b) to reflect her climate?
 d. Is a sinusoidal function necessarily a good model? Explain. (*Hint:* Think about the *rates* at which the temperature increases and decreases.)

9. Ocean waves move in a roughly sinusoidal pattern. As a rule of thumb, the length of a wave (crest to crest, say) on the open seas is about 20 times the height of the wave (trough to crest). (This rule doesn't apply near coastlines where waves are much choppier and their intervals shorter.)

 a. Write a formula for ocean waves that are 4 feet high in moderately calm seas.
 b. Write a formula for ocean waves that are 15 feet high in rough seas.

10. Meryl is a normal individual with a pulse rate of 72 beats per minute and a blood pressure of 120 over 80. Thus her heart is beating 72 times each minute and her blood pressure is oscillating between a low (diastolic) reading of 80 and a high (systolic) reading of 120. Assume that the oscillation in Meryl's blood pressure can be modeled by a sinusoidal function.

 a. What is the period of this sinusoid?
 b. What is the frequency of this sinusoid?
 c. What is the equation of this sinusoid?

11. Your Thanksgiving turkey is taken from a refrigerator at 40°F and placed in an oven set at 350°. Suppose that the temperature of the bird is 130° after 60 minutes. You know that an oven cycles on and off as some of the heat escapes. Suppose that the cycle occurs every 10 minutes and that the actual temperatures inside the oven oscillate between 340° and 360°.

 a. Use this information to construct a sinusoidal function to model the temperature of the oven as a function of time t.
 b. Use Newton's law of heating from Section 5.4 to estimate how large a variation is possible in the temperature of the turkey after 60 minutes, and after 100 minutes. (*Hint:* Solve the problem with the minimum and maximum oven temperatures.)

12. A standard radio has two bands—the AM (*amplitude modulation* band) and the FM (*frequency modulation* band). In one case, the amplitude of a sinusoidal wave is modulated (varied) to produce the desired output sounds; in the other, the frequency of a sinusoidal wave is modulated. Which of the following represents an AM sound and which represents an FM sound?

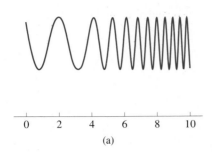

(a)

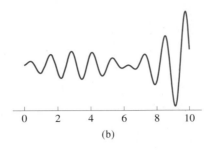

(b)

13. Two successive turning points of a sinusoidal function are at $(8, 72)$ and $(20, 30)$.

    a. Write a possible formula for this function, using a sine function.
    b. Write a possible formula for this function, using a cosine function.

14. Two successive inflection points of a sinusoidal function are at $(6, 20)$ and $(18, 20)$; the maximum attained by the function is 43.

    a. Write a possible formula for this function, using a sine function.
    b. Write a possible formula for this function, using a cosine function.

15. Suppose that the historical average daytime high temperature in Fairbanks ranges from a low of $-20°F$ to a high of $64°F$ and that the coldest day of the year, historically, is the 40th day. Write a formula for a sinusoidal function that can be used to model the average daytime high temperature in Fairbanks as a function of the day of the year.

16. Sketch by hand the graph of each function. Draw the basic curve $y = \sin x$ or $y = \cos x$ on the same set of axes for comparison. (Do not use your function grapher.)

    a. $y = 3 \sin 4x$ b. $y = 3 \sin\left(\frac{1}{2}x\right)$
    c. $y = 2 \sin 3x$ d. $y = 4 \cos 2x$
    e. $y = -3 \cos 2x$ f. $y = 4 + 2 \sin x$
    g. $y = \sin\left(x - \frac{\pi}{4}\right)$ h. $y = 3 \sin\left(2x - \frac{\pi}{6}\right)$
    i. $y = 4 + 2 \cos\left(x + \frac{\pi}{3}\right)$

17. Write a possible formula for each sinusoidal function $(a)-(l)$ from its graph.

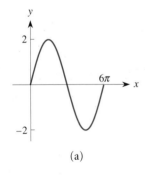

(a)

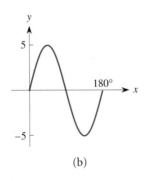

(b)

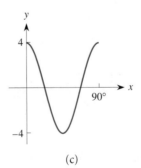

(c)

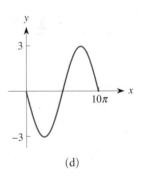

(d)

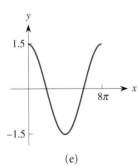

(e)

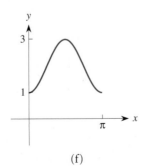

(f)

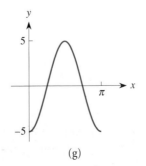

(g)

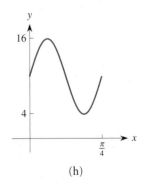

(h)

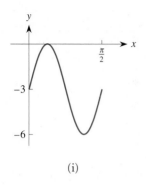
(i)

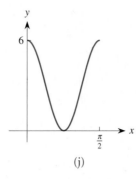
(j)

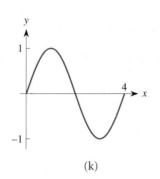
(k)

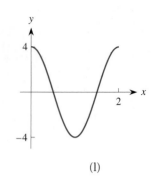
(l)

18. Identify a possible formula for each of the six sinu-soidal functions $f_1, f_2, \ldots, f_6$ whose values are given in the table. Note that there are many possible correct answers.

x	f_1	f_2	f_3	f_4	f_5	f_6
-6	0.279	2.279	0.559	0.537	-0.141	0.721
-5	0.959	2.959	1.918	0.544	-0.598	0.041
-4	0.757	2.757	1.514	-0.989	-0.909	0.243
-3	-0.141	1.859	-0.282	0.279	-0.997	1.141
-2	-0.909	1.091	-1.819	0.757	-0.841	1.909
-1	-0.841	1.159	-1.683	-0.909	-0.479	1.841
0	0.000	2.000	0.000	0.000	0.000	1.000
1	0.841	2.841	1.683	0.909	0.479	0.159
2	0.909	2.909	1.819	-0.757	0.841	0.091
3	0.141	2.141	0.282	-0.279	0.997	0.859
4	-0.757	1.243	-1.514	0.989	0.909	1.757
5	-0.959	1.041	-1.918	-0.544	0.598	1.959
6	-0.279	1.721	-0.559	-0.537	0.141	1.279

19. The table gives the outdoor temperatures in °F in Chicago during one 24-hour period.

Midnight	2 A.M.	4 A.M.	6 A.M.	8 A.M.	10 A.M.	
53	48	47	49	53	59	
Noon	2 P.M.	4 P.M.	6 P.M.	8 P.M.	10 P.M.	Midnight
66	71	68	65	58	54	53

If you were to fit a sinusoidal function to this set of data, what is the vertical shift? the amplitude? the period? the frequency? What is the equation of the resulting sinusoidal function?

Month	Jan	Feb	Mar	Apr	May	June	July	Aug	Sept	Oct	Nov	Dec
Avg. daily high temp. (°F)	65.2	64.4	65.9	67.8	68.6	71.3	75.6	77.6	76.8	74.6	69.9	66.1

20. The table above shows the average daytime high temperature each month in San Diego.
    a. Construct a sinusoidal function that best fits these data.
    b. How does the phase shift for this function compare to the phase shift used in the text for the number of hours of daylight in San Diego? In particular, explain in practical terms why the sinusoidal function for air temperature lags behind the function for hours of daylight.

21. The table gives the average daytime high temperature in Dallas on different days of the year (roughly every 2 weeks), based on historical weather records.

a. Assuming that the temperature behavior in Dallas is periodic from year to year, determine a sinusoidal function that models the average daytime high temperature in Dallas.
b. The values shown in the table are temperatures roughly every 2 weeks, but two entries are missing. Use your model from part (a) to predict the average daytime high temperature in Dallas on the missing dates.

Day	1	15	32	46	60	74	91	105	121	135	152
Avg. daily high temp. (°F)	55	53	56	59	63	67	72	77	81	84	89
Day	196	213	227	244	258	274	288	305	319	335	349
Avg. daily high temp. (°F)	98	99	98	94	90	85	80	72	66	61	58

22. The table shows the average number of tornados reported in the United States per month based on historical records.

Month	Jan	Feb	Mar	Apr	May	June	July	Aug	Sept	Oct	Nov	Dec
Tornados	16	24	60	111	191	179	96	66	41	26	31	22

Determine a sinusoidal function that models the monthly number of tornados as a function of time.

23. For a normal adult at rest, the rate R, in liters per second, at which air flows in and out of the lungs can be modeled by the function $R(t) = 0.85 \sin[(2\pi/5)t]$, where t is measured in seconds. The person is inhaling when $R > 0$ and exhaling when $R < 0$. How many times does the person breathe per minute?

24. Astronomers recently reported the discovery of the first known planets outside the solar system. They found three worlds orbiting around a pulsar, a rotating star that emits radiation with constant frequency. For this pulsar, the astronomers detected slight variations in the intensity of the radiation, as shown in the accompanying figure. This variation would be the effect of a planet in orbit about the pulsar.

a. From the figure, estimate the length of the year for the planet.
b. Use Kepler's law from Example 4 in Section 3.6 (assuming that the same coefficient applies) to calculate the distance from this planet to its star.
c. Assuming that the orbit of this new planet is circular, how fast is it moving in its orbit about the pulsar?
d. For comparison, Earth takes 365 days to complete one revolution about the sun at a distance of about 93 million miles. How fast is Earth moving in its orbit about the sun?

Variations (milliseconds)

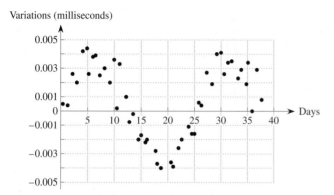

25. You are taking a ride on a Ferris wheel that is 200 feet in diameter and has a bottom point 10 feet above the ground. Suppose that the Ferris wheel rotates twice every minute and, from your friend's viewpoint on the ground, is rotating clockwise.

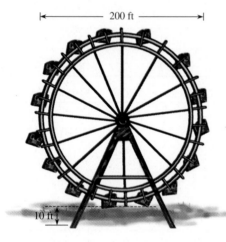

|← ——— 200 ft ——— →|

10 ft

a. Sketch your height y above the ground as a function of time t.
b. Find a formula for your height y as a function of t. Does it agree with your rough sketch in part (a)?
c. Find a formula for the horizontal distance x from the center of the wheel as a function of t.
d. Find all intervals of t values for which you are moving forward. Indicate these intervals on the graph of the function in part (c). What do you observe?
e. Suppose that the Ferris wheel rotates in the opposite direction, so it is now moving counterclockwise. How do your answers to parts (c) and (d) change?
f. Find a formula relating your height y above the ground and the horizontal distance x from the vertical axis through the center of the wheel.

26. A Ferris wheel is 12 meters in diameter and completes one full revolution every 20 seconds. If the bottom of the Ferris wheel is 2 meters above the ground, write a formula for the height above ground of a person on the Ferris wheel as a function of time.

27. Certain stars are called *variable stars* because their brightness increases and decreases in a periodic manner. The brightest variable star that can be observed from Earth is Delta Cephei, whose brightness varies between a minimum brightness (or magnitude) level of 3.65 and a maximum brightness of 4.35, with a cycle of 5.4 days. Write an equation that represents the brightness of Delta Cephei as a function of time t based on $t = 0$ at an instant when the star has minimal brightness.

28. Many people believe that a person's life is determined by three independent cycles, called *biorhythms*. One cycle, with a period of 23 days, represents the physical or health dimension of a person, $H(t) = \sin(2\pi t/23)$, where time t is measured in days starting at birth. A second cycle, with a period of 28 days, represents the emotional or sensitivity aspects of a person, $E(t) = \sin(2\pi t/28)$. A third cycle, with a period of 33 days, represents the mental or intellectual aspects of an individual, $M(t) = \sin(2\pi t/33)$.

a. Suppose that Tony was born on January 1. Consider the 60-day period immediately following his 20th birthday. What set of values for t are appropriate?
b. Which days would you recommend as being suitable for Tony to compete in a track-and-field meet?
c. Which days would you recommend as being good days for Tony to ask his girlfriend to marry him?
d. Which days could you suggest as days on which Tony could hope to have a major exam at school?
e. Are there any days when you would recommend that Tony simply not get out of bed?
f. Are there any days when all the signs are highly positive?

29. As part of a study on the possibility of global warming at a National Science Foundation math modeling workshop at Pellissippi State College, the accompanying scatterplot was produced. It suggests that the average global temperature values appear to oscillate about the regression line,

$T = 0.0042t + 14.67$, where t represents years since 1880.

a. Use the scatterplot to estimate the parameters for a sinusoidal function that oscillates above and below the indicated line. What is the equation of the resulting function?

b. Use your function grapher to draw the graph of that function. Does it have the correct shape?

c. What is your prediction for the average global temperature in 2005, based on the combination of the given linear function and the sinusoidal function you created?

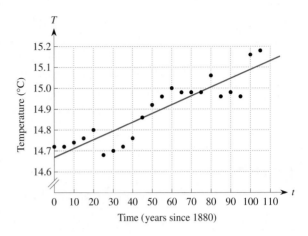

7.3 Solving Equations with Sine and Cosine: The Inverse Functions

We have shown that the number of hours H of daylight in San Diego as a function of the day of the year t is given by

$$H(t) = 12 + 2.4 \sin\left[\frac{2\pi}{365}(t - 80)\right].$$

Suppose that we now ask: When will there be 13 hours of daylight? That is, on which day t of the year will $H = 13$? To find this date, as illustrated in Figure 7.31, we must solve for the independent variable t in the equation

$$H(t) = 12 + 2.4 \sin\left[\frac{2\pi}{365}(t - 80)\right] = 13.$$

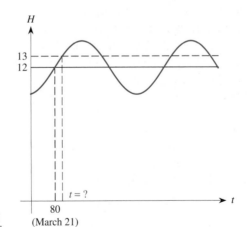

FIGURE 7.31

We can solve such an equation in a variety of ways, as we illustrate in Examples 1 and 2.

◆ **EXAMPLE 1** ⋯⋯

Determine graphically all days on which there will be 13 hours of daylight in San Diego.

Solution The function $H(t)$ oscillates between 9.6 and 14.4, so there are 13 hours of daylight on two different days each year, as indicated in Figure 7.31. One of these days occurs in the spring when the days are lengthening; the other occurs during the fall when the days are shortening. When we trace the graph of this function, we find that the two solutions are $t \approx 105$ and $t \approx 238$. The 105th day of the year is April 15 (31 days in January + 28 in February + 31 in March + 15 in April = 105). The 238th day of the year is August 26. Moreover, these same values will occur *every* year because the function is periodic. Check this result on your calculator by evaluating $H(105)$ and $H(238)$.

◆

EXAMPLE 2

Determine algebraically when there will be 13 hours of daylight in San Diego.

Solution   We start with the equation

$$H(t) = 12 + 2.4 \sin\left[\frac{2\pi}{365}(t - 80)\right] = 13.$$

To solve algebraically for t, we first subtract 12 from both sides:

$$2.4 \sin\left[\frac{2\pi}{365}(t - 80)\right] = 1.$$

We next divide both sides by 2.4:

$$\sin\left[\frac{2\pi}{365}(t - 80)\right] = \frac{1}{2.4} = 0.417.$$

Our task now is to extract the variable t from the argument of the sine function.

Compare this problem to the situation we repeatedly faced of extracting the variable from an exponential function, such as 10^x. We solved that problem by using logarithms to undo the exponential function. The reason that this method works is because the exponential and logarithmic functions are inverse functions of one another.

Similarly, we can undo the sine function by using the *inverse sine function.* You can do so on your calculator by pressing either INV or 2nd followed by SIN to get the arcsine function. When you do this in radian mode, you will find that

$$\arcsin 0.417 \approx 0.430;$$

that is, the value whose sine is 0.417 is 0.430 radians. Therefore

$$\frac{2\pi}{365}(t - 80) = \arcsin 0.417 \approx 0.430.$$

To solve for t, we now multiply both sides by 365 and get

$$2\pi(t - 80) = (0.430)(365) = 156.95.$$

Dividing both sides by 2π yields

$$t - 80 = \frac{156.95}{2\pi} = 24.98 \approx 25.$$

Hence

$$t = 25 + 80 = 105.$$

That is, there will be 13 hours of daylight in San Diego on approximately the 105th day of the year (April 15), which is the same answer we obtained graphically in Example 1.

◆

Actually, the result in Example 2 is not complete because it is only one of the two possible days each year on which the sinusoidal function H passes across the 13-hour level. However, this is the only solution that you can get directly from a calculator or computer when you use the inverse sine function. You can determine the other day when 13 hours of daylight occurs by using the following line of reasoning, based on some key facts about the sine function. We found that the solution $t = 105$ corresponds to April 15, which is 25 days *after* the spring equinox on March 21. Using the symmetry of the sine curve, we should expect that there will also be 13 hours of daylight 25 days *before* the fall equinox on September 21. But 25 days before September 21 is August 27, which is roughly the other solution we found in Example 1. Finally, because of the periodicity of the sine function, there will be 13 hours of daylight in San Diego on April 15 and August 27 every year.

The Inverse Sine Function

Let's now examine more carefully what the inverse sine function is all about and the reason for the limitation in Example 2. Recall that a continuous function f has an inverse f^{-1} when it is either strictly increasing or strictly decreasing or, equivalently, if it satisfies the horizontal line test. Obviously, the sine function does not fulfill either of these conditions because of its shape. The only way to obtain an inverse for the sine function is to restrict its domain, as we did for the parabola in Section 2.9 when we considered only the right side of the parabola $y = x^2$. We thus use only a small portion of the sine curve $y = \sin\theta$—where the function is strictly increasing. By convention, the restricted domain for the sine function is from $\theta = -\pi/2$ to $\theta = \pi/2$, as depicted in Figure 7.32.

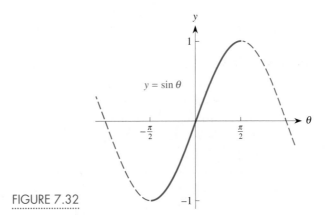

FIGURE 7.32

Let's work temporarily in degrees. Suppose that

$$\sin\theta = 0.825$$

and θ is some unknown value. From the graph of the sine function, we expect that θ will be closer to 90° than to 0°. The inverse sine function, arcsin y or $\mathrm{Sin}^{-1}y$, allows us to solve for the correct value of θ. That is, if

$$\sin\theta = 0.825,$$

then

$$\theta = \arcsin(0.825) = 55.59°.$$

Equivalently, if we use radians instead of degrees, then

$$\theta = \arcsin(0.825) = 0.9702 \text{ radians}.$$

Thus the inverse sine function undoes the sine function to extract any value θ between $-90°$ and $90°$ or, equivalently, between $-\pi/2$ and $\pi/2$ radians. Check this result by taking the sine of $55.59°$ or 0.9702 radians. It may be helpful to think of $\arcsin y$ as *the angle whose sine is y.*

In general, for any function f, if $y = f(x)$ and $x = g(y) = f^{-1}(y)$ are inverse functions, we know that

$$f^{-1}(f(x)) = x, \qquad \text{for all } x \text{ in the domain of } f;$$
$$f(f^{-1}(y)) = y, \qquad \text{for all } y \text{ in the domain of } f^{-1}.$$

In the case of the sine function and its inverse, we have

$$\arcsin(\sin \theta) = \theta, \qquad -\frac{\pi}{2} \le \theta \le \frac{\pi}{2};$$
$$\sin(\arcsin y) = y, \qquad -1 \le y \le 1.$$

As we found previously, when we wanted to find the day on which there would be 13 hours of daylight in San Diego, the inverse sine function returned only $t = 105$. This is the only answer we get because we must restrict the domain in order to have an inverse function.

Let's try a few values for y to see the effects of the inverse sine function.

$$\text{If } y = 0.2, \quad \arcsin 0.2 = 11.5°.$$
$$\text{If } y = 0.6, \quad \arcsin 0.6 = 36.9°.$$
$$\text{If } y = 0.95, \quad \arcsin 0.95 = 71.8°.$$
$$\text{If } y = -0.4, \quad \arcsin(-0.4) = -23.6°.$$
$$\text{If } y = -0.88, \quad \arcsin(-0.88) = -61.6°.$$

What if you try to find $\arcsin(2.5)$? Most calculators will return an error message, usually indicating a problem with the domain. The reason is that you are trying to find a number whose sine is 2.5. But the only permissible values for the sine function are between -1 and 1. Thus, if you try to use any value outside this interval, the inverse sine function is not defined and you will get an error message on most calculators. (Some models return a complex number instead of an error message, but that is a topic for considerably more advanced mathematics courses.)

You have seen that when you use any value for y between -1 and 1, the inverse sine function returns an answer between $-90°$ and $90°$ in degree mode or between $-\pi/2$ and $\pi/2$ in radian mode. These values are called the *principal values* of the inverse sine. The inverse sine function does not give any values larger than $90°$ (or $\pi/2$) or smaller than $-90°$ (or $-\pi/2$). It is up to you to realize that, for any real number k between -1 and 1, there are infinitely many values for θ whose sine is k. You can find these values by visualizing the graphs of the sine function and the horizontal line $y = k$ and determining all points where they intersect, that is, where $\sin \theta = k$. You can also use what you know about the symmetry of each arch of the sine curve. Finally, you can always estimate these values graphically, an approach that is usually straightforward. In Examples 3–5 we demonstrate the algebraic approach and the associated reasoning, because it is more difficult than the graphical approach.

EXAMPLE 3

Find all values of θ, in degrees, for which $\sin \theta = 0.6$.

Solution   If $y = \sin \theta = 0.6$,

$$\arcsin 0.6 = 36.9° \approx 37°,$$

which occurs while the sine curve is increasing. We have to find the second solution which occurs while the sine curve is decreasing. We know that the sine curve reaches its maximum height of 1 at 90° and is symmetric about 90°. Therefore, because 37° is 53° *before* 90°, a second value at which the sine function reaches the same height of 0.6 occurs 53° *after* 90°, or at $\theta = 143°$, as shown in Figure 7.33. (Verify that $\sin 143° \approx 0.6$ with your calculator.)

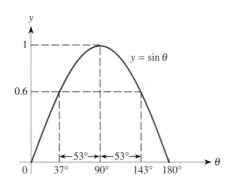

FIGURE 7.33

Further, because of the periodicity of the sine curve, we know that the same behavior pattern will repeat every 360°. Therefore, other values of θ whose sine is 0.6 are at $37° + 360° = 397°$, at $143° + 360° = 503°$, at $37° + 2(360°) = 757°$, at $143° + 2(360°) = 863°$, and so on. Verify some of these values with your calculator as well.

◆

In Example 5 of Section 7.2, we created the sinusoidal function

$$T = 71 + 3\sin\left(\frac{\pi}{10}t\right)$$

to model the temperature in a house where the air conditioning control is set to turn on the air conditioner when the temperature rises to 74°F and to turn it off when the temperature drops to 68°F, a cycle that repeats every 20 minutes. Also, we were told that, at noon, the temperature was 71°F and rising. We now consider some inverse predictions based on this model.

EXAMPLE 4

Use the sinusoidal model to determine all times between noon and 1 P.M. when the temperature in the house is 70°F.

Solution   We need to solve the equation

$$T = 71 + 3\sin\left(\frac{\pi}{10}t\right) = 70.$$

We show an algebraic solution. We first subtract 71 from both sides and get

$$3\sin\left(\frac{\pi}{10}t\right) = -1 \quad \text{so that} \quad \sin\left(\frac{\pi}{10}t\right) = -\frac{1}{3}.$$

Consequently, in radians

$$\frac{\pi}{10}t = \arcsin\left(-\frac{1}{3}\right) \approx -0.3398,$$

so that

$$t \approx \frac{10}{\pi}(-0.3398) \approx -1.08,$$

or about 1 minute *before* noon. The air conditioning cycle takes 20 minutes, so we conclude that the temperature will be 70° again about one minute *before* 12:20, or at about 12:19, again at about 12:39, and once more at about 12:59. All these values occur while the sine curve is increasing, as shown in Figure 7.34.

To find the times that the temperature is 70° while the sine curve is decreasing, we reason as follows. A complete cycle takes 20 minutes, so a half cycle takes 10 minutes. The first time the temperature reaches 70° while the curve is increasing is at about $t = 19$ minutes, or 1 minute *before* the end of the cycle. Therefore, from the symmetry of the sine curve over the first 20 minutes, as illustrated in Figure 7.35, the first time the sine curve passes the 70° level while the curve is decreasing must occur about 1 minute after the middle ($t = 10$ minutes) of the cycle. That is, the other solution is at about $t = 11$ minutes, or at about 12:11. Because of the periodic nature of the sine function, the 70° temperature will also occur at about 12:31 and at about 12:51.

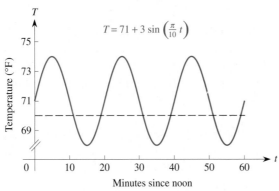

FIGURE 7.34

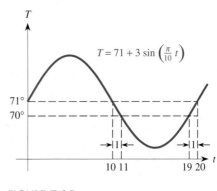

FIGURE 7.35

◆ **EXAMPLE 5**

In Example 6 of Section 7.2, we created the sinusoidal function

$$C = 4800 + 2400 \sin\left(\frac{\pi}{12}t\right)$$

to model a cricket's chirp rate C in chirps per hour, where t is measured in hours since 9 A.M. At what times, if any, does the cricket chirp at a rate of 6000 times per hour?

Solution   We first solve this problem graphically. The graph of C over a 24-hour period starting at 9 A.M. (when $t = 0$) is shown in Figure 7.36. Note the horizontal line at a height of 6000. The times when the cricket chirps at a rate of 6000 times per hour are the points at which the curve intersects the line. If we zoom in on the graph about these two points, we find that t is approximately $t = 2$ hours after 9 A.M. (or about 11 A.M.) and $t = 10$ hours after 9 A.M. (or about 7 P.M.).

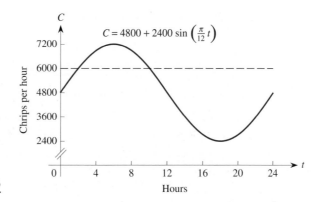

FIGURE 7.36

Alternatively, let's solve for t algebraically from

$$C = 4800 + 2400 \sin\left(\frac{\pi}{12}t\right) = 6000.$$

We subtract 4800 from both sides to get

$$2400 \sin\left(\frac{\pi}{12}t\right) = 1200$$

so that

$$\sin\left(\frac{\pi}{12}t\right) = \frac{1200}{2400} = 0.5.$$

Using the inverse sine function in radian mode, we get

$$\frac{\pi}{12}t = \arcsin(0.5) \approx 0.5236$$

so that

$$t \approx \frac{12}{\pi}(0.5236) \approx 2.0000,$$

or 2 hours after 9 A.M., which is 11 A.M., as we found graphically.

To find the other time of day when the cricket is chirping 6000 times per hour, we note that the period of the sinusoidal function is 24 hours, starting at 9 A.M., so a half cycle takes 12 hours. The symmetry of the sinusoidal function, as depicted in Figure 7.37, shows that the other time that the curve passes a height of 6000 must be 2 hours before the 12-hour mark at 9 P.M. (or 10 hours after 9 A.M.), which is at 7 P.M.

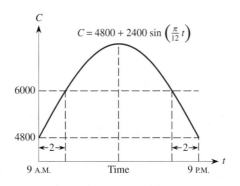

FIGURE 7.37

We summarize the important properties of the inverse sine function as follows.

> ## Properties of the Inverse Sine Function $y = \arcsin x$
>
> 1. $\arcsin x$ is defined only for values of x (the domain) between -1 and 1.
> 2. The principal values for y (the range) lie between $-\pi/2$ and $\pi/2$ radians.
> 3. $\arcsin(\sin y) = y,\qquad$ for $-\dfrac{\pi}{2} \le y \le \dfrac{\pi}{2}$.
> 4. $\sin(\arcsin x) = x,\qquad$ for $-1 \le x \le 1$.

In general, suppose that f^{-1} is the inverse of a function f. If we write the two functions in terms of the same independent variable x so that $y = f(x)$ and $y = f^{-1}(x)$, their graphs are mirror images of each other about the line $y = x$, as we showed in Section 2.9. (Visualize the graphs of the exponential and logarithmic functions.) Using this fact, we can easily construct the graph of the inverse sine function from our knowledge of the graph of the sine function. We begin with the sine curve shown in Figure 7.38(a), in which the portion corresponding to the domain from $-\pi/2$ to $\pi/2$ is highlighted. When we reflect the highlighted portion of the sine curve about the line $y = x$, we get the graph of the inverse sine function; both graphs are shown in Figure 7.38(b). Finally, the graph of the inverse sine function alone is shown in Figure 7.38(c). Note that $y = \arcsin x$ exists only for values of x between -1 and 1 and that the corresponding heights range from $-\pi/2$ to $\pi/2$.

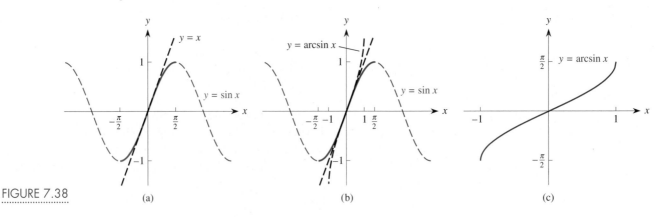

FIGURE 7.38   (a)   (b)   (c)

The Inverse Cosine Function

Suppose that

$$\cos \theta = 0.3$$

and we want to find a number θ whose cosine has this value. To solve this equation, we introduce the *inverse cosine function,* $y = \arccos x$, using ideas that are similar to those that led to the inverse sine function. We use the inverse cosine (press 2nd COS or INV COS on your calculator, in Degree mode) to get

$$\theta = \arccos 0.3 \approx 72.5°.$$

Verify that $\cos 72.5° \approx 0.3$.

We know that the only possible values for $y = \cos\theta$ lie between -1 and $+1$. For the inverse cosine function $\theta = \arccos y$, we must restrict the domain to a suitable portion of the cosine curve where the cosine function is either strictly increasing or strictly decreasing. By convention, we consider only values of θ between $0°$ and $180°$ (or equivalently between 0 and π radians) where the cosine function is strictly decreasing, as shown in Figure 7.39. These are the principal values for the inverse cosine and they are the only values that your calculator will return when you use the inverse cosine function. As with the inverse sine function, you will have to use what you know about the behavior of the cosine graph, including the symmetry of the arches on the graph, if you want to determine all other numbers having the specified cosine value.

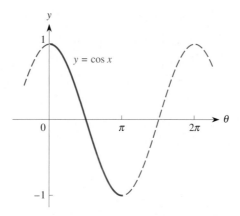

FIGURE 7.39

EXAMPLE 6

Find all values of θ in both degrees and radians for which $\cos\theta = 0.92$.

Solution   Although we can solve this problem graphically, we illustrate the details of the algebraic solution. Because $\cos\theta = 0.92$,

$$\theta = \arccos(0.92) \approx 23°$$

in degree mode. Because of the periodicity of the cosine function, we also know that the value of 0.92 will repeat every 360°, so the solutions include

$$\theta = 23°, \quad 23° + 360°, \quad 23° + 2(360°), \quad 23° + 3(360°), \ldots,$$

or, in general,

$$\theta = 23° + n \cdot (360°), \qquad \text{for any integer } n \geq 0.$$

Moreover the graph of the cosine function shown in Figure 7.40 indicates that there must be another value of θ just before 360° whose cosine is also 0.92. In particular, because our first solution is $\theta = 23°$, the other value must be 23° before 360°, or $360° - 23° = 337°$. Thus the solutions also include

$$\theta = 337°, \quad 337° + 360°, \quad 337° + 2(360°), \quad 337° + 3(360°), \ldots,$$

or, in general,

$$\theta = 337° + n \cdot (360°), \qquad \text{for any integer } n \geq 0.$$

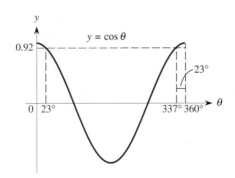

FIGURE 7.40

Using radian measure gives the equivalent solutions

$$\theta = 0.401 + 2n\pi \text{ radians} \quad \text{and} \quad \theta = 5.882 + 2n\pi \text{ radians,} \quad \text{for any integer } n \geq 0.$$

Moreover, the cosine function is symmetric about the vertical axis $\theta = 0$ (it is an even function; see Appendix D). Therefore we know that the same patterns repeat with negative values for θ, so all the other solutions are

$$\theta = -23° - n \cdot (360°) \quad \text{and} \quad \theta = -337° - n \cdot (360°), \quad \text{for any integer } n \geq 0.$$

Equivalently, with radian measure, $\theta = -0.401 - 2n\pi$ radians and $\theta = -5.882 - 2n\pi$, for any integer $n \geq 0$.

EXAMPLE 7

A clock is mounted on the wall with its center 7 feet (or 84 inches) above the floor. Suppose that the minute hand is 5 inches long.

a. Write a formula for the vertical height y of the arrowhead on the minute hand above or below the horizontal line through the center of the clock as a function of the time t in minutes from the instant that the minute hand is pointing vertically upward to the 12.

b. Determine all times during the first hour when the arrowhead on the minute hand is 2 inches above that horizontal line.

Solution

a. The midline, or vertical shift, is 84 inches above floor level. The minute hand is 5 inches long, so the height of the arrowhead on the hand oscillates between $84 - 5 = 79$ and $84 + 5 = 89$ inches above the floor, giving an amplitude of 5. This cycle repeats every 60 minutes, so the period is 60 and the frequency is $2\pi/60 = \pi/30$. Because $t = 0$ corresponds to the instant when the hand is pointing vertically upward, the initial height for the arrowhead is $y = 84 + 5 = 89$ inches. This suggests that we use a cosine function with a phase shift of 0 as our model. The resulting function is

$$y = 84 + 5\cos\left(\frac{\pi}{30}t\right).$$

b. To find all times when the arrowhead is 2 inches above the midline (equivalently, when the arrowhead is $84 + 2 = 86$ inches above the floor), we need to solve the equation

$$84 + 5\cos\left(\frac{\pi}{30}t\right) = 86.$$

Proceeding algebraically, we have

$$5 \cos\left(\frac{\pi}{30}t\right) = 2 \quad \text{so that} \quad \cos\left(\frac{\pi}{30}t\right) = \frac{2}{5} = 0.4.$$

Therefore, in radians,

$$\frac{\pi}{30}t = \arccos 0.4 \approx 1.1593$$

so that

$$t \approx \frac{30}{\pi}(1.1593) \approx 11.07,$$

or about 11 minutes after the hour, as illustrated in Figure 7.41.

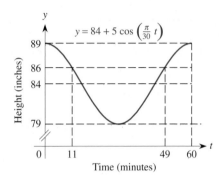

FIGURE 7.41

Because the first solution occurs at about 11 minutes after the hour, we know from the symmetry of the cosine curve that the same height must occur at about 11 minutes before the next hour, or at $t = 49$.

◆

We summarize the important properties of the inverse cosine function as follows.

Properties of the Inverse Cosine Function $y = \arccos x$

1. $\arccos x$ is defined only for values of x (the domain) between -1 and 1.
2. The principal values for y (the range) lie between 0 and π radians.
3. $\arccos(\cos y) = y$, for $0 \leq y \leq \pi$.
4. $\cos(\arccos x) = x$, for $-1 \leq x \leq 1$.

Finally, as with the graph of the inverse sine function, the graph of the inverse cosine function $y = \arccos x$ is the mirror image of the cosine graph about the line $y = x$, as shown in Figure 7.42. Note that the inverse cosine is defined only for x between -1 and 1 and that the inverse cosine values lie between 0 and π.

As a final note, you may find the names arcsine and arccosine to be rather strange. To see where they come from, think about how we defined radians. In a unit circle, we measured a length of 1 and defined the corresponding angle to be 1 radian. The same is true for any angle—its measure in radians equals the length of arc along the unit circle. So, to solve $\sin \theta = a$, say, we find the angle θ that equals the length of an arc on the circle corresponding to the value of a. Thus we have arcsin a.

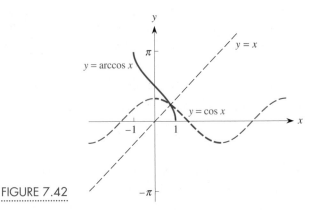

FIGURE 7.42

Problems

1. On what days of the year will San Diego have 11 hours of daylight? 10 hours of daylight? 9 hours of daylight?

2. The height of water at a dock is given by the formula $h(t) = 10 + 4\sin(\pi t/6)$, where t is measured in hours since midnight.

   a. When does high tide occur?
   b. When does low tide occur?
   c. When does the water level reach 8 feet? 10 feet? 11 feet? 12 feet?

3. An air conditioner is being used to cool a room. The temperature T oscillates according to the formula $T(t) = 69 + 3\sin(\pi t/10)$, where t is measured in minutes after 9 A.M.

   a. At about what temperature is the thermostat set? (i.e., when does the air conditioner kick in?)
   b. At about what temperature does the air conditioner kick out?
   c. When does the room temperature reach 70°F? 67°F?

4. The thermostat in an apartment is set to turn the heat on when the temperature falls to 64°F and to turn it off when the temperature rises to 70°F. This cycle takes 15 minutes.

   a. Write a formula for the temperature T as a function of time t, where t is the number of minutes after noon. Assume that the temperature at noon is 70°.
   b. Determine all times between noon and 1 P.M. when the temperature is 66°.
   c. Suppose that the temperature at noon is 67°. Repeat parts (a) and (b).

   d. Suppose that the temperature at noon is 68°. Repeat parts (a) and (b).

5. One of the dangers at places that have very high tides, such as Canada's Bay of Fundy, is the rate at which the tide can come in and potentially trap unwary visitors. Use the formula you devised for a sinusoidal function that models the heights of the tides at the Bay of Fundy in Problem 7 of Section 7.2 to determine how long it takes for the water level to rise 5 feet

   a. from a point of low tide.
   b. from a point at the average tide level.

6. A Ferris wheel is 12 meters in diameter and completes one full revolution every 20 seconds. The bottom of the Ferris wheel is 2 meters above the ground. In Problem 26 of Section 7.2, you were asked to write a formula for the height above ground of a person on the Ferris wheel as a function of time. Use that model to determine the times at which a person is 10 meters above the ground.

7. The historical average daytime high temperature in Fairbanks ranges from a low of -20°F to a high of 64°F, and the coldest day of the year, historically, is the 40th day. In Problem 15 of Section 7.2, you were asked to write a formula for a sinusoidal function that can be used to model the average daytime high temperature in Fairbanks as a function of the day of the year. Use this model to determine the days on which the high temperature in Fairbanks will be 0°.

8. The table below gives the outdoor temperatures in Chicago during one 24-hour period:

Time	Midnight	2 A.M.	4 A.M.	6 A.M.	8 A.M.	10 A.M.	Noon	2 P.M.	4 P.M.	6 P.M.	8 P.M.	10 P.M.	Midnight
Temp.(°F)	53	48	47	49	53	59	66	71	68	65	58	54	53

In Problem 19 of Section 7.2, you were asked to create the equation of the sinusoidal function that fits these data. Use that model to determine the times at which the temperature in Chicago will be (**a**) 50° and (**b**) 60°.

9. The table below shows the average daytime high temperature each month in San Diego. In Problem 20 of Section 7.2, you were asked to construct a sinusoidal function that fits these data. Use that model to determine the months on which the average daytime high temperature in San Diego will be (**a**) 65°, (**b**) 70°, and (**c**) 80°.

Month	Jan	Feb	Mar	Apr	May	June	July	Aug	Sept	Oct	Nov	Dec
Avg. daily high temp. (°F)	65.2	64.4	65.9	67.8	68.6	71.3	75.6	77.6	76.8	74.6	69.9	66.1

10. A 25-foot ladder is leaning against the side of a building and begins to slip. Write a formula for the angle θ that the ladder makes with the ground as a function of the distance x from the foot of the ladder to the building. Use your function grapher to draw the graph of this function. What are appropriate values for the domain of the function? Is the graph concave up or concave down? When is it maximum?

Exercising Your Algebra Skills

Several of the following equations do not have solutions. By inspection, decide which ones do not (give reasons) and then find the solutions to the remaining equations.

1. $\sin \theta = 4$
2. $\sin \theta = 0.4$
3. $3 \sin \theta = 4$
4. $4 \sin \theta = 3$
5. $4 \sin \theta = -3$
6. $5 \sin 2x = 3$
7. $5 \cos 2x = -3$
8. $3 \cos x = 2$
9. $5 \sin 2x = 3 \cos x$

7.4 The Tangent Function

We have considered many situations that can be modeled with either the sine or cosine function. In this section we return to the third trigonometric function, the *tangent*, and consider its properties and some applications. Recall from Section 6.1 that the tangent is defined by the ratio

$$\tan \theta = \frac{\text{opposite}}{\text{adjacent}}$$

of sides in a right triangle in terms of an angle θ, as shown in Figure 7.43.

The Graph of $y = \tan x$

As with any function, our first concern is to determine the graph of the tangent function to help us understand its behavior. The graph of the tangent function is shown in Figure 7.44, from which we observe the following characteristics.

1. The tangent function is periodic with period π; the tangent graph completes one full cycle between $x = -3\pi/2$ and $-\pi/2$, another full cycle between $-\pi/2$ and $\pi/2$, a third cycle between $\pi/2$ and $3\pi/2$, and so on. Each segment is called a *branch* of the graph.

2. The tangent function has zeros at $x = 0, \pm\pi, \pm2\pi, \ldots$

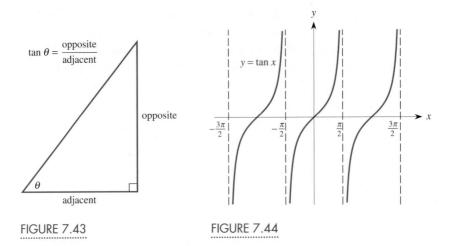

$$\tan \theta = \frac{\text{opposite}}{\text{adjacent}}$$

opposite

θ

adjacent

$y = \tan x$

FIGURE 7.43

FIGURE 7.44

3. The tangent graph has vertical asymptotes at $x = \pm\pi/2$, $\pm 3\pi/2$, $\pm 5\pi/2, \ldots$, so the tangent function is not defined at these points (which separate the branches).

4. The tangent is an increasing function for all intervals between successive vertical asymptotes.

5. The tangent curve is first concave down and then concave up on each branch. The point of inflection on each branch occurs at the point where the tangent curve crosses the x-axis: at $x = 0$, $\pm\pi$, $\pm 2\pi, \ldots$

To understand the behavior of the tangent function, we consider the fundamental relationship.

$$\tan x = \frac{\sin x}{\cos x}, \qquad \text{for any value of } x \text{ for which } \cos x \neq 0.$$

Because the tangent is the quotient of two functions, we can analyze this relationship in the same way that we analyzed the behavior of rational functions in Section 4.6. First, the tangent function must have a zero wherever the numerator, $\sin x$, is 0. This corresponds to $x = 0$, $\pm\pi$, $\pm 2\pi, \ldots$, which clearly agrees with what the graph of the tangent function shows. Second, the tangent function is undefined and therefore has a vertical asymptote wherever the denominator, $\cos x$, is 0. This occurs at $x = \pm\pi/2$, $\pm 3\pi/2$, $\pm 5\pi/2, \ldots$, which again agrees with what the graph of the tangent function shows.

Now let's see how these ideas help in understanding the graph of the tangent function. Consider what happens between $x = 0$ and $\pi/2$. The sine function (the numerator for $y = \tan x$) is positive and increasing toward 1, whereas the cosine function (the denominator for $y = \tan x$) is positive and decreasing toward 0. Because both are positive, $\tan x$ must be positive between 0 and $\pi/2$. Also, the ratio involves a numerator that is getting larger and a denominator that is getting smaller and approaching 0, so there is a vertical asymptote at $x = \pi/2$. The tangent is a positive function that increases toward ∞ as x approaches $\pi/2$ from the left.

Similarly, between $x = -\pi/2$ and $x = 0$, the sine function is negative and increasing toward 0, whereas the cosine function is positive and increasing toward 1.

Thus their ratio is negative and increases toward 0 as x increases toward 0. Moreover, as x approaches $-\pi/2$ from the right, the tangent ratio becomes ever more negative and eventually approaches $-\infty$.

Next, why is the period of the tangent function π when the periods for the sine and the cosine are both 2π? Visualize the sine curve from 0 to π and then from π to 2π, as shown in Figure 7.45. If you flip the second half of the curve over the x-axis, you get a curve identical to the first half. So the values for sin x between π and 2π are the same as those between 0 and π, but with the signs reversed. The same is true for the cosine between π and 2π—its values repeat those for the cosine between 0 and π, but with the signs reversed, as shown in Figure 7.46.

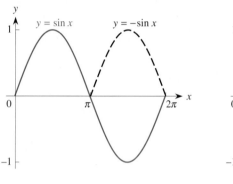

FIGURE 7.45

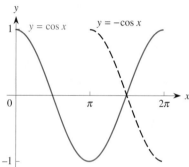

FIGURE 7.46

So, when we take the ratio sin x/cos x for x between π and 2π ($x \neq 3\pi/2$), the numerator and the denominator have the same numerical values as the ratio of sin x/cos x for x between 0 and π ($x \neq \pi/2$), but the signs of both the numerator and the denominator are reversed. In the quotient, these reversed signs cancel, so that the values for tan x from π to 2π match the corresponding values for tan x from 0 to π. But, if the same values are repeated, the function is periodic and therefore its period is π.

Finally, because the tangent function is periodic with period π, it repeats the behavior that we've outlined here, leading to the graph previously shown in Figure 7.44. For reference purposes, you should know that

$$\tan 0 = \frac{\sin 0}{\cos 0} = \frac{0}{1} = 0 \quad \text{and} \quad \tan \frac{\pi}{4} = \tan 45° = 1.$$

EXAMPLE 1

A video cameraman is taping a 100 meter dash down a straight track. He is positioned halfway along the track 40 meters from the inside lane where the race's favorite is running. He plans to focus his camera on the favorite throughout the race.

a. Write a formula for a function that models the distance d from the runner to the point A on the track as a function of the angle θ, as illustrated in Figure 7.47.

b. What is the runner's distance from the line extending from the cameraman to the middle of the track when the angle $\theta = 30°$?

c. What is the runner's distance from that line when $\theta = 0.6$ radian?

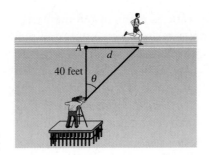

Solution

a. From Figure 7.47,

$$\tan \theta = \frac{d}{40} \quad \text{so that} \quad d = 40 \tan \theta = f(\theta).$$

b. When $\theta = 30°$,

$$d = f(30°) = 40 \tan 30° \approx 23.09 \text{ meters.}$$

c. When $\theta = 0.6$ radian,

$$d = f(0.6) = 40 \tan 0.6 \approx 27.37 \text{ meters.}$$

We can write more general tangent functions of the form

$$y = D + A \tan[B(x - C)],$$

where D is the vertical shift or midline, A is the amplitude, B is the frequency, and C is the phase shift. These ideas are the same as those that we encountered with general sinusoidal functions in Section 7.2. We explore some of these ideas for the tangent function in the Problems at the end of this section.

The Inverse Tangent

Suppose that we have an equation such as $\tan \theta = 1.5$. To find a value of θ that satisfies this equation, we use the *inverse tangent function,* arctan x, that gives the number whose tangent value is x. Using a calculator, we find

$$\theta = \arctan 1.5 \approx 0.9828 \text{ radian} \approx 56.31°.$$

As with the inverse sine and inverse cosine functions, we have to restrict the domain of the tangent function in order to define the inverse tangent function. By convention, the principal values for the tangent function are from $-\pi/2$ to $\pi/2$ where the tangent function is strictly increasing. Accordingly, a calculator returns a value only between $-\pi/2$ and $\pi/2$ (or between $-90°$ and $90°$) for the inverse tangent.

We summarize the important properties of the inverse tangent function as follows.

Properties of the Inverse Tangent Function $y = \arctan x$

1. $\arctan x$ is defined for values of x (the domain) between $-\infty$ and ∞.
2. The principal values for y (the range) lie between $-\pi/2$ and $\pi/2$ radians.
3. $\arctan(\tan y) = y$, for $-\pi/2 < y < \pi/2$.
4. $\tan(\arctan x) = x$, for $-\infty < x < \infty$.

Finally, as with the graphs of the other inverse functions, the graph of the inverse tangent function $y = \arctan x$ is the mirror image of the tangent graph about the line $y = x$, as shown in Figure 7.48, where x goes from -15 radians to 15 radians. Note how the curve levels off to the right at a height of $\pi/2 \approx 1.57$ and to the left at a height of about $-\pi/2 \approx -1.57$. These are a pair of horizontal asymptotes.

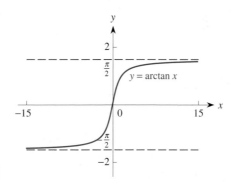

FIGURE 7.48

If you need other values of θ outside the interval from $-\pi/2$ to $\pi/2$, you will have to determine them by using what you know about the symmetry of the graph of the tangent function.

EXAMPLE 2

You enter a movie theater that has a screen 20 feet high positioned 5 feet above your eye level. If you sit too far back in the theater, the screen appears too small because your viewing angle is too small. If you sit too close to the screen, the picture will seem distorted because your viewing angle is again too small.

a. Find a formula giving the viewing angle θ as a function of your distance d from the screen, as illustrated in Figure 7.49.

b. What is your viewing angle θ if you sit 40 feet back from the screen?

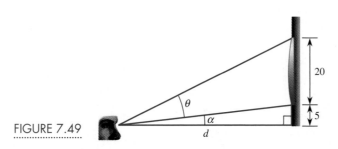

FIGURE 7.49

c. Does the viewing angle increase or decrease if you move farther back? In particular, estimate the distance you should sit from the screen to get the largest possible viewing angle.

Solution

a. Your viewing angle θ is the angle subtended by the screen. There is no direct way to get an expression for θ, because the associated triangle shown in Figure 7.49 is not a right triangle. Thus you have to get an expression for θ in a somewhat indirect manner. To do so, introduce the angle α shown in Figure 7.49 representing the angle from your eye level vertically upward to the bottom of the screen. This gives you two right triangles. In the smaller triangle,

$$\tan \alpha = \frac{5}{d} \quad \text{so that} \quad \alpha = \arctan \frac{5}{d}.$$

In the larger triangle,

$$\tan(\theta + \alpha) = \frac{25}{d} \quad \text{so that} \quad \theta + \alpha = \arctan \frac{25}{d}.$$

Consequently, the desired expression for θ is

$$\theta = (\theta + \alpha) - \alpha = \arctan \frac{25}{d} - \arctan \frac{5}{d} = f(d).$$

(Note that these two terms *cannot* be combined algebraically.)

b. For $d = 40$ feet,

$$\theta = f(40) = \arctan \frac{25}{40} - \arctan \frac{5}{40} \approx 0.4342 \text{ radians,}$$

or about 24.9°.

c. To determine what happens to this viewing angle θ as you move farther back, just replace the 40-foot distance with somewhat larger values—say, 41 or 45 feet. Alternatively, graph the function f that gives the angle θ as a function of the distance d. If you graph this function on the interval from 0 to 50, say, as shown in Figure 7.50, you can determine the behavior of this function for θ more thoroughly. The function increases rapidly, starting at $d = 0$, and rises to a maximum viewing angle when d is approximately 11 feet. You can verify this result on your calculator. Then the function slowly decreases as d increases thereafter. Therefore if you move farther back from the screen, the viewing angle will decrease.

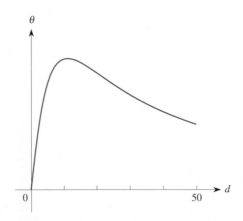

FIGURE 7.50

Finally, we note that it is possible to consider three other trigonometric functions—the *cotangent,* the *secant,* and the *cosecant*—which are just reciprocals of the tangent, the cosine, and the sine functions, respectively. These functions are defined as

$$\cot \theta = \frac{1}{\tan \theta}, \quad \sec \theta = \frac{1}{\cos \theta}, \quad \text{and} \quad \csc \theta = \frac{1}{\sin \theta}.$$

Be sure that you understand that these are reciprocals of the basic trigonometric functions. They have nothing to do with the associated inverse functions.

The cosecant, secant, and cotangent functions have been useful in the past because they simplified the hand calculations required in working with certain trigonometric problems. However, with technology, working with the actual reciprocals is just as easy as using $\csc \theta$, $\sec \theta$, and $\cot \theta$. Consequently, the cotangent, secant, and cosecant functions gradually are being laid to rest, and we don't consider them further.

Problems

1. Each of the figures (a)–(f) shows the graph of $y = \tan x$, where x is in degrees.

   a. Write an equation for a tangent function with frequency 2 and sketch its graph superimposed over the graph of $y = \tan x$ in Figure (a).

   b. Write an equation for a tangent function with frequency $\frac{1}{2}$ and sketch its graph superimposed over the graph of $y = \tan x$ in Figure (b).

   c. Write an equation for a tangent function with amplitude 3 and sketch its graph superimposed over the graph of $y = \tan x$ in Figure (c).

   d. Write an equation for a tangent function with amplitude -2 and sketch its graph superimposed over the graph of $y = \tan x$ in Figure (d).

   e. Write an equation for a tangent function with phase shift of 30° and sketch its graph superimposed over the graph of $y = \tan x$ in Figure (e).

   f. Write an equation for a tangent function with vertical shift of -10 and sketch its graph superimposed over the graph of $y = \tan x$ in Figure (f).

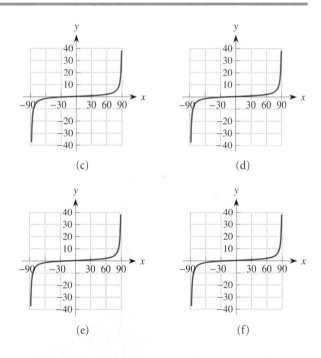

(c)

(d)

(e)

(f)

2. Write a possible formula involving tangent functions for each function (a)–(c) shown.

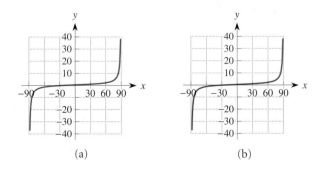

(a)

(b)

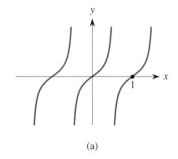

(a)

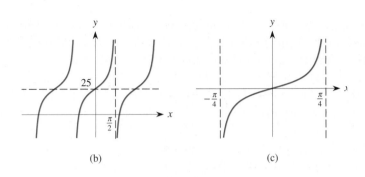

(b)   (c)

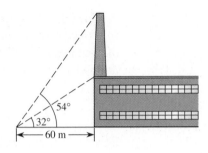

3. Write a possible formula involving a tangent function for the function whose values are given in the table.

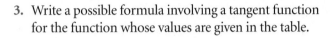

θ	$-\pi/3$	$-\pi/4$	$-\pi/6$	0	$\pi/6$	$\pi/4$	$\pi/3$
$f(\theta)$	UNDEF	-2.414	-1	0	1	2.414	UNDEF

4. The Statue of Liberty is 46 meters tall and stands on a base that is also 46 meters tall. Find an expression for the angle subtended by the statue from ground level as a function of distance from the base of the statue. Use this function to estimate graphically the distance when the angle is maximum. Approximately what is this maximum angle?

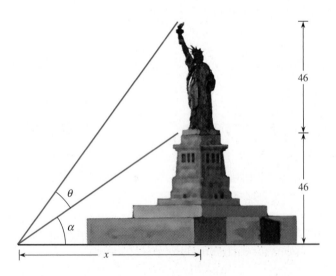

5. A tall smokestack extends from the roof of a large industrial plant. At a point 60 meters from the base of the building, the angle of elevation to the roof (the bottom of the smokestack) is 32°, and the angle of elevation to the top of the smokestack is 54°. Find both the height of the building and the height of the smokestack.

6. The lines $y = x$, $y = 2x$, $y = 3x$, and $y = 4x$ all pass through the origin. Find the angle each line makes with the x-axis.

7. **a.** For the general equation of a line through the origin $y = mx$, interpret the meaning of the slope m in terms of trig functions.
   b. What is the significance of the slope m in $y = mx + b$ from this point of view?

8. Use the graph of $y = \sin x$ to sketch the graph of its reciprocal function $y = 1/\sin x$. (This is the cosecant function.)

9. Use the graph of $y = \cos x$ to help you sketch the graph of its reciprocal function $y = 1/\cos x$. (This is the secant function.)

10. Use the graph of $y = \tan x$ to sketch the graph of $y = 1/\tan x$. (This is the cotangent function.)

11. A 5-foot high painting is hanging on the wall of an art museum when a photographer takes a picture of it. The lens of his camera is 1 foot below the bottom of the painting when he snaps the picture.

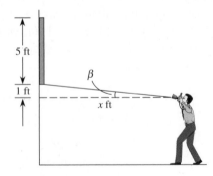

   a. Find a formula for the angle β subtended by the painting at the camera's lens at a distance of x feet from the wall.
   b. Using your function grapher, estimate the distance x from the wall at which the photographer

should position his camera to subtend the greatest possible angle with the painting.

12. A TV cameraman is videotaping the liftoff of the space shuttle. The cameraman is positioned at ground level 500 meters from the launch pad and is tracking the shuttle as it rises.

   a. Write a formula for the angle of inclination α to the shuttle as a function of the height y of the shuttle above the ground.

   b. Find the angle of inclination α when the shuttle is 1000 meters high.

   c. Find the angle of inclination α when the shuttle is 2000 meters high.

13. According to Einstein's theory of relativity, the mass M of an object increases as its speed v increases according to the formula

$$M = f(v) = \frac{M_0}{\sqrt{1 - \dfrac{v^2}{c^2}}} = M_0 \cdot \left(1 - \frac{v^2}{c^2}\right)^{-1/2},$$

where M_0 is the mass of the object at rest ($v = 0$) and c is the speed of light (about 186,282 miles per second). Suppose that an object has a rest mass of $M_0 = 1$ unit.

   a. Construct a table of values for the mass of the object for each of the following speeds expressed as a fraction of the speed of light: $v = 0$, $0.5c$, $0.9c$, $0.95c$, $0.99c$, and $0.999c$.

   b. Sketch a graph showing the behavior of the mass M of an object as its speed approaches the speed of light.

   c. The speed of light is the physical equivalent of a vertical asymptote. Write the formula for a function involving the tangent that can be used to model the mass of an object as a function of its speed expressed as a fraction of the speed of light.

Exercising Your Algebra Skills

Solve each trigonometric equation.

1. $4 \tan \theta = 5$
2. $5 \tan \theta = 4$
3. $\cos \theta = -\sin \theta$
4. $2 \cos \theta = \sin \theta$
5. $5 \sin \theta = 4 \cos \theta$
6. $4 \sin \theta = 5 \cos \theta$
7. $\sin \theta + \cos \theta = 0$
8. $4 \sin \theta - 3 \cos \theta = 0$

Chapter Summary

In this chapter, we introduced the use of the sine and cosine functions for modeling periodic phenomena and the tangent function. In particular, we discussed the following:

◆ The behavior of the sine and cosine functions.

◆ How to convert between radian measure and degree measure.

◆ What the *vertical shift* or *midline* means for the sine and cosine functions.

◆ What *amplitude* means for the sine and cosine functions.

◆ What *frequency* means for the sine and cosine functions.

◆ What *period* means for the sine and cosine functions.

◆ What *phase shift* means for the sine and cosine functions.

◆ How to use the sine and cosine functions to model periodic behavior.

◆ How to fit sine and cosine functions to data.

◆ The behavior of the inverse sine and inverse cosine functions.

◆ How to solve trigonometric equations, using the inverse sine and inverse cosine functions.

◆ The behavior of the tangent function.

◆ The behavior of the inverse tangent function.

◆ How to solve trigonometric equations, using the inverse tangent function.

Review Problems

1. The student with whom you are working finishes a problem and announces her answer is $\cos(5.70)$. You get an answer in the form $\sin(7.2708)$. Under what circumstances are these answers the same?

2. Suppose that θ is 60°.

   a. Find two positive angles and two negative angles that have the same sine as θ.

   b. Write the angles from part (a) in radian form.

3. Let $\theta = 45°$.

   a. Find two positive angles and two negative angles with the same cosine as θ.

   b. Write the radian form of the angles from part (a).

For each sinusoidal function in Problems 4–11, identify the vertical shift, the amplitude, the frequency, the period, and the phase shift.

4. $y = 325 + 10 \sin\left(\dfrac{2\pi}{9}t\right)$

5. $y = 63 + 3 \sin\left(\dfrac{2\pi}{25}t\right)$

6. $y = 71 + 2 \cos\left(\dfrac{2\pi}{15}t\right)$

7. $y = 80 + 13 \cos\left[\dfrac{2\pi}{24}(t - 15)\right]$

8. $y = 38 + 8 \sin\left[\dfrac{2\pi}{24}(t - 5)\right]$

9. $y = 100 + 25 \sin\left(\dfrac{2\pi}{72}t\right)$

10. $y = 100 + 25 \sin\left(\dfrac{2\pi}{97}t\right)$

11. $y = 145 + 40 \sin\left(\dfrac{2\pi}{83}t\right)$

12–19. Each of the functions in Problems 4–11 can be a model for a common periodic phenomenon. For each function,

   a. describe a phenomenon that each function could model.

   b. What do the variables represent?

   c. What are the units?

   d. What are possible values for the domain and range?

20. Bernice is swinging on a playground swing whose supporting crossbar is 11 feet above the ground and the length of the chain to her seat is 8 feet. At the end of each swing, she makes an angle of 60° with the vertical and it takes her 3 seconds to complete each full cycle.

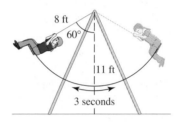

   a. Write a sinusoidal function that can be used to model the height of the seat above the ground as a function of time t.

   b. Write a sinusoidal function that can be used to model the horizontal displacement from directly under the crossbar as a function of time t.

21. A bungee jumper dives off a bridge that spans across a deep gorge. The bungee cord initially stretches to a maximum length of 200 feet before the jumper begins her first rebound. Over the course of the next

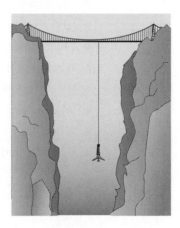

60 seconds, she bounces up and down with ever-diminishing oscillations, each lasting about 6 seconds, until she comes to rest about 160 feet below the bridge. Write the equation of a decaying oscillatory function that models the height of the bungee jumper as a function of time as measured from the instant the cord is extended to its maximum stretch.

22. If a car's engine is operating at 2000 rpm, its pistons are moving up and down 2000 times per minute. Thus, in a four-cylinder engine, each piston moves up and down 500 times per minute. Suppose that the total vertical distance that a piston moves is 3 inches.

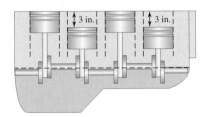

    a. Write a sinusoidal function that models the height of the piston as a function of time in minutes, based on the midline for the height.

    b. Write a sinusoidal function that models the height of the piston as a function of time in minutes, based on the lowest height of the piston.

23. A pogo stick consists of a spring in a vertical tube with two fixed pedals on which a person stands and jumps up and down. Suppose that a child on a pogo stick hops up and down every 3 seconds and that the height of the pedals varies from 4 inches above the ground to 14 inches above the ground. Write a sinusoidal function to model the height of the pedals above ground level as a function of time. (*Hint:* Assume that the pedals are at the midline level at the start.)

24. The child in Problem 23 is also moving forward 20 inches with each bounce of the pogo stick.

    a. Write a sinusoidal function to model the path of the child's feet—that is, the height y above the ground as a function of the horizontal distance x covered.

    b. By comparing the graph of the function you created in part (a) to your image of what is actually happening, explain why the sinusoidal model may not make sense.

    c. Look at the graph of the absolute value of the function you created in part (a). Is it a better or worse model for the behavior you envision?

To solve Problems 25–27, use the fact that, if an arc of length s on a circle of radius r subtends an angle of θ radians, then $s = r\theta$.

25. The distance between two points P and Q on the Earth is measured as the distance along the arc of the circle through P and Q and centered at the center of the Earth O. The radius of the Earth is about 4000 miles. Find the distance from P to Q if the angle POQ has the following measurements.

    a. $\dfrac{\pi}{4}$ **b.** $\dfrac{\pi}{3}$

    c. $\dfrac{5\pi}{6}$ **d.** $15°$

26. A wheel of radius 2 feet rotates at a constant rate of 180 revolutions per minute.

    a. How many radians per minute are swept by the wheel?

    b. How far does a point on the rim of the wheel travel in 1 minute?

27. Find the diameter of the tires on your car. Assume that the car is traveling at 60 mph and determine the number of revolutions the tire makes every minute.

28. On the same set of axes, graph the functions

$$S(x) = 2 \sin x, \qquad R(x) = 2 \sin 3x$$
$$\text{and} \quad T(x) = 2 \sin 0.5x.$$

Clearly mark the zeros of each function.

29. For each function give the frequency, period, amplitude, and phase shift.

    a. $y = 5 + 2 \cos\left(\dfrac{3}{4}x\right)$

    b. $y = 5 - 2 \cos\left(\dfrac{3}{4}x + \pi\right)$

c. $y = 5 - 2 \cos \left(\pi x + \dfrac{3}{4} \right)$

d. $y = 5 - 2 \cos \left[\dfrac{3}{4}(\pi x - 1) \right]$

30. Determine the values of x for which each function in Problem 29 equals 6.

31. a. Graph the function $y = \arcsin \left[\sin(x) \right]$ for x between -10 and 10 radians. Explain why you get the pattern you do.

    b. Repeat part (a) with the function $y = \sin \left[\arcsin(x) \right]$ for x between -1 and 1 radians.

32. Solve for θ.

    a. $-4 \sin \theta = 6 \cos \theta$

    b. $2 \cos \theta = \sin \theta$

    c. $3 \tan \theta - 21 = 0$

33. Solve for x.

    a. $\arctan x = 1.35$

    b. $\arcsin x = 0.5$

8

More About the Trigonometric Functions

Relationships Among Trigonometric Functions

In many applications of trigonometry, particularly in calculus, it often is necessary to transform one trigonometric function into another using an appropriate *trigonometric identity*. Recall that an identity is a relationship that is true for *all* values of the variable. For instance, the Pythagorean identity

$$\sin^2 x + \cos^2 x = 1 \qquad\qquad \textbf{(1)}$$

that we discussed in Section 6.4 holds for every value of *x*.

However, suppose that we ask whether $\sin x + \cos x$ equals 1. Figure 8.1 shows a portion (one complete cycle) of the graph of $y = \sin x + \cos x$. Note that the function is not identically equal to 1 because its graph is not a horizontal line of height 1. Although there are several specific values of *x* for which $\sin x + \cos x$ equals 1 (such as $x = 0$, $x = \pi/2$ and $x = 2\pi$), the relationship does not hold for *every* value of *x*. So $\sin x + \cos x = 1$ is not an identity, but simply an equation that holds for some specific values of the variable.

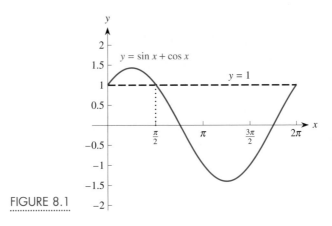

FIGURE 8.1

Identities Involving the Sine and Cosine

Consider again the Pythagorean identity (1). We can use it to transform sines to cosines with

$$\cos^2 x = 1 - \sin^2 x$$

so that

$$\cos x = \pm\sqrt{1 - \sin^2 x}.$$

Similarly, we can transform cosines to sines by using

$$\sin^2 x = 1 - \cos^2 x$$

so that

$$\sin x = \pm\sqrt{1 - \cos^2 x}.$$

Each of these equations holds for all values of the variable x, so each is an identity.

The Reflection Identities

We explore several other useful relationships among the trigonometric functions here. Two properties of the sine and cosine functions are

$$\sin(-x) = -\sin x \qquad \text{(2)}$$
$$\cos(-x) = \cos x, \qquad \text{(3)}$$

for any x. These two relationships, known as the **reflection identities,** are easy to see graphically. The graph of the cosine function is symmetric about the vertical y-axis, as illustrated in Figure 8.2. That is, for any positive value of x, the height of the cosine function is the same to the left of the y-axis (at $-x$) as it is at the same distance to the right of the y-axis (at x). Thus

$$\cos(-x) = \cos x$$

for any value of x. We discussed this same type of behavior in Section 2.7 for power functions with even powers such as $f(x) = x^2$ and $g(x) = x^4$. For this reason, the cosine function is called an *even function.*

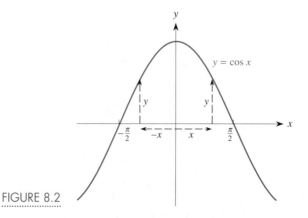

FIGURE 8.2

However, the sine curve is not symmetric about the y-axis. Rather, if you move a distance of x to the left of the y-axis and consider the height to the sine curve, it is equivalent, but opposite in sign, to the height you get if you move the same distance x to the right of the y-axis, as shown in Figure 8.3. Thus

$$\sin(-x) = -\sin x,$$

for any value of x. We encountered this type of behavior with power functions such as $g(x) = x^3$ when the power is odd. As a result, the sine function is called an *odd function*.

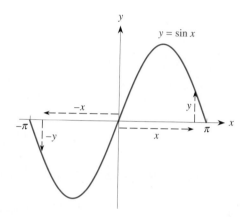

FIGURE 8.3

We discuss even and odd functions again in Section 8.2 when we describe connections between polynomial functions and trigonometric functions.

Think About This   Write a reflection identity for the tangent function. ▭

The Double-Angle Identities

We next consider some additional relationships involving the sine and cosine. The Pythagorean identity says that $\sin^2 x + \cos^2 x = 1$, which is equivalent to $\cos^2 x + \sin^2 x = 1$. What happens if we take the difference instead of the sum? Figure 8.4 shows the graph of $y = \cos^2 x - \sin^2 x$, for x between 0 and 2π. It is a sinusoidal curve that oscillates between -1 and 1 and completes two full cycles between 0 and 2π, so it has a period of π and a frequency of 2. But these features exactly describe the function $y = \cos 2x$. Therefore it seems that

$$\cos^2 x - \sin^2 x = \cos 2x,$$

or, equivalently,

$$\cos 2x = \cos^2 x - \sin^2 x. \qquad (4)$$

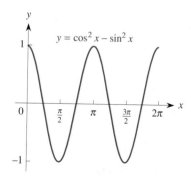

FIGURE 8.4

You can verify this relationship numerically by substituting any value of x into Equation (4). Alternatively, you can verify this relationship graphically by examining the graphs of the functions $y = \cos 2x$ and $y = \cos^2 x - \sin^2 x$.

We can rewrite Equation (4) in several alternative, but equivalent, forms by making use of the Pythagorean identity (1). Thus

$$\cos 2x = \cos^2 x - \sin^2 x = (1 - \sin^2 x) - \sin^2 x$$

so that

$$\cos 2x = 1 - 2 \sin^2 x. \tag{4a}$$

Similarly, we can rewrite Equation (4) as

$$\cos 2x = \cos^2 x - \sin^2 x = \cos^2 x - (1 - \cos^2 x)$$

so that

$$\cos 2x = 2 \cos^2 x - 1. \tag{4b}$$

Verify the identities in Equations (4a) and (4b) visually by using your function grapher and numerically by substituting several different values for x into the equations.

We next consider $\sin 2x$. Suppose that we want to express $\sin 2x$ in an equivalent form that does not show the frequency 2 explicitly. Is it possible that $\sin 2x$ and $2 \sin x$ are equivalent? Graph the two functions and you'll see that they cannot be the same. The first, $y = \sin 2x$, is a sinusoidal curve with an amplitude of 1 and a frequency of 2, so its values oscillate between -1 and 1 and it completes two full cycles between $x = 0$ and $x = 2\pi$. The second function, $y = 2 \sin x$, is a sinusoidal curve with an amplitude of 2 and a frequency of 1, so its values oscillate between -2 and 2 and it completes one full cycle between 0 and 2π.

The actual relationship for $\sin 2x$ is

$$\sin 2x = 2 \sin x \cos x. \tag{5}$$

You can verify Equation (5) graphically on your function grapher. When you graph the two functions $y = \sin 2x$ and $y = 2 \sin x \cos x$ simultaneously, you will see only one graph—the second traces precisely over the first. You can also verify this result numerically: Pick any value for x and evaluate $\sin 2x$ and $2 \sin x \cos x$. The results will be identical for every value of x, thus supporting the fact that Equation (5) is an identity.

The identities in Equations (4), (4a), (4b), and (5) are known as the **double-angle identities** for the sine and cosine.

The Sum and Difference Identities

The double-angle identities in Equations (4) and (5) actually are special cases of more general identities known as the sum and difference identities for sine and cosine that are formally derived in any trigonometry text. The **sum identities** are

$$\sin(x + y) = \sin x \cos y + \cos x \sin y \tag{6}$$

$$\cos(x + y) = \cos x \cos y - \sin x \sin y. \tag{7}$$

To show how the double-angle identities are derived from these formulas, we set $y = x$ in Equations (6) and (7). For instance, in Equation (6),

$$\sin(x + x) = \sin x \cos x + \cos x \sin x = 2 \sin x \cos x,$$

giving

$$\sin(2x) = 2 \sin x \cos x.$$

The same process in Equation (7) produces the double-angle formula for the cosine.

Similarly, we can replace y with $-y$ in the two sum identities, Equations (6) and (7), and then use the reflection identities to derive the **difference identities** for the sine and cosine:

$$\sin(x - y) = \sin x \cos(-y) + \cos x \sin(-y)$$
$$= \sin x \cos y - \cos x \sin y \qquad \textbf{(8)}$$

$$\cos(x - y) = \cos x \cos(-y) - \sin x \sin(-y)$$
$$= \cos x \cos y + \sin x \sin y. \qquad \textbf{(9)}$$

EXAMPLE 1

Show that $\cos(x - \pi/2) = \sin x$ for all x by using the difference identity for the cosine.

Solution   Using the difference identity in Equation (9), we have

$$\cos\left(x - \frac{\pi}{2}\right) = \cos x \cos \frac{\pi}{2} + \sin x \sin \frac{\pi}{2}$$
$$= \cos x \cdot (0) + \sin x \cdot (1) = \sin x.$$

EXAMPLE 2

Reduce $\sin 3x$ to an equivalent expression involving only sines and not including any multiple angles.

Solution   We write

$$\sin 3x = \sin(2x + x) = \sin 2x \cos x + \cos 2x \sin x \qquad \text{Sum identity}$$

$$= (2 \sin x \cos x)\cos x + (1 - 2 \sin^2 x)\sin x \qquad \text{Double angle identity}$$

$$= 2 \sin x \cos^2 x + \sin x - 2 \sin^3 x$$

$$= 2 \sin x \cdot (1 - \sin^2 x) + \sin x - 2 \sin^3 x \qquad \text{Pythagorean identity}$$

$$= 2 \sin x - 2 \sin^3 x + \sin x - 2 \sin^3 x$$

$$= 3 \sin x - 4 \sin^3 x.$$

EXAMPLE 3

Reduce $\sin 4x$ to an equivalent expression involving sines and cosines that has no multiple angles.

Solution   Following the approach in Example 2, we write

$$\sin 4x = \sin(3x + x)$$

$$= \sin 3x \cos x + \cos 3x \sin x. \qquad \text{Sum identity}$$

But, this expression involves expanding cos 3x, and we haven't worked that out yet. (You are asked to do so in a problem at the end of this section.) Alternatively, we could start with

$$\sin 4x = \sin(2x + 2x) = \sin[2(2x)]$$

$$= 2\sin 2x \cos 2x \qquad\qquad \text{Double angle identity}$$

$$= 2(2\sin x \cos x)(\cos^2 x - \sin^2 x) \qquad \text{Double angle identities}$$

$$= 4\sin x \cos^3 x - 4\sin^3 x \cos x.$$

◆

The Half-Angle Identities

Occasionally we face the reverse problem of starting with powers of the sine or cosine—say, $\sin^3 x$ or $\cos^4 x$—and having to eliminate all powers by rewriting the expression in terms of sines and cosines with multiple angles. To eliminate the powers, we make use of two additional identities. Starting with the double-angle identity in Equation (4a),

$$\cos 2x = 1 - 2\sin^2 x,$$

we have

$$2\sin^2 x = 1 - \cos 2x$$

so that

$$\sin^2 x = \frac{1}{2}(1 - \cos 2x). \qquad\qquad \textbf{(10)}$$

Similarly, if we start with the double-angle identity in Equation (4b),

$$\cos 2x = 2\cos^2 x - 1,$$

we get

$$2\cos^2 x = 1 + \cos 2x$$

or

$$\cos^2 x = \frac{1}{2}(1 + \cos 2x). \qquad\qquad \textbf{(11)}$$

The identities in Equations (10) and (11) are the **half-angle identities.** Verify them graphically on your function grapher. We illustrate their use in Example 4.

◆**EXAMPLE 4** ....

Rewrite $\cos^4 x$ in terms of cosines of multiple angles by eliminating all exponents.

Solution   Using Equation (11), we have

$$\cos^4 x = (\cos^2 x)^2 = \left[\frac{1}{2}(1 + \cos 2x)\right]^2$$

$$= \frac{1}{4}[1 + 2\cos 2x + \cos^2(2x)].$$

This expression involves $\cos^2(2x)$, so we apply Equation (11) again to get

$$\cos^4 x = \frac{1}{4}\left\{1 + 2\cos 2x + \frac{1}{2}[1 + \cos 2(2x)]\right\}$$

$$= \frac{1}{4}\left(1 + 2\cos 2x + \frac{1}{2} + \frac{1}{2}\cos 4x\right)$$

$$= \frac{1}{4}\left(\frac{3}{2} + 2\cos 2x + \frac{1}{2}\cos 4x\right)$$

$$= \frac{3}{8} + \frac{1}{2}\cos 2x + \frac{1}{8}\cos 4x.$$

For easy reference, we list all the fundamental trigonometric identities involving the sine and cosine functions. These identities reappear both in this course and in later mathematics and associated courses.

Trigonometric Identities

Pythagorean identity:	$\sin^2 x + \cos^2 x = 1$	(1)
Reflection identities:	$\sin(-x) = -\sin x$	(2)
	$\cos(-x) = \cos x$	(3)
Double-angle identities:	$\cos 2x = \cos^2 x - \sin^2 x$	(4)
	$= 1 - 2\sin^2 x = 2\cos^2 x - 1$	(4a,b)
	$\sin 2x = 2\sin x \cos x$	(5)
Sum identities:	$\sin(x + y) = \sin x \cos y + \cos x \sin y$	(6)
	$\cos(x + y) = \cos x \cos y - \sin x \sin y$	(7)
Difference identities:	$\sin(x - y) = \sin x \cos y - \cos x \sin y$	(8)
	$\cos(x - y) = \cos x \cos y + \sin x \sin y$	(9)
Half-angle identities:	$\sin^2 x = \frac{1}{2}(1 - \cos 2x)$	(10)
	$\cos^2 x = \frac{1}{2}(1 + \cos 2x)$	(11)

Using the Trigonometric Identities

Suppose that a projectile, such as a cannonball or a high-pressure stream of water, is shot off with an initial velocity v_0 at an angle θ with the horizontal, as shown in Figure 8.5. The distance R that the cannonball or the water travels—its *range*—depends on the angle θ. For very small angles, the range is minimal because gravity pulls the object to the ground quickly. For very large angles (close to 90°), the object is shot almost vertically upward and comes back to the ground fairly near the point at which it was released. For moderately sized angles, the range is considerably larger.

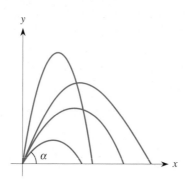

FIGURE 8.5

It is shown in physics that the range R is given by

$$R = \frac{2v_0{}^2 \sin \theta \cos \theta}{g},$$

where g is the acceleration due to gravity.

EXAMPLE 5

Use a trigonometric identity to simplify the formula for the range of a projectile and use the result to determine the angle that leads to the maximum range for any initial velocity.

Solution   The range is

$$R = \frac{2v_0{}^2 \sin \theta \cos \theta}{g}.$$

Because $\sin 2\theta = 2 \sin \theta \cos \theta$, this expression for the range reduces to

$$R = \frac{v_0{}^2 \sin 2\theta}{g}.$$

Because g and v_0 are fixed, the range is maximal when $\sin 2\theta$ is maximal and the largest value of the sine function is 1, which occurs when $2\theta = 90°$ or $\theta = 45°$. Therefore a projectile subject only to the force of gravity has a maximum range when the initial angle $\theta = 45°$.

◆

In most derivations in physics and engineering involving wave phenomena such as electromagnetic waves (e.g., radio signals or electric currents in a circuit), sound waves, or water waves, the height y of the wave as a function of time t is usually given in the form

$$y = A \sin kt + B \cos kt,$$

where A and B are constants and k is the frequency. In Example 6, we show how this type of expression can be simplified by using a trigonometric identity to give far more insight into the behavior of the wave than this fairly complicated expression provides.

EXAMPLE 6

The equation of a wave is $y = 4 \sin t - 3 \cos t$. Use a trigonometric identity to explain the behavior of this wave.

Solution   The graph of this function between $t = 0$ and $t = 2\pi$ is shown in Figure 8.6. It looks like a sine wave shifted horizontally to the right by about $\pi/6$, or 30°. Also, the amplitude of this wave seems to be about 5, compared to the amplitudes of 4 and 3 in the two terms of the function. Finally, the period of this wave seems to be about 2π. As a result, the equation of the wave $y = 4 \sin t - 3 \cos t$ appears to be equivalent to $y = C \sin(t - D)$, where the amplitude $C \approx 5$ and the phase shift D is about $\pi/6$. Let's see why.

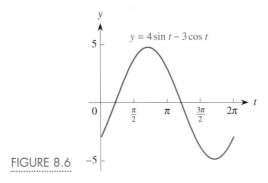

FIGURE 8.6

The seemingly equivalent form for the wave $y = C \sin(t - D)$ suggests using the difference formula for the sine, or

$$C \sin(t - D) = C \cdot (\sin t \cos D - \cos t \sin D). \qquad \textbf{(12)}$$

Also, the fact that the individual amplitudes in the original formula are 4 and 3 and the apparent amplitude we observe for the wave is about $C = 5$ suggests the Pythagorean theorem

$$\sqrt{4^2 + 3^2} = 5.$$

Factoring 5 out of the original formula for the wave yields

$$y = 4 \sin t - 3 \cos t = 5\left(\frac{4}{5} \sin t - \frac{3}{5} \cos t\right).$$

Comparing this expression to Equation (12) suggests that we make the association

$$C = 5, \qquad \cos D = \frac{4}{5}, \quad \text{and} \quad \sin D = \frac{3}{5}.$$

If $\cos D = 4/5$,

$$D = \arccos \frac{4}{5} \approx 36.87°,$$

which is close to what we predicted for the phase shift, based on the graph shown in Figure 8.6. To be sure that this result is consistent with the third condition, we see that if $\sin D = 3/5$,

$$D = \arcsin \frac{3}{5} \approx 36.87°.$$

Consequently, the wave formula

$$y = 4 \sin t - 3 \cos t = 5(\cos D \sin t - \sin D \cos t)$$
$$= 5 \sin(t - 36.87°)$$

and the original wave is the same as a pure sinusoidal function centered about $y = 0$ with amplitude 5, period $2\pi = 360°$, and a phase shift of about 36.87°, or 0.6435 radian.

◆

Identities Involving the Tangent

Just as there are trigonometric identities relating the sine and cosine, there are identities involving the tangent function. We encountered two of them in Section 6.4. The first is the key identity relating the tangent function to the sine and the cosine:

$$\tan x = \frac{\sin x}{\cos x}.$$

We also derived an analog to the Pythagorean identity in Section 6.4:

$$\tan^2 x + 1 = \frac{1}{\cos^2 x},$$

provided that $\cos x \neq 0$.

Likewise, there are double-angle, sum, and difference formulas, and so on, for the tangent function. We investigate a double-angle identity and a sum identity in the Problems. If you're interested, you can find more details about them in any trigonometry textbook.

Problems

1. Using ideas on amplitude and frequency, explain why $\cos 3x$ cannot be identically equal to $3 \cos x$.

2. Using ideas on amplitude, explain why $\cos 2x = 2 \cos^2 x - 1$ is reasonable. (Recognize that such an argument is not a proof.)

Examine each equation in Problems 3–14 graphically to see if the relationship may be an identity. If it is not an identity, attempt to locate graphically or numerically at least one point that lies on both curves. If it seems to be an identity, prove it algebraically.

3. $\sin^3 x + \cos^3 x = 1$

4. $\cos 3x = \cos^3 x - \sin^3 x$

5. $\dfrac{\sin 2x}{\sin x} = 2 \cos x$

6. $(1 - \cos \theta)(1 + \cos \theta) = \sin^2 \theta$

7. $\sin 3x = 3 \sin x$

8. $\dfrac{\cos^2 \theta}{1 + \sin \theta} = 1 - \sin \theta$

9. $\dfrac{1 - \cos \alpha}{\sin \alpha} = \dfrac{\sin \alpha}{1 + \cos \alpha}$

10. $\cos 3\beta = 3 \cos^3 \beta - 1$

11. $\sin^2 3x + \cos 6x = \cos^2 3x$

12. $\cos^2 2x = 3(1 - \sin 2x)$

13. $\sin(\cos x) = \cos(\sin x)$

14. $\sin(\cos x) = \sin x \cos x$

15. Express $\cos 3x$ in terms of powers of $\sin x$ and $\cos x$, but with no multiple angles.

16. Express $\cos 4x$ in terms of powers of $\sin x$ and $\cos x$, but with no multiple angles.

17. Express $\cos 5x$ in terms of powers of $\sin x$ and $\cos x$, but with no multiple angles.

18. Examine the results of Problems 15–17 and the formula for $\cos 2x$. Are there any patterns in the terms? If so, what are they?

19. By setting $y = x$ in the sum identity in Equation (7), show that you get the double-angle identity in Equation (4).

20. Rewrite $\sin^4 x$ in terms of multiple angles by eliminating all exponents.

21. Rewrite $\sin^2 x \cos^2 x$ in terms of multiple angles by eliminating all exponents.

22. **a.** Sketch the graph of $\sin(x + \pi/2)$. What familiar function do you get from this phase shift?
    b. Use the sum identity for the sine function to show that $\sin(x + \pi/2)$ actually equals that function.

23. **a.** Repeat Problem 22 for $\sin(x + \pi)$.
    b. Repeat Problem 22 for $\cos(x + \pi/2)$.

24. In the half-angle identities in Equations (10) and (11), let $y = 2x$, so that $x = \frac{1}{2}y$. Rewrite each identity in terms of y to see why they are called half-angle identities.

25. William Tell is about to shoot the most important arrow of his life. His son is standing 250 feet away. Tell releases the arrow at a height of 5 feet above the ground with an initial speed of 180 feet per second. The height of the center of the apple on his son's head is also 5 feet above the ground. Find algebraically two different angles α at which Tell should release the arrow in order to have it pass through the apple without hitting the boy.

26. Suppose that William Tell's son is actually a foot shorter than in Problem 26 so that the center of the apple is now 4 feet above the ground and that the arrow comes off the bow string at a height of 5 feet. Estimate, graphically, two different angles α at which Tell should release the arrow in order for it to pass through the apple without hitting the boy.

27. In Example 6, we converted the wave $y = 4 \sin t - 3 \cos t$ to the equivalent pure sinusoidal expression $y = 5 \sin(t - 36.87°)$.
    a. Convert this formula to a pure cosine curve by an appropriate horizontal shift.
    b. Repeat the derivation in Example 6 by using the sum or difference identity for cosines to derive the equivalent formula as a cosine wave.
    c. How does the result in part (b) compare to the result in part (a)?

28. A baseball player hits a ball with an initial velocity of 120 feet per second at a height of 5 feet above the ground. The ball is caught 320 feet from home plate by an outfielder whose glove is also 5 feet above the ground. Use the formula for the range of a projectile to determine the angle of inclination of the ball as it comes off the bat.

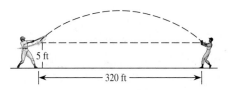

29. The accompanying figure shows the graph of the function $y = \sin^4 x + \cos^4 x$ from $x = 0$ to 2π. The curve suggests that the function is equivalent to some sinusoidal function.

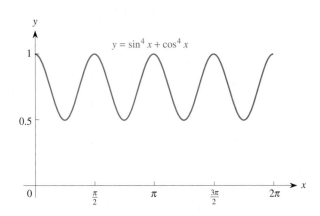

    a. By examining the graph carefully on your function grapher, estimate values for each parameter to find a sinusoidal function that seems to have the matching behavior pattern. (*Hint*: The parameters should be simple fractions or whole numbers.)
    b. Superimpose the graph of your function over the graph of $y = \sin^4 x + \cos^4 x$ to verify that they do appear to be the same.
    c. Use the half-angle identities for sine and cosine repeatedly to prove that $y = \sin^4 x + \cos^4 x$ does reduce to the expression you conjectured.

30. Repeat Problem 29 with the function $y = \sin^6 x + \cos^6 x$.

31. Refer to the functions shown in Problem 1 of Section 7.2 and decide which are odd, even, or neither.

32. **a.** Use some ideas from Section 5.5 on the sum of the terms in an exponential sequence to explain why you can calculate the value of
$$1 + \sin x + \sin^2 x + \sin^3 x + \sin^4 x + \cdots$$
    as $\dfrac{1}{1 - \sin x}$.

Are there any values of x for which this approach does not work?

b. What formula would you get for the sum of the terms
$$1 + \sin x + \sin^2 x + \cdots + \sin^n x$$
for any given positive integer n?

33. Use the result of Problem 32(b) with different values of n to calculate the value of
$$1 + \sin x + \sin^2 x + \sin^3 x + \sin^4 x + \cdots,$$
for $x = \pi/6$ correct to three decimal places. Now suppose that you want to do so for $x = \pi/3$ instead. Will you need approximately the same number of terms, more terms, or fewer terms to get the same three decimal place accuracy? Explain.

34. a. Verify graphically that
$$\tan \theta + \frac{1}{\tan \theta} = \frac{1}{\sin \theta \cos \theta},$$
for all θ for which the denominators are nonzero.
    b. Show algebraically that the expression in part (a) is an identity. (*Hint*: Transform $\tan \theta$ to equivalent expressions in $\sin \theta$ and $\cos \theta$.)

35. Use appropriate trigonometric identities to show that
$$\frac{1}{\tan x} - \tan x = \frac{2}{\tan 2x}.$$

36. Use the identity in Problem 35 to derive a double-angle formula for the tangent function.

37. a. Derive the double-angle identity
$$\tan 2x = \frac{2 \tan x}{1 - \tan^2 x}$$

by using the double-angle identities for sine and cosine. (*Hint*: Divide both the numerator and the denominator by $\cos^2 x$.)

b. Derive the addition identity for the tangent,
$$\tan(x + y) = \frac{\tan x + \tan y}{1 - \tan x \tan y}$$
by using the addition formulas for sine and cosine. (*Hint*: Divide both the numerator and the denominator by $\cos x \cos y$.)

Examine each equation in Problems 38–46 graphically to see whether the relationship may be an identity. If it is not an identity, attempt to locate graphically or numerically at least one point that lies on both curves. If it seems to be an identity, prove it algebraically.

38. $1 + \dfrac{1}{\tan^2 \theta} = \dfrac{1}{\sin^2 \theta}$

39. $\tan 2\theta = 2 \tan \theta$

40. $\tan^2 x - \sin^2 x = (\tan x \sin x)^2$

41. $\tan 2x = \dfrac{2 \tan x}{1 - \tan^2 x}$

42. $\tan \dfrac{\alpha}{2} = \dfrac{1 - \cos \alpha}{\sin \alpha}$ (*Hint*: Let $\dfrac{\alpha}{2} = \theta$.)

43. $\tan^2 x = 1 + 2 \tan x$

44. $1 - \cos 2x = \tan x \sin 2x$

45. $\tan(\sin x) = \tan x \sin x$

46. $\cos(\tan x) = \tan(\cos x)$

47. What is wrong with the following "proof"?
$$\cos(\tan x) = \cos\left(\frac{\sin x}{\cos x}\right) = \sin x.$$

8.2 Approximating Sine and Cosine with Polynomials

Have you ever wondered what happens when you press either the SIN or COS key on your calculator and the value for the function appears? How does the calculator actually find the values of these functions?

Approximating the Sine Function

In this section, we consider one approach that has been used to compute function values. We begin by examining the graph of the sine function, with x measured in radians, as shown in Figure 8.7(a). We zoom in on the portion of the curve close to the origin, as marked by the box; the corresponding curve is shown in Figure 8.7(b). If we zoom in still further about the origin, as marked by the box in Figure 8.7(b), we get the portion of the sine curve shown in Figure 8.7(c). This final graph looks like a straight line rather than a portion of a curve. (In fact, if you zoom in sufficiently on any smooth curve, it will eventually look like a straight line.)

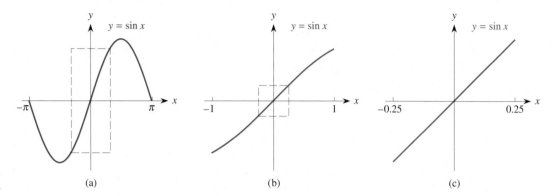

FIGURE 8.7

(a)   (b)   (c)

Linear Approximation to the Sine   When x is very close to the origin, the sine curve looks like a line. Let's find the equation of this "line." Because it passes through the origin, the vertical intercept must be 0. To find the slope, we need a second point. If we trace along the sine curve very close to the origin, we find that $x = 0.001$ and $y = \sin 0.001 = 0.0009999998$ is a point on the sine curve. The slope of the line through this point and the origin is

$$m = \frac{0.0009999998 - 0}{0.001 - 0} = 0.9999998 \approx 1.$$

Therefore the equation of a line that very closely hugs the sine curve near the origin is $y = x$. We show the graph of this line, along with the sine curve, in Figure 8.8.

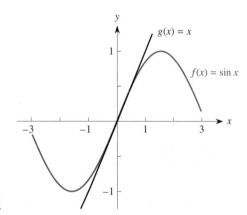

FIGURE 8.8

Observe that, when x is very close to 0, the graphs of $f(x) = \sin x$ and $g(x) = x$ are very close to one another. In fact, when x is very close to 0, the two graphs are virtually indistinguishable. That is,

$$\sin x \approx x, \qquad \text{if } x \text{ is very close to } 0$$

Of course, as the value of x gets farther from 0, the sine curve eventually bends away from the line $y = x$.

x	$\sin x$	x	$\sin x$
0	0		
0.1	0.100	-0.1	-0.100
0.2	0.199	-0.2	-0.199
0.3	0.296	-0.3	-0.296
0.4	0.389	-0.4	-0.389
0.5	0.479	-0.5	-0.479
0.6	0.565	-0.6	-0.565
0.7	0.644	-0.7	-0.644

To show this result numerically, we look at some values of x to see how close the values along the line match the values of the sine function.

From the table at the left, we see that, when x is extremely close to 0, the value of $\sin x$ is almost identical to x itself, but the farther that x is from the origin, the less accurate the approximation. Thus, whenever x is very close to 0, we can replace $\sin x$ with x for the purposes of approximating the value of $\sin x$. For instance, to approximate $\sin(0.00243)$, we could say that

$$\sin(0.00243) \approx 0.00243.$$

Using a calculator gives $\sin(0.00243) = 0.0024299976$, so the approximation is accurate to five decimal places.

We can approximate the sine function with a linear function in a different way by using methods of linear regression. We use a set of points that lie on the sine curve $y = \sin x$ very close to the origin. They are shown rounded to 6 decimal places in Table 8.1. The line that best fits these "data" is $y = 0.9999258x$ with a correlation coefficient $r = 1.000000000$, which tells us that a line with slope of about 1 is virtually perfect. Thus we again see that when x is very close to 0, $\sin x \approx x$. However, if we move too far away from $x = 0$, the accuracy of the approximation breaks down. For instance, we would not want to approximate $\sin(0.75)$ with the value $x = 0.75$ because $\sin(0.75) = 0.6816$; the value $x = 0.75$ is too far from $x = 0$ for the approximation to be good.

TABLE 8.1

x	-0.025	-0.02	-0.015	-0.01	-0.005	0	0.005	0.01	0.015	0.02	0.025
$y = \sin x$	-0.024997	-0.019999	-0.014999	-0.01	-0.005	0	0.005	0.01	0.014999	0.019999	0.024997

This idea of approximating a function such as $y = \sin x$ with a simpler function (often a linear function) is an essential principle in mathematics. We use this principle to approximate the values of trigonometric functions because it is *impossible* to calculate them directly with algebraic methods.

Improving on the Linear Approximation to the Sine

Unfortunately, as we have noted, the linear approximation to the sine curve is only accurate if x is very close to the origin. As we take values of x farther and farther from the origin, the sine curve bends ever more sharply and eventually bends away from the line. Let's see how we can improve on the linear approximation $\sin x \approx x$ when x is somewhat farther from 0. To do so, we need a simple curve (at least one that is simpler to work with than the sine function) that bends in a similar manner. For computational purposes, the simplest curves are usually polynomials.

In Figure 8.7, we zoomed in on the sine curve very close to the origin so that the curve looked like a line. Now we zoom out a bit to see what happens for values of x from -3 to 3, as previously shown in Figure 8.8. Although the line is indistinguishable from the sine curve near the origin (roughly from $x = -0.6$ to $x = 0.6$), the sine curve bends away from the line as the first pair of turning points in the sine curve come into view. In fact, the overall shape of this portion of the sine curve is quite suggestive of a cubic polynomial with a negative leading coefficient. (Recognize that, if you zoom out a bit farther, more turning points appear and the cubic-like appearance disappears.) This result suggests that we try to approximate this portion of the sine curve with a cubic curve. We use the data in Table 8.1 that we used for the linear fit, but now fit a cubic polynomial instead. We then get the cubic

$$y = -0.1666601x^3 + 0x^2 + 0.999999999x + 0.$$

Note that (1) the constant term is 0, which assures us that the cubic passes through the origin; (2) the coefficient of the linear term is essentially 1; (3) the coefficient of the quadratic term is 0; (4) the leading coefficient is negative, which is what we expected; and (5) the value of the leading coefficient, -0.1666601, is quite close to $-1/6 = -0.16666667$. Thus a cubic polynomial that approximates the sine function is

$$T_3(x) = -\frac{x^3}{6} + x,$$

or, equivalently,

$$\sin x \approx x - \frac{x^3}{6}$$

when x is fairly close to 0.

Figure 8.9 shows both the sine curve and the cubic polynomial for x from -3.5 to 3.5. The two curves are indistinguishable from about $x = -1.2$ to $x = 1.2$, which extends over a considerably larger interval than the linear approximation, which is accurate only from about $x = -0.6$ to $x = 0.6$.

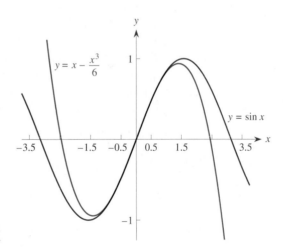

FIGURE 8.9

To illustrate the accuracy of the approximation for $\sin x$ using the cubic polynomial for values of x near 0, we try $x = 0.125$, say, and find that

$$\sin(0.125) \approx (0.125) - \frac{(0.125)^3}{6} \approx 0.1246744,$$

which agrees with the true value of $\sin(0.125) = 0.1246747$ to six decimal places. If we move farther from 0 and try $x = 0.7$, we find that

$$\sin(0.7) \approx (0.7) - \frac{(0.7)^3}{6} \approx 0.643,$$

compared to the actual value of $\sin(0.7) = 0.644$, which is correct to the nearest hundredth, so the approximation is still fairly accurate. However, the graphs in Figure 8.9 show that the two curves eventually diverge. Thus if we take x too far from 0, the accuracy of the approximation diminishes. Moreover, the farther from $x = 0$ we go, the worse the approximation is. For instance, if $x = 1$ radian $\approx 57°$,

$$\sin 1 \approx 1 - \frac{1^3}{6} \approx 0.83333,$$

compared to the correct value of $\sin 1 = 0.84147$. If $x = 1.5$ radians,

$$\sin(1.5) \approx (1.5) - \frac{(1.5)^3}{6} \approx 0.93750,$$

compared to the correct value of $\sin(1.5) = 0.99749$. If $x = 2$ radians,

$$\sin 2 \approx 2 - \frac{2^3}{6} \approx 0.66667,$$

compared to the correct value of $\sin 2 = 0.90930$. If $x = \pi$ radians,

$$\sin \pi \approx \pi - \frac{\pi^3}{6} \approx -2.02612,$$

compared to the correct value $\sin \pi = 0$. In fact, this last approximation is so bad that it gives us a value, -2.02612, outside the range of the sine function.

What if we wanted to improve on the approximation still further so that we could use it to estimate values for $\sin x$ when x is still farther from the origin? Consider the graph of the sine curve from $x = -6$ to $x = 6$ shown in Figure 8.10. It has four turning points and three inflection points, which suggests that the sine curve looks like a polynomial of degree 5. Although graphing calculators don't fit a fifth degree polynomial to a set of data, that task can be accomplished by many software packages. Using a spreadsheet, we find that the fifth degree polynomial that fits the data in Table 8.1 is

$$T_5(x) = 0.0083x^5 + 0x^4 - 0.1667x^3 + 0x^2 + 0.999999999x + 0,$$

or essentially

$$T_5(x) = 0.0083x^5 - \frac{x^3}{6} + x.$$

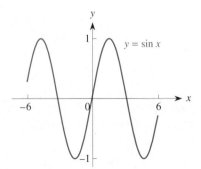

FIGURE 8.10

Note that this polynomial has a positive leading coefficient, so it increases toward the right, as we want. It also has a 0 constant coefficient, so it passes through the origin. Note also that the only change from the cubic polynomial to this fifth degree polynomial is the fifth degree term—all other terms remained the same.

Because $0.0083 \approx 1/120$, we can write this polynomial as

$$T_5(x) = \frac{x^5}{120} - \frac{x^3}{6} + x \quad \text{or} \quad T_5(x) = x - \frac{x^3}{6} + \frac{x^5}{120}.$$

A simple and interesting pattern is developing here with the coefficients: $6 = 3 \times 2 \times 1 = 3!$ and $120 = 5 \times 4 \times 3 \times 2 \times 1 = 5!$. (See Appendix A2 for a discussion of factorial notation.) So we can rewrite the approximation formula for the sine function as

$$\sin x \approx x - \frac{x^3}{3!} + \frac{x^5}{5!}.$$

Figure 8.11 shows the graphs of the sine function and the fifth degree polynomial for x between -4 and 4. The two curves are indistinguishable for x between roughly -2 and 2, so we have achieved a considerable improvement over the cubic approximation, which was a good match for x between roughly -1.2 and 1.2.

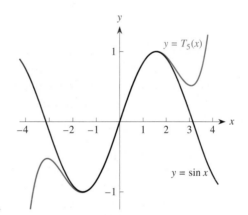

FIGURE 8.11

To verify the accuracy of this approximation, let's see how much improvement we get compared to the previous values. The results are shown in the following table.

x	$\sin x$	$T_3(x)$	$T_5(x)$
0.7	0.64422	0.643	0.64423
1	0.84147	0.83333	0.84167
1.5	0.99749	0.93750	1.00078
2	0.90930	0.66667	0.93333
π	0	-2.02612	0.52404

The fifth degree approximation is better still because we get more accurate estimates for the values of sin x over larger intervals of x-values centered at 0.

We can continue this process, using higher degree polynomials, and get even better approximations. However, before doing so, let's examine the sequence of polynomial approximations we have so far. They are

$$\sin x \approx x,$$

$$\sin x \approx x - \frac{x^3}{3!},$$

$$\sin x \approx x - \frac{x^3}{3!} + \frac{x^5}{5!}.$$

First, each successive polynomial involves just one additional term, compared to the preceding polynomial. Second, each polynomial involves only *odd* powers—and we know that the sine function is an *odd* function. This means that both the sine function and the approximating polynomials are symmetric about the origin. Third, the signs of successive coefficients alternate. Fourth, there is a definite pattern involving factorials in the coefficients. These polynomials are known as **Taylor polynomial approximations** after English mathematician Brook Taylor, who investigated them in the early 1700s.

Think About This ....

Predict the next higher degree polynomial approximation to sin x. How accurate is this approximation for the values $x = 0.7, 1, 1.5, 2$, and π? ▭

Improving the Approximation Using the Behavior of sin x   We could continue this process and construct Taylor polynomial approximations of higher and higher degree. However, that isn't necessary if we cleverly use some of the basic behavioral properties of the sine function. First, recall the reflection identity

$$\sin(-x) = -\sin x.$$

It allows us to approximate sin x when x is negative simply by using the corresponding positive value for x and reversing the sign of the estimate.

Second, we know that the sine function is periodic with period 2π. Therefore, if x is any number greater than $2\pi \approx 6.28$, the value of sin x is the same as the value of sin x_0, where x_0 is the corresponding number between 0 and 2π radians. Consequently, we need only obtain an approximation that is accurate as far out as 2π. We can handle anything beyond that by reducing the value of x to an appropriate value x_0 between 0 and 2π by "removing" all multiples of 2π, as illustrated in Figure 8.12.

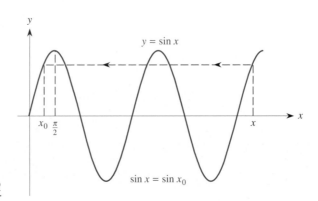

FIGURE 8.12

Now visualize the portion of the sine curve between $x = \pi$ and $x = 2\pi$, as shown in Figure 8.13. It has the same shape as the portion from $x = 0$ to $x = \pi$, only "flipped" across the x-axis. Thus, if we have a value of x between π and 2π (where $\sin x$ is negative), there is a point between 0 and π, namely at $x - \pi$, where the sine function has the same value, but with a positive sign. That is,

$$\sin x = -\sin(x - \pi).$$

So, all we need is an approximation that is sufficiently accurate for x between 0 and π.

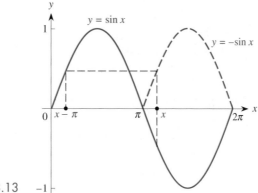

FIGURE 8.13

Think About This   Use an appropriate trigonometric identity to show that $\sin x = -\sin(x - \pi)$. ▭

Now visualize the sine curve from $x = 0$ to $x = \pi$. The two halves are symmetric, as shown in Figure 8.14. Therefore, for any point x between $\pi/2$ and π, the value of $\sin x$ is the same as that at a corresponding point between 0 and $\pi/2$. So, all we need is an approximation to $\sin x$ that is sufficiently accurate for x between 0 and $\pi/2$. The previous fifth degree polynomial $T_5(x)$ gives two-decimal accuracy for any value of x in this interval. If we want more than two-decimal accuracy, we have to use a higher degree polynomial—say, the seventh degree Taylor polynomial that we asked you to produce in a previous *Think About This* exercise.

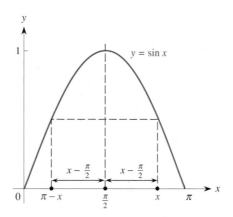

FIGURE 8.14

Think About This   Try the seventh degree polynomial

$$\sin x \approx x - \frac{x^3}{3!} + \frac{x^5}{5!} - \frac{x^7}{7!}$$

for various values of x in the interval from 0 to $\pi/2$. Does it provide four-decimal accuracy? Is that adequate? If not, what would you do? ⬜

Approximating the Cosine Function

We now consider the comparable problem of approximating the cosine function by using a polynomial. If you zoom in on the cosine curve very close to $x = 0$, it appears indistinguishable from a horizontal line. In fact, because $\cos 0 = 1$, that line must be $y = 1$. So, for x very close to 0,

$$\cos x \approx 1.$$

However, once you move away from $x = 0$, the cosine curve bends away from the line $y = 1$.

Let's now look at the cosine curve in a somewhat wider interval about $x = 0$—say, from $x = -2$ to $x = 2$, as shown in Figure 8.15. Its overall shape suggests a parabola opening downward. Therefore we try to approximate the cosine function with a quadratic function so long as x remains fairly close to 0.

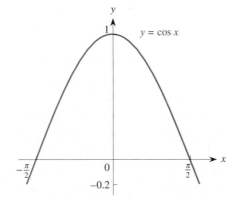

FIGURE 8.15

Approximating $y = \cos x$ Using Data Analysis   There are several ways to find an equation for such a quadratic. One way is to fit a quadratic function to some set of values for $\cos x$ when x is relatively close to 0. Consider the values in Table 8.2.

TABLE 8.2

x	-0.08	-0.06	-0.04	-0.02	0	0.02	0.04	0.06	0.08
$y = \cos x$	0.9968	0.9982	0.9992	0.9998	1	0.9998	0.9992	0.9982	0.9968

Using a calculator, we find that the quadratic function that fits these data is

$$y = -0.49973x^2 + 0x + 0.99999.$$

The constant term is essentially 1 and the leading coefficient is approximately -0.5. Hence we have the following approximation to the cosine function near $x = 0$.

$$\cos x \approx 1 - \frac{x^2}{2}$$

Figure 8.16 shows the graphs of the cosine function and this quadratic Taylor approximating polynomial for x between -2 and 2. The two are virtually indistinguishable for x between about -0.8 and 0.8. For instance, $\cos(0.5) = 0.87758$ compared to the value of the approximating quadratic,

$$\cos(0.5) \approx 1 - \frac{(0.5)^2}{2} = 0.875,$$

so we have two decimal place accuracy.

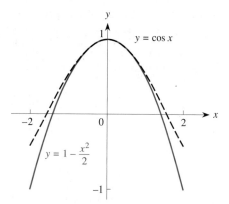

FIGURE 8.16

If we zoom out somewhat farther on the graph of the cosine curve—say, for x between -5 and 5, as shown in Figure 8.17—the cosine function no longer suggests a quadratic function. This portion of the cosine curve has four real roots, three turning points, and two inflection points, which suggest a polynomial of degree 4. Using a calculator, we find a quartic function that fits the data values in Table 8.2 is

$$y = 0.041653x^4 + 0x^3 - 0.499999x^2 + 0x + 0.999999.$$

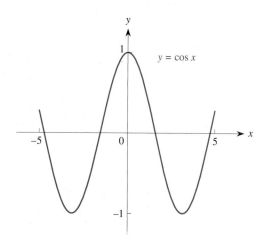

FIGURE 8.17

Note that the coefficients of both odd-powered terms are 0, the constant coefficient is essentially 1, and the quadratic coefficient is essentially $-\frac{1}{2}$. As for the leading coefficient, 0.041653,

$$\frac{1}{0.041653} = 24.00787,$$

so the leading coefficient is essentially $1/24 = 1/4!$. Therefore we have the fourth degree Taylor polynomial approximation

$$\cos x \approx 1 - \frac{x^2}{2} + \frac{x^4}{4!}.$$

Figure 8.18 shows the graph of this polynomial $T_4(x) = 1 - x^2/2 + x^4/4!$ and the cosine function for x between -5 and 5. Observe that $T_4(x)$ tries hard to capture the pattern in the cosine curve. In fact, the polynomial is an excellent match to the cosine for x between roughly -1.5 and 1.5. For instance, if $x = 0.5$, then $T_4(0.5) = 0.87760$, compared to $\cos(0.5) = 0.87758$; similarly, $T_4(1) = 0.541666$, compared to $\cos(1) = 0.54030$. If we choose a value of x too far from 0, the approximation breaks down. Thus $T_4(1.5) = 0.08594$, compared to $\cos(1.5) = 0.07074$.

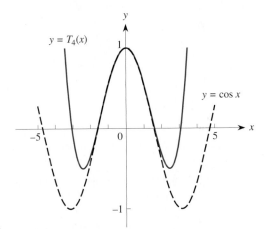

FIGURE 8.18

If we zoom out still farther on the cosine curve—say, from $x = -8$ to $x = 8$—two more turning points come into view, which suggests that we could get a better approximation to the cosine with a sixth degree polynomial. Using the values in Table 8.2 and a spreadsheet, we find, after rounding the coefficients, that

$$\cos x \approx T_6(x) = 1 - \frac{x^2}{2} + \frac{x^4}{4!} - \frac{x^6}{6!}.$$

Summarizing these results, the successive Taylor polynomial approximations to the cosine function are:

$$\cos x \approx 1 - \frac{x^2}{2},$$

$$\cos x \approx 1 - \frac{x^2}{2} + \frac{x^4}{4!}$$

$$\cos x \approx 1 - \frac{x^2}{2} + \frac{x^4}{4!} - \frac{x^6}{6!}.$$

As with the sine approximations, (1) each successive polynomial involves just one additional term; (2) each polynomial involves only *even* powers, and we know that the cosine function is an *even* function; (3) the signs of successive coefficients alternate; and (4) there is a clear pattern in the coefficients involving factorials.

The graphs of these polynomials, as well as that of the cosine curve, are shown in Figure 8.19. Note that each successive polynomial fits the cosine curve more accurately over a larger and larger interval centered at $x = 0$. You should examine these successive approximations using your function grapher.

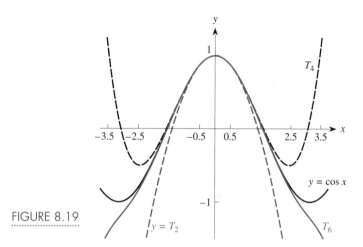

FIGURE 8.19

Think About This   How could you improve on the sixth degree polynomial approximation to the cosine? By eye, over what interval does it appear to be a good fit to the cosine curve? ▢

Think About This   Devise a scheme to reduce any value of x to an equivalent value that allows you to use the smallest possible interval of x-values. How accurate is the fourth degree Taylor polynomial on this interval (i.e., what is the largest error between the cosine and the polynomial)? How accurate is the sixth degree polynomial? ▢

Approximating sin x and cos x Using Trigonometric Identities

We now approach the problem of approximating the sine and cosine from a different viewpoint using several trigonometric identities. Recall the double-angle identity

$$\cos 2\theta = 1 - 2\sin^2\theta$$

from Equation (4a) of Section 8.1. If we let $x = 2\theta$ so that $x/2 = \theta$, the expression for $\cos 2\theta = \cos x$ becomes

$$\cos x = 1 - 2\sin^2\left(\frac{x}{2}\right) = 1 - 2\left[\sin\left(\frac{x}{2}\right)\right]^2.$$

When θ is close to 0, we have $\sin\theta \approx \theta$, so that

$$\sin\left(\frac{x}{2}\right) \approx \frac{x}{2}.$$

Consequently we can approximate $\cos x$ by

$$\cos x \approx 1 - 2\left(\frac{x}{2}\right)^2 = 1 - \frac{x^2}{2}$$

when x is close to 0. Note that this approximation is identical to the quadratic Taylor polynomial that we obtained previously by using data analysis techniques.

We now use the double-angle formula for the sine,

$$\sin 2\theta = 2 \sin \theta \cos \theta$$

with $x = 2\theta$ so that $x/2 = \theta$, which gives

$$\sin x = 2 \sin \left(\frac{x}{2}\right) \cos \left(\frac{x}{2}\right). \tag{13}$$

We use the linear approximation for $\sin x \approx x$ and the quadratic approximation $\cos x \approx 1 - x^2/2$ to get

$$\sin \left(\frac{x}{2}\right) \approx \frac{x}{2} \quad \text{and} \quad \cos \left(\frac{x}{2}\right) \approx 1 - \frac{(x/2)^2}{2} = 1 - \frac{x^2}{8}.$$

Substituting these expressions into Equation (13), we obtain an approximation for $\sin x$ that improves on $\sin x \approx x$:

$$\sin x \approx 2\left(\frac{x}{2}\right)\left(1 - \frac{x^2}{8}\right) = x\left(1 - \frac{x^2}{8}\right)$$

$$= x - \frac{x^3}{8}.$$

This is a cubic function that approximates the sine function near $x = 0$, but it is slightly different from the third degree Taylor approximation, $\sin x \approx x - x^3/6$. We compare these two cubic approximations in the Problems at the end of this section.

Figure 8.20 shows the graph of this cubic along with the sine function on the interval from -3 to 3. The two curves seem almost identical for x between -1 and 1; they are reasonably close between -2.5 and -1 and again between 1 and 2.5; but they apparently begin to diverge farther from 0. Note how much better this cubic function seems to approximate $\sin x$ than our linear approximation $\sin x \approx x$.

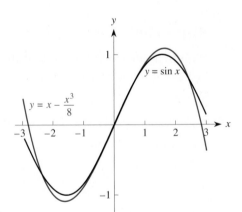

FIGURE 8.20

Think About This   Check numerically on your calculator how close the cubic is to the sine function at $x = 0.5$, at $x = 1$, and at $x = 1.5$. ▭

We can continue this process to produce still better approximations to both the cosine and the sine functions by using the same trigonometric identities.

For instance, using the double-angle formula $\cos 2\theta = 1 - 2\sin^2\theta$ and our new approximation

$$\sin x \approx x - \frac{x^3}{8},$$

we get

$$\cos x = 1 - 2\sin^2\left(\frac{x}{2}\right) = 1 - 2\left[\sin\left(\frac{x}{2}\right)\right]^2$$

$$\approx 1 - 2\left[(x/2) - \frac{(x/2)^3}{8}\right]^2.$$

After some algebraic simplification, we eventually get

$$\cos x \approx 1 - \frac{x^2}{2} + \frac{x^4}{32} - \frac{x^6}{2048}.$$

The two graphs in Figure 8.21 illustrate that this sixth degree polynomial $P_6(x)$ is an almost perfect match to the cosine function from about $x = -1.5$ to about $x = 1.5$. It is quite accurate from about $x = -2$ to about $x = -1.5$ and from about $x = 1.5$ to about $x = 2$; thereafter its accuracy diminishes. For comparison, Figure 8.22 shows three graphs: the basic cosine curve, the initial quadratic approximation $P_2(x) = 1 - x^2/2$, and this sixth degree polynomial approximation $P_6(x)$. The higher degree polynomial is clearly a much better fit. It follows the bends of the cosine curve and stays close to it over a wider interval of x-values.

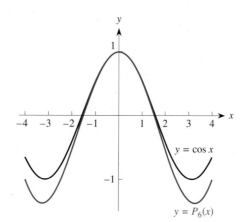

FIGURE 8.21

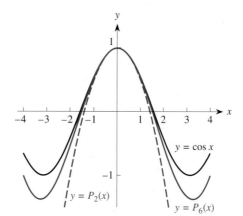

FIGURE 8.22

Approximating sin x and cos x Using Taylor Polynomials

We could continue this process to get better polynomial approximations to both sin x and cos x by using the trigonometric identities to construct polynomials of still higher degree. Unfortunately, each successive improvement is based on a series of approximations—we used sin $x \approx x$ to generate the approximation

$$\cos x \approx 1 - \frac{x^2}{2},$$

and so on. The approximation errors in this process mount up and give less than the best possible approximation at each successive stage. For instance, we first found

$$\sin x \approx x = T_1(x)$$

and then used it to find

$$\sin x \approx x - \frac{x^3}{8} = Q_3(x).$$

Actually, as you will learn in calculus, the *best* possible cubic curve to approximate the sine curve near $x = 0$ is the Taylor polynomial of degree 3.

$$\sin x \approx x - \frac{x^3}{6} = T_3(x).$$

It is identical to the cubic polynomial we obtained earlier based on fitting a cubic function to a set of values of the sine function near 0. Figure 8.23 shows the graph of this cubic and the underlying sine curve. Figure 8.24 shows the sine curve and the two different cubic approximations: the Taylor approximation of degree three, $T_3(x) = x - x^3/6$, and the polynomial of degree three based on the trig identities, $Q_3(x) = x - x^3/8$. Note that $T_3(x)$ remains closer to the sine curve over a wider interval than $Q_3(x)$ does. Note also that $T_3(x)$ bends in such a way that it remains very close to the sine curve over a relatively large portion of its first arch.

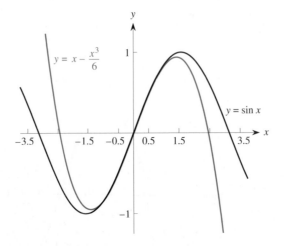

FIGURE 8.23

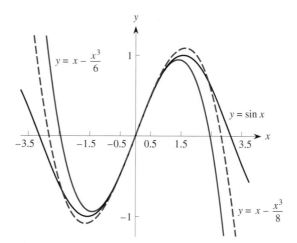

FIGURE 8.24

Use your function grapher to view what happens in a more dynamic way. In particular, examine the three curves near $x = 0$ to see that T_3 actually is closer to the sine curve than Q_3 is.

As a final note, let's look at the ideas we have developed in this section from a somewhat different perspective. Until now, we have interpreted Taylor polynomials as a means of *approximating* one function by a polynomial. An alternative interpretation is that we have been *constructing* a function (or a portion of a function) from simpler functions. That is, we have been constructing the trigonometric functions by using polynomials as the fundamental building blocks. More specifically, we have used *linear combinations* of power terms (i.e., sums of constant multiples of power terms) as these fundamental building blocks. This idea of using linear combinations of basic mathematical elements to construct more complicated mathematical structures is a continuing theme throughout mathematics.

Problems

1. Use the Taylor polynomial approximation to $f(x) = \cos x$ of degree 2 to estimate the value of the cosine function for $x = 0, 0.1, 0.2, 0.3, 0.4, 0.5,$ and 0.6. Compare each estimate to the correct value.

2. Use the values you calculated in Problem 1 to construct a table that has a column containing the error in the approximation (the difference between the estimate and the correct value). Analyze the column of errors. Do they appear to grow approximately linearly? exponentially? quadratically? cubically?

3. Use the Taylor polynomial approximation to $f(x) = \sin x$ of degree 3 to construct a table of estimates for the values of the sine function when $x = 0,$ $0.1, 0.2, 0.3, 0.4, 0.5,$ and 0.6. Calculate the errors and analyze them the same way you did in Problem 2.

4. **a.** Construct a table containing Taylor polynomial approximations of degrees $n = 1$ and $n = 3$ to $f(x) = \sin x$ for $x = -\pi/3, -\pi/4, -\pi/6, 0,$ $\pi/6, \pi/4, \pi/3$.

   b. Add 2 columns to the table, one for $\sin x - x$ and another for $\sin x - (x - x^3/6)$, to compare the linear and cubic approximations to the correct value for each x.

   c. Use your function grapher to graph $y = \sin x - x$. What type of function does it appear to be?

   d. Use polynomial regression to find an appropriate polynomial to fit the data values of $\sin x - x$ versus x.

   e. Repeat part (c) with $y = \sin x - (x - x^3/6)$.

5. Construct a table of values of $\sin x$ for $x = -4\pi/25,$ $-3\pi/25, -2\pi/25, \ldots, 4\pi/25$. Use your calculator to find the cubic polynomial that fits this set of sine values. How close does it come to the cubic Taylor polynomial approximation $\sin x \approx x - x^3/6$?

6. Use the Taylor polynomial approximation of degree $n = 5$ to $f(x) = \sin x$ to find a polynomial approximation of degree $n = 5$ to $g(x) = \sin(-x)$. Is the result surprising? Explain.

7. Use the Taylor polynomial approximation of degree $n = 5$ to $f(x) = \sin x$ to find a polynomial approximation of degree $n = 10$ to $g(x) = \sin(x^2)$. Graph both $g(x)$ and your approximation to it on the interval from $-\pi$ to π. Based on this graph, over what interval does your polynomial seem to be a good approximation to $g(x)$?

8. **a.** Use the Taylor polynomial approximation of degree $n = 5$ to $f(x) = \sin x$ to find a polynomial approximation of degree $n = 5$ to $h(x) = \sin 2x$.
   b. What do you get if you multiply the polynomial approximation of degree $n = 3$ to $f(x) = \sin x$ by the polynomial approximation of degree $n = 4$ to $g(x) = \cos x$?
   c. Graph $h(x)$ and twice the product of the two approximations found in part (b). What do you observe? Explain why.

9. Write the Taylor polynomial approximation of degree 3 to $f(x) = \sin x$ and the approximation of degree 4 to $g(x) = \cos x$. Square each expression and add them. What do you get? What do you think will happen if you use higher degree approximations? Explain.

10. In the text, we used the double-angle identity $\cos 2\theta = 1 - 2\sin^2\theta$ with $\theta = x/2$ to construct the approximation
$$\cos x \approx 1 - \frac{x^2}{2} + \frac{x^4}{32} - \frac{x^6}{2048}.$$
Instead, use the alternative form of the identity $\cos 2\theta = \cos^2\theta - \sin^2\theta$ and any lower degree polynomials to find a different approximation to $\cos x$.

11. Repeat Problem 10 with the third form of the double-angle identity $\cos 2\theta = 2\cos^2\theta - 1$ to construct still another approximation formula for $\cos x$.

12. The function $f(x) = (\sin x)/x$ is not defined at $x = 0$.
   a. Use values of $x = 0.1, 0.01, 0.001, 0.0001, 0.00001, \ldots$ to investigate the behavior of this function close to $x = 0$. What limiting value does this function appear to approach?
   b. Use the linear Taylor polynomial approximation to $\sin x$ to explain why the limiting value you found in part (a) appears to make sense.

13. In calculus, you will have to determine the value of
$$\frac{\sin(x + \Delta x) - \sin x}{\Delta x},$$
where Δx is a very small quantity.
   a. Estimate the value of this quotient by using linear approximations to both sine expressions.
   b. Estimate the value of this quotient by using a cubic approximation to both $\sin x$ and $\sin(x + \Delta x)$.
   c. With the cubic approximation, suppose that Δx is actually 0. What does the resulting expression suggest?

14. The exponential function $f(x) = e^x$ with base $e = 2.71828\ldots$ is used extensively in mathematics and the sciences. As with the trig functions, its values are calculated using Taylor polynomial approximations:
$$e^x \approx 1 + x,$$
$$e^x \approx 1 + x + \frac{x^2}{2!},$$
$$e^x \approx 1 + x + \frac{x^2}{2!} + \frac{x^3}{3!},$$
and so on. Use these and any further approximations that you need to approximate the values of
   a. $e^{0.1}$
   b. $e^{-0.1}$
   c. Use the given polynomials and any additional approximations to e^x that you need to estimate the value of e reasonably accurately. What degree polynomial will produce two-decimal accuracy? three-decimal accuracy? four-decimal accuracy?

8.3 Properties of Complex Numbers

One of the most amazing developments in the history of mathematics was the introduction of complex numbers to solve quadratic equations. For example, if $x^2 + 4 = 0$, then $x^2 = -4$, so that $x = \pm\sqrt{-4} = \pm 2i$, and the two roots are $x = 2i$ and $x = -2i$, where $i = \sqrt{-1}$. Similarly, from the quadratic formula, the roots of $x^2 - 2x + 10 = 0$ are
$$x = \frac{-(-2) \pm \sqrt{(-2)^2 - 4(10)}}{2} = \frac{2 \pm \sqrt{-36}}{2} = \frac{2 \pm (6i)}{2} = 1 \pm 3i,$$

or $x = 1 - 3i$ and $x = 1 + 3i$.

In our exploration of the nature of the roots of polynomials in Section 4.4, we demonstrated that quadratic, cubic, and higher degree polynomials have a surprisingly high proportion of complex zeros. We now develop a way to visualize complex numbers that gives a deeper understanding of the processes that lead to such polynomial equations.

Any complex number $z = a + bi$ is composed of two parts, a *real part, a*, and an *imaginary part, b*. For instance, in $z = 4 + 7i$, the real part is 4 and the imaginary part is 7. We occasionally write $a = \text{Re}(z)$ and $b = \text{Im}(z)$, respectively. Note that a and b are both real numbers; it is the combination $a + bi$ that is a complex number. In the special case when $b = 0$, the complex number $z = a + bi$ reduces to a real number. In another special case where $a = 0$, the complex number z reduces to a pure imaginary number, bi.

The arithmetic of complex numbers, for the most part, is quite straightforward, and we review it briefly in Appendix E. Because $i = \sqrt{-1}$, it follows that

$$i^2 = (\sqrt{-1})^2 = -1$$
$$i^3 = (i^2)(i) = (-1)(i) = -i$$
$$i^4 = (i^2)(i^2) = (-1)(-1) = 1$$
$$i^5 = (i^4)(i) = 1(i) = i.$$

In fact, all higher powers of i simply cycle through the four "values" $i, -1, -i$, and 1. That is, $i^6 = i^2 = -1, i^7 = i^3 = -i, i^8 = i^4 = 1, i^9 = i^5 = i$, and so on.

Visualizing complex numbers geometrically is extremely helpful. We do so by using the **complex plane,** which is a two-dimensional coordinate system designed to display a complex number $z = a + bi$. We measure the real part a horizontally and the imaginary part b vertically. In Figure 8.25 we plot the complex number $z = 2 + 5i$. Note that it lies 2 units to the right and 5 units up from the origin. Similarly, the complex numbers $1 - 3i$, and $-2 + i$ are also plotted in Figure 8.25. Any purely real number, such as 4 (which is $4 + 0i$) or -6 (which is $-6 + 0i$), lies on the horizontal axis. Any purely imaginary number, such as $4i$ (which is $0 + 4i$) or $-3i$ (which is $0 - 3i$) lies on the vertical axis.

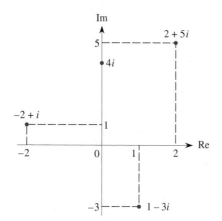

FIGURE 8.25

Suppose that $z = a + bi$ is an arbitrary complex number that we plot as a point in the complex plane. We connect the point to the origin with a line segment, which is the hypotenuse of a right triangle, as shown in Figure 8.26. The base of the triangle is a, the real part of z, and the height of the triangle is b, the size of the imaginary part of z. The Pythagorean theorem gives the length of the hypotenuse as $\sqrt{a^2 + b^2}$, which we interpret as the size of the complex number $z = a + bi$. We call it the **modulus** of the complex number and write it as

$$\|z\| = \sqrt{a^2 + b^2}.$$

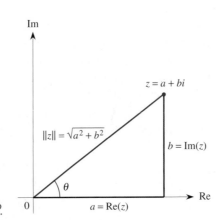

FIGURE 8.26

For instance, if $z = 4 + 3i$, then its modulus is

$$\|z\| = \sqrt{4^2 + 3^2} = \sqrt{25} = 5.$$

The complex numbers $4 - 3i$, $-4 + 3i$, and $-4 - 3i$ all have the same modulus of 5. Sketch them to verify that this is indeed the case.

Think About This   Are there any other points in the complex plane that also have a modulus of 5? What can you say about all such complex numbers? ▭

We again consider the complex number $z = a + bi$ and the associated right triangle in the complex plane. We now focus on the angle θ shown in Figure 8.26. By convention, θ is measured counterclockwise from the horizontal, or real, axis. Thus

$$\tan \theta = \frac{b}{a}.$$

We also have the two further relations

$$\cos \theta = \frac{a}{\|z\|} \quad \text{and} \quad \sin \theta = \frac{b}{\|z\|},$$

which lead to

$$a = \|z\| \cos \theta \quad \text{and} \quad b = \|z\| \sin \theta.$$

Consequently, we can write the original complex number z in the equivalent *trigonometric form*

$$z = a + bi = \|z\|\cos \theta + i\|z\|\sin \theta$$
$$= \|z\|(\cos \theta + i \sin \theta).$$

The *trigonometric form* for the complex number $z = a + bi$ is

$$z = \|z\|(\cos \theta + i \sin \theta),$$

where

$$\|z\| = \sqrt{a^2 + b^2} \quad \text{and} \quad \tan \theta = \frac{b}{a}, a \neq 0.$$

EXAMPLE 1

Find the trigonometric form for the complex number $z = 4 + 3i$.

Solution   For $z = 4 + 3i$, we have $\|z\| = 5$, so that

$$z = 4 + 3i = 5(\cos \theta + i \sin \theta),$$

where $\tan \theta = 3/4$, so that $\theta = \arctan 3/4 = 0.6435$ radian or $36.87°$.

Think About This   Use the value for the angle θ in Example 1 to show that the trigonometric form for the complex number z is identical to the original expression $4 + 3i$. ▱

Powers of Complex Numbers

The trigonometric form for a complex number $z = a + bi$ allows us to interpret z as being located at a certain distance, the modulus, from the origin and rotated through an angle θ from the horizontal. This model gives us a way to gain some special insights into powers of complex numbers.

EXAMPLE 2

For $z = 4 + 3i$, (**a**) find z^2 algebraically and (**b**) interpret z^2 geometrically in the complex plane.

Solution

a. If $z = 4 + 3i$,

$$z^2 = (4 + 3i)^2 = 4^2 + 2(4)(3i) + (3i)^2 \qquad (u + v)^2 = u^2 + 2uv + v^2$$
$$= 16 + 24i + 9(i^2)$$
$$= 16 + 24i - 9 \qquad\qquad i^2 = -1$$
$$= 7 + 24i.$$

This algebraic result provides no special insight into how z^2 is related to z.

b. We look at the trigonometric form for $z = 4 + 3i$. The modulus is $\|z\| = 5$, and the associated angle is $\theta = \arctan(3/4) = 0.6435$ radians, or $36.87°$, as in Example 1. Now consider the trigonometric form for z^2. Its modulus is

$$\|z^2\| = \sqrt{7^2 + 24^2} = \sqrt{49 + 576} = \sqrt{625} = 25,$$

which is the square of the modulus of the original complex number z. Next, the angle ϕ associated with z^2 is defined by

$$\tan \phi = \frac{24}{7} \quad \text{so that} \quad \phi = \arctan \frac{24}{7} \approx 1.2870 \text{ radians or } 73.74°,$$

which is exactly twice the angle θ (=0.6435 radians or 36.87°) associated with z, as illustrated in Figure 8.27.

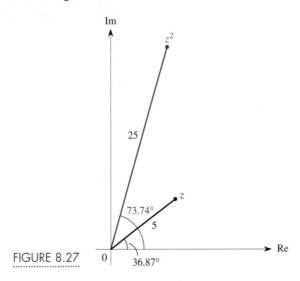

FIGURE 8.27

For this particular complex number $z = 4 + 3i$, z^2 is related to z by the process of squaring the modulus and doubling the angle. Does this rule hold in general? Let's look at two other simple cases.

EXAMPLE 3

Find the modulus and angle associated with $z^2 = (2i)^2$ and relate them to the modulus and angle associated with $z = 2i$.

Solution We know that $z = 2i = 0 + 2i$ is located at a distance of $\|z\| = 2$ from the origin with an associated angle of $\theta = \pi/2$ measured in the usual positive direction from the horizontal axis. We now consider $z^2 = (2i)^2$, which is

$$z^2 = 4i^2 = -4 = -4 + 0i.$$

This complex number has modulus 4 and associated angle π because it is on the negative real axis. That is, the modulus of $z^2 = (2i)^2$ is the square of the modulus of $z = 2i$, and the associated angle π is twice the angle $\pi/2$ associated with $z = 2i$.

EXAMPLE 4

Find the modulus and angle associated with $z^2 = (1 + i)^2$, where $z = 1 + i$, and relate them to the corresponding modulus and angle for z.

Solution For $z = 1 + i$, the modulus is

$$\|z\| = \sqrt{1^2 + 1^2} = \sqrt{2},$$

and the associated angle is $\theta = \pi/4$. Further,

$$z^2 = (1 + i)^2 = 1 + 2i + i^2 = 1 + 2i - 1 = 2i,$$

so $\|z^2\| = \sqrt{4} = 2$, the square of the modulus of z. The associated angle is $\pi/2$, or double the angle associated with z. So, again, when we square a complex number, the modulus is squared and the angle is doubled.

Let's now consider any complex number $z = a + bi$ in the equivalent trigonometric form

$$z = \|z\|(\cos\theta + i\sin\theta).$$

Squaring z gives

$$z^2 = \|z\|^2(\cos\theta + i\sin\theta)^2$$
$$= \|z\|^2(\cos^2\theta + 2i\cos\theta\sin\theta + i^2\sin^2\theta).$$

Using $i^2 = -1$ and collecting the real and imaginary terms yields

$$z^2 = \|z\|^2[(\cos^2\theta - \sin^2\theta) + 2i\cos\theta\sin\theta].$$

Now recall the double-angle identities:

$$\cos 2\theta = \cos^2\theta - \sin^2\theta$$
$$\sin 2\theta = 2\sin\theta\cos\theta.$$

Examining the real part of the previous expression for z^2, we see that it equals $\cos 2\theta$, whereas the imaginary part equals $\sin 2\theta$. Thus we have

$$z^2 = \|z\|^2(\cos 2\theta + i\sin 2\theta).$$

Geometrically, squaring any complex number always produces a new complex number whose modulus is the square of the original modulus and whose angle is double the original angle. If the modulus of the original number is greater than 1, z^2 is a "larger" complex number, as shown in Figure 8.28(a). If $\|z\|$ is smaller than 1, z^2 is a "smaller" complex number, as shown in Figure 8.28(b).

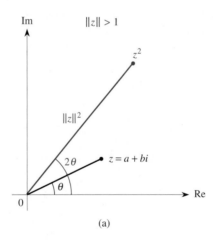
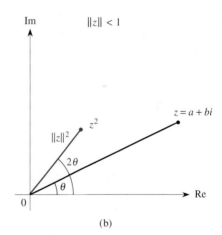

FIGURE 8.28

(a)   (b)

What about other powers of $z = a + bi$? Is there any pattern for z^n when $n > 2$?

EXAMPLE 5

Find the modulus and angle associated with z^3 when $z = 2i$.

Solution   The complex number $z = 2i$ is located at a distance of 2 from the origin and at an angle of $\pi/2$. Now consider

$$z^3 = (2i)^3 = 8i^3 = -8i = 0 - 8i.$$

It is located at a distance of 8 from the origin and is rotated through an angle of $3\pi/2$, which is triple $\pi/2$. Thus the modulus of $(2i)^3$ is the cube of the modulus of $2i$, and the associated angle $3\pi/2$ is three times the angle $\pi/2$ associated with $2i$.

Let's find out whether the same pattern holds when we cube any complex number $z = a + bi$. We do so by using the trigonometric form. Because

$$z^2 = \|z\|^2(\cos 2\theta + i \sin 2\theta),$$

we have

$$z^3 = z^2(z)$$
$$= [\|z\|^2(\cos 2\theta + i \sin 2\theta)][\|z\|(\cos \theta + i \sin \theta)]$$
$$= \|z\|^3[\cos 2\theta \cos \theta + i \cos 2\theta \sin \theta + i \sin 2\theta \cos \theta + i^2\sin 2\theta \sin \theta].$$

Using $i^2 = -1$ and collecting the real and imaginary terms, we get

$$z^3 = \|z\|^3[(\cos 2\theta \cos \theta - \sin 2\theta \sin \theta) + i(\cos 2\theta \sin \theta + \sin 2\theta \cos \theta)].$$

We now use the sum identities

$$\cos(x + y) = \cos x \cos y - \sin x \sin y$$
$$\sin(x + y) = \sin x \cos y + \cos x \sin y$$

with $x = 2\theta$ and $y = \theta$ to get

$$z^3 = \|z\|^3[\cos(2\theta + \theta) + i \sin(2\theta + \theta)]$$
$$= \|z\|^3(\cos 3\theta + i \sin 3\theta).$$

Thus cubing any complex number always results in cubing the modulus and tripling the rotation of the original complex number. It either "lengthens" the complex number if the original modulus is greater than 1, as illustrated in Figure 8.29(a), or "contracts" it if the modulus is less than 1, as illustrated in Figure 8.29(b). If the modulus equals 1 and $\theta \neq 0$, all that happens is a rotation.

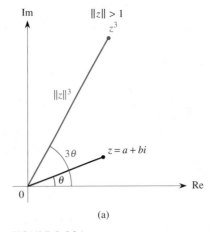

(a)

FIGURE 8.29A

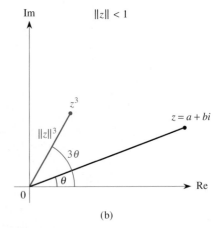

(b)

FIGURE 8.29B

In the Problems at the end of this section, we ask you to show that

$$z^4 = z^3(z) = \|z\|^4[\cos 4\theta + i \sin 4\theta]$$

and that, in general for any positive integer power n,

$$z^n = z^{n-1}(z)$$
$$= \|z\|^n(\cos \theta + i \sin \theta)^n$$
$$= \|z\|^n(\cos n\theta + i \sin n\theta).$$

This important and extremely useful result is known as **DeMoivre's theorem** after French mathematician Abraham DeMoivre who first discovered it.

> **DeMoivre's Theorem**
> If
> $$z = a + bi = \|z\|(\cos \theta + i \sin \theta)$$
> then
> $$z^n = \|z\|^n(\cos n\theta + i \sin n\theta)$$
> for any positive integer n.

Complex Conjugates

We know that complex numbers occur in complex conjugate pairs, such as $z = 3 + 5i$ and $z = 3 - 5i$ when we use the quadratic formula. If $z = a + bi$ is any complex number, we write its conjugate as $\bar{z} = a - bi$, which is shown geometrically in Figure 8.30. Clearly, z and $\bar{z}$ have the same modulus, $\sqrt{a^2 + b^2}$, so $\|\bar{z}\| = \|z\|$. Also, if the angle associated with z is θ, the angle associated with $\bar{z}$ is $-\theta$.

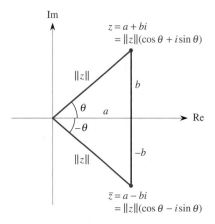

FIGURE 8.30

Using the reflection identities
$$\cos(-\theta) = \cos \theta \quad \text{and} \quad \sin(-\theta) = -\sin \theta,$$
we find that
$$\bar{z} = a - bi = \|z\|[\cos(-\theta) + i \sin(-\theta)] = \|z\|(\cos \theta - i \sin \theta).$$
Applying DeMoivre's theorem to $\bar{z}$ gives
$$(\bar{z})^n = \|z\|^n(\cos \theta - i \sin \theta)^n$$
$$= \|z\|^n(\cos n\theta - i \sin n\theta).$$

A simple extension of these ideas provides a way of visualizing both the product and the quotient of any two complex numbers. We explore this approach in the Problems at the end of this section.

Problems

In Problems 1–9, find the modulus and the associated angle for each complex number.

1. $z = 4 - 3i$
2. $z = 5 + 12i$
3. $z = 12 - 5i$
4. $z = -15 + 20i$
5. $z = 64 - 36i$
6. $z = 8 - 3i$
7. $z = -5 + 7i$
8. $z = 3 + \sqrt{8}i$
9. $z = -8 - \sqrt{3}i$

10–18. Find the trigonometric form for each complex number in Problems 1–9.

19–22. For each complex number in Problems 1–4, find z^2 algebraically.

23–31 For each complex number in Problems 1–9, find z^2 by using DeMoivre's theorem.

32–35. For each complex number in Problems 1–4, find z^3 algebraically.

36–44. For each complex number in Problems 1–9, find z^3 by using DeMoivre's theorem.

45. For $z = 1 + 2i$, calculate and plot z^n, for $n = 0, 1, 2, 3,$ and 4.

46. Repeat Problem 45 for $z = 0.6 + 0.8i$. What difference do you observe about the behavior of the two sets of points?

47. Show that $z^4 = z^3 \cdot z = \|z\|^4(\cos 4\theta + i \sin 4\theta)$ for any complex number z.

48. Prove DeMoivre's theorem for any integer power n:
$$z^n = \|z\|^n(\cos n\theta + i \sin n\theta).$$
Hint: Write $z^n = z^{n-1} \cdot z$ and assume that
$z^{n-1} = \|z\|^{n-1}[\cos((n-1)\theta) + i \sin((n-1)\theta)].$

49. Suppose that you have two complex numbers $z = a + bi$ and $w = c + di$.
   a. What is the product of z and w algebraically?
   b. What is the product of z and w using the trigonometric forms of z and w?

c. Hypothesize and prove an extension of DeMoivre's theorem that will allow you to multiply any two complex numbers in trigonometric form. (*Hint*: Your extension should reduce to DeMoivre's theorem for z^2 when $w = z$.)
d. Apply the rule that you discovered in part (b) to find the product of
   i. $z = 1 + 2i$ and $w = 1 - 2i$
   ii. $z = \dfrac{1}{2} + \dfrac{\sqrt{3}}{2}i$ and $w = \dfrac{\sqrt{3}}{2} - \dfrac{1}{2}i.$

50. a. Hypothesize an extension of DeMoivre's theorem that will allow you to divide one complex number by another in trigonometric form.
   b. Apply the rule that you proposed in part (a) to find the quotient of
   i. $z = 1 + 2i$ and $w = 1 - 2i$
   ii. $z = \dfrac{1}{2} + \dfrac{\sqrt{3}}{2}i$ and $w = \dfrac{\sqrt{3}}{2} - \dfrac{1}{2}i.$

51. a. Hypothesize an extension of DeMoivre's theorem that will allow you to determine the square root of a complex number z.
   b. Apply the rule that you proposed in part (a) to find the square root of
   $$z = \dfrac{1}{2} + \dfrac{\sqrt{3}}{2}i.$$
   c. Algebraically square the complex number that you obtained in part (b) to verify that it actually is the square root of the original number in part (a).
   d. Can you hypothesize a further extension of DeMoivre's theorem to extract any desired root of a complex number? any desired rational power of a complex number? Explain.

52. A negative real number can be thought of as being produced by rotating the corresponding positive real number (which is located on the horizontal axis) through an angle π in the complex plane. Use this interpretation to explain why the product of two negative numbers is positive.

53. Show that, for any pair of complex conjugates $z = a + bi$ and $\bar{z} = a - bi$, $z \cdot \bar{z} = \|z\|^2$.

8.4 ······ The Road to Chaos

In this section we investigate some fascinating results that arise from iteration processes applied to complex numbers. Let's begin with any complex number in trigonometric form—say, $z_0 = \|z_0\|(\cos \theta + i \sin \theta)$—and square it to produce $z_1 = z_0{}^2$. Using

DeMoivre's theorem, we know that the geometric result is a complex number whose associated angle is 2θ and whose modulus is $\|z_0\|^2$. Recall that, if $\|z_0\| > 1$, we get a rotation and an expansion to a "larger" complex number; if $\|z_0\| < 1$, we get a rotation and a contraction to a "smaller" complex number; if $\|z_0\| = 1$, we get only a rotation.

Suppose that we next square z_1 to produce $z_2 = z_1^2 = z_0^4$. If $\|z_0\| > 1$, we get a further rotation (to the angle $2 \times 2\theta = 4\theta$) and a further expansion. If $\|z_0\| < 1$, we get the same further rotation (to 4θ) and a further contraction. If $\|z_0\| = 1$, we get only the rotation (to 4θ).

What happens if we continue this process indefinitely to produce a sequence of complex numbers $z_0, z_1 = z_0^2, z_2 = z_1^2, z_3 = z_2^2, \ldots$? The geometric behavior of the terms of this sequence can be predicted easily by extending the reasoning we just used. If the modulus of the initial value z_0 is greater than 1, each successive iterate is farther from the origin in the complex plane, at a larger angle, and the sequence clearly diverges in a counterclockwise spiral pattern for $\theta > 0$, as shown in Figure 8.31(a). If $\|z_0\| < 1$, each successive term is closer to the origin; the successive iterates converge to 0 in a counterclockwise spiral pattern as each one is a further rotation of the original angle $\theta > 0$, as shown in Figure 8.31(b). Finally, if $\|z_0\| = 1$, all successive iterates fall on the boundary of the unit circle centered at the origin in the complex plane.

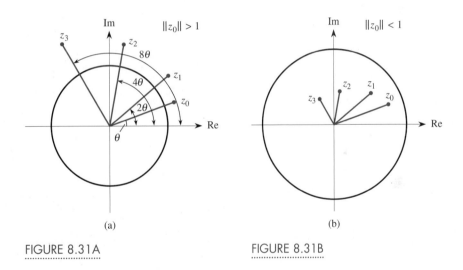

(a)

(b)

FIGURE 8.31A

FIGURE 8.31B

The Julia Set

Let's focus on the possible initial values for z_0. Any initial point inside the unit circle starts a sequence that spirals in to the origin; any initial point on the circle itself starts a sequence that remains on the unit circle; and any initial point outside the unit circle starts a sequence that spirals away toward infinity.

We can display this graphically in the following way. Visualize the unit circle centered at the origin in the complex plane, as shown in Figure 8.32. The circle is drawn in heavy black, the interior is shaded, and the region outside the circle is unshaded. Think of the unshaded region as indicating any point that begins a sequence that diverges, the shaded region as indicating those initial points for which the sequence converges to 0, and the black as indicating those initial points for which the sequence remains *on* the circle forever. The set of initial points for which the resulting sequences do not diverge to infinity is known as the **Julia set** associated with the function $f(z) = z^2$. (It is named after French mathematician Gaston

Julia, who discovered the properties of these sets in the 1920s.) The Julia set associated with $f(z) = z^2$ consists of the unit circle and all points inside it.

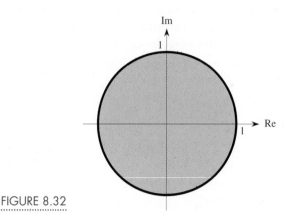

FIGURE 8.32

A relatively small change in what we have just done puts us on the road to chaos. Instead of using $f(z) = z^2$, let's see what happens if we use $f(z) = z^2 + C$, where C is any constant, either real or complex. (You may want to think of this as a family of functions for different values of C.) We take $z_1 = f(z_0) = z_0^2 + C$, so that $z_2 = f(z_1) = z_1^2 + C$, $z_3 = f(z_2) = z_2^2 + C, \ldots$ We now consider a variety of cases with different values for C and with different starting values z_0.

Let's begin with $C = 2$. If the initial value is $z_0 = 3$,

$$z_1 = z_0^2 + C = 9 + 2 = 11,$$
$$z_2 = z_1^2 + C = 121 + 2 = 123,$$
$$z_3 = z_2^2 + C = 123^2 + 2 = 15{,}131, \ldots,$$

and the sequence clearly diverges. Using the same $C = 2$ with some other starting values, we get

if $z_0 = 0.2$,   then $z_1 = 2.04$, $z_2 = 6.1616$, $z_3 = 39.9653, \ldots$;

if $z_0 = 1 + i$,   then $z_1 = 2 + 2i$, $z_2 = 2 + 8i$, $z_3 = -58 + 32i, \ldots$;

if $z_0 = 0.5 + 0.2i$,   then $z_1 = 2.21 + 0.2i$, $z_2 = 6.844 + 0.884i$, $z_3 = 48.059 + 12.10i \ldots$

All three sequences seem to diverge to infinity. Of course, we can't reach such a conclusion based on just a few examples; they can, at best, suggest what may happen.

Let's use DeMoivre's theorem to analyze the behavior of the successive iterates. Suppose that z_0 is any initial value inside the unit circle, so its modulus is less than 1. When we square it, the modulus for z_0^2 is smaller still. However, when we add 2 to it, the point is shifted 2 units to the right, so that z_1 must be outside and to the right of the unit circle.

Now suppose that z_0 (or some subsequent iterate) is outside the unit circle. Its modulus is greater than 1, so the modulus for z_0^2 is larger still. When we add 2 to it, the point is again shifted 2 units to the right. For almost all possible values of z_0, the resulting z_1 will be outside the unit circle.

There are some exceptions—say, $z_0 = 1.1i$ so that

$$z_1 = (1.1i)^2 + 2 = -1.21 + 2 = 0.79.$$

However, it can be shown that, eventually all subsequent iterates will land outside the circle and ultimately diverge to infinity. (Because each iteration involves a rotation, at some stage one of the successive iterates will eventually land near the horizontal axis to the right and the following iterate will be outside and to the right of the unit circle.) Thus it turns out that, with $f(z) = z^2 + 2$, for every initial point in the complex plane, the resulting sequence diverges. The Julia set associated with the function $f(z) = z^2 + C$, when $C = 2$, will be completely empty because all initial points give rise to sequences that eventually diverge. Our diagram of this Julia set will be entirely unshaded because there are no initial points that start convergent sequences.

Similarly, if $C = 2i$, all sequences will diverge regardless of the initial value for z_0. The additive constant $2i$ results in a shift upward of 2 units in the imaginary direction. Pick several initial values for z_0 (real, imaginary, or complex) and see what happens when you calculate the successive iterates.

However, if $C = 0.2i$ and we start with $z_0 = 0.5 + 0.2i$, we obtain

$$z_1 = (0.5 + 0.2i)^2 + 0.2i = 0.21 + 0.4i;$$

$$z_2 = -0.1159 + 0.368i; \qquad z_6 = -0.0394 + 0.1881i;$$

$$z_3 = -0.1220 + 0.1147i; \qquad z_7 = -0.0338 + 0.1852i;$$

$$z_4 = 0.0017 + 0.1720i; \qquad z_8 = -0.0332 + 0.1875i.$$

$$z_5 = -0.0296 + 0.2006i; \qquad z_9 = -0.0341 + 0.1876i.$$

The sequence apparently converges to some point in the complex plane.

Unfortunately, repeating this process for every possible starting value z_0 is not practical. Instead, we use a computer to perform such calculations for a large number of points in a grid to give a representative picture of what happens. As with the previous cases, we leave any initial point that starts a sequence that diverges to infinity unmarked to become part of the unshaded region. We put a small dot at any initial point that starts a sequence that converges to some point in the complex plane so that it will be part of the shaded Julia set.

The resulting picture of the Julia set for the function $f(z) = z^2 + 0.2i$ is shown in Figure 8.33.

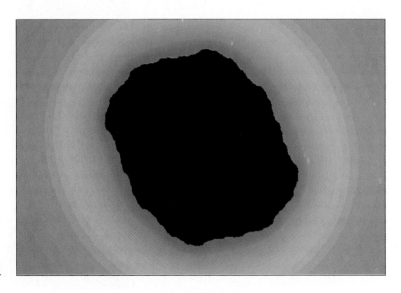

FIGURE 8.33

This picture does not indicate the limits of any of the sequences; it shows only those points that start sequences that have limits. Typically, if you take points in the interior of the shaded region, it turns out that nearby starting points tend to converge to limits that are relatively close to one another. However, if you take points near the boundary, very different results can occur. Initial points that are extremely close together can produce sequences that converge to radically different limits. The result is an instance of mathematical chaos because the behavior has no predictable patterns. Points that are very close together, provided that they are both near the boundary of the Julia set, may well lead to sequences that behave very differently. If you were to zoom in on the portion of the Julia set near these boundaries, you would see an ever more intricate design illustrating how nearby points can start sequences that either diverge or converge. They occur in a totally chaotic and unpredictable manner.

A striking illustration of this outcome is shown in Figure 8.34, which is the Julia set corresponding to $C = -0.2 + 0.7i$. Note how intricate the boundary appears. Figure 8.35 shows the result of zooming in on the upper left corner of the Julia set shown in Figure 8.34. Observe how roughly similar patterns repeat; this kind of repetition is typical of what happens when you zoom in repeatedly on Julia sets associated with most values for C. Also, note how much more jagged the boundary looks as more details appear in the magnified image in Figure 8.35, which also is typical.

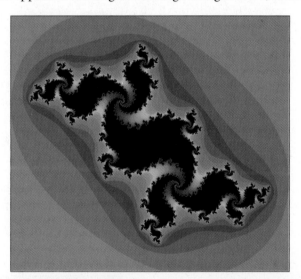

FIGURE 8.34

FIGURE 8.35

The Julia set associated with a complex constant C can be far more intricate than we have shown so far; it can, for instance, consist of a large variety of disconnected pieces. It may even consist of nothing but a collection of isolated points like a set of dust particles. You may want to experiment with some of these ideas yourself, using any of the many computer programs available for displaying Julia sets for iterated functions.

The Mandelbrot Set

There is a completely different way of looking at these ideas. In the discussion of Julia sets, we considered the function $f(z) = z^2 + C$, selected a particular value for C, and then examined points in the complex plane as starting points z_0 for iterated sequences. Now let's reverse this.

Suppose instead that we select a particular starting point z_0 and examine the effects of using different values for the complex constant C in $f(z) = z^2 + C$. Thus our view of the complex plane has shifted—it now represents all different constants rather than all different starting points. In particular, suppose that we select $z_0 = 0$ as the starting point for all sequences. Then, for any constant C, $z_1 = 0^2 + C = C$, $z_2 = z_1^2 + C = C^2 + C$, and so on. Clearly, if C is large (far from the origin in the complex plane), all successive iterates will be larger still and the successive points of the sequence will diverge. However, if C is fairly small, the successive iterates may remain close to the origin and the sequence may converge to some finite complex value.

The Julia set associated with the function $f(z) = z^2 + C$ consists of all initial points z_0 for which the sequences converge for a given constant C. Similarly, the **Mandelbrot set** associated with the function $f(z) = z^2 + C$ (named after French mathematician Benoit Mandelbrot) consists of all constants C for which the sequences starting from $z_0 = 0$ fail to diverge. For this initial point $z_0 = 0$, the Mandelbrot set illustrated in Figure 8.36 shows those constants C for which the corresponding sequences remain close to the origin. As with a Julia set, the boundary of the Mandelbrot set is an incredibly intricate structure. If you zoom in on it, as shown in Figure 8.37, you will see remarkable shapes with no predictable patterns; however, the original overall shape shown in Figure 8.36 appears to repeat at all levels of magnification. The main heart-shaped portion of the Mandelbrot set is called a *cardioid*, which we discuss in Chapter 9; the portion to the left of the cardioid is actually a circle.

FIGURE 8.36

FIGURE 8.37A

FIGURE 8.37B

FIGURE 8.37C

These displays show the Mandelbrot set with different shadings to indicate how quickly different sequences diverge from the starting value $z_0 = 0$. When different colors are used, the results are even more dramatic. You may want to examine the Mandelbrot set, using one of the programs available for displaying it. All such programs allow you to see the details at different levels of magnification as you zoom in on the boundary. In theory, there is no limit to the degree of complexity of the boundary. Such a shape is known as a *fractal*.

Many shareware programs are available (one of the most popular is called FracInt) that will let you investigate both Julia and Mandelbrot sets. This subject is one of the most exciting areas of current mathematical research, and many new and important theorems have been proven in the last few years. These ideas have also formed the basis for many of the computer graphics images that you have undoubtedly seen in today's movies.

Problems

1. **a.** Use the quadratic formula to find a condition on those values of C for which the sequence of iterates $x = f(x) = x^2 + C$ has a real limiting value.
   b. Verify your condition in part (a) by using $C = 0.1$, starting with $x_0 = 0.5$ and performing enough iterations to see the eventual behavior.
   c. Repeat part (b), using $C = 0.4$.

2. **a.** What is the limiting value you expect if $C = 5/4$ for the sequence of function iterations based on $x = f(x) = x^2 + C$?
   b. Start the iteration process at $x_0 = 0.5 + 0.5i$ and perform enough iterations to verify that the process seems to be converging to your answer for part (a).
   c. Start the iteration process at $x_0 = 1 + i$ and perform enough iterations to determine the eventual behavior of the sequence of iterates. How could you have anticipated the result without performing the actual calculations?

3. You can think of the iteration scheme for $x = f(x) = x^2 + C$ as the difference equation $x_{n+1} = f(x_n) = x_n^2 + C$. What are the equilibrium levels for the solutions to the difference equation? Under what conditions on C will the equilibrium values be real?

4. Explain graphically the significance of C in determining whether the iteration process based on $x = f(x) = x^2 + C$ has a real limiting value by looking at the graphs of $y = x^2 + C$ and $y = x$.

5. Explain graphically why the iteration process based on the function $x = f(x) = x^3 + C$ must have at least one real limiting value.

6. Consider iterations $x = f(x)$ based on the function $f(x) = x + \sin x$.
   a. Begin the iteration process at $x_0 = 2$ and perform enough iterations to allow you to recognize the limit of the resulting sequence.
   b. Repeat part (a), starting with $x_0 = 5$.
   c. Repeat part (a), starting with $x_0 = 8$. How does the limiting value compare to π?
   d. Repeat part (c), starting with $x_0 = 15$.
   e. Based on the function f, explain why all limits will be some multiple of π.

7. Consider iterations $x = f(x)$ based on the function $f(x) = x + \cos x$. Predict the possible values that can arise for the limits based on the function f. Verify whether your predictions are correct if you start with initial values $x_0 = 1, 3, 7$, and -12.

Chapter Summary

In this chapter, we continued our discussion of trigometric functions. In particular, we discussed the following:

◆ The fundamental identities that relate the sine and cosine functions.

◆ Some identities involving the tangent function.

◆ How to approximate the sine and cosine functions with polynomial functions.

◆ How the accuracy of a polynomial approximation depends on the degree of the polynomial.

◆ How to convert a complex number to its equivalent trigonometric form.

◆ How to construct powers of complex numbers with DeMoivre's theorem.

◆ The Julia set that is associated with a function $f(z)$ and the idea of chaos.

◆ The Mandelbrot set that is associated with a function $f(z)$.

Review Problems

Determine graphically which of the relationships in Problems 1–9 might be identities and which clearly are not identities. For those that appear to be identities, prove them algebraically.

1. $\sin x \cos^2 x + \sin^3 x = \cos x$

2. $\sin x \cos^2 x + \sin^3 x = \sin x$

3. $\sin x \cos^2 x + \sin^3 x = \cos^2 x$

4. $\sin x \cos^2 x + \sin^3 x = \sin^2 x$

5. $(\sin x + \cos x)^2 = 1 + \sin 2x$

6. $\dfrac{\sin \theta}{1 + \cos \theta} + \dfrac{\cos \theta}{\sin \theta} = \dfrac{1}{\sin \theta}$

7. $\dfrac{1}{1 + \cos t} + \dfrac{1}{1 - \cos t} = \dfrac{2}{\sin^2 t}$

8. $\cos^4 \theta - \sin^4 \theta = \cos 2\theta$

9. $\cos^6 \theta - \sin^6 \theta = \cos 2\theta$

10. Prove each identity.

   a. $\cos(x + y) + \cos(x - y) = 2 \cos x \cos y$
   b. $\sin(x + y) + \sin(x - y) = 2 \sin x \cos y$

11. Use the Taylor polynomial approximation of degree $n = 3$ for the sine function to estimate the value of the function $f(x) = \sin 3x$ at $x = 0.2$. Sketch the graph of the function and the approximating polynomial on the same set of axes.

12. Repeat Problem 11 with degree $n = 5$. Discuss any differences you observe.

13. You know that $\sin x \approx T_3(x) = x - x^3/6$ and $\cos x \approx T_2(x) = 1 - x^2/2$ when x is reasonably close to 0, so $[T_3(x)]^2 + [T_2(x)]^2$ should be fairly close to 1. Using your function grapher, estimate how far the expression $[T_3(x)]^2 + [T_2(x)]^2$ is from 1 for any value of x between -1 and 1 radian.

14. a. Convert the complex numbers $z = -6 + 8i$ and $w = 5 - 2i$ to trigonometric form.
   b. Use the results from part (a) to find $z \cdot w$, z/w, and w/z.

15. A complex number z has modulus 3 and an associated angle of 52°.

   a. Write the complex number in trigonometric form.
   b. Write the complex number in the usual form $z = a + bi$.
   c. Find the fifth power of this complex number z.
   d. Find the square root of this complex number z.

Use your function grapher to estimate the period for each function. Express your answers as multiples of π.

16. $f(x) = \sin 3x + \cos 2x$

17. $f(x) = \sin 3x + \cos 4x$

18. $f(x) = \sin 4x + \cos 2x$

19. $f(x) = \sin 2x + \cos 4x$

20. $f(x) = \sin \left(\dfrac{1}{2}x\right) + \cos 2x$

21. $f(x) = \sin 3x + \cos \left(\dfrac{1}{2}x\right)$

22. Based on your answers to Problems 16–21, conjecture a general rule for the period of the function $f(x) = \sin mx + \cos nx$, for any m and n.

Geometric Models

9

9.1 Introduction to Coordinate Systems

What is a coordinate system? In simple terms, a coordinate system provides a way to *locate* and *identify* points in the plane. In the usual *rectangular* or *Cartesian coordinate system*, every point can be pictured in either of two ways. First, a point P with coordinates (x_0, y_0) can be thought of as lying at the corner of a unique rectangle whose opposite corner is at the origin and two of whose sides lie along the two coordinate axes, as illustrated in Figure 9.1(a). The base of this rectangle is x_0, and its height is y_0. Second, the point P can be thought of as the intersection of two perpendicular lines, one parallel to the y-axis at a distance of x_0 from it and the other parallel to the x-axis at a height of y_0 from it, as illustrated in Figure 9.1(b).

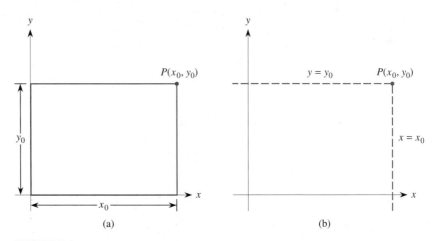

(a)   (b)

FIGURE 9.1

Mathematicians have found that, in many situations, rectangular coordinates are not the most natural or the most effective way to locate points and have developed alternative coordinate systems. One such approach involves the use of two axes that are not perpendicular, but rather meet at the origin at some angle other than a right angle. Points in such a slanted coordinate system can be located at the opposing vertex of a parallelogram, as illustrated in Figure 9.2.

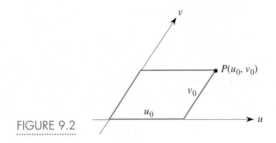

FIGURE 9.2

Another approach is to locate a point by using a circle centered at the origin O instead of a rectangle. To do so requires specifying both the radius of the circle and an angle θ to indicate where on the circle the point is located. This approach leads to the **polar coordinate system,** which is illustrated in Figure 9.3. We investigate this coordinate system in Sections 9.6 and 9.7.

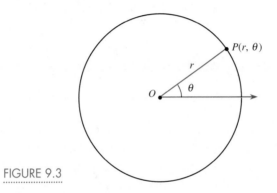

FIGURE 9.3

Other approaches are used for particular applications that involve locating points lying on some ellipse centered at the origin (an elliptic coordinate system), on some parabola (a parabolic coordinate system), or on a hyperbola (a hyperbolic coordinate system), as illustrated in Figures 9.4(a–c), respectively. In fact, the long range navigation (LORAN) system used by navigators in ships and planes to locate their positions is based on the fact that every point in a plane can be interpreted as lying at the intersection of two hyperbolas in a hyperbolic coordinate system.

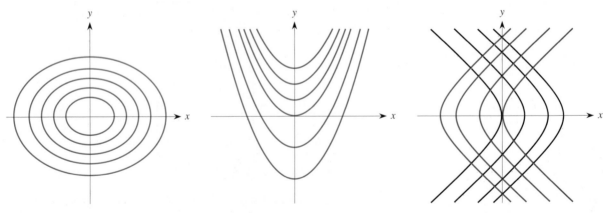

FIGURE 9.4

In this chapter, we first develop several special characteristics of the rectangular coordinate system and then show how to represent some extremely important curves. Later we explore other ways to represent functions.

9.2 ···· **Analytic Geometry**

One of the most useful and far-reaching developments in mathematics is Rene Descartes's idea of representing algebraic concepts geometrically. This approach, known as **analytic geometry,** lets you visualize the mathematics graphically to complement the algebraic approach that is based on symbols. Everything we have done involving graphs of functions is an outgrowth of Descartes's ideas. In this section we examine some additional ideas involving points, lines, and circles in the plane.

We begin by considering the two points A at (x_0, y_0) and B at (x_1, y_1) in the plane. You already know how to find an equation of the line through them by using either the point–slope form

$$y - y_0 = m(x - x_0)$$

or the slope–intercept form

$$y = mx + b,$$

where the slope of the line is

$$m = \frac{y_1 - y_0}{x_1 - x_0}.$$

Alternatively, we have the implicit form for the equation of a line,

$$ax + by = c,$$

where c/a and c/b represent the x- and the y-intercepts of the line, respectively.

Distance Between Points

We now ask: What is the distance between the points A at (x_0, y_0) and B at (x_1, y_1)? We write this distance as $|AB|$. Figure 9.5 shows that the points A and B determine a right triangle ABC; the coordinates of point C are (x_1, y_0) because C is at the same horizontal distance as B (measured from the y-axis) and at the same vertical height as A (measured from the x-axis). Moreover, the horizontal distance from A to C is $x_1 - x_0$; it is the change, or difference, in the x-coordinates. Similarly, the vertical distance from C to B is $y_1 - y_0$; it is the change in the y-coordinates. Consequently, the distance from A to B is the length of the hypotenuse of this right triangle. The Pythagorean theorem therefore gives us the **distance formula.**

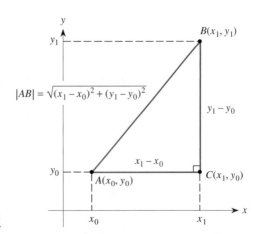

FIGURE 9.5

Distance Formula

The distance from point A at (x_0, y_0) to point B at (x_1, y_1) is

$$|AB| = \sqrt{(x_1 - x_0)^2 + (y_1 - y_0)^2}.$$

◆ **EXAMPLE 1**

Find the distance from the point A at $(2, 5)$ to the point B at $(6, 8)$.

Solution   Applying the distance formula gives

$$|AB| = \sqrt{(6 - 2)^2 + (8 - 5)^2}$$
$$= \sqrt{16 + 19} = \sqrt{25} = 5 \text{ units.}$$

Consider again the two points A at (x_0, y_0) and B at (x_1, y_1) in the plane. Suppose that we want to determine the *midpoint* M of the line segment connecting A to B. Figure 9.6 shows that the points A and B determine a right triangle ABC and that the points A and M determine a smaller right triangle AMD. These two right triangles

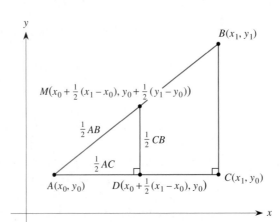

FIGURE 9.6

are similar, and hence their corresponding sides are proportional (see Appendix A4). Because M is halfway from A to B, we see that D is halfway from A to C, and the height DM is half the height CB. Thus the x-coordinate at D (and hence also at M) is

$$x = x_0 + \frac{1}{2}(x_1 - x_0).$$

Similarly, because the height DM is half the height CB, the y-coordinate at M is

$$y = y_0 + \frac{1}{2}(y_1 - y_0).$$

We can rewrite these expressions as

$$x_0 + \frac{1}{2}(x_1 - x_0) = x_0 + \frac{1}{2}x_1 - \frac{1}{2}x_0 = \frac{1}{2}(x_1 + x_0)$$

and

$$y_0 + \frac{1}{2}(y_1 - y_0) = y_0 + \frac{1}{2}y_1 - \frac{1}{2}y_0 = \frac{1}{2}(y_1 + y_0).$$

Thus the coordinates of the midpoint M of a line segment are simply the averages of the x-coordinates and the y-coordinates of the endpoints, respectively.

> ## Midpoint Formula
>
> The midpoint M of the line segment from A at (x_0, y_0) to B at (x_1, y_1) is at
>
> $$x = x_0 + \frac{1}{2}(x_1 - x_0), \qquad y = y_0 + \frac{1}{2}(y_1 - y_0)$$
>
> or
>
> $$x = \frac{x_1 + x_0}{2}, \quad y = \frac{y_1 + y_0}{2}.$$

◆ **EXAMPLE 2**

Find the midpoint of the line segment joining A at $(1, 11)$ and B at $(3, 7)$.

Solution   The coordinates of the midpoint are

$$x = x_0 + \frac{1}{2}(x_1 - x_0)$$

$$= 1 + \frac{1}{2}(3 - 1) = 1 + 1 = 2$$

and

$$y = y_0 + \frac{1}{2}(y_1 - y_0)$$

$$= 11 + \frac{1}{2}(7 - 11) = 11 + \frac{1}{2}(-4) = 9.$$

Alternatively,

$$x = \frac{x_1 + x_2}{2} = \frac{3 + 1}{2} = 2 \quad \text{and} \quad y = \frac{y_1 + y_2}{2} = \frac{7 + 11}{2} = 9.$$

Figure 9.7 shows the solution.

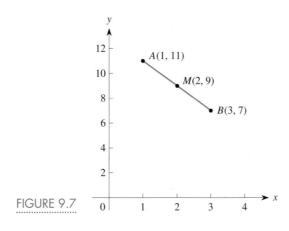

FIGURE 9.7

We might also want to determine a point at some other fraction of the distance from A to B. To do so, we simply extend the preceding argument to determine a point P at any distance from A to B. Suppose that we want the point one-quarter of the way from A to B. We then have

$$x = x_0 + \frac{1}{4}(x_1 - x_0) \quad \text{and} \quad y = y_0 + \frac{1}{4}(y_1 - y_0).$$

Think About This   Verify that this quarter-distance formula is correct by using an argument comparable to the one used for the midpoint formula. ▭

In general, if we want the point P at a fraction t of the distance from A to B, it will be located at

$$x = x_0 + t \cdot (x_1 - x_0)$$
$$y = y_0 + t \cdot (y_1 - y_0),$$

as shown in Figure 9.8. Incidentally, if $t > 1$, we get a point on the line *beyond B*, and if $t < 0$, we get a point on the line *before A*.

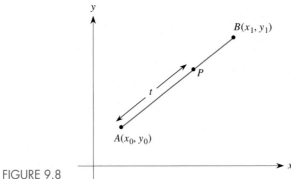

FIGURE 9.8

EXAMPLE 3

a. Find the point P located three-fifths of the way from A at $(-1, 3)$ to B at $(4, 13)$.

b. Find the point Q located seven-fifths of the way from A to B.

Solution

a. For P with $t = 3/5$,

$$x = x_0 + t \cdot (x_1 - x_0) = -1 + \frac{3}{5}[4 - (-1)] = 2 \quad \text{and}$$

$$y = y_0 + t \cdot (y_1 - y_0) = 3 + \frac{3}{5}(13 - 3) = 9.$$

Verify by plotting the points $(-1, 3)$, $(4, 13)$, and $(2, 9)$ that P is located three-fifths the way from A to B.

b. Similarly, for Q with $t = 7/5$,

$$x = x_0 + t \cdot (x_1 - x_0) = -1 + \frac{7}{5}[4 - (-1)] = 6 \quad \text{and}$$

$$y = y_0 + t \cdot (y_1 - y_0) = 3 + \frac{7}{5}(13 - 3) = 17.$$

Plot Q at $(6, 17)$ and note that this point lies on the line through A and B and is beyond B.

We could continue this example with different values of t to find other points on the line. In fact, every value of t determines a unique point on the line joining A at (x_0, y_0) and B at (x_1, y_1). Therefore the *two* equations for x and y give us a different way of representing the line. They are known as a *parametric representation* or *parametric equations* of the line, and the quantity t is called a *parameter*.

Parametric equations of the line through (x_0, y_0) and (x_1, y_1) are

$$x = x_0 + (x_1 - x_0)t \quad \text{and}$$
$$y = y_0 + (y_1 - y_0)t.$$

Note that this parametric form involves two interrelated equations for the line, not a single equation as in the point–slope form. It is possible to eliminate the parameter t to produce a single equation for the line. We ask you to do so in the Problems at the end of this section. However, the parametric form can provide valuable information.

The Equation of a Circle

We apply the concept of distance between two points in the plane to define a circle. A **circle** is the set of all points in the plane at a fixed distance from a fixed point. The fixed distance is called the **radius,** and the fixed point is called the **center.**

This definition allows us to find a general equation for any circle. Let r be the radius and let point C with coordinates (x_0, y_0) be the center of a circle. A point P with coordinates (x, y) lies on this circle provided that its distance from the center C is r, as shown in Figure 9.9. The distance formula gives the equivalent expression

$$|CP| = \sqrt{(x - x_0)^2 + (y - y_0)^2} = r.$$

We can eliminate the square root in this equation by squaring both sides and thus obtain the *standard form* for the equation of a circle.

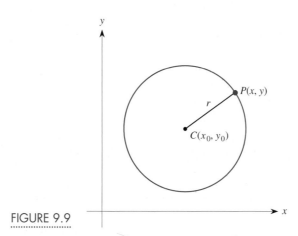

FIGURE 9.9

The equation of the circle with radius r centered at (x_0, y_0) is

$$(x - x_0)^2 + (y - y_0)^2 = r^2.$$

For instance, the circle of radius 8 centered at $(5, 2)$ has the equation

$$(x - 5)^2 + (y - 2)^2 = 8^2 = 64,$$

whereas the equation of the circle of radius 3 centered at $(-5, 0)$ is

$$(x + 5)^2 + y^2 = 9.$$

As a special case, the equation of a circle of radius r centered at the origin is

$$x^2 + y^2 = r^2.$$

Note that the equation of a circle does not represent a function. Picture any vertical line that passes through the circle but is not tangent to the circle—it intersects the circle twice and so the circle fails the vertical line test. That is, each such value of x has two corresponding values of y, which violates the definition of a function, as shown in Figure 9.10.

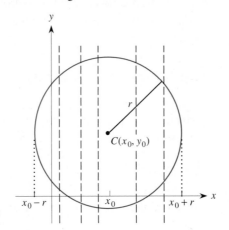

FIGURE 9.10

We get a similar result algebraically. For example, the circle of radius 10 centered at the origin has the equation

$$x^2 + y^2 = 100.$$

If we select $x = 6$, say, then

$$36 + y^2 = 100 \quad \text{so that} \quad y^2 = 64,$$

which has the solutions

$$y = 8 \quad \text{or} \quad y = -8.$$

Again, two values of y correspond to one value of x. Even though a circle does not represent a function, it is nonetheless a very important curve. We discuss several other curves that do not represent functions in Sections 9.3 and 9.4.

Let's start with the equation of the circle of radius 8 centered at $(5, 2)$:

$$(x - 5)^2 + (y - 2)^2 = 8^2 = 64.$$

We expand the left-hand side and combine like terms to get

$$x^2 - 10x + 25 + y^2 - 4y + 4 = 64$$

or

$$x^2 + y^2 - 10x - 4y - 35 = 0,$$

which is an equivalent, although different, representation for the same circle. Clearly, we could do the same with the equation of any circle,

$$(x - x_0)^2 + (y - y_0)^2 = r^2,$$

which is centered at (x_0, y_0) with radius r. Expanding the left-hand side, we get

$$x^2 - 2x_0x + x_0{}^2 + y^2 - 2y_0y + y_0{}^2 = r^2$$

or, equivalently,

$$x^2 + y^2 - 2x_0x - 2y_0y + x_0{}^2 + y_0{}^2 - r^2 = 0.$$

Because x_0, y_0, and r are constants, we can write this equation in the alternative form

$$x^2 + y^2 + Cx + Dy + E = 0,$$

where we have introduced the new constants

$$C = -2x_0, \quad D = -2y_0, \quad \text{and} \quad E = x_0{}^2 + y_0{}^2 - r^2.$$

Such an equation, known as the **general equation of the circle,** represents a circle for any choice of constants C, D, and E such that the radius of the circle is positive.

Suppose that we start with an equation such as

$$x^2 + y^2 - 10x - 4y - 35 = 0,$$

which we know from the preceding derivation is the equation of a circle. Examples 4 and 5 demonstrate how to work backward from this equation to determine the center and radius of the circle.

EXAMPLE 4

Show that

$$x^2 + y^2 - 10x - 4y - 35 = 0$$

is the equation of a circle by finding its center and radius.

Solution To solve this problem, we use the technique of *completing the square* (see Appendix A8) in both the x- and y-terms on the left-hand side and obtain

$$
\begin{aligned}
x^2 + y^2 - 10x - 4y - 35 &= (x^2 - 10x) + (y^2 - 4y) - 35 \\
&= [(x^2 - 10x + 25) - 25] + [(y^2 - 4y + 4) - 4] - 35 \\
&= [(x - 5)^2 - 25] + [(y - 2)^2 - 4] - 35 \\
&= (x - 5)^2 + (y - 2)^2 - 64 = 0.
\end{aligned}
$$

Adding 64 to both sides, we get

$$(x - 5)^2 + (y - 2)^2 = 64 = 8^2.$$

This is the equation of the circle with radius 8 and center at $(5, 2)$.

EXAMPLE 5

Find the radius and the center of the circle whose equation is

$$x^2 + y^2 + 8x - 10y - 8 = 0.$$

Solution To solve this problem, we again complete the square in both the x- and y-terms on the left-hand side and obtain

$$
\begin{aligned}
x^2 + y^2 + 8x - 10y - 8 &= (x^2 + 8x) + (y^2 - 10y) - 8 \\
&= [(x^2 + 8x + 16) - 16] + [(y^2 - 10y + 25) - 25] - 8 \\
&= [(x + 4)^2 - 16] + [(y - 5)^2 - 25] - 8 \\
&= (x + 4)^2 + (y - 5)^2 - 49.
\end{aligned}
$$

Therefore the original equation becomes

$$(x + 4)^2 + (y - 5)^2 - 49 = 0.$$

Adding 49 to both sides, we get

$$(x + 4)^2 + (y - 5)^2 = 49 = 7^2,$$

which is the equation of a circle with radius 7 and center at $(-4, 5)$. ◆

Note that not every equation of the form $x^2 + y^2 + Cx + Dy + E = 0$ represents the equation of a circle. In Examples 4 and 5, the constant term we ended up with on the right, either 64 or 49, was a positive number, so we could take the square root to get a radius. However, the constant on the right can be a negative number, in which case the equation does not represent a circle. In fact, even if the constant on the right is 0, we would not have a circle because the radius would be 0. We ask you to investigate such cases in the Problems at the end of this section.

EXAMPLE 6

In Example 4 of Section 4.6, we found two points on the line through the Earth and the moon at which the gravitational forces of the Earth and the moon on a spacecraft are exactly equal in size numerically. One point is 216 thousand miles from the Earth toward the moon and the other is 270 thousand miles from the Earth, which is 30 thousand miles beyond the moon. Find all such points in the plane containing the Earth, the moon, and the sun.

Solution We set up a coordinate system with the Earth at the origin and the moon on the horizontal axis, 240 thousand miles to the right. Suppose that the two gravitational forces are numerically equal at some other point P in the plane with coordinates (x, y), as shown in Figure 9.11. (Technically, we should consider not only the size of the two forces but also the directions in which they are exerted; that requires the notion of a vector quantity, which we discuss in Chapter 10.)

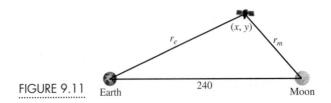

FIGURE 9.11

The distance formula gives the distance r_e from the Earth to the point P as

$$r_e = \sqrt{x^2 + y^2} \quad \text{so that} \quad r_e^2 = x^2 + y^2.$$

Similarly, the distance r_m from the moon to the point P is

$$r_m = \sqrt{(x - 240)^2 + y^2} \quad \text{so that} \quad r_m^2 = (x - 240)^2 + y^2.$$

Let m_0 be the mass of the spacecraft, m_1 be the mass of the Earth, and m_2 be the mass of the moon. Because the Earth is 81 times as massive as the moon, $m_1 = 81m_2$. Based on the universal law of gravitation, the size of the gravitational force F_e that the Earth exerts on the spacecraft is

$$F_e = \frac{Gm_0 m_1}{r_e^2} = \frac{Gm_0 \cdot (81m_2)}{r_e^2} = \frac{81Gm_0 m_2}{r_e^2}$$

and the size of the gravitational force F_m that the moon exerts on the spacecraft is

$$F_m = \frac{Gm_0 m_2}{r_m^2}.$$

Equating these two expressions yields

$$\frac{81Gm_0 m_2}{r_e^2} = \frac{Gm_0 m_2}{r_m^2}.$$

Dividing both sides by the constant $Gm_0 m_2$ gives

$$\frac{81}{r_e^2} = \frac{1}{r_m^2},$$

or cross-multiplying,

$$81r_m^2 = r_e^2.$$

We substitute the expressions for the two distances to get

$$81[(x - 240)^2 + y^2] = x^2 + y^2,$$

or

$$81[x^2 - 480x + (240)^2 + y^2] = x^2 + y^2$$
$$81x^2 - 81(480)x + 81(240)^2 + 81y^2 = x^2 + y^2.$$

Collecting like terms gives

$$80x^2 - 38{,}880x + 81(240)^2 + 80y^2 = 0.$$

Dividing through by the common factor 80 yields

$$x^2 - 486x + 58{,}320 + y^2 = 0,$$

which suggests the equation of a circle. We complete the square on the x terms on the left-hand side and get

$$x^2 - 486x + 58{,}320 + y^2 = x^2 - 486x + \left(\frac{486}{2}\right)^2 - \left(\frac{486}{2}\right)^2 + 58{,}320 + y^2 = 0.$$

Therefore

$$(x - 243)^2 - (243)^2 + 58{,}320 + y^2 = 0,$$

or

$$(x - 243)^2 + y^2 = (243)^2 - 58{,}320 = 729.$$

That is, in the plane formed by the Earth, the moon, and the sun, the size of the gravitational forces from the Earth and the moon are numerically equal at every point on a circle of radius $\sqrt{729} = 27$ thousand miles centered at a distance of 243 thousand miles from the Earth on a line through the moon. In fact, the center of this circle is just beyond the moon.

Incidentally, even though a circle does not fulfill the requirements of a function, if we restrict our attention to either its upper half or its lower half, the resulting semicircle does represent a function. For the circle $x^2 + y^2 = r^2$, we obtain the equations for the semicircles by solving for y:

$$y^2 = r^2 - x^2 \quad \text{so that} \quad y = \pm\sqrt{r^2 - x^2}.$$

Thus the upper semicircle is the graph of the function $y = f(x) = \sqrt{r^2 - x^2}$, and the lower semicircle is the graph of the function $y = g(x) = -\sqrt{r^2 - x^2}$.

Problems

In Problems 1–6, find the distance between each pair of points.

1. $(2, 4)$ and $(5, 8)$

2. $(2, 4)$ and $(7, 16)$

3. $(4, -1)$ and $(0,4)$

4. $(2, -5)$ and $(0, 4)$

5. $(-1, 5)$ and $(3, 7)$

6. $(3, 1)$ and $(-5, -4)$

7. Find the midpoint of the line segment joining the points in Problem 1.

8. Find the midpoint of the line segment joining the points in Problem 2.

9. Find the point one-third the way from the first point to the second point in Problem 3.

10. Find the point three-fourths the way from the first point to the second point in Problem 4.

11. Find the equation of the circle that has center $(5, 2)$ and passes through the point P at $(8, -2)$.

12. Find the equation of the circle that has center $(-3, 7)$ and passes through the point P at $(2, -5)$.

13. Find the equation of the circle that has $(2, 4)$ and $(10, 4)$ as the endpoints of a diameter.

14. Find the equation of the circle that has $(-2, 3)$ and $(4, 11)$ as the endpoints of a diameter.

15. Repeat Problems 13 and 14, using the facts that any angle inscribed in a semicircle is a right angle and that perpendicular lines have slopes that are negative reciprocals.

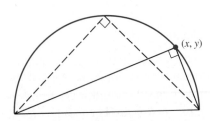

In Problems 16–21, complete the square in both the x- and y-terms for each equation to obtain the standard

form for the equation of a circle. Use it to determine the radius and center of the circle. Then draw the graph of the circle.

16. $x^2 + y^2 + 4x + 6y = 3$

17. $x^2 + y^2 + 4x + 6y = 12$

18. $x^2 + y^2 + 10x - 4y = 7$

19. $x^2 + y^2 + 10x - 4y = 71$

20. $x^2 + y^2 - 2x + 6y = -9$

21. $x^2 + y^2 - 2x + 6y + 6 = 0$

22. Determine which of the following equations represent a circle and which do not. Explain.

 a. $x^2 + y^2 - 4x - 6y + 15 = 0$
 b. $x^2 + y^2 - 4x - 6y + 13 = 0$
 c. $x^2 + y^2 - 4x - 6y + 12 = 0$

23. The equations $x = 3 + 2t$, $y = 4 - 5t$ form a parametric representation of a line.

 a. Construct a table of values for x and y corresponding to $t = -2, -1, 0, 1, \ldots, 5$.
 b. Plot these points and verify that they do seem to lie on a line.
 c. What is the slope of this line?
 d. What is a point–slope form for the equation of this line using $t = 1$?
 e. Use the midpoint formula to find the midpoint of each of the consecutive line segments determined by the entries in your table from part (a). Then use the parametric representation of the line with $t = -1.5, -0.5, 0.5, 1.5, 2.5, 3.5$, and 4.5. How do the results compare? Explain.

24. Consider again the parametric representation of the line $x = 3 + 2t$, $y = 4 - 5t$ in Problem 23. Eliminate the parameter t from the two equations by first solving the first equation for t in terms of x and then substituting the result into the second equation. How does this result compare to the result obtained in part (d) of Problem 23?

25. Start with the general parametric equations of a line

$$x = x_0 + (x_1 - x_0)t, \quad y = y_0 + (y_1 - y_0)t$$

and algebraically eliminate the parameter t. Identify the equation you produce.

26. a. Find the slope of the line having the parametric representation
$$x = 1 + 2t, \quad y = 2 - 3t.$$
   b. Sketch the graph of this line.

27. Find the points at which the line $x = 1 + 2t$, $y = 2 - t$ intersects the circle $x^2 + y^2 = 25$. (*Hint:* First find values of the parameter t that satisfy the equation of the circle.)

28. The three points P at $(0, 2)$, Q at $(2, 4)$, and R at $(4, 0)$ are noncollinear and as such determine a circle. Find an equation of this circle. (*Hint:* Substitute the coordinates of each point into the general equation of the circle, $x^2 + y^2 + Cx + Dy + E = 0$, and then solve the resulting system of three equations in three unknowns.)

Exercising Your Algebra Skills

In Problems 1–8, complete the square for each expression.

1. $x^2 + 8x + 25$ **2.** $x^2 - 8x + 25$

3. $x^2 - 6x + 5$ **4.** $x^2 + 6x + 5$

5. $y^2 + 10y + 26$ **6.** $y^2 - 10y + 26$

7. $y^2 + 4y - 12$ **8.** $y^2 - 4y - 12$

9.3   Conic Sections: The Ellipse

When we introduced functions and their graphs in Chapter 1, we said that not every graph represents a function. In Section 9.2 we pointed out that a circle is not a function. Several other important curves that have useful and interesting properties similarly are not functions. We investigate some of these curves in this section and Section 9.4.

Consider any equation of the form

$$Ax^2 + By^2 + Cx + Dy + E = 0,$$

where A, B, C, D, and E are constants, provided that at least one of A and B is not 0. In particular, if $A = B$, we can divide through by this constant, so that the result is the equation of a circle if the coefficients lead to a positive radius. Let's see what happens when $A \neq B$.

The graph of any equation of the form $Ax^2 + By^2 + Cx + Dy + E = 0$ is known as a **conic section.** To see why, consider a slice through the double right circular cone shown in Figure 9.12. If the slicing plane is horizontal, each slice is a *circle.* However, if the slicing plane is inclined slightly from the horizontal, the curve produced is oval in shape, rather than circular, and is an *ellipse.* (Imagine a diagonal slice through a round salami.) In fact, the sharper the angle of the slice, the more elongated the ellipse will be, as shown in Figure 9.13. If the angle of slicing is increased further so that it is parallel to the "edge" of the cone, the resulting curve is a *parabola,* as shown in Figure 9.14. If the angle of the slice is increased still further, the slicing plane intersects both the upper and lower parts of the cone and produces a pair of separated curves, known as a *hyperbola,* as shown in Figure 9.15.

In summary, there are three types of conic sections: (1) the ellipse, (2) the parabola, and (3) the hyperbola. The circle is a special case of the ellipse.

The conic sections occur often in various applications of mathematics and science. For instance, the orbits of the planets about the sun are ellipses. The paths of many comets and meteoroids are hyperbolic. A cross section of the metallic reflector inside a flashlight or an automobile headlight is a parabola. The path of a

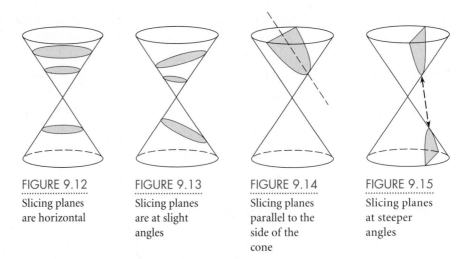

FIGURE 9.12
Slicing planes
are horizontal

FIGURE 9.13
Slicing planes
are at slight
angles

FIGURE 9.14
Slicing planes
parallel to the
side of the
cone

FIGURE 9.15
Slicing planes
at steeper
angles

thrown object, such as a perfect "spiral" pass in football or a "line drive" in baseball, is also a parabola.

Although we typically use formulas when working with the conic sections, we define them formally from a purely geometric perspective. This approach is analogous to the way we defined a circle in Section 9.2 as the set of all points at a fixed distance from a single fixed point, its center. In this section, we study the ellipse and consider the hyperbola and the parabola in Section 9.4.

The Ellipse

An **ellipse** is defined as the set of all points in the plane for which the sum of the distances to two fixed points is a constant. The two fixed points are called the **foci** (the plural of *focus*) of the ellipse. The midpoint of the line segment joining the foci is the **center** of the ellipse.

When the two foci are far apart, the resulting ellipse is very elongated. When the two foci are close together, the ellipse is close to circular and, in fact, when the two foci merge into a single point, the ellipse is a circle.

For convenience, we assume that the center of an ellipse is at the origin and that the two foci lie on the x-axis. Suppose that the foci are at F_1 with coordinates $(c, 0)$ and at F_2 with coordinates $(-c, 0)$, as illustrated in Figure 9.16. A point P

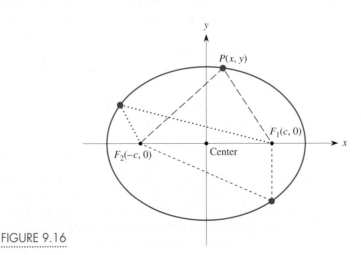

FIGURE 9.16

with coordinates (x, y) lies on the ellipse if the sum of the two distances $|F_1P|$ and $|F_2P|$ is some constant k. To make things easier, we write $k = 2a$. That is,

$$|F_1P| + |F_2P| = 2a,$$

or, equivalently,

$$\sqrt{(x - c)^2 + (y - 0)^2} + \sqrt{(x + c)^2 + (y - 0)^2} = 2a$$

or

$$\sqrt{(x - c)^2 + y^2} + \sqrt{(x + c)^2 + y^2} = 2a.$$

When we simplify this equation by eliminating both square roots (we leave the actual simplification for you to do as a problem at the end of this section), we eventually obtain

$$\frac{x^2}{a^2} + \frac{y^2}{b^2} = 1$$

as the equations of the ellipse, where $b^2 = a^2 - c^2$ is a new constant. The three constants a, b, and c are related by the equation

$$a^2 = b^2 + c^2, \qquad \text{where } a > b.$$

We investigate the meaning of the constants a and b below.

To determine where the ellipse intersects the x-axis, we set $y = 0$. The equation of the ellipse then reduces to

$$\frac{x^2}{a^2} = 1 \quad \text{so that} \quad x^2 = a^2$$

from which we find that either

$$x = a \quad \text{or} \quad x = -a.$$

This result indicates that a represents the distance from the center of the ellipse to the two points where the ellipse crosses the x-axis, as illustrated in Figure 9.17. Similarly, if $x = 0$, the equation of the ellipse yields $y^2 = b^2$, from which either

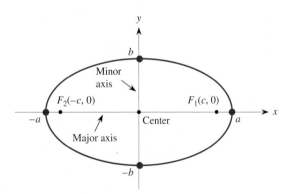

FIGURE 9.17

$$y = b \quad \text{or} \quad y = -b.$$

Thus b represents the distance from the center of the ellipse to the two points where the ellipse crosses the y-axis. The four points, $(a, 0)$, $(-a, 0)$, $(0, b)$, and $(0, -b)$, are the **vertices** of the ellipse; any one of them is a *vertex*. The lines connecting opposite vertices are called the **axes** of the ellipse. The longer axis, whether horizontal or vertical, is called the *major axis* and always contains the two foci; the shorter axis is called the *minor axis*.

In summary, a represents the distance from the center to either of the two more distant vertices along the major axis of the ellipse; b represents the distance from the center to either of the two closer vertices along the minor axis; and c represents the distance from the center to either focus of the ellipse. Because a is half the length of the major axis, we sometimes call a the length of a *semi-major axis*. Similarly, b is sometimes called the length of a *semi-minor axis*.

EXAMPLE 1

Describe and sketch the graph of the ellipse

$$\frac{x^2}{16} + \frac{y^2}{9} = 1.$$

Solution   This ellipse is centered at the origin. Its vertices occur when $y = 0$, so $x = \pm 4$, or when $x = 0$, so $y = \pm 3$. Therefore $a = 4$ and $b = 3$, so that the major axis extends horizontally from $x = -4$ to $x = 4$, and the minor axis extends vertically from $y = -3$ to $y = 3$. See Figure 9.18. Because a, b, and c are related by $a^2 = b^2 + c^2$ we have

$$c^2 = a^2 - b^2 = 16 - 9 = 7,$$

so that $c = \pm \sqrt{7}$. Therefore the foci are located at $(\sqrt{7}, 0)$ and $(-\sqrt{7}, 0)$, giving the graph shown in Figure 9.18.

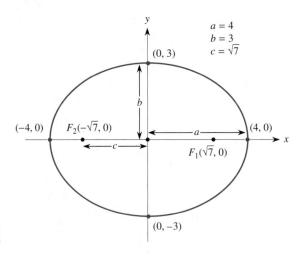

FIGURE 9.18

So far, we have considered an ellipse centered at the origin with foci along the x-axis. If we consider the analogous situation where the foci lie along the y-axis, the resulting equation for such an ellipse is

$$\frac{x^2}{b^2} + \frac{y^2}{a^2} = 1, \qquad a > b.$$

Note that the constant a is still measured along the major axis and b is measured along the minor axis of the ellipse, so $a > b$. From the equation of an ellipse, we can identify its major axis immediately by observing which of the two denominators is larger.

We have only considered ellipses that are centered at the origin. In fact, an ellipse can be centered at any point (x_0, y_0). In such a case, we get the following **standard forms for the equation of an ellipse.**

The equation of an ellipse centered at (x_0, y_0) with its major axis parallel to the x-axis is

$$\frac{(x - x_0)^2}{a^2} + \frac{(y - y_0)^2}{b^2} = 1.$$

The equation of an ellipse centered at (x_0, y_0) with its major axis parallel to the y-axis is

$$\frac{(x - x_0)^2}{b^2} + \frac{(y - y_0)^2}{a^2} = 1.$$

In each case,

$$a^2 = b^2 + c^2,$$

where c is the distance from the center to either focus.

EXAMPLE 2

Describe and sketch the ellipse whose equation is

$$\frac{(x - 2)^2}{4} + \frac{(y - 7)^2}{25} = 1.$$

Solution   The center of this ellipse is at the point $(2, 7)$. Because $25 > 4$, the major axis is parallel to the y-axis. In particular, the major axis is on the vertical line $x = 2$, and the foci also lie on this line. The minor axis is on the horizontal line $y = 7$. Also, because $a^2 = 25$ and $b^2 = 4$, 2. Thus the length of the major axis is, $2a$ 10, and the length of the minor axis is $2b$ 4. Consequently, the maximum horizontal distance from the center is 2 on either side of $x = 2$, and the maximum vertical distance from the center is 5 above and below $y = 7$. The ellipse therefore extends horizontally from $x = 0$ to $x = 4$ and extends vertically from $y = 2$ to $y = 12$. Figure 9.19 shows the graph of the ellipse.

To locate the foci, we use

$$c^2 = a^2 - b^2 = 25 - 4 = 21,$$

which gives $c = \sqrt{21}$. Because the foci are on the major axis of the ellipse, they are on the vertical line $x = 2$. Thus the foci are at the points $(2, 7 + \sqrt{21})$ and $(2, 7 - \sqrt{21})$.

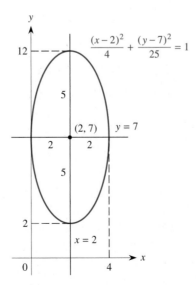

FIGURE 9.19

Suppose that we again consider the ellipse in Example 2,

$$\frac{(x-2)^2}{4} + \frac{(y-7)^2}{25} = 1.$$

We now multiply both sides of the equation by 100 to eliminate the fractions, so that

$$25(x-2)^2 + 4(y-7)^2 = 100.$$

Expanding the terms, we get

$$25(x^2 - 4x + 4) + 4(y^2 - 14y + 49) = 100,$$
$$25x^2 - 100x + 100 + 4y^2 - 56y + 196 = 100,$$
$$25x^2 + 4y^2 - 100x - 56y + 196 = 0.$$

This last equation is an equivalent equation for the same ellipse. Often we start with such an equation and have to rewrite it algebraically to uncover the key information about the ellipse. We illustrate how to do so in Example 3.

EXAMPLE 3

Verify that the equation

$$25x^2 + 9y^2 - 50x - 36y - 164 = 0$$

represents an ellipse, and find its center, vertices, and foci. Use this information to sketch the ellipse.

Solution We first collect the terms in x and y separately and then factor out the coefficients of x^2 and y^2:

$$25x^2 + 9y^2 - 50x - 36y - 164 = 25x^2 - 50x + 9y^2 - 36y - 164$$

$$= \left[25(x^2 - 2x)\right] + \left[9(y^2 - 4y)\right] - 164.$$

Finally, we complete the squares on both x and y to obtain

$$25\left[(x^2 - 2x + 1) - 1\right] + 9\left[(y^2 - 4y + 4) - 4\right] - 164$$

$$= 25\left[(x-1)^2 - 1\right] + 9\left[(y-2)^2 - 4\right] - 164$$
$$= 25(x-1)^2 - 25 + 9(y-2)^2 - 36 - 164$$
$$= 25(x-1)^2 + 9(y-2)^2 - 225.$$

Therefore the original equation is equivalent to

$$25(x-1)^2 + 9(y-2)^2 - 225 = 0 \quad \text{or} \quad 25(x-1)^2 + 9(y-2)^2 = 225.$$

Dividing both sides by 225 yields

$$\frac{(x-1)^2}{9} + \frac{(y-2)^2}{25} = 1,$$

which is the standard form for the equation of an ellipse. The center is at $(1, 2)$. The major axis is vertical because $25 > 9$. Moreover, because $a = 5$, the major axis extends from $y = 2 - 5 = -3$ to $y = 2 + 5 = 7$. The minor axis is horizontal with $b = 3$, so it extends from $x = 1 - 3 = -2$ to $x = 1 + 3 = 4$. To find the foci, we solve

$$c^2 = a^2 - b^2 = 25 - 9 = 16,$$

which gives $c = 4$. Therefore the foci are 4 units above and below the center $(1, 2)$, so they are at $(1, -2)$ and $(1, 6)$. The graph of this ellipse is shown in Figure 9.20.

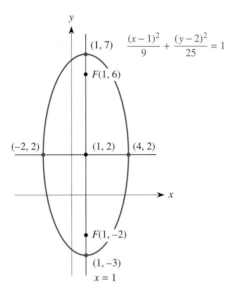

FIGURE 9.20

At the beginning of the section, we pointed out that the orbits of the planets about the sun are ellipses. More specifically, these elliptical orbits all have the sun as one of their two foci. A natural question to ask is: What is the equation of the ellipse for the orbit of the Earth? To answer it, we need two pieces of data used by astronomers to describe the orbits of the planets. The *perihelion* is the smallest distance from a planet to the sun, and the *aphelion* is the greatest distance, as depicted in Figure 9.21. For the Earth, the perihelion is approximately 147.1 million kilometers, or 91.38 million miles, and the aphelion is approximately 152.1 million kilometers, or 94.54 million miles. These two distances help identify the location of the sun on the major axis of Earth's elliptical orbit.

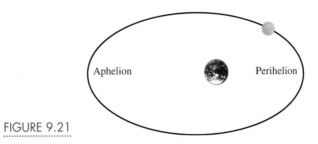

FIGURE 9.21

EXAMPLE 4

Find an equation of the Earth's orbit about the sun.

Solution   We first set up a coordinate system with the sun, the other (phantom) focus, and the major axis on the x-axis, as shown in Figure 9.22. Because the perihelion and aphelion distances are almost the same, the two foci are quite close together and the orbit of the Earth is nearly circular. From Figure 9.22, we see that the distance from one vertex to the other, or $2a$, is $91.38 + 94.54 = 185.92$ million miles, so

$$a = 185.92/2 = 92.96 \text{ million miles.}$$

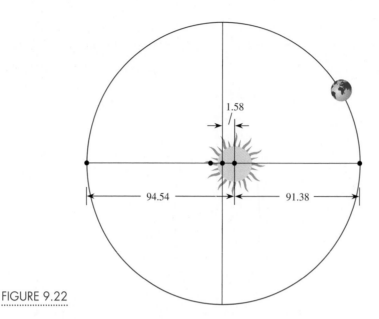

FIGURE 9.22

To find b, we first have to determine c, using the following reasoning. At perihelion, the Earth is 91.38 million miles from the sun, so the distance from the center of the ellipse to the sun (a focus) must be

$$c = 92.96 - 91.38 = 1.58 \text{ million miles.}$$

From $a^2 = b^2 + c^2$, we have

$$b^2 = a^2 - c^2$$
$$= (92.96)^2 - (1.58)^2 = 8639.07,$$

so that

$$b = \sqrt{8639.07} = 92.95 \text{ million miles.}$$

Consequently, the equation of the Earth's orbit about the sun is

$$\frac{x^2}{(92.96)^2} + \frac{y^2}{(92.95)^2} = 1.$$

As we observed previously, the Earth's orbit is very nearly circular.

The table of planetary data on the following page lists the perihelion and aphelion distances, in millions of miles, for the planets in the solar system. You can use it to compare the Earth's orbit to that of the other planets. You will use some of these entries for the Problems at the end of the section.

Reflection Property of the Ellipse

One of the most fascinating properties of an ellipse is known as the **reflection property.** Consider any line segment emanating from one of the two foci—say, F_1—as shown in Figure 9.23. It eventually intersects the ellipse and then reflects.

Planet	Perihelion	Aphelion
Mercury	28.56	43.38
Venus	66.74	67.68
Earth	91.38	94.54
Mars	128.49	154.83
Jupiter	460.43	506.87
Saturn	837.05	936.37
Uranus	1700.07	1867.76
Neptune	2771.72	2816.42
Pluto	2749.57	4582.61

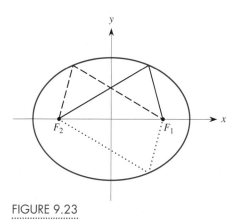

FIGURE 9.23

(According to physical principles, the angle of incidence with a tangent line equals the angle of reflection.) *Any* such reflected line segment will pass through the second focus F_2.

This property is significant because many physical phenomena, such as light and sound, travel in straight lines and reflect off solid surfaces. Thus, if a light-bulb is placed at one focus of a three-dimensional shell whose cross sections containing the major axis are all ellipses, all its light rays will bounce off the inside surface of the shell and reflect back through the other focus. The effect is similar with sound waves. Probably the best known example of this is the whispering gallery effect in the U.S. Capitol in Washington, D.C. The dome of the Capitol has the approximate shape of a three-dimensional ellipse, and there are two foci near floor level. If you stand at one of the foci and whisper, your voice is carried to the second focus across the hall and can be heard clearly by anyone standing there.

◆**EXAMPLE 5**

The distance between the foci in the "whispering gallery" of the Capitol is 38.5 feet, and the maximum height of the ceiling above ear level is 37 feet. Find the equation of an elliptical cross section of the gallery under the Capitol dome.

Solution   We set up axes as shown in Figure 9.24. Because the distance between foci is 38.5 feet, $c = \frac{1}{2}(38.5) = 19.25$ feet. Also, from the maximum height of the dome, $b = 37$ feet. For an ellipse, we know that

$$a^2 = b^2 + c^2 = (37)^2 + (19.25)^2 = 1739.5625,$$

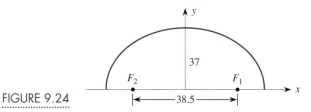

FIGURE 9.24

so $a \approx 41.7$ feet. Therefore the equation of an elliptical cross section of the Capitol whispering gallery is

$$\frac{x^2}{(41.7)^2} + \frac{y^2}{(37)^2} = 1.$$

The Average Distance from the Sun

We have stated that the orbit of each planet is an ellipse with one focus at the sun. A natural question to ask is: What is the *average* distance of a planet from the sun during a full orbit? The answer is particularly simple, yet surprising, as we demonstrate in Example 6.

EXAMPLE 6

Show that the average distance from all points on any ellipse

$$\frac{x^2}{a^2} + \frac{y^2}{b^2} = 1$$

is precisely equal to a, the length of the semi-major axis.

Solution   We begin with the ellipse shown in Figure 9.25, with foci at F_1 and F_2. Let P_1 be any point on the right half of the ellipse. From the geometric definition of the ellipse, we know that the sum of the two distances $|F_1P_1|$ and $|F_2P_1|$ must equal the constant $2a$, or

$$|F_1P_1| + |F_2P_1| = 2a.$$

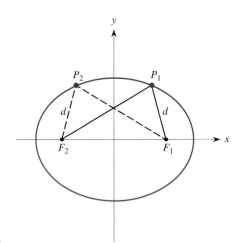

FIGURE 9.25

Using the symmetry of the ellipse, there is a comparable point P_2 on the left half of the ellipse so that

$$|F_2P_2| = |F_1P_1| \quad \text{and} \quad |F_2P_1| = |F_1P_2|.$$

Therefore

$$|F_1P_1| + |F_1P_2| = |F_1P_1| + |F_2P_1| = 2a.$$

Hence the average of these two distances from F_1 to P_1 and from F_1 to P_2 is simply a. This argument can be applied to every possible pair of matching points on the ellipse, so the average distance from F_1 to *all* points on the ellipse must be a.

◆

Now let's apply this result to the orbits of the planets. For instance, the aphelion distance for the Earth is 94.54 million miles and the perihelion distance is 91.38 million miles. These distances can be expressed as

$$\text{Perihelion} = a - c \quad \text{and} \quad \text{Aphelion} = a + c.$$

Their arithmetic average is

$$\frac{1}{2}(\text{perihelion} + \text{aphelion}) = \frac{1}{2}(a - c + a + c) = a;$$

that is, the average distance of the Earth (or any other planet) from the sun is just the average of its perihelion and aphelion distances.

Problems

1. For the satellite whose elliptic orbit about the Earth is shown in the accompanying diagram, indicate the location of the following points and give reasons for your answers.

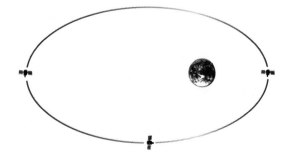

   a. The point at which the gravitational force F exerted by the Earth on the satellite is greatest; where it is least. (Recall Newton's law of universal gravitation: $F = f(r) = GmM/r^2$, where r is the distance between the two objects.)
   b. The point at which the speed of the satellite is greatest. (*Hint*: Think of the satellite as always "falling" toward the Earth.)
   c. The point at which the speed of the satellite is least.

2. Suppose that the satellite in Problem 1 fires its retrorockets to slow down somewhat at the point in its orbit where it is closest to the Earth. Compare the graph of the new orbit to the graph of the old one in a sketch.

3. Suppose that the satellite in Problem 1 is in a relatively low orbit about the Earth so that it encounters the upper fringe of the Earth's atmosphere. What will be the atmosphere's effect on the satellite's path? Sketch the graph of the resulting trajectory. What will happen eventually?

4. Suppose that the satellite in Problem 1 fires its booster rocket to speed up at the point in its orbit where it is closest to the Earth. Compare the graph of the new orbit to the graph of the old one in a sketch. What happens to the orbit if the booster rockets are extremely strong or continue firing for a long time?

5. Which of the nine planets in the solar system has the most circular orbit? the least circular orbit? Explain.

6. During the next few years, Pluto's orbit takes it inside the orbit of Neptune. Use the values in the table of planetary data in the text to explain why this situation can occur.

7. Use the fact that the perihelion and aphelion distances for Mercury are 46.0 and 69.8 million kilometers respectively to find the equation of the orbit of Mercury about the sun.

8. A salami is 4 inches in diameter. When the deli clerk slices it, however, the slices are at an angle of 65° to the main axis of the salami. Consequently, each slice will be in the shape of an ellipse with a minor axis of length 4 inches, as shown on the following page. Find the length of the major axis of each slice.

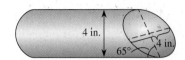

9. The Roman Coliseum is in the shape of an ellipse whose major axis measures 620 feet and whose minor axis measures 513 feet.

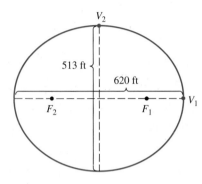

   a. What is the equation of this ellipse?
   b. How far apart are the foci?
   c. How far apart are two adjacent vertices?

10. Write formulas expressing the perihelion and aphelion of an elliptic orbit in terms of the *semimajor axis* length a and the focal distance c in an ellipse.

In Problems 11 and 12, each equation represents an ellipse. In each case, identify the center, the vertices, and the foci and use the pertinent information to draw its graph.

11. $4x^2 + 9y^2 = 1$

12. $25x^2 + 4y^2 = 100$

In Problems 13–18, each equation represents an ellipse. Complete the square for x and y in each case to obtain the standard form for an ellipse. Then identify the center, the vertices, and the foci of the ellipse and use the pertinent information to draw its graph.

13. $x^2 + 4y^2 + 2x + 8y = -1$

14. $x^2 + 4y^2 + 2x + 8y = 11$

15. $x^2 + 4y^2 + 20x - 40y + 100 = 0$

16. $4x^2 + y^2 + 24x - 2y + 4 = 0$

17. $9x^2 + y^2 - 54x + 4y = -76$

18. $x^2 + 9y^2 - 6x + 36y = -36$

19. Complete the derivation of the equation of the ellipse by simplifying the equation in the text by eliminating the two square roots. (*Hint*: First isolate one of the radicals, then square both sides, and finally eliminate the remaining radical.)

9.4   Conic Sections: The Hyperbola and the Parabola

We now turn our attention to the two remaining conic sections, the hyperbola and the parabola. We begin this section by investigating the properties and some applications of the hyperbola.

The Hyperbola

We defined an ellipse geometrically as the set of all points for which the *sum* of the distances to two fixed foci is constant. In an analogous way, we define a hyperbola in terms of the *difference* of the distances to two fixed points being constant. A **hyperbola** is the set of all points for which the *difference* between the distances to two fixed points is a constant. The two points are the **foci** of the hyperbola. The midpoint of the line segment joining the foci is the **center**.

◤ **EXAMPLE 1** ....

During a severe thunderstorm, two lightning bolts appear to strike simultaneously. You hear the thunderclap from one lightning bolt exactly 1 second after the lightning strikes at point P, and you hear the thunderclap from the second lightning bolt

2 seconds after it hits at point Q, as depicted in Figure 9.26. Sound travels at a speed of about 1100 feet per second.

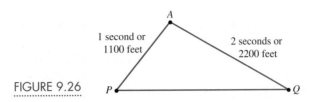

FIGURE 9.26

a. Based on this information what can you conclude about the point A, where you are?

b. Two friends of yours also see the same two lightning strikes. From Becky's location at point B, the thunder from the lightning bolt at point P takes 2 seconds to reach her and the thunder from the lightning bolt at point Q takes 3 seconds. From Carl's location at point C, the times are 4 and 5 seconds, respectively. What can you conclude about the three points A, B, and C?

Solution

a. The lightning bolts hit at points P and Q, and you're located at point A. Because it takes 1 second for the sound of the lightning bolt at P to reach A and sound travels at 1100 feet per second, the distance from P to A must be 1100 feet. Similarly, it takes 2 seconds for the sound of the strike at Q to reach A so that distance must be 2 seconds × 1100 feet/second = 2200 feet.

b. Figure 9.27 shows the points A, B, and C. Reasoning as in part (a), you can conclude that Becky is 2 seconds × 1100 feet/second = 2200 feet from P and 3 seconds × 1100 feet/second = 3300 feet from Q. Similarly, Carl is 4400 feet from P and 5500 feet from Q.

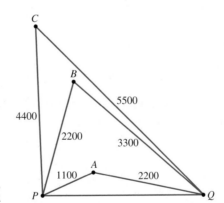

FIGURE 9.27

However, you can deduce one more piece of information: For all three, the *difference in time* between the two thunderclaps is 1 second. That is, the *difference in distance* from each of the three points A, B, and C to the points P and Q is a constant equal to 1100 feet. But if the differences in the distances from these three points to the fixed points where the lightning bolts hit are all equal, the three points A, B, and C must lie on a hyperbola whose foci are at P and Q.

Think About This

Explain why you cannot determine the distance from P to Q in Example 1 based on the triangle APQ by using trigonometry. What additional information would you need to be able to find that distance?

The Equation of a Hyperbola We now determine the equation of a hyperbola from the geometric definition. For convenience, we place the center of a hyperbola at the origin and the foci on the x-axis at the point F_1 with coordinates $(c, 0)$ and F_2 with coordinates $(-c, 0)$, as shown in Figure 9.28. A point P with coordinates (x, y) lies on the hyperbola if, for some constant k,

$$|F_2P| - |F_1P| = k.$$

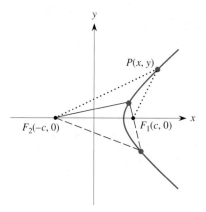

FIGURE 9.28

As with the equation of an ellipse, we let the constant $k = 2a$ for convenience. Thus

$$|F_2P| - |F_1P| = 2a,$$

so that

$$\sqrt{(x + c)^2 + y^2} - \sqrt{(x - c)^2 + y^2} = 2a.$$

We simplify this equation by eliminating both square roots, as was done for the equation of an ellipse (see Problem 19, Section 9.3) and eventually obtain

$$\frac{x^2}{a^2} - \frac{y^2}{b^2} = 1,$$

where

$$c^2 = a^2 + b^2.$$

The graph of this hyperbola is shown in Figure 9.29. Note that the hyperbola has two distinct *branches,* which is what we should expect from the discussion in Section 9.3 of slicing through a double right circular cone. The two points where this hyperbola crosses the x-axis are called its **vertices;** they correspond to the points where $y = 0$ and thus represent the points where the two branches are closest.

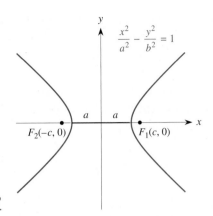

FIGURE 9.29

At the vertices, we have $y = 0$, so that

$$\frac{x^2}{a^2} = 1,$$

and therefore $x = \pm a$. Thus a represents the distance from the center to a vertex, whereas c represents the distance from the center to a focus. The line containing the foci is the **axis** of the hyperbola.

Alternatively, we can place the foci for a hyperbola on the vertical axis, as shown in Figure 9.30. The equation of such a hyperbola is

$$\frac{y^2}{a^2} - \frac{x^2}{b^2} = 1.$$

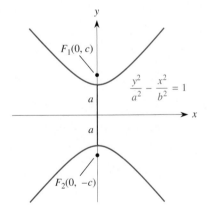

FIGURE 9.30

EXAMPLE 3

Describe the hyperbola whose equation is

$$\frac{x^2}{16} - \frac{y^2}{9} = 1.$$

Solution   The form of the equation indicates that the hyperbola is centered at the origin and that its axis is horizontal. Because $a^2 = 16$ and $b^2 = 9$, we have $a = 4$ and $b = 3$, so that

$$c^2 = a^2 + b^2 = 16 + 9 = 25,$$

and so $c = 5$. Thus the vertices are at $x = -4$ and $x = 4$, or the points $(-4, 0)$ and $(4, 0)$. The foci are the points $(-5, 0)$ and $(5, 0)$.

◆

Think About This

Use your function grapher to see what the graph of the hyperbola in Example 3 looks like. To do so, you have to rewrite the equation by solving for y as a function of x. In particular,

$$\frac{y^2}{9} = \frac{x^2}{16} - 1 \quad \text{so that} \quad y^2 = 9\left(\frac{x^2}{16} - 1\right).$$

The upper and lower halves of the hyperbola are therefore given separately by the two functions

$$y = 3\sqrt{\frac{x^2}{16} - 1} \quad \text{and} \quad y = -3\sqrt{\frac{x^2}{16} - 1}.$$

What are the domains of the two functions? ⌐

More generally, we can consider a hyperbola as being shifted horizontally and/or vertically so that its center is at the point P with coordinates (x_0, y_0) rather than at the origin. We then have the following **standard forms for the equation of a hyperbola.**

The equation of a hyperbola centered at (x_0, y_0) with its axis parallel to the x-axis is

$$\frac{(x - x_0)^2}{a^2} - \frac{(y - y_0)^2}{b^2} = 1.$$

The equation of a hyperbola centered at (x_0, y_0) with its axis parallel to the y-axis is

$$\frac{(y - y_0)^2}{a^2} - \frac{(x - x_0)^2}{b^2} = 1.$$

In each case,

$$c^2 = a^2 + b^2.$$

Note that, in the equation of a hyperbola in standard form, the term with the positive coefficient determines the orientation. If the x^2-term is positive, the two branches open about the x-axis; if the y^2-term is positive, the two branches open about the y-axis. Also, be sure to distinguish between the equation $c^2 = a^2 + b^2$ relating the constants for a hyperbola and the equation $a^2 = b^2 + c^2$ relating the constants for an ellipse.

◆ **EXAMPLE 4**

Verify that

$$x^2 - y^2 + 8x - 6y = 2$$

is an equation of a hyperbola. Find the center, vertices, and foci of the hyperbola and sketch its graph.

Solution   We complete the square on both x and y so that the left-hand side becomes

$$x^2 - y^2 + 8x - 6y = [(x^2 + 8x + 16) - 16] - [(y^2 + 6y + 9) - 9]$$
$$= (x + 4)^2 - 16 - (y + 3)^2 + 9$$
$$= (x + 4)^2 - (y + 3)^2 - 7.$$

The original equation therefore becomes

$$(x + 4)^2 - (y + 3)^2 - 7 = 2 \quad \text{or} \quad (x + 4)^2 - (y + 3)^2 = 9.$$

Dividing by 9, we obtain

$$\frac{(x + 4)^2}{9} - \frac{(y + 3)^2}{9} = 1.$$

Consequently, the center of the hyperbola is $(-4, -3)$ and $a = b = 3$. Furthermore, since the x^2-term is the positive one, the axis of the hyperbola is parallel to the x-axis. That is, the vertices and the foci lie on the horizontal line $y = -3$ through the center, and the hyperbola opens to the left and the right. Because $a = 3$, the vertices are 3 units left and right of the center, or at $(-7, -3)$ and at $(-1, -3)$. Also,

$$c^2 = a^2 + b^2 = 9 + 9 = 18,$$

so $c = \sqrt{18} = 3\sqrt{2}$. Thus the foci are located at F_1 at $(-4 - 3\sqrt{2}, -3)$ and F_2 at $(-4 + 3\sqrt{2}, -3)$, as shown in Figure 9.31.

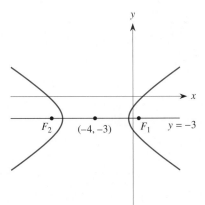

FIGURE 9.31

Based on their graphs and the vertical line test, hyperbolas (like ellipses) cannot be functions because any x-value can lead to two distinct y-values. Of course, if an x-value is beyond the limits of the ellipse, there is no corresponding y-value. Similarly, if an x-value is between the two branches of a hyperbola whose axis is horizontal, there is no corresponding y-value.

Applications of the Hyperbola   Probably the most significant application of the hyperbola has been the long range navigation (LORAN) system used around the world by sailors to locate their positions before the advent of the global positioning system (GPS) that makes use of orbiting satellites. With LORAN, radio transmitters along coastal waters emit simultaneous radio signals that are picked up by electronic equipment on ships. As we demonstrated in Example 1 with sound waves, there will usually be a difference in the times at which a ship

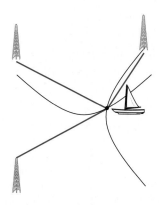

FIGURE 9.32

receives radio signals from different stations, and this difference is used to "place" the vessel on a specific hyperbola. When the same procedure is used with other radio transmitters in the LORAN network, the ship is simultaneously "placed" on a second hyperbola. Finding a point of intersection of the two hyperbolas and locating the position of the vessel is then a relatively simple matter, as illustrated in Figure 9.32.

In Example 5 we demonstrate the actual use of these ideas. To do so, we use the fact that any radio wave travels at the speed of light, or about 186,300 miles per second, or 300,000 kilometers per second.

◆ **EXAMPLE 5** ⋯⋯⋯

A sailboat is out on Long Island Sound when a heavy fog moves in. To the south of the sailboat, on Long Island's shore, are two LORAN radio transmitters 60 km apart at points P and Q. A third transmitter is to the north on Connecticut's shore at point R, which is 40 km directly north of P. Figure 9.33 depicts this situation.

a. The receiver on the boat receives signals from P and Q that arrive 0.00016 second apart. Find an equation of the hyperbola having foci at P and Q on which the boat is located.

b. The receiver on the boat receives signals from P and R that arrive 0.0001067 second apart. Find an equation of the hyperbola having foci at P and R on which the boat is located, based on the same coordinate system used in part (a).

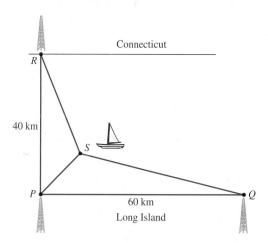

FIGURE 9.33

c. Estimate the location of the boat based on the results of parts (a) and (b).

Solution

a. Suppose that we set up a coordinate system with a horizontal axis through P and Q and the origin midway between them, as shown in Figure 9.34. In this system, the coordinates of P are $(-30, 0)$ and the coordinates of Q are $(30, 0)$. One focus of the hyperbola is at Q, so we have $c = 30$. Suppose that the sailboat is at S with coordinates (x, y) on this hyperbola. The difference in times between receipt of the two signals is 0.00016 second; using the speed of light as 300,000 km/second, this difference in times is equivalent to a difference in distance of 300,000 km/second $\times$ 0.00016 seconds = 48 km. That is, $2a = 48$, so $a = 24$.

For a hyperbola $c^2 = a^2 + b^2$, so that

$$b^2 = c^2 - a^2 = (30)^2 - (24)^2 = 324 \quad \text{or} \quad b = 18.$$

Consequently, the equation of this hyperbola, which opens to the left and the right, is

$$\frac{x^2}{(24)^2} - \frac{y^2}{(18)^2} = 1.$$

b. We now consider the hyperbola with foci at P and R and center at the point $(-30, 20)$, as shown in Figure 9.35. We use a_1, b_1, and c_1 to represent the parameters for this hyperbola. Because the hyperbola opens upward and downward, its equation has the form

$$\frac{(y - 20)^2}{a_1^{\,2}} - \frac{(x + 30)^2}{b_1^{\,2}} = 1.$$

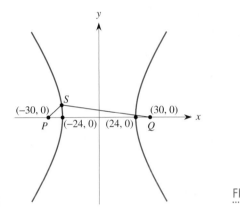

FIGURE 9.34

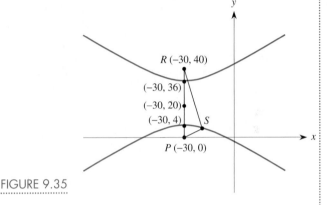

FIGURE 9.35

The distance between the foci is 40 km, so we have $c_1 = 20$.

The difference in time for receipt of the two signals is 0.0001067 second, which is equivalent to a difference in distance of $300{,}000 \times 0.0001067 = 32$ km, so $2a_1 = 32$ and $a_1 = 16$. Therefore

$$b_1^{\,2} = c_1^{\,2} - a_1^{\,2} = (20)^2 - (16)^2 = 144 \quad \text{or} \quad b_1 = 12,$$

and the equation of this hyperbola is

$$\frac{(y - 20)^2}{(16)^2} - \frac{(x + 30)^2}{(12)^2} = 1.$$

c. To locate the position of the sailboat, we have to find the point of intersection of the two hyperbolas. (Note that there can be as many as four points of intersection, but in practice we would know which branch of each hyperbola the boat is on, based on the strength of the signal, so the problem reduces to finding a single point of intersection.) Although we can find the point of intersection algebraically, to do so is rather complicated, so instead we estimate the point graphically. Using the equation of the first hyperbola from part (a), we have

$$\frac{y^2}{(18)^2} = \frac{x^2}{(24)^2} - 1 \quad \text{or} \quad \frac{y^2}{324} = \frac{x^2}{576} - 1$$

so that

$$y^2 = 324\left(\frac{x^2}{576} - 1\right).$$

Figure 9.36 shows that the sailboat is above the line on which the center and the foci lie, so y must be positive (otherwise the boat would be on land). We therefore take the positive square root and get

$$y = 18\sqrt{\frac{x^2}{576} - 1}.$$

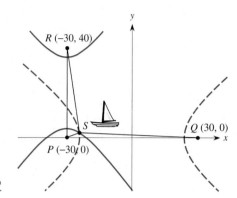

FIGURE 9.36

Similarly, starting with the equation of the second hyperbola, we have

$$\frac{(y-20)^2}{(16)^2} = \frac{(x+30)^2}{(12)^2} + 1 \quad \text{or} \quad \frac{(y-20)^2}{256} = \frac{(x+30)^2}{144} + 1$$

so that

$$(y-20)^2 = 256\left[\frac{(x+30)^2}{144} + 1\right].$$

Because the sailboat, as shown in Figure 9.35, is closer to the transmitter at P than the one at R, we need the lower branch of the hyperbola. Taking the negative square root gives

$$y - 20 = -16\sqrt{\frac{(x+30)^2}{144} + 1} \quad \text{or} \quad y = 20 - 16\sqrt{\frac{(x+30)^2}{144} + 1}.$$

To locate the sailboat, we need to find the point where the curves corresponding to these two equations intersect. From Figure 9.36 we estimate graphically that the point of intersection occurs at about $x \approx -24.18$ and $y \approx 2.21$. That is, the sailboat is located about 2.21 km off the north coast of Long Island at a position about $30 - 24.18 = 5.82$ km east of the transmitter at point P.

The Parabola

We have shown that the graph of any quadratic function $y = ax^2 + bx + c$ is a parabola opening upward or downward. However, the same parabolic shape can open to the left or the right, as shown in Figure 9.37, although neither of these graphs represents a function. To unify these ideas about parabolas, we consider the parabola from a somewhat different perspective in terms of its geometric definition as a conic section. We define a **parabola** as the set of all points in the plane for which the distance to a single fixed point is equal to the distance to a fixed line, as depicted in Figure 9.38. The fixed point is called the **focus** of the parabola. The fixed line is the **directrix** of the parabola.

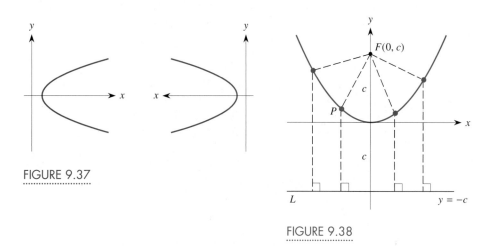

FIGURE 9.37

FIGURE 9.38

The Equation of a Parabola   For convenience, we place the focus of the parabola at the point F on the y-axis with coordinates $(0, c)$ and let the directrix be the horizontal line $y = -c$, as shown in Figure 9.38. The graph shown corresponds to the case where $c > 0$. The parabola consists of all points P having the property that the distance from P to the focus F is equal to the vertical distance from P to the directrix line L. Thus a point P with coordinates (x, y) lies on the parabola if the distance from P to F, $\sqrt{x^2 + (y - c)^2}$, equals the distance from P to the line L, which is $y + c$; that is,

$$\sqrt{x^2 + (y - c)^2} = y + c.$$

We square both sides of this equation and get

$$x^2 + (y - c)^2 = (y + c)^2,$$

or, equivalently, when we expand the equation, we have

$$x^2 + y^2 - 2cy + c^2 = y^2 + 2cy + c^2.$$

We subtract y^2 and c^2 from both sides of this equation and obtain

$$x^2 - 2cy = 2cy.$$

Finally, we add $2cy$ to both sides and solve for y;

$$y = \frac{x^2}{4c}.$$

This is the equation of a parabola with vertex (or turning point) at the origin. If $c > 0$, the parabola opens upward. If $c < 0$, the parabola opens downward. The vertical line through the vertex is called the **axis of symmetry** of the parabola.

Alternatively, had we positioned the focus on the x-axis at F with coordinates $(c, 0)$ with a vertical directrix at $x = -c$, then we would have obtained

$$x = \frac{y^2}{4c}$$

as the equation for the parabola. This parabola likewise has its vertex at the origin; it opens to the right if $c > 0$ and opens to the left if $c < 0$. Finally, its axis of symmetry is now the horizontal line through the vertex, as depicted in Figure 9.39.

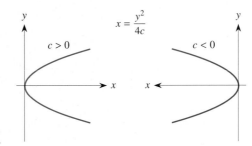

FIGURE 9.39

More generally, we can describe a parabola whose vertex is at (x_0, y_0) with the following **standard forms of the equation of a parabola.**

The equation of a parabola with its vertex at (x_0, y_0) and opening vertically is

$$y - y_0 = \frac{(x - x_0)^2}{4c}.$$

The equation of a parabola with its vertex at (x_0, y_0) and opening horizontally is

$$x - x_0 = \frac{(y - y_0)^2}{4c}.$$

Reflection Property of the Parabola   Just as the ellipse has a remarkable—and useful—reflection property, the parabola has one that is even more commonly encountered. It can be shown that any ray coming into a parabola along a line parallel to the axis of symmetry of the parabola will "reflect" off the curve and pass through the focus, as shown in Figure 9.40. Alternatively, any ray emanating from the focus will reflect off the parabola and continue on a path parallel to the axis of symmetry. This reflection property is used, for example, in flashlights and in the headlights of an automobile, where the light source is located at the focus and the beams of light bounce off a parabolic reflector to concentrate more light in a particular direction. The reflection property is also used by satellite TV dishes, which are constructed in such a way that every cross section containing the axis of symmetry of the dish is a parabola, as illustrated in Figure 9.41. The TV signals coming from a satellite relay in orbit arrive at the dish along rays parallel to the axis of the dish and its parabolic cross sections. They reflect off the dish and pass through a receptor unit positioned at the focus. There the signal is collected and then transmitted to the television set.

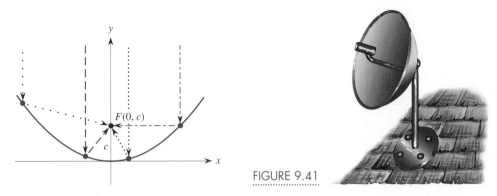

FIGURE 9.40

FIGURE 9.41

Conic Sections in General

The general equation of a conic section has the form

$$Ax^2 + By^2 + Cx + Dy + E = 0,$$

where A, B, C, D, and E are constants and at least one of A and B is nonzero. When $B = A \neq 0$, we divide both sides by A to get

$$x^2 + y^2 + \frac{C}{A}x + \frac{D}{A}y + \frac{E}{A} = 0,$$

which is the equation of a circle, provided that certain conditions are satisfied that lead to a positive radius. When $B = 0$ and $A \neq 0$, the resulting equation is quadratic in x but only linear in y and so gives the equation of a parabola opening either upward or downward. Similarly, when $A = 0$ and $B \neq 0$, we get a parabola opening either left or right. If A and B have the same sign—say, both are positive—the resulting curve is an ellipse, provided that certain conditions are satisfied. If A and B have opposite signs, the curve is a hyperbola. Thus, for instance,

$$4x^2 + 9y^2 + 8x - 36y - 5 = 0$$

represents an ellipse, whereas

$$4x^2 - 9y^2 + 8x - 36y - 5 = 0 \quad \text{and} \quad 25y^2 - 16x^2 + 10y + 8x + 3 = 0$$

both represent hyperbolas (one of which opens left and right and the other opens up and down). You have to be able to identify the type of curve from the given equation.

Finally, our discussions of conic sections have been restricted to their being in *standard position*—that is, their axes are parallel to the x- and y-axes. However, the same shapes can be rotated through some angle θ about the x-axis. When that occurs, the equation for the conic section—whether an ellipse, hyperbola, or parabola—includes a term of the form xy. For the most part, we aren't concerned with such situations here except for the case

$$xy = k,$$

where k is any constant. If we solve for y, this equation is equivalent to the power function

$$y = \frac{k}{x} = kx^{-1}.$$

This function is not defined at $x = 0$ and has two branches, the more familiar one in the first quadrant when $x > 0$ and a symmetric one in the third quadrant when $x < 0$. Together, they form the graph of a hyperbola that has been rotated from standard position through an angle of 45° or $\pi/4$ (when $k > 0$), as shown in Figure 9.42.

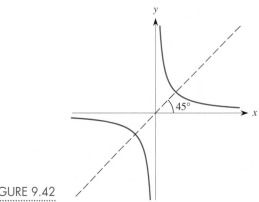

FIGURE 9.42

Problems

1. The small satellite TV dishes now on the market for home use have parabolic cross sections containing the axis of the dish. The focus is located at a point about 6 inches from the vertex of the parabola.

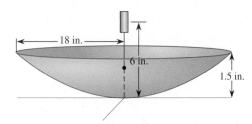

   a. Find an equation of a parabolic cross section. Assume that the dish is aimed directly upward—which makes sense only at the equator because the communications satellites are in orbit over the equator.
   b. The rim of the dish is a circle with a diameter of 18 inches at a height of about 1.5 inches above the vertex. If the dish were extended, the rim would enlarge. Find the diameter of the rim if the rim reached the height, 6 inches, of the focus.

2. Suppose that a satellite receiver is 36 inches across and 16 inches deep (vertex to plane of the rim). How far from the vertex must the receptor unit be located to ensure that it is at the focus of the parabolic cross sections?

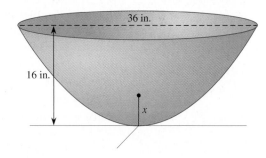

3. In this problem, we ask you to investigate the significance of the quantities a and b in the equation of a hyperbola

$$\frac{x^2}{a^2} - \frac{y^2}{b^2} = 1.$$

Suppose that you zoom out far enough on the graph of the hyperbola so that what you see appears to be a pair of lines that intersect at the origin.

   a. Explain why—when x and y both are very large, either positive or negative—you can ignore the number 1 in the equation.
   b. Ignore the number 1 and solve for y in terms of x to find the equations of the two lines described.
   c. What are the slopes of the two lines that the branches of the hyperbola approach?

In Problems 4–13 complete the square for x and y in each equation to obtain the standard form for a conic section. In each case, identify the conic section and use the pertinent information to draw its graph.

4. $x^2 + 4y^2 + 2x + 8y = -1$

5. $x^2 + 4y^2 + 2x + 8y = 11$

6. $x^2 - 4y^2 + 2x - 8y = 7$

7. $x^2 - 4y^2 + 2x - 8y = 19$

8. $4x^2 + y^2 + 24x - 2y + 4 = 0$

9. $4x^2 - 9y^2 - 16x - 18y = 31$

10. $9x^2 - 16y^2 - 90x + 64y = -17$

11. $4x^2 + 4y^2 - 24x + 16y + 43 = 0$

12. $9x^2 - 4y^2 + 18x - 16y = 8$

13. $9x^2 - 4y^2 + 18x - 16y = 6$

14. Explain why $x^2 + y^2 - 2x - 2y = -4$ is not the equation of a conic section.

15. In this problem, we ask you to look at the mathematics of string and wire art designs. Start with the hyperbola $xy = 1$ or $y = 1/x$ and construct a series of lines tangent to the curve, as shown in the accompanying figure. For instance, the line tangent to the curve when $x = 1$ crosses the x-axis at $x = 2$ and crosses the y-axis at $y = 2$. The line tangent to the hyperbola when $x = 2$ has an x-intercept of 4

and a y-intercept of 1. If you were to erase the curve, you would still see its outline from the tangent lines. String and wire art designers use this idea to suggest a variety of curves by using line segments made of the string or wire. The points on the axes are selected so as to follow the outline of a desired curve, such as the hyperbola.

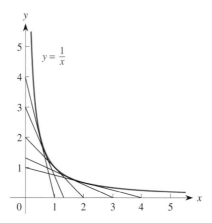

a. Find the slope m of each of the five tangent lines shown in the figure.

b. Find a formula for the slope m as a function of the point of tangency x.

9.5 · Parametric Curves

Throughout our discussion of functions, we have almost always considered expressions for which the dependent variable is given in terms of the independent variable. In some cases, however, introducing an additional variable, a *parameter,* can provide more insight into what is happening.

Parametric Representations of a Line

There are many ways to write an equation of a line, including the point–slope form, the slope–intercept form, and the normal form. However, certain questions about a line can't be answered with any of these forms. For instance, as we demonstrated in Section 9.2, if we want to locate the point that is a certain fraction of the way from the point P at (x_0, y_0) to the point Q at (x_1, y_1), it is essential to use the parametric form

$$x = x_0 + (x_1 - x_0)t$$
$$y = y_0 + (y_1 - y_0)t$$

for the line, where the parameter t takes on any value. With this form, each possible value of t gives a corresponding point on the line through P and Q.

EXAMPLE 1

Consider the parametric equations $x = 2 + 4t$ and $y = 5 - 3t$.

a. Construct a table of values for x and y corresponding to $t = -1, 0, 1, \ldots, 4$.
b. Use the table to explain why the two equations give a linear function.
c. Plot the points and draw the line that passes through them.

Solution

a. When $t = -1$, the parametric equations yield $x = 2 + 4(-1) = -2$ and $y = 5 - 3(-1) = 8$. Similarly, when $t = 0$, the equations give $x = 2 + 4(0) = 2$ and $y = 5 - 3(0) = 5$. We show the results for the given values of t in the following table:

t	-1	0	1	2	3	4
x	-2	2	6	10	14	18
y	8	5	2	-1	-4	-7

b. Note that each value of the parameter t gives rise to a pair of values x and y, which in turn produces a point (x, y). To show that these points lie on a line, we consider just the x and y values in the table. Each successive x value increases by 4 units and, simultaneously, each corresponding y value decreases by 3. Because there is a constant change in y when the x's are uniformly spaced, we conclude that these points lie on a line and so the parametric equations represent a line. In particular, the slope of this line is $\Delta y / \Delta x = -\frac{3}{4}$.

c. Figure 9.43 shows the plot of these points and the line through them.

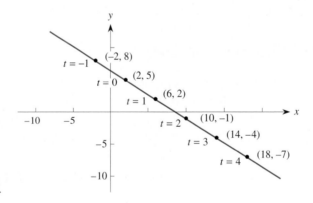

FIGURE 9.43

The pair of equations

$$x = 2 + 4t \quad \text{and} \quad y = 5 - 3t$$

giving x and y in terms of the parameter t is called a *parametric representation* or the *parametric equations* of the line. More generally, if we have a curve instead of a line, the pair of equations for x and y in terms of a parameter t is a parametric representation of the curve.

EXAMPLE 2

Eliminate the parameter t from the pair of parametric equations $x = 2 + 4t$ and $y = 5 - 3t$ and so find the slope–intercept form for the equation of this line.

Solution   We start with the parametric equation $x = 2 + 4t$ and solve for t:

$$4t = x - 2 \quad \text{so that} \quad t = \frac{1}{4}(x - 2).$$

When we substitute this expression into the parametric equation for y, we get

$$y = 5 - 3t = 5 - 3\left(\frac{1}{4}\right)(x - 2) = 5 - \frac{3}{4}(x - 2),$$

or

$$y - 5 = -\frac{3}{4}(x - 2),$$

which is the point–slope form for the equation of a line with slope $-\frac{3}{4}$ that passes through the point $(2, 5)$. This value for the slope is the same value we found in Example 1. Figure 9.43 shows that the line clearly passes through the point $(2, 5)$, which is also a point in the table we created in Example 1.   ◆

The Path of a Projectile

Another case of a parametric representation of a function is the path of a thrown object, such as a football. The path, or trajectory, is a parabola of the form

$$y = ax^2 + bx + c.$$

However, for most real-world applications, the equation of the parabola by itself is of little value. Far more important is knowing *when* the ball, or other object, will reach a particular point. Therefore introducing time as a variable is necessary, and we do so by writing both x and y, the coordinates of each point along the parabola, in terms of a parameter t that represents time. In particular, if the object is released at time $t = 0$ from an initial height y_0 with an initial velocity v_0 at an initial angle α, as shown in Figure 9.44, then at any time t thereafter, it turns out that

$$x = (v_0 \cos \alpha)t \quad \text{and} \quad y = -\frac{1}{2}gt^2 + (v_0 \sin \alpha)t + y_0.$$

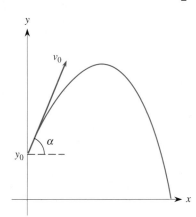

FIGURE 9.44

Each of these expressions can be thought of as a function of the parameter t, so we rewrite them as

$$x(t) = (v_0 \cos \alpha)t \quad \text{and} \quad y(t) = -\frac{1}{2}gt^2 + (v_0 \sin \alpha)t + y_0.$$

Each value of t determines a corresponding value for x (the horizontal distance) and a value for y (the vertical height), which produces a point (x, y) on the parabola. Again, the pair of equations for x and y as a function of the parameter t is a parametric representation of the curve, and the two equations are the parametric equations of the curve.

Parametric Representation of a Circle

The fact that the parametric representations of a line and of a parabola are so valuable suggests that parametric representations of other curves might also be useful. There are two key steps: (1) to decide on an appropriate parameter t, and (2) to find a way to express the usual variables x and y in terms of t.

We begin with a circle of radius r centered at the origin:

$$x^2 + y^2 = r^2.$$

Recall that we can express both x and y in terms of an angle θ drawn from the center of the circle, as shown in Figure 9.45. We write

$$x = r \cos \theta \quad \text{and} \quad y = r \sin \theta.$$

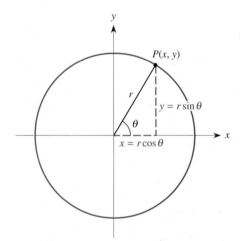

FIGURE 9.45

In retrospect, this equation is a parametric representation of the circle with the angle θ as the parameter. For each value of θ, we can calculate x and y and so get the point (x, y) on the circle. In fact, we can use any other letter, such as t, and get

$$x = r \cos t \quad \text{and} \quad y = r \sin t$$

as a parametric representation of the circle.

If we start with a parametric representation of a curve, we can sometimes *eliminate the parameter* to construct a single equation of the curve, as we did for the line in Example 2. For the circle $x = r \cos t$ and $y = r \sin t$, we eliminate the parameter t as follows:

$$x^2 + y^2 = (r \cos t)^2 + (r \sin t)^2 = r^2(\cos^2 t + \sin^2 t) = r^2.$$

Thus we are left with $x^2 + y^2 = r^2$, the usual equation for the circle.

Parametric Representation of an Ellipse

Now let's consider the ellipse centered at the origin with its major axis along the x-axis:

$$\frac{x^2}{a^2} + \frac{y^2}{b^2} = 1.$$

Its graph is shown in Figure 9.46. The question is: What might be an appropriate parameter to introduce to help describe this ellipse?

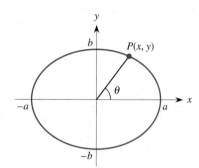

FIGURE 9.46

EXAMPLE 3

Find a parametric representation of the ellipse.

Solution   Visualize a point moving around the ellipse shown in Figure 9.46. Although we can locate each point P in terms of its x- and y-coordinates, we may also be able to locate it by using the angle θ determined by P and the positive x-axis. How do we express x and y as functions of θ?

At first thought, you might be tempted to create a right triangle by dropping a perpendicular from P to the x-axis, as we did for the circle. The problem with this approach is that the length of the hypotenuse would change along with θ as the point P moves around the ellipse, unlike a circle in which the lengths of the line segments from O to P remain constant. Thus the angle θ is not a good choice for the parameter.

Nevertheless, our experience with the circle can provide some guidance. The parametric representation of a circle of radius r is

$$x = r \cos t \quad \text{and} \quad y = r \sin t.$$

If we think of the ellipse as having a "radius" of a associated with x and a "radius" of b associated with y, we might write

$$x = a \cos t \quad \text{and} \quad y = b \sin t.$$

Let's see if doing so makes sense. Suppose that (x, y) is any point on the ellipse so that x and y must satisfy

$$\frac{x^2}{a^2} + \frac{y^2}{b^2} = 1.$$

If we substitute our conjectured expressions for x and y into this equation, we find that

$$\frac{(a \cos t)^2}{a^2} + \frac{(b \sin t)^2}{b^2} = \frac{a^2 \cos^2 t}{a^2} + \frac{b^2 \sin^2 t}{b^2}$$

$$= \cos^2 t + \sin^2 t = 1.$$

Because these expressions for x and y as functions of t satisfy the equation of the ellipse, we conclude that $x = a \cos t$ and $y = b \sin t$ form a parametric representation of the ellipse with parameter t.

◆

Using a Calculator

One of the options available on all graphing calculators is a `Parametric` mode. To use it, you need to supply an expression for x in terms of the parameter t and an expression for y in terms of t.[1] You then have to define a window not only in terms of x and y, but also in terms of an interval of values for the parameter t. Enter the parametric representation

$$x = 5 \cos t$$
$$y = 3 \sin t$$

for the ellipse

$$\frac{x^2}{25} + \frac{y^2}{9} = 1$$

with a range of values for the parameter from 0 to 2π in radians. Verify that the graph is indeed that of an ellipse.

To use this parametric representation, suppose that we want to know the point on the ellipse

$$\frac{x^2}{25} + \frac{y^2}{9} = 1$$

corresponding to a value of the parameter—say, $t = \pi/6$. We find that

$$x = 5 \cos\left(\frac{\pi}{6}\right) = 4.330$$

$$y = 3 \sin\left(\frac{\pi}{6}\right) = 1.5.$$

Alternatively, suppose you are told that the point $(4, 9/5)$ lies on the ellipse. (Verify that it does.) To find the value of the parameter t for this point, we consider

$$x = 5 \cos t = 4$$

$$y = 3 \sin t = \frac{9}{5}.$$

The first of these equations gives

$$\cos t = \frac{4}{5},$$

from which

$$t = \arccos\left(\frac{4}{5}\right) = 0.6435 \text{ radian.}$$

Verify that this value of t also satisfies the second equation $y = 3 \sin t = 9/5$.

---

[1]Note that in `Parametric` mode, different calculators use `t` or `T` as the "generic" variable just as different models use `x` or `X` as the "generic" variable in the usual `Function` mode.

Parametric Representations of a Hyperbola

We now consider how to write a parametric representation of the hyperbola.

EXAMPLE 4

Show that the equations

$$x = b \tan t \quad \text{and} \quad y = \frac{a}{\cos t}$$

are a parametric representation of the hyperbola

$$\frac{y^2}{a^2} - \frac{x^2}{b^2} = 1.$$

Solution   If we substitute the expressions for x and y into the equation of the hyperbola, we get

$$\frac{(a/\cos t)^2}{a^2} - \frac{(b \tan t)^2}{b^2} = \frac{a^2}{a^2 \cos^2 t} - \frac{b^2 \tan^2 t}{b^2}$$

$$= \frac{1}{\cos^2 t} - \tan^2 t.$$

But, recalling the trigonometric identity

$$1 + \tan^2 \theta = \frac{1}{\cos^2 \theta} \quad \text{or} \quad \frac{1}{\cos^2 \theta} - \tan^2 \theta = 1,$$

we see that the previous expression equals 1. Thus the two equations for x and y satisfy the equation of the hyperbola and therefore represent a pair of parametric equations of the hyperbola. ◆

An alternative way to develop a parametric representation of the hyperbola requires introducing two new functions known as the *hyperbolic sine* and the *hyperbolic cosine*. We discuss them briefly in the Problems at the end of this section.

Parametric Representations of a Parabola

At the beginning of this section, we described how to find a parametric representation of the parabolic path of a projectile. We now consider the same situation geometrically. Actually, we can introduce a parameter in an extremely simple way. If the equation of the parabola is

$$y = ax^2 + bx + c,$$

we can let $x = t$, so that

$$y = at^2 + bt + c.$$

This approach may strike you as somewhat unfair (too easy!), but it is effective. In fact, it can be used with any function $y = f(x)$. Let's look at one of the advantages of doing so.

If we restrict our attention to the right-hand side of the parabola, we know that the curve is strictly increasing (if $a > 0$) and so has an inverse f^{-1}. We know that the graph of the inverse function is the mirror image of the graph of f about the diagonal line $y = x$. However, in all but the simplest cases, finding an explicit, or closed form, expression for the inverse f^{-1} is not easy, or even possible. Without such a formula for f^{-1}, constructing the graph of the inverse function would normally be almost impossible.

With parametric functions, however, this becomes a simple chore. Recall the definition of a function and its inverse function. If $y = f(x)$, for each value of x, the function determines a single corresponding value for y. The inverse function undoes this process in the sense that, for each value of y, f^{-1} returns the value of x that led to y. We can draw the graph of f in the parametric form

$$x = t \quad \text{and} \quad y = f(t)$$

by using the `Parametric` mode of the graphing calculator. To produce the graph of the inverse function, all we need do is reverse the roles of x and y. That is, if we set

$$x = f(t) \quad \text{and} \quad y = t,$$

so that

$$y = t = f^{-1}(x),$$

the calculator will draw the graph of the inverse function! Try it with, say, the right side of a parabola or with an exponential function, where you know what the function and its inverse should look like.

Other Parametric Curves

Many curves that cannot be represented simply, if at all, with y as a function of x can be represented fairly readily with parametric equations. Suppose that your friend has a reflector attached to the rim of her bicycle tire. As she rides past you at a constant speed, you observe that the path of the reflector is a curve such as the one shown in Figure 9.47. This curve, showing y as a function of x, is called a **cycloid.** If the radius of the tire is a, the parametric representation of the cycloid is

$$x = at - a \sin t \quad \text{and} \quad y = a - a \cos t,$$

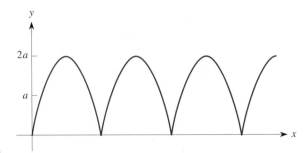

FIGURE 9.47

where the parameter t represents time. We simply cite these equations, which are typically derived in calculus, and only discuss their reasonableness here. Since the variable is time, be sure that you graph all such curves in radian mode.

Let's begin with the expression for the height $y = a - a \cos t$ of the reflector as a function of time t. The constant term a is the vertical shift, so y oscillates above and below it as the midline. The amplitude also equals a, so the height y actually oscillates between 0 and $2a$, which makes sense in terms of the physical phenomenon.

What about the expression $x = at - a \sin t$ for the horizontal distance? Note that this expression involves a sine term, which oscillates between $-a$ and a. This term is subtracted from at, which grows linearly, which again should make sense. The bicycle wheel is rolling along, so the horizontal distance traveled by the center of the wheel is simply at. Because the reflector is rotating about the rim of the tire, there must be an oscillatory adjustment to the linear distance covered.

In Problem 26 of Section 2.5, we raised the question about the shape of a water slide along which a person would slide most rapidly from one point to another; it is called the *brachistochrone* problem. From physical principles, the curve should be decreasing and concave up, so that the person gains the greatest speed at the beginning of the slide. It turns out that the specific curve along which the time needed is a minimum is an upside-down cycloid.

Let's consider another application involving a parametric representation of a curve. You have likely seen a spirograph, a toy with which you can draw intricate shapes by tracing curves as one plastic wheel rotates about another plastic wheel. Suppose that you have a large wheel of radius b and a smaller wheel of radius a that is rolling on the outside of the larger wheel, as shown in Figure 9.48. A fixed point on the outer (rolling) circle describes a curve that is known as an **epicycloid.** A parametric representation of the epicycloid is

$$x = (a + b)\cos\left(\frac{at}{b}\right) - a\cos\left[\left(\frac{a + b}{b}\right)t\right]$$

$$y = (a + b)\sin\left(\frac{at}{b}\right) - a\sin\left[\left(\frac{a + b}{b}\right)t\right].$$

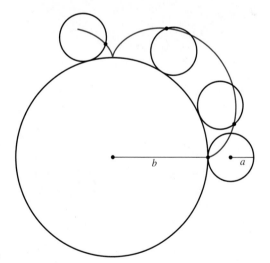

FIGURE 9.48

Let's see what the path of the fixed point on the rolling circle looks like. As the outer circle rolls on the fixed inner circle, the point on it moves back and forth, getting closer to and farther from the origin. It is closest to the origin—at a distance b—at the points where the two circles touch. It is farthest from the origin when the point is at the farthest possible position on the rolling circle, a distance of $b + 2a$. At any other time, the point is at an intermediate distance between b and $b + 2a$.

The actual shapes of epicycloids are often visually surprising and striking, as shown in Figure 9.49 for $a = 11$, $b = 28$, and t between 0 and 421π. A much simpler case is when the fixed inner circle has a radius of $b = 4$ and the rolling circle has a radius of $a = 1$, which gives the epicycloid shown in Figure 9.50 for t between 0 and $8\pi \approx 25.13$; the same curve thereafter repeats with period 8π because the identical points are repeatedly traced out. We also superimpose the inner fixed circle to indicate how the epicycloid is traced out by the fixed point as the outer (unseen) circle rolls around on the inner circle.

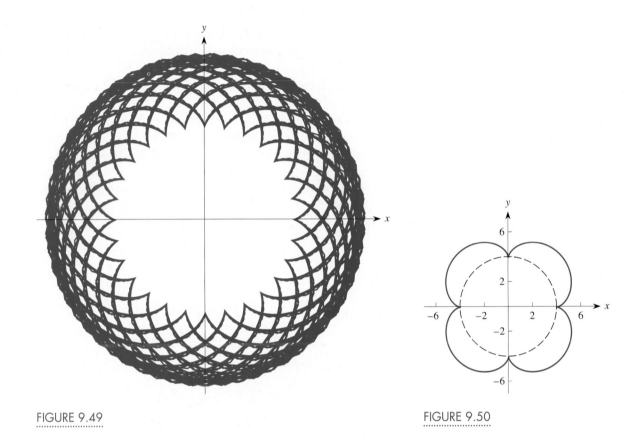

FIGURE 9.49   FIGURE 9.50

In the Problems for this section we ask you to experiment with the epicycloid and other curves by using parametric equations. You will see some surprising shapes if you simply try interesting combinations of functions. A favorite parametric curve that you can try is the "snowman" function whose parametric representation is

$$x = t - \frac{1}{2}\sin 10t \quad \text{and} \quad y = 5\sin t - \frac{1}{2}\cos 10t.$$

Problems

1. Consider the parametric representation of the line $x = 4 - 3t$, $y = 2 - 5t$.

   a. Construct a table of points that lie on this line and find the slope of the line from the table.

   b. Use the slope and a point on the line to write an equation of the line with y as a function of x.

   c. How does the slope of the line relate to the coefficients in the parametric representation?

   d. Eliminate the parameter t algebraically by solving for t from the first equation and substituting the result into the second.

e. Eliminate the parameter t algebraically by solving for t from the second equation and substituting the result into the first.

2. Consider the parametric representation of the line: $x = 7 - 3t$, $y = 4 + 5t$.

 a. Based on the results of part (c) of Problem 1, what do you expect the slope of this line to be?

 b. Create a table of values for this line and use the entries to sketch the graph of the line.

 c. Find a point–slope form of the equation of the line.

 d. Find an equation of the line by eliminating the parameter t algebraically.

3. Consider the curve given in parametric form $x = t^3 + 1$, $y = t^2 - 2$.

 a. Create a table of values for this function by using $t = -2, -1.5, -1, \ldots, 2$ and plot the points to construct a rough sketch of the graph.

 b. Draw the curve using the `Parametric` mode on your function grapher. How does the result compare to that in part (a)?

 c. Eliminate the parameter t algebraically by first solving for t in terms of x.

 d. Graph the function you obtained in part (c) by using the `Function` mode on your function grapher. How does it compare to your graph in part (b)?

4. Consider the curve with the parametric representation $x = t^2 + 1$, $y = t^2 - 2$.

 a. Create a table of values for this function, using $t = -1, 0, 1, \ldots$, and plot the points to construct a rough sketch of the graph. What surprising result do you get?

 b. Use your function grapher in `Parametric` mode to verify that the result you obtained in part (a) is correct.

 c. In terms of x and y, what are the domain and range for the curve you found in part (a)?

 d. Eliminate the parameter t algebraically and explain why you got the shape you did.

5. Use the `Parametric` mode on your function grapher to draw the graph of the epicycloid with $a = 1$ and $b = 3$. Repeat with $b = 4$, $b = 5$, and $b = 6$ while keeping $a = 1$. Do you see any pattern in the periods of these curves? Do you observe any pattern in the number of loops that you get? Explain these patterns.

6. Repeat Problem 5 with $a = 1$ and $b = 6, 8, 10,$ and 12. How do the curves you get compare to the ones you obtained before?

7. Repeat Problem 5 with $a = 2$ and $b = 3, 5, 7,$ and 9.

8. Figure 9.47 shows the graph of a cycloid, which is the path of a reflector mounted on the rim of a tire of radius a. Determine the coordinates of the points where the curve touches the horizontal axis.

9. A **hypocycloid** is the curve generated by a fixed point on a circle of radius a that rolls around the inside of a larger circle of radius b. The parametric equations for a hypocycloid are

$$x = (b - a)\cos t + a \cos\left[\left(\frac{b-a}{a}\right)t\right]$$

$$y = (b - a)\sin t - a \sin\left[\left(\frac{b-a}{a}\right)t\right].$$

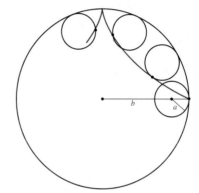

Use your function grapher to see the shapes that result when **(a)** $a = 1$, $b = 2$; **(b)** $a = 1$, $b = 3$; **(c)** $a = 1$, $b = 4$; and **(d)** $a = 2$, $b = 3$.

10. Suppose that a bicycle reflector is mounted partway along one of the spokes in a wheel at a distance $b < a$ from the center of the wheel. How should the parametric equations for the cycloid be modified to reflect this new position?

11. In the equations for the cycloid, $x = at - a \sin t$ and $y = a - a \cos t$, use the second equation to solve for t and then substitute the result into the first equation to eliminate the parameter and obtain x as a function of y. What does the resulting equation tell you about the path of the reflector?

9.6 ····· **The Polar Coordinate System**

As we discussed in Section 9.1, the *polar coordinate system* is based on the idea that every point in the plane must lie on some circle centered at the origin. In this coordinate system, the origin is known as the **pole.** To locate a point P in such a system, we must indicate the radius r of the particular circle on which P lies, as shown in Figure 9.51. That is, does P lie on a circle of radius $r = 3$ or a circle of radius $r = 4$ or a circle of radius $r = 4.2689$? Knowing that P lies on a specific circle still does not locate the point exactly. We must also specify where the point lies on that circle. Is it at an angle of 30° or an angle of 45° or an angle of 263° measured counterclockwise from the horizontal?

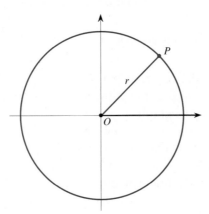

FIGURE 9.51

To formulate these ideas more precisely, let's develop some appropriate terminology. We first introduce a horizontal axis starting at the pole and pointing to the right. It is called the **polar axis** and serves as a reference. The distance from the pole to the point P, which is equivalent to the radius of a circle centered at the pole, is denoted by the coordinate r. To locate a specific point P on this circle of radius r, we must indicate how far around the circle P lies, starting from the polar axis. We measure this distance around the circle in terms of an angle coordinate θ drawn counterclockwise, or in a positive direction, from the polar axis, as shown in Figure 9.52. Thus we can locate any point in the plane if we know its distance r from the pole (to determine a circle) and the angle θ around this circle. The **polar coordinates** of the point P consist of r and θ, so we write the point as (r, θ).

For example, a point that lies 5 units from the pole at an angle of 60°, or $\pi/3$ radians, with the polar axis has polar coordinates $(5, 60°)$ or $(5, \pi/3)$, as shown in Figure 9.53. Similarly, the point Q that lies 3 units from the pole at an angle of $2\pi/3$, or 120°, has coordinates $(3, 2\pi/3)$ or $(3, 120°)$.

We can visualize the polar coordinates of a point P in the following alternative way. Any point is located on a line passing through the pole, as measured by the angle of inclination θ, which is known as the **polar angle.** The particular location of the point P along that line is determined by its distance r from the pole. Thus P can be visualized as lying at the intersection of a line through the pole and a circle centered at the pole.

This approach has an added geometric advantage. Recall from geometry that the radius drawn to any point on a circle is perpendicular to the tangent line at that point. Therefore the polar coordinates of a point are determined by the intersection

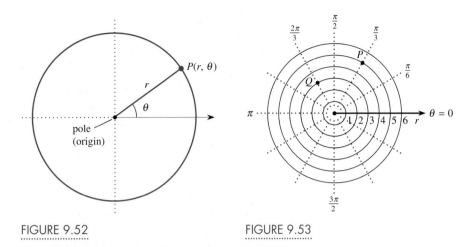

FIGURE 9.52   FIGURE 9.53

of two curves that we can think of as being "perpendicular" at the point. This property is analogous to what we do in rectangular coordinates where the vertical and horizontal lines that determine a point are perpendicular to each other.

Although there are many advantages to working with a polar coordinate system, it does have one disadvantage. In rectangular coordinates, every point has a unique pair of coordinates. However, every point in polar coordinates has more than one address. Consider the point 1 unit to the right of the pole on the polar axis. According to our discussion so far, you might conclude that its polar coordinates are $r = 1$ and $\theta = 0$. However, with a little thought, it should be evident that the address for this point could also be $r = 1$ and $\theta = 2\pi$, or $r = 1$ and $\theta = 4\pi$, and so on. Thus there are infinitely many polar coordinate representations of the same point.

In fact, there are still other ways to give the polar address of this point. In general, any angle θ measured counterclockwise from the polar axis is considered positive; any angle θ measured clockwise from the polar axis is considered negative, as illustrated in Figure 9.54. Thus our point on the polar axis could also be written as $(1, -2\pi)$, for instance.

Furthermore, we encounter some situations in Section 9.7 in which an angle θ gives rise to a negative value for r. Let's see what this means because a negative value of r cannot represent the radius of a circle. If $\theta = \pi/4$ and $r = 3$, we simply measure a distance of 3 units along the terminal side of the angle $\theta = \pi/4$. However, if $\theta = \pi/4$ and $r = -3$, we can locate the corresponding point by extending the terminal side of the angle backward through the pole and measuring 3 units along this extension, as illustrated in Figure 9.55.

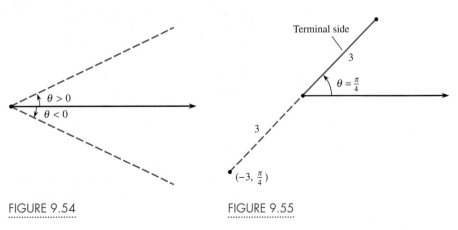

FIGURE 9.54   FIGURE 9.55

Let's look at this issue a bit more formally. When we draw any angle θ, it determines a terminal side OP from the pole through some point P, as illustrated in Figure 9.56. The obvious polar coordinate representation for this point is (r, θ), where r is positive because the distance is measured along the terminal side. However, we can also represent that point by considering the angle $\theta + \pi$ (corresponding to an additional rotation of π radians or 180°) and measuring a distance r from the pole in the opposite direction. In such a case, we think of r as negative and the polar coordinates of the point as $(-r, \theta + \pi)$. Thus, if a point P is located at $\theta = \pi/4$ and $r = 3$, we can consider the associated angle $\pi/4 + \pi = 5\pi/4$ and assign the coordinates $(-3, 5\pi/4)$ to the point P as well.

With these ideas in mind, we can find even more ways to write our earlier point with coordinates $r = 1$ and $\theta = 0$. For instance, we can obtain this point when $r = -1$ and $\theta = \pi$ or when $r = 1$ and $\theta = -2\pi$ or when $r = -1$ and $\theta = -\pi$, as illustrated in Figure 9.57.

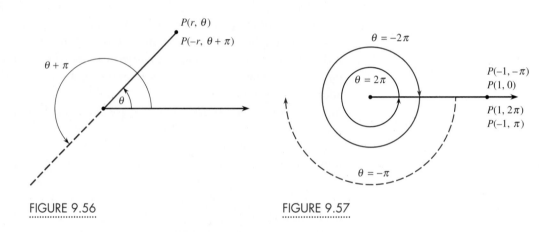

FIGURE 9.56   FIGURE 9.57

Think About This Can you think of any other coordinates for this point when r is negative? when θ is negative? ☐

Hence any point in the polar coordinate system has infinitely many pairs of coordinates. Even the pole, where $r = 0$, has infinitely many representations because it can be thought of as corresponding to *any* possible angle θ.

Transforming Between Polar and Rectangular Coordinates

Often, it is useful to think of the two coordinate systems, polar and rectangular, as being superimposed. In such a case, the pole and the origin are the same point; the polar axis and the positive x-axis coincide. The question then is: How do the coordinates of a point P in one system relate to the coordinates of the same point in the other system? That is, how do we transform the rectangular coordinates (x, y) of a point into the equivalent polar coordinates (r, θ) and vice versa?

Suppose that we start with a point P having polar coordinates (r, θ) and we want to determine the corresponding rectangular coordinates x and y. From the right triangle shown in Figure 9.58, it is clear that

$$x = r \cos \theta \quad \text{and} \quad y = r \sin \theta.$$

EXAMPLE 1

The point P has polar coordinates $r = 5$ and $\theta = \pi/3$. Find the corresponding rectangular coordinates x and y.

Solution   Using the preceding two equations, we find that the rectangular coordinates are

$$x = r \cos \theta = 5 \cos \frac{\pi}{3} = 5\left(\frac{1}{2}\right) = 2.5$$

and

$$y = r \sin \theta = 5 \sin \frac{\pi}{3} = 5\left(\frac{\sqrt{3}}{2}\right) = 4.33.$$

For the reverse problem, suppose that we start with a point P whose rectangular coordinates are (x, y), as shown in Figure 9.59. We now want to find the corresponding polar coordinates r and θ. First, we observe that r is the distance from the pole (origin) to P. The Pythagorean theorem gives

$$r^2 = x^2 + y^2 \quad \text{so that} \quad r = \pm \sqrt{x^2 + y^2}.$$

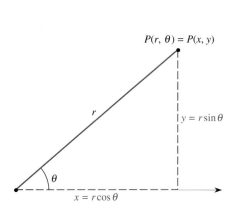

FIGURE 9.58

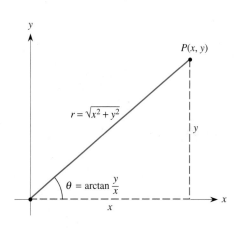

FIGURE 9.59

Next, observe that

$$\tan \theta = \frac{y}{x} \quad \text{so that} \quad \theta = \arctan \frac{y}{x}.$$

Thus, given the rectangular coordinates (x, y) of a point, we can find the polar coordinates (r, θ) by using

$$r = \pm \sqrt{x^2 + y^2} \quad \text{and} \quad \theta = \arctan \frac{y}{x}.$$

However, these formulas have to be used with great care, as demonstrated in Example 2.

EXAMPLE 2

If the rectangular coordinates of a point are $x = 3$ and $y = 4$, find one set of polar coordinates for that point.

Solution We find one set of polar coordinates using

$$r = \pm\sqrt{x^2 + y^2} = \pm\sqrt{3^2 + 4^2} = \pm\sqrt{25} = \pm 5$$

and

$$\theta = \arctan\frac{4}{3} = 0.927 \text{ radian,}$$

or about 53.13°. These values give rise to infinitely many possible pairs of coordinates, but not all of them are appropriate for the point P. The point P lies in the first quadrant, as shown in Figure 9.60, so one possible polar representation is $(r, \theta) = (5, 0.927)$. But the coordinates $(-5, 0.927)$ are not correct because they lie in the third quadrant. However, $(-5, \pi + 0.927)$, which is $(-5, 4.069)$ in radians or $(-5, 233.13°)$, is another possible pair of coordinates for P. But the polar coordinates $(5, \pi + 0.927)$ in radians or $(5, 233.13°)$ is not correct because it also is a point in the third quadrant. Be sure to plot the point in order to decide which value of r to match with which value of θ.

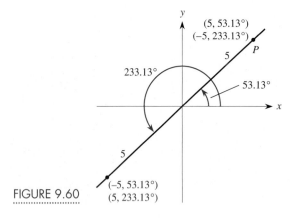

FIGURE 9.60

Think About This   Explain why only even multiples of π are used for the θ-coordinate of the different representations for the point P in Example 2.

Polar coordinates are particularly useful in representing situations in which there is a single special point and all other ideas of interest are centered at that point. For instance, in physics, the total mass of a body is often assumed to be at a single point corresponding to the pole. Thus satellites can often be thought of as moving in circular orbits about a planet located at the pole of a polar coordinate system. Similarly, the magnetic field associated with a magnet can be thought of as being centered at the pole of a polar coordinate system, and all related phenomena are often best expressed in terms of polar coordinates.

Problems

1. For each of the points P, Q, R, and S shown, do the following.

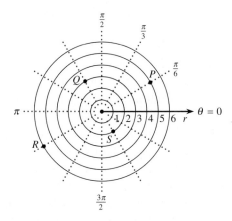

  a. Write a polar coordinate representation with r and θ both positive.
  b. Write a polar coordinate representation with r positive and θ negative.
  c. Write a polar coordinate representation with r negative and θ positive.
  d. Write a polar coordinate representation with r and θ both negative.

2. A merry-go-round at an amusement park has an inner radius of 9 feet and an outer radius of 26 feet. On the merry-go-round are five concentric circles of horses, 3 feet apart, starting with the innermost circle 10 feet from the center. Let the polar axis extend from the center of the merry-go-round to the entrance gate of the ride.

  a. What are the polar coordinates of the horse in the outer, or fifth, circle that is one-third of the way around the merry-go-round to the right from the gate?
  b. What are the polar coordinates of the horse in the second circle that is one-fifth of the way around to the left from the gate?

3. Transform the rectangular coordinates (x, y) in (a)–(h) to the equivalent polar coordinates. Sketch the location of each point in the polar coordinate plane.

  a. $(4, 4)$
  b. $(-4, 4)$
  c. $(-4, -4)$
  d. $(4, -4)$
  e. $(3, -4)$
  f. $(-3, 4)$
  g. $(8, 3)$

  h. $(3, 8)$

4. Transform the polar coordinates (r, θ) in (a)–(i) to equivalent rectangular coordinates. Indicate the location of each point graphically in the polar plane.

  a. $(5, 0)$
  b. $(5, \pi/2)$
  c. $(-5, \pi/2)$
  d. $(-5, 0)$
  e. $(3, \pi/3)$
  f. $(-3, -\pi/3)$
  g. $(2, 3\pi/2)$
  h. $(2, 5\pi/4)$
  i. $(2, -5\pi/3)$

5. A satellite is in a circular orbit about the equator at a height of 22,800 miles above the surface of the Earth. The radius of the Earth is about 4000 miles. The longitude line running north–south from the north pole to the south pole through Greenwich, England, serves as the 0° reference. Because the circumference of the Earth is about 24,000 miles, each 15° of longitude corresponds to about 1000 miles along the equator.

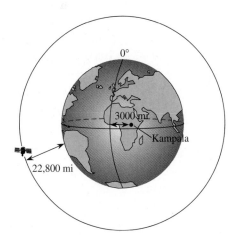

  a. What are the polar coordinates of the Ugandan capital Kampala, which is 3000 miles east (positive direction) of the Greenwich baseline?
  b. What are the polar coordinates of the capital of Borneo, which is about 7500 miles east of the Greenwich baseline?
  c. What are the polar coordinates of the satellite when it passes over Quito, Ecuador, which is 5200 miles west of the Greenwich baseline?

9.7 Families of Curves in Polar Coordinates

In Section 9.6, we introduced the notion of polar coordinates and considered coordinates (r, θ) for individual points in such a system. A far more interesting and useful question is: How do we represent curves and families of curves in polar coordinates?

Recall that, in rectangular coordinates, the curve associated with an equation $y = f(x)$ consists of all points (x, y) whose coordinates satisfy the equation. We use the comparable notion when working with polar coordinates but with the understanding that it is only necessary that one representation of a point in polar coordinates satisfies the equation.

To begin, it is usually much simpler to think of the angle θ as the independent variable and the distance r from the pole as the dependent variable. Thus for most functions in polar coordinates, we write $r = f(\theta)$ for some set of values of the angle θ. Then for each allowable value of θ, the function determines a corresponding value for r and the pair (r, θ) represents the polar coordinates of a point P in the plane. The totality of all such points determined by the equation constitutes the graph of the function. Note that writing the polar coordinates of a point as (r, θ) reverses our usual notation of writing the independent variable first and the dependent variable second, as is done in rectangular coordinates with (x, y).

Let's begin with some particularly simple cases. First, consider the equation

$$r = f(\theta) = c,$$

where c is a constant. Thus, no matter what the angle θ is, the distance r from the pole is the constant c. The set of all points that satisfy this condition forms the circle of radius c centered at the pole, as shown in Figure 9.61.

Next, consider the equation $\theta = \alpha$, where α is a constant; note that the distance r is not explicitly mentioned. No matter what distance we use, the corresponding point is always located at the angle α as measured from the polar axis. The set of all such points forms a line, inclined at the angle α which passes through the pole, as depicted in Figure 9.62. For instance, $\theta = \pi/4$ represents the line through the pole inclined at a 45° angle.

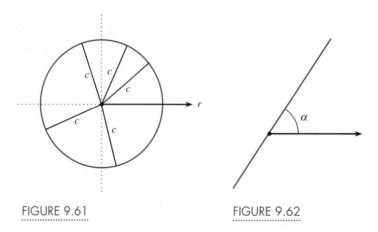

FIGURE 9.61   FIGURE 9.62

Shapes that are far more interesting and intricate than a circle and a line arise from relatively simple polar equations. We investigate some types of shapes and their underlying patterns for various families of polar coordinate curves. In the rest of this section, you should use your graphing calculator set in `Polar` mode or a polar graphing program for a computer.

Working with polar coordinates often has a special advantage over working with rectangular coordinates. Consider the simple curve shown in Figure 9.63, which is known as an Archimedean spiral. Its equation in polar coordinates is $r = \theta$. When $\theta = 0$, we have $r = 0$, so the curve starts at the pole. As θ increases, the distance r from the pole likewise increases, and as θ loops repeatedly around the pole, so does the curve to form the spiral shape shown.

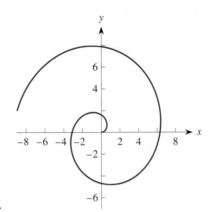

FIGURE 9.63

Now, let's find an equivalent equation in rectangular coordinates, using the transforming equations we derived in Section 9.6:

$$r^2 = x^2 + y^2 \quad \text{and} \quad \theta = \arctan \frac{y}{x},$$

and

$$x = r \cos \theta \quad \text{and} \quad y = r \sin \theta.$$

Substituting the first pair of these expressions into the equation $r = \theta$, we get the rectangular equation

$$\sqrt{x^2 + y^2} = \arctan \frac{y}{x},$$

which is not particularly attractive. We can simplify this expression slightly by taking the tangent of both sides to eliminate the arctangent function:

$$\tan \left(\sqrt{x^2 + y^2} \right) = \frac{y}{x}.$$

Or, if we multiply through by x,

$$x \tan \left(\sqrt{x^2 + y^2} \right) = y.$$

Neither of these expressions is any more attractive. Moreover, we can't simplify any of these expressions to write y as a function of x or to write x as a function of y. (In fact, recall that such a curve does not represent a function.) Furthermore, none of these rectangular expressions gives any insight into the behavior of the curve, whereas the polar representation $r = \theta$ was very helpful in understanding the spinal curve shown in Figure 9.63.

Think About This   What happens to the spiral if $\theta < 0$? ⌐

EXAMPLE 1

Consider the polar function $r = f(\theta) = \cos\theta$, whose graph is shown in Figure 9.64. By eye, the curve appears to be circular, appears to pass through the pole, and appears to be symmetrical about the polar axis. Show that this curve is a circle.

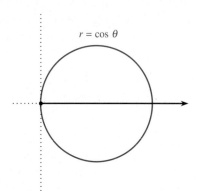

$r = \cos\theta$

FIGURE 9.64

Solution   To prove that the curve is a circle, we can try to express it in rectangular coordinates where the equation of a circle would be recognizable. Although we could attempt to substitute the transforming expressions for r and θ into the equation $r = \cos\theta$, using a little trick is much easier. We multiply both sides of the given equation $r = \cos\theta$ by r to get

$$r^2 = r\cos\theta,$$

which is equivalent to the rectangular equation

$$x^2 + y^2 = x \quad\text{or}\quad x^2 - x + y^2 = 0.$$

To determine whether this is the equation of a circle, we complete the square in the x-terms:

$$(x^2 - x) + y^2 = \left[x^2 - x + \left(-\frac{1}{2}\right)^2 - \left(-\frac{1}{2}\right)^2\right] + y^2$$

$$= \left(x^2 - x + \frac{1}{4}\right) - \frac{1}{4} + y^2$$

$$= \left(x - \frac{1}{2}\right)^2 + y^2 - \frac{1}{4} = 0,$$

so that

$$\left(x - \frac{1}{2}\right)^2 + y^2 = \frac{1}{4}.$$

This is the equation of a circle with radius $\frac{1}{2}$ centered at the (rectangular) point $(\frac{1}{2}, 0)$. This circle is indeed symmetrical about the horizontal axis and does pass through the pole.   ◆

Think About This

1. Describe the graph of $r = 5\cos\theta$.

2. Describe the graph of $r = a\cos\theta$, for any multiple $a > 0$.

3. Describe the graph of $r = a\sin\theta$, for any multiple $a > 0$.   ▭

The Family of Rose Curves

Let's consider some related curves in polar coordinates. When we graph the equation $r = \cos 2\theta$, we obtain the result shown in Figure 9.65. This graph corresponds to angles θ ranging from 0 to 2π. If we extend the values beyond this interval, in either direction, the same points repeat, so the result is a periodic function with period 2π. If you experiment with your polar function grapher, you will notice that the graph shown is traced repeatedly when you take a large range of values for the angle θ.

Don't just look at the completed shape, but rather consider this curve and other polar coordinate curves we discuss in a dynamic manner. How are the curves produced or traced? Think of the cursor on the calculator or the computer screen as a moving point that traces the curve and observe carefully how the curve is generated.

Note that the graph shown in Figure 9.65 for $r = \cos 2\theta$ consists of four loops of equal size. (Actually, depending on the calculator or computer graphics package you use, there may be some distortion and the loops may not appear to be precisely the same size even though they are.) To get a better feel for how the particular shape evolves, watch carefully as the curve is traced in Figure 9.66. Note that it starts at the far right (corresponding to $\theta = 0$ where $r = 1$) and then loops around (portion ①) until it passes through the pole (corresponding to $\theta = \pi/4$, where $r = \cos 2(\pi/4) = \cos(\pi/2) = 0$). It then starts to form a second loop (portion ②) as r takes on negative values. Eventually, it completes the loop (portion ③) before it again passes through the pole, this time at an angle of $\theta = 3\pi/4$ so that again $r = 0$. It then begins to form the third loop (portion ④) and completes that loop (portion ⑤) when it passes through the pole, where $\theta = 5\pi/4$. It then forms a fourth loop (portions ⑥ and ⑦), for θ between $5\pi/4$ and $7\pi/4$. It finally completes the original loop (portion ⑧) as θ progresses to 2π. This curve is known as a *four leaf rose*.

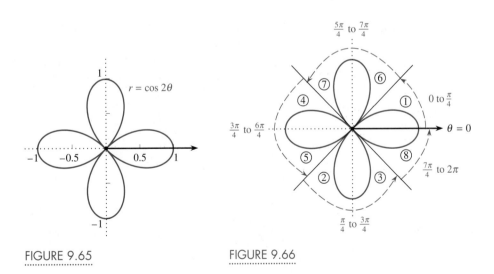

FIGURE 9.65   FIGURE 9.66

Think About This

1. What shape is produced if you graph $r = a \cos 2\theta$ for any multiple a?

2. Describe the graph corresponding to $r = a \sin 2\theta$ for any multiple a. How does it compare to the graph of the cosine function in part (1)? ▭

Let's now make a relatively simple change and consider $r = \cos 3\theta$ instead of the four-leaf rose $r = \cos 2\theta$.

EXAMPLE 2

Describe the graph of $r = \cos 3\theta$.

Solution The resulting graph is shown in Figure 9.67, but we need to observe carefully how the curve is traced. First, we observe that the curve now consists of only three loops, and they are traced for values of θ between 0 and π. For any angles outside the interval $[0, \pi]$, the same points are produced, so the polar curve is periodic with period π, even though the function $f(\theta) = \cos 3\theta$ is periodic with period $2\pi/3$. Next, we observe that the curve starts when $\theta = 0$ and $r = 1$ to produce the point at the far right. It then forms a half loop and passes through the pole when $\theta = \pi/6$. The lower left full loop is traced for values of θ between $\pi/6$ and $\pi/2$. The upper left full loop is traced as θ ranges from $\pi/2$ to $5\pi/6$. The bottom half of the right-hand loop is completed as θ ranges from $5\pi/6$ to π. This curve is known as a *three-leaf rose*.

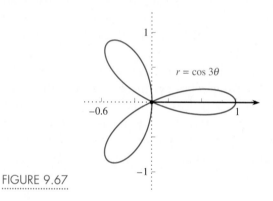

FIGURE 9.67

In general, the family of curves given by $r = \cos n\theta$ or $r = \sin n\theta$ for any positive integer n are called **rose curves.** Figures 9.68(a) and (b) show the graphs of $r = \cos 4\theta$ and $r = \cos 5\theta$; note that they contain eight and five loops, or *petals,* respectively.

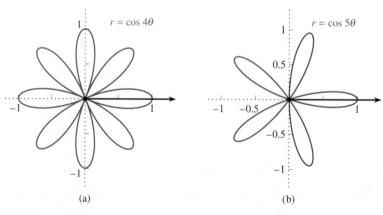

(a)   (b)

FIGURE 9.68

1. Investigate some other cases using your polar function grapher until you can devise a rule to predict the number of petals in the rose curve $r = \cos n\theta$ for any positive integer n. Are there any numbers of petals that cannot occur in this family of rose curves? If so, what are they?

2. What can you conclude about the number of petals in the related family of rose curves given by $r = \sin n\theta$?

The Family of Cardioids

Let's consider another family of polar coordinate curves, those given in the form $r = a(1 \pm \cos \theta)$. Figure 9.69 shows the graph of $r = 1 + \cos \theta$, with $a = 1$. The heart-shaped appearance of this curve suggested its name, a *cardioid.*

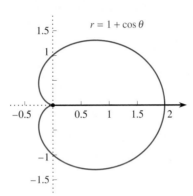

FIGURE 9.69

EXAMPLE 3

Describe how the graph of the cardioid $r = 1 + \cos \theta$ is traced.

Solution   We start with $\theta = 0$ so that $r = 2$. The curve begins at the point at the far right. As θ increases to $\pi/2$, the curve arches upward. For θ between $\pi/2$ and π, the curve bends downward and eventually inward to the pole; the resulting point at $\theta = \pi$ is called a *cusp.* As the angle θ increases from π to 2π, the curve traces the mirror image of the upper half of the cardioid; this cardioid is symmetric about the polar axis with period 2π. ◆

Those of you who have read Section 8.4 on chaos have seen that the primary central portion of the Mandelbrot set is a cardioid.

a. What is the effect of a multiple a on the shape of the curve

$$r = a(1 + \cos \theta)?$$

b. Describe the graph of the related equation $r = 1 - \cos \theta$. How does it compare with the cardioid $r = 1 + \cos \theta$?

Sketch some graphs of the related equations $r = 1 \pm \sin \theta$. Identify an axis of symmetry for them.

Suppose that you combine the ideas on the rose curves and the cardioids to consider the class of polar equations of the form $r = 1 \pm \cos n\theta$ for different positive integers n. Determine a pattern regarding their shapes.

The Family of Limaçons

An extension of the cardioid known as the *limaçon* is defined by the equations $r = a \pm b \cos \theta$ and $r = a \pm b \sin \theta$. In particular, we consider two cases: $a > b$ and $a < b$. When $a = b$, both equations reduce to that of a cardioid. Figure 9.70 shows the graph of $r = 3 + 4 \cos \theta$.

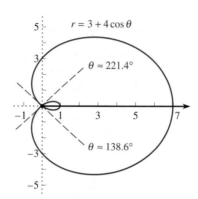

FIGURE 9.70

The curve starts at the far right, where $\theta = 0$ and $r = 7$. The curve then traces around the upper arch and eventually bends inward to pass through the pole. After passing through the pole, the curve traces the small inner loop and then passes through the pole again. It then traces the large outer loop below the polar axis, which is a mirror image of the large loop above the polar axis. The resulting curve is called a *limaçon with a loop*. (It comes from the Greek word *limax*, for snail, because the first half of the curve traced from $\theta = 0$ to $\theta = \pi$ resembles a snail-like shape.) For $\theta > 2\pi$, the curve precisely repeats this behavior.

◆ **EXAMPLE 4**

At what angles does the limaçon curve $r = 3 + 4 \cos \theta$ pass through the pole?

Solution The graph of the limaçon in Figure 9.70 shows two such angles—one in the "second quadrant" and the other in the "third quadrant". To find these angles, we use the fact that the pole corresponds to $r = 0$. Therefore, if we set $r = 0$, we get the equation

$$3 + 4 \cos \theta = 0 \quad \text{so that} \quad \cos \theta = -\frac{3}{4}.$$

Thus one angle at which the limaçon passes through the pole must satisfy is

$$\theta = \arccos\left(-\frac{3}{4}\right) = 2.419 \text{ radians}, \quad \text{or} \quad 138.59°.$$

We use the symmetry of the cosine function to find the second solution at $\theta = 3.864$ radians, or 221.41°. These values agree with the visual estimates that can be made by looking at Figure 9.70.

◆

Figure 9.71 shows the graph of the polar curve $r = 5 + 4 \cos \theta$. This curve is known as a *limaçon without a loop*, or a *dimpled limaçon*. It is similar in appearance to a cardioid, but it does not reach the pole.

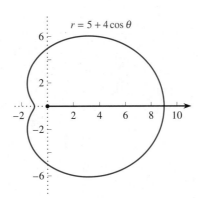

FIGURE 9.71

Think About This

Can you account for the fact that the curve $r = 5 + 4 \cos \theta$ never passes through the pole (where $r = 0$) and so never produces a loop?

Think About This

Devise criteria based on the values of a and b in $r = a + b \cos \theta$ so that you can determine whether there is a loop. Be sure that you graph a variety of limaçons using your polar function grapher to collect enough information to know that you are correct.

Think About This

Describe the shape of limaçons given by $r = a - b \cos \theta$.

Think About This

What happens in the related family of limaçons given by $r = a \pm b \sin \theta$?

We urge you to experiment with the curves generated by polar coordinate equations. You can get some incredibly striking effects just by creating strange combinations of different functions. For instance, the graph of the polar function

$$r = \sin^5 \theta + 8 \sin \theta \cos^3 \theta$$

is the butterfly shape shown in Figure 9.72.

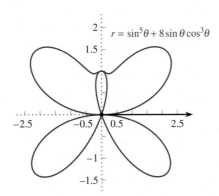

FIGURE 9.72

Think About This

Explore systematically some family of polar functions—say, $r = \sin^n \theta$ for various integers n. You may well discover some fascinating new patterns and add some new items to the literature of mathematics.

Problems

In Problems 1–11, graph each polar curve using your polar function grapher. For each, use a variety of intervals for the angle θ until you obtain a "good" picture of the graph.

1. $r = \dfrac{4 \sin^2\theta}{\cos\theta}$ (*Cissoid of Diocles*)

2. $r = \dfrac{1}{\sin\theta} - 2$ (*Conchoid of Nicomedes*)

3. $r = 4 \sin\theta \cos^2\theta$ (*Bifolium*)

4. $r = 5\left(4 \cos\theta - \dfrac{1}{\cos\theta}\right)$ (*Trisectrix*)

5. $r = \dfrac{3 \sin\theta}{\theta}$ (*Cochleoid*)

6. $r = \dfrac{4}{\sqrt{\theta}}$ (*Lituus*)

7. $r = \dfrac{8}{\sin 2\theta}$ (*Cruciform*)

8. $r = \dfrac{10}{3 + 2 \cos\theta}$ (*Ellipse*)

9. $r^2 = 4 \cos 2\theta$ (*Lemniscate of Bernoulli*)

(*Caution*: Be sure to restrict your attention to values of θ that cause the right-hand side to be positive.)

10. $r^3 = 4 \cos 3\theta$ (*Generalized Lemniscate*)

(*Caution*: Some programs and calculators are not able to evaluate the cube root of a negative number.)

11. $r = \dfrac{4}{\sin\theta}$

12–22. Repeat Problems 1–11 by changing some of the terms. What happens to the shape you produced if you use different values for the coefficients? What happens if you interchange sines and cosines? What happens if you change the multiples of θ? Keep a record of what you do and of your findings.

23. Consider the family of "hybrid rose curves"[1] given by $r = \cos(\frac{a}{b}\theta)$ for any rational number a/b.

a. By experimenting with different combinations of a and b, can you determine any rules for predicting the number of (overlapping) loops that will result? If so, state them.

b. Can you determine any rules for predicting the interval of angles θ needed to trace one complete petal of this curve? If so, state them.

c. Can you determine any rules for predicting the interval of angles θ needed to trace the entire curve? If so, state them. (*Hint*: Consider different cases, depending on whether a and b are odd or even.)

24. Consider the family of polar curves given by $r = \sin^n\theta$.

a. After graphing the curves corresponding to $n = 1$ and $n = 2$, what shape do you expect for $n = 3$? for $n = 4$?

b. Account for the fact that the shapes are not what you expected.

c. Determine a pattern for the number of loops that will correspond to any value of n.

d. What interval of angles corresponds to a complete curve? Do the same conclusions apply to $r = \cos^n\theta$?

25. Consider the family of generalized lemniscates given by $r^2 = \cos n\theta$. Can you find any pattern for the number and location of the loops that will result for any n? (*Caution*: When you try to graph these curves, you must take into account intervals of angles for which the function is well defined.)

26. Consider the family of generalized lemniscates given by $r^n = \cos n\theta$. Can you determine a pattern for the number and location of the loops that will result for any n? If so, what is it?

27. Consider the family of generalized limaçons given by $r = a + b \cos n\theta$. Can you find a pattern for the number and location of the loops that will result for any n? If so, what is it?

Chapter Summary

In this chapter we introduced and discussed a variety of topics related to coordinate systems in general and several specific coordinate systems in particular. This includes:

[1] These curves were studied in detail by a student, Kenneth Gordon, in the article, Investigating the petals of hybrid roses, *Mathematics and Computer Education*, 1992, 26, 66–73.

◆ What a coordinate system is.

◆ How to find the distance between points in the plane.

◆ How to find the midpoint of a line segment.

◆ How to find a point at any given distance along the line through two points.

◆ The parametric equations of a line.

◆ The equation of a circle.

◆ The equation of an ellipse, including finding its center, vertices, and foci.

◆ The reflection property of an ellipse and its applications.

◆ The equation of a hyperbola, including finding its center, vertices, and foci.

◆ Applications of the hyperbola.

◆ The equation of a parabola, including finding its vertex, focus, and directrix.

◆ The reflection property of a parabola and its applications.

◆ The parametric representation of curves in the plane.

◆ What the polar coordinate system is and how to transform between polar and rectangular coordinates.

◆ The behavior of families of curves in polar coordinates.

Review Problems

1. Find an equation of each ellipse shown.

2. Find an equation of each hyperbola shown.

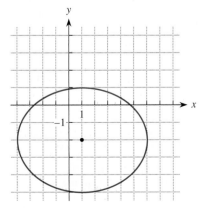

a.

a.

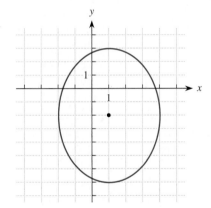

b.

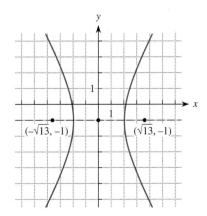

b.

3. Find an equation of each parabola shown.

a.

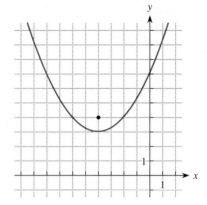

b.

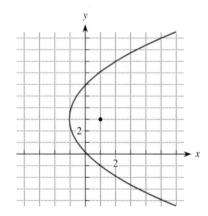

4. Identify the conic whose equation is $xy = 5$ and sketch the curve.

In Problems 5–9, determine the equation for the standard form of the conic section. Identify the conic and sketch the curve. Wherever applicable, give the focus (foci), vertex (vertices), and center for each conic.

5. $x^2 - 6x + y - 34 = 0$
6. $x^2 + y^2 - 8x + 6y + 9 = 0$
7. $2x^2 + 3y^2 + 20x - 12y + 28 = 0$
8. $3x^2 - 4y^2 - 6x - 24y - 45 = 0$
9. $3y^2 - 2x^2 - 12y + 12x - 24 = 0$

10. Find the equation of the ellipse with foci $(-8, 0)$ and $(8, 0)$ and with vertices $(-10, 0)$ and $(10, 0)$.

11. Find the equation of the ellipse centered at $(-6, 3)$ with one focus at $(0, 3)$ and the minor axis 4 units long.

12. Find the equation of the hyperbola centered at $(2, 3)$ with one focus at $(2, 7)$ and the corresponding vertex at $(2, 6)$.

13. Find the equation of the hyperbola that has vertices $(0, \pm 4)$ and passes through the point $(6, \sqrt{80})$.

14. Find the equation of the set of points P with coordinates (x, y) in the plane such that the sum of the distance from $(12, 0)$ to P and the distance from $(-12, 0)$ to P is 30.

15. An ellipse passes through the point P at $(\sqrt{3}/2, 2)$ and has foci at $(-3, 0)$ and $(3, 0)$. Use the geometric definition of an ellipse to find the equation of this ellipse.

16. The ceiling of a whispering gallery is built so that the highest point of the structure is 16 feet above the floor. The floor has vertices 40 feet apart. Where along the axes should each person stand to be able to get the "whispering effect"? Ignore the height of the two people.

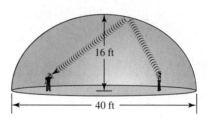

17. A lithotripter is a medical device used by doctors to break up kidney stones by bombarding them with intense bursts of sound waves, using the reflection property of an ellipse. The device is situated so that the sound waves emanate from one focus, reflect off an elliptic-shaped bowl, and come together to strike the kidney stone at the other focus. The distance between the two foci is 23 cm, and the distance from the source focus to the vertex on the elliptic reflector bowl is 3 cm. Find the equation of an elliptic cross section of the lithotripter bowl.

18. Let $x = 4 - t$ and $y = 2 + 3t$. Graph the points (x, y) for $t = -2, -1, 0, 1,$ and 2. Find the function determined by the parametric equations.

19. Sketch the parametric curve given by
$$x = 3t \quad \text{and} \quad y = t^2 + 1, \quad \text{for } -2 \le t \le 2.$$

20. Let $x = t^2 + 3$ and $y = t^3 - 1$.
   a. Graph the curve for $-4 \le t \le 4$.
   b. Eliminate t and write an expression for the curve in x and y.
   c. At what value of x is $y = 0$?

21. Sketch the curve

$$x = 1 - \log t \quad \text{and} \quad y = \log t, \qquad \text{for } 1 \le t \le 10.$$

   a. Eliminate the parameter to find the expression for y as a function of x.
   b. What is the largest possible domain for this function?

22. Graph the equations

$$x = \sin 2t \quad \text{and} \quad y = \cos \frac{1}{2} t,$$

   for $-2\pi \le t \le 2\pi$.

23. Compare the graph in Problem 22 to the graphs of

   a. $x = \sin 4t$, $y = \cos 2t$;
   b. $x = \sin 6t$, $y = \cos 2t$;
   c. $x = \sin 6t$, $y = \cos t$.
   d. Determine the period of each graph in parts (a)–(c).

24. Use appropriate trigonometric identities to eliminate t and write the following expressions in terms of x and y:

$$x = \cos 2t \quad \text{and} \quad y = \sin t.$$

25. Transform each point from rectangular coordinates (x, y) to an equivalent point in polar coordinates.

   a. $(3, 3)$
   b. $(-1, 3)$
   c. $(4, -1)$
   d. $(0, 6)$.

26. Transform each point from polar coordinates (r, θ) to rectangular coordinates.

   a. $(3, \pi/3)$
   b. $(3, \pi/4)$
   c. $(4, 3\pi/2)$
   d. $(4, 5\pi/4)$
   e. $(5, 5\pi/6)$
   f. $(5, 2)$

27. Using polar coordinates, sketch the curve

$$r = \frac{1}{1 + \cos \theta}.$$

   Convert the polar expression to rectangular coordinates and find the equation of the conic.

28. The polar equation of a well-known family of curves is

$$r = \frac{1}{\sqrt{\dfrac{\cos^2\theta}{a^2} + \dfrac{\sin^2\theta}{b^2}}}.$$

   What are these curves?

In Problems 29–32, compare the graphs of each set of equations.

29. $r = \cos \theta$, $r = \cos 2\theta$, and $r = \cos 4\theta$.

30. $r = \cos \theta$, $r = \cos 3\theta$, and $r = \cos 5\theta$.

31. $r = \sin \theta$, $r = \sin 2\theta$, and $r = \sin 4\theta$.

32. $r = \sin \theta$, $r = \sin 3\theta$, and $r = \sin 5\theta$.

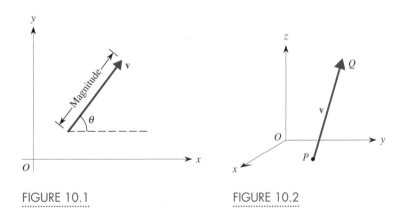

Matrix Algebra and Its Applications

10.1 ⸱⸱⸱Geometric Vectors

In the early 1800s, physicists found that most physical quantities could be categorized in one of two ways: those that have only size, such as length, time, or mass, and those that have both size and direction, such as force, velocity, or acceleration. Quantities that have only size, or *magnitude,* are called **scalars;** those that have both magnitude and direction are called **vectors.** We use lightface, italic lowercase letters, such as *a*, *m*, or *x* to denote scalars; we use boldface, roman lowercase letters such as **b, v,** or **x** to denote vectors.

Although vectors may represent physical (as well as other) quantities, we will think of them geometrically in this section. In two dimensions, we visualize a vector as an arrow connecting two points, as shown in Figure 10.1. The length of the arrow represents the magnitude of the vector. The slope of the line through any two points on the arrow, along with the arrowhead, gives the direction of the vector **v**. In three dimensions, we likewise visualize a vector as an arrow connecting two points, as shown in Figure 10.2.

FIGURE 10.1

FIGURE 10.2

Any vector starting at the origin is known as a *position vector* because it gives the position of the arrowhead with respect to the origin. A vector connecting two points *P* and *Q*, sometimes written **v** = *PQ*, is called a *displacement vector;* it indicates how to get from P to Q by moving a given distance from P in the desired direction. Obviously, a position vector is also a displacement vector, indicating how to move from the origin to point Q. But a displacement vector is not a position vector if it starts at any point P other than the origin.

If we know the coordinates of the initial point and the final point of the arrow, we can write the vector simply. First, we consider the position vector $\mathbf{v}$ from the origin to the point $(3, 4)$ shown in Figure 10.3. It involves moving 3 units to the right and 4 units upward from the origin, and we write the vector $\mathbf{v}$ as either a *row vector* $\mathbf{v} = \begin{bmatrix} 3 & 4 \end{bmatrix}$ or as a *column vector* $\mathbf{v} = \begin{bmatrix} 3 \\ 4 \end{bmatrix}$. The numerical entries 3 and 4 in the vector are called its *components*. The decision to write the vector as a row vector or as a column vector is usually a matter of choice, so long as you are consistent. We cover several specific cases later in this chapter in which the choice of column vectors is essential; in this section we primarily use row vectors for convenience.

The magnitude of the vector $\mathbf{v} = \begin{bmatrix} 3 & 4 \end{bmatrix}$ is the length of the arrow, as shown in Figure 10.3. It is the distance from the origin to the point $(3, 4)$, and so is 5, using the Pythagorean theorem.

Next, we consider the displacement vector $\mathbf{w}$ from the point $(6, 15)$ to the point $(11, 3)$, as shown in Figure 10.4. It involves a move of $5 (=11 - 6)$ to the right and a move of $-12 (=3 - 15)$ vertically. We therefore write this vector either as the row vector $\mathbf{w} = \begin{bmatrix} 5 & -12 \end{bmatrix}$ or as the column vector $\mathbf{w} = \begin{bmatrix} 5 \\ -12 \end{bmatrix}$. Note that, in a displacement vector, the components are the differences in the coordinates of the points defining the vector. The magnitude of the vector $\mathbf{w} = \begin{bmatrix} 5 & -12 \end{bmatrix}$ equals the distance from one point to the other, or

$$\text{Magnitude} = \sqrt{(11 - 6)^2 + (3 - 15)^2} = \sqrt{5^2 + (-12)^2}$$
$$= \sqrt{25 + 144} = \sqrt{169} = 13.$$

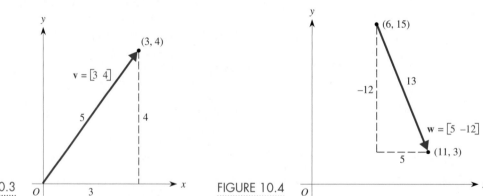

FIGURE 10.3 FIGURE 10.4

We write the magnitude of a vector $\mathbf{v}$ as $\|\mathbf{v}\|$. In general, in two dimensions, we have the following.

If $\mathbf{v} = \begin{bmatrix} a & b \end{bmatrix}$, then

$$\|\mathbf{v}\| = \sqrt{a^2 + b^2}.$$

For example, if $\mathbf{v} = [7 \quad -4]$, then $\|\mathbf{v}\| = \sqrt{7^2 + (-4)^2} = \sqrt{65}$.

Similarly, in three dimensions we can write a vector $\mathbf{v}$ in terms of three components a, b, and c as $\mathbf{v} = [a \quad b \quad c]$. The magnitude of such a vector is defined analogously with the Pythagorean theorem.

If $\mathbf{v} = [a \quad b \quad c]$, then
$$\|\mathbf{v}\| = \sqrt{a^2 + b^2 + c^2}.$$

However, specifying the direction of a vector in space is considerably harder than in the plane, and we don't go into it here.

EXAMPLE 1

Find the magnitude of the vector from the point $(1, 2, 4)$ to the point $(3, -3, 8)$.

Solution   We first write this vector in terms of its components, which are the differences in each of the three coordinates. Therefore $\mathbf{v} = [3 - 1 \quad -3 - 2 \quad 8 - 4] = [2 \quad -5 \quad 4]$, and its magnitude is
$$\|\mathbf{v}\| = \sqrt{2^2 + (-5)^2 + 4^2} = \sqrt{4 + 25 + 16} = \sqrt{45}.$$

We say that two vectors $\mathbf{v}$ and $\mathbf{w}$ are equal, written $\mathbf{v} = \mathbf{w}$, if all their corresponding components are equal. For instance, $[3 \quad 4 \quad 7] = [3 \quad \sqrt{16} \quad 7]$, but
$$\begin{bmatrix} 1 \\ 8 \\ 2 \end{bmatrix} \neq \begin{bmatrix} 1 \\ 0 \\ 2 \end{bmatrix}.$$

Geometrically, two vectors are equal if they have the same magnitude and the same direction. Figure 10.5 shows that $\mathbf{v} = \mathbf{y}$ (they have the same magnitude and the same direction); but $\mathbf{v} \neq \mathbf{x}$ (they are parallel and have the same direction, but have different magnitudes) and $\mathbf{v} \neq \mathbf{w}$ (they have the same magnitude, but do not have the same direction because they are not parallel).

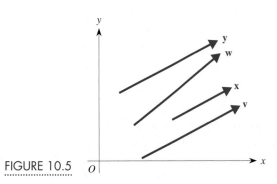

FIGURE 10.5

A Constant Multiple of a Vector

If the vector $\mathbf{v} = \begin{bmatrix} 3 & 4 \end{bmatrix}$, it seems reasonable to assume that two times $\mathbf{v}$ is just

$$2\mathbf{v} = 2 \cdot \begin{bmatrix} 3 & 4 \end{bmatrix} = \begin{bmatrix} 6 & 8 \end{bmatrix}.$$

Does this make sense geometrically? Figure 10.6 shows $\mathbf{v}$ as the displacement vector from an arbitrary point (x, y) to the point $(x + 3, y + 4)$. Note that the magnitude of $\mathbf{v}$ is 5. We also show the vector $\mathbf{w} = \begin{bmatrix} 6 & 8 \end{bmatrix}$ starting from the same point (x, y); it extends 6 units to the right and 8 units up, so it ends at the point $(x + 6, y + 8)$. From the Pythagorean theorem, the magnitude of $\mathbf{w}$ is

$$\|\mathbf{w}\| = \sqrt{6^2 + 8^2} = \sqrt{100} = 10.$$

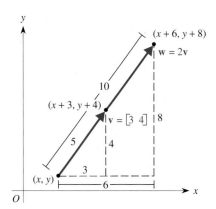

FIGURE 10.6

Thus the magnitude of $\mathbf{w}$ is twice the magnitude of $\mathbf{v}$. So multiplying $\mathbf{v}$ by 2 produces a vector $\mathbf{w} = 2\mathbf{v}$ that is twice the length of $\mathbf{v}$, but in the same direction.

In general, for any vector $\mathbf{v} = \begin{bmatrix} a & b \end{bmatrix}$ and any scalar multiple m,

$$m \cdot \mathbf{v} = \begin{bmatrix} ma & mb \end{bmatrix}$$

is a vector m times as long as $\mathbf{v}$ (shorter if $0 < m < 1$) that points in the same direction if $m > 0$. If the multiple $m < 0$, the resulting vector is parallel to $\mathbf{v}$, but it points in the opposite direction.

Moreover, this definition of the multiple of a vector suggests the following important and useful fact.

> Two vectors $\mathbf{v}$ and $\mathbf{w}$ are parallel if and only if one is a multiple of the other.

Unit Vectors

EXAMPLE 2

Find a vector $\mathbf{u}$ of length 1 that is in the same direction as the vector $\mathbf{v} = \begin{bmatrix} -6 & 8 \end{bmatrix}$.

Solution Because the vector $\mathbf{u}$ we want to find is in the same direction as $\mathbf{v}$, it will be parallel to $\mathbf{v}$ and so must be some multiple of $\mathbf{v}$, $\mathbf{u} = m \cdot \mathbf{v}$, as shown in Figure 10.7. The problem is to find the appropriate multiple m. The magnitude of $\mathbf{v}$ is

$$\|\mathbf{v}\| = \sqrt{(-6)^2 + 8^2} = \sqrt{36 + 64} = \sqrt{100} = 10.$$

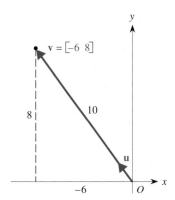

FIGURE 10.7

The vector **u** we seek is to have length 1, so it must be one-tenth of **v**. That is,

$$\mathbf{u} = \left(\frac{1}{10}\right)\mathbf{v} = \left[-\frac{6}{10}\quad\frac{8}{10}\right].$$

Any vector whose length is 1 is called a **unit vector.** In general, if **v** is any nonzero vector, a unit vector **u** in the same direction as **v** is

$$\mathbf{u} = \frac{\mathbf{v}}{\|\mathbf{v}\|}.$$

In two dimensions, the two most important unit vectors are the *coordinate vectors* along the horizontal and vertical axes. The unit coordinate vector pointing to the right is denoted by $\mathbf{i} = \begin{bmatrix}1 & 0\end{bmatrix}$, and the unit coordinate vector pointing upward is denoted by $\mathbf{j} = \begin{bmatrix}0 & 1\end{bmatrix}$.

The Sum of Two Vectors

We add vectors—whether they are two row vectors or two column vectors—by adding the corresponding components. For instance, if $\mathbf{v} = \begin{bmatrix}4 & -9\end{bmatrix}$ and $\mathbf{w} = \begin{bmatrix}7 & 3\end{bmatrix}$ are row vectors, their sum is the row vector

$$\mathbf{v} + \mathbf{w} = \begin{bmatrix}4 & -9\end{bmatrix} + \begin{bmatrix}7 & 3\end{bmatrix} = \begin{bmatrix}4+7 & -9+3\end{bmatrix} = \begin{bmatrix}11 & -6\end{bmatrix}.$$

In general, if $\mathbf{v} = \begin{bmatrix}v_1 & v_2\end{bmatrix}$ and $\mathbf{w} = \begin{bmatrix}w_1 & w_2\end{bmatrix}$ are any two row vectors,

$$\mathbf{v} + \mathbf{w} = \begin{bmatrix}v_1 + w_1 & v_2 + w_2\end{bmatrix}.$$

Geometrically, adding vectors involves "adding" the arrows, which can be thought of in two ways. First, in Figure 10.8 vector **w** is "moved" so that it starts at the end of vector **v** and still points in the same direction. (Equivalently, vector **w** is replaced by an equal vector that starts at the end of vector **v**.) Then **v** + **w** is the vector from the start of **v** to the end of **w**. The sum of the two vectors is the third side of the triangle formed by the vector **v** and the shifted vector **w**.

Alternatively, in Figure 10.9, **v** and **w** form adjacent sides of a parallelogram. The upper side has the same length as **v** and is parallel to **v**; therefore it equals **v**. Similarly, the right side of the parallelogram has the same magnitude as **w** and is parallel to **w**, so it equals **w**. The sum **v** + **w** then is the long diagonal in the parallelogram.

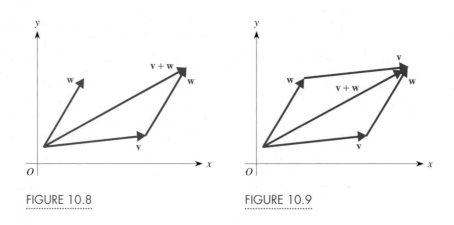

FIGURE 10.8

FIGURE 10.9

The Coordinate Vectors i and j

One of the advantages of the coordinate vectors $\mathbf{i} = \begin{bmatrix} 1 & 0 \end{bmatrix}$ and $\mathbf{j} = \begin{bmatrix} 0 & 1 \end{bmatrix}$ is that any vector in the plane $\mathbf{v} = \begin{bmatrix} a & b \end{bmatrix}$ can be written in terms of $\mathbf{i}$ and $\mathbf{j}$. In particular, as shown in Figure 10.10, the vector $\mathbf{v}$ can be thought of as the sum of a horizontal vector with magnitude a and a vertical vector with magnitude b. We write the horizontal vector with magnitude a as $a\mathbf{i}$ and the vertical vector with magnitude b as $b\mathbf{j}$. Consequently, $\mathbf{v} = \begin{bmatrix} a & b \end{bmatrix} = a\mathbf{i} + b\mathbf{j}$.

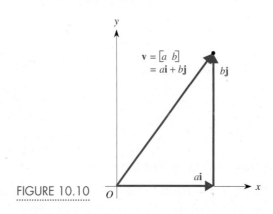

FIGURE 10.10

All operations with vectors, such as addition, can be done in terms of $\mathbf{i}$ and $\mathbf{j}$.

EXAMPLE 3

Given the vectors $\mathbf{v} = 3\mathbf{i} + 5\mathbf{j}$ and $\mathbf{w} = 7\mathbf{i} - 4\mathbf{j}$, find **(a)** their sum and **(b)** 4 times the first vector.

Solution

a. The sum of the two vectors is $\mathbf{v} + \mathbf{w} = (3\mathbf{i} + 5\mathbf{j}) + (7\mathbf{i} - 4\mathbf{j}) = 10\mathbf{i} + \mathbf{j}$.

b. $4\mathbf{v} = 4(3\mathbf{i} + 5\mathbf{j}) = 12\mathbf{i} + 20\mathbf{j}$.

Applications of Vectors

We next look at several examples involving physical situations that use vector addition.

EXAMPLE 4

Tom is trying to open a window that is stuck. He exerts a force of 30 lb at an angle of 20° with the wall. What is the effective vertical force that he exerts upward against the window?

Solution   We start by drawing a sketch of the situation, as shown in Figure 10.11. The force that Tom exerts is a vector **F** whose magnitude is 30 and whose direction is at a 20° angle with the wall. The force actually consists of two components—one vertically upward (**F**$_y$), which represents the effective force that he exerts to raise the window, and the other (**F**$_x$) perpendicular to the window, which doesn't have any effect on moving the window vertically. Thus the total force **F** is just the sum of the two vectors **F**$_x$ and **F**$_y$, and what we seek is the vertical vector **F**$_y$.

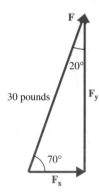

FIGURE 10.11

The angle at the upper vertex of the triangle is 20°, so the angle at the starting point of vector **F** is 70°. The length of the hypotenuse of the triangle is just the magnitude of the force vector, or $\|\mathbf{F}\| = 30$. Using trigonometry, we have

$$\sin 70° = \frac{\text{opposite}}{\text{hypotenuse}},$$

where the length of the hypotenuse is $\|\mathbf{F}\| = 30$. Therefore the length of the vertical side of the triangle is $\|\mathbf{F}\| \sin 70° = 30 \sin 70° \approx 28.19$, and the corresponding vertical vector is $\mathbf{F}_y = \begin{bmatrix} 0 & 28.19 \end{bmatrix}$. Consequently, the effective force that Tom exerts upward to move the window actually is $\|\mathbf{F}_y\| = 30 \sin 70° \approx 28.19$ pounds.

EXAMPLE 5

A flock of Canadian geese is trying to fly due south for the winter with a constant velocity of 12 mph. A stiff wind is blowing at a constant rate of 20 mph from a direction 35° west of north. Find the actual direction that the geese end up flying and their actual speed with respect to the ground.

Solution   We begin with a sketch of the situation, as shown in Figure 10.12. Each goose is trying to fly due south, so there is one velocity vector, **g**, for the goose having magnitude 12 and pointing vertically downward. In addition, each goose is pushed by the wind, which is coming from a northwesterly direction. The wind is represented by a second velocity vector, **w**, having magnitude 20 and pointing from a direction 35° west of north.

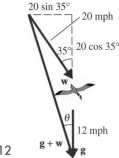

FIGURE 10.12

The actual velocity vector for the goose is the sum of these two vectors. To find the direction that a goose actually flies—and then the speed at which it flies—we need to find the components of $\mathbf{g} + \mathbf{w}$. Because the goose is trying to fly due south, $\mathbf{g}$ has only a vertical component of -12, so $\mathbf{g} = \begin{bmatrix} 0 & -12 \end{bmatrix}$. The velocity vector for the wind has both a horizontal and a vertical component. Using trigonometry, we see that the horizontal component of $\mathbf{w}$ is $20 \sin 35° \approx 11.47$. Similarly, the vertical component of $\mathbf{w}$ is $-20 \cos 35° \approx -16.38$ (it is negative because it is directed downward). Consequently, the wind vector $\mathbf{w}$ is

$$\mathbf{w} = \begin{bmatrix} 11.47 & -16.38 \end{bmatrix}.$$

Thus the sum of the two vectors is

$$\mathbf{g} + \mathbf{w} = \begin{bmatrix} 0 & -12 \end{bmatrix} + \begin{bmatrix} 11.47 & -16.38 \end{bmatrix} = \begin{bmatrix} 11.47 & -28.38 \end{bmatrix}.$$

The actual speed with which the goose flies is the magnitude of this vector $\mathbf{g} + \mathbf{w}$, which is

$$\text{Speed} = \sqrt{(11.47)^2 + (-28.38)^2} = \sqrt{131.56 + 805.42} = \sqrt{936.90} \approx 30.61.$$

Next, to find the direction in which the goose flies, we need to find the angle θ that the vector $\mathbf{g} + \mathbf{w}$ makes with the vertical. From the large right trangle in Figure 10.12, we find

$$\tan \theta = \frac{20 \sin 35°}{20 \cos 35° + 12}$$

$$= \frac{11.47}{28.38} \approx 0.404$$

so that

$$\theta = \arctan 0.404 \approx 22°.$$

Thus the geese actually end up flying in a direction 22° east of south instead of due south.

The Difference of Two Vectors

We define the difference of two vectors $\mathbf{v} = \begin{bmatrix} v_1 & v_2 \end{bmatrix}$ and $\mathbf{w} = \begin{bmatrix} w_1 & w_2 \end{bmatrix}$ to be

$$\mathbf{v} - \mathbf{w} = \begin{bmatrix} v_1 - w_1 & v_2 - w_2 \end{bmatrix};$$

that is, we simply take the difference of corresponding components.

For instance, if $\mathbf{v} = \begin{bmatrix} 14 & 3 \end{bmatrix}$ and $\mathbf{w} = \begin{bmatrix} 6 & 10 \end{bmatrix}$, then $\mathbf{v} - \mathbf{w} = \begin{bmatrix} 8 & -7 \end{bmatrix}$. Alternatively, in terms of the unit vectors $\mathbf{i}$ and $\mathbf{j}$, $\mathbf{v} = 14\mathbf{i} + 3\mathbf{j}$ and $\mathbf{w} = 6\mathbf{i} + 10\mathbf{j}$, so that

$$\mathbf{v} - \mathbf{w} = (14\mathbf{i} + 3\mathbf{j}) - (6\mathbf{i} + 10\mathbf{j}) = 8\mathbf{i} - 7\mathbf{j}.$$

Now let's interpret the difference of two vectors geometrically. Consider the vector $\mathbf{x}$ in Figure 10.13, which connects the end of $\mathbf{v}$ to the end of $\mathbf{w}$. We know from the sum of two vectors that

$$\mathbf{v} + \mathbf{x} = \mathbf{w} \quad \text{so that} \quad \mathbf{x} = \mathbf{w} - \mathbf{v}.$$

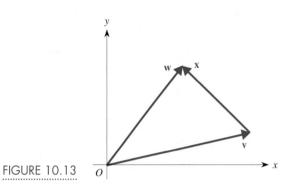

FIGURE 10.13

In general, the difference of two vectors always connects the end of the second vector to the end of the first. The only question is: Which way does the difference vector point? The easiest way to decide that is to draw a sketch such as Figure 10.13.

Another way of looking at the difference $\mathbf{v} - \mathbf{w}$ is to think of it as $\mathbf{v} - \mathbf{w} = \mathbf{v} + (-\mathbf{w})$, where

$$-\mathbf{w} = (-1)\begin{bmatrix} w_1 & w_2 \end{bmatrix} = \begin{bmatrix} -w_1 & -w_2 \end{bmatrix},$$

a vector with the same length as $\mathbf{w}$ but pointing in the opposite direction.

Problems

1. Plot each position vector as an arrow in the xy-plane from the origin to the point having the appropriate coordinates.

   a. $\mathbf{r} = \begin{bmatrix} 4 \\ 0 \end{bmatrix}$ **b.** $\mathbf{s} = \begin{bmatrix} 2 \\ 4 \end{bmatrix}$

   c. $\mathbf{t} = \begin{bmatrix} -2 \\ 4 \end{bmatrix}$ **d.** $\mathbf{u} = \begin{bmatrix} 3 \\ -4 \end{bmatrix}$

   e. $\mathbf{v} = \begin{bmatrix} -1 \\ -2 \end{bmatrix}$

2. Using the vectors in Problem 1, plot the result of:

   a. Adding vector $\mathbf{u}$ to vector $\mathbf{r}$.

   b. Adding vector $\mathbf{r}$ to $\mathbf{u}$. Compare your answer to the vector obtained in part (a).

   c. Adding $\mathbf{t}$ to $\mathbf{s}$.

   d. Adding $\mathbf{u}$ to $\mathbf{v}$.

3. Using the vectors in Problem 1, plot each vector in a–c on the same graph.

   a. The result of adding one-half of $\mathbf{r}$ to one-half of $\mathbf{s}$.

   b. The result of adding one-quarter of $\mathbf{r}$ to three-quarters of $\mathbf{s}$.

   c. The result of adding three-quarters of $\mathbf{r}$ to one-quarter of $\mathbf{s}$.

   d. Plot $\mathbf{r}$ and $\mathbf{s}$. Draw a straight line joining these two vectors. Which of the vectors drawn in parts (a), (b), and (c) lie on this line?

4. Determine the magnitude of the position vectors from the origin to the following points.

   a. $(3, 4)$　　　　　**b.** $(12, 5)$
   c. $(3, -2)$　　　　**d.** $(-7, -3)$
   e. $(1, 2, 2)$　　　　**f.** $(2, -3, 4)$

5. Determine the magnitude of the displacement vector from point A to point B for each pair of points.

   a. $A = (1, 2), B = (5, 5)$
   b. $A = (-2, -1), B = (4, 1)$
   c. $A = (1, -3), B = (-3, -4)$
   d. $A = (1, 2, 3), B = (4, 5, 3)$
   e. $A = (-1, 6, 3), B = (3, -1, 2)$

6. Determine the vector that is the given multiple of the vector $\begin{bmatrix} 1 & 2 \end{bmatrix}$.

   a. 2　　　　　　　**b.** 7
   c. -2　　　　　　**d.** 0

7. Determine the vector that is the given multiple of the vector $\begin{bmatrix} 3 & -1 & 2 \end{bmatrix}$.

   a. 3　　　　　　　**b.** 10
   c. -7　　　　　　**d.** $\frac{1}{2}$

8. Find the unit vector that points in the same direction as the given vector.

   a. $\begin{bmatrix} 3 & 4 \end{bmatrix}$　　　　**b.** $\begin{bmatrix} 1 & 1 \end{bmatrix}$
   c. $\begin{bmatrix} 0 & 5 \end{bmatrix}$　　　　**d.** $\begin{bmatrix} 1 & -1 \end{bmatrix}$
   e. $\begin{bmatrix} 1 & 2 & 2 \end{bmatrix}$　**f.** $\begin{bmatrix} 1 & -1 & 1 \end{bmatrix}$

9. Express each vector as a sum of multiples of the co-ordinate vectors $\mathbf{i} = \begin{bmatrix} 1 & 0 \end{bmatrix}$ and $\mathbf{j} = \begin{bmatrix} 0 & 1 \end{bmatrix}$.

   a. $\begin{bmatrix} 2 & 1 \end{bmatrix}$　　　　**b.** $\begin{bmatrix} -1 & 3 \end{bmatrix}$
   c. $\begin{bmatrix} \frac{1}{2} & \frac{1}{2} \end{bmatrix}$　　　　**d.** $\begin{bmatrix} 3 & 0 \end{bmatrix}$

10. Express each vector as a sum of multiples of the three coordinate vectors $\mathbf{i} = \begin{bmatrix} 1 & 0 & 0 \end{bmatrix}$, $\mathbf{j} = \begin{bmatrix} 0 & 1 & 0 \end{bmatrix}$, and $\mathbf{k} = \begin{bmatrix} 0 & 0 & 1 \end{bmatrix}$.

    a. $\begin{bmatrix} 1 & 2 & 3 \end{bmatrix}$　　　**b.** $\begin{bmatrix} 2 & 0 & 3 \end{bmatrix}$
    c. $\begin{bmatrix} -1 & -3 & 1 \end{bmatrix}$　**d.** $\begin{bmatrix} 0 & \frac{1}{2} & 0 \end{bmatrix}$

11. Refer to Example 4 in the text. What is the upward force on the window if Tom exerts

    a. a force of 30 lb at an angle of 25°?
    b. a force of 20 lb at an angle of 30°?
    c. a force of 40 lb at an angle of 15°?

12. **a.** A sliding door is difficult to open. If Claire exerts a horizontal force of 30 lb at an angle of 25° to the sliding door, what is the effective force on the door in the direction in which the door slides?
    b. Repeat part (a) with a force of 25 lb at an angle of 40°.

c. Repeat part (a) with a force of 40 lb at an angle of 20°.

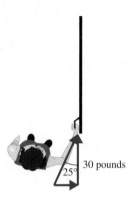

30 pounds

13. **a.** If a jet plane is flying on a heading of due east at 600 mph and the wind is blowing due south at 100 mph, what are the actual direction and speed of the plane?
    b. Repeat part (a) if the plane is flying due east at 600 mph and the wind is blowing in a direction that is 40° south of east at 100 mph.
    c. Repeat part (a) if the plane is flying southwest (45° south of west) at 300 mph and the wind is blowing in a direction 50° south of west at 100 mph.

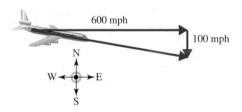

600 mph

100 mph

14. Suppose that a boat is moving at 10 mph in the direction of 20° north of east across a bay and the tide is moving the water in the bay at 4 mph in the direction of 40° west of south. What are the actual direction and speed of the boat? (*Hint:* Express both the boat's vector and the tide's vector in terms of the two coordinate vectors $\mathbf{i}$ and $\mathbf{j}$).

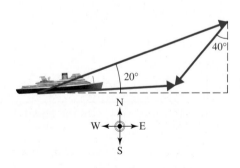

10.2 Linear Models

The real-world problems to which people apply mathematical models often involve large and very complex situations. For instance, one might want to analyze the effect that imposing a 50¢ per gallon tax on gasoline would have on the national economy with its thousands of interdependent businesses and industries. An airline must have a reservations system that takes into account all its aircraft, the cities it serves, flight schedules, dates, and different fare structures in effect. A company may need a battery of tests that can predict how well applicants will perform at a given job. The data in such problems usually come in the form of a rectangular array of numbers, called a **matrix.** For example, if four students, Ann, Bob, Carol, and Dan, take exams in French, mathematics, and sociology, the set of exam results could be displayed in the matrix

$$\begin{array}{c} \\ \text{French} \\ \text{Mathematics} \\ \text{Sociology} \end{array} \begin{array}{cccc} \text{Ann} & \text{Bob} & \text{Carol} & \text{Dan} \\ \begin{bmatrix} 84 & 73 & 82 & 85 \\ 88 & 78 & 94 & 92 \\ 76 & 81 & 83 & 78 \end{bmatrix} \end{array}$$

Thus, for instance, Bob received an 81 in sociology and Carol a 94 in math.

Matrix algebra provides a systematic way of working with such arrays of numbers. In this chapter, we develop the basic language and methods of matrix algebra that will allow you to use matrices to solve various problems involving systems of linear equations.

We refer to a rectangular array of numbers as an $m \times n$ matrix when it has m rows horizontally and it has n columns vertically. A 3×4 matrix thus has 3 horizontal rows and 4 vertical columns, as in the preceding matrix of exam scores. We use boldface capital letters, such as **A**, to denote matrices in print. (When writing matrices by hand, you may find it convenient to use a wavy line under the letter, as in $\underset{\sim}{A}$.)

Two more examples of matrices are

$$\mathbf{A} = \begin{bmatrix} 5 & 1 & -1 \\ 1 & 7 & 2 \\ 6 & 5 & 0 \end{bmatrix} \quad \text{and} \quad \mathbf{N} = \begin{bmatrix} 1 & 0 & 3 \\ 1 & 7 & 4 \end{bmatrix}.$$

Here **A** is a 3×3 matrix and **N** is a 2×3 matrix because it has two rows across and three columns vertically.

We denote the entry in row i and column j of matrix **A** by a_{ij}. Thus in matrix **A**, $a_{13} = -1$ because -1 is the entry in the first row and the third column, whereas $a_{31} = 6$ because 6 is the entry in the third row and first column. Similarly, in matrix **N**, $n_{23} = 4$ because 4 is the entry in the second row and the third column.

When a matrix has only one row or only one column, we call it a **row vector** or a **column vector,** respectively, or simply a vector. A vector having three numbers is called a **3-vector,** whereas a vector consisting of n numbers is called an **n-vector.** As noted in Section 10.1, we use boldface lowercase letters, such as **b** or **x**, to denote vectors. Note that any vector is also a matrix. Some examples of vectors are

$$\mathbf{b} = \begin{bmatrix} 1 & 3 & 0 & 5 \end{bmatrix} \quad \text{and} \quad \mathbf{x} = \begin{bmatrix} 2 \\ 4 \\ 1 \end{bmatrix}.$$

Here **b** is a row 4-vector and **x** is a column 3-vector. We write b_1 for the first entry in vector **b**, b_2 for the second entry in **b**, and b_i for the ith entry in **b**. Thus for the vectors **b** and **x**, we have $b_2 = 3$ and $x_2 = 4$.

Recall from Section 10.1 that two vectors **v** and **w** are *equal*, written **v** = **w**, if all their corresponding entries, or *components*, are equal. Similarly, two matrices are *equal* if all their corresponding components are the same.

Any list of numbers can be thought of as a column vector or a row vector. Whether we choose a column or row format if only vectors are involved usually doesn't matter, but when a vector and a matrix are multiplied, it is important to distinguish clearly whether the vector is a row vector or a column vector. For reasons that will be clear shortly, we *usually treat most vectors as column vectors*. Note that a column n-vector is an $n \times 1$ matrix and a row n-vector is a $1 \times n$ matrix.

An $m \times n$ matrix **A** can be thought of as a set of n column m-vectors or as a set of m row n-vectors. In the case of students and their test results, each column vector of the matrix gives the scores for one student in all these courses, whereas each row vector gives the scores in one course for all these students.

We use the following notation to refer to rows and columns in a matrix:

$\mathbf{a}_j$ denotes the jth column vector in **A**; and

$\mathbf{a}_i'$ denotes the ith row vector in **A**.

For instance, in the matrix

$$\mathbf{A} = \begin{bmatrix} 5 & 1 & -1 \\ 1 & 7 & 2 \\ 6 & 5 & 0 \end{bmatrix}, \quad \mathbf{a}_2 = \begin{bmatrix} 1 \\ 7 \\ 5 \end{bmatrix} \quad \text{and} \quad \mathbf{a}_1' = \begin{bmatrix} 5 & 1 & -1 \end{bmatrix}.$$

A Geometric View of Vectors

In Section 10.1, we presented vectors geometrically as positions and displacements in coordinate space. As we pointed out there, vectors can be used to represent points in space. In two-dimensional space, we use a 2-vector; in three-dimensional space, we use a 3-vector. The point $(5, 2)$ in the plane can be thought of as the 2-vector $\begin{bmatrix} 5 \\ 2 \end{bmatrix}$. Similarly, the point $(3, 2, 7)$ in three-dimensional space with coordinates $x = 3$, $y = 2$, $z = 7$, or equivalently, $x_1 = 3$, $x_2 = 2$, $x_3 = 7$, can be written as the 3-vector

$$\begin{bmatrix} 3 \\ 2 \\ 7 \end{bmatrix}.$$

Thus the coordinates of a point become the components of a position vector.

We next consider how matrices in general and vectors in particular occur in applied problems from many different fields.

A Clothes Production Model

A textile company runs three clothing factories. Each factory produces three types of women's clothing: vests, pants, and coats. For simplicity, we assume that one size fits all. Suppose that the first factory produces 20 vests, 10 pants, and 5 coats from each roll of cloth. The second and third factories produce different amounts of these three products, as described in matrix **A**.

	Factory 1	Factory 2	Factory 3
Vests	20	4	4
Pants	10	14	5
Coats	5	5	12

$= \mathbf{A}$

Each column of **A** is a vector of clothing produced by a factory from one roll of cloth. For instance, Factory 3 has the output vector

$$\mathbf{a}_3 = \begin{bmatrix} 4 \\ 5 \\ 12 \end{bmatrix},$$

which indicates that it makes 4 vests, 5 pants, and 12 coats from each roll. Each row of **A** is a vector of factory production of one particular type of clothing from one roll of cloth. The row vector for coats is $\mathbf{a}_3' = \begin{bmatrix} 5 & 5 & 12 \end{bmatrix}$, which indicates that Factory 1 produces 5 coats, Factory 2 produces 5 coats, and Factory 3 produces 12 coats from each roll.

Let x_1 denote the number of rolls of cloth used by the first factory; similarly, x_2 and x_3 denote the numbers of rolls used by the second and third factories, respectively. Suppose that the company gets an order for 500 vests, 850 pants, and 1000 coats. This triple of numbers is called the *demand,* which we write as a column vector

$$\begin{bmatrix} 500 \\ 850 \\ 1000 \end{bmatrix}.$$

Then x_1, x_2, and x_3 need to satisfy the system of linear equations

$$\begin{aligned} \text{vests:} \quad 20x_1 + 4x_2 + 4x_3 &= 500 \\ \text{pants:} \quad 10x_1 + 14x_2 + 5x_3 &= 850 \\ \text{coats:} \quad 5x_1 + 5x_2 + 12x_3 &= 1000. \end{aligned}$$

In words, the vests equation says: The number of vests produced by Factory 1, $20x_1$ (this expression is 20 vests per roll times the x_1 rolls used by Factory 1), plus the number of vests produced by Factory 2, which is $4x_2$, plus the number of vests produced by Factory 3, which is $4x_3$, must equal the demand of 500 vests.

As we demonstrate in Section 10.3, we can write this system of linear equations as the *matrix–vector equation*

$$\begin{bmatrix} 20 & 4 & 4 \\ 10 & 14 & 5 \\ 5 & 5 & 12 \end{bmatrix} \begin{bmatrix} x_1 \\ x_2 \\ x_3 \end{bmatrix} = \begin{bmatrix} 500 \\ 850 \\ 1000 \end{bmatrix}.$$

We can also write this as a single *vector equation* in the column vectors of the matrix as

$$x_1 \begin{bmatrix} 20 \\ 10 \\ 5 \end{bmatrix} + x_2 \begin{bmatrix} 4 \\ 14 \\ 5 \end{bmatrix} + x_3 \begin{bmatrix} 4 \\ 5 \\ 12 \end{bmatrix} = \begin{bmatrix} 500 \\ 850 \\ 1000 \end{bmatrix}.$$

If **b** is the demand vector on the right side of the matrix–vector equation and **x** is a (column) vector of the x_i's, matrix algebra gives us a way to write the system of linear equations concisely in terms of **A**, **b**, and **x** as $\mathbf{Ax} = \mathbf{b}$. We discuss how to do this in Section 10.3. In Section 10.5, we extend these ideas to solve any system of three equations in three variables by using matrix algebra techniques.

The expression $2x_1 + 4x_2 - \frac{1}{2}x_3$ is a *linear expression* in three variables; it involves only the first power of the variables x_1, x_2, and x_3. Other cases are the expressions on the left side of the preceding system of linear equations. More formally, a linear expression is one that involves a sum of terms made up of constants multiplying individual variables that are raised only to the first power. In contrast, a nonlinear expression involves one or more variables that are raised to various powers (different from 1), or exponential, logarithmic, trigonometric, or other more complex expressions. (This terminology is similar to that used to describe linear difference equations.) The term *linear* is used to indicate that a "line-like" graph is associated with each variable in the expression. For example, the vest expression $20x_1 + 4x_2 + 4x_3$ is a linear expression; if x_2 and x_3 are fixed—say, $x_2 = x_3 = 3$—with only x_1 remaining as a free variable, the resulting expression $20x_1 + 4(3) + 4(3)$, or $20x_1 + 24$, defines a function $y = 20x_1 + 24$ whose graph is a line.

The clothing production equations form what is called a **linear model** because the equations involve only linear expressions (linear equations). In the following sections, we will return to this and other models introduced here as we develop the mathematical methods needed to analyze linear models.

A Markov Chain Model for the Stock Market

We next develop a linear model for the behavior of the stock market. Here we show how matrix methods can be used to represent a situation in which the values of a number of variables at one stage of a process are related to their values at the preceding stage.

Each business day, the stock market goes up, goes down, or stays the same. Suppose that historical studies show that if the market goes up one day—say today—the probability is $\frac{1}{4}$ that it will go up tomorrow, the probability is $\frac{1}{2}$ that it will go down tomorrow, and the probability is $\frac{1}{4}$ that it will stay the same tomorrow. If the market goes down today, there are three other observed probabilities for tomorrow's market performance. Similarly, if the market stays the same today, there is a third set of three probabilities for what will happen tomorrow. We can conveniently display all nine of these probabilities in a matrix **A**:

$$
\begin{array}{cc}
 & \text{Market Today} \\
 & \begin{array}{ccc} \text{Up} & \text{Down} & \text{Same} \end{array} \\
\begin{array}{c} \\ \textbf{Market} \\ \textbf{Tomorrow} \end{array}
\begin{array}{c} \text{Up} \\ \text{Down} \\ \text{Same} \end{array}
\left[\begin{array}{ccc}
\frac{1}{4} & \frac{1}{2} & \frac{1}{4} \\
\frac{1}{2} & \frac{1}{4} & \frac{1}{2} \\
\frac{1}{4} & \frac{1}{4} & \frac{1}{4}
\end{array} \right] = \mathbf{A}.
\end{array}
$$

The probabilities in this matrix are called *transition probabilities* because they give us information about how to relate one stage of a process to the next. The matrix **A** is called a **transition matrix.** Each column corresponds to a type of market movement today, and each row corresponds to a type of market movement tomorrow. The matrix entry a_{23}, which equals $\frac{1}{2}$, is in the "down tomorrow" row and in the "same today" column. The value $\frac{1}{2}$ represents the probability that the market will go down tomorrow given that it stays the same today.

Note that the probabilities in each column of the transition matrix *must add to 1* because they include all possible outcomes for tomorrow given a particular type of market behavior today. A mathematical model such as this with given transition probabilities is known as a **Markov process,** or **Markov chain** (named after Russian mathematician Andrei Markov, who first developed these ideas). A convenient way to display the information in a Markov chain is with a *transition diagram,* such as the one shown in Figure 10.14. In this diagram, there are three *nodes,* one for each type of market movement: go up (U), go down (D), or stay the same (S). These are the possible *states* for the system. Note also that we indicate each transition probability with an arrow.

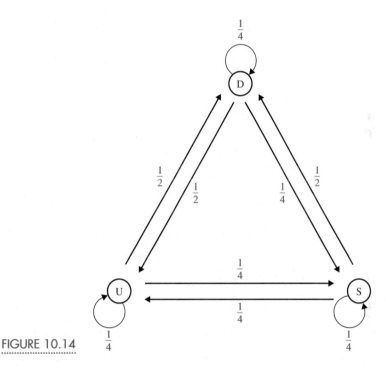

FIGURE 10.14

EXAMPLE 1

Suppose that, before the stock market opens today, we believe that there is a 50–50 chance of the market going down or staying the same, but no chance of its going up. Use the values in the preceding transition matrix **A** to compute the probabilities of the market being in any of the three states tomorrow—up, down, or the same—based on the probabilities of the market being up, down, or the same today.

Solution   Let p_1, p_2, and p_3 denote today's probabilities of the market being up, down, and the same, respectively, and let p_1^+, p_2^+, and p_3^+ denote tomorrow's probabilities of being up, down, and the same, respectively. Let's see how to compute p_1^+. First, to compute the probability of two successive events—such as (i) being in State 1 today (probability p_1) followed by (ii) switching from State 1 today to State 1 tomorrow (probability $\frac{1}{4}$)—we multiply the probabilities of the two events and get $\frac{1}{4}p_1$. Similarly, the probability of (i) being in State 2 today (probability p_2) followed by (ii) switching from State 2 to State 1 tomorrow (probability $\frac{1}{2}$) is the product $\frac{1}{2}p_2$. Also, the probability of (i) being in State 3 today (probability p_3) followed by (ii) switching from State 3 to State 1 tomorrow (probability $\frac{1}{4}$)is $\frac{1}{4}p_3$.

To get the total probability p_1^+ that the stock market will go up tomorrow, we add these three values to get

$$p_1^+ = \frac{1}{4}p_1 + \frac{1}{2}p_2 + \frac{1}{4}p_3.$$

In the same way, we calculate the probabilities p_2^+ and p_3^+ that the stock market goes down or stays the same tomorrow and so obtain a set of three equations based on the transition matrix **A**:

$$
\begin{aligned}
p_1^+ &= \frac{1}{4}p_1 + \frac{1}{2}p_2 + \frac{1}{4}p_3 \\
p_2^+ &= \frac{1}{2}p_1 + \frac{1}{4}p_2 + \frac{1}{2}p_3 \\
p_3^+ &= \frac{1}{4}p_1 + \frac{1}{4}p_2 + \frac{1}{4}p_3.
\end{aligned}
\qquad (1)
$$

Note that the coefficients in this system of linear equations come directly from the entries in the transition matrix **A**.

Believing that $p_1 = 0$, $p_2 = \frac{1}{2}$, and $p_3 = \frac{1}{2}$ are today's probabilities of the market being up, down, and the same, respectively, we calculate the probability p_1^+ that the market goes up tomorrow, using the first of Equations (1), as follows:

$$p_1^+ = \frac{1}{4}p_1 + \frac{1}{2}p_2 + \frac{1}{4}p_3 = \frac{1}{4}\cdot 0 + \frac{1}{2}\cdot\frac{1}{2} + \frac{1}{4}\cdot\frac{1}{2} = \frac{3}{8}.$$

In the same way, using Equations (1), we obtain the probabilities for the other two market outcomes tomorrow. Thus tomorrow's probabilities p_1^+, p_2^+, and p_3^+ are

$$
\begin{aligned}
p_1^+ &= \frac{1}{4}p_1 + \frac{1}{2}p_2 + \frac{1}{4}p_3 = \frac{1}{4}\cdot 0 + \frac{1}{2}\cdot\frac{1}{2} + \frac{1}{4}\cdot\frac{1}{2} = \frac{3}{8}, \\
p_2^+ &= \frac{1}{2}p_1 + \frac{1}{4}p_2 + \frac{1}{2}p_3 = \frac{1}{4}\cdot 0 + \frac{1}{4}\cdot\frac{1}{2} + \frac{1}{2}\cdot\frac{1}{2} = \frac{3}{8}, \\
p_3^+ &= \frac{1}{4}p_1 + \frac{1}{4}p_2 + \frac{1}{4}p_3 = \frac{1}{4}\cdot 0 + \frac{1}{4}\cdot\frac{1}{2} + \frac{1}{4}\cdot\frac{1}{2} = \frac{1}{4}.
\end{aligned}
$$

EXAMPLE 2

Use the equations for p_1^+, p_2^+, and p_3^+ to predict the market probabilities p_1^{++}, p_2^{++}, and p_3^{++} two days ahead. Then predict the market probabilities farther into the future.

Solution   We repeat the process in Example 1, using $p_1^+ = \frac{3}{8}$, $p_2^+ = \frac{3}{8}$, and $p_3^+ = \frac{1}{4}$ to obtain

$$
\begin{aligned}
p_1^{++} &= \frac{1}{4}p_1^+ + \frac{1}{2}p_2^+ + \frac{1}{4}p_3^+ = \frac{1}{4}\cdot\frac{3}{8} + \frac{1}{2}\cdot\frac{3}{8} + \frac{1}{4}\cdot\frac{1}{4} = \frac{11}{32}, \\
p_2^{++} &= \frac{1}{2}p_1^+ + \frac{1}{4}p_2^+ + \frac{1}{2}p_3^+ = \frac{1}{2}\cdot\frac{3}{8} + \frac{1}{4}\cdot\frac{3}{8} + \frac{1}{2}\cdot\frac{1}{4} = \frac{13}{32}, \\
p_3^{++} &= \frac{1}{4}p_1^+ + \frac{1}{4}p_2^+ + \frac{1}{4}p_3^+ = \frac{1}{4}\cdot\frac{3}{8} + \frac{1}{4}\cdot\frac{3}{8} + \frac{1}{4}\cdot\frac{1}{4} = \frac{8}{32} = \frac{1}{4}.
\end{aligned}
$$

From these probabilities for 2 days hence, we can predict the market 3 days ahead, and so on indefinitely, so long as the probabilities of the market going up, going down, or staying the same continue to hold. In Section 10.4, we introduce a far simpler way to perform these calculations based on matrix algebra. For now, we simply indicate the results in the following table, assuming that today's market probabilities are 0 for going up, $\frac{1}{2}$ for going down, and $\frac{1}{2}$ for staying the same.

	Up	**Down**	**Same**
Today	0	$\frac{1}{2}$	$\frac{1}{2}$
Tomorrow	$\frac{3}{8}$	$\frac{3}{8}$	$\frac{1}{4}$
2 days ahead	$\frac{11}{32}$	$\frac{13}{32}$	$\frac{1}{4}$
3 days ahead	$\frac{45}{128}$	$\frac{51}{128}$	$\frac{1}{4}$
5 days ahead	≈ 0.35	≈ 0.40	0.25
10 days ahead	≈ 0.35	≈ 0.40	0.25
100 days ahead	0.35	0.40	0.25

Each triple of probabilities for any day—say p_1, p_2, p_3 for the first day— can be thought of as the components of a 3-vector of probabilities for that day. Also, note that on any given day, the sum of the probabilities is always 1 because one of the possibilities (up, down, or the same) must occur.

The sequence of all the successive vectors $\mathbf{p}, \mathbf{p}^+, \mathbf{p}^{++}, \mathbf{p}^{+++}, \ldots$ associated with any transition matrix $\mathbf{A}$ is called a *Markov chain* because the successive vectors are linked by the matrix $\mathbf{A}$. Eventually, the successive probabilities in this Markov chain stabilize at 0.35 for the market going up, 0.40 for the market going down, and 0.25 for the market staying the same. That is, the probabilities converge over time to these limiting values. This behavior occurs regardless of the initial values we used for today's probabilities. Later, we formulate a system of three linear equations in three variables and solve it to determine these stable probabilities directly.

A Population Growth Model

The following model relates populations of hares and wolves from one week to the next. To make the numbers work out conveniently, we measure the hare population in groups of 10 hares and the wolf population in single wolves. Suppose that the number of groups of hares H grows by 20% per week when no wolves are present, so the population of hares next week H^+ would be $1.2H$. But W wolves *are* present and the wolves eat the hares at the rate of each wolf eating 3 hares each week, which is $30\% = 0.3$ of a group of 10 hares. Thus the hare population is reduced by $0.3W$ per week, giving $H^+ = 1.2H - 0.3W$ as next week's hare population.

Next, without hares present, the wolf population decreases at a rate of 30% per week, so the wolf population next week W^+ would be $0.7W$. But the wolf population grows each week when hares are present at the rate of one wolf for every 50 hares or every 5 groups of 10 hares, which is $\frac{1}{5} = 0.2$ wolf per group of 10 hares. Hence $W^+ = 0.2H + 0.7W$.

Together, these two equations are our model for the hare and wolf population over time:

$$H^+ = 1.2H - 0.3W$$
$$W^+ = 0.2H + 0.7W.$$

EXAMPLE 3

Suppose that we start with 1000 groups of hares and 800 wolves. Use the preceding expressions for H^+ and W^+ to calculate the populations of hares and wolves over time.

Solution   If we start with $H = 1000$ and $W = 800$, the hare–wolf model predicts that

$$H^+ = 1.2(1000) - 0.3(800) = 960$$
$$W^+ = 0.2(1000) + 0.7(800) = 760$$

as the populations after the first week. We now use these values to predict the two populations after the second week:

$$H^{++} = 1.2H^+ - 0.3W^+ = 1.2(960) - 0.3(760) = 924$$
$$W^{++} = 0.2H^+ + 0.7W^+ = 0.2(960) + 0.7(760) = 724.$$

Extending these values from one week to the next, we obtain the following table for the sizes of the hare and wolf populations over time.

Weeks	Groups of hares	Wolves
0	1000	800
1	960	760
2	924	724
3	892	692
10	739	539
20	649	449
50	602	402
100	600	400

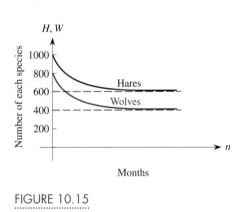

FIGURE 10.15

Note that over time the populations converge to 600 groups of 10 hares and 400 wolves. Figure 10.15 shows the graphs of both populations as functions of time.

We can visualize this situation another way: We can think one population depends on the other. That is, the number of wolves W can be viewed as a function of the number of hares H. If we plot the number of wolves versus groups of hares, we find that they fall in a straight line. In particular, the linear function that fits the points in the preceding table is $W = H - 200$. The graph in Figure 10.16 shows several *trajectories* for the populations (in hundreds) of groups of 10 hares and wolves. One trajectory shown starts from the initial point $(10, 8)$ and leads to the point $(6, 4)$.

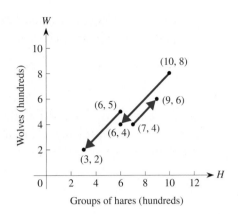

FIGURE 10.16

Suppose we start with a different set of initial values for the two populations—say, $H_0 = 700$ and $W_0 = 400$. Then

$$H^+ = 1.2(700) - 0.3(400) = 720$$
$$W^+ = 0.2(700) + 0.7(400) = 420$$

and so on. The resulting points all lie on a line starting at the point $(700, 400)$ and converge to the point $(900, 600)$. Similarly, if we start with $H_0 = 600$ and $W_0 = 500$, the resulting points all lie on a line and converge to $(300, 200)$, as also shown in Figure 10.16. In each case, the limiting values for H and W satisfy $W = \frac{2}{3}H$. That is, the limiting points lie on the line $W = \frac{2}{3}H$, as shown in Figure 10.17. In fact, for *any* initial pair of population values (H_0, W_0), the points of successive pairs (H_n, W_n) all lie on some line, and in each case the successive points are converging (as indicated by the arrows) toward a limiting point on the line $W = \frac{2}{3}H$, as illustrated in Figure 10.18. Thus under this model, all populations, regardless of the initial values, converge over time to populations in which the number of wolves is two-thirds the number of groups of 10 hares.

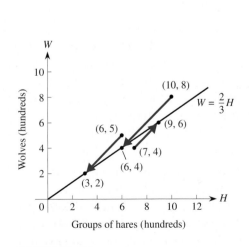

FIGURE 10.17

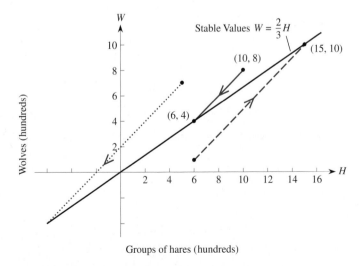

FIGURE 10.18

Note that the pair of equations we used to define this model can be rewritten as a pair of difference equations:

$$H_{n+1} = 1.2H_n - 0.3W_n$$
$$W_{n+1} = 0.2H_n + 0.7W_n.$$

We present a more detailed analysis of population models based on systems of difference equations, including a more sophisticated predator–prey model, in supplementary Section 12.6.

Problems

1. Ted, Carol, and Alice took tests in German, physics, theater, and politics. Ted's test scores in these subjects were 64, 73, 86, 85; Carol's scores were 82, 69, 77, 91; Alice's were 82, 84, 81, 83. Construct a matrix of these test results. Label the columns and rows.

2. For the matrix

$$\mathbf{A} = \begin{bmatrix} 1 & 5 & 3 \\ 6 & 1 & 7 \\ 6 & 9 & 5 \\ 0 & 2 & 8 \end{bmatrix},$$

write the following row and column vectors and entries.

a. $\mathbf{a}_1'$
b. $\mathbf{a}_3$
c. $\mathbf{a}_4$
d. a_{12}
e. a_{23}
f. a_{41}

3. In the matrix of letters

$$\mathbf{A} = \begin{bmatrix} H & R & B & I \\ N & S & O & A \\ E & T & Y & L \\ M & G & D & I \end{bmatrix},$$

spell the words represented by the following sequences of entries.

a. $a_{11}\, a_{23}\, a_{12}\, a_{21}$
b. $a_{41}\, a_{23}\, a_{32}\, a_{11}\, a_{31}\, a_{12}$
c. $a_{31}\, a_{24}\, a_{12}\, a_{34}\, a_{33}$
d. $a_{33}\, a_{31}\, a_{22}\, a_{32}\, a_{31}\, a_{12}\, a_{43}\, a_{24}\, a_{33}$

4. A clothing company's three factories (1, 2, and 3) produce the following numbers of vests, pants, and coats from each roll of cloth.

	Factory 1	Factory 2	Factory 3
Vests	6	4	2
Pants	4	8	4
Coats	3	2	8

Suppose that the company has a demand for 400 vests, 800 pants, and 500 coats. Write a system of equations whose solution would determine production levels to yield the desired numbers of vests, pants, and coats. As in the clothes production model, let x_i be the number of rolls of cloth processed by the ith factory.

5. Three oil refineries (1, 2, and 3) produce the following amounts, in thousands of gallons, of heating oil, diesel oil, and gasoline from each shipment of crude petroleum.

	Refinery 1	Refinery 2	Refinery 3
Heating Oil	8	5	3
Diesel Oil	2	5	5
Gasoline	3	7	6

Suppose that demand is for 6200 thousand gallons of heating oil, 4000 thousand gallons of diesel oil, and 4700 thousand gallons of gasoline. Write a system of equations whose solution would determine production levels to yield the desired amounts of heating oil, diesel oil, and gasoline. Let x_i be the number of shipments processed by the ith refinery.

6. The staff dietitian at the California Institute of Trigonometry has to make up a meal with 600 calories, 20 grams of protein, and 200 mg of vitamin C. The three food types that the dietitian can choose from are gelatin, fish sticks, and mystery meat. They have the following nutritional content per unit.

	Gelatin	Fish Sticks	Mystery Meat
Calories	10	50	200
Protein	1	3	0.2
Vitamin C	30	10	0

Construct a mathematical model for this situation, based on a system of three linear equations.

7. A company has a budget of $280,000 for computing equipment. The types of equipment available are microcomputers at $2000 each, terminals at $500 each, and workstations at $5000 each. There should be five times as many terminals as microcomputers and twice as many microcomputers as workstations. Write a system of three linear equations to describe this situation.

8. In the clothes production model in the text, suppose that Factory 1 processes 15 rolls of cloth, Factory 2 processes 20 rolls, and Factory 3 processes 60 rolls. For which product, vests, pants, or coats, does production deviate the most from the demand for 600, 800, 1000?

9. Refer to the stock market Markov chain in Example 1. Determine the set of probabilities for tomorrow's market for each set of probabilities that the market will be up, down, or the same today.

   a. $p_1 = 1, p_2 = 0, p_3 = 0$
   b. $p_1 = 0, p_2 = \frac{1}{2}, p_3 = \frac{1}{2}$
   c. $p_1 = \frac{1}{2}, p_2 = 0, p_3 = \frac{1}{2}$
   d. $p_1 = \frac{1}{4}, p_2 = \frac{1}{2}, p_3 = \frac{1}{4}$
   e. $p_1 = 0.35, p_2 = 0.40, p_3 = 0.25$

10. The copy machine at the student union breaks down according to the following pattern. If it is working today, it has a 70% chance of working tomorrow (and a 30% chance of breaking down). If the copy machine is broken today, it has a 50% chance of working tomorrow (and a 50% chance of being broken again).

    a. Construct a Markov chain for this situation; give the matrix of transition probabilities and draw the transition diagram.
    b. If there is a 50–50 chance of the copy machine's working today, what is the chance of its working tomorrow?
    c. Based on the situation in part (b), what is the chance that the copy machine is working the day after tomorrow?
    d. If the copy machine is working today, what is the chance that it is working the day after tomorrow?

11. The Pins, a bowling team, plays in a bowling league each week. If they win this week's game, they have a $\frac{2}{3}$ chance of winning next week's game. If they lose this week's game, they have a $\frac{1}{2}$ chance of winning next week's game.

    a. Construct a Markov chain for this situation; give the matrix of transition probabilities and draw the transition diagram.
    b. If there is a 50–50 chance of the Pins' winning this week's game, what is their chance of winning next week's game?
    c. If they won this week, what is their chance of winning the game 2 weeks from now?

12. Consider a weather Markov chain having two states: sunny and cloudy. If today is sunny, there is a $\frac{3}{4}$ probability that tomorrow will be sunny and a $\frac{1}{4}$ probability that tomorrow will be cloudy. If today is cloudy, there is a $\frac{1}{4}$ probability that tomorrow will be sunny and a $\frac{3}{4}$ probability that tomorrow will be cloudy.

    a. Write the transition matrix for this Markov chain and draw its transition diagram.
    b. In this weather Markov chain, starting with the vector of probabilities $\begin{bmatrix} 1 \\ 0 \end{bmatrix}$ (a sunny day), compute and plot the vectors of probabilities for four successive days.
    c. Repeat the process starting with the probability vector $\begin{bmatrix} 0 \\ 1 \end{bmatrix}$ (a cloudy day). Can you guess the values of the equilibrium state to which your probability vectors are converging?

13. The following model for learning a concept over a set of lessons identifies four states of learning: I = ignorance, E = exploratory thinking, S = superficial understanding, and M = mastery. If you are now in state I, after one lesson you have a probability of $\frac{1}{2}$ of still being in I and a probability of $\frac{1}{2}$ of being in E. If you are now in state E, after one lesson you have a probability of $\frac{1}{4}$ of being in I, $\frac{1}{2}$ in E, and $\frac{1}{4}$ in S. If you are now in state S, after one lesson you have a probability of $\frac{1}{4}$ of being in E, $\frac{1}{2}$ in S, and $\frac{1}{4}$ in M. If you are in M, you always stay in M (with a probability of 1).

    a. Construct a Markov chain for this learning model.
    b. If you start in state I, what is your probability vector after two lessons? After three lessons?

14. a. In Example 3, if initially there were 800 groups of 10 hares and 300 wolves, how many hares and wolves would there be after 1 month, 3 months,

5 months, and 10 months? What do the limiting values for the two populations appear to be?

b. Repeat part (a) with an initial population of 600 groups of 10 hares and 800 wolves.

c. Repeat part (a) with an initial population of 900 groups of 10 hares and 600 wolves.

15. Consider the following cattle–sheep models in which the two species compete for common grazing land. In each case, compute the populations after 1 month, 2 months, and 3 months if the initial populations are 50 cattle and 100 sheep.

a. $C^+ = 1.2C - 0.3S$
$S^+ = -0.2C + 1.2S$

b. $C^+ = 1.2C - 0.1S$
$S^+ = 0.5C + 1.4S$

16. Consider the rabbit–fox model

$$R^+ = 1.1R - 0.2F$$
$$F^+ = 0.2R + 0.6F.$$

Plot the following on the same graph.

a. The trajectory of populations starting from $(10, 15)$.

b. The trajectory of populations starting from $(10, 30)$.

c. The trajectory of populations starting from $(20, 10)$.

17. In Example 3 we found that the solution to the system

$$H^+ = 1.2H - 0.3W$$
$$W^+ = 0.2H + 0.7W$$

converged to a point on the line $W = \frac{2}{3}H$ for any starting values H_0 and W_0.

a. To find the equation of this limiting line, assume that $H^+ = H$ and $W^+ = W$ in the two equations defining the system and solve these two equations in two unknowns.

b. For any given starting populations—say, $H_0 = 1000$ and $W_0 = 800$—calculate the next point (H^+, W^+) on the trajectory. What is the equation of the line through (H_0, W_0) and (H^+, W^+)?

c. You now have the equation of the limiting line and the equation of the trajectory. Describe how you would use them to find the final population values for H and W. What are the values for initial population values of $H_0 = 1000$ and $W_0 = 800$?

18. **a.** The population models in Example 3 and Problems 15–17 all involve linear expressions in H and W. Suppose that the equations for H^+ and W^+ contained nonlinear expressions in H and W. How might such expressions affect the trajectory?

b. We develop a more sophisticated mathematical model for two species, known as the predator–prey model, in supplementary Section 12.6. It is based on equations such as

$$H^+ = 1.2H - 0.003HW$$
$$W^+ = 0.2W + 0.005HW.$$

If the initial population values are $H_0 = 1000$ and $W_0 = 800$, calculate and plot the population values over the first 3 months. What do you observe about the trajectory?

10.3   Scalar Products

In this section, we explain how to multiply two vectors in what is called a scalar product. For instance, suppose that a family normally eats three vegetables—asparagus, beans, and corn. Suppose that the costs per pound for these vegetables are $0.80 for asparagus, $1.00 for beans, and $0.60 for corn. We can then form a vector $\mathbf{p} = \begin{bmatrix} 0.80 & 1.00 & 0.60 \end{bmatrix}$ for the prices of these three vegetables. Suppose further that the family consumes 2 lb of asparagus, 5 lb of beans, and 3 lb of corn each week, so we can also form a vector $\mathbf{d} = \begin{bmatrix} 2 & 5 & 3 \end{bmatrix}$ of the family's weekly demand for these vegetables. The total cost of the family's weekly demand for the vegetables is

$$2 \times 0.80 + 5 \times 1.00 + 3 \times 0.60 = 1.60 + 5.00 + 1.80 = \$8.40$$

because the cost for asparagus is 2 × $0.80, the cost for beans is 5 × $1.00, and the cost for corn is 3 × $0.60. This result suggests a natural way to define the product of the two vectors **d** and **p**. We write this product as **d · p** and call it the **scalar product** of **d** and **p**. Here the scalar product **d · p** is

$$\mathbf{d} \cdot \mathbf{p} = \begin{bmatrix} 2 & 5 & 3 \end{bmatrix} \cdot \begin{bmatrix} 0.80 & 1.00 & 0.60 \end{bmatrix}$$
$$= 2(0.80) + 5(1.00) + 3(0.60)$$
$$= 1.60 + 5.00 + 1.80 = 8.40,$$

or \$8.40. Note that the scalar product involves multiplying the corresponding entries in each position of the vectors and adding the results. Each vector in a scalar product can be either a row or a column vector, but multiplication makes sense only when the two vectors have the same size (that is, they have the same number of entries). The scalar product of two vectors of different sizes can't be formed because there would be terms that do not match. For instance, the scalar product of $\begin{bmatrix} 1 & 2 & 3 & 4 & 5 \end{bmatrix}$ and $\begin{bmatrix} 20 & 40 & 60 \end{bmatrix}$ can't be formed.

Recall that the word *scalar* means a single number, as opposed to a vector or matrix. The scalar product of two vectors is so named because its result is a single number—a scalar. The scalar product also is known as the *dot product* and the *inner product,* but we will not use either of these terms.

The product of two vectors can also be defined in a different way, known as the *vector product,* which produces a vector instead of a scalar, or number, as the result. However, we don't consider it here.

More formally we have the following definition of the scalar product.

Scalar Product

Let

$$\mathbf{a} = \begin{bmatrix} a_1 & a_2 & \cdots & a_n \end{bmatrix} \quad \text{and} \quad \mathbf{b} = \begin{bmatrix} b_1 & b_2 & \cdots & b_n \end{bmatrix}$$

be vectors of the same size n. Then the *scalar product* $\mathbf{a} \cdot \mathbf{b}$ of **a** and **b** is the single number (a scalar) equal to the sum of the products,

$$\mathbf{a} \cdot \mathbf{b} = a_1 b_1 + a_2 b_2 + \cdots + a_n b_n.$$

Figure 10.19 will help you visualize how to calculate the scalar product of two vectors **a** and **b**.

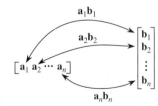

FIGURE 10.19

For instance, the scalar product of the vectors $\mathbf{a} = \begin{bmatrix} 3 & 2 & -5 & 4 \end{bmatrix}$ and $\mathbf{b} = \begin{bmatrix} 6 & -3 & 2 & 0 \end{bmatrix}$ is

$$\mathbf{a} \cdot \mathbf{b} = 3(6) + 2(-3) + (-5)(2) + 4(0) = 18 - 6 - 10 + 0 = 2.$$

We now see how the scalar product arises in a variety of situations.

EXAMPLE 1

Suppose that peaches cost 30¢ each, pears cost 20¢ each, apples cost 35¢ each, and grape-fruits cost 50¢ each. Amy wants to get 5 peaches, 3 pears, 2 apples, and 2 grapefruits, and Bill wants to get 3 peaches, 4 pears, 3 apples, and 3 grapefruits.

a. Write vectors to represent the prices of the fruits and the amount of each fruit that Amy and Bill will purchase.

b. Write the total costs of their fruit purchases, using vector methods.

Solution

a. We form the price vector $\mathbf{p} = \begin{bmatrix} 0.30 & 0.20 & 0.35 & 0.50 \end{bmatrix}$ for the prices of the respec-tive fruits and the two demand vectors, $\mathbf{a} = \begin{bmatrix} 5 & 3 & 2 & 2 \end{bmatrix}$ for Amy and $\mathbf{b} = \begin{bmatrix} 3 & 4 & 3 & 3 \end{bmatrix}$ for Bill.

b. The scalar products $\mathbf{a} \cdot \mathbf{p}$ and $\mathbf{b} \cdot \mathbf{p}$ give the total costs of Amy's and Bill's purchases:

$$\mathbf{a} \cdot \mathbf{p} = \begin{bmatrix} 5 & 3 & 2 & 2 \end{bmatrix} \cdot \begin{bmatrix} 0.30 & 0.20 & 0.35 & 0.50 \end{bmatrix}$$
$$= 5(0.30) + 3(0.20) + 2(0.35) + 2(0.50)$$
$$= 1.50 + 0.60 + 0.70 + 1.00 = 3.80;$$
$$\mathbf{b} \cdot \mathbf{p} = \begin{bmatrix} 3 & 4 & 3 & 3 \end{bmatrix} \cdot \begin{bmatrix} 0.30 & 0.20 & 0.35 & 0.50 \end{bmatrix}$$
$$= 3(0.30) + 4(0.20) + 3(0.35) + 3(0.50)$$
$$= 0.90 + 0.80 + 1.05 + 1.50 = 4.25.$$

Thus the cost of fruit was $3.80 for Amy and $4.25 for Bill.

A Geometric View of Scalar Products

An interesting special case of the scalar product involves the use of coordinate vectors, which we introduced in Section 10.1. In two dimensions, they are the unit vectors $\begin{bmatrix} 1 & 0 \end{bmatrix}$ along the horizontal axis and $\begin{bmatrix} 0 & 1 \end{bmatrix}$ along the vertical axis. See Figure 10.20. In three dimensions, the coordinate vectors are $\begin{bmatrix} 1 & 0 & 0 \end{bmatrix}$, $\begin{bmatrix} 0 & 1 & 0 \end{bmatrix}$, and $\begin{bmatrix} 0 & 0 & 1 \end{bmatrix}$. They lie along three mutually perpendicular axes, as illustrated in Fig-ure 10.21.

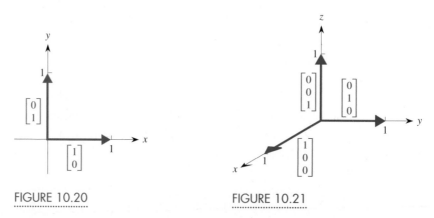

FIGURE 10.20

FIGURE 10.21

Any vector can be built from its components. For instance, for the 3-vector $\begin{bmatrix} 2 & 5 & 3 \end{bmatrix}$, we use the components 2, 5, and 3 and the coordinate vectors to write

$$\begin{bmatrix} 2 & 5 & 3 \end{bmatrix} = 2\begin{bmatrix} 1 & 0 & 0 \end{bmatrix} + 5\begin{bmatrix} 0 & 1 & 0 \end{bmatrix} + 3\begin{bmatrix} 0 & 0 & 1 \end{bmatrix}.$$

In general, for any 3-vector $\mathbf{a} = \begin{bmatrix} a_1 & a_2 & a_3 \end{bmatrix}$, we can write

$$\mathbf{a} = a_1\begin{bmatrix} 1 & 0 & 0 \end{bmatrix} + a_2\begin{bmatrix} 0 & 1 & 0 \end{bmatrix} + a_3\begin{bmatrix} 0 & 0 & 1 \end{bmatrix}.$$

In many geometric uses of vectors, we need to know whether two vectors "point" in the same general direction or in opposite directions. When two vectors point in approximately the same general direction, as shown by the arrows in Figure 10.22, their scalar product will be positive. When two vectors point in approximately opposite directions, as shown in Figure 10.23, their scalar product will be negative. Most interestingly, when two vectors form a right angle, as shown in Figure 10.24, their scalar product will be zero.

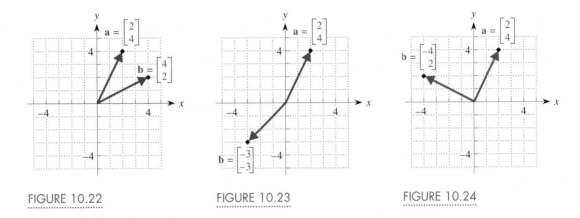

FIGURE 10.22   FIGURE 10.23   FIGURE 10.24

These assertions follow from the fact that, if θ is the angle between the vectors $\mathbf{a}$ and $\mathbf{b}$, as shown in Figure 10.25, we have the following relationship.

$$\cos\theta = \frac{\mathbf{a}\cdot\mathbf{b}}{\sqrt{\mathbf{a}\cdot\mathbf{a}}\sqrt{\mathbf{b}\cdot\mathbf{b}}}$$

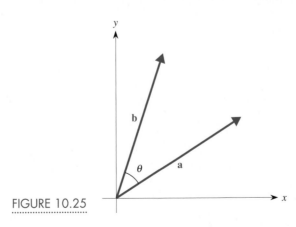

FIGURE 10.25

(This formula is based on the Law of Cosines introduced in Section 6.5; we ask you to derive this formula in a problem at the end of the section.) Note that, if $\mathbf{a} = \begin{bmatrix} a_1 & a_2 \end{bmatrix}$ is a 2-vector, $\mathbf{a}\cdot\mathbf{a} = a_1{}^2 + a_2{}^2$. We interpret $\mathbf{a}\cdot\mathbf{a}$ geometrically by using the Pythagorean theorem to represent the square of the hypotenuse in a right triangle with sides a_1 and a_2, as shown in Figure 10.26. Thus, in the formula for $\cos \theta$, $\sqrt{\mathbf{a} \cdot \mathbf{a}}$ is the length of the vector $\mathbf{a}$. Similarly, if $\mathbf{a}$ is a 3-vector, $\mathbf{a} = \begin{bmatrix} a_1 & a_2 & a_3 \end{bmatrix}$, then

$$\mathbf{a} \cdot \mathbf{a} = a_1{}^2 + a_2{}^2 + a_3{}^2$$

and so $\sqrt{\mathbf{a} \cdot \mathbf{a}}$ is the length of $\mathbf{a}$. Similarly, $\sqrt{\mathbf{b} \cdot \mathbf{b}}$ is the length of $\mathbf{b}$. Therefore we can rewrite the formula for $\cos \theta$ as

$$\cos \theta = \frac{\mathbf{a} \cdot \mathbf{b}}{(\text{length of } \mathbf{a})(\text{length of } \mathbf{b})}$$

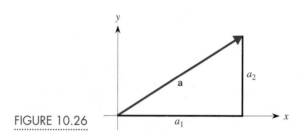

FIGURE 10.26

Clearly the length of $\mathbf{a}$ and the length of $\mathbf{b}$ are both positive. Thus, when the vectors point in roughly the same direction and the angle θ is between 0° and 90°, $\cos \theta$ is positive and so is $\mathbf{a} \cdot \mathbf{b}$. When the vectors point in roughly opposite directions and θ is between 90° and 180°, $\cos \theta$ is negative and so is $\mathbf{a} \cdot \mathbf{b}$. When θ is 90°, so that $\mathbf{a}$ and $\mathbf{b}$ are perpendicular, $\cos \theta$ is zero and so $\mathbf{a} \cdot \mathbf{b} = 0$.

◆ **EXAMPLE 2** ···

Find the angle between the lines $y = 2x$ and $y = 3x$.

Solution   We know that both lines pass through the origin, as illustrated in Figure 10.27. To find the angle θ between the lines at the origin, we need to find vectors along each line. Suppose that we arbitrarily choose $x = 1$. On the first line $y = 2x$, the corresponding value of y is 2, so the point $(1, 2)$ is on that line and the vector from the origin to that

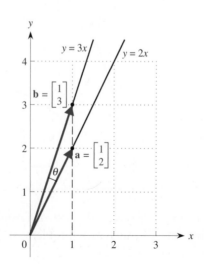

FIGURE 10.27

point is $\mathbf{a} = \begin{bmatrix} 1 & 2 \end{bmatrix}$. Similarly, using $x = 1$, we find the corresponding point $(1, 3)$ on the second line $y = 3x$, so the vector from the origin to that point is $\mathbf{b} = \begin{bmatrix} 1 & 3 \end{bmatrix}$. The angle θ between the lines is the same as the angle between the vectors, so

$$\cos \theta = \frac{\mathbf{a} \cdot \mathbf{b}}{\sqrt{\mathbf{a} \cdot \mathbf{a}} \sqrt{\mathbf{b} \cdot \mathbf{b}}} = \frac{(1)(1) + (2)(3)}{\sqrt{(1)(1) + (2)(2)} \sqrt{(1)(1) + (3)(3)}}$$

$$= \frac{7}{\sqrt{5} \sqrt{10}} \approx 0.98995.$$

Consequently, the angle between the two lines is

$$\theta = \arccos(0.98995) \approx 8.13° \approx 0.1419 \text{ radian.}$$

The Clothes Production Model Using Scalar Products

In Section 10.2 we introduced a mathematical model for production in three clothing factories to meet a demand for 500 vests, 850 pants, and 1000 coats. If x_i denotes the number of rolls of cloth used by the ith factory ($i = 1, 2, 3$), the x_i's must satisfy the system of linear equations

$$\begin{array}{llll} \text{vests:} & 20x_1 & + \ \ 4x_2 & + \ \ 4x_3 & = 500 \\ \text{pants:} & 10x_1 & + 14x_2 & + \ \ 5x_3 & = 850 \\ \text{coats:} & \ \ 5x_1 & + \ \ 5x_2 & + 12x_3 & = 1000. \end{array}$$

A key property of scalar products is that $\mathbf{a} \cdot \mathbf{b}$ is a linear combination of the entries in each vector. (Similarly, you can think of a polynomial as being a linear combination of power functions.) Conversely, any linear combination of variables or numbers always can be interpreted as a scalar product of two vectors.

EXAMPLE 3

Rewrite the three linear equations for the clothes production model using scalar products of vectors.

Solution   Consider the first linear equation of the clothes production model

$$20x_1 + 4x_2 + 4x_3 = 500.$$

The left-hand side of this equation is a linear combination of the three variables. If

$$\mathbf{a} = \begin{bmatrix} 20 & 4 & 4 \end{bmatrix} \quad \text{and} \quad \mathbf{x} = \begin{bmatrix} x_1 \\ x_2 \\ x_3 \end{bmatrix},$$

(we explain shortly the reason for writing $\mathbf{a}$ as a row vector and $\mathbf{x}$ as a column vector), we can write the left-hand side of this equation as the scalar product

$$\mathbf{a} \cdot \mathbf{x} = \begin{bmatrix} 20 & 4 & 4 \end{bmatrix} \cdot \begin{bmatrix} x_1 \\ x_2 \\ x_3 \end{bmatrix} = 20x_1 + 4x_2 + 4x_3.$$

The first equation is then $\mathbf{a} \cdot \mathbf{x} = 500$. Note that the vector $\mathbf{a}$ consists of the entries in the first row of the matrix

$$A = \begin{bmatrix} 20 & 4 & 4 \\ 10 & 14 & 5 \\ 5 & 5 & 12 \end{bmatrix}$$

that we used in the clothes production model of Section 10.2.

Using vectors $\mathbf{b}$ and $\mathbf{c}$ to represent the second and third rows of matrix $\mathbf{A}$, we can write the other two linear equations in the same way as $\mathbf{b} \cdot \mathbf{x} = 850$ and $\mathbf{c} \cdot \mathbf{x} = 1000$.

◆

Any system of linear equations can be written in terms of a system of scalar products. For example, the left-hand sides of the equations for the clothes production model are

$$\text{vests: } 20x_1 + 4x_2 + 4x_3 = \begin{bmatrix} 20 & 4 & 4 \end{bmatrix} \cdot \begin{bmatrix} x_1 \\ x_2 \\ x_3 \end{bmatrix} = \mathbf{a}_1' \cdot \mathbf{x},$$

$$\text{pants: } 10x_1 + 14x_2 + 5x_3 = \begin{bmatrix} 10 & 14 & 5 \end{bmatrix} \cdot \begin{bmatrix} x_1 \\ x_2 \\ x_3 \end{bmatrix} = \mathbf{a}_2' \cdot \mathbf{x},$$

$$\text{coats: } 5x_1 + 5x_2 + 12x_3 = \begin{bmatrix} 5 & 5 & 12 \end{bmatrix} \cdot \begin{bmatrix} x_1 \\ x_2 \\ x_3 \end{bmatrix} = \mathbf{a}_3' \cdot \mathbf{x},$$

where $\mathbf{a}_i'$ is the ith row of the clothes production coefficient matrix

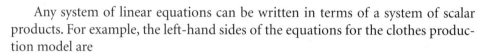

$$\begin{array}{c} \\ \text{Shirts} \\ \text{Pants} \\ \text{Coats} \end{array} \begin{array}{ccc} \text{Factory 1} & \text{Factory 2} & \text{Factory 3} \\ \begin{bmatrix} 20 & 4 & 4 \\ 10 & 14 & 5 \\ 5 & 5 & 12 \end{bmatrix} & & \end{array} = \mathbf{A}.$$

The Matrix–Vector Product

Although we encounter many important uses of single scalar products in matrix algebra, their most important use is as a building block for defining the product of a matrix and a vector and, in Section 10.4, the product of two matrices. We define the matrix–vector product as follows.

Matrix–Vector Product

The product of an $m \times n$ matrix $\mathbf{A}$ and a column n-vector $\mathbf{c}$ is a column m-vector of scalar products

$$\mathbf{A}\mathbf{c} = \mathbf{a}_i' \cdot \mathbf{c} \quad (\text{each } row \ \mathbf{a}_i' \text{ of } \mathbf{A} \text{ multiplies } \mathbf{c}).$$

The diagram in Figure 10.28 will help you visualize this definition. Think of each row in the first matrix **A** as a row vector (or think of **A** as consisting of a collection of row vectors). The scalar product of each row vector in **A** and the column vector **c** creates a new column vector. In order for this definition to make sense, the entries in each row of the matrix **A** (equivalently, the number of columns in matrix **A**) must equal the size of the vector **c**.

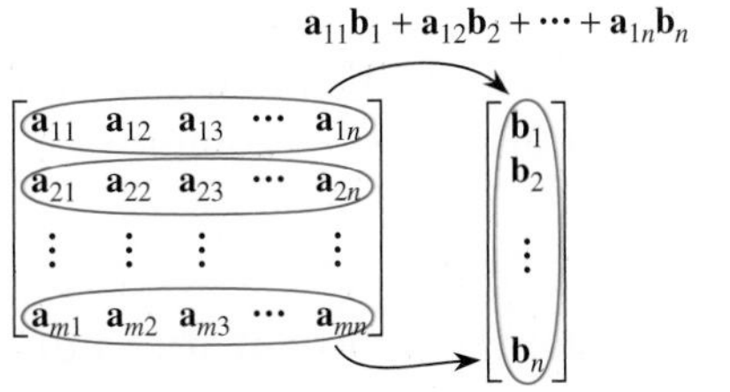

FIGURE 10.28

For instance, if **A** is a 2×3 matrix and **c** is a column 3-vector,

$$\mathbf{A} = \begin{bmatrix} 2 & 1 & 0 \\ 5 & 3 & 6 \end{bmatrix} \quad \text{and} \quad \mathbf{c} = \begin{bmatrix} 3 \\ 4 \\ 2 \end{bmatrix},$$

then

$$\mathbf{Ac} = \begin{bmatrix} 2 & 1 & 0 \\ 5 & 3 & 6 \end{bmatrix}\begin{bmatrix} 3 \\ 4 \\ 2 \end{bmatrix} = \begin{bmatrix} 2(3) + 1(4) + 0(2) \\ 5(3) + 3(4) + 6(2) \end{bmatrix} = \begin{bmatrix} 10 \\ 39 \end{bmatrix}.$$

As we said previously, each row of matrix **A** is treated as if it were a row vector, and it is used to form a scalar product with the column vector **c**. However, we cannot multiply

$$A = \begin{bmatrix} 2 & 1 & 0 \\ 5 & 3 & 6 \end{bmatrix} \quad \text{and} \quad c = \begin{bmatrix} 3 \\ 4 \\ 2 \\ 5 \end{bmatrix}$$

because the number of entries in each row of **A** does not match the number of entries in **C**.

We can recast this illustration symbolically as follows. If

$$\mathbf{A} = \begin{bmatrix} a_{11} & a_{12} & a_{13} \\ a_{21} & a_{22} & a_{23} \end{bmatrix} \quad \text{and} \quad \mathbf{c} = \begin{bmatrix} c_1 \\ c_2 \\ c_3 \end{bmatrix},$$

then

$$\mathbf{Ac} = \begin{bmatrix} \mathbf{a}_1' \cdot \mathbf{c} \\ \mathbf{a}_2' \cdot \mathbf{c} \end{bmatrix} = \begin{bmatrix} a_{11}c_1 + a_{12}c_2 + a_{13}c_3 \\ a_{21}c_1 + a_{22}c_2 + a_{23}c_3 \end{bmatrix}.$$

The only other allowable way to multiply a vector and a matrix is to multiply a row vector by a matrix. In that case, the matrix must have the same number of rows as the row vector has entries. That is, we can multiply

$$\mathbf{c} = \begin{bmatrix} 1 & 4 & -2 & 3 \end{bmatrix} \quad \text{and} \quad \mathbf{A} = \begin{bmatrix} 2 & 0 \\ 6 & -2 \\ -3 & 5 \\ 1 & -5 \end{bmatrix}$$

because there are four entries in the row vector, which matches the number of rows (vertical entries per column) in the matrix $\mathbf{A}$.

EXAMPLE 4

Consider again the set of equations discussed in Example 3 for the clothes production model:

$$\text{vests:} \quad 20x_1 + 4x_2 + 4x_3 = 500$$
$$\text{pants:} \quad 10x_1 + 14x_2 + 5x_3 = 850$$
$$\text{coats:} \quad 5x_1 + 5x_2 + 12x_3 = 1000.$$

Rewrite the system of equations as a matrix–vector equation.

Solution If we make vectors of the left-hand sides of the three equations, we have

$$\begin{bmatrix} 20x_1 + 4x_2 + 4x_3 \\ 10x_1 + 14x_2 + 5x_3 \\ 5x_1 + 5x_2 + 12x_3 \end{bmatrix} = \begin{bmatrix} \begin{bmatrix} 20 & 4 & 4 \end{bmatrix} \cdot \mathbf{x} \\ \begin{bmatrix} 10 & 14 & 5 \end{bmatrix} \cdot \mathbf{x} \\ \begin{bmatrix} 5 & 5 & 12 \end{bmatrix} \cdot \mathbf{x} \end{bmatrix} = \begin{bmatrix} \mathbf{a}_1' \cdot \mathbf{x} \\ \mathbf{a}_2' \cdot \mathbf{x} \\ \mathbf{a}_3' \cdot \mathbf{x} \end{bmatrix} = \mathbf{Ax}.$$

If we let

$$\mathbf{b} = \begin{bmatrix} 500 \\ 850 \\ 1000 \end{bmatrix}$$

represent the column vector of demands for the different products, the system of equations becomes

$$\mathbf{Ax} = \mathbf{b} \quad \text{or} \quad \begin{bmatrix} 20x_1 + 4x_2 + 4x_3 \\ 10x_1 + 14x_2 + 5x_3 \\ 5x_1 + 5x_2 + 12x_3 \end{bmatrix} = \begin{bmatrix} 500 \\ 850 \\ 1000 \end{bmatrix}.$$

◆

As Example 4 suggests, *any* system of linear equations can be written as a matrix–vector equation. As we develop the tools for working with and solving such equations, we demonstrate that having such a formulation has significant advantages, especially for systems that are larger than three equations in three unknowns.

The Fruit Purchase Model Revisited In Example 1, we computed the scalar products for the costs of fruit purchased by Amy and Bill. Recall that Amy wanted 5 peaches, 3 pears, 2 apples, and 2 grapefruits, whereas Bill wanted 3 peaches, 4 pears, 3 apples, and 3 grapefruits. Also, we know that peaches cost 30¢ each, pears 20¢ each, apples 35¢ each, and grapefruits 50¢ each.

EXAMPLE 5

Write a matrix–vector equation to represent the costs of the fruit purchases by Amy and Bill.

Solution   Let's make a matrix **A** of fruit purchases. The columns represent the different fruits, the first row gives Amy's fruit shopping list, and the second row gives Bill's list. We also make a (column) vector **p** of the costs, in cents.

$$\begin{array}{cccc}\text{Peaches} & \text{Pears} & \text{Apples} & \text{Grapefruits}\end{array}$$

$$\mathbf{A} = \begin{bmatrix} 5 & 3 & 2 & 2 \\ 3 & 4 & 3 & 3 \end{bmatrix}\begin{array}{l}\text{Amy} \\ \text{Bill}\end{array} \qquad \mathbf{p} = \begin{bmatrix} 30 \\ 20 \\ 35 \\ 50 \end{bmatrix}\begin{array}{l}\text{Peaches} \\ \text{Pears} \\ \text{Apples} \\ \text{Grapefruits}\end{array}$$

We can now write the costs of Amy's and Bill's fruit purchases as the matrix–vector product

$$\mathbf{Ap} = \begin{bmatrix} 5 & 3 & 2 & 2 \\ 3 & 4 & 3 & 3 \end{bmatrix}\begin{bmatrix} 30 \\ 20 \\ 35 \\ 50 \end{bmatrix} = \begin{bmatrix} 5(30) + 3(20) + 2(35) + 2(50) \\ 3(30) + 4(20) + 3(35) + 3(50) \end{bmatrix}$$

$$= \begin{bmatrix} 380 \\ 425 \end{bmatrix} = \begin{bmatrix} \$3.80 \\ \$4.25 \end{bmatrix}.$$

This result is the same as we obtained in Example 1.

Markov Chain for the Stock Market Revisited   Recall the Markov chain introduced in Example 1 of Section 10.2. The equations for determining the probabilities $p_1{}^+$, $p_2{}^+$, and $p_3{}^+$ that the stock market goes up, goes down, or stays the same tomorrow given the probabilities p_1, p_2, and p_3 of its going up, going down, or staying the same today were

$$p_1{}^+ = \frac{1}{4}p_1 + \frac{1}{2}p_2 + \frac{1}{4}p_3$$

$$p_2{}^+ = \frac{1}{2}p_1 + \frac{1}{4}p_2 + \frac{1}{2}p_3$$

$$p_3{}^+ = \frac{1}{4}p_1 + \frac{1}{4}p_2 + \frac{1}{4}p_3.$$

EXAMPLE 6

Write the preceding Markov chain model for the stock market as a matrix–vector equation.

Solution   Let

$$\mathbf{p} = \begin{bmatrix} p_1 \\ p_2 \\ p_3 \end{bmatrix} \quad \text{and} \quad \mathbf{p}^+ = \begin{bmatrix} p_1{}^+ \\ p_2{}^+ \\ p_3{}^+ \end{bmatrix},$$

respectively, be the vectors of today's and tomorrow's probabilities and let the matrix of transition probabilities be **A**:

Market Today

		Up	Down	Same	
Market Tomorrow	Up	$\frac{1}{4}$	$\frac{1}{2}$	$\frac{1}{4}$	
	Down	$\frac{1}{2}$	$\frac{1}{4}$	$\frac{1}{2}$	$=\mathbf{A}.$
	Same	$\frac{1}{4}$	$\frac{1}{4}$	$\frac{1}{4}$	

Then the preceding set of probability equations can be written simply as $\mathbf{p}^+ = \mathbf{Ap}$ because

$$\mathbf{Ap} = \begin{bmatrix} \frac{1}{4} & \frac{1}{2} & \frac{1}{4} \\ \frac{1}{2} & \frac{1}{4} & \frac{1}{2} \\ \frac{1}{4} & \frac{1}{4} & \frac{1}{4} \end{bmatrix} \begin{bmatrix} p_1 \\ p_2 \\ p_3 \end{bmatrix} = \begin{bmatrix} \frac{1}{4}p_1 + \frac{1}{2}p_2 + \frac{1}{4}p_3 \\ \frac{1}{2}p_1 + \frac{1}{4}p_2 + \frac{1}{2}p_3 \\ \frac{1}{4}p_1 + \frac{1}{4}p_2 + \frac{1}{4}p_3 \end{bmatrix}.$$

◆

Similarly, the probabilities $\mathbf{p}^{++}$ for the day after tomorrow can be found from

$$\mathbf{p}^{++} = \mathbf{Ap}^+$$

and so on.

In Section 10.4 we present a concise way of writing these probability vectors.

A Geometric View of Matrix–Vector Products

In geometric terms, when we multiply a vector $\mathbf{v}$ by some matrix $\mathbf{A}$, the vector $\mathbf{v}$ is transformed by the multiplication into another vector $\mathbf{w} = \mathbf{Av}$. Because of the dimensions of $\mathbf{A}$ and $\mathbf{v}$, this product makes sense only in the order $\mathbf{Av}$ (rather than $\mathbf{vA}$); we thus call this operation *pre-multiplication* of $\mathbf{v}$ by $\mathbf{A}$. We can view premultiplication by $\mathbf{A}$ as defining a function: $\mathbf{w} = f(\mathbf{v})$, where $f(\mathbf{v}) = \mathbf{Av}$. This type of transformation is used in computer graphics, for instance, to produce creative lettering and moving images (animation) on the screen. It also provides a way to visualize the effect on a vector of multiplying it by a matrix.

Rather than using arrows, here we simply represent 2-vectors as points in the plane.

EXAMPLE 7

Consider the vectors

$$\mathbf{v}_1 = \begin{bmatrix} 0 \\ 0 \end{bmatrix} \qquad \mathbf{v}_2 = \begin{bmatrix} 0 \\ 2 \end{bmatrix} \qquad \mathbf{v}_3 = \begin{bmatrix} 2 \\ 2 \end{bmatrix} \qquad \mathbf{v}_4 = \begin{bmatrix} 2 \\ 0 \end{bmatrix}$$

$$\mathbf{v}_5 = \begin{bmatrix} 0 \\ 1 \end{bmatrix} \qquad \mathbf{v}_6 = \begin{bmatrix} 1 \\ 2 \end{bmatrix} \qquad \mathbf{v}_7 = \begin{bmatrix} 2 \\ 1 \end{bmatrix} \qquad \mathbf{v}_8 = \begin{bmatrix} 1 \\ 0 \end{bmatrix}.$$

Here, $\mathbf{v}_1, \mathbf{v}_2, \mathbf{v}_3,$ and $\mathbf{v}_4$ point to the corners of a square whose sides are of length 2 and $\mathbf{v}_5, \mathbf{v}_6, \mathbf{v}_7,$ and $\mathbf{v}_8$ point to the midpoints of the sides of this square, as shown in Figure 10.29(a). Describe the effects of pre-multiplying each of these eight vectors by the matrix $\mathbf{A} = \begin{bmatrix} 1 & 1 \\ -1 & 1 \end{bmatrix}.$

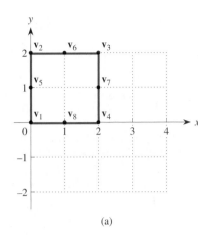

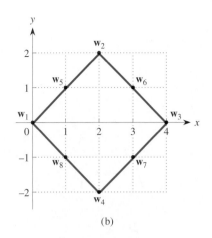

FIGURE 10.29   (a)   (b)

Solution   When we pre-multiply the $\mathbf{v}$'s by the matrix $\mathbf{A}$, we obtain the following eight vectors $\mathbf{w}_i = \mathbf{A}\mathbf{v}_i$.

$$\mathbf{w}_1 = \begin{bmatrix} 1 & 1 \\ -1 & 1 \end{bmatrix}\begin{bmatrix} 0 \\ 0 \end{bmatrix} = \begin{bmatrix} 0 \\ 0 \end{bmatrix} \qquad \mathbf{w}_2 = \begin{bmatrix} 1 & 1 \\ -1 & 1 \end{bmatrix}\begin{bmatrix} 0 \\ 2 \end{bmatrix} = \begin{bmatrix} 2 \\ 2 \end{bmatrix}$$

$$\mathbf{w}_3 = \begin{bmatrix} 1 & 1 \\ -1 & 1 \end{bmatrix}\begin{bmatrix} 2 \\ 2 \end{bmatrix} = \begin{bmatrix} 4 \\ 0 \end{bmatrix} \qquad \mathbf{w}_4 = \begin{bmatrix} 1 & 1 \\ -1 & 1 \end{bmatrix}\begin{bmatrix} 2 \\ 0 \end{bmatrix} = \begin{bmatrix} 2 \\ -2 \end{bmatrix}$$

$$\mathbf{w}_5 = \begin{bmatrix} 1 & 1 \\ -1 & 1 \end{bmatrix}\begin{bmatrix} 0 \\ 1 \end{bmatrix} = \begin{bmatrix} 1 \\ 1 \end{bmatrix} \qquad \mathbf{w}_6 = \begin{bmatrix} 1 & 1 \\ -1 & 1 \end{bmatrix}\begin{bmatrix} 1 \\ 2 \end{bmatrix} = \begin{bmatrix} 3 \\ 1 \end{bmatrix}$$

$$\mathbf{w}_7 = \begin{bmatrix} 1 & 1 \\ -1 & 1 \end{bmatrix}\begin{bmatrix} 2 \\ 1 \end{bmatrix} = \begin{bmatrix} 3 \\ -1 \end{bmatrix} \qquad \mathbf{w}_8 = \begin{bmatrix} 1 & 1 \\ -1 & 1 \end{bmatrix}\begin{bmatrix} 1 \\ 0 \end{bmatrix} = \begin{bmatrix} 1 \\ -1 \end{bmatrix}.$$

When we examine these eight new vectors, shown in Figure 10.29(b), we observe that they also form a square, but it has a different orientation and size. Each of the four vertex vectors $\mathbf{v}_1$, $\mathbf{v}_2$, $\mathbf{v}_3$, and $\mathbf{v}_4$ of the original square was transformed into a vertex vector $\mathbf{w}_1$, $\mathbf{w}_2$, $\mathbf{w}_3$, and $\mathbf{w}_4$ of the new square. Each of the midpoint vectors $\mathbf{v}_5$, $\mathbf{v}_6$, $\mathbf{v}_7$, and $\mathbf{v}_8$ in the original square was transformed into a corresponding midpoint vector $\mathbf{w}_5$, $\mathbf{w}_6$, $\mathbf{w}_7$, and $\mathbf{w}_8$ in the new square. Also, in the original square, the sides were of length 2 and each diagonal was $2\sqrt{2}$, from the Pythagorean theorem. In the transformed square, each diagonal has length 4, so that the sides have length $2\sqrt{2}$. Thus pre-multiplying the vectors $\mathbf{v}_1$, $\mathbf{v}_2$, $\mathbf{v}_3$, and $\mathbf{v}_4$ forming the original square by $\mathbf{A}$ does three things to the square.

1. It rotates the square clockwise through $45°$ or $\pi/4$.

2. It increases the length of each side by a factor of $\sqrt{2}$.

3. It moves the center of the square from $\begin{bmatrix} 1 \\ 1 \end{bmatrix}$ to $\begin{bmatrix} 2 \\ 0 \end{bmatrix}$.

Example 7 demonstrates that one square can be transformed into another by multiplying a set of vectors by an appropriate matrix. The same principle applies to any shape whose corners are determined by a set of vectors.

If we pre-multiplied the vectors forming the original square by other (appropriately chosen) matrices, we could get rotations of the square through any desired angle, increase or decrease the lengths of the sides by any desired multiple, and place the center in any desired location. However, the origin will always be unaffected because $\begin{bmatrix} 0 \\ 0 \end{bmatrix}$ multiplied by any matrix yields $\begin{bmatrix} 0 \\ 0 \end{bmatrix}$.

For example, using $\mathbf{B} = \begin{bmatrix} 0 & -1 \\ 1 & 0 \end{bmatrix}$ rotates the square counterclockwise through an angle of $\pi/2$ and transforms the center to $\begin{bmatrix} -1 \\ 1 \end{bmatrix}$. We ask you to verify this result in the Problems at the end of this section.

In general, pre-multiplying by the matrix

$$\mathbf{R} = \begin{bmatrix} \cos\theta & -\sin\theta \\ \sin\theta & \cos\theta \end{bmatrix}$$

rotates the square counterclockwise through an angle of θ.

EXAMPLE 8

Show that the effect of pre-multiplying any vector $\mathbf{v} = \begin{bmatrix} a \\ b \end{bmatrix}$ by the matrix $\mathbf{R}$ is to rotate $\mathbf{v}$ through an angle θ.

Solution We start with an arbitrary vector $\mathbf{v}$, as shown in Figure 10.30, that is inclined at an angle α from the horizontal, so that

$$\tan\alpha = \frac{b}{a}.$$

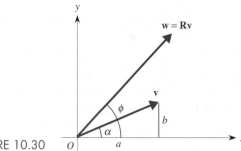

FIGURE 10.30

Also, we let the angle associated with the vector $\mathbf{w} = \mathbf{R}\mathbf{v}$ be ϕ. We want to show that the difference $\phi - \alpha$ in the two angles must equal θ.

When we pre-multiply vector $\mathbf{v}$ by $\mathbf{R}$ to form vector $\mathbf{w}$, we get

$$\mathbf{w} = \mathbf{R}\mathbf{v} = \begin{bmatrix} \cos\theta & -\sin\theta \\ \sin\theta & \cos\theta \end{bmatrix} \begin{bmatrix} a \\ b \end{bmatrix} = \begin{bmatrix} a\cos\theta - b\sin\theta \\ a\sin\theta + b\cos\theta \end{bmatrix}.$$

Consequently, the tangent of angle ϕ is

$$\tan\phi = \frac{a\sin\theta + b\cos\theta}{a\cos\theta - b\sin\theta}.$$

We factor a out of both the numerator and denominator and divide both the numerator and denominator by $\cos\theta$:

$$\tan\phi = \frac{a\cdot\left(\dfrac{\sin\theta}{\cos\theta}+\dfrac{b}{a}\right)}{a\cdot\left(1-\dfrac{b}{a}\dfrac{\sin\theta}{\cos\theta}\right)} = \frac{\tan\theta+\dfrac{b}{a}}{1-\dfrac{b}{a}\tan\theta}.$$

However, we know that $\tan\alpha = b/a$, so that this expression reduces to

$$\tan\phi = \frac{\tan\theta + \tan\alpha}{1 - \tan\alpha\,\tan\theta}.$$

When we compare this expression to the sum identity for the tangent from Problem 37 of Section 8.1, we see that

$$\tan\phi = \tan(\theta+\alpha)\quad\text{so that}\quad \phi = \theta + \alpha.$$

Therefore the effect of pre-multiplying any vector $\mathbf{v}$ by the matrix $\mathbf{R}$ is to rotate the resulting vector through an angle of θ.

What if we want to change the length of a vector by using matrix multiplication? We note that the lengths of the sides of the square in Example 8 are unchanged when we use this rotation matrix $\mathbf{R}$. If we want to enlarge or contract the square, we must use an appropriate *scalar multiple* of the rotation matrix, so that we multiply every entry in the matrix by a constant amount. Thus, if

$$\mathbf{A} = \begin{bmatrix} 1 & 1 \\ -1 & 1 \end{bmatrix},\quad\text{then}\quad 2\mathbf{A} = 2\begin{bmatrix} 1 & 1 \\ -1 & 1 \end{bmatrix} = \begin{bmatrix} 2 & 2 \\ -2 & 2 \end{bmatrix}.$$

In Example 7, to make each side of the square grow by a factor of 2 instead of a factor of $\sqrt{2}$, we would multiply $\mathbf{A}$ by $\sqrt{2}$ because $\sqrt{2}\cdot\sqrt{2} = 2$. Incidentally, not every matrix can be a rotation matrix; a special form is necessary.

Any transformation that takes a 2-vector $\mathbf{v}$ into the 2-vector $\mathbf{w} = \mathbf{Av,}$ for any 2×2 matrix $\mathbf{A}$, always *transforms* or *maps* lines into lines. Because of this linearity property, a mapping of the form $\mathbf{v}\to\mathbf{w} = \mathbf{Av}$ is called a *linear transformation*.

Suppose that we start with any figure comprising (very short) line segments. Each segment can be interpreted as a vector, and appropriate matrices can be constructed to create any kind of transformation—a shift, a stretch, or a rotation—that we desire. This method is the mathematical foundation of the computer graphics animation that appears in movies, on television, and on computer screens.

Problems

1. Let

$$\mathbf{a} = \begin{bmatrix} 3 \\ 5 \end{bmatrix},\quad \mathbf{b} = \begin{bmatrix} 0 \\ 2 \end{bmatrix},\quad \mathbf{c} = \begin{bmatrix} 5 \\ -1 \end{bmatrix},$$

$$\mathbf{d} = \begin{bmatrix} -1 \\ 0 \end{bmatrix},\quad \mathbf{e} = \begin{bmatrix} 4 \\ 5 \end{bmatrix}.$$

Compute the scalar products.

a. $\mathbf{a}\cdot\mathbf{c}$ b. $\mathbf{b}\cdot\mathbf{c}$ c. $\mathbf{b}\cdot\mathbf{d}$

d. $\mathbf{a}\cdot\mathbf{d}$ e. $\mathbf{c}\cdot\mathbf{d}$ f. $\mathbf{b}\cdot\mathbf{e}$

g. $\mathbf{c}\cdot\mathbf{e}$

2. Let

$$\mathbf{a} = \begin{bmatrix} 4 \\ 2 \end{bmatrix}, \qquad \mathbf{b} = \begin{bmatrix} -2 \\ 4 \end{bmatrix}, \qquad \mathbf{c} = \begin{bmatrix} -4 \\ 2 \end{bmatrix},$$

$$\mathbf{d} = \begin{bmatrix} 4 \\ -2 \end{bmatrix}, \quad \text{and} \quad \mathbf{e} = \begin{bmatrix} 2 \\ 4 \end{bmatrix}.$$

Plot the pairs of vectors and compute their scalar products.

a. **a, b**   b. **a, c**   c. **a, d**
d. **b, d**   e. **c, d**   f. **d, e**

What geometric pattern do you find in pairs of vectors whose scalar products are zero? What numerical pattern do you find in the ratios of components in these pairs?

3. Let

$$\mathbf{a} = \begin{bmatrix} 2 \\ 5 \\ 1 \end{bmatrix}, \qquad \mathbf{b} = \begin{bmatrix} 0 \\ 1 \\ 0 \end{bmatrix},$$

$$\mathbf{c} = \begin{bmatrix} 7 \\ -2 \\ -1 \end{bmatrix}, \quad \mathbf{d} = \begin{bmatrix} -2 \\ 3 \\ 1 \end{bmatrix}.$$

Compute the scalar products.

a. **a · c**   b. **b · c**   c. **b · d**
d. **a · d**   e. **a · a**   f. **c · d**

4. Let

$$\mathbf{a} = \begin{bmatrix} 3 \\ 1 \\ 2 \end{bmatrix}, \qquad \mathbf{b} = \begin{bmatrix} 0 \\ 2 \\ -2 \end{bmatrix},$$

$$\mathbf{c} = \begin{bmatrix} 1 \\ 4 \\ 8 \end{bmatrix}, \quad \mathbf{d} = \begin{bmatrix} 2 \\ 1 \\ 5 \end{bmatrix}.$$

Compute the scalar products.

a. **a · c**   b. **b · d**   c. **c · d**
d. **(a + c) · a**   e. **a · (c + d)**
f. **(b − a) · (a + c)**

5. Let **a, b, c,** and **d** be as in Problem 1 and let

$$\mathbf{A} = \begin{bmatrix} 1 & 7 \\ 4 & 2 \end{bmatrix}, \qquad \mathbf{B} = \begin{bmatrix} 5 & 2 \\ -1 & 0 \end{bmatrix}, \qquad \mathbf{C} = \begin{bmatrix} 2 & 1 \\ 4 & -2 \\ 3 & 1 \end{bmatrix}.$$

Compute the matrix–vector products.

a. **Ab**   b. **Ac**   c. **Ba**
d. **Bb**   e. **Cc**   f. **Ca**

6. Let **a, b, c,** and **d** be as in Problem 3 and let

$$\mathbf{A} = \begin{bmatrix} 3 & 1 & 2 \\ 0 & -1 & 5 \\ 0 & 4 & 3 \end{bmatrix}, \qquad \mathbf{B} = \begin{bmatrix} 0 & 1 & 3 \\ -1 & 5 & 1 \\ 4 & -1 & 6 \end{bmatrix},$$

$$\mathbf{C} = \begin{bmatrix} 0 & 1 & 2 \\ 3 & 4 & 5 \\ 6 & 7 & 8 \\ 9 & 10 & 11 \end{bmatrix}.$$

Find the matrix–vector products.

a. **Aa**   b. **Ab**   c. **Bc**
d. **Bd**   e. **Cb**   f. **Cc**

7. Let **a, b, c,** and **d** be as in Problem 3 and let

$$\mathbf{A} = \begin{bmatrix} 1 & 0 & 0 \\ 0 & 1 & 0 \\ 0 & 0 & 1 \end{bmatrix}, \qquad \mathbf{B} = \begin{bmatrix} 0 & 0 & 1 \\ 0 & 1 & 0 \\ 1 & 0 & 0 \end{bmatrix},$$

$$\mathbf{C} = \begin{bmatrix} 1 & 0 & 0 \\ 0 & -1 & 0 \\ 0 & 0 & 2 \end{bmatrix}.$$

Compute the matrix–vector products and describe in words the effect of multiplying a vector by the particular matrix.

a. **Ac**   b. **Ad**   c. **Ba**
d. **Bd**   e. **Cb**   f. **Ca**

8. Explain why it is not possible to multiply the matrices

$$\mathbf{A} = \begin{bmatrix} 1 & 2 & 3 \\ 4 & 5 & 6 \end{bmatrix} \quad \text{and} \quad \mathbf{b} = \begin{bmatrix} 4 \\ 7 \end{bmatrix}.$$

9. Write each vector–matrix equation as a system of equations. Here **x** denotes a column vector of variables $x_1, x_2, \ldots$, where the number of variables equals the number of columns in **A**.

a. **Ax = b**, where $\mathbf{A} = \begin{bmatrix} 5 & 1 \\ 4 & 3 \end{bmatrix}$ and $\mathbf{b} = \begin{bmatrix} 2 \\ 5 \end{bmatrix}$

b. **Ax = b**, where

$$\mathbf{A} = \begin{bmatrix} 1 & 4 \\ 2 & -3 \end{bmatrix} \quad \text{and} \quad \mathbf{b} = \begin{bmatrix} 4 \\ 9 \end{bmatrix}$$

c. **Ax = b**, where

$$\mathbf{A} = \begin{bmatrix} 5 & 2 & 1 \\ 4 & 1 & 6 \\ 3 & 1 & 0 \end{bmatrix} \quad \text{and} \quad \mathbf{b} = \begin{bmatrix} 1 \\ 5 \\ 2 \end{bmatrix}$$

d. $\mathbf{Ax} = \mathbf{b}$, where

$$\mathbf{A} = \begin{bmatrix} 2 & -1 & 5 \\ 3 & 1 & 2 \\ 5 & 1 & -3 \end{bmatrix} \quad \text{and} \quad \mathbf{b} = \begin{bmatrix} 0 \\ 0 \\ 0 \end{bmatrix}$$

10. Write each system of equations in matrix notation. Define any matrix or vector that you use.

a. $x_1 + 2x_2 = 6$
$2x_1 - 6x_2 = 4$

b. $5x_1 + 2x_2 + 4x_3 = 6$
$x_1 + 3x_2 - 2x_3 = 2$
$2x_1 - 5x_2 + 5x_3 = 5$

c. $2x_1 + 5x_2 - 2x_3 = 0$
$3x_1 + 8x_2 + 4x_3 = 0$
$x_1 + x_2 - 7x_3 = 0$

11. Write in matrix notation the systems of linear equations obtained in the following Problems in Section 10.2. Define any matrix or vector that you use.

a. Problem 4 **b.** Problem 5
c. Problem 6 **d.** Problem 7

12. Write in matrix notation the systems of linear equations obtained in the hare–wolf population model in Example 3 of Section 10.2. Define any matrix or vector that you use.

13. Suppose that you will need 10 hero sandwiches, 6 quarts of fruit punch, 3 pounds of potato salad, and 2 plates of hors d'oeuvres for a party. The matrix shows the cost per unit of these supplies from three different caterers.

	Caterer A	Caterer B	Caterer C
Hero sandwich	$4	$6	$5
Fruit punch	$2	$1	$0.85
Potato salad	$1.50	$2	$2.50
Hors d'oeuvres	$6	$5	$7

a. Express the cost of catering the party by each caterer as a matrix–vector product.
b. Determine the cost of each caterer.

14. Plot a square with corners determined by the vectors $\mathbf{v}_1, \mathbf{v}_2, \mathbf{v}_3$, and $\mathbf{v}_4$ in (a) and (b). Then, for the matrix $\mathbf{A} = \begin{bmatrix} 1 & 1 \\ -1 & 1 \end{bmatrix}$, plot the transformed corners $\mathbf{w}_1 = \mathbf{Av}_1$, $\mathbf{w}_2 = \mathbf{Av}_2$, $\mathbf{w}_3 = \mathbf{Av}_3$, and $\mathbf{w}_4 = \mathbf{Av}_4$. Confirm that the midpoints of the sides of the original square are mapped to the midpoints of the sides of the transformed square.

a. $\mathbf{v}_1 = \begin{bmatrix} 1 \\ 1 \end{bmatrix}$, $\mathbf{v}_2 = \begin{bmatrix} 1 \\ 3 \end{bmatrix}$, $\mathbf{v}_3 = \begin{bmatrix} 3 \\ 3 \end{bmatrix}$, $\mathbf{v}_4 = \begin{bmatrix} 3 \\ 1 \end{bmatrix}$

b. $\mathbf{v}_1 = \begin{bmatrix} 0 \\ 0 \end{bmatrix}$, $\mathbf{v}_2 = \begin{bmatrix} -1 \\ 1 \end{bmatrix}$, $\mathbf{v}_3 = \begin{bmatrix} 0 \\ 2 \end{bmatrix}$, $\mathbf{v}_4 = \begin{bmatrix} 1 \\ 1 \end{bmatrix}$

15. Repeat Problem 14 with the matrix $\mathbf{B} = \begin{bmatrix} 0 & -1 \\ 1 & 0 \end{bmatrix}$.

In Example 8, we proved that the mapping of $\mathbf{v}$ to $\mathbf{w} = \mathbf{Bv}$ acts to rotate a square counterclockwise through an angle of $\pi/2$. Do your results confirm this outcome?

16. Transform each square in Problem 14 by using the matrix $\mathbf{C} = \begin{bmatrix} 1 & 2 \\ 3 & 4 \end{bmatrix}$. Does the mapping of $\mathbf{v}$ to $\mathbf{w} = \mathbf{Cv}$ transform the square into a square?

17. Consider the rotation matrix $\mathbf{R} = \begin{bmatrix} \cos\theta & -\sin\theta \\ \sin\theta & \cos\theta \end{bmatrix}$. Show that, for any vector $\mathbf{v} = \begin{bmatrix} a \\ b \end{bmatrix}$, the vector $\mathbf{w} = \mathbf{Rv}$ has the same length as $\mathbf{v}$.

18. Use the matrix $\mathbf{R}$ from Problem 17 to determine the effect that the matrix $2\mathbf{R}$ has on the vector $\mathbf{v} = \begin{bmatrix} a \\ b \end{bmatrix}$.

19. The rotation matrix $\mathbf{R}$ in Problem 17 acts to rotate any nonzero vector counterclockwise through an angle θ.

a. Modify matrix $\mathbf{R}$ to produce a matrix that rotates any nonzero vector clockwise through an angle θ.
b. Write a matrix that will rotate any nonzero vector counterclockwise through an angle of 30°.
c. Plot the position vector $\mathbf{v} = \begin{bmatrix} 5 \\ 2 \end{bmatrix}$, pre-multiply it by the matrix you created in part (b), and then plot the resulting vector $\mathbf{w}$ on the same graph.
d. Pre-multiply the vector $\mathbf{w}$ by the matrix you created in part (a) with $\theta = 30°$. What is the result of this operation?

20. The five points A, B, C, D, and E at $(3, 2)$, $(7, -1)$, $(9, 1), (10, 3)$, and $(6, 6)$ determine a five sided figure having two right angles. Use vectors to determine which of the five angles in the figure are the right angles.

21. Find the acute angle between each pair of vectors.

    a. $\begin{bmatrix} 3 & 5 \end{bmatrix}$ and $\begin{bmatrix} -2 & 4 \end{bmatrix}$
    b. $\begin{bmatrix} 1 & 4 \end{bmatrix}$ and $\begin{bmatrix} -2 & 5 \end{bmatrix}$
    c. $\begin{bmatrix} 1 & 4 & 5 \end{bmatrix}$ and $\begin{bmatrix} 2 & 3 & -2 \end{bmatrix}$
    d. $\begin{bmatrix} 6 & 4 & -1 \end{bmatrix}$ and $\begin{bmatrix} 5 & -3 & 2 \end{bmatrix}$

22. Rewrite each polynomial as the scalar product of a vector of numbers and a vector of power function terms of the form $\mathbf{x} = \begin{bmatrix} 1 & x & x^2 \end{bmatrix}$ or $\mathbf{x}' = \begin{bmatrix} 1 & x & x^2 & x^3 \end{bmatrix}$.

    a. $5 + 3x + 2x^2$
    b. $8 - 7x + 3x^2$
    c. $4 - 2x - 6x^2 + 5x^3$

23. When Susan was applying to college, she was turned down by her top choice, Ivy Tech, but she was accepted by State Tech and the Hawaii Institute of Technology. To decide on which school to attend, she rated each school (on a scale of 0 to 10) on five important criteria and then selected the one that was closer to Ivy Tech. She used her knowledge of vectors to decide how to interpret *closer*—the two

	Ivy	State	Hawaii
Location	8	7	10
Size	9	6	7
Campus	7	9	6
Faculty	7	6	4
Programs	5	7	6

vectors with the smallest angle between them. Which college did she choose?

24. (Continuation of Problem 23) Suggest some other ways that Susan could have decided which school was closer to Ivy Tech? Would you necessarily make the same decision as to which school to choose? Explain.

25. (Derivation of the formula for $\cos \theta$.) In the triangle shown, let $\mathbf{a} = \begin{bmatrix} a_1 & a_2 \end{bmatrix}$ and $\mathbf{b} = \begin{bmatrix} b_1 & b_2 \end{bmatrix}$. The lengths of $\mathbf{a}$ and $\mathbf{b}$ can be written $\|\mathbf{a}\| = \sqrt{a_1^2 + a_2^2}$ and $\|\mathbf{b}\| = \sqrt{b_1^2 + b_2^2}$, so that $\|\mathbf{a}\|^2 = \mathbf{a} \cdot \mathbf{a}$ and $\|\mathbf{b}\|^2 = \mathbf{b} \cdot \mathbf{b}$.

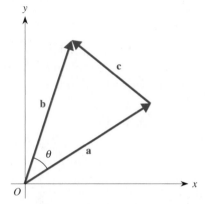

a. Write the vector $\mathbf{c}$ in terms of the vectors $\mathbf{a}$ and $\mathbf{b}$.
b. Use the expression from part (a) to form the scalar product $\mathbf{c} \cdot \mathbf{c}$.
c. Use the law of cosines to write an equation giving $\|\mathbf{c}\|^2$.
d. Compare the expressions from part (b) and part (c) and use the facts that $\mathbf{a} \cdot \mathbf{a} = \|\mathbf{a}\|^2$, $\mathbf{b} \cdot \mathbf{b} = \|\mathbf{b}\|^2$, and $\mathbf{c} \cdot \mathbf{c} = \|\mathbf{c}\|^2$ to show that

$$\cos \theta = \frac{\mathbf{a} \cdot \mathbf{b}}{\|\mathbf{a}\| \|\mathbf{b}\|}.$$

10.4 Matrix Multiplication

In Section 10.3, we introduced the concept of the scalar product $\mathbf{a} \cdot \mathbf{b}$ of two vectors $\mathbf{a}$ and $\mathbf{b}$. For example, if $\mathbf{a} = \begin{bmatrix} 2 & 1 \end{bmatrix}$ and $\mathbf{b} = \begin{bmatrix} 4 \\ 6 \end{bmatrix}$, then $\mathbf{a} \cdot \mathbf{b} = 2(4) + 1(6) = 14$.

We used this scalar product of vectors to define the matrix–vector product $\mathbf{Ab}$ of a matrix $\mathbf{A}$ times a vector $\mathbf{b}$. For instance, if

$$\mathbf{A} = \begin{bmatrix} 2 & 1 \\ 0 & 3 \end{bmatrix} \quad \text{and again} \quad \mathbf{b} = \begin{bmatrix} 4 \\ 6 \end{bmatrix},$$

then

$$\mathbf{Ab} = \begin{bmatrix} 2 & 1 \\ 0 & 3 \end{bmatrix} \begin{bmatrix} 4 \\ 6 \end{bmatrix} = \begin{bmatrix} 2(4) + 1(6) \\ 0(4) + 3(6) \end{bmatrix} = \begin{bmatrix} 14 \\ 18 \end{bmatrix}.$$

Thus $\mathbf{Ab}$ is a column vector consisting of the scalar products of each row of $\mathbf{A}$ with $\mathbf{b}$.

In this section, we extend this process to define the product of two matrices $\mathbf{A}$ and $\mathbf{B}$. In particular, in the product $\mathbf{AB}$ we think of the second matrix $\mathbf{B}$ as consisting of a series of column vectors and pre-multiply each of them by the matrix $\mathbf{A}$. That is, we multiply the first column of $\mathbf{B}$ by $\mathbf{A}$, then multiply the second column of $\mathbf{B}$ by $\mathbf{A}$, and so on, as illustrated in Figure 10.31. The result of each product is a column vector, so the product $\mathbf{AB}$ will be a matrix having the same number of columns as there are in $\mathbf{B}$. We can think of the product $\mathbf{AB}$ as a matrix whose columns are a sequence of matrix–vector products $\begin{bmatrix} \mathbf{Ab}_1 & \mathbf{Ab}_2 & \ldots & \mathbf{Ab}_n \end{bmatrix}$.

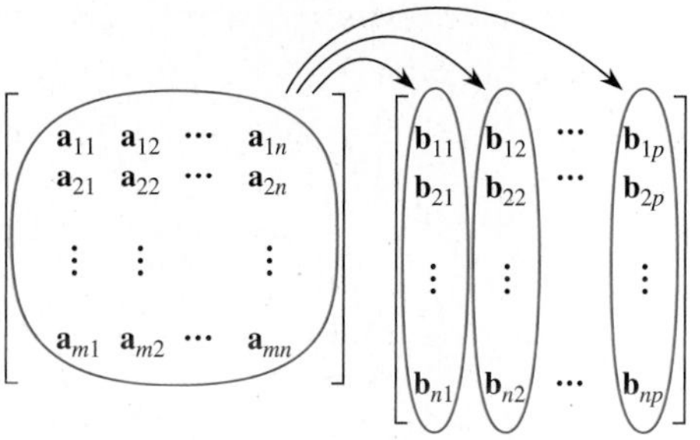

FIGURE 10.31

Think About This

In the matrix product $\mathbf{AB}$, you can also think of the first matrix $\mathbf{A}$ as consisting of a series of row vectors and the second matrix $\mathbf{B}$ as consisting of a series of column vectors and then take the scalar product of each of the row vectors making up $\mathbf{A}$ with each of the column vectors making up $\mathbf{B}$. Draw a sketch comparable to Figure 10.31 to illustrate this interpretation. ◻

To demonstrate the product of two matrices, consider again the matrix

$$\mathbf{A} = \begin{bmatrix} 2 & 1 \\ 0 & 3 \end{bmatrix} \quad \text{and let} \quad \mathbf{B} = \begin{bmatrix} 4 & 7 & -1 \\ 6 & 5 & 9 \end{bmatrix}.$$

We think of each of the three columns of matrix $\mathbf{B}$ as a column vector. Note that the first column $\mathbf{b}_1 = \begin{bmatrix} 4 \\ 6 \end{bmatrix}$ of $\mathbf{B}$ is precisely the column vector $\mathbf{b} = \begin{bmatrix} 4 \\ 6 \end{bmatrix}$ we used above. The corresponding first column in the product matrix $\mathbf{AB}$ is then $\mathbf{Ab}_1$, which we computed above to be $\begin{bmatrix} 14 \\ 18 \end{bmatrix}$.

Similarly, we take the matrix–vector product of the matrix $\mathbf{A}$ with the second column of $\mathbf{B}$ thinking of it as the vector $\mathbf{b}_2 = \begin{bmatrix} 7 \\ 5 \end{bmatrix}$ to produce the second column of the product $\mathbf{AB}$. That gives

$$\mathbf{Ab}_2 = \begin{bmatrix} 2 & 1 \\ 0 & 3 \end{bmatrix} \begin{bmatrix} 7 \\ 5 \end{bmatrix} = \begin{bmatrix} 2(7) + 1(5) \\ 0(7) + 3(5) \end{bmatrix} = \begin{bmatrix} 19 \\ 15 \end{bmatrix}.$$

Finally, we take the matrix–vector product of **A** with the third column of **B**, which is the vector $\mathbf{b}_3 = \begin{bmatrix} -1 \\ 9 \end{bmatrix}$, to produce the third column of the product **AB**. The complete product **AB** is therefore

$$\mathbf{AB} = \begin{bmatrix} 2 & 1 \\ 0 & 3 \end{bmatrix} \begin{bmatrix} 4 & 7 & -1 \\ 6 & 5 & 9 \end{bmatrix} = \begin{bmatrix} 2(4) + 1(6) & 2(7) + 1(5) & 2(-1) + 1(9) \\ 0(4) + 3(6) & 0(7) + 3(5) & 0(-1) + 3(9) \end{bmatrix}$$

$$= \begin{bmatrix} 14 & 19 & 7 \\ 18 & 15 & 27 \end{bmatrix}.$$

In general, we have the following definition for the product of two matrices.

Matrix Multiplication

Let **A** be an $m \times r$ matrix and **B** be an $r \times n$ matrix. The number of columns in **A** must equal the number of rows in **B**.

The matrix product **AB** is the $m \times n$ matrix obtained by forming the scalar product of each row $\mathbf{a}_i'$ in **A** with each column $\mathbf{b}_j$ in **B**. That is, the (i, j)th entry in **AB** is $\mathbf{a}_i'\mathbf{b}_j$. Thus

$$\mathbf{AB} = \begin{bmatrix} \mathbf{a}_1'\mathbf{b}_1 & \mathbf{a}_1'\mathbf{b}_2 & \cdots & \mathbf{a}_1'\mathbf{b}_n \\ \mathbf{a}_2'\mathbf{b}_1 & \mathbf{a}_2'\mathbf{b}_2 & \cdots & \mathbf{a}_2'\mathbf{b}_n \\ \vdots & \vdots & \vdots & \vdots \\ \mathbf{a}_m'\mathbf{b}_1 & \mathbf{a}_m'\mathbf{b}_2 & \cdots & \mathbf{a}_m'\mathbf{b}_n \end{bmatrix}.$$

In summary, there are three ways to interpret matrix multiplication.

1. We can think of **AB** as the scalar product of each row of **A** with each column of **B**. Figure 10.32 illustrates this interpretation. The product requires that the number of entries in each row of **A** must equal the number of entries in each column of **B**.

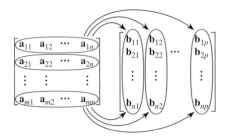

FIGURE 10.32

2. We can think of **AB** as a sequence of matrix–vector products—that is, the product of **A** with each of the column vectors making up **B**, so that

$$\mathbf{AB} = \mathbf{A}[\mathbf{b}_1 \quad \mathbf{b}_2 \quad \ldots \quad \mathbf{b}_n] = [\mathbf{Ab}_1 \quad \mathbf{Ab}_2 \quad \ldots \quad \mathbf{Ab}_n].$$

For example, for the matrices **A** and **B** above, we verify that the first column of **AB** is the matrix–vector product $\mathbf{Ab}_1$. Then

$$\mathbf{Ab}_1 = \begin{bmatrix} 2 & 1 \\ 0 & 3 \end{bmatrix}\begin{bmatrix} 4 \\ 6 \end{bmatrix} = \begin{bmatrix} 2(4) + 1(6) \\ 0(4) + 3(6) \end{bmatrix} = \begin{bmatrix} 14 \\ 18 \end{bmatrix}.$$

3. We can view **AB** as a set of vector-matrix products $\mathbf{a}_i'\mathbf{B}$ (defined analogously to matrix–vector products), where each row of **A**, considered as a row vector, pre-multiplies the matrix **B**. Thus

$$\mathbf{AB} = \begin{bmatrix} \mathbf{a}_1' \\ \mathbf{a}_2' \\ \vdots \\ \mathbf{a}_m' \end{bmatrix}\mathbf{B} = \begin{bmatrix} \mathbf{a}_1'\mathbf{B} \\ \mathbf{a}_2'\mathbf{B} \\ \vdots \\ \mathbf{a}_m'\mathbf{B} \end{bmatrix}.$$

For instance, for the matrices **A** and **B,** we verify that the first row of **AB** is the vector–matrix product $\mathbf{a}_1'\mathbf{B}$:

$$\mathbf{a}_1'\mathbf{B} = \begin{bmatrix} 2 & 1 \end{bmatrix}\begin{bmatrix} 4 & 7 & -1 \\ 6 & 5 & 9 \end{bmatrix}$$
$$= \begin{bmatrix} 2(4) + 1(6) & 2(7) + 1(5) & 2(-1) + 1(9) \end{bmatrix} = \begin{bmatrix} 14 & 19 & 7 \end{bmatrix}.$$

For the matrix product **AB** to make sense, the number of columns in **A** must equal the number of rows in **B**. Thus, if **A** is $m \times n$ (m rows and n columns) and **B** is $n \times k$ (n rows and k columns), the product **AB** is $m \times k$. For instance, the product of a 3×5 matrix and a 5×8 matrix will be a 3×8 matrix. But, the product of a 5×8 matrix and a 3×5 matrix is not defined—the numbers of columns and rows do not match, so it is not possible to perform the multiplication.

Note that our definition of a matrix–vector product in Section 10.3 is a special case of matrix multiplication because, in the matrix–vector product **Ab,** the column vector **b** can be interpreted as an $n \times 1$ matrix. Then **Ab** is the matrix product of an $m \times n$ matrix **A** and an $n \times 1$ matrix **b.** The result **Ab** is an $m \times 1$ matrix (a column m-vector).

In general, except in rather unusual circumstances, matrix multiplication is not commutative; that is, $\mathbf{AB} \neq \mathbf{BA}$. In fact, unless **A** and **B** are both square matrices with the same size, only one at most of **AB** and **BA** is defined. For instance, if **A** is 3×5 and **B** is 5×2, we can form **AB**, but not **BA**.

EXAMPLE 1

Given

$$\mathbf{A} = \begin{bmatrix} 2 & 1 \\ 0 & 3 \end{bmatrix} \quad \text{and} \quad \mathbf{B} = \begin{bmatrix} 3 & -1 \\ 1 & 4 \end{bmatrix},$$

find **AB** and **BA** and decide whether matrix multiplication is commutative.

Solution   Using the given matrices, we find that

$$\mathbf{AB} = \begin{bmatrix} 2 & 1 \\ 0 & 3 \end{bmatrix}\begin{bmatrix} 3 & -1 \\ 1 & 4 \end{bmatrix} = \begin{bmatrix} 2(3) + 1(1) & 2(-1) + 1(4) \\ 0(3) + 3(1) & 0(-1) + 3(4) \end{bmatrix} = \begin{bmatrix} 7 & 2 \\ 3 & 12 \end{bmatrix},$$

whereas

$$\mathbf{BA} = \begin{bmatrix} 3 & -1 \\ 1 & 4 \end{bmatrix} \begin{bmatrix} 2 & 1 \\ 0 & 3 \end{bmatrix} = \begin{bmatrix} 3(2) + (-1)(0) & 3(1) + (-1)(3) \\ 1(2) + 4(0) & 1(1) + 4(3) \end{bmatrix} = \begin{bmatrix} 6 & 0 \\ 2 & 13 \end{bmatrix}.$$

Thus, for these two matrices, $\mathbf{AB} \neq \mathbf{BA}$, so matrix multiplication is not, in general, a commutative operation.

We also need a way to add two matrices. Recall how we added vectors in Example 3 of Section 10.1: We simply added the entries in the corresponding positions. For instance, the sum of the row vectors $\mathbf{c} = \begin{bmatrix} 2 & 1 & 5 \end{bmatrix}$ and $\mathbf{d} = \begin{bmatrix} 4 & 3 & 0 \end{bmatrix}$ is

$$\mathbf{c} + \mathbf{d} = \begin{bmatrix} 2 & 1 & 5 \end{bmatrix} + \begin{bmatrix} 4 & 3 & 0 \end{bmatrix} = \begin{bmatrix} 2+4 & 1+3 & 5+0 \end{bmatrix} = \begin{bmatrix} 6 & 4 & 5 \end{bmatrix}.$$

We add matrices in the same way by simply adding all corresponding entries. However, for addition to make sense, the two matrices or the two vectors must be of the same size.

◆ **EXAMPLE 2**

Find the sum of the two matrices

$$\mathbf{A} = \begin{bmatrix} 2 & 1 \\ 0 & 3 \end{bmatrix} \quad \text{and} \quad \mathbf{B} = \begin{bmatrix} 3 & -1 \\ 1 & 4 \end{bmatrix}.$$

Solution Matrix addition involves adding the entries in the corresponding positions, so we have

$$\mathbf{A} + \mathbf{B} = \begin{bmatrix} 2 & 1 \\ 0 & 3 \end{bmatrix} + \begin{bmatrix} 3 & -1 \\ 1 & 4 \end{bmatrix} = \begin{bmatrix} 2+3 & 1+(-1) \\ 0+1 & 3+4 \end{bmatrix} = \begin{bmatrix} 5 & 0 \\ 1 & 7 \end{bmatrix}.$$

◆

◆ **EXAMPLE 3**

Find the sum of the 3×3 matrices

$$\mathbf{E} = \begin{bmatrix} 4 & 2 & 3 \\ 2 & 1 & 3 \\ 0 & 5 & 4 \end{bmatrix} \quad \text{and} \quad \mathbf{F} = \begin{bmatrix} 2 & 6 & -1 \\ 4 & 0 & 2 \\ 7 & 3 & 8 \end{bmatrix}.$$

Solution Again, by adding the entries in the corresponding positions, we find that

$$\mathbf{E} + \mathbf{F} = \begin{bmatrix} 4 & 2 & 3 \\ 2 & 1 & 3 \\ 0 & 5 & 4 \end{bmatrix} + \begin{bmatrix} 2 & 6 & -1 \\ 4 & 0 & 2 \\ 7 & 3 & 8 \end{bmatrix} = \begin{bmatrix} 4+2 & 2+6 & 3+(-1) \\ 2+4 & 1+0 & 3+2 \\ 0+7 & 5+3 & 4+8 \end{bmatrix} = \begin{bmatrix} 6 & 8 & 2 \\ 6 & 1 & 5 \\ 7 & 8 & 12 \end{bmatrix}.$$

◆

Subtraction of matrices is defined in the analogous way—we simply take the difference of the corresponding entries in each position. However, the quotient of two matrices can't be defined.

We now summarize the laws of matrix algebra for matrix addition and multiplication. In each case we assume that the matrices have the appropriate sizes so that the operations make sense.

Basic Laws of Matrix Algebra

Associative Law

Matrix addition and matrix multiplication are associative:

$$(A + B) + C = A + (B + C) \quad \text{and} \quad (AB)C = A(BC).$$

Commutative Law

Matrix addition is commutative:

$$A + B = B + A.$$

Matrix multiplication is *not* commutative (except in special cases):

$$AB \neq BA.$$

Distributive Law

$$A(B + C) = AB + AC \quad \text{and} \quad (B + C)A = BA + CA$$

Law of Scalar Factoring

$$k(AB) = (kA)B = A(kB),$$

where k is a scalar constant.

With the exception that matrix multiplication is not commutative, these laws are basically the same as the laws used in algebra for working with real numbers. However, because the objects now are arrays and the operation of matrix multiplication is much more complicated than real-number multiplication, it is not at all obvious that these matrix laws should be true. Some effort is required to verify them (but this is beyond the scope of this chapter).

Because a vector is just a $1 \times n$ matrix or an $n \times 1$ matrix, these laws also apply to vectors. However, scalar products of vectors are commutative.

Matrix calculations are standard features on most calculators and in many software packages. Typically, you have to give a name for a matrix, such as **A**, then specify the *dimensions* of the matrix (the number of rows by the number of columns)—say, 3×4—and then enter the values in the appropriate positions. Once you have entered the matrices, you can use the calculator to perform any of the allowable operations such as sums, differences, and products. See your instruction manual for details.

The Fruit Purchase Model Revisited

In Example 1 of Section 10.3, we computed the scalar products of fruit costs and quantities of fruit to be purchased by Amy and Bill. Recall that Amy wanted 5 peaches, 3 pears, 2 apples, and 2 grapefruits, whereas Bill wanted 3 peaches, 4 pears, 3 apples, and 3 grapefruits. Also, peaches cost 30¢ each, pears 20¢ each, apples 35¢ each, and grapefruits 50¢ each. In Example 6 of Section 10.2, we constructed a matrix **A** of the fruit shopping lists of Amy and Bill and made a column vector **p** of the fruit costs (in cents).

$$\begin{array}{cccc} & \text{Peaches} & \text{Pears} & \text{Apples} & \text{Grapefruits} \\ \mathbf{A} = & \begin{bmatrix} 5 & 3 & 2 & 2 \\ 3 & 4 & 3 & 3 \end{bmatrix} & \begin{array}{l} \text{Amy} \\ \text{Bill} \end{array} \end{array} \qquad \mathbf{p} = \begin{bmatrix} 30 \\ 20 \\ 35 \\ 50 \end{bmatrix} \begin{array}{l} \text{Peaches} \\ \text{Pears} \\ \text{Apples} \\ \text{Grapefruits} \end{array}$$

The matrix-vector product $\mathbf{Ap} = \begin{bmatrix} 380 \\ 425 \end{bmatrix} = \begin{bmatrix} \$3.80 \\ \$4.25 \end{bmatrix}$ gave the cost of the fruit purchases of Amy and Bill.

EXAMPLE 4

Suppose now that there are two other stores at which Amy and Bill can shop for fruit. Instead of a vector of fruit prices, we now have a matrix **P** of fruit prices (whose first column is the original store's set of prices).

$$\begin{array}{cccc} & \text{Store 1} & \text{Store 2} & \text{Store 3} \\ \mathbf{P} = & \begin{bmatrix} 30 & 25 & 30 \\ 20 & 25 & 25 \\ 35 & 40 & 30 \\ 50 & 60 & 45 \end{bmatrix} & \begin{array}{l} \text{Peaches} \\ \text{Pears} \\ \text{Apples} \\ \text{Grapefruits} \end{array} \end{array}$$

Construct a matrix giving the cost to Amy and Bill of their fruit purchases at each store.

Solution   We need to compute the matrix product **AP**, which is well defined becasue **A** is a 2 × 4 matrix and **P** is a 4 × 3 matrix:

$$\mathbf{AP} = \begin{bmatrix} 5 & 3 & 2 & 2 \\ 3 & 4 & 3 & 3 \end{bmatrix} \begin{bmatrix} 30 & 25 & 30 \\ 20 & 25 & 25 \\ 35 & 40 & 30 \\ 50 & 60 & 45 \end{bmatrix}$$

$$= \begin{bmatrix} 5(30) + 3(20) + 2(35) + 2(50) & 5(25) + 3(25) + 2(40) + 2(60) & 5(30) + 3(25) + 2(30) + 2(45) \\ 3(30) + 4(20) + 3(35) + 3(50) & 3(25) + 4(25) + 3(40) + 3(60) & 3(30) + 4(25) + 3(30) + 3(45) \end{bmatrix}$$

$$= \begin{bmatrix} 380 & 400 & 375 \\ 425 & 475 & 415 \end{bmatrix}.$$

Alternatively, had we entered the cost of fruit in the form $0.30 instead of 30¢, say, we would get

$$\mathbf{AP} = \begin{bmatrix} \$3.80 & \$4.00 & \$3.75 \\ \$4.25 & \$4.75 & \$4.15 \end{bmatrix}.$$

Note that we would have gotten the same results using a calculator to multiply the two matrices.

Powers of Markov Chain Transition Matrices

Next, we consider a more substantial use of matrix multiplication—one that greatly expands the power of the Markov chain model for the stock market introduced in Section 10.2.

In the Markov chain model, the equations for determining the probabilities p_1^+, p_2^+, and p_3^+ of the market going up, going down, or staying the same tomorrow given the probabilities p_1, p_2, and p_3 of the market going up, going down, or staying the same today were

$$p_1^+ = \frac{1}{4}p_1 + \frac{1}{2}p_2 + \frac{1}{4}p_3$$

$$p_2^+ = \frac{1}{2}p_1 + \frac{1}{4}p_2 + \frac{1}{2}p_3$$

$$p_3^+ = \frac{1}{4}p_1 + \frac{1}{4}p_2 + \frac{1}{4}p_3.$$

If

$$\mathbf{p} = \begin{bmatrix} p_1 \\ p_2 \\ p_3 \end{bmatrix} \quad \text{and} \quad \mathbf{p}^+ = \begin{bmatrix} p_1^+ \\ p_2^+ \\ p_3^+ \end{bmatrix}$$

are the vector of today's probabilities and the vector of tomorrow's probabilities, respectively, and the matrix $\mathbf{A}$ of transition probabilities is

Market Today

		Up	Down	Same
Market Tomorrow	Up	$\frac{1}{4}$	$\frac{1}{2}$	$\frac{1}{4}$
	Down	$\frac{1}{2}$	$\frac{1}{4}$	$\frac{1}{2}$
	Same	$\frac{1}{4}$	$\frac{1}{4}$	$\frac{1}{4}$

then the given system of transition probabilities can be written simply as $\mathbf{p}^+ = \mathbf{Ap}$, or, equivalently, as

$$\begin{bmatrix} p_1^+ \\ p_2^+ \\ p_3^+ \end{bmatrix} = \begin{bmatrix} \frac{1}{4} & \frac{1}{2} & \frac{1}{4} \\ \frac{1}{2} & \frac{1}{4} & \frac{1}{2} \\ \frac{1}{4} & \frac{1}{4} & \frac{1}{4} \end{bmatrix} \begin{bmatrix} p_1 \\ p_2 \\ p_3 \end{bmatrix} = \begin{bmatrix} \frac{1}{4}p_1 + \frac{1}{2}p_2 + \frac{1}{4}p_3 \\ \frac{1}{2}p_1 + \frac{1}{4}p_2 + \frac{1}{2}p_3 \\ \frac{1}{4}p_1 + \frac{1}{4}p_2 + \frac{1}{4}p_3 \end{bmatrix}.$$

EXAMPLE 4

Use matrices to find the probabilities of the stock market going up, going down, or staying the same tomorrow if

$$\mathbf{p} = \begin{bmatrix} 0 \\ \frac{1}{2} \\ \frac{1}{2} \end{bmatrix}$$

is the vector of probabilities of the market going up, going down, or staying the same today, as in Example 1 of Section 10.2.

Solution   The vector of probabilities for the stock market tomorrow is $\mathbf{p}^+ = \mathbf{Ap}$, so we get

$$\mathbf{p}^+ = \mathbf{Ap} = \begin{bmatrix} \frac{1}{4} & \frac{1}{2} & \frac{1}{4} \\ \frac{1}{2} & \frac{1}{4} & \frac{1}{2} \\ \frac{1}{4} & \frac{1}{4} & \frac{1}{4} \end{bmatrix} \begin{bmatrix} 0 \\ \frac{1}{2} \\ \frac{1}{2} \end{bmatrix} = \begin{bmatrix} \frac{3}{8} \\ \frac{3}{8} \\ \frac{2}{8} \end{bmatrix}.$$

Alternatively, using a calculator where the entries in the matrix **A** and the vector **p** are given as decimals instead of fractions, we get

$$\mathbf{p}^+ = \mathbf{Ap} = \begin{bmatrix} 0.25 & 0.5 & 0.25 \\ 0.5 & 0.25 & 0.5 \\ 0.25 & 0.25 & 0.25 \end{bmatrix} \begin{bmatrix} 0 \\ 0.5 \\ 0.5 \end{bmatrix} = \begin{bmatrix} 0.375 \\ 0.375 \\ 0.25 \end{bmatrix}.$$

◆

Recall that a Markov chain can be extended farther into the future. Just as tomorrow's probability vector **p**$^+$ can be computed by the matrix expression **p**$^+$ = **Ap**, so too the probability vector **p**$^{++}$ for the day after tomorrow is given by

$$\mathbf{p}^{++} = \mathbf{Ap}^+ = \mathbf{A}(\mathbf{Ap}) = (\mathbf{AA})\mathbf{p} = \mathbf{A}^2\mathbf{p},$$

where we have written **A**^{2} to represent **AA**, the *square* of the matrix **A.** Note that **A**^{2} = **AA** is possible provided that **A** is a square matrix. (Don't confuse "square of a matrix" and "square matrix", where the number of rows equals the number of columns.) Similarly, the vector of probabilities **p**$^{+++}$ three days hence is given by

$$\mathbf{p}^{+++} = \mathbf{Ap}^{++} = \mathbf{A}(\mathbf{A}^2\mathbf{p}) = \mathbf{A}^3\mathbf{p}.$$

In writing these two matrix equations for **p**$^{++}$ and **p**$^{+++}$, we made use of the fact that matrix algebra is an associative operation, so **A**(**Ap**) = (**AA**)**p** = **A**^{2}**p**, and **A**(**A**^{2}**p**) = (**AA**^{2})**p** = **A**^{3}**p**, where **A**^{3} can be thought of as **AAA.**

◆ **EXAMPLE 5**

Compute **A**^{2} and **A**^{3} for the stock market Markov transition matrix **A.**

Solution   We first do the calculations by hand:

$$\mathbf{A}^2 = \begin{bmatrix} \frac{1}{4} & \frac{1}{2} & \frac{1}{4} \\ \frac{1}{2} & \frac{1}{4} & \frac{1}{2} \\ \frac{1}{4} & \frac{1}{4} & \frac{1}{4} \end{bmatrix} \begin{bmatrix} \frac{1}{4} & \frac{1}{2} & \frac{1}{4} \\ \frac{1}{2} & \frac{1}{4} & \frac{1}{2} \\ \frac{1}{4} & \frac{1}{4} & \frac{1}{4} \end{bmatrix}$$

$$= \begin{bmatrix} \frac{1}{4}\cdot\frac{1}{4} + \frac{1}{2}\cdot\frac{1}{2} + \frac{1}{4}\cdot\frac{1}{4} & \frac{1}{4}\cdot\frac{1}{2} + \frac{1}{2}\cdot\frac{1}{4} + \frac{1}{4}\cdot\frac{1}{4} & \frac{1}{4}\cdot\frac{1}{4} + \frac{1}{2}\cdot\frac{1}{2} + \frac{1}{4}\cdot\frac{1}{4} \\ \frac{1}{2}\cdot\frac{1}{4} + \frac{1}{4}\cdot\frac{1}{2} + \frac{1}{2}\cdot\frac{1}{4} & \frac{1}{2}\cdot\frac{1}{2} + \frac{1}{4}\cdot\frac{1}{4} + \frac{1}{2}\cdot\frac{1}{4} & \frac{1}{2}\cdot\frac{1}{4} + \frac{1}{4}\cdot\frac{1}{2} + \frac{1}{2}\cdot\frac{1}{4} \\ \frac{1}{4}\cdot\frac{1}{4} + \frac{1}{4}\cdot\frac{1}{2} + \frac{1}{4}\cdot\frac{1}{4} & \frac{1}{4}\cdot\frac{1}{2} + \frac{1}{4}\cdot\frac{1}{4} + \frac{1}{4}\cdot\frac{1}{4} & \frac{1}{4}\cdot\frac{1}{4} + \frac{1}{4}\cdot\frac{1}{2} + \frac{1}{4}\cdot\frac{1}{4} \end{bmatrix}$$

$$= \begin{bmatrix} \frac{3}{8} & \frac{5}{16} & \frac{3}{8} \\ \frac{3}{8} & \frac{7}{16} & \frac{3}{8} \\ \frac{1}{4} & \frac{1}{4} & \frac{1}{4} \end{bmatrix}.$$

Similarly, we compute **A**^{3} as

$$\mathbf{A}^3 = \mathbf{AA}^2 = \begin{bmatrix} \frac{1}{4} & \frac{1}{2} & \frac{1}{4} \\ \frac{1}{2} & \frac{1}{4} & \frac{1}{2} \\ \frac{1}{4} & \frac{1}{4} & \frac{1}{4} \end{bmatrix} \begin{bmatrix} \frac{3}{8} & \frac{5}{16} & \frac{3}{8} \\ \frac{3}{8} & \frac{7}{16} & \frac{3}{8} \\ \frac{1}{4} & \frac{1}{4} & \frac{1}{4} \end{bmatrix}$$

$$= \begin{bmatrix} \frac{1}{4}\cdot\frac{3}{8} + \frac{1}{2}\cdot\frac{3}{8} + \frac{1}{4}\cdot\frac{1}{4} & \frac{1}{4}\cdot\frac{5}{16} + \frac{1}{2}\cdot\frac{7}{16} + \frac{1}{4}\cdot\frac{1}{4} & \frac{1}{4}\cdot\frac{3}{8} + \frac{1}{2}\cdot\frac{3}{8} + \frac{1}{4}\cdot\frac{1}{4} \\ \frac{1}{2}\cdot\frac{3}{8} + \frac{1}{4}\cdot\frac{3}{8} + \frac{1}{2}\cdot\frac{1}{4} & \frac{1}{2}\cdot\frac{5}{16} + \frac{1}{4}\cdot\frac{7}{16} + \frac{1}{2}\cdot\frac{1}{4} & \frac{1}{2}\cdot\frac{3}{8} + \frac{1}{4}\cdot\frac{3}{8} + \frac{1}{2}\cdot\frac{1}{4} \\ \frac{1}{4}\cdot\frac{3}{8} + \frac{1}{4}\cdot\frac{3}{8} + \frac{1}{4}\cdot\frac{1}{4} & \frac{1}{4}\cdot\frac{5}{16} + \frac{1}{4}\cdot\frac{7}{16} + \frac{1}{4}\cdot\frac{1}{4} & \frac{1}{4}\cdot\frac{3}{8} + \frac{1}{4}\cdot\frac{3}{8} + \frac{1}{4}\cdot\frac{1}{4} \end{bmatrix}$$

$$= \begin{bmatrix} \frac{11}{32} & \frac{23}{64} & \frac{11}{32} \\ \frac{13}{32} & \frac{25}{64} & \frac{13}{32} \\ \frac{1}{4} & \frac{1}{4} & \frac{1}{4} \end{bmatrix}.$$

Alternatively, we can find these matrices by using the matrix capabilities of a calculator or computer package, though likely with the entries in decimal form. In that case, we would define the 3×3 matrix $\mathbf{A}$ and then refer to it by name to form

$$\mathbf{A}^2 = \mathbf{A}\wedge 2 = \begin{bmatrix} 0.375 & 0.3125 & 0.375 \\ 0.375 & 0.4375 & 0.375 \\ 0.25 & 0.25 & 0.25 \end{bmatrix}$$

and

$$\mathbf{A}^3 = \mathbf{A}\wedge 3 = \begin{bmatrix} 0.34375 & 0.359375 & 0.34375 \\ 0.40625 & 0.390625 & 0.40625 \\ 0.25 & 0.25 & 0.25 \end{bmatrix}.$$

You can verify that these decimal entries are equivalent to the fractions we calculated by hand initially.

◆

The entries in $\mathbf{A}^2$ are the transition probabilities for 2 days from now and the entries in $\mathbf{A}^3$ are the transition probabilities for 3 days from now. For instance, the value $\frac{3}{8} = 0.375$ in position $(2, 1)$ of $\mathbf{A}^2$ (the entry in the second row, first column) means that, if we are now in State 1 (Up—first column), the chance is $\frac{3}{8}$ that in 2 days we will be in State 2 (Down—second row). Similarly, the value of $13/32 = 0.40625$ in entry $(2, 1)$ of $\mathbf{A}^3$ indicates that, if the market is now up, the probability is $13/32$ that in 3 days it will go down.

The values we obtained in computing $\mathbf{A}^2$ and $\mathbf{A}^3$ look reasonable. In particular, the numbers in each column of $\mathbf{A}^2$ and $\mathbf{A}^3$ sum to 1 (the sum of all probabilities for the market on any given day must be 1). Note that all the entries in each of the three rows of $\mathbf{A}^3$ have roughly the same numerical value; the entries in the first row are slightly larger than $\frac{1}{3}$, those in the second row are all about 0.40, and all the entries in the last row are 0.25.

EXAMPLE 6

If the probabilities of the stock market going up, going down, or staying the same today are given by the vector

$$\mathbf{p} = \begin{bmatrix} 0 \\ \frac{1}{2} \\ \frac{1}{2} \end{bmatrix},$$

use matrices to find the probabilities $\mathbf{p}^{++}$ of the market going up, going down, or staying the same the day after tomorrow and the probabilities $\mathbf{p}^{+++}$ for the day after that.

Solution   We know that $\mathbf{p}^{++} = \mathbf{A}^2\mathbf{p}$ and $\mathbf{p}^{+++} = \mathbf{A}^3\mathbf{p}$. Multiplying, we get

$$\mathbf{p}^{++} = \mathbf{A}^2\mathbf{p} = \begin{bmatrix} 0.34375 \\ 0.40625 \\ 0.25 \end{bmatrix} \quad \text{and} \quad \mathbf{p}^{+++} = \mathbf{A}^3\mathbf{p} = \begin{bmatrix} 0.3515625 \\ 0.3984375 \\ 0.25 \end{bmatrix}.$$

You can easily verify that these entries are the same values we obtained in Section 10.2.

◆

If you refer back to the table that gave the market probabilities over many days near the end of Example 2 in Section 10.2, you will observe that the probabilities shown there are similar to the entries *in the columns* of A^3. That is, the probabilities after 3 days are actually fairly close to the long-term probabilities, so this Markov chain converges to a limiting state rather quickly.

Example 6 illustrates how, with concise notation, matrix algebra allows us to express quite complex expressions.

We can visualize what happens in a Markov chain graphically in the case of a two-by-two transition matrix—say, $A = \begin{bmatrix} 0.6 & 0.3 \\ 0.4 & 0.7 \end{bmatrix}$. (Again, notice that the sum of the entries in each column is 1.) Let's start with an initial vector of probabilities $\mathbf{p} = \mathbf{p}_0 = \begin{bmatrix} 0.75 \\ 0.25 \end{bmatrix}$. Any vector, including this vector of probabilities, can be interpreted geometrically, as shown in Figure 10.33. When we pre-multiply the vector $\mathbf{p}_0$ by the transition matrix A, we obtain another vector of probabilities $\mathbf{p}^+ = \mathbf{p}_1 = \begin{bmatrix} 0.525 \\ 0.475 \end{bmatrix}$, which we can interpret geometrically, as shown in Figure 10.33. Note that the resulting vector $\mathbf{p}_1$ points in a direction quite different from that of the initial vector $\mathbf{p}_0$. We can think of the matrix A as transforming the probability vector $\mathbf{p}_0$ into the vector $\mathbf{p}_1 = A\mathbf{p}_0$.

When we multiply the vector $\mathbf{p}_1$ by the matrix A to form the next probability vector $\mathbf{p}_2 = A\mathbf{p}_1 = \begin{bmatrix} 0.4575 \\ 0.5425 \end{bmatrix}$, we get a vector $\mathbf{p}_2 = A^2\mathbf{p}_0$ that points in still another direction. What happens when we continue the Markov process to get $\mathbf{p}_3 = A^3\mathbf{p}_0$, $\mathbf{p}_4 = A^4\mathbf{p}_0, \ldots$? (We have moved from using the notation $\mathbf{p}^+$, $\mathbf{p}^{++}$, and so on, to use subscript notation because the use of multiple $+$'s quickly becomes too unwieldy.) In the following table, we show the results of continuing this process numerically for the two components p_1 and p_2 for each successive probability vector $\mathbf{p}$. As in Example 6, the probabilities seem to converge to a pair of limiting values—approximately 0.42857, which is about 3/7, and approximately 0.57143, or 4/7.

n	0	1	2	3	4	5	6	7	...	
p_1	0.75	0.525	0.4575	0.43725	0.43118	0.42935	0.42881	0.42864	...	0.42857
p_2	0.25	0.475	0.5425	0.56275	0.56883	0.57065	0.57119	0.57136	...	0.57143

Figure 10.34 shows the corresponding geometric behavior. Note how the sequence of vectors also converges, getting closer and closer to a single limiting vector.

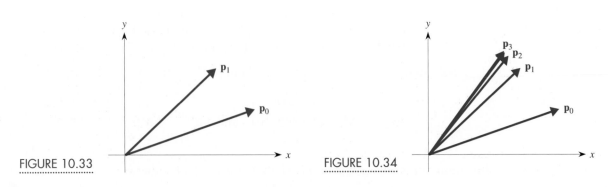

FIGURE 10.33

FIGURE 10.34

If we start with any other vector of probabilities, say, $\mathbf{p}_0 = \begin{bmatrix} 0.2 \\ 0.8 \end{bmatrix}$, then the corresponding sequence of vectors will also converge to this same limiting vector $\begin{bmatrix} \frac{3}{7} \\ \frac{4}{7} \end{bmatrix}$. This special limiting vector is called an **eigenvector** of the matrix **A**. A similar type of convergence occurs with most other transition matrices, whether a two-by-two matrix or larger.

A Geometric View of Powers of a Matrix

We have shown that any matrix of the form $\mathbf{R} = \begin{bmatrix} \cos\theta & -\sin\theta \\ \sin\theta & \cos\theta \end{bmatrix}$ acts as a rotation matrix to rotate a vector **v** through an angle θ. We now examine the effect of applying $\mathbf{R}^2$ to a vector.

EXAMPLE 7

Show that the matrix $\mathbf{R}^2$ is a rotation matrix that will rotate any vector through an angle 2θ.

Solution   We have

$$\mathbf{R}^2 = \begin{bmatrix} \cos\theta & -\sin\theta \\ \sin\theta & \cos\theta \end{bmatrix}\begin{bmatrix} \cos\theta & -\sin\theta \\ \sin\theta & \cos\theta \end{bmatrix}$$

$$= \begin{bmatrix} \cos^2\theta - \sin^2\theta & -\sin\theta\cos\theta - \sin\theta\cos\theta \\ \sin\theta\cos\theta + \sin\theta\cos\theta & -\sin^2\theta + \cos^2\theta \end{bmatrix}$$

$$= \begin{bmatrix} \cos^2\theta - \sin^2\theta & -2\sin\theta\cos\theta \\ 2\sin\theta\cos\theta & \cos^2\theta - \sin^2\theta \end{bmatrix}.$$

If we now apply the double angle identities for the sine and cosine from Chapter 8, we get $\mathbf{R}^2 = \begin{bmatrix} \cos 2\theta & -\sin 2\theta \\ \sin 2\theta & \cos 2\theta \end{bmatrix}$, so that $\mathbf{R}^2$ indeed is a rotation matrix with an associated angle 2θ.

Problems

1. Let

$$\mathbf{A} = \begin{bmatrix} 0 & 2 \\ 1 & 4 \end{bmatrix}, \qquad \mathbf{B} = \begin{bmatrix} 4 & 2 \\ 1 & 1 \end{bmatrix},$$

$$\mathbf{C} = \begin{bmatrix} 3 & 1 \\ -1 & -3 \end{bmatrix}.$$

Compute the matrix products.

a. **AB**   b. **BA**
c. **CB**   d. **BC**
e. **AC**   f. $\mathbf{A}^2$
g. $\mathbf{B}^2$

2. For the matrices in Problem 1, compute the sums and linear combinations of matrices.

a. **A + B**   b. **B + C**
c. **A + B + C**   d. **2A + 3C**

3. Let

$$\mathbf{A} = \begin{bmatrix} 5 & 0 \\ 1 & 4 \end{bmatrix}, \qquad \mathbf{B} = \begin{bmatrix} 5 & -1 \\ 0 & 2 \end{bmatrix},$$

$$\mathbf{C} = \begin{bmatrix} 1 & 2 \\ 3 & 4 \\ 5 & 6 \end{bmatrix}, \qquad \mathbf{D} = \begin{bmatrix} 1 & 5 & -2 \\ 3 & 0 & 2 \end{bmatrix}.$$

Compute the products, if possible.

a. **AB**   b. **CB**   c. **BC**
d. **AD**   e. **DA**   f. **CD**
g. **DC**   h. $\mathbf{C}^2$

4. For **A**, **B**, and **C** in Problem 1,

   a. compute **AB** and then compute **(AB)C**.
   b. compute **BC** and then compute **A(BC)**. Does **(AB)C** equal **A(BC)**?

5. For **B**, **C**, and **D** in Problem 3,

   a. compute **CB** and then compute **(CB)D**.
   b. compute **BD** and then compute **C(BD)**. Does **(CB)D** equal **C(BD)**?

6. Let

$$\mathbf{A} = \begin{bmatrix} -1 & 3 & -2 \\ 3 & 4 & -1 \\ 4 & 0 & 1 \\ 4 & 0 & 1 \end{bmatrix}, \quad \mathbf{B} = \begin{bmatrix} 0 & 1 & 0 \\ 1 & 0 & 1 \\ 0 & 1 & 0 \\ 1 & 0 & 1 \end{bmatrix},$$

$$\mathbf{C} = \begin{bmatrix} 4 & 1 & 0 & 3 \\ -2 & 1 & 6 & 3 \\ 3 & 0 & 2 & 0 \end{bmatrix}.$$

   Compute the products, if possible. If not possible, explain why.

   a. **AB**   b. **BA**   c. **AC**
   d. **CA**   e. **BC**   f. **CB**
   g. $\mathbf{A}^2$

7. Pre-multiply each matrix by the matrix **A** in Problem 6 and in each case describe how the columns of the product matrix compare to the columns of **A**.

   a. $\begin{bmatrix} 0 & 0 & 1 \\ 0 & 1 & 0 \\ 1 & 0 & 0 \end{bmatrix}$ b. $\begin{bmatrix} 0 & 1 & 0 \\ 0 & 0 & 1 \\ 1 & 0 & 0 \end{bmatrix}$

   c. $\begin{bmatrix} 1 & 0 & 0 \\ 0 & 3 & 0 \\ 0 & 0 & 2 \end{bmatrix}$

8. For **A**, **B**, and **C** in Problem 6, compute the entry in the third row, third column in **(BC)A**. Explain why you do not have to multiply **BC** fully to determine this entry in **(BC)A**.

9. Perform the matrix multiplication for Amy's and Bill's fruit purchases at the three different stores if the matrix of Amy's and Bill's needs are

	Peaches	Pears	Apples	Grapefruits
Amy	7	2	4	5
Bill	2	5	1	8

10. Suppose that you have four robots: Supremo, Ultramatic, Maximus, and Gandalf, and three types of jobs: Job 1 (washing clothes), Job 2 (walking the dog), and Job 3 (doing a student's homework assignment). There are three families, the Joneses, the Smiths, and the Madonnas. Matrix **A** gives the times in hours it takes each robot to do each job. Matrix **B** tells how many jobs of each type are required by each family. Compute with **A** and **B** to find a matrix showing how long it will take each robot to do each family's set of jobs.

Matrix of Times

	Job 1	Job 2	Job 3	
A =	3	4	2	Supremo
	5	7	3	Ultramatic
	1	2	1	Maxiumus
	3	3	3	Gandalf

Matrix of Jobs

	Jones	Smith	Madonna	
B =	6	2	4	Job 1
	8	5	4	Job 2
	10	5	4	Job 3

11. Suppose that you are given the following matrices involving the costs of fruit at different stores, the amounts of fruit that professors and engineers typically want, and the number of each type of person in two towns.

	Store 1	Store 2	
A =	0.15	0.20	Bananas
	0.25	0.15	Peaches
	0.20	0.25	Pears

	Bananas	Peaches	Pears	
B =	6	12	4	Professors
	6	8	5	Engineers

	Professors	Engineers	
C =	2000	800	Town 1
	1500	1200	Town 2

   a. Compute a matrix product to find how much each type of person's fruit purchases cost at each store.
   b. Compute a matrix product to find how many of each fruit will be purchased in each town.
   c. Compute a matrix product to find how much was spent by each town at each store.

12. Consider the following population model for the numbers of goats (G) and sheep (S) from year to year.

$$G^+ = 2G + 2S$$
$$S^+ = G + 3S$$

Let $\mathbf{x}_0 = \begin{bmatrix} G_0 \\ S_0 \end{bmatrix}$ be the initial vector and $\mathbf{x}_n = \begin{bmatrix} G_n \\ S_n \end{bmatrix}$ be the vector of goats and sheep after n years. Let $\mathbf{A}$ be the matrix of coefficients in this system. (a) Write an expression for $\mathbf{x}_n$ in terms of $\mathbf{A}$ and $\mathbf{x}_0$. (b) Find $\mathbf{A}^2$ and $\mathbf{A}^3$, and use them and your formula to determine $\mathbf{x}_2$ and $\mathbf{x}_3$, if the starting population, in thousands, is $\mathbf{x}_0 = \begin{bmatrix} 2 \\ 5 \end{bmatrix}$. (Check your answer by applying the model equations for 2 and 3 years.) (c) What does the model predict for the two populations in 10 years?

13. Consider the following population model for the numbers of goats (G), sheep (S), and bears (B) from year to year.

$$G^+ = G + S + B$$
$$S^+ = G + 2S - B$$
$$B^+ = 2G - S + B$$

Let

$$\mathbf{x_0} = \begin{bmatrix} G_0 \\ S_0 \\ B_0 \end{bmatrix}$$

be the initial vector and let $\mathbf{x}_n$ denote the vector of goats, sheep, and bears after n years. Let $\mathbf{A}$ be the matrix of coefficients in this system. (a) Write an expression for $\mathbf{x}_n$ in terms of $\mathbf{A}$ and $\mathbf{x}_0$. (b) Find $\mathbf{A}^2$ and $\mathbf{A}^3$. (c) Determine $\mathbf{x}_2$ and $\mathbf{x}_3$, if the starting population, in hundreds, is

$$\mathbf{x_0} = \begin{bmatrix} 1 \\ 1 \\ 2 \end{bmatrix}.$$

(Check your answer by applying the model equations for 2 and 3 years.) (d) How long does it take for one of the populations to die out?

14. The copy machine at the student union breaks down as follows. If it is working today, it has a 70% chance of working tomorrow (and a 30% chance of breaking down). If the copy machine is broken today, it has a 50% chance of working tomorrow (and a 50% chance of being broken again). Write the Markov chain transition matrix $\mathbf{A}$ for this scenario.

a. Compute $\mathbf{A}^2$. What probability does the entry in position $(1, 2)$ in $\mathbf{A}^2$ represent?

b. Compute $\mathbf{A}^3$. If the copy machine is working today, what is the probability that it will be working in 3 days?

c. If it is working today, what is the probability that it will be working a week from today?

15. Consider a weather Markov chain with 2 states, sunny and cloudy. If today is sunny, there is a $\frac{3}{4}$ probability that tomorrow will be sunny and a $\frac{1}{4}$ probability that tomorrow will be cloudy. If today is cloudy, there is a $\frac{1}{4}$ probability that tomorrow will be sunny and a $\frac{3}{4}$ probability that tomorrow will be cloudy. Write the transition matrix $\mathbf{A}$ for this Markov chain.

a. Compute $\mathbf{A}^2$. What probability does the entry in position $(1, 2)$ in $\mathbf{A}^2$ represent?

b. Compute $\mathbf{A}^3$. If today is cloudy, what is the probability that it will be sunny in three days?

c. If today is cloudy, what is the probability that it will be sunny a week from today?

16. Consider the transition matrix $\mathbf{A} = \begin{bmatrix} 0.2 & 0.4 \\ 0.8 & 0.6 \end{bmatrix}$ and an initial vector $\mathbf{p} = \begin{bmatrix} 0.7 \\ 0.3 \end{bmatrix}$.

a. Calculate $\mathbf{p}^+$ and $\mathbf{p}^{++}$ by hand.

b. Use the matrix features of your calculator to calculate the next four iterates $\mathbf{p}^{+++}$, $\mathbf{p}^{++++}$, and so on, corresponding to $\mathbf{A}^3$, $\mathbf{A}^4$, $\mathbf{A}^5$, and $\mathbf{A}^6$. Create a table listing the entries in the vectors that result. Do they appear to be converging?

c. Plot the vectors you found in parts (a) and (b). Do they appear to be converging?

d. Repeat parts (a)–(c) if the initial vector is $\mathbf{p} = \begin{bmatrix} 0.5 \\ 0.5 \end{bmatrix}$.

e. Repeat parts (a)–(c) if the initial vector is $\mathbf{p} = \begin{bmatrix} 0.25 \\ 0.75 \end{bmatrix}$.

f. What do you observe about the three sequences of vectors from parts (b), (d), and (e)?

17. In Problem 22 of Section 10.3, we described how to write a polynomial as a scalar product of a vector of coefficients and a vector of power functions. Suppose that you now have two cubic polynomials $P(x) = x^3 + 4x^2 - 7x + 2$ and $Q(x) = 4x^3 - 5x^2 + 8x + 17$.

a. Express the sum of the two polynomials in terms of vectors.

b. Express the difference of the two polynomials in terms of vectors.

18. Explain why you can't have a power of a non-square matrix **A**.

19. Prove that, if **R** is the 2×2 rotation matrix associated with an angle θ, then $\mathbf{R}^3$ is a rotation matrix associated with an angle 3θ. (*Hint*: Write $\mathbf{R}^3 = \mathbf{R}\mathbf{R}^2$ and use appropriate trigonometric identities to simplify the result of the multiplication.)

20. The matrix $\mathbf{R} = \begin{bmatrix} 0.9272 & -0.3746 \\ 0.3746 & 0.9272 \end{bmatrix}$ is a rotation matrix. What is the associated angle of rotation?

21. Consider the matrix $\mathbf{R} = \begin{bmatrix} \cos 1° & -\sin 1° \\ \sin 1° & \cos 1° \end{bmatrix}$.

Describe the effect of successively applying **R**, $\mathbf{R}^2$, $\mathbf{R}^3$, $\mathbf{R}^4, \ldots, \mathbf{R}^{90}$ to any nonzero vector **v**.

10.5 Gaussian Elimination

In this section we develop a procedure known as **Gaussian elimination** for solving any system of linear equations. The elimination method was devised by Carl Friedrich Gauss in about 1820 to solve systems of linear equations in astronomical and land-surveying computations. For our purposes here, when we speak of a system of linear equations, we assume that *the number of equations equals the number of variables*.

Suppose that we start with the system of linear equations

$$3x - 5y + 4z = 10$$
$$3x - 2y + 5z = 11$$
$$6x + 2y - 2z = 10.$$

Gaussian elimination involves two stages. The first stage transforms the given system of equations, or equivalently the associated matrix of coefficients, into *upper triangular form*, with only 0's below the *main diagonal* of the matrix:

$$3x - 5y + 4z = 10$$
$$3y + z = 1$$
$$z = 1.$$

(We show how to do this shortly.) Once we have transformed the original system of equations to upper triangular form, solving it is quite simple. The second stage of the process uses "back substitution" to obtain values for the variables. That is, knowing from the third equation that $z = 1$, we can solve for y in the second equation:

$$3y + (1) = 1 \quad \text{so that} \quad y = 0.$$

Now, knowing y and z, we can solve for x from the first equation:

$$3x - 5(0) + 4(1) = 10 \quad \text{so that} \quad x = 2.$$

We can simplify this procedure by using vectors and matrices. Instead of using x, y, and z as the variables, we use x_1, x_2, and x_3. We then write the upper triangular form for the preceding system of equations in matrix form as $\mathbf{A}\mathbf{x} = \mathbf{b}$, where

$$\mathbf{A} = \begin{bmatrix} 3 & -5 & 4 \\ 0 & 3 & 1 \\ 0 & 0 & 1 \end{bmatrix}, \quad \mathbf{x} = \begin{bmatrix} x_1 \\ x_2 \\ x_3 \end{bmatrix}, \quad \text{and} \quad \mathbf{b} = \begin{bmatrix} 10 \\ 1 \\ 1 \end{bmatrix}.$$

To apply Gaussian elimination to any system of linear equations, we transform the original coefficient matrix into upper triangular form by applying the following three *elementary row operations* repeatedly.

Elementary Row Operations

♦ Multiply or divide any row of a matrix (or an equation) by a nonzero number.

♦ Add a multiple of one row (or equation) to another row (or equation).

♦ Interchange two rows (or two equations).

Realize that performing any of these operations on the rows of a matrix is equivalent to performing the same operation on the original equations. Becasue the operations do not materially change the equations, the equivalent operations do not materially change the matrix, so we will get the same solution.

Because the numbers, not the variables, are what matter in the equations, we work with the coefficient matrix **A** extended by a column for the right-hand side vector **b.** This matrix $[A \,|\, b]$ of coefficients along with the right-hand side vector is called the *augmented coefficient matrix.* For instance, if the system of equations is

$$2x + 3y = 7$$
$$4x - 5y = 3$$

so that

$$A = \begin{bmatrix} 2 & 3 \\ 4 & -5 \end{bmatrix} \quad \text{and} \quad b = \begin{bmatrix} 7 \\ 3 \end{bmatrix},$$

then the augmented matrix is

$$[A \,|\, b] = \begin{bmatrix} 2 & 3 & | & 7 \\ 4 & -5 & | & 3 \end{bmatrix}.$$

In Example 1 we show the steps involved in applying Gaussian elimination to both the system of linear equations and the associated augmented matrix.

◆ **EXAMPLE 1**

Use Gaussian elimination to solve the following system of two equations in two unknowns.

$$2x + y = 7 \tag{1}$$
$$x - 2y = -4 \tag{2}$$

Solution This system is equivalent to the augmented matrix

$$\begin{bmatrix} 2 & 1 & | & 7 \\ 1 & -2 & | & -4 \end{bmatrix}.$$

To eliminate the *x*-term from Equation (2), we add $-\frac{1}{2}$ times Equation (1) to Equation (2) (the second elementary row operation). The result is a new second Equation (2′):

(2)	$x - 2y = -4$
$-\frac{1}{2}(1)$	$-x - 0.5y = -3.5$
$(2') = (2) - \frac{1}{2}(1)$	$0 - 2.5y = -7.5$

Equivalently, we perform the identical operation on the corresponding rows of the augmented matrix: Add $-\frac{1}{2}$ times the first row to the second row. Our new system of equations and the new augmented matrix become

$$2x + y = 7 \tag{1}$$
$$-2.5y = -7.5, \tag{2'}$$

and

$$\begin{bmatrix} 2 & 1 & \bigm| & 7 \\ 0 & -2.5 & \bigm| & -7.5 \end{bmatrix}.$$

Note that any solution to Equations (1) and (2) is also a solution to Equations (1) and (2') because we can reverse the step that created Equation (2'). That is, $(2') = (2) - \frac{1}{2}(1)$ implies that $(2) = (2') + \frac{1}{2}(1)$. Thus Equation (2) is formed from Equation (2') and a multiple of Equation (1), and so any solution to Equations (1) and (2') is also a solution to Equations (1) and (2). But Equation (2') is trivial to solve, and gives $y = 3$. Substituting $y = 3$ into Equation (1), we get

$$2x + 3 = 7, \quad \text{so} \quad x = \frac{(7 - 3)}{2} = 2.$$

Verify that $x = 2$ and $y = 3$ satisfy the *two* original Equations (1) and (2). ◆

When solving a system of linear equations, you should always check your result by substituting the values for the variables into the original equations, not the transformed equations, in case you made a mathematical error along the way.

Note also that the work we did on the system of equations exactly parallels what happens with the augmented matrix. Eventually, we will dispense with the equations altogether and work exclusively with the augmented matrix because it eliminates writing and keeping track of the variables at every step.

A Geometric Interpretation

If there are only two equations in two unknowns, you can also solve the system graphically by plotting the two lines. The first equation represents all points on one line and the second equation represents all points on the second line. Thus the solution to the system of equations corresponds to the point of intersection and you can approximate this point with your graphing calculator.

There is a comparable geometric interpretation for a system of three linear equations in three unknowns. Just as an equation of the form $ax + by = d$ represents a line in the two-dimensional coordinate plane, an equation of the form $ax + by + cz = d$ represents a plane in three-dimensional space. When you have three equations in three unknowns, you actually have three different planes in space, as suggested in Figure 10.35. Visualize, for instance, the ceiling, the wall in front of you, and the wall to your left. These three planes intersect at a single point in the upper corner of the room to your left. This point of intersection also is the solution of the system of three equations. Unfortunately, graphing calculators cannot yet use this geometric interpretation to solve such a system of equations.

Also, there can be some complications: You know that two lines can be parallel, so the resulting system of two equations in two unknowns will not have a solution. Similarly, three planes don't necessarily have a common point of intersection. Visualize

FIGURE 10.35

the ceiling, the floor, and the wall in front of you—they don't meet at a single point. Or, picture a long triangular prism made up of three flat sides that likewise have no single point of intersection. Alternatively, three planes can all pass through a common line of intersection, as with a revolving door. In that case the system of linear equations has infinitely many solutions. We consider such cases later in Example 3.

The Clothes Production Model and Gaussian Elimination

Recall the clothes production model introduced in Section 10.2 with three clothing factories whose raw material production levels had to be chosen to meet the demands for vests, pants, and coats.

$$\text{Vests:} \quad 20x_1 + 4x_2 + 4x_3 = 500 \tag{3}$$

$$\text{Pants:} \quad 10x_1 + 14x_2 + 5x_3 = 850 \tag{4}$$

$$\text{Coats:} \quad 5x_1 + 5x_2 + 12x_3 = 1000 \tag{5}$$

EXAMPLE 2

Set up the corresponding augmented matrix and use Gaussian elimination to solve the system of equations.

Solution The augmented matrix is

$$\begin{bmatrix} 20 & 4 & 4 & | & 500 \\ 10 & 14 & 5 & | & 850 \\ 5 & 5 & 12 & | & 1000 \end{bmatrix}.$$

To solve this system, we first use multiples of Equation (3) to eliminate x_1 from Equations (4) and (5). Because $10x_1$ is half of $20x_1$ in Equation (3), we add $-\frac{1}{2}$ times Equation (3) to Equation (4) to cancel the terms $-\frac{1}{2}(20x_1)$ and $10x_1$ and so get a new second Equation $(4')$.

$$
\begin{array}{ll}
(4) & 10x_1 + 14x_2 + 5x_3 = 850 \\
-\frac{1}{2}(3) & \underline{-10x_1 - 2x_2 - 2x_3 = -250} \\
(4') = (4) + (-\frac{1}{2})(3) &   0 + 12x_2 + 3x_3 = 600
\end{array}
$$

Similarly, we add $-\frac{1}{4}$ times Equation (3) to Equation (5) to eliminate the x_1-term from Equation (5) and get a new Equation (5′).

$$
\begin{array}{ll}
(5) & 5x_1 + 5x_2 + 12x_3 = 1000 \\
-\frac{1}{4}(3) & -5x_1 - 1x_2 - 1x_3 = -125 \\
\hline
(5') = (5) + (-\frac{1}{4})(3) & 4x_2 + 11x_3 = 875
\end{array}
$$

Our new system of equations is now

$$
\begin{array}{ll}
20x_1 + 4x_2 + 4x_3 = 500 & \textbf{(3)} \\
12x_2 + 3x_3 = 600 & \textbf{(4')} \\
4x_2 + 11x_3 = 875, & \textbf{(5')}
\end{array}
$$

and the corresponding augmented matrix is

$$
\left[\begin{array}{ccc|c}
20 & 4 & 4 & 500 \\
0 & 12 & 3 & 600 \\
0 & 4 & 11 & 875
\end{array}\right].
$$

Next we use Equation $(4')$ to eliminate the x_2-term from Equation $(5')$ by adding $-\frac{1}{3}$ times Equation $(4')$ to Equation $(5')$. We obtain a new Equation $(5'')$.

$$
\begin{array}{ll}
(5') & 4x_2 + 11x_3 = 875 \\
-\frac{1}{3}(4') & -4x_2 - 1x_3 = -200 \\
\hline
(5'') = (5') + (-\frac{1}{3})(4') & 10x_3 = 675
\end{array}
$$

Our final system of equations in upper triangular form then is

$$
\begin{array}{ll}
20x_1 + 4x_2 + 4x_3 = 500 & \textbf{(3)} \\
12x_2 + 3x_3 = 600 & \textbf{(4')} \\
10x_3 = 675 & \textbf{(5'')}
\end{array}
$$

with the associated augmented matrix

$$
\left[\begin{array}{ccc|c}
20 & 4 & 4 & 500 \\
0 & 12 & 3 & 600 \\
0 & 0 & 10 & 675
\end{array}\right].
$$

Any solution to the original system is also a solution to the new system. Furthermore, reversing the steps used in going from the original system to the final system (so that the original system is formed from linear combinations of the equations in the final system), we see that any solution to the new system is also a solution to the original system. The final system is in upper triangular form, so we can solve it using back substitution. From Equation $(5'')$, we have

$$
x_3 = \frac{675}{10} = 67.5.
$$

Substituting this value into Equation $(4')$ yields

$$
12x_2 + 3(67.5) = 600, \quad \text{so} \quad 12x_2 = 600 - 202.5, \quad \text{or} \quad x_2 = 33.125.
$$

Next substituting $x_3 = 67.5$ and $x_2 = 33.125$ into Equation (3) gives

$$20x_1 + 4(33.125) + 4(67.5) = 500$$

so that

$$x_1 = \frac{500 - 132.5 - 270}{20} = 4.875.$$

Therefore the vector of the number of rolls of cloth needed by each of the three clothing factories is

$$\begin{bmatrix} x_1 \\ x_2 \\ x_3 \end{bmatrix} = \begin{bmatrix} 4.875 \\ 3.125 \\ 67.5 \end{bmatrix} = \begin{bmatrix} 4\frac{7}{8} \\ 33\frac{1}{8} \\ 67\frac{1}{2} \end{bmatrix}.$$

Clearly in practice, cloths come in full rolls, so a more realistic solution would involve rounding up to the next full roll of cloth.

Incidentally, you can apply the three elementary row operations in many different ways to solve a particular system of equations. However, although the steps you use might differ from those of others solving the same problem, you should all obtain the same solution in the end, if none of you have made any mathematical errors.

Most graphing calculators have a built-in routine for applying Gaussian elimination to any set of linear equations. Typically, you would enter the coefficients, press the SOLVE key, and the calculator will respond with the solution or will indicate that either there is not a unique solution or no solution exists. (The specific key operations differ from one machine to another, so check your manual for details.) From one point of view, this approach is extremely simple because you get the answer instantly. However, it does have the drawback of not letting you see *how* the method works or understand what went wrong when the method fails, as we discuss below.

Systems of Linear Equations with Multiple Solutions

In our discussion regarding the geometric interpretation of systems of three equations in three unknowns, we mentioned that it is possible that three planes can have many points of intersection, as in a revolving door. We illustrate this type of situation in Example 3 where we change the number on the right-hand side of the third equation in Example 2 from 1000 to 325 and the coefficient of x_3 from 12 to 2.

EXAMPLE 3

Apply Gaussian elimination to the system of linear equations

$$20x_1 + 4x_2 + 4x_3 = 500 \tag{6}$$
$$10x_1 + 14x_2 + 5x_3 = 850 \tag{7}$$
$$5x_1 + 5x_2 + 2x_3 = 325. \tag{8}$$

Solution   The corresponding augmented matrix is

$$\left[\begin{array}{ccc|c} 20 & 4 & 4 & 500 \\ 10 & 14 & 5 & 850 \\ 5 & 5 & 2 & 325 \end{array} \right].$$

After eliminating x_1 from Equations (7) and (8), we have

$$
\begin{array}{ll}
(6) & 20x_1 + 4x_2 + 4x_3 = 500 \\
(7') = (7) - \frac{1}{2}(6) & 12x_2 + 3x_3 = 600 \\
(8') = (8) - \frac{1}{4}(6) & 4x_2 + 1x_3 = 200,
\end{array}
$$

along with the associated augmented matrix

$$
\left[
\begin{array}{ccc|c}
20 & 4 & 4 & 500 \\
0 & 12 & 3 & 600 \\
0 & 4 & 1 & 200
\end{array}
\right].
$$

Next we add $-\frac{1}{3}$ times Equation $(7')$ to Equation $(8')$ to eliminate the x_2-term:

$$
\begin{array}{ll}
(6) & 20x_1 + 4x_2 + 4x_3 = 500 \\
(7') & 12x_2 + 3x_3 = 600 \\
(8'') = (8') - \frac{1}{3}(7') & 0x_3 = 0
\end{array}
$$

The corresponding augmented matrix is

$$
\left[
\begin{array}{ccc|c}
20 & 4 & 4 & 500 \\
0 & 12 & 3 & 600 \\
0 & 0 & 0 & 0
\end{array}
\right].
$$

But this process eliminates the x_3 term, and simultaneously the constant term on the right becomes 0. Also, note that Equation $(8')$ is equivalent to $\frac{1}{3}$ times Equation $(7')$. As a consequence, we actually have only two equations in three unknowns. This system has infinitely many solutions because we can choose *any* value for x_3 and then use back substitution to determine x_2 and x_1, based on our choice for x_3. For instance, if we choose $x_3 = 0$, Equation $(7')$ gives $12x_2 + 3(0) = 600$, so $x_2 = 50$, and then Equation (6) gives $20x_1 + 4(50) + 4(0) = 500$, so $x_1 = 15$. Alternatively, if we choose $x_3 = 40$, eventually Equation $(7')$ gives $x_2 = 40$, and so Equation (6) gives $x_1 = 9$. That is, we have obtained two very different solutions to the same system of equations. In fact, had we selected any other positive value for x_3, we would have obtained still another solution to the system (but not necessarily to the original problem). Thus, geometrically, the corresponding planes in three dimensions all pass through a common line of intersection, and every point on this line is a solution to the system.

◆

Incidentally, if you attempted to solve the system of equations in Example 3 on a graphing calculator, say, you would get an error message of the form SINGULAR MATRIX. We discuss what this message means later in this section.

Systems of Linear Equations with No Solutions

We pointed out in our discussion of the geometry of systems of equations that three planes may have no common point of intersection, as with the floor, ceiling and one wall in a room. We illustrate this situation in Example 4 by making another minor change in the value on the right-hand side of the third equation in the original system in Examples 2 and 3.

EXAMPLE 4

Apply Gaussian elimination to the system of linear equations

$$20x_1 + 4x_2 + 4x_3 = 500 \qquad \textbf{(9)}$$
$$10x_1 + 14x_2 + 5x_3 = 850 \qquad \textbf{(10)}$$
$$5x_1 + 5x_2 + 2x_3 = 1000. \qquad \textbf{(11)}$$

Solution The corresponding augmented matrix is

$$\begin{bmatrix} 20 & 4 & 4 & \bigm| & 500 \\ 10 & 14 & 5 & \bigm| & 850 \\ 5 & 5 & 2 & \bigm| & 1000 \end{bmatrix}.$$

Eliminating x_1 as before, we get

$$20x_1 + 4x_2 + 4x_3 = 500 \qquad \textbf{(9)}$$
$$12x_2 + 3x_3 = 600 \qquad \textbf{(10')}$$
$$4x_2 + 1x_3 = 875. \qquad \textbf{(11')}$$

The associated augmented matrix is

$$\begin{bmatrix} 20 & 4 & 4 & \bigm| & 500 \\ 0 & 12 & 3 & \bigm| & 600 \\ 0 & 4 & 1 & \bigm| & 875 \end{bmatrix}.$$

We now use Equation $(10')$ to eliminate the x_2-term in Equation $(11')$ to get

$$(9) \qquad\qquad\qquad 20x_1 + 4x_2 + 4x_3 = 500$$
$$(10') \qquad\qquad\qquad 12x_2 + 3x_3 = 600$$
$$(11'') = (11') + (-\tfrac{1}{3})(10') \qquad\qquad\qquad 0x_3 = 675,$$

along with the augmented matrix

$$\begin{bmatrix} 20 & 4 & 4 & \bigm| & 500 \\ 0 & 12 & 3 & \bigm| & 600 \\ 0 & 0 & 0 & \bigm| & 675 \end{bmatrix}.$$

The two Equations $(10')$ and $(11')$ are called *inconsistent equations* because they lead to the impossible Equation $(11'')$: $0 = 675$. Hence the original system has no solution. Geometrically, this outcome indicates that the corresponding planes in three dimensions do not have a common point of intersection. ◆

If you attempted to solve this system of equations on a graphing calculator, say, you would again get an error message about a `SINGULAR MATRIX`.

In the real world, the inconsistency that occurred in Example 4 would be resolved by increasing one of the right-hand side demands (thus producing an excess amount of one type of clothing).

We now summarize the steps of Gaussian elimination.

> ## Gaussian Elimination
>
> 1. Add multiples of the ith equation, for $i = 1, 2, \ldots, n - 1$, to the remaining equations to eliminate the ith variable from the other equations.
> 2. Solve the resulting upper triangular system of equations, using back substitution.

If, in the process, the coefficient of x_i in the ith equation $(i = 1, 2, \ldots, n - 1)$ is zero and the coefficient of x_i in one of the following equations is nonzero, simply interchange the two equations.

Using the Inverse Matrix

We can use a related approach with matrices to solve systems of equations. Let's begin with the matrices

$$\mathbf{A} = \begin{bmatrix} 1 & 0 \\ 4 & 2 \end{bmatrix} \quad \text{and} \quad \mathbf{B} = \begin{bmatrix} 1 & 0 \\ -2 & \frac{1}{2} \end{bmatrix}.$$

Their product is

$$\mathbf{AB} = \begin{bmatrix} 1 & 0 \\ 4 & 2 \end{bmatrix} \begin{bmatrix} 1 & 0 \\ -2 & \frac{1}{2} \end{bmatrix} = \begin{bmatrix} 1 & 0 \\ 0 & 1 \end{bmatrix}.$$

The matrix $\begin{bmatrix} 1 & 0 \\ 0 & 1 \end{bmatrix}$ is called the 2×2 **identity matrix** and is denoted by $\mathbf{I}_2$. For any 2×2 matrix $\mathbf{A}$,

$$\mathbf{AI}_2 = \mathbf{I}_2\mathbf{A} = \mathbf{A},$$

so $\mathbf{I}_2$ plays the same role for 2×2 matricies as the number 1 plays in arithmetic: $a \cdot 1 = 1 \cdot a = a$. Similarly, the identity matrix for 3×3 matrices is

$$\mathbf{I}_3 = \begin{bmatrix} 1 & 0 & 0 \\ 0 & 1 & 0 \\ 0 & 0 & 1 \end{bmatrix},$$

and so on. The preceding matrices $\mathbf{A}$ and $\mathbf{B}$ are said to be **inverses** of each other. We write $\mathbf{B} = \mathbf{A}^{-1}$, and the product of the matrix and its inverse is the identity matrix

$$\mathbf{AA}^{-1} = \mathbf{I}_2 = \mathbf{A}^{-1}\mathbf{A}.$$

The inverse of a matrix can be extremely useful in solving systems of linear equations.

Not every matrix has an inverse. A matrix that does not have an inverse is called a **singular matrix.** (Note that this is the same term that is used in the error message on graphing calculators when a system of equations has no solution or multiple solutions.) A square matrix $\mathbf{A}$ has an inverse if, when we use Gaussian elimination to reduce it to upper triangular form, all the diagonal elements are nonzero. However, if any of the diagonal elements are zero, no inverse exists and the matrix is singular.

Suppose now that we have the matrix-vector equation $\mathbf{Ax} = \mathbf{b}$ and that the $n \times n$ matrix $\mathbf{A}$ is not singular, so its inverse $\mathbf{A}^{-1}$ exists. Matrix $\mathbf{A}$ can be any size, so we denote the corresponding identity matrix by $\mathbf{I}$ without designating its size. Then we can left-multiply the equation by this inverse matrix and obtain

$$\mathbf{A}^{-1}\mathbf{Ax} = \mathbf{A}^{-1}\mathbf{b}$$
$$\mathbf{Ix} = \mathbf{A}^{-1}\mathbf{b}$$
$$\mathbf{x} = \mathbf{A}^{-1}\mathbf{b},$$

which automatically gives the desired solution vector $\mathbf{x}$.

Note that we haven't discussed *how* to calculate the inverse of a matrix $\mathbf{A}$. We can do so by using an extension of the Gaussian elimination process, but that is somewhat outside the scope of what we want to focus on here. Instead, note that your calculator will do this for you. Enter a square matrix $\mathbf{A}$, call it up and press the x^{-1} key, and your calculator will give you the inverse matrix $\mathbf{A}^{-1}$ if it exists. If the inverse does not exist, you will get an error message. Multiply this inverse matrix and the vector of constants $\mathbf{b}$ to get the desired solution vector $\mathbf{x} = \mathbf{A}^{-1}\mathbf{b}$.

Some of you may have seen a method called *Cramer's rule*, which uses *determinants* for solving a system of linear equations. Although fairly effective for solving systems of two or three equations, Cramer's rule is very inefficient for larger systems. Suppose, for instance, that you use a relatively slow computer that can perform only 1 million operations per second. To solve a system of 20 equations in 20 unknowns with Cramer's rule would take it about 77,000 years! The same computer could solve that system in about 0.003 second using Gaussian elimination or matrix methods.

Applications of Gaussian Elimination

We now consider a series of further applications involving systems of equations. We begin by using Gaussian elimination to investigate the long-term behavior of Markov chains.

Steady State of the Markov Chain for the Stock Market   Consider again the stock market Markov chain introduced in Section 10.2 with the transition matrix

$$\begin{array}{c} \textbf{Market} \\ \textbf{Tomorrow} \end{array} \begin{array}{c} \text{Up} \\ \text{Down} \\ \text{Same} \end{array} \begin{bmatrix} \frac{1}{4} & \frac{1}{2} & \frac{1}{4} \\ \frac{1}{2} & \frac{1}{4} & \frac{1}{2} \\ \frac{1}{4} & \frac{1}{4} & \frac{1}{4} \end{bmatrix} = \mathbf{A}.$$

(Market Today: Up, Down, Same)

We noted that over many time periods, the successive probability vectors converged to 0.35 for the market going up, 0.40 for the market going down, and 0.25 for the market staying the same. We confirm the fact that

$$\mathbf{p} = \begin{bmatrix} 0.35 \\ 0.40 \\ 0.25 \end{bmatrix}$$

is the **steady state** or the **equilibrium state** of the Markov chain by showing that if the market ever reaches this state on one day, the vector $\mathbf{p}^+ = \mathbf{Ap}$ for the following day will be the same as $\mathbf{p}$. That is, at the steady state, there is no subsequent change in the values for the probabilities. Using a calculator to perform the matrix–vector product gives

$$\mathbf{p}^+ = \mathbf{A}\mathbf{p} = \begin{bmatrix} \frac{1}{4} & \frac{1}{2} & \frac{1}{4} \\ \frac{1}{2} & \frac{1}{4} & \frac{1}{2} \\ \frac{1}{4} & \frac{1}{4} & \frac{1}{4} \end{bmatrix} \begin{bmatrix} 0.35 \\ 0.40 \\ 0.25 \end{bmatrix} = \begin{bmatrix} 0.35 \\ 0.40 \\ 0.25 \end{bmatrix} = \mathbf{p}.$$

In Example 5, we show how to find the equilibrium state, denoted by $\mathbf{p}^\star$, for a Markov chain exactly.

EXAMPLE 5

Find the equilibrium state $\mathbf{p}^\star$ for the preceding matrix $\mathbf{A}$.

Solution We want a vector $\mathbf{p}^\star$ that satisfies the matrix-vector equation $\mathbf{p}^\star = \mathbf{A}\mathbf{p}^\star$. Let the components of $\mathbf{p}^\star$ be denoted by $p_1{}^\star$, $p_2{}^\star$, and $p_3{}^\star$. Note that $p_1{}^\star + p_2{}^\star + p_3{}^\star = 1$ because the sum of all the probabilities associated with an event must sum to 1.

Because $\mathbf{p}^\star = \mathbf{A}\mathbf{p}^\star$, we have

$$\begin{bmatrix} p_1{}^\star \\ p_2{}^\star \\ p_3{}^\star \end{bmatrix} = \begin{bmatrix} \frac{1}{4}p_1{}^\star + \frac{1}{2}p_2{}^\star + \frac{1}{4}p_3{}^\star \\ \frac{1}{2}p_1{}^\star + \frac{1}{4}p_2{}^\star + \frac{1}{2}p_3{}^\star \\ \frac{1}{4}p_1{}^\star + \frac{1}{4}p_2{}^\star + \frac{1}{4}p_3{}^\star \end{bmatrix}.$$

Collecting the variables on the left, we get the following system of three equations in the three unknowns $p_1{}^\star$, $p_2{}^\star$, and $p_3{}^\star$:

$$\frac{3}{4}p_1{}^\star - \frac{1}{2}p_2{}^\star - \frac{1}{4}p_3{}^\star = 0$$
$$-\frac{1}{2}p_1{}^\star + \frac{3}{4}p_2{}^\star - \frac{1}{2}p_3{}^\star = 0$$
$$-\frac{1}{4}p_1{}^\star - \frac{1}{4}p_2{}^\star + \frac{3}{4}p_3{}^\star = 0.$$

Solving this system by Gaussian elimination applied to the augmented coefficient matrix, we obtain

$$\begin{bmatrix} \frac{3}{4} & -\frac{1}{2} & -\frac{1}{4} & 0 \\ -\frac{1}{2} & \frac{3}{4} & -\frac{1}{2} & 0 \\ -\frac{1}{4} & -\frac{1}{4} & \frac{3}{4} & 0 \end{bmatrix} \Rightarrow \begin{bmatrix} \frac{3}{4} & -\frac{1}{2} & -\frac{1}{4} & 0 \\ 0 & \frac{5}{12} & -\frac{2}{3} & 0 \\ 0 & -\frac{5}{12} & \frac{2}{3} & 0 \end{bmatrix} \Rightarrow \begin{bmatrix} \frac{3}{4} & -\frac{1}{2} & -\frac{1}{4} & 0 \\ 0 & \frac{5}{12} & -\frac{2}{3} & 0 \\ 0 & 0 & 0 & 0 \end{bmatrix}.$$

To eliminate the fractions, we multiply the first row by 4 and the second row by 12:

$$\begin{bmatrix} 3 & -2 & -1 & 0 \\ 0 & 5 & -8 & 0 \\ 0 & 0 & 0 & 0 \end{bmatrix}.$$

This augmented matrix is equivalent to the reduced system of equations

$$3p_1{}^\star - 2p_2{}^\star - p_3{}^\star = 0$$
$$5p_2{}^\star - 8p_3{}^\star = 0.$$

Because there are only two equations in the three unknowns, there is no unique solution. Instead, we can solve for any two of the variables in terms of the third—say, $p_3{}^\star$. The second equation then gives

$$p_2{}^\star = \frac{8}{5}p_3{}^\star$$

and, when we substitute it into the first equation and solve for $p_1{}^*$, we have

$$p_1{}^* = \frac{2p_2{}^* + p_3{}^*}{3} = \frac{2(8/5)p_3{}^* + p_3{}^*}{3} = \frac{7}{5}p_3{}^*.$$

Where is our unique vector of stable probabilities? In this solution we obtained infinitely many solutions—one for each value of $p_3{}^*$. We now use the fact that the sum of the three probabilities must equal 1, or $p_1{}^* + p_2{}^* + p_3{}^* = 1$. We substitute the expressions for $p_1{}^*$ and $p_2{}^*$ in terms of $p_3{}^*$ to obtain

$$p_1{}^* + p_2{}^* + p_3{}^* = \frac{7}{5}p_3{}^* + \frac{8}{5}p_3{}^* + p_3{}^* = \frac{20}{5}p_3{}^* = 4p_3{}^* = 1,$$

so that $p_3{}^* = 0.25$. Therefore

$$p_2{}^* = \frac{8}{5}(0.25) = 0.40 \quad \text{and} \quad p_1{}^* = \frac{7}{5}(0.25) = 0.35.$$

These probabilities are identical to those in the equilibrium state vector of the market Markov chain given previously.

Systems of linear equations of the form $\mathbf{Ax} = \mathbf{0}$ with a zero vector, $\mathbf{0}$, on the right-hand side, such as the one in Example 5, occur frequently in matrix algebra. They are called *homogeneous systems* of linear equations. In comparison, systems of equations of the form $\mathbf{Ax} = \mathbf{b}$, where $\mathbf{b} \neq \mathbf{0}$, are called *nonhomogeneous systems*.

When solving a homogeneous system, we usually are interested in a nonzero solution, as was the case here. Note, however, that $\mathbf{x} = \mathbf{0}$ is always a solution to any homogeneous system $\mathbf{Ax} = \mathbf{0}$. Thus, if we are to get a nonzero solution, we must have a case of multiple solutions.

Applications to Analytic Geometry   Next, we apply Gaussian elimination to solve some systems of linear equations that arise in analytic geometry.

One of the most basic objects in geometry is a line, whose equation can be written as $y = ax + b$. Suppose that we have two points, P with coordinates $(3, 1)$ and Q with coordinates $(6, 7)$, and we want to find the equation of the line through these points. We could solve this problem by finding the slope of the line segment joining the two points and then using the point–slope formula. Here, however, we solve this problem using systems of equations to illustrate some important mathematical ideas that can be extended to answer considerably more complicated questions. The results either way are the same.

EXAMPLE 6

Use matrix methods to find the equation of the line through the points $(3, 1)$ and $(6, 7)$.

Solution   First, let's be clear about what we need to find. The equation of the line $y = ax + b$ is determined by two constants—the slope a and the vertical intercept b. Although a and b are constants for a particular line, we can think of them as parameters that distinguish one line from another. We need to determine appropriate values for a and b so that the line $y = ax + b$ passes through the points $(3, 1)$ and $(6, 7)$. Accordingly, a and b must satisfy the following two equations.

$$\text{For point } (3, 1): \quad 1 = a(3) + b$$
$$\text{For point } (6, 7): \quad 7 = a(6) + b$$

We rewrite these two equations in standard form to get

$$3a + b = 1$$
$$6a + b = 7.$$

Subtracting twice the first equation from the second equation yields

$$-b = 5, \quad \text{so that} \quad b = -5.$$

Substituting back into the second equation gives

$$6a + (-5) = 7, \quad \text{so} \quad 6a = 12, \quad \text{and} \quad a = 2.$$

Therefore the equation of the desired line is

$$y = 2x - 5.$$

We now extend the ideas in Example 6 to find the equation of a parabola $y = ax^2 + bx + c$. We need to determine the values of the parameters a, b, and c. To solve for these three unknowns, we need three linear equations. As with the line, we get a linear equation associated with each point the parabola goes through. For three equations, we need three points. Suppose that the three points are $(2, -1)$, $(4, 3)$ and $(-1, 8)$. Note that the three points must be noncollinear (i.e., they cannot lie on a common line). Each point determines an equation when we substitute its coordinates, x and y, into the parabola's formula $y = ax^2 + bx + c$.

$$\text{For point } (2, -1): \quad -1 = a(2)^2 + b(2) + c$$
$$\text{For point } (4, 3): \quad 3 = a(4)^2 + b(4) + c$$
$$\text{For point } (-1, 8): \quad 8 = a(-1)^2 + b(-1) + c$$

Hence we have the system of linear equations

$$4a + 2b + c = -1$$
$$16a + 4b + c = 3$$
$$a - b + c = 8.$$

In Problem 17 we ask you to solve these three equations to determine a, b, and c.

Next, we apply this curve-fitting method to a more complicated situation—namely, finding the equation of a circle. The familiar form of the equation of a circle is

$$(x - a)^2 + (y - b)^2 = c^2,$$

where (a, b) is the center of the circle and c is the radius. It turns out that, just as three noncollinerar points determine a parabola, any three noncollinear points also determine a circle. We now have three parameters a, b, and c to determine, which should lead to three equations in the three unknowns. But the equations arising from giving particular values to x and y in the above equation for the circle will not be linear (there will be a^2 and b^2 terms), so we cannot use Gaussian elimination to determine a, b, and c based on the equation $(x - a)^2 + (y - b)^2 = c^2$. However, if we expand this expression algebraically, we get

$$(x - a)^2 + (y - b)^2 = (x^2 - 2ax + a^2) + (y^2 - 2by + b^2)$$
$$= x^2 + y^2 - 2ax - 2by + a^2 + b^2 = c^2.$$

Therefore we can transform the equation of the circle to the form

$$x^2 + y^2 + Cx + Dy + E = 0,$$

where $C = -2a$, $D = -2b$, and $E = a^2 + b^2 - c^2$ are three other parameters. Using this form, we can set up three linear equations in these three unknowns and solve for them with Gaussian elimination. Once we have their values, we can rewrite the equation of the circle in the more familiar form and read the coordinates of the center and the radius. We illustrate these ideas in Example 7.

EXAMPLE 7

Find the equation of the circle passing through the points $(4, -2)$, $(6, 2)$ and $(5, 5)$.

Solution Using the equation

$$x^2 + y^2 + Cx + Dy + E = 0,$$

we create three linear equations that C, D, and E must satisfy.

For the point $(4, -2)$: $4^2 + (-2)^2 + C(4) + D(-2) + E = 0$
For the point $(6, 2)$: $6^2 + 2^2 + C(6) + D(2) + E = 0$
For the point $(5, 5)$: $5^2 + 5^2 + C(5) + D(5) + E = 0$

Moving the constant terms to the right-hand side in each equation, we have

$$4C - 2D + E = -20 \tag{12}$$
$$6C + 2D + E = -40 \tag{13}$$
$$5C + 5D + E = -50. \tag{14}$$

We now apply Gaussian elimination to this (nonhomogeneous) system of three linear equations in three unknowns. First, to eliminate C from Equations (13) and (14), we subtract $\frac{3}{2}$ of Equation (12) from Equation (13) and subtract $\frac{5}{4}$ of Equation (12) from Equation (14).

(12) $\qquad\qquad 4C - 2D + E = -20$

$(13') = (13) - \frac{3}{2}(12)$ $\qquad\qquad 5D - \dfrac{1}{2}E = -10$

$(14') = (14) - \frac{5}{4}(12)$ $\qquad\qquad \dfrac{15}{2}D - \dfrac{1}{4}E = -25$

Next we eliminate D from the last equation by subtracting $\frac{3}{2}$ of Equation $(13')$ from Equation $(14')$.

(12) $\qquad\qquad 4C - 2D + E = -20$

$(13')$ $\qquad\qquad 5D - \dfrac{1}{2}E = -10$

$(14'') = (14') - \frac{3}{2}(13')$ $\qquad\qquad \dfrac{1}{2}E = -10$

From Equation $(14'')$, we have $E = -20$. Substituting back into Equation $(13')$ yields

$$5D - \frac{1}{2}(-20) = -10, \quad \text{so} \quad 5D = -20 \quad \text{and} \quad D = -4.$$

Then substituting back into Equation (12) yields

$$4C - 2(-4) + (-20) = -20, \quad \text{so} \quad 4C = -8 \quad \text{and} \quad C = -2.$$

Therefore one form of an equation for our circle is

$$x^2 + y^2 - 2x - 4y - 20 = 0.$$

We rewrite this equation in the more familiar form by completing the square on both x and y (see Appendix A8):

$$[(x^2 - 2x)] + [(y^2 - 4y)] - 20 = 0$$
$$[(x^2 - 2x + 1) - 1] + [(y^2 - 4y + 4) - 4] - 20 = 0$$
$$[(x - 1)^2 - 1] + [(y - 2)^2 - 4] - 20 = 0$$
$$(x - 1)^2 + (y - 2)^2 = 1 + 4 + 20 = 25.$$

Thus our circle is centered at the point $(1, 2)$ and has radius 5, as shown in Figure 10.36.

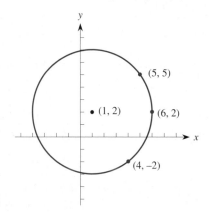

FIGURE 10.36

Think About This   Substitute the coordinates of the three points in Example 7 into the usual equation for a circle, $(x - a)^2 + (y - b)^2 = c^2$, to see the types of equations involving the three parameters a, b and c that would result. ▭

Problems

1. Solve each system of equations using Gaussian elimination.

a. $\begin{aligned} x + y &= 8 \\ 2x - 3y &= 4 \end{aligned}$

b. $\begin{aligned} x - 4y &= 3 \\ -x + 2y &= 9 \end{aligned}$

c. $\begin{aligned} 2x - 3y &= 4 \\ -4x + y &= 1 \end{aligned}$

d. $\begin{aligned} 3x - 2y &= 3 \\ -2x + 5y &= 2 \end{aligned}$

e. $\begin{aligned} 3x - 2y &= 3 \\ -4x + 3y &= -2 \end{aligned}$

2. Use Gaussian elimination to solve each variation on the clothes production model in Example 2. Some variations may have no solution, some may have multiple solutions (express such an infinite family of solutions in terms of x_3), and some may have an unrealistic solution involving negative values.

a. $\begin{aligned} 20x_1 + 4x_2 + 4x_3 &= 500 \\ 8x_1 + 3x_2 + 5x_3 &= 850 \\ 4x_1 + 5x_2 + 11x_3 &= 2050 \end{aligned}$

b. $\begin{aligned} 6x_1 + 5x_2 + 6x_3 &= 500 \\ 10x_1 + 10x_2 &= 850 \\ 2x_1 + 12x_3 &= 1000 \end{aligned}$

c. $6x_1 + 2x_2 + 2x_3 = 500$
$3x_1 + 6x_2 + 3x_3 = 300$
$3x_1 + 2x_2 + 6x_3 = 1000$

d. $8x_1 + 4x_2 + 3x_3 = 500$
$4x_1 + 8x_2 + 5x_3 = 500$
$12x_2 + 6x_3 = 500$

3. Solve each system of equations using Gaussian elimination.

a. $x_1 + 2x_2 + x_3 = 3$
$-x_1 + 2x_2 + 3x_3 = 1$
$2x_1 - x_2 + 2x_3 = 2$

b. $2x_1 - x_2 + x_3 = 2$
$-x_1 - 2x_2 + 2x_3 = -3$
$3x_1 + 2x_2 - x_3 = 0$

c. $-x_1 - 3x_2 + 2x_3 = -1$
$5x_1 + 4x_2 + 6x_3 = 12$
$2x_1 + x_2 + 3x_3 = 4$

d. $2x_1 + 4x_2 - 2x_3 = 4$
$-2x_1 + 2x_2 - 3x_3 = -4$
$x_1 - x_2 - 2x_3 = -1$

e. $x_1 + 2x_2 + 2x_3 = 3$
$x_1 + 2x_2 + 3x_3 = 11$
$2x_1 + x_2 + 2x_3 = 5$

f. $x_1 - x_2 - 2x_3 = 2$
$3x_1 + 2x_2 + 4x_3 = 2$
$3x_1 + x_2 - 2x_3 = -3$

4. In each set of three equations, show that the third equation equals some multiple r of the first equation added to the second equation: $(3) = r \cdot (1) + (2)$.

a. $(1)\, x + 2y = 4$
$(2)\, 3x + y = 9$
$(3)\, x - 3y = 1$

b. $(1)\, x - y + z = 2$
$(2)\, x + y - z = 3$
$(3)\, -2x + 4y - 4z = -3$

c. $(1)\, 2x + y - 2z = -5$
$(2)\, 3x - y + z = 8$
$(3)\, 6x + 0.5y - 2z = 0.5$

5. Determine whether each system of equations has a unique solution, multiple solutions, or is inconsistent (has no solution).

a. $x - 2y = -3$
$-3x + 6y = 9$

b. $x - 2y = -3$
$-3x + 6y = 4$

c. $2x - 5y = 3$
$6x - 15y = 9$

d. $x_1 + 2x_2 + 3x_3 = 10$
$2x_1 - x_2 + 4x_3 = 20$
$5x_2 + 2x_3 = 0$

e. $x_1 + 5x_2 + 4x_3 = 10$
$x_1 + 3x_2 + 6x_3 = 9$
$x_1 - x_2 + x_3 = 5$

f. $-x_1 + 2x_2 + x_3 = 0$
$2x_1 + x_2 - 3x_3 = 0$
$x_1 + x_2 + 2x_3 = 0$

6. For each augmented matrix, state whether the associated system of equations has a unique solution, multiple solutions, or no solution.

a. $\begin{bmatrix} 3 & 3 & 1 & | & 0 \\ 0 & 1 & 5 & | & 1 \\ 0 & 0 & 1 & | & 2 \end{bmatrix}$
b. $\begin{bmatrix} 1 & 2 & 2 & | & 3 \\ 0 & 5 & 1 & | & 6 \\ 0 & 0 & 0 & | & 2 \end{bmatrix}$

c. $\begin{bmatrix} 1 & 2 & 7 & | & 5 \\ 0 & 4 & 2 & | & 1 \\ 0 & 0 & 0 & | & 0 \end{bmatrix}$
d. $\begin{bmatrix} 4 & 2 & 3 & | & 3 \\ 0 & 4 & 2 & | & 6 \\ 0 & 2 & 1 & | & 3 \end{bmatrix}$

e. $\begin{bmatrix} 2 & 1 & 3 & | & 0 \\ 0 & 1 & 0 & | & 0 \\ 0 & 0 & 1 & | & 0 \end{bmatrix}$

7. (Continuation of Problem 4 of Section 10.2) Consider the following clothes production model. There are three clothing factories (1, 2, and 3) and from each roll of cloth, the different factories produce the following numbers of vests, pants, and coats.

	Factory 1	Factory 2	Factory 3
Vests	6	4	2
Pants	4	8	4
Coats	3	2	8

Suppose that the demand is for 400 vests, 800 pants, and 500 coats. Write a system of equations whose solution would determine production levels (rolls of cloth needed by each factory) to yield the desired numbers of vests, pants, and coats. Find the solution using Gaussian elimination.

8. (Continuation of Problem 5 of Section 10.2) From each shipment of crude petroleum, Refineries 1, 2, and 3 produce the following amounts (in thousands of gallons) of heating oil, diesel oil, and gasoline.

$$
\begin{array}{c}
\phantom{\text{Heating Oil}}\text{Refinery 1}\quad\text{Refinery 2}\quad\text{Refinery 3}\\
\begin{array}{r}
\text{Heating Oil}\\
\text{Diesel Oil}\\
\text{Gasoline}
\end{array}
\left[
\begin{array}{ccc}
8 & 5 & 3\\
2 & 5 & 5\\
3 & 7 & 6
\end{array}
\right]
\end{array}
$$

Suppose that the demand is for 6200 thousand gallons of heating oil, 4000 thousand gallons of diesel oil, and 4700 thousand gallons of gasoline. Write a system of equations whose solution would determine production levels to yield the desired amounts of heating oil, diesel oil, and gasoline. Find the solution using Gaussian elimination.

9. (Continuation of Problem 6 of Section 10.2) The staff dietitian at the California Institute of Trigonometry has to make up a meal with 600 calories, 20 grams of protein, and 200 mg of vitamin C. The three food types that the dietitian can choose from are gelatin, fish sticks, and mystery meat. They have the following nutritional content per ounce.

$$
\begin{array}{c}
\phantom{\text{Calories}}\text{Gelatin}\quad\text{Fish Sticks}\quad\text{Mystery Meat}\\
\begin{array}{r}
\text{Calories}\\
\text{Protein}\\
\text{Vitamin C}
\end{array}
\left[
\begin{array}{ccc}
10 & 50 & 200\\
1 & 3 & 0.2\\
30 & 10 & 0
\end{array}
\right]
\end{array}
$$

Describe the dietitian's problem and create a mathematical model for it with a system of three linear equations. Find the solution using Gaussian elimination.

10. (Continuation of Problem 7 of Section 10.2) A company has a budget of $280,000 for computing equipment. The types of equipment available are microcomputers at $2000 each, terminals at $500 each, and workstations at $5000 each. There should be five times as many terminals as microcomputers and twice as many microcomputers as workstations. Set up a system of three linear equations for this situation. Find the solution using Gaussian elimination.

11. Find the equilibrium state for the Markov chains with the following transition matrices.

a. $\begin{bmatrix} \frac{2}{3} & \frac{1}{3} \\ \frac{1}{3} & \frac{2}{3} \end{bmatrix}$ b. $\begin{bmatrix} \frac{3}{4} & \frac{1}{4} \\ \frac{1}{4} & \frac{3}{4} \end{bmatrix}$ c. $\begin{bmatrix} 1 & \frac{1}{2} \\ 0 & \frac{1}{2} \end{bmatrix}$

12. The copy machine at the student union breaks down with the following pattern. If it is working today, there is a 70% chance that it works tomorrow (and a 30% chance of breaking down). If it is broken today, there is a 50% chance that it works tomorrow. Construct a Markov chain for this problem and find the equilibrium state.

13. From past experience, the Pins bowling team knows that if they win this week's game, they have a $\frac{2}{3}$ chance of winning next week's game. If they lose this week's game, they have a $\frac{1}{2}$ chance of winning next week's game. Construct a Markov chain for this problem and find the stable distribution.

14. Find the equilibrium state for the Markov chains with the following transition matrices.

a. $\begin{bmatrix} \frac{1}{3} & \frac{2}{3} & \frac{1}{3} \\ 0 & 0 & \frac{1}{3} \\ \frac{2}{3} & \frac{1}{3} & \frac{1}{3} \end{bmatrix}$ b. $\begin{bmatrix} \frac{1}{4} & \frac{1}{4} & \frac{1}{2} \\ \frac{1}{4} & \frac{1}{2} & \frac{1}{4} \\ \frac{1}{2} & \frac{1}{4} & \frac{1}{4} \end{bmatrix}$

c. $\begin{bmatrix} 0 & \frac{1}{2} & \frac{1}{2} \\ \frac{1}{2} & 0 & \frac{1}{2} \\ \frac{1}{2} & \frac{1}{2} & 0 \end{bmatrix}$

15. The following model for learning a concept over a set of lessons identifies four states of learning: I = Ignorance, E = Exploratory thinking, S = superficial understanding, and M = Mastery. If you are now in State I, after one lesson you have probability $\frac{1}{2}$ of still being in I and probability $\frac{1}{2}$ of being in E. If you are now in State E, you have probability $\frac{1}{4}$ of being in I, $\frac{1}{2}$ in E, and $\frac{1}{4}$ in S. If you are now in State S, you have probability $\frac{1}{4}$ of being in E, $\frac{1}{2}$ in S, and $\frac{1}{4}$ in M. If you are in State M, you always stay in M (with probability 1). Construct a Markov chain for this problem and find the equilibrium state.

16. Find the equation of the line passing through each pairs of points, using Gaussian elimination.
    a. $(1, 1), (2, 3)$ b. $(1, 0), (0, 2)$
    c. $(2, 3), (-1, 1)$ d. $(2, 1), (3, 1)$
    e. $(-1, 1), (2, -5)$ f. $(4, 1), (-1, -1)$

17. Solve the three equations for determining the parameters of the parabola $y = ax^2 + bx + c$ that passes through the three points $(2, -1), (4, 3)$ and $(-1, 8)$.

18. A parabola passes through the points $(-1, 4), (1, 2)$ and $(4, 7)$. Set up a system of linear equations in a, b, and c and then solve it, using Gaussian elimination, to find the equation of the parabola.

19. Find an equation of the circle passing through each triple of points, using Gaussian elimination.
    a. $(1, 0), (0, 1), (1, 1)$
    b. $(2, 1), (2, -3), (0, 1)$
    c. $(3, 1), (2, 5), (-3, 6)$

d. $(-1, -1), (1, 3), (2, -1)$
e. $(0, 0), (4, 1), (-4, 1)$

20. Just as two distinct points determine a line uniquely, three noncollinear points determine a parabola. Moreover, four noncollinear points, all of which do not lie on a parabola, determine a cubic polynomial.

a. A parabola $y = ax^2 + bx + c$ passes through the points $(-1, 5)$, $(1, -2)$, and $(4, 6)$. Set up a system of linear equations in a, b, and c and then solve it, using Gaussian elimination, to find the equation of the parabola.

b. A cubic polynomial $y = ax^3 + bx^2 + cx + d$ passes through the four points $(-1, 4)$, $(1, -2)$, $(4, 7)$, and $(5, 2)$. Set up a system of linear equations in a, b, c, and d that could be used to find the equation of the cubic. Then solve the system using Gaussian elimination.

c. How does your answer to part (b) compare to the result you can obtain with your calculator using the routine for fitting a cubic function to a set of data?

21. Find the inverse of the matrix $\mathbf{A} = \begin{bmatrix} 1 & 0 \\ 4 & 2 \end{bmatrix}$ algebraically. (*Hint:* Write the inverse as $\mathbf{A}^{-1} = \begin{bmatrix} a & b \\ c & d \end{bmatrix}$ and use the fact that $\mathbf{AA}^{-1} = \mathbf{I}_2$ to solve for a, b, c, and d.) How does your answer compare with what your calculator shows using its matrix features?

22. a. Consider the rotation matrix

$$\mathbf{R} = \begin{bmatrix} \cos\theta & -\sin\theta \\ \sin\theta & \cos\theta \end{bmatrix}.$$

Based on geometric principles, predict what the inverse matrix $\mathbf{R}^{-1}$ has to be.

b. Prove algebraically that the matrix you constructed in part (a) is indeed the inverse matrix for **R**.

23. The method of linear regression discussed in Chapter 3, in which the best fit line $y = ax + b$ is constructed for a set of (x, y) data, is based on solving a system of linear equations for the unknown parameters a and b. It can be shown that, if the set of n data points is $(x_1, y_1), (x_2, y_2), (x_3, y_3), \ldots, (x_n, y_n)$, then a and b must satisfy the system of linear equations

$$\left(\sum_{i=1}^{n} x_i\right)a + nb = \left(\sum_{i=1}^{n} y_i\right)$$

$$\left(\sum_{i=1}^{n} x_i^2\right)a + \left(\sum_{i=1}^{n} x_i\right)b = \left(\sum_{i=1}^{n} x_i y_i\right).$$

Recall that the summations indicate adding up all the x-values, adding up all the y-values, adding up the squares of all the x-values, and adding up all the products of the x and y-values. Note that all the summation terms are known constants once the x's and y's have been given.

a. For the set of data $(1, 11)$, $(2, 25)$, $(3, 33)$, $(4, 45)$, $(5, 57)$, and $(6, 65)$, find the equation of the regression line using your graphing calculator or appropriate software package.

b. Calculate the sums needed in the two preceding equations by completing the entries in the table.

x	y	x^2	xy
1	11		
2	25		
3	33		
4	45		
5	57		
6	65		
$\Sigma x =$	$\Sigma y =$	$\Sigma x^2 =$	$\Sigma xy =$

c. Use the results of part (b) to write the system of linear equations in a and b that you can use to determine the values for the unknown coefficients a and b in the regression equation.

d. Solve the system of equations in part (c) using Gaussian elimination. How do your results compare to what you obtained directly in part (a)?

e. In Chapter 3 we suggested that you scale down large numbers, such as the full years 2000, 2001, 2002, . . . , in a set of data. Based on your calculations in part (b), explain why doing so is desirable. In particular, what might happen if there were many data values, if the x' s, say, consisted of full years and the y' s were also large numbers?

24. Repeat parts (a)–(d) of Problem 23 for $(0, 24)$, $(5, 21)$, $(10, 17)$, $(15, 14)$, $(20, 12)$, and $(25, 9)$.

Chapter Summary

In this chapter we introduced vectors and matrices and some of their properties and applications. In particular, we emphasized

◆ What a vector is geometrically, algebraically, and as an ordered pair or ordered triple of numbers.

◆ What a matrix is algebraically and as an array of numbers.

◆ How to use vectors and matrices to construct mathematical models, including Markov chains and growth models, of a wide variety of real-world problems.

◆ How to use vectors and matrices to rotate geometric figures.

◆ How to add and subtract vectors and matrices.

◆ How to compute the scalar product of two vectors.

◆ How to multiply matrices and vectors, and matrices and other matrices.

◆ How to use matrix multiplication to solve applied problems, including Markov chains and linear growth models.

◆ How to solve a system of n linear equations in n unknowns, using Gaussian elimination.

◆ How to apply Gaussian elimination to solve various real-world problems with a system of linear equations.

◆ How to find an equation of various types of curves, such as parabolas and circles, that pass through a given set of points.

Chapter Review Problems

1. Determine the magnitude of the displacement vectors from point A to point B for each pair of points.

   a. $A = (6, 6), B = (2, 3)$
   b. $A = (-1, 1), B = (4, -1)$
   c. $A = (1, 2, -3), B = (3, 3, -5)$

2. a. If a plane is flying due south at 200 mph and the wind is blowing *in* a direction that is 40° north of west at 50 mph, what are the actual direction and speed of the plane?

   b. Suppose instead that the wind is blowing *from* a direction that is 40° north of west at 50 mph. How do your answers to part (a) change?

3. A furniture manufacturer makes tables, chairs, and sofas. In one month, the company has available 1300 units of wood, 2300 units of labor, and 1300 units of upholstery. The manufacturer wants a production schedule for the month that uses all these resources. The different products require the following amounts of the resources.

	Table	Chair	Sofa
Wood	4	1	3
Labor	3	2	5
Upholstery	0	2	4

Write a system of equations whose solution would determine production levels to yield the desired numbers of tables, chairs, and sofas.

4. If the stock market goes up today, historical data show that for tomorrow it has a 60% chance of going up, a 20% chance of staying the same, and a 20% chance of going down. If the market is unchanged today, it has a 20% chance of being unchanged, a 40% chance of going up, and a 40% chance of going down tomorrow. If the market goes down today, it has a 20% chance of going up, a 20% chance of being unchanged, and a 60% chance of going down tomorrow.

   a. Construct a Markov chain for this problem. Give **A**, the matrix of transition probabilities, and draw the transition diagram.

b. If there is a 30% chance of the market going up today, a 10% chance of being unchanged, and a 60% chance of going down, what is the probability distribution for the market tomorrow?

5. Let $\mathbf{a} = \begin{bmatrix} 1 \\ 2 \\ 3 \end{bmatrix}$, $\mathbf{b} = \begin{bmatrix} -1 \\ 3 \\ -1 \end{bmatrix}$, and $\mathbf{c} = \begin{bmatrix} 2 \\ 5 \\ 8 \end{bmatrix}$.

Compute: **(a)** $\mathbf{a} \cdot \mathbf{b}$ **(b)** $\mathbf{b} \cdot \mathbf{c}$ **(c)** $\mathbf{a} \cdot (\mathbf{b} + \mathbf{c})$
(d) $\mathbf{a} \cdot \mathbf{a}$

6. Let **a**, **b**, and **c** be as in Problem 5. Let

$$A = \begin{bmatrix} 1 & 2 & 3 & 4 \\ 2 & 4 & 6 & 8 \\ 3 & 5 & 7 & 9 \end{bmatrix},$$

$$B = \begin{bmatrix} 1 & 0 & -1 \\ 2 & -2 & 0 \\ 0 & 1 & 1 \end{bmatrix},$$

$$C = \begin{bmatrix} 5 & 4 & 1 \\ 1 & 0 & 2 \\ 3 & 2 & 1 \\ 0 & 1 & 3 \end{bmatrix}.$$

Which of the following matrix calculations are well defined (the sizes match)? If the computation makes sense, perform it. If necessary, **a, b,** or **c** may be changed to row vectors.

a. aA **b.** bB **c.** cC
d. Aa **e.** Bb **f.** Cc

7. Three different types of computers need varying amounts of four different types of integrated circuits. Matrix **A** gives the number of each circuit needed by each computer.

$$\begin{array}{c} \\ \\ \text{Computers} \end{array}\begin{array}{c} \\ A \\ B \\ C \end{array}\overset{\begin{array}{cccc} 1 & 2 & 3 & 4 \end{array}}{\begin{bmatrix} 2 & 3 & 2 & 1 \\ 5 & 1 & 3 & 2 \\ 3 & 2 & 2 & 2 \end{bmatrix}} = A$$

Let $\mathbf{d} = \begin{bmatrix} 10 & 20 & 30 \end{bmatrix}$ be the computer demand vector. Let

$$\mathbf{p} = \begin{bmatrix} 2 \\ 5 \\ 1 \\ 10 \end{bmatrix}$$

be the price vector for the circuits (the cost in dollars of each type of circuit).

a. Write an expression in terms of **A**, **d**, and **p** for the total cost of the circuits needed to produce the set of computers demanded; indicate where the matrix–vector product occurs and where the scalar product occurs.
b. Compute the total cost.

8. Let

$$A = \begin{bmatrix} 1 & 2 & 3 & 4 \\ 2 & 4 & 6 & 8 \\ 3 & 5 & 7 & 9 \end{bmatrix},$$

$$B = \begin{bmatrix} 1 & 0 & -1 \\ 2 & -2 & 0 \\ 0 & 1 & 1 \end{bmatrix},$$

$$C = \begin{bmatrix} 5 & 4 & 1 \\ 1 & 0 & 2 \\ 3 & 2 & 1 \\ 0 & 1 & 3 \end{bmatrix}.$$

Compute each matrix product (if possible).

a. AB **b.** BA **c.** AC
d. CA **e.** CB

9. Suppose that you are given the following matrices involving the costs of fruits at different stores, the amounts of fruit different types of people want, and the numbers of people of different types in different towns

	Store 1	Store 2
Apples	0.10	0.15
Oranges	0.15	0.20
Pears	0.10	0.10

	Apples	Oranges	Pears
Person 1	5	10	3
Person 2	4	5	5

	Person 1	Person 2
Town 1	1000	500
Town 2	2000	1000

a. Compute a matrix that represents the cost of each person's fruit purchases at each store.
b. Compute a matrix that represents the quantity of each fruit to be purchased in each town.

10. a. For the Markov chain matrix $\mathbf{A}$ in Problem 4, compute $\mathbf{A}^2$, $\mathbf{A}^3$, and $\mathbf{A}^5$.

b. What vectors do the columns of the powers of $\mathbf{A}$ appear to be approaching?

11. Solve each system of equations, using Gaussian elimination.

a.
$$2x_1 - 3x_2 + 2x_3 = 0$$
$$x_1 - x_2 + x_3 = 7$$
$$-x_1 + 5x_2 + 4x_3 = 4$$

b.
$$-x_1 - x_2 + x_3 = 2$$
$$2x_1 + 2x_2 - 4x_3 = -4$$
$$x_1 - 2x_2 + 3x_3 = 5$$

12. Solve the system of equations obained for the furniture model in Problem 3.

13. Find the stable distribution for the Markov chain in Problem 4.

14. Use matrix methods to find the equation of the parabola that passes through the three points $(1, 1)$, $(2, 2)$, and $(3, 5)$.

Appendices

A   Some Mathematical Moments to Remember

A.1   Absolute Value

Absolute value is used to transform any number, positive or negative, to the corresponding positive value. We write the absolute value of a number x as $|x|$. Thus

$$|5| = 5, \qquad |-6| = 6, \quad \text{and} \quad |-2.3| = 2.3.$$

In general, for any number $x \geq 0$, $|x| = x$ and for any number $x < 0$, $|x| = -x$. We can write this as

$$|x| = \begin{cases} x & \text{if } x \geq 0 \\ -x & \text{if } x < 0. \end{cases}$$

A.2   Factorial Notation $n!$

Expressions of the form $4 \times 3 \times 2 \times 1$ involving the product of consecutive positive integers starting with 1 are written using *factorial notation*. We write such products as $n!$ (and read it as *n factorial*) for any positive integer n. For instance,

$$3! = (3)(2)(1) = 6,$$
$$4! = (4)(3)(2)(1) = 24,$$
$$5! = (5)(4)(3)(2)(1) = 120,$$

and so on. In general, for any positive integer n,

$$n! = n(n-1)(n-2)\cdots(3)(2)(1).$$

Note that $10! = 10 \times 9 \times 8 \times \cdots \times 1 = 10(9 \cdot 8 \cdots 1) = 10(9!)$ and, in general,

$$n! = n \cdot (n-1)!,$$

for any n. For completeness, it is necessary to define $0! = 1$.

721

A.3 Summation or Sigma Notation

Summation notation is used as a shorthand for writing expressions such as

$$1 + 2 + 3 + \cdots + 100,$$

$$1 + 2 + 2^2 + 2^3 + \cdots + 2^{50},$$

and

$$1 + \frac{1}{2} + \frac{1}{3} + \cdots + \frac{1}{n}.$$

The Greek letter sigma (Σ) is used to denote *summation* and an *index of summation*—say, k—is used to indicate which specific terms are included in the sum. Thus, to add the squares of all the integers between $k = 1$ and $k = 100$, we write

$$\sum_{k=1}^{100} k^2$$

because when $k = 1$ we have 1^2, when $k = 2$ we have 2^2, when $k = 3$ we have 3^2, and so on until $k = 100$ when we have 100^2. Therefore

$$\sum_{k=1}^{100} k^2 = 1^2 + 2^2 + 3^2 + \cdots + 100^2.$$

To add all the integers between 25 and 60, we write

$$\sum_{k=25}^{60} k = 25 + 26 + 27 + \cdots + 60.$$

Similarly, we write the sum of the powers of 2 from 2^0 to 2^{50} as

$$\sum_{k=0}^{50} 2^k = 1 + 2 + 2^2 + 2^3 + \cdots + 2^{50}.$$

Also, to add the *reciprocals* of all the integers between $k = 1$ and some unspecified upper limit $k = n$, we write

$$\sum_{k=1}^{n} \frac{1}{k} = 1 + \frac{1}{2} + \frac{1}{3} + \frac{1}{4} + \cdots + \frac{1}{n}.$$

The letter we use for the index of summation is immaterial; we could equivalently write

$$\sum_{i=1}^{n} \frac{1}{i} = 1 + \frac{1}{2} + \frac{1}{3} + \frac{1}{4} + \cdots + \frac{1}{n} \quad \text{or}$$

$$\sum_{j=1}^{n} \frac{1}{j} = 1 + \frac{1}{2} + \frac{1}{3} + \frac{1}{4} + \cdots + \frac{1}{n}.$$

The numerical result of these summations is the same regardless of the letter used as the index.

A.4 Similar Triangles

Two triangles are *similar* if all three angles in one triangle are the same size as all three angles in the other. Of course, the lengths of the sides of the two triangles may be quite different. However, the key fact about similar triangles is:

Corresponding sides of similar triangles are proportional.

The right triangles ABC and ADE shown in Figure A.1 are similar because the angle θ is common to both, the angle ϕ is the same in both, and both triangles have a right angle. Therefore, we have a variety of ratios that are equal, including

$$\frac{AB}{AD} = \frac{BC}{DE} = \frac{AC}{AE} \quad \text{and} \quad \frac{AB}{AC} = \frac{AD}{AE} \quad \text{and} \quad \frac{BC}{AC} = \frac{DE}{AE}.$$

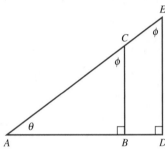

FIGURE A.1

A.5 Distance Between Points in the Plane

The distance from a point A with coordinates (x_0, y_0) to a point B with coordinates (x_1, y_1) is given by the *distance formula*

$$|AB| = \sqrt{(x_1 - x_0)^2 + (y_1 - y_0)^2}.$$

It is based on the Pythagorean theorem, as illustrated in Figure A.2. The distance $|AB|$ is the hypotenuse of the right triangle formed by a base of $x_1 - x_0$ (the horizontal change) and a height of $y_1 - y_0$ (the vertical change).

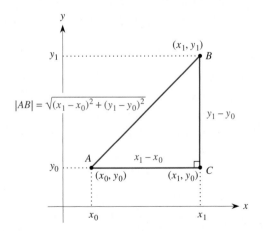

FIGURE A.2

For instance, the distance between the two points A at $(2, 5)$ and B at $(6, 8)$ is

$$|AB| = \sqrt{(6 - 2)^2 + (8 - 5)^2} = \sqrt{16 + 9} = \sqrt{25} = 5.$$

A.6 The Equation of a Circle

The equation of the circle with radius r and center at (x_0, y_0) is

$$(x - x_0)^2 + (y - y_0)^2 = r^2.$$

For instance, the equation of the circle with radius 7 and center at $(2, -5)$ is

$$(x - 2)^2 + (y + 5)^2 = 7^2 = 49,$$

as shown in Figure A.3. The equation of a circle is discussed in detail in Section 9.2.

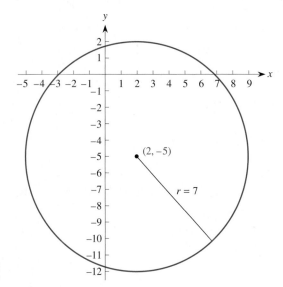

FIGURE A.3

A.7 The Equation of an Ellipse

The equation of the ellipse with center at (x_0, y_0) having its horizontal *major* (or longer) *axis* of length $2a$ and its vertical *minor* (or shorter) *axis* of length $2b$ is

$$\frac{(x - x_0)^2}{a^2} + \frac{(y - y_0)^2}{b^2} = 1,$$

as shown in Figure A.4.

For instance, the equation of the ellipse with center at $(2, -5)$, whose major axis is horizontal and has length 12 $(=2a)$ and whose minor axis is vertical and has length 8 $(=2b)$ is

$$\frac{(x - 2)^2}{6^2} + \frac{(y + 5)^2}{4^2} = 1 \quad \text{or}$$

$$\frac{(x - 2)^2}{36} + \frac{(y + 5)^2}{16} = 1,$$

as shown in Figure A.5. The ellipse is discussed in detail in Section 9.3.

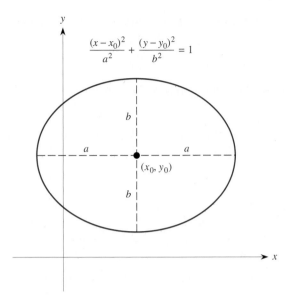

FIGURE A.4

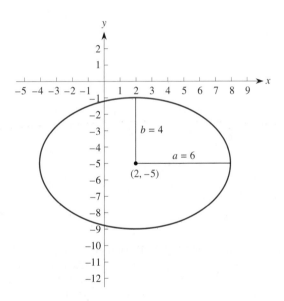

FIGURE A.5

A.8 Completing the Square

The two quadratic functions $y = x^2 - 6x + 13$ and $y = (x - 3)^2 + 4$ are algebraically equivalent because the second can be expanded to give

$$y = (x - 3)^2 + 4 = (x^2 - 6x + 9) + 4 = x^2 - 6x + 13.$$

You can also check this result by graphing the two functions on your function grapher. Because their graphs are the same parabola, they are indeed the same function. Although the first representation is more common, the second gives more information about the behavior of the corresponding parabola: It is shifted 3 units to the right and 4 units up compared to the basic parabola $y = x^2$.

The process of transforming the first expression to the second is called *completing the square*. We illustrate this procedure as follows.

- ◆ Start with $y = x^2 - 6x + 13$.
- ◆ Take one-half of the coefficient of x: $\frac{1}{2}(-6) = -3$.
- ◆ Square this number: $(-3)^2 = 9$.
- ◆ *Add* and immediately *subtract* the resulting number 9 (we are actually adding 0 and thus still have the equivalent of the original expression):

$$y = (x^2 - 6x + 9) - 9 + 13 = (x^2 - 6x + 9) + 4.$$

- ◆ Recognize that the first three terms, $(x^2 - 6x + 9)$, form a perfect square—the square of $(x - 3)$. Therefore we have

$$y = (x - 3)^2 + 4.$$

Note that the -3 in this expression is the same as half of the original coefficient of x.

EXAMPLE 1

Complete the square on the quadratic $y = x^2 + 10x - 11$ and compare its graph to that of the parabola $y = x^2$.

Solution We have

$$y = [x^2 + 10x] - 11$$

$$= [(x^2 + 10x + 5^2) - 5^2] - 11 \qquad \frac{1}{2}(10) = 5$$

$$= (x + 5)^2 - 25 - 11 = (x + 5)^2 - 36.$$

Graphically, the corresponding parabola is obtained by shifting $y = x^2$ to the left by 5 and down by 36.

◆

If the original quadratic expression has a leading coefficient other than 1, that coefficient must first be factored out. We illustrate how to do so in Example 2.

EXAMPLE 2

Complete the square on the quadratic $y = 2x^2 - 16x + 42$ and compare its graph to that of the parabola $y = x^2$.

Solution We begin by factoring out the leading coefficient 2 so that

$$2x^2 - 16x + 42 = 2[(x^2 - 8x) + 21]$$

$$= 2[(x^2 - 8x + (-4)^2) - (-4)^2 + 21] \qquad \frac{1}{2}(-8) = -4$$

$$= 2[(x - 4)^2 + 5] = 2(x - 4)^2 + 10.$$

The corresponding parabola is obtained by doubling $y = x^2$ and then shifting it 4 units to the right and 10 units up.

◆

Problems

In Problems 1–5, evaluate each number.

1. $|9 + 12|$

2. $|6 - 10|$

3. $|-7 - 3|$

4. $|-5| - |4|$

5. $|-5| - |-4|$

6. Calculate the value of $y = |x|$ for $x = -4, -3, -2, \ldots, 4$. Plot the points and then connect them. How would you describe the graph of $y = |x|$?

7. Repeat Problem 6, using $y = |x + 3|$ for $x = -7, -6, -5, \ldots, 1$. How does the graph $y = |x + 3|$ compare to the graph of $y = |x|$?

In Problems 8 and 9, evaluate the expression.

8. $\dfrac{5!}{3!}$

9. $\dfrac{20!}{3!17!}$

10. Rewrite the expression $1 + 4 + 9 + 16 + \cdots + 100$ in summation notation.

11. Rewrite the expression $1/5 + 1/6 + 1/7 + \cdots + 1/40$ in summation notation.

12. Find the value of $\sum\limits_{k=1}^{4} k^3$.

13. Find the distance between the points $(2, 4)$ and $(5, 8)$.

14. Find the equation of the circle that has $(2, 4)$ and $(5, 8)$ as endpoints on a diameter.

15. In the accompanying figure, triangle ABC is similar to triangle ADE.

    a. Find DE.
    b. Find AE.

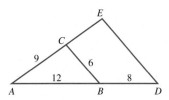

In Problems 16 and 17, complete the square to identify the center and radius of the circle.

16. $x^2 + y^2 - 2x - 10y = 55$

17. $x^2 + y^2 + 8x - 6y - 11 = 0$

18. Write an equation for the ellipse shown.

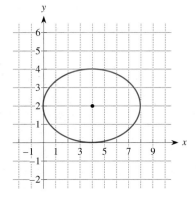

B   Solving Systems of Linear Equations Algebraically

A pair of linear equations such as

$$x + 5y = -3 \tag{1}$$
$$4x - 3y = 11 \tag{2}$$

is a *system of linear equations.* Its solution is a pair of values, one for x and the other for y, that satisfy both equations simultaneously. Geometrically, each of the two equations represents a line and every pair (x, y) that satisfies each individual equation is a point on that line. A pair of values for x and y that simultaneously satisfy both equations must be the point of intersection of the two lines, as shown in Figure A.6. The two lines seem to intersect at the point $(2, -1)$. We verify that this

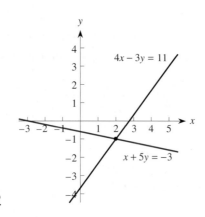

FIGURE A.6

point is indeed the solution by substituting $x = 2$ and $y = -1$ into the two original equations. For the first, we have

$$(2) + 5(-1) = -3$$

and for the second,

$$4(2) - 3(-1) = 11.$$

We can always use this kind of geometric approach to solve systems of two equations in two unknowns, but if the solutions are not simple numbers, the best we can get is a reasonably accurate estimate. Alternatively, we can solve such a system algebraically using either of two methods.

1. The **method of substitution:** (a) Solve for one variable in terms of the other, using one of the two equations. (b) Next, substitute the expression for that variable into the other equation to eliminate that variable. (This step is usually straightforward if the coefficient of one of the variables is 1 or -1.) (c) Then solve for the remaining variable. (d) Finally, substitute its value back into one of the equations to find the value of the other variable.

2. The **method of elimination:** Add or subtract an appropriate multiple of one of the equations to the other equation to eliminate one of the variables. (This method is discussed in detail in Section 10.5).

EXAMPLE 1

Solve the system of Equations (1) and (2) by using the method of substitution.

Solution The coefficient of x in Equation (1) is 1, so we use Equation (1) to solve for x in terms of y:

$$x = -5y - 3. \tag{3}$$

We substitute this expression into Equation (2) to get

$$4(-5y - 3) - 3y = 11 \quad \text{so that} \quad -20y - 12 - 3y = 11.$$

Note that we eliminated the variable x and now have a single equation in y only. Next, we collect like terms:

$$-23y = 23 \quad \text{so that} \quad y = \frac{23}{-23} = -1.$$

When we substitute $y = -1$ back into Equation (3), we find that

$$x = -5(-1) - 3 = 5 - 3 = 2.$$

This solution is the same, $x = 2$ and $y = -1$, that we obtained graphically in Figure A.6.

EXAMPLE 2

Solve the same system of linear equations

$$x + 5y = -3 \qquad \textbf{(1)}$$
$$4x - 3y = 11, \qquad \textbf{(2)}$$

using the method of elimination.

Solution   To eliminate the variable x from the two equations, we multiply Equation (1) by -4:

$$-4x - 20y = 12 \qquad \textbf{(4)}$$

while

$$4x - 3y = 11 \qquad \textbf{(2)}$$

Note that the coefficients of x are numerically equal but of opposite sign. If we add Equations (2) and (4), the x terms cancel, leaving

$$-23y = 23 \quad \text{so that} \quad y = -1.$$

If we now substitute this value of y into either Equation (1) or (2)—say, Equation (1)—we get

$$x + 5(-1) = -3 \quad \text{or} \quad x = 5 - 3 = 2,$$

which again is the same solution.

Alternatively, we could eliminate the variable y from Equations (1) and (2). To do so, we multiply Equation (1) by 3 and Equation (2) by 5 to get:

$$3\,\text{Eqn}\,(1) \qquad 3x + 15y = -9$$
$$5\,\text{Eqn}\,(2) \qquad 20x - 15y = 55.$$

We eliminate y by adding the two equations to get

$$23x = 46,$$

which again leads to $x = 2$ and hence $y = -1$.

◆

If we have a system of three linear equations in three unknowns—say, x, y, and z or x_1, x_2, and x_3 or a, b, and c—or an even larger system of linear equations, the method of substitution quickly becomes unworkable. The method of elimination is almost always preferable. However, in practice, for systems larger than two-by-two, calculators and computers are typically the method of choice instead of attempting to solve the systems by hand.

EXAMPLE 3

Find the solution to the system of linear equations

$$2x + 5y + 4z = -7 \qquad \textbf{(5)}$$
$$3x - y + 5z = 14 \qquad \textbf{(6)}$$
$$6x + 2y - 2z = 4, \qquad \textbf{(7)}$$

using the method of elimination.

Solution   We use Equation (7) to eliminate the variable x from the remaining two equations. To eliminate x from Equation (5), we first add -3 times Equation (5) to Equation (7):

$$
\begin{array}{ll}
\text{Eqn (7)} & 6x + 2y - 2z = 4 \\
-3 \text{ Eqn (5)} & \underline{-6x - 15y - 12z = 21} \\
& \qquad\; -13y - 14z = 25.
\end{array} \tag{8}
$$

Similarly, to eliminate the variable x from Equation (6), we add -2 times Equation (6) to Equation (7) to get

$$
\begin{array}{ll}
\text{Eqn (7)} & 6x + 2y - 2z = 4 \\
-2 \text{ Eqn (6)} & \underline{-6x + 2y - 10z = -28} \\
& \qquad\;\; 4y - 12z = -24.
\end{array} \tag{9}
$$

Equations (8) and (9) together are a system of two linear equations in the two unknowns y and z:

$$-13y - 14z = 25 \tag{8}$$
$$4y - 12z = -24 \tag{9}$$

We can solve this reduced system as before.

For instance, to eliminate the variable y, we multiply Equation (8) by 4 and multiply Equation (9) by 13, so that

$$
\begin{array}{ll}
4 \text{ Eqn (8)} & -52y - 56z = 100 \\
13 \text{ Eqn (9)} & 52y - 156z = -312.
\end{array}
$$

We add these two equations to get

$$-212z = -212 \quad \text{so that} \quad z = 1.$$

Substituting $z = 1$ into Equation (9), say, gives

$$4y - 12(1) = -24,$$

so that

$$4y = -24 + 12 = -12 \quad \text{and so } y = -3.$$

Substituting both of these values into Equation (5), say, gives

$$
\begin{aligned}
2x + 5(-3) + 4(1) &= -7 \\
2x - 15 + 4 &= -7 \\
2x = 4 \quad &\text{and so} \quad x = 2.
\end{aligned}
$$

Thus, the solution to the original system of three equations in three unknowns is $x = 2$, $y = -3$, and $z = 1$. Substitute these values into the three original equations to verify that they satisfy all three equations. ◆

Graphing calculators have the capability of solving systems of up to 99 equations in 99 unknowns at the push of a button. On some calculators, there is a `SIMULT` key for simultaneous equations; enter the number of linear equations, then enter the coefficients and the constant terms, and finally press `Solve` to get the solutions. On other calculators, you can solve a system of linear equations with the `Solve` command; enter the list of equations and the list of variables and then press enter. On still other calculators, you can solve systems of linear equations by using matrix methods, as discussed briefly in Appendix C and in detail in Chapter 10. (Actually, all calculators use matrix methods for solving systems of linear equations.)

C   Solving Systems of Linear Equations using Matrices

Briefly, a matrix is any rectangular array of numbers, such as

$$\begin{bmatrix} 4 & 0 & 5 & 1 \\ 7 & -2 & 8 & 6 \\ -3 & 1 & 2 & 3 \end{bmatrix}.$$

The size, or *dimension,* of a matrix is measured by the number of rows (horizontally across) and the number of columns (vertically down) in the array. The dimension of the preceding matrix is 3 by 4, which we write as 3×4.

Let's look at the system of linear equations:

$$2x + 5y + 4z = -7$$
$$3x - y + 5z = 14$$
$$6x + 2y - 2z = 4.$$

(We considered this system in Example 3 of Appendix B, where we found the solution algebraically to be $x = 2$, $y = -3$, and $z = 1$.) We first construct the *coefficient matrix* **A,** which consists of the coefficients from each of the equations:

$$\mathbf{A} = \begin{bmatrix} 2 & 5 & 4 \\ 3 & -1 & 5 \\ 6 & 2 & -2 \end{bmatrix}.$$

This matrix has 3 rows and 3 columns, so its size is 3 by 3, or 3×3. We also construct the matrix **B** of constants, consisting of the constants on the right-hand side of the equations:

$$\mathbf{B} = \begin{bmatrix} -7 \\ 14 \\ 4 \end{bmatrix}.$$

This matrix (also known as a vector) has three rows and one column, so its size is 3×1. Finally, we construct a 3×1 matrix **X** of variables:

$$\mathbf{X} = \begin{bmatrix} x \\ y \\ z \end{bmatrix};$$

these are the unknowns that we want to determine.

The system of three linear equations in three unknowns is then equivalent to the simple matrix equation

$$\mathbf{AX} = \mathbf{B}.$$

The solution to this matrix equation is found in terms of the **inverse matrix A^{-1}** (if it exists) of the matrix **A**:

$$\mathbf{X} = \mathbf{A}^{-1}\mathbf{B}.$$

To solve this system on the calculator, you must "name" each of the matrices in turn by giving its size: 3×3 for **A** and 3×1 for **B**. Then enter the values for each position. Finally, by selecting the appropriate names of the matrices, you have the calculator find **A^{-1} ∗ B**. The calculator displays the entries in the column matrix **X**:

$$\begin{bmatrix} 2 \\ -3 \\ 1 \end{bmatrix}.$$

See the instruction manual for your calculator for details on how to use these matrix features.

EXAMPLE

Use matrices to solve the system of linear equations

$$5x - 7y + 4z = 6$$
$$2x + 4y - 8z = 13$$
$$3x - 5y - 9z = -2.$$

Solution The coefficient matrix and the matrix of constants are

$$\mathbf{A} = \begin{bmatrix} 5 & -7 & 4 \\ 2 & 4 & -8 \\ 3 & -5 & -9 \end{bmatrix} \quad \text{and} \quad \mathbf{B} = \begin{bmatrix} 6 \\ 13 \\ -2 \end{bmatrix}.$$

After entering these matrices in the calculator and forming the expression **A^{-1} ∗ B**, we find the corresponding matrix of variables is

$$\mathbf{x} = \begin{bmatrix} x \\ y \\ z \end{bmatrix} = \mathbf{A}^{-1}\mathbf{B} = \begin{bmatrix} 3.791079812 \\ 2.049295775 \\ 0.347178404 \end{bmatrix}.$$

That is, the solution to the system of equations is $x \approx 3.791$, $y \approx 2.049$, and $z \approx 0.347$. The identical values are obtained on calculators having simultaneous equations (SIMULT or SOLVE) capabilities. ◆

D Symmetry

The notion of **symmetry** arises throughout mathematics in a variety of ways. We use symmetry to describe the behavior of functions and other geometric objects when one portion is a mirror image of another portion. We may describe a curve (whether or not it represents a function) as being *symmetric about a line*, or *symmetric with respect to an axis*, or *symmetric with respect to the origin*, or *symmetric about a point P*.

The ellipse shown in Figure A.7(a) is symmetric with respect to the x-axis because the lower half is the mirror image of the upper half. Similarly, as shown in

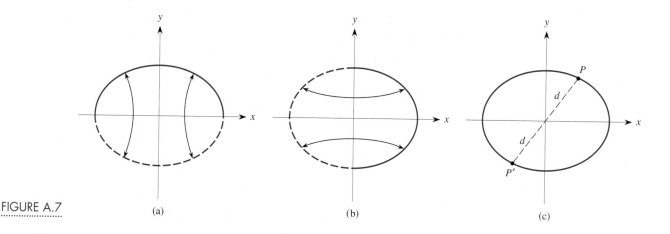

FIGURE A.7   (a)   (b)   (c)

Figure A.7(b), the ellipse is also symmetric with respect to the y-axis because the left half is the mirror image of the right half. The ellipse is also symmetric with respect to the origin because, for any point P on the ellipse, we can find the mirror image P' through the origin on the ellipse, as shown in Figure A.7(c).

The parabola $y = x^2$ is symmetric about the y-axis because the left and right sides are mirror images of one another. In fact, every parabola of the form $y = ax^2 + bx + c$ is symmetric about the vertical line through its turning point, or vertex. However, the cubic $y = x^3$ shown in Figure A.8 is not symmetric about the x-axis nor is it symmetric about the y-axis because the two portions of the curve are not mirror images of each other about either axis. However, the curve *is* symmetric with respect to the origin; if any point P with coordinates (a, b) is on the curve, so is the point P' with coordinates $(-a, -b)$, as shown in Figure A.8. This condition is equivalent to one portion of the curve being rotated through an angle of 180° to produce the other portion.

We summarize the key information about symmetry as follows.

1. A curve is symmetric about the x-axis if, when a point P at (a, b) lies on the curve, the point P' at $(a, -b)$ also lies on the curve, as shown in Figure A.9(a).

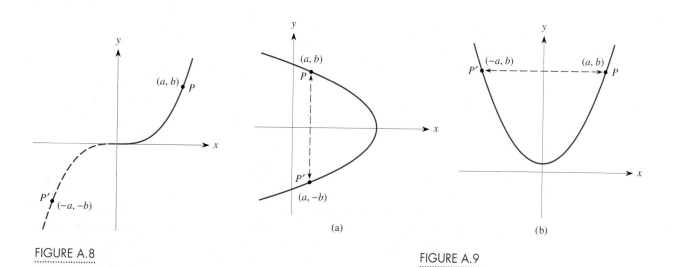

FIGURE A.8   (a)   (b)

FIGURE A.9

2. A curve is symmetric about the y-axis if, when a point P at (a, b) lies on the curve, the point P' at $(-a, b)$ also lies on the curve, as shown in Figure A.9(b).

3. A curve is symmetric about the origin if, when a point P at (a, b) lies on the curve, the point P' at $(-a, -b)$ also lies on the curve, as previously shown in Figure A.8.

◆EXAMPLE

Show algebraically that the curve $y = x^3$ is (a) not symmetric about the x-axis, (b) not symmetric about the y-axis, and (c) symmetric about the origin.

Solution

a. It is obvious from the graph of $y = x^3$ shown in Figure A.10 that the curve is not symmetric about the x-axis; however, we illustrate how to prove this fact by applying the principles of symmetry. Suppose that a point (a, b) lies on the curve so that $b = a^3$. We now consider the point $(a, -b)$. When we substitute $x = a$, we again get $a^3 = b$, not $-b$, so the point $(a, -b)$ is not on the curve and therefore the curve is not symmetric about the x-axis.

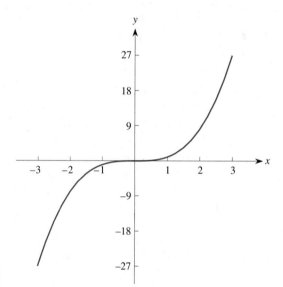

FIGURE A.10

b. Again, suppose that a point (a, b) lies on the curve so that $b = a^3$, and consider the point $(-a, b)$. When we substitute $x = -a$, we get $(-a)^3 = -a^3 = -b$, not b, so the curve is not symmetric about the y-axis.

c. Again, suppose that (a, b) lies on the curve, so that $b = a^3$, and consider the point $(-a, -b)$. When we substitute $x = -a$, we get $(-a)^3 = -a^3 = -b$, so that $(-a, -b)$ also satisfies the equation of the curve and the curve is symmetric about the origin.

◆

E Arithmetic of Complex Numbers

Complex numbers were originally introduced to allow people to solve quadratic equations such as $x^2 + 4 = 0$, which is equivalent to $x^2 = -4$. When we try to solve this equation by taking square roots, we obtain

$$x = \pm\sqrt{-4},$$

which has no solution among the real numbers.

To work with numbers such as these that involve the square root of a negative number, we introduce the **imaginary number** i:

$$i = \sqrt{-1}.$$

Using i, we can take the square root of any negative number. For instance,

$$\sqrt{-25} = \sqrt{25(-1)} = \sqrt{25}\sqrt{-1} = 5i.$$

Similarly, the roots of the equation $x^2 + 4 = 0$ are therefore

$$x = \pm\sqrt{-4} = \pm\sqrt{4}\sqrt{-1} = \pm 2i,$$

which is equivalent to $x = 2i$ and $x = -2i$.

◆ **EXAMPLE**

Find the roots of $x^2 - 2x + 10 = 0$, using the quadratic formula

$$x = \frac{-b \pm \sqrt{b^2 - 4ac}}{2a}.$$

Solution Substituting $a = 1$, $b = -2$, and $c = 10$ into the quadratic formula yields

$$x = \frac{-(-2) \pm \sqrt{(-2)^2 - 4(1)(10)}}{2(1)}$$

$$= \frac{2 \pm \sqrt{4 - 40}}{2} = \frac{2 \pm \sqrt{-36}}{2}$$

$$= \frac{2 \pm 6i}{2} = 1 \pm 3i.$$

Therefore the roots are $x = 1 + 3i$ and $x = 1 - 3i$. ◆

Each of these roots is called a *complex number*. In general, we write a **complex number** in the form $z = a + bi$, where a and b are both real numbers. Thus z consists of a real number a and an imaginary number bi. We call a the *real part* of z and b the *imaginary part* of z. Note that an imaginary number is just a multiple of i, such as $4i$; a complex number is the sum of a real number and an imaginary number, such as $3 - 7i$.

In the equation $x^2 + 4 = 0$, the roots were $x = 2i$ and $x = -2i$. In the above example, the roots were $x = 1 + 3i$ and $x = 1 - 3i$. As in both cases, complex numbers typically arise in pairs of the form $a + bi$ and $a - bi$, called **complex conjugates**.

Powers of i

Because $i = \sqrt{-1}$, we have

$$i^2 = (\sqrt{-1})^2$$
$$= -1;$$

that is,

$$i^2 = -1.$$

From this relation we can find other powers of i:

$$i^3 = (i^2)(i) = (-1)(i) = -i,$$
$$i^4 = (i^2)(i^2) = (-1)(-1) = 1,$$
$$i^5 = (i^4)(i) = (1)(i) = i,$$

and so on. All higher powers simply cycle through these four values: i, -1, $-i$, and 1.

Complex Arithmetic

The arithmetic of complex numbers, for the most part, is straightforward. Consider the two complex numbers $z = 5 + 4i$ and $w = 3 - 11i$.

Addition   To add complex numbers, we add the real parts and the imaginary parts separately. For instance,

$$z + w = (5 + 4i) + (3 - 11i) = 8 - 7i.$$

This is totally analogous to how we collect like terms in algebra. In general,

If $z = a + bi$ and $w = c + di$, then
$$z + w = (a + c) + (b + d)i.$$

Subtraction   To subtract complex numbers, we subtract the real parts and the imaginary parts separately. For our numbers $z = 5 + 4i$ and $w = 3 - 11i$,

$$z - w = (5 + 4i) - (3 - 11i)$$
$$= (5 - 3) + (4i - (-11i)) = 2 + 15i.$$

In general,

If $z = a + bi$ and $w = c + di$, then
$$z - w = (a - c) + (b - d)i.$$

Multiplication   To multiply complex numbers, we multiply them algebraically, use the fact that $i^2 = -1$ to simplify any power of i, and collect like terms consisting of the real part and the imaginary part. For instance,

$$z \times w = (5 + 4i)(3 - 11i)$$
$$= (5)(3) + (5)(-11i) + (4i)(3) + (4i)(-11i)$$
$$= 15 - 55i + 12i - 44i^2$$
$$= 15 - 43i + 44 \qquad\qquad i^2 = -1$$
$$= 59 - 43i.$$

In general,

> If $z = a + bi$ and $w = c + di$, then
> $$z \times w = (ac - bd) + (ad + bc)i.$$

Note that, in the particular case where z and w are complex conjugates (say, $z = 6 + 8i$ and $w = 6 - 8i$) we have

$$z \times w = (6 + 8i)(6 - 8i)$$
$$= 6(6) + 6(-8i) + (8i)(6) - (8i)(8i)$$
$$= 36 - 48i + 48i - 64i^2$$
$$= 36 - 64(-1) = 100. \qquad i^2 = -1$$

In general,

> If $z = a + bi$ and $w = a - bi$ are complex conjugates, then
> $$z \times w = a^2 + b^2.$$

Consequently, the product of complex conjugates is always a real number.

Division of complex numbers is somewhat more difficult, but the ideas we developed in Section 8.3 provide a simple way to do it.

Problems

In Problems 1–4, simplify each expression.

1. i^{23}
2. i^{72}
3. i^{58}
4. i^{45}

In Problems 5–12, perform each operation. Write each answer in the form $a + bi$, where a and b are real numbers.

5. $(8 + i) + (-6 + 3i)$
6. $(5 - 2i) + (7 + 6i)$
7. $(10 - i) - (-1 + i)$
8. $(4 + 5i) - 8i$
9. $(1 - 3i)(2 + i)$
10. $(5 + 6i)^2$
11. $(15 + 2i)(15 - 2i)$
12. $(7 - 4i)(7 + 4i)$

In Problems 13–18, find the roots of each equation.

13. $x^2 + 25 = 0$
14. $4x^2 + 9 = 0$
15. $5x^2 + 2x + 1 = 0$
16. $x^2 - 4x + 7 = 0$
17. $3x^2 + 4x + 8 = 0$
18. $5x^2 - 2x + 4 = 0$

F   Introduction to Data Analysis

In all the cases presented in this book, we have used data relating two quantities and sought to identify a relationship—linear or nonlinear—between them. Situations also are common in which we have data only on a single quantity. For instance, your professor may want to analyze the scores of all students in your class on an exam. Alternatively, she may want to compare the set of scores that each student has on all the exams in a course. In such cases, methods are needed to analyze the data to extract useful information.

Measuring the Center of a Set of Data

Given a set of data on a single quantity, it is usually most important to find the *center* of the data to give a single number that describes the entire set of values. Probably the most common method is to use the **mean,** or arithmetic average, of the data. Suppose that we have n data values, $x_1, x_2, x_3, \ldots, x_n$. Their mean is given by

$$\bar{x} = \frac{x_1 + x_2 + \cdots + x_n}{n} = \frac{\left(\sum_{k=1}^{n} x_k\right)}{n}.$$

For instance, if Amy's grades on her four math tests are 85, 91, 98, and 94, her mean grade for the course is $\bar{X} = (85 + 91 + 98 + 94)/4 = 92$.

The mean also gives a simple way to compare one set of data to another. If Bret had a mean of 87 on the same four tests, clearly he didn't do quite as well as Amy did. This type of comparison is usually applied by a professor in assigning final grades in a course.

However, there are situations in which the mean gives misleading information about a set of data, and other measures for the center for the data must be used. The accompanying set of data shows the number of home runs for the 2001 season for each of the starting players in the San Francisco Giants lineup. The mean number of home runs for the eight players was

$$\bar{x} = \frac{37 + 22 + 6 + 73 + 15 + 5 + 6 + 8}{8} = 21.5.$$

However, this value is unrepresentative of the team as a whole because Bobby Bonds hit so many home runs that season. His record-breaking 73 home runs has a disproportionate effect on the mean because his contribution is so far away from the values for the other players.

Player	Position	Home runs
R. Aurillia	SS	37
J. Kent	2B	22
B. Santiago	C	6
B. Bonds	OF	73
M. Benard	OF	15
R. Martinez	3B	5
C. Murray	OF	6
J. Snow	1B	8

Think About This   Recalculuate the mean number of home runs of the Giants starting players when you remove Bobby Bonds's 73 home runs. ▭

For this set of data, the **median** is a far better measure of the center. To find the median, you must list the data values in either increasing or decreasing order, as in 5, 6, 6, 8, 15, 22, 37, 73. If there are an odd number of entries, the median is defined as the middle value when the data are in order. If there are an even number of entries, the median is defined as the average of the middle two values. In this way, the median is always located at the center of the ordered list of data—there are just as many values below it as there are above it. For the number of home runs hit by the Giants during 2001, there are 8 entries, so the median is the average of the middle two, 8 and 15. Thus the median is $(8 + 15)/2 = 11.5$ home runs. This value is more representative of *all* of the players; Bonds's 73 home runs does not have the disproportionate effect on the median that it has on the mean. In fact, if Bonds had hit 103 home runs, instead of 73, the median would not change, but the mean certainly would change.

Think About This   Recalculuate the median number of home runs of the Giants players when you remove Bobby Bonds's 73 home runs. Did the value change as much as it did when you recalculated the mean in the previous Think About This exercise? ◻

In general, the median is more representative of a set of data when there is large variation among the data entries, especially if the data contain extreme values. In such a case, a handful of very large or very small values has a disproportionate effect on the mean, but not on the median.

Measuring the Spread in a Set of Data

Locating the center in a set of data often isn't sufficient to give a full picture of the data. For instance, consider two students, Carol and Doug, who had the following scores on five tests in a course:

Carol: 78, 80, 85, 75, 82

Doug: 92, 75, 88, 63, 82.

Both have means of 80, but there is a clear difference in the overall scores of the two students. Carol's results are consistent; they are all quite close to the mean of 80. But Doug's results are very scattered, ranging from a minimum of 63 to a maximum of 92.

We therefore need a way to measure the amount of scatter, or *spread,* in a set of data. One way to do so is to use the mean as the center of the data and calculate the amount of spread about the mean. The result is a quantity called the **standard deviation.**

There are two ways to calculate this quantity, depending on whether the data represents a *population* (*all* possible values associated with a quantity) or just a *sample* drawn from a much larger population. For a population, the standard deviation is denoted by σ (lowercase Greek letter sigma) and is found from

$$\sigma = \sqrt{\frac{(x - \bar{x})^2}{n}},$$

where n is the number of data points and $\bar{x}$ is the mean of the population. (In statistics texts, it is standard to write the mean of a population as μ (the lowercase Greek letter mu) instead of $\bar{x}$. For a sample, the standard deviation is denoted by s and is found from

$$s = \sqrt{\frac{(x - \bar{x})^2}{n - 1}},$$

where n is the number of data points and $\bar{x}$ is the mean of the sample. (We won't go into the reason for the difference in the two formulas here; any introductory statistics text gives a full discussion of the reason.)

All these quantities, as well as several others that we discuss shortly, are routinely obtained by using the statistical features of all calculators under the `one-variable` option in the `Statistics` menu, as well as many software packages. In particular, for Carol's test scores, the calculator gives the values

$$\overline{X} = 80, \quad s = 3.808, \quad \text{and} \quad \sigma = 3.406;$$

and for Doug's scores

$$\overline{X} = 80, \quad s = 11.467, \quad \text{and}$$

(We write $\overline{X}$ as the calculator output; whether it represents a population mean μ or a sample mean $\bar{x}$ depends on the context.) Note that the value for the standard deviation for Doug is about three times as large as that for Carol, which indicates that Doug's test scores are much more widely spread about the mean than Carol's are.

There is also a way to measure the spread in a set of data about the median as the center of the data. We do so by partitioning the data values into four groups— the top 25%, the next 25%, the 25% below the median, and the bottom 25%. These are known as the **quartiles.** For instance, suppose that we have a set of data with 60 values that have been arranged in either increasing or decreasing order. Then the number that separates the bottom 15 data values from the rest of the data values is called the *first quartile, Q_1*. The *second quartile, Q_2*, separates the bottom 30 values from the top 30 values and is the median. The number that separates the top 15 values from the rest is the *third quartile, Q_3*. In addition, it is standard to report the minimum and maximum values in the data.

As with the standard deviation, these values are also included in the output of all calculators with statistical features and in many software packages. In particular, the values corresponding to Carol's test grades are

$$\text{Minimum} = 75, \quad Q_1 = 76.5, \quad \text{median} = 80,$$
$$Q_3 = 83.5, \quad \text{and} \quad \text{maximum} = 85.$$

Those corresponding to Doug's grades are

$$\text{Minimum} = 63, \quad Q_1 = 69, \quad \text{median} = 82,$$
$$Q_3 = 90, \quad \text{and} \quad \text{maximum} = 92.$$

These values give a clear picture of the way that the different sets of data values are spread about the respective medians.

G ···· 2002 World Population Data

	Population Mid-2002 (millions)	Births per 1,000 Pop.	Deaths per 1,000 Pop.	Growth Rate (%)	Doubling Time	Projected Population (millions) 2025	Projected Population (millions) 2050	Life Expectancy at Birth (years)	Percent of pop. 15–49 with HIV/AIDS
WORLD	6,215	21	9	1.3	53.7	7,859	9,104	67	1.2
NORTH AMERICA	319	14	9	0.6	115.9	382	450	77	0.6
Canada	31.3	11	7	0.3	231.4	36.0	36.6	79	0.3
United States	287.4	15	9	0.6	115.9	346.0	413.5	77	0.6

	Population Mid-2002 (millions)	Births per 1,000 Pop.	Deaths per 1,000 Pop.	Growth Rate (%)	Doubling Time	Projected Population (millions) 2025	2050	Life Expectancy at Birth (years)	Percent of pop. 15–49 with HIV/AIDS
LATIN AMERICA & THE CARIBBEAN	531	23	6	1.7	41.1	697	815	71	0.7
CENTRAL AMERICA	140	27	5	2.2	31.9	188	225	74	0.5
Costa Rica	3.9	21	4	1.7	41.1	5.2	5.9	77	0.6
El Salvador	6.6	30	7	2.3	30.5	9.3	12.4	70	0.6
Guatemala	12.1	36	7	2.9	24.2	19.8	27.2	66	1.0
Honduras	6.7	33	6	2.8	25.1	9.6	12.2	66	1.6
Mexico	101.7	26	5	2.1	33.4	131.7	150.7	75	0.3
Nicaragua	5.4	34	5	2.8	25.1	8.6	11.6	68	0.2
Panama	2.9	23	4	1.9	36.8	3.8	4.3	74	1.5
CARIBBEAN	37	21	8	1.3	53.7	45	50	69	2.4
Antigua and Barbuda	0.1	22	6	1.6	43.7	0.1	0.1	71	—
Bahamas	0.3	18	5	1.3	53.7	0.4	0.5	72	3.5
Barbados	0.3	15	8	0.6	115.9	0.3	0.3	73	1.3
Cuba	11.3	12	7	0.5	139.0	11.8	11.1	76	0.1
Dominican Republic	8.8	26	5	2.1	33.4	12.1	14.9	69	2.7
Grenada	0.1	19	7	1.2	58.1	0.1	0.1	71	—
Guadeloupe	0.5	17	6	1.2	58.1	0.5	0.6	77	—
Haiti	7.1	33	15	1.7	41.1	9.6	11.9	49	6.1
Jamaica	2.6	20	5	1.5	46.6	3.3	3.8	75	1.2
Martinique	0.4	14	6	0.8	87.0	0.4	0.4	79	—
Netherlands Antilles	0.2	14	6	0.7	99.4	0.2	0.3	76	—
Puerto Rico	3.9	15	7	0.8	87.0	4.1	4.1	76	—
St. Kitts-Nevis	0.04	19	9	1.0	69.7	0.05	0.1	71	—
Saint Lucia	0.2	18	6	1.2	58.1	0.2	0.2	71	—
Trinidad and Tobago	1.3	14	8	0.7	99.4	1.4	1.4	71	2.5
SOUTH AMERICA	354	22	6	1.5	46.6	463	540	70	0.6
Argentina	36.5	19	8	1.1	63.4	47.2	54.5	74	0.7
Bolivia	8.8	32	9	2.3	30.5	13.2	17.1	63	0.1
Brazil	173.8	20	7	1.3	53.7	219.0	247.2	69	0.7
Chile	15.6	18	6	1.2	58.1	19.5	22.2	77	0.3
Colombia	43.8	22	6	1.7	41.1	59.7	71.5	71	0.4
Ecuador	13.0	28	6	2.2	31.9	18.5	22.9	71	0.3
Guyana	0.8	24	8	1.5	46.6	0.7	0.5	63	2.7
Paraguay	6.0	31	5	2.7	26.0	10.1	15.0	71	0.1
Peru	26.7	26	7	2.0	35.0	35.7	42.8	69	0.4
Suriname	0.4	24	7	1.7	41.1	0.5	0.4	71	1.2
Uruguay	3.4	16	10	0.7	99.4	3.8	4.2	75	0.3
Venezuela	25.1	24	5	1.9	36.8	34.8	41.0	73	0.5
EUROPE	728	10	11	−0.1	—	718	651	74	0.4
NORTHERN EUROPE									
Denmark	5.4	12	11	0.1	693.5	5.9	6.4	77	0.2

	Population Mid-2002 (millions)	Births per 1,000 Pop.	Deaths per 1,000 Pop.	Growth Rate (%)	Doubling Time	Projected Population (millions) 2025	2050	Life Expectancy at Birth (years)	Percent of pop. 15–49 with HIV/AIDS
Estonia	1.4	9	14	−0.4	—	1.2	0.9	71	1.0
Finland	5.2	11	10	0.2	346.9	5.3	4.8	78	0.1
Iceland	0.3	15	6	0.9	77.4	0.3	0.4	79	0.2
Ireland	3.8	14	8	0.6	115.9	4.5	4.5	77	0.1
Latvia	2.3	8	14	−0.6	—	2.2	1.8	71	0.4
Lithuania	3.5	9	12	−0.3	—	3.5	3.1	73	0.1
Norway	4.5	13	10	0.3	231.4	5.0	5.2	79	0.1
Sweden	8.9	10	11	0.0	—	9.5	9.8	80	0.1
United Kingdom	60.2	11	10	0.1	693.5	64.8	65.4	78	0.1

WESTERN EUROPE

Austria	8.1	9	9	0.0	—	8.4	8.2	78	0.2
Belgium	10.3	11	10	0.1	693.5	10.8	11.0	78	0.2
France	59.5	13	9	0.4	173.6	64.2	65.1	79	0.3
Germany	82.4	9	10	−0.1	—	78.1	67.7	78	0.1
Liechtenstein	0.03	12	7	0.5	139.0	0.04	0.04	—	—
Luxembourg	0.5	13	9	0.5	139.0	0.6	0.6	78	0.2
Monaco	0.03	23	16	0.6	115.9	0.04	0.04	—	—
Netherlands	16.1	13	9	0.4	173.6	17.7	18.0	78	0.2
Switzerland	7.3	10	8	0.2	346.9	7.6	7.4	80	0.5

EASTERN EUROPE

Belarus	9.9	9	14	−0.5	—	9.4	8.5	69	0.3
Bulgaria	7.8	9	14	−0.5	—	6.6	5.3	72	z[1]
Czech Republic	10.3	9	11	−0.2	—	10.3	9.4	75	z[1]
Hungary	10.1	10	13	−0.4	—	9.2	8.1	72	0.1
Moldova	4.3	9	10	−0.1	—	4.5	4.2	68	0.2
Poland	38.6	10	10	0.0	—	38.6	33.9	74	0.1
Romania	22.4	10	12	−0.2	—	20.6	17.1	71	z
Russia	143.5	9	16	−0.7	—	129.1	101.7	65	0.9
Slovakia	5.4	10	10	0.0	—	5.2	4.7	73	z
Ukraine	48.2	8	15	−0.8	—	45.1	38.4	68	1.0

SOUTHERN EUROPE

Albania	3.1	17	5	1.2	58.1	4.1	4.7	74	z
Bosnia-Herzegovina	3.4	12	8	0.4	173.6	3.6	3.4	68	z
Croatia	4.3	10	12	−0.2	—	4.1	3.6	74	z
Greece	11.0	10	10	0.0	—	10.4	9.7	78	0.2
Italy	58.1	9	9	0.0	—	57.5	52.2	80	0.4
Macedonia	2.0	15	9	0.6	115.9	2.2	2.1	73	z
Malta	0.4	11	8	0.3	231.4	0.4	0.4	77	0.1
Portugal	10.4	12	10	0.2	346.9	9.7	8.6	76	0.5
Slovenia	2.0	9	9	0.0	—	2.0	1.7	76	z
Spain	41.3	10	9	0.1	693.5	44.3	42.1	79	0.5
Yugoslavia	10.7	12	11	0.2	346.9	10.7	10.2	72	0.2

[1]z = less than $\frac{1}{2}$%

	Population Mid-2002 (millions)	Births per 1,000 Pop.	Deaths per 1,000 Pop.	Growth Rate (%)	Doubling Time	Projected Population (millions)		Life Expectancy at Birth (years)	Percent of pop. 15–49 with HIV/AIDS
						2025	2050		
AFRICA	**840**	**38**	**14**	**2.4**	**29.2**	**1,281**	**1,845**	**53**	**6.6**
NORTHERN AFRICA									
Algeria	31.4	23	5	1.8	38.9	43.0	51.3	70	0.1
Egypt	71.2	27	7	2.0	35.0	96.1	115.4	66	z
Libya	5.4	28	4	2.4	29.2	8.3	10.8	75	0.2
Morocco	29.7	25	6	1.9	36.8	40.5	48.4	69	0.1
Sudan	32.6	36	12	2.4	29.2	49.6	63.5	56	2.6
Tunisia	9.8	17	6	1.2	58.1	11.6	12.2	72	z
WESTERN AFRICA									
Benin	6.6	41	12	2.9	24.2	12.0	18.1	54	3.6
Burkina Faso	12.6	47	17	3.0	23.4	21.6	34.3	47	6.5
Cote d'Ivoire	16.8	36	16	2.0	35.0	25.6	35.7	45	9.7
Gambia	1.5	42	13	2.9	24.2	2.7	4.2	53	1.6
Ghana	20.2	32	10	2.2	31.9	26.5	32.0	58	3.0
Guinea	8.4	45	18	2.7	26.0	14.1	20.7	48	1.5
Guinea-Bissau	1.3	45	20	2.5	28.1	2.2	3.3	45	2.8
Liberia	3.3	49	17	3.1	22.7	6.0	10.0	50	2.8
Mali	11.3	49	19	3.0	23.4	21.6	36.4	47	1.7
Mauritania	2.6	34	14	2.0	35.0	5.1	7.2	53	0.5
Niger	11.6	55	20	3.5	20.1	25.7	51.9	45	1.4
Nigeria	129.9	41	14	2.7	26.0	204.5	303.6	52	5.8
Senegal	9.9	38	12	2.6	27.0	16.5	22.7	53	0.5
Sierra Leone	5.6	49	25	2.4	29.2	10.6	14.9	39	7.0
Togo	5.3	40	11	2.9	24.2	7.6	9.7	55	6.0
EASTERN AFRICA									
Burundi	6.7	43	21	2.2	31.9	12.4	20.2	41	8.3
Eritrea	4.5	43	12	3.0	23.4	8.3	13.3	56	2.8
Ethiopia	67.7	40	15	2.5	28.1	117.6	172.7	52	6.4
Kenya	31.1	34	14	2.0	35.0	33.3	37.4	48	15.0
Madagascar	16.9	43	13	3.0	23.4	30.8	47.0	55	0.3
Malawi	10.9	46	22	2.4	29.2	12.8	15.0	38	15.0
Mauritius	1.2	16	7	1.0	69.7	1.4	1.5	72	0.1
Mozambique	19.6	43	23	2.0	35.0	20.6	22.9	38	13.0
Rwanda	7.4	42	21	2.2	31.9	8.0	8.9	39	8.9
Somalia	7.8	48	19	2.9	24.2	14.9	25.5	47	1.0
Tanzania	37.2	40	13	2.7	26.0	59.8	88.3	52	7.8
Uganda	24.7	48	18	3.0	23.4	48.0	84.1	43	5.0
Zambia	10.0	42	22	2.0	35.0	14.3	20.3	37	21.5
Zimbabwe	12.3	29	20	0.9	77.4	10.3	10.1	38	33.7
MIDDLE AFRICA									
Angola	12.7	48	20	2.9	24.2	28.2	53.3	45	5.5

	Population Mid-2002 (millions)	Births per 1,000 Pop.	Deaths per 1,000 Pop.	Growth Rate (%)	Doubling Time	Projected Population (millions)		Life Expectancy at Birth (years)	Percent of pop. 15–49 with HIV/AIDS
						2025	2050		
Cameroon	16.2	37	12	2.5	28.1	24.7	34.7	55	11.8
Central African Republic	3.6	38	18	2.0	35.0	4.9	6.4	44	12.9
Chad	9.0	49	16	3.3	21.3	18.2	33.3	51	3.6
Congo	3.2	44	14	3.0	23.0	6.3	10.7	51	7.2
Congo, Dem. Rep. of	55.2	46	15	3.1	22.7	106.0	181.9	49	4.9
Gabon	1.2	32	16	1.6	43.7	1.4	1.8	50	4.2
SOUTHERN AFRICA									
Botswana	1.6	31	22	0.8	87.0	1.2	1.2	39	38.8
Lesotho	2.2	33	15	1.8	38.9	2.4	2.8	51	31.0
Namibia	1.8	35	20	1.6	43.7	2.0	2.5	43	22.5
South Africa	43.6	25	15	1.1	63.4	35.1	32.5	51	20.1
Swaziland	1.1	41	20	2.0	35.0	1.4	2.0	40	33.4
ASIA	3,776	20	7	1.3	53.7	4,741	5,297	67	0.4
WESTERN ASIA									
Armenia	3.8	8	6	0.2	346.9	3.7	3.2	72	0.2
Azerbaijan	8.2	14	6	0.8	87.0	10.2	13.0	72	z
Bahrain	0.7	22	3	1.9	36.8	1.7	2.9	74	0.3
Cyprus	0.9	12	7	0.6	115.9	1.0	1.0	77	0.3
Georgia	4.4	9	9	0.0	—	3.6	2.5	73	z
Iraq	23.6	35	10	2.5	28.1	41.2	60.1	58	z
Israel	6.6	21	6	1.5	46.6	9.3	11.0	78	0.1
Jordan	5.3	28	5	2.3	30.5	8.7	11.8	70	z
Kuwait	2.3	32	3	2.9	24.2	3.9	5.5	76	0.1
Lebanon	4.3	21	7	1.4	49.9	5.4	5.8	73	0.1
Oman	2.6	33	4	2.9	24.2	5.1	7.4	73	0.1
Palestinian Territory	3.5	40	4	3.5	20.1	7.4	11.2	72	—
Qatar	0.6	31	4	2.7	26.0	0.8	0.9	72	0.1
Saudi Arabia	24.0	35	6	2.9	24.2	40.9	60.3	72	z
Syria	17.2	31	6	2.6	27.0	26.5	34.4	70	z
Turkey	67.3	22	7	1.5	46.6	85.0	96.9	69	z
United Arab Emirates	3.5	17	2	1.5	46.6	4.5	5.1	74	0.2
Yemen	18.6	44	11	3.3	21.3	39.6	71.1	59	0.1
SOUTH CENTRAL ASIA									
Afghanistan	27.8	43	19	2.4	29.2	45.9	67.2	45	z
Bangladesh	133.6	30	8	2.2	31.9	177.8	205.4	59	z
Bhutan	0.9	34	9	2.5	28.1	1.4	2.0	66	z
India	1049.5	26	9	1.7	41.1	1363.0	1628.0	63	0.8
Iran	65.6	18	6	1.2	58.1	84.7	96.5	69	0.1
Kazakhstan	14.8	15	10	0.5	139.0	14.7	14.0	66	0.1
Kyrgyzstan	5.0	20	7	1.3	53.7	6.5	7.5	69	z[1]

	Population Mid-2002 (millions)	Births per 1,000 Pop.	Deaths per 1,000 Pop.	Growth Rate (%)	Doubling Time	Projected Population (millions)		Life Expectancy at Birth (years)	Percent of pop. 15–49 with HIV/AIDS
						2025	2050		
Nepal	23.9	31	11	2.1	33.4	36.1	43.4	58	0.5
Pakistan	143.5	30	9	2.1	33.4	242.1	332.0	63	0.1
Sri Lanka	18.9	18	6	1.2	58.1	22.1	22.7	72	z
Tajikistan	6.3	19	4	1.4	49.9	7.8	8.5	68	z
Turkmenistan	5.6	19	5	1.3	53.7	7.2	7.9	67	z
Uzbekistan	25.4	22	5	1.7	41.1	37.2	38.6	70	z
SOUTHEAST ASIA									
Brunei	0.4	22	3	2.0	35.0	0.5	0.6	74	0.2
Cambodia	12.3	28	11	1.7	41.1	18.4	21.9	56	2.7
East Timor	0.8	29	15	1.5	46.6	1.2	1.4	48	—
Indonesia	217.0	22	6	1.6	43.7	281.9	315.8	68	0.1
Laos	5.5	36	13	2.3	30.5	8.6	11.3	54	0.1
Malaysia	24.4	23	4	1.9	36.8	35.6	46.4	73	0.4
Myanmar	49.0	25	12	1.3	53.7	60.2	68.5	56	2.0
Philippines	80.0	28	6	2.2	319	115.5	145.7	68	z
Singapore	4.2	12	4	0.8	87.0	8.0	10.4	78	0.2
Thailand	62.6	14	6	0.8	87.0	72.1	71.9	72	1.8
Vietnam	79.7	19	5	1.4	49.9	104.1	117.2	68	0.3
EAST ASIA									
China	1,280.7	13	6	0.7	99.4	1,454.7	1,393.6	71	0.1
China, Hong Kong	6.8	7	5	0.2	346.9	8.4	7.5	79	0.1
China, Macao	0.4	7	3	0.4	173.6	0.6	0.8	77	—
Japan	127.4	9	8	0.2	346.9	121.1	100.6	81	z
Korea, North	23.2	18	10	0.7	99.4	25.7	26.4	64	z
Korea, South	48.4	13	5	0.8	87.0	50.5	50.0	76	z
Mongolia	2.4	23	8	1.5	46.6	3.3	3.9	63	z
Taiwan	22.5	11	6	0.6	115.9	25.3	25.2	75	—
OCEANIA	**32**	**18**	**7**	**1.0**	**69.7**	**40**	**46**	**75**	**0.2**
Australia	19.7	13	7	0.6	115.9	23.2	25.0	80	0.1
Fiji	0.9	25	6	1.9	36.8	1.0	0.9	67	0.1
French Polynesia	0.2	21	5	1.6	43.7	0.3	0.4	72	—
Guam	0.2	24	4	2.0	35.0	0.2	0.3	77	—
New Caledonia	0.2	21	6	1.6	43.7	0.3	0.4	73	—
New Zealand	3.9	14	7	0.7	99.4	4.6	5.0	78	0.1
Papua-New Guinea	5.0	34	11	2.3	30.5	8.0	10.9	57	0.7
Samoa	0.2	30	6	2.4	29.2	0.2	0.2	68	—
Solomon Islands	0.5	41	7	3.4	20.7	0.9	1.5	67	—

Selected Answers

Chapter 1

Section 1.1

1. **a.** Function, number of miles depends on number of gallons
   b. Function, price depends on number of carats
   e. Function, amount of rain depends on the day
   c, d, and **f** are not functions.

3. **a.** (ii)   **c.** (i)   **d.** (iii)

Section 1.2

1. **e** is strictly increasing.
   g and **i** are strictly decreasing
   a, b, c, d, f, and **h** are neither

3.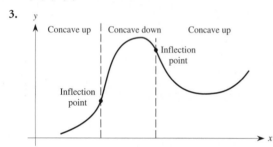

5.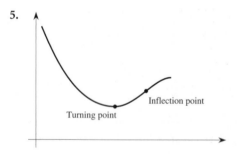

9.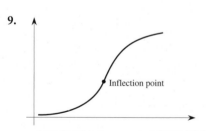

15. **a, b, c, d, e, h, i,** and **j** are periodic. **f** and **g** are not periodic.

17.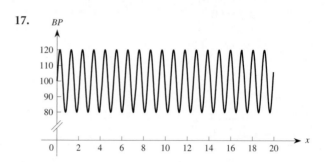

19. **a.** about $10\frac{1}{2}$ years
    b. 2000 and 2011

Section 1.3

3. **a.** A 10-year-old child is 50 inches tall.
   b. Increasing
   c. Generally concave down because growth starts quickly but usually slows down

5. $P(T) = kT$ for some constant k

7. $F(d) = \dfrac{k}{d^2}$ for some constant k

9. **a.** 156   **b.** 210; 243; 270; 330
   c. 150

11. **a.** 348
    b. 16 grams of peanut butter and 20 grams of jelly, or 25 grams of peanut butter and no jelly
    c. Peanut butter

13. **b.** 164 feet; 104 feet
    c. $H(2) = 176$ feet is the height of the ball after 2 seconds, and $H(3) = 156$ feet is the height of the ball after 3 seconds.
    d. 1.875 seconds; 176.25 feet
    e. 5.19 seconds
    f. Domain: $0 \le t \le 5.19$; range: $0 \le H \le 176.25$

15. $-\dfrac{1}{4}; -\dfrac{1}{3}; \dfrac{1}{5}; \dfrac{1}{12}; \dfrac{1}{21}; x = 2$ would make the denominator $0; x = -2$; domain: $x \ne \pm 2$

17. 6, 20, 30, 110; no; nonnegative; domain: $s \ge 0$

Section 1.4

1.(i) Function, with domain $-3.5 \le x \le 4$ and range $0 \le y \le 2$; decreasing from -3 to -2.5 and from -1 to 1.7; increasing elsewhere; concave down from -1.8 to 0 and concave up from -3 to -2 and from 0 to 4

746

(ii) Function, with domain $0 \leq x \leq 5$ and range $-1 \leq y \leq 2$; increasing and concave down everywhere

(iii) Not a function

(iv) Not a function

(v) Function, with domain $-1 \leq x \leq 5$ and range $-1 \leq y \leq 2$; decreasing and concave up everywhere

5.

x	$f(x) = x^2 - 3x + 2$
-3	20
-2	12
-1	6
0	2
1	0
2	0
3	2
4	6
5	12

$$f\left(\frac{1}{2}\right) = \frac{3}{4}; \quad f\left(\frac{3}{2}\right) = -\frac{1}{4}; \quad f\left(-\frac{5}{2}\right) = \frac{63}{4}$$

9. $f(5)$ can be any number between 10 and 15.

11. **a.** From about $x = -2.2$ to $x = 0.6$ and from $x = 3.9$ to infinity

   b. From $-\infty$ to about $x = -2.2$ and from $x = 0.6$ to $x = 3.9$

   c. Roughly $(-2.2, -3), (0.6, 3.5), (3.9, -7)$

   d. Roughly $(0.6, 3.5)$

   e. Roughly $(-2.2, -3)$ and $(3.9, -7)$

   f. $(-1, 0)$ and $(2.5, -2.5)$

   g. From $-\infty$ to roughly $x = -1$ and from $x = 2.5$ to ∞

   h. From $x = -1$ to $x = 2.5$

   i. As $x \to \infty$

   j. As $x \to -\infty$

   k. $x = -3, -1, 2, 5$

13. **a.** $g(x)$ **b.** $f(x)$ **c.** $h(x)$

17. **a.** Increasing   **b.** Concave up

19.

x	$f(x) = \sqrt{x}$
0	0
1	1
2	$\sqrt{2} \approx 1.414$
3	$\sqrt{3} \approx 1.732$
4	2
5	$\sqrt{5} \approx 2.236$
6	$\sqrt{6} \approx 2.449$

$$f\left(\frac{1}{2}\right) = \sqrt{\frac{1}{2}} \approx 0.707; \quad f\left(\frac{3}{2}\right) = \sqrt{\frac{3}{2}} \approx 1.225;$$

$$f\left(\frac{5}{2}\right) = \sqrt{\frac{5}{2}} \approx 1.581; \quad \text{domain: } x \geq 0$$

Section 1.5

5. **a.** 38 mph   **b.** 49 mph   **c.** 58 mph
   d. 69 mph   **e.** 150 ft

7. **a.** $y = f(t) = 80t - 16t^2$
   b. $y = f(t) = v_0 t - 16t^2$

Review Problems

1. Independent variable: depth of tumor; dependent variable: amount of radiation

3.

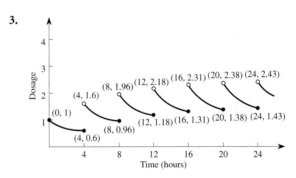

4.

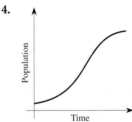

5. **a.** Function   **b.** Not a function

6. Overall, zoos with larger budgets have greater attendance.

7. **b.** 1990
   c. Most rapid change is between 1994 and 1995; slowest change is between 1993 and 1994.

8. 1, 2, 2.43, 2.0403, 34, $f(a) = 3a^2 - 2a + 1$

9. **a.** \$18,896.50   **b.** \$24,174.10   **c.** about 6.7 years

10. **a.** 5.6 years   **b.** 3.9 to 24.8 years
    c. Increasing and concave down
    d. 232 days   **e.** 16.2 years

11. **a.** $(-0.375, -5.5625)$ **b.** $y \geq -5.5625$

12. **a.** $(0, -9)$ **b.** All real numbers

13. **a.** $x \geq -5$ **b.** $x \leq -4$ or $x \geq 4$
    c. $x \neq \pm 3$ **d.** All real numbers

14. **a.**

Weight W (oz)	Cost (\$)
$0 < W \leq 1$	0.34
$1 < W \leq 2$	0.57
$2 < W \leq 3$	0.80
$3 < W \leq 4$	1.03
$4 < W \leq 5$	1.26

15. a. From A to B the track is increasing and concave up.
From B to C the track is increasing and concave down.
From C to D the track is decreasing and concave down.
From D to E the track is decreasing and concave up.
From E to F the track is increasing and concave up.

b. From A to B the car's speed is decreasing at an increasing rate.
From B to C the car's speed is decreasing at a decreasing rate.
From C to D the car's speed is increasing at an increasing rate.
From D to E the car's speed is increasing at a decreasing rate.
From E to F the car's speed is decreasing at an increasing rate.

Chapter 2

Section 2.2

1. a. (iii)   **b.** (i)   **c.** (v)   **d.** (vi)
   e. (iv)   **f.** (ii)

3. a. $y - 5 = 7(x - 2)$, or $y = 7x - 9$
   b. $y = -2$
   c. $y = -0.626x + 7.164$

5. a. $CT = -35.4625t + 442.2$
   b. $CD = 32.325t + 865.7$, where $t = 0$ in 1990
   c. The sales of cassette tapes are falling by 35.4625 million per year the sales of CDs are rising by 32.325 million per year.
   d. Late 1983
   e. $B = -3.1375t + 1307.9$

7. a. $C = 0.30t + 0.40$
   b. Each minute costs 30¢, and it costs 40¢ to place the call.
   c. $8.20
   d. $C = 0.21t + 0.28$; each minute costs 21¢, and it costs 28¢ to place the call; $5.74

9. a. DJ1: $C = 60t + 120$; DJ2: $C = 75t + 100$
   c. DJ1 costs less than DJ2 if she is hired for longer than $1\frac{1}{3}$ hours or 1 hour 20 minutes.

11. a. $T = 3217.50 + 0.28(I - 21,450)$ for $21,450 \leq I \leq 51,900$ with a range of $3217.50 \leq T \leq 11743.50$
   b. There is a 28% tax on income for single taxpayers with taxable income between $21,450 and $51,900.

13. a. $d = 15x + 300$ **b.** about 1880
   c. no

15. b. 30,300 feet

17. a. $y = 5x - 26$ **b.** $y = -\frac{1}{5}x + \frac{26}{5}$

19. b. $(3, 2)$ **c.** $7x - 2y = 17$
   d. $9x - y = 25$ **e.** $x = 3$

21. a. (i) and (iii)   **b.** $\left(2, -\frac{3}{2}\right)$

Exercising Your Algebra Skills

1. $x = \dfrac{19}{5}$ **3.** $x = -\dfrac{1}{2}$ **5.** $y = \dfrac{29}{18}$ **7.** $x = 1$

9. $q = \dfrac{194}{47} \approx 4.1$ **11.** $k = 3$

13. $x = \dfrac{19}{6}$ **15.** $t = 2511.25$

17. a. $\dfrac{2}{3}$ **b.** $4, -\dfrac{8}{3}$

19. a. $-\dfrac{4}{7}$ **b.** $-\dfrac{5}{4}, -\dfrac{5}{7}$

Section 2.3

1. a. Yes, 47   **b.** No   **c.** Yes, 137.1

3. $y = 0.057x + 1.329$

5. a. $W = 0.66t + 26.5$ **b.** 41.02 billion
   c. $W - 26.5 = 0.66(t - 80)$, 41.02 billion

11. a. $3N + 2G = 30$
   c. Domain: $0 \leq N \leq 10$; range: $0 \leq G \leq 15$
   d. $3N + 2G = 60$ **e.** $3N + G = 30$
   f. $6N + 2G = 30$

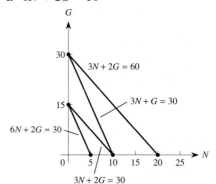

13. a. For $3y - 2x = 12$, the slope is $\frac{2}{3}$ and the vertical intercept is at 4; for $4x + 5y = 20$, the slope is $-\frac{4}{5}$ and the vertical intercept is at 4.
   c. $(0, 4)$

15. Possible: **a, f,** and **i**; Impossible: **b, c, d, e, g** and **h**

17. a. PQ, PR, QR **b.** QR, PR, PQ

Exercising Your Algebra Skills

1. $h = \dfrac{A}{b}$ **3.** $r = \sqrt{\dfrac{A}{\pi}}$ **5.** $v = \sqrt{\dfrac{2K}{m}}$

7. $l = g \cdot \left(\dfrac{T}{2\pi}\right)^2$ **9.** $v = \sqrt{\dfrac{rF}{m}}$

11. $y = -\dfrac{6}{5}x + 6; -\dfrac{6}{5}$ **13.** $y = -\dfrac{2}{7}x + \dfrac{9}{7}; -\dfrac{2}{7}$

Section 2.4

1. a. A   **b.** C   **c.** B, C   **d.** D
   e. B, D

5. a. $I = 495(1.0794)^t$, where $t = 0$ in 1990
   b. $I = 0.511(1.0794)^t$, where $t = 0$ in 1900
   c. $I = (4.59 \times 10^{-64})(1.0794)^t$, where $t = 0$ in year 0
   e. \$1557.19 billion; \$1558.02 billion; \$1557.95 billion

7. a. $S = 119(1.0614)^t$, where $t = 0$ in 1980
   b. \$215.94 trillion   **c.** \$497.33 trillion   **d.** During 2004

9. a. $P(t) = 21.8(1.026)^t$, where $t = 0$ in 1995
   b. 28.2 million   **c.** 27 years

11. a. $C = 1.36(1.0201)^t$, where $t = 0$ in 1990
   b. 1.80 billion metric tons

13. a. 7.5 billion   **b.** about 47 years

15. \$106.14; \$181.40

17. 1.24×10^{11}, or \$124 billion

19. a. 1.125   **b.** 12.5%   **c.** 36.4
   d. $f(x) = 11.2(1.125)^x$

21. $D = 964(1.1386)^t$, where $t = 0$ in 1981; 19,081

23. **a** possible; **b** and **c** impossible

25. In mid 1997

Exercising Your Algebra Skills

1. x^8 **3.** a^{12} **5.** x^{-2} **7.** r^{12}

9. w^3 **11.** $x^{1/4}$ **13.** z^{-1} **15.** x^{15}

17. $a^{12}b^{20}$ **19.** $a^6 - 2a^3b + b^2$

Section 2.5

1. a. Exponential: 1, 3, 6; not exponential: 2, 4, and 5

3. a.

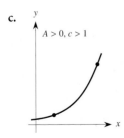

     $A > 0, c < 1$

   c.

     $A > 0, c > 1$

   d.

     $A < 0, c < 1$

5. a. (vi)   **b.** (v)   **c.** (iv)   **d.** (ii)

7. a. $C = 27,700(0.9807)^t$, where $t = 0$ in 1980
   b. 17,352 cases   **c.** 52.3 years

9. a. $N = 0.4(0.65)^t$ **b.** 10.2 hours

11. a. $M = 3(0.69)^t$ **b.** 0.68 mg   **c.** 7.3 hours

13. About 96.3 years

15. a. $I = I_0 \cdot (0.7)^n$, where I_0 is the original level
   b. $0.16807 I_0$, or about 16.8%

19. a. 64%   **b.** $S = 0.956^t$ **c.** 15.4 years

21. a. $T = 442.2(0.8796)^t$ **b.** $C = 865.7(1.0332)^t$
   c. Sales of cassette tapes are falling by 12.04% a year; sales of CDs are rising by 3.32% a year.
   d. In late 1985

25. $A = 10$, $B = -10$, c $= 0.5$

Exercising Your Algebra Skills

1. 2^{m+n} **3.** 5^{3+x} **5.** 3^{5-2a} **7.** 10^{-5x}

9. 2^{5x} **11.** $9(3^x)$ **13.** $10(10^{3x}) = 10(1000^x)$

Section 2.6

3. a. $P = 34.6(1.013)^t$ **b.** 39.4 million   **c.** 53.7 years

5. 6.6 weeks

7. a. 23.45 years, 17.67 years, 14.21 years, 11.9 years, 10.24 years

9. In mid 2051

11. 2137 B.C.

13. 780 years old

15. a. 78.5%   **b.** 34 months, or 2 years 10 months
   c. 114 months, or 9.5 years

17. 1000 times as strong

19. a. 10^6, or 1,000,000 times as loud
   b. 10^{15} times more intense   **c.** 10^{12} times more intense
   d. 60 decibels   **e.** 110 decibels

21. $P(t) = 6(10^{0.0065t})$; $P(t) = 6e^{0.0149t}$

Exercising Your Algebra Skills

1. $\log x^6 = 6 \log x$ **3.** $\log (xy)$ **5.** x^2

7. $\dfrac{\log 11}{\log 7} \approx 1.23$ **9.** $\dfrac{\log 0.6}{\log 0.4} \approx 0.56$

11. $\dfrac{\log 0.25}{\log 0.86} \approx 9.19$ **13.** $\dfrac{\log(5/4)}{\log(1.05/1.04)} \approx 23.32$

15. 100   **17.** 6   **19.** -4

Section 2.7

1. a. Power function; $k > 0, p < 0$
  c. Power function; $k > 0, 0 < p < 1$
  d. Power function; $k > 0, p > 1$
  e. Power function; $k < 0, p < 0$
  b and **f** not power functions

3. Exponential functions: **b, c, g,** and **n**; power functions: **a, d, e, h, j,** and **k**; neither: **f, i, l,** and **m**

5. a. $y = ax^2$ **b.** exponential   **c.** $y = bx^3$

9. 603.6 lb

11. a. $f(x) = 3x^{1/2}$ **b.** $f(x) \approx 3x^{0.70752}$
  c. $f(x) \approx 3x^{0.86848}$ **d.** $f(x) \approx 0.55794x^{2.2239}$
  e. $f(x) = 10x^{-1/2}$ **f.** $f(x) = 40x^{-1}$

13. a. 15 watts   **b.** 120 watts
  c. 8   **d.** 1.875 watts; yes
  e. 9,645 windmills   **f.** about 19 mph

17. a. $T = 442.2t^{-0.467}$, where $t = 1$ in 1990
  b. $C = 865.7t^{0.119}$, where $t = 1$ in 1990
  c. In early 1989

19. About 43 miles

21. The range is an additional 4 miles; the area increases by 1246 square miles.

23. About 7058 mph

25.

H	$D = 89\sqrt{H}$	$D(H) = \sqrt{H^2 + 7920H}$
0.1 miles	28.14427	28.1427
1 mile	89	89
10 miles	281.44	281.60
100 miles	890	895.54

27. Slope of PQ is 1, slope of QR is 3, and slope of PR is 2.

29. Slope of PQ is $2a + h$, slope of QR is $2a + 3h$, and slope of PR is $2a + 2h$.

Exercising Your Algebra Skills

1. 3   **3.** 16   **5.** x^7 **7.** r^4
9. x^2 **11.** $x^{-1/2}$ **13.** $a^{7/3}$

Section 2.8

3. $x \approx 1.098$

5. a. 4   **b.** 4   **c.** -3 **d.** -3
  e. $\dfrac{4}{3}$

7. a. 1   **b.** 2   **c.** 3

9. b. 1   **c.** About 0.414   **d.** About 0.707

Section 2.9

1. a. Inverse   **b.** Inverse
  c. No inverse, assuming that different students are the same height
  d. No inverse   **e.** Inverse

f. No inverse   **g.** No inverse
h. No inverse

3. a. Domain: 0, 1, 2, 3, 4, 5; range: 1.12, 1.44, 1.84, 2.05, 2.48, 2.94
  b.

x	1.12	1.44	1.84	2.05	2.48	2.94
$f(x)$	0	1	2	3	4	5

  Domain: 1.12, 1.44, 1.84, 2.05, 2.48, 2.94; range: 0, 1, 2, 3, 4, 5

7. $p^{-1}(t) = \dfrac{\log t}{\log 1.04} \approx 58.7084 \log t$

17. a. Each letter of the alphabet is matched with only one letter of the alphabet.
  b. IS THIS MATH?

Exercising Your Algebra Skills

1. $\sqrt[25]{14} \approx 1.1113$ **3.** $\dfrac{\log 0.20}{\log 0.84} \approx 9.2309$

5. $\sqrt[8]{\dfrac{32}{17}} \approx 1.0823$ **7.** $\dfrac{\log 1.75}{\log 1.02} \approx 28.2597$

9. $\left(\dfrac{y}{12}\right)^{2/7}$ **11.** $\left(\dfrac{Q}{27}\right)^{-4/3} = \dfrac{81}{Q^{4/3}}$

Review Problems

1. a. Exponential, $c < 1$
  b. Power function, $0 < p < 1$
  c. Exponential, $c > 1$

2. a. Exponential   **b.** Power   **c.** Logarithmic
  d. Power   **e.** Power   **f.** Exponential
  g. Exponential   **h.** Power   **i.** Exponential
  j. Power   **k.** Power   **l.** Linear
  m. Linear   **n.** Exponential

3. a. A   **b.** D   **c.** C   **d.** B
  e. E   **f.** I   **g.** H   **h.** G
  i. F   **j.** J   **k.** L   **l.** K

4. a. (6)   **b.** (5)   **c.** (4)   **d.** (3)
  e. (1)

6. $F(t) = 5(1.1935)^t$; 1.1935

7. a. $P(t) = 1.5 + 0.1t$; 1995

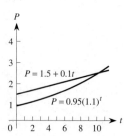

b. $P(t) = 0.95(1.10)^t$; during 1997
c. During 2000

8. **a.** About 8.1 years   **b.** About 20,424 eagles
   c. during 2004

9. About 50.5°F

10. $W \approx 2t + 141$; 191 million

11. $D \approx 382t + 2048$; deaths are increasing by 382 deaths per year; 10,834 deaths

12. **a.** $F(x) \approx 8(0.7249)^x$ **b.** $G(x) = \left(-\dfrac{13}{3}\right)x + 8$
    c. $H(x) \approx 2.7222(1.60357)^x$

13. **a.** $f(t) = 103.6t + 1055$ **b.** $2401.8 billion
    c. $F(t) = 1055(1.0832)^t$ **d.** $2981.7 billion
    e. Linear: $3127 billion; exponential: $5217 billion

14. **a.** $I = 1.135t + 17.7$ **b.** $I = 17.7(1.044)^t$

c.

t	Linear Model	Exponential Model
0	17.7	17.7
5	23.375	21.952
10	29.05	27.226
15	34.725	33.766
17	37	37
25	46.075	51.938
30	51.75	64.415

   d. Linear: $43,812; exponential: $47,998
   e. Power function

15. **a.** Linear   **b.** Linear

16. **a.**
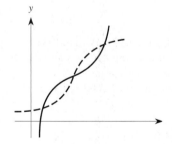

   b.
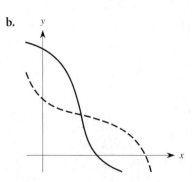

   c. No inverse

17. **a.** Domain: $t > 2$; $f^{-1}(t) = \dfrac{10^{2t} + 4}{2}$

   b. Domain: all real numbers; $g^{-1}(x) = \sqrt[3]{x - 6}$

18. Exponential; $F(x) = 4(1.5)^x$

19. **a.** $g(x)$ **b.** $h(x)$ **c.** $f(x)$

20. **a.** (i) $f(t) = 0.20(1.0565)^t$, where $t = 0$ in 1960 (ii) $2.37
    b. Ice cream   **c.** 2093
    d. $291.04   **e.** No

21. 1166; 45 years

22. 5.6 hours; 8.4 hours

Chapter 3

Section 3.2

1. **a.** (iii)   **b.** (i)   **c.** (iv)
   d. (ii)   **e.** (v)

3. **a.** Correlation coefficient must be between -1 and 1.
   b. Slope and correlation coefficient must have the same sign.
   c. The smaller x is, the larger y will be.

5. $r = 0.9897$; $T = 0.231s - 4.06$; 6.3 sec, 16.7 sec; the 45-mph estimate

7. $r = 0.992$; $P = 0.254t - 10.682$, where $t = 0$ in 1900

9. **a.** $H = 1.78S + 50.863$; $r = 0.95$.
   b. same as (a)
   c. $H = 1.96S + 49.051$

11. **a.** $r = 0.999$; yes   **b.** $S = 0.817T + 67.508$
    c. 102.3

13. $v = -964.97t - 0.32$; cm/sec/sec

15. **a.** Length depends on mass
    b. $r = 0.99998$; yes
    c. $L = 0.0404m - 0.1$
    d. cm/gram; length increases by 0.0404 cm for each additional gram of mass.
    e. $L = 0.0404m$

17. **a.** $A = 28.23t + 963.82$, where $t = 0$ in 1965
    b. 2,234.17 million tons
    c. In late 2001

19. **a.** $C = 0.022t + 13.707$
    b. Cost per minute is 2.2¢; charge for service is $13.71
    c. $57.71

21. **a.** 33
    b. Increase slope and decrease vertical intercept

Section 3.3

1. **i.** Exponential   **ii.** None   **iii.** Power
   iv. Exponential   **v.** Power   **vi.** None
   vii. Logarithmic   **viii.** Power

3. **a.** 9.3; 21.5 **b.** 0.3; 1.7 **c.** 151.2; 348.6
   d. −45.5; −197.9 **e.** Slower

5. 881 square miles

7. **a.** 345 nanograms per milliliter
   b. 335.5 minutes

9. **a.** $N = 1.506A^{0.3293}$ **b.** 1075.2 square kilometers

11. **a.** Linear: $C = 2.55t + 11$; exponential: $C = 15.29(1.0735)^t$; power: $C = 6.76t^{0.744}$, $t = 0$ in 1975
    b. 87.50, 128.37, 84.90

13. **a.** Linear: $N = 16.619t + 16.726$; exponential: $N = 133.392(1.0428)^t$
    b. Linear: 848; exponential: 1,084
    c. Linear: in early 2009; exponential: in early 1998
    d. 16.5 years

15. **a.** Linear: $N = 12.9203t − 272.9816$
    exponential: $N = 26.0174(1.0405)^t$
    power: $N = 0.3347 t^{1.7269}$
    logarithmic: $N = −1455.3584 + 501.7958 t$
    b. Linear: 1213
    exponential: 2498
    power: 1211
    logarithmic: 926
    d. Linear: in late 2065
    exponential: in mid 1999
    power: in late 2043
    logarithmic: in mid 2868
    e. 17.5 years

17. **b.** Linear: $C = −40.3864 + 0.335H$
    exponential: $C = 52.7295(1.0011)^H$
    power: $C = 0.2568H^{1.014}$
    logarithmic: $C = −1402.8353 + 271.786 \ln H$
    c. College degrees increase by 335 for every 1000 additional high school diplomas.
    d. The number of college degrees is increasing with respect to the number of high school diplomas.

19. **a.** $N = 0.3072(1.32996)^t$, where $t = 0$ in 1979
    b. 1.32996; growth rate is about 33% per year.
    c. 509.6 million
    d. Slower from 1984 to 1995 and then faster
    e. $N = 0.0657t^{2.1432}$, where $t = 0$ in 1979
    f. 70.8 million
    g. Matches very well from 1980 to 1995 and then is slower
    h. Power

21. **a.** $P = 0.3378t + 28.1705$, where $t = 0$ in 1945
    b. 53.5%
    c. $P = 29.0314(1.0090)^t$, where $t = 0$ in 1945
    d. 56.8%
    e. $P = 20.5929 t^{0.1907}$, where $t = 0$ in 1945
    f. 46.9%

23. **a.** $T = 4.4258 + 19.528 \ln P$; $r = 0.9950$, so it seems a good fit
    b. 40.1°C **c.** 371.8 kiloPascals

Section 3.4

1. **a.** $P = 3.0003(1.3303)^t$, where $t = 0$ in 1780
   b. Yes, since $r = 0.9989$, whereas previously $r = 0.9982$

3. **a.** $D = 216.97(1.0464)^t$, where $t = 0$ in 1940
   b. $4137.6 billion

5.

Year	t	Average Debt
1940	10	4.9431
1950	20	1.7054
1960	30	1.6230
1970	40	1.8741
1980	50	4.0132
1990	60	12.8951
2000	70	20.2061

$A = 0.2520t − 3.3284$, where $t = 0$ in 1930; $15.6 thousand

11. **a.** $W = 52.4972(1.3730)^t$, where $t = 0$ in 1980
    b. 2.2 years; amount of wind energy generating capacity doubles every 2.2 years.
    c. 708,357 megawatts

13. **a.** $C = 314.2542(1.0039)^t$, where $t = 0$ in 1960
    b. 178.1 years; it will take 178.1 years for the carbon dioxide concentration to double.
    c. 381.8 parts per million
    d. In late 2021

15. **a.** $N = 91.1092(1.0602)^t$, where $t = 0$ in 1960
    b. 11.9 years; the number of telephones will double in 11.9 years.
    c. 3040 million
    d. In late 2000
    e. During 2040.

17. **a.** $N = 7.1195(2.0315)^t$, where $t = 0$ in 1985
    b. 0.98; the number of computers connected to the Internet doubles in less than a year.
    c. 4.2×10^{11} thousand, or 420,000,000,000,000
    d. In late 1999

Exercising Your Algebra Skills

1. $y = 6.0007(1.0517)^x$ 3. $y = 12.0005(0.9499)^x$
5. $y = 2.2501(12.1423)^x$ 7. $y = 7.1203(0.0533)^x$

Section 3.5

1. **a.** 166.3 feet **b.** 234.8 feet
   c. 569.4 thousand pounds

3. **a.** $D = 0.000385n^{1.4865}$
   b. 9.2 cm; 15.8 cm **c.** 661

5. **b.** $S = 18.4490L^{1.1526}$ **c.** 45.4 cm
   d. 7470 cm/sec, or 245 ft/sec

7. a. $S = 10.2782L^{0.8600}$ **b.** 297 cm/sec
 c. 205.0 cm

9. a. The number of species depends on area.
 b. $N = 63.7256A^{0.3521}$
 c. 109 species
 d. 25.7 thousand square miles

11. a. Angle depends on radius.
 b. 0° to 90°
 c. $A = 31.4584R^{-0.8102}$
 d. 2.49 meters
 e. 22.6°

13. a. $T = 0.0836R^{0.6216}$
 b. 8.7 seconds
 c. 2200 meters
 d. 12.1 seconds

Exercising Your Algebra Skills

1. $y = 8.0002x^{1.5}$ **3.** $y = 12.0005x^{-1.5}$
5. $y = 2.2501x^{1.0843}$ **7.** $y = 7.1203x^{-1.2733}$

Section 3.6

1. a. (ii)   **b.** (i)
 c. (ii)   **d.** (i)

3. a. Residuals are 0, 0, 1, 1, 1; sum of squares is 3.
 b. Sum of squares is 1.5, which is one-half the sum of squares for $y = 2x + 1$.
 c. $y = 2.3x + 1$; sum of squares is 0.3, or much less than for the preceding two models.

5. a. $S = 5602.99E - 43477.62$; each additional year of education increases salary by $5,602.99.
 c. 82,788,080.84

7. a. $S = 111.5539E^{2.1533}$
 b. 29,045,904.48

9. a.

Planet	Period t	Distance D	Average speed $S = 2\pi D/t$
Mercury	88	36.0	2.5704
Venus	225	67.2	1.8766
Earth	365	92.9	1.5992
Mars	687	141.5	1.2941
Jupiter	4,329	483.3	0.7015
Saturn	10,753	886.2	0.5178
Uranus	30,660	1782.3	0.3653
Neptune	60,150	2792.6	0.2917
Pluto	90,670	3668.2	0.2542

 b. $S = 11.4233t^{-0.3333}$

Section 3.7

1. a. 0.9961; 0.9980; 99.61% of the variation
 b. $y = 6.284x_1 - 0.435x_2 + 34.561$
 c. 173.1 cm; within 1.1 cm

3. a. 0.9157; 0.9569; 91.57% of the variation
 b. $y = 0.7301x_1 + 0.0547x_2 - 0.1624$
 c. 1.9 liters per second
 d. Vital lung capacity has a greater effect because its coefficient is larger than that of total lung capacity.

5. a. 0.3877; 0.6226; 38.77% of the variation
 b. $y = 0.1620x_1 + 0.7117x_2 - 5.891$
 c. 78.2, or close to 78.
 d. 0.7386; 0.8594; 73.86% of the variation; $y = 0.1882x_1 - 2.9607x_2 + 1.9798x_3 - 141.501$; 86.266
 e. 0.8518; 0.9229; 85.18% of the variation; $y = 0.1702x_1 - 1.2035x_2 + 1.4929x_3 + 1.8644x_4 - 195.431$; 72.8187

Review Problems

1. $A = 0.1678B^{0.8757}$; $r = 0.7793$

2. For all models, let $t = 0$ in 1979; linear: $H = 51.79t + 159.97$; exponential: $H = 251.92(1.089)^t$; power: $H = 218.35t^{0.532}$; logarithmic: $H = 124.91 + 300.89 \ln t$; $1454.7 billion; $2123.0 billion; $1210.2 billion; $1093.4 billion

3. For all models, let $t = 0$ in 1979; linear: $H = 13.90t + 320.47$; exponential: $H = 332.11(1.0311)^t$; power: $H = 322.05t^{0.1824}$; logarithmic: $H = 310.59 + 80.945 \ln t$; $668.0 billion; $714.2 billion; $579.3 billion; $571.1 billion

4. a. Health expenditures are increasing by $51.8 billion per year whereas public education expenditures are increasing by $13.9 billion per year.
 b. During 1983

5. a. Health expenditures are increasing by 8.9% a year, whereas public education expenditures are increasing by 3.11% a year.
 b. In early 1984

6. $H = 3.68E - 1012.33$; Health expenditures increase by $3.68 billion for every $1 billion increase in education expenditures.

7. a. $L = 0.5432G^{0.6471}$

8. $P = 0.23w + 0.14$; an 8.5-oz letter would cost $2.10 by the model, but in reality it would cost $2.21.

9. a. $C = 38.3991t + 45.8474$, where $t = 0$ in 1970
 b. $890.6 billion
 c. In late 2007

10. a. $N = 5.491(1.0298)^t$, where $t = 0$ in 1983
 c. In mid 2003

11. a. $H = 12.5877n + 20.8676$
 b. Height of the building increases by 12.5877 feet for each story.
 c. About $12\frac{1}{2}$ feet

12. f is linear, g is exponential, and h is logarithmic.

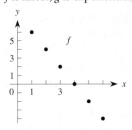

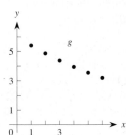

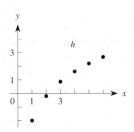

13. a. $t = 1.03n + 0.0186$; each bounce will take about 1.03 seconds to return to the floor.
b. $H = -1.0889n + 8.3482$
c. $H = 9.0002(0.80007)^n$; the height of each bounce is decreasing by 20%.
d. $H = -1.007t + 8.695$
e. The next bounce will reach a maximum height of 1.89 feet at 7.79 seconds and hit the floor at 8.26 seconds.

14. Yes.

15. If $W = f(S)$, $W \approx 45.8$ lbs; If $S = f(W)$, $W \approx 57.4$ lbs

Chapter 4

Section 4.1

1. a. 3   **b.** 4   **c.** 2
    d. 8   **e.** 3   **f.** 6

3. $1, 0, -3$

5. $-3, -1, 2, 4$

7. 2.165

9. $(0.59, -1.73)$ and $(1.41, -2.27)$ are the two turning points; one inflection point

Exercising Your Algebra Skills

1. $6x^3 - 10x^2 + 6x + 8$

3. $5x^4 + 2x^3 - 9x^2 + 7x + 1$

5. $8x^4 + 11x^3 + 3x - 1$ **7.** $3x^2 - 5x$

9. $7x + 3x^2$ **11.** $x^2 - 4x + 3$

13. $x^2 + x - 6$ **15.** $x^2 - 25$

17. $x^2 - 4$ **19.** $x^2 - 2x + 1$

21. $x^2 + 4x + 4$ **23.** $4x^2 - 24x + 36$

Section 4.2

3. a. (i) 3   (ii) 4   (iii) 4   (iv) 5
    b. (i) Positive   (ii) Positive   (iii) Negative   (iv) Negative

5. a. (iv)   **b.** (vi)   **c.** (v)
    d. (i)   **e.** (ii)   **f.** (iii)

7.

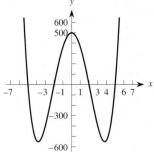

9.

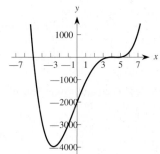

11. a. $f(x) = \dfrac{2}{9}(x - 1)(x - 3)(x + 3)$

    b. $f(x) = -\dfrac{1}{3}(x + 1)^2(x - 3)$

13. a. (i) 2 real, 2 complex   (ii) 3 real, 2 complex
      (iii) 2 real, 4 complex   (iv) 4 real, 2 complex
    b. (i) $-3, 3$ (ii) $-2.386, 0.929, 5.108$ (iii) $0.586, 3.414$
      (iv) $0.209, 0.586, 3.414, 4.791$

15. a. $-0.558, 0.769,$ and 2.040
    b. Increasing on $-0.558 < x < 0.769$ and $2.040 < x < \infty$; decreasing on $-\infty < x < -0.558$ and $0.769 < x < 2.040$
    c. 0 and 1.5
    d. concave up on $-\infty < x < 0$ and $1.5 < x < \infty$; concave down on $0 < x < 1.5$
    e. -1.305 and 2.780

17. $P(x) = -5x^2 + 10x$

19. a. $f(x) = x^2 - 4x - 12$ **b.** $x = 2$
    c. $f(x) = -\dfrac{5}{4}x^2 + 5x + 15$ **d.** $f(x) = \dfrac{5}{4}x^2 - 5x - 15$

21. a. 0 **b.** $\dfrac{2v_0}{g}$ seconds

c. $\dfrac{v_0}{g}$ seconds   **d.** $\dfrac{v_0^2}{2g}$ cm

23. $P(x) = (x + 0.87)(x - 2.21)(x - 3.66)$

25. a. 3   **b.** Average value = 3.
c. Average value = 8   **d.** Average value = 9

27. a. $p(x) = 1$ **b.** $p(x) = \dfrac{1}{2}x^2 - \dfrac{1}{2}x + 1$
c. $p(x) = ax^2 - ax + 1$, a any real number
d. $p(x) = ax^2 - ax + c$, a and c any real numbers

Exercising Your Algebra Skills

1. $(x + 4)(x + 3)$ **3.** $(x - 4)(x - 3)$
5. $(x + 4)(x - 3)$ **7.** $(x - 3)^2$
9. $(x - 10)(x + 10)$ **11.** $x(x + 5)(x - 4)$
13. $x(x + 5)^2$

Section 4.3

1. a. 806,991 cases.
c. 895,619 cases   **d.** In early 2001

3. Answer depends on estimates.

7. $T = -0.00001873R^2 + 0.0622R + 323.5359$

9. $P = -0.03316t^2 + 0.7830t + 25.1921$, $t = 0$ in 1970;
18.84 million, 11.98 million

11. $h = 0.0001047d^2 - 0.4466d + 496.9888$

13. a. $A = -0.1437t^2 + 5.9982t + 266.0148$, $t = 0$ in 1965
b. 328.6 kg   **c.** 244.9 kg

15. a. $D = -0.000003644T^2 - 0.00005986T + 1.0004$
b. 0.97835   **c.** 43.2°C

17. 3.4 seconds

Section 4.4

1. 50%

3. The discriminant becomes $k^2(b^2 - 4ac)$; $k \ne 0$ gives the same roots.

5. a. 0.77 and -3.44 **b.** -0.14 and 4.81
c. -0.46 and 0.86   **d.** -0.44 and 0.94
e. None

7. a. Stable   **b.** Unstable   **c.**   Stable
d. Unstable   **e.** Stable

Section 4.5

3. The 344 should be 334.

5. $P(x) = x^2 + 2x + 1$; $P(0.5) = \dfrac{9}{4} = 2.25$; $P(3) = 16$

7. a. 325   **b.** 5050   **c.** 500,500

9. 13 complete layers use only 819 grapefruit; 14 complete layers use 1015 grapefruit.

11. 6565 inches

13. a. $\Delta^2 y = 2a$ **b.** $\Delta^3 y = 6a$
c. $\Delta^4 y = 24a$ **d.** $\Delta^5 y = 120a$

17. 105,625

Section 4.6

1. a. $56/5$ **b.** $54/5$ **c.** $11/5$ **d.** 55
e. $-17/5$ **f.** $1/11$ **g.** 29 **h.** 5
i. $3x - 4 + 1/x$ **j.** $3x - 4 - 1/x$ **k.** $3 - 4/x$
l. $3x^2 - 4x$ **m.** $3/x - 4$ **n.** $1/(3x - 4)$
o. $9x - 16$ **p.** x

3. a. $10^5 + \log 5$ **b.** $10^5 - \log 5$ **c.** $10^5 \log 5$
d. $10^5/\log 5$ **e.** $10^{\log 5} = 5$ **f.** $\log(10^5) = 5$
g. $10^{10^5} = 10^{100,000}$ **h.** $\log(\log 5)$ **i.** $10^x + \log x$
j. $10^x - \log x$ **k.** $10^x \log x$ **l.** $10^x/\log x$
m. x **n.** x **o.** 10^{10^x}
p. $\log(\log x)$

5. c, d, and **g** are correct; while **a, b, e,** and **f** are incorrect.

13. a. 2   **b.** 0
c. -2 **d.** 0

15. $F(x) = x + 5$, $G(x) = x^4$

17. $F(x) = \log x$, $G(x) = x + 3$

19. a. $t = 4.8$ seconds   **b.** $t = 5.6$ seconds
c. Domain: between 15 mph and 55 mph, which is between 22 ft/sec and 81 ft/sec; range: $0 < t < 7$ seconds
d. decrease the 20 and/or increase the 70
e. $t = \dfrac{s^2 + 20s + 1400}{20s}$

25. a. 2   **b.** $\dfrac{3}{2}$ **c.** $\dfrac{5}{3}$
e. Yes, 1.618034

27. a. $V = \dfrac{1}{6\sqrt{\pi}}S^{3/2}$ **b.** $S = 6^{2/3}\pi^{1/3}V^{2/3}$

29. a. G   **b.** L   **c.** Q
d. X   **e.** L

Section 4.7

1.

x	$f(x)$	$5f(x)$	$f(x) + 3$	$f(x - 1)$	$[f(x)]^2$
3	5	25	8	UNDEF	25
4	2	10	5	5	4
5	-1	-5	2	2	1
6	3	15	6	-1	9
7	8	40	11	3	64

5. a. $y - 12 = m \cdot (x - 5)$
b. $y - y_0 = m \cdot (x - x_0)$, the point–slope form of a line

7. a.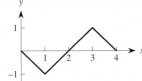

b.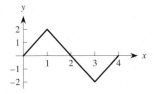

c.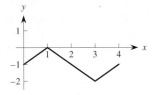

d.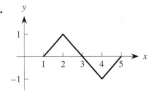

e.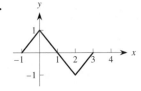

f.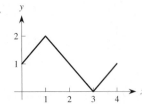

9.

x	3	4	5	6	7	8	9	10	11
y	31	23	18	15	16	22	34	46	42

11. a. $x^2 - 3x + 4 + h$
   b. $x^2 + 2xh + h^2 - 3x - 3h + 4$
   c. $2xh + h^2 - 3h$ **d.** $2x + h - 3$
   e. 7.1; 7.01; 7.0001

13. $T = 350 - 310(0.9947)^t$, where t is in seconds

15.

n	1	2	3	4	5
Turning point for $x > 0$	1.4427	2.8854	4.3281	5.7708	7.2135

Regression equation: $y = 1.4427n$; the linear fit is appropriate; 2.1641

Exercising Your Algebra Skills

   1. $f(2x) = 4x^2 - 10x + 3$
   3. $f(4x) = 16x^2 - 20x + 3$
   5. $f(x + 1) = x^2 - 3x - 1$
   7. $f(2x - 1) = 4x^2 - 14x + 9$

Section 4.8

   1. $T = 8.65 + 45.2015(0.8292)^t$; worse
   3. $T = 70 + 128.1068(0.9395)^t$
   5. a. $T = 110 - 89.69(0.978)^P$
      b. 31.87°C   **c.** 129.77 kiloPascals

Section 4.9

   1. a. Males: 176.8 cm; Females: 163.7 cm
      b. Males: $H = \dfrac{192.91}{1 + 1.876e^{-0.166t}}$,
         Females: $H = \dfrac{171.52}{1 + 1.73e^{-0.211t}}$
      c. Males: 192.91 cm; Females: 171.52 cm.

   3. a. Turning point is farther to the right and higher; function eventually decays at the same rate.
      b. Turning point is farther to the left and lower; function eventually decays at the same rate.
      c. Turning point is farther to the left and lower; function eventually decays at a faster rate.
      d. Turning point is farther to the right and higher; function eventually decays at a slower rate.

Review Problems

   1.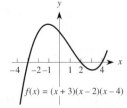

$f(x) = (x + 3)(x - 2)(x - 4)$

   2.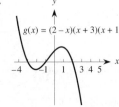

$g(x) = (2 - x)(x + 3)(x + 1)$

3.

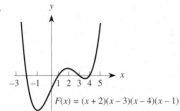

$$F(x) = (x + 2)(x - 3)(x - 4)(x - 1)$$

4.

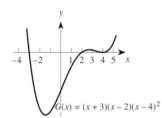

$$G(x) = (x + 3)(x - 2)(x - 4)^2$$

5. $P(x) = (x - 2)(x + 3)$; the roots are $x = 2$ and -3.

6. $Q(x) = (x + 5)(2x - 1)$; the roots are $x = -5$ and $x = \frac{1}{2}$.

7. $R(x) = x(x - 2)(x - 1)$; the roots are $x = 0, 1,$ and 2

9. 4   **10.** -20

11. 0 and -2.667

12. a.

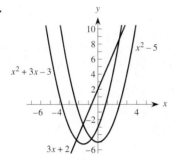

b.

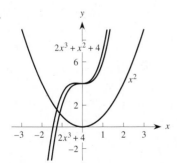

13. a. Zeros at $x = -2$, $x = 2$; turning point at $x = 0$; no vertical asymptotes; end behavior as x approaches ∞ is 1 and end behavior as x approaches $-\infty$ is 1.

  b. Zeros at $x = -2$, $x = 2$; turning point at $x = 0$; vertical asymptotes at $x = -3$, $x = 3$; end behavior as x approaches ∞ is 1 and end behavior as x approaches $-\infty$ is 1.

c. No zeros; turning point at $x = 0$; no vertical asymptotes; end behavior as x approaches ∞ is 1 and end behavior as x approaches $-\infty$ is 1.

d. No zeros; turning point at $x = 0$; vertical asymptotes at $x = -3$, $x = 3$; end behavior as x approaches ∞ is 1 and end behavior as x approaches $-\infty$ is 1.

14. a. (i)

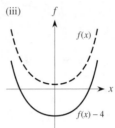

(ii)

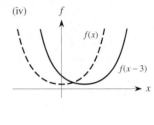

(iii) (iv)

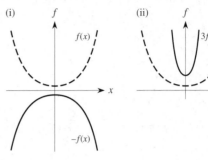

(v)

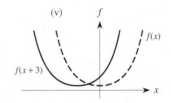

(vi)

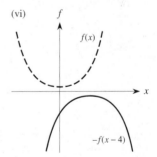

b. (i)

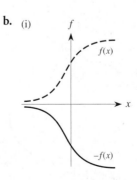

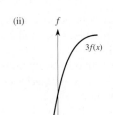

(ii)

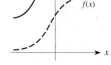

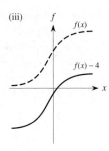

(iii)

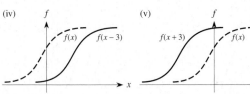

(iv)   (v)

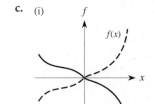

(vi)

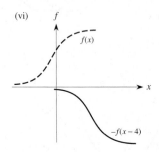

c. (i)

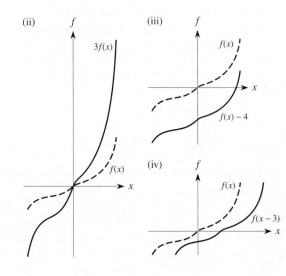

(ii)   (iii)

(iv)

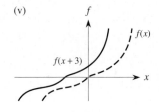

(v)   (vi)

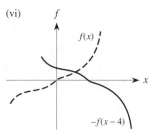

15. a. 19.4   **b.** 723   **c.** $\dfrac{6}{7}$

d. $-\dfrac{1}{4}$ **e.** 7.6   **f.** 47.5

g. $\dfrac{3(x^2 + 2)}{(x + 2)^2}$ **h.** $8x^4 + 8x^2 + 3$ **i.** $\dfrac{2x^2}{2x^2 + 3}$

j. $\dfrac{-1}{x + 1}$ **k.** $\dfrac{2x^3 - 2x^2 + x - 1}{x + 2}$

l. $\dfrac{2x^3 + 4x^2 + x + 2}{x - 1}$

16. a. $f(g(0)) = 2,\ f(g(1)) = 2,\ f(g(2)) = 3,\ f(g(3)) = 0$
   b. $g(f(0)) = 2,\ g(f(1)) = 2,\ g(f(2)) = 3,\ g(f(3)) = 1$
   c. $f(0) + g(0) = 3,\ f(1) + g(1) = 2,\ f(2) + g(2) = 5,$
      $f(3) + g(3) = 3$
   d. $\dfrac{f(0)}{g(0)} = 2,\ \dfrac{f(1)}{g(1)}$ is not defined, $\dfrac{f(2)}{g(2)} = \dfrac{3}{2},\ \dfrac{f(3)}{g(3)} = 0$

17. a. $f(g(1)) = 3,\ f(g(2)) = 3,\ f(g(3)) = 1,\ f(g(4)) = 1$
   b. $g(f(1)) = 3,\ g(f(2)) = 1,\ g(f(3)) = 1,\ g(f(4)) = 3$
   c. $f(1) + g(1) = 4,\ f(2) + g(2) = 5,\ f(3) + g(3) = 4,$
      $f(4) + g(4) = 2$
   d. $\dfrac{f(1)}{g(1)} = \dfrac{1}{3},\ \dfrac{f(2)}{g(2)} = \dfrac{3}{2},\ \dfrac{f(3)}{g(3)} = 3,\ \dfrac{f(4)}{g(4)} = 1$

18. $0.106262 < t < 3.29374$

19. 15150

20. a. 3   **b.** 5   **c.** 1   **d.** 5
   e. Yes, 3

21. $y = -x^4 + 6x^2 - x + 4$

22. a.

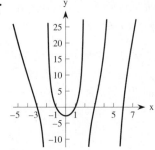

   b. $R(x) = \dfrac{(x + 3)(x + 1)(x - 1)(x - 3)(x - 6)}{(x + 2)(x - 2)(x - 5)}$

23. a. (i) and (iv) are exponential; (ii), (v), and (vi) are power;
      (iii) is logarithmic.
   b. There are many possible answers.
      (i) $f(x) = 2^x - 11$ (ii) $f(x) = \sqrt{x} + 5$

(iii) $f(x) = \log(x + 4)$ (iv) $f(x) = \left(\dfrac{1}{3}\right)^x + 7$

(v) $f(x) = x^{-1} + 9, x > 0$

(vi) $f(x) = (x - 5)^{-3}, x > 5$

24. a. $N = 4.8196(1.3840)^t$, where $t = 0$ in 1990

b. $B = 82.7650(0.9170)^t$, where $t = 0$ in 1990

c. $R = 398.8942(1.2691)^t$, where $t = 0$ in 1990

d. $R = 398.8901(1.2691)^t$, where $t = 0$ in 1990; almost identical

Chapter 5

Section 5.1

1. 5.5 mL; 1.9 mL; about 31.1 hours; about 82.7 hours

3. a. 320, 320, 320, 320, 320, 320; same

7. 16.67 mg   **9.** No

11. 325 mg; virtually all washed out in 24 hours

13. $r_3 < r_1 < r_2$

15. $\{0, 4, 8, 12, 20, \ldots\}$ **17.** $\{0, 0.5, 1, 1.5, 2, 2.5, \ldots\}$

19. $\{-9, -2, 17, 54, 115, 206, \ldots\}$

21. $\left\{1, \dfrac{2}{3}, \dfrac{4}{9}, \dfrac{8}{27}, \dfrac{16}{81}, \dfrac{32}{243}, \ldots\right\}$

23. $1, \dfrac{1}{2}, \dfrac{1}{3}, \dfrac{1}{4}, \dfrac{1}{5}, \dfrac{1}{6}, \ldots$

25. $\{0, 0.8, 0.96, 0.992, 0.9984, 0.99968, \ldots\}$

27. Diverges   **29.** Diverges   **31.** Diverges

33. Converges to 0   **35.** Converges to 0

37. Converges to 1   **39.** Strictly increasing, no concavity

41. Strictly increasing, no concavity

43. Strictly increasing, concave up

45. Strictly decreasing, concave up

47. Strictly decreasing, concave up

49. Strictly increasing, concave down

51. a. 2, 2.25, 2.3704, 2.4414, 2.4883, 2.5216, 2.5465, 2.5658, 2.5812, 2.5937; yes

b. $e_{100} = 2.704814$, $e_{500} = 2.715569$, $e_{1000} = 2.716923$, $e_{10,000} = 2.718146$, $e_{100,000} = 2.718268$, $e_{1,000,000} = 2.718280$; $e \approx 2.718281828$

53. $\dfrac{1}{e}$ **55.** 1

Section 5.2

1. a. \$383,519   **b.** \$506,055

3. About 7.6 months

5. a. $b_{n+1} = (1.06)b_n$, $b_0 = 2000$

b. $b_{n+1} = (1.06)b_n + 1000$, $b_0 = 2000$

c. $b_{n+1} = (1.06)b_n + 2000 + 1000(n + 1)$, $b_0 = 2000$

d. $b_{n+1} = (1.06)b_n + 2000(1.1)^{n+1}$, $b_0 = 2000$

7. a. $W_{n+1} = 1.03W_n - 50$, $W_0 = 8000$, $n = 0$ in 1992

b. $W_0 = 8000$, $W_1 = 8190$, $W_2 = 8386$, $W_3 = 8587$, $W_4 = 8795$, $W_5 = 9009$, $W_6 = 9229$, $W_7 = 9456$, $W_8 = 9690$, $W_9 = 9930$, $W_{10} = 10{,}178$, $W_{11} = 10{,}433$, $W_{12} = 10{,}696$, $W_{13} = 10{,}967$

c. 240

9. a. $W_{n+1} = W_n + r \cdot (400 - W_n)$, where r is the proportion learned

b. $0 < r < 1$

11. 41 rabbits

Section 5.3

1. 10.15, 10.30, 10.45, 10.61, 10.76, 10.92, 11.08, 11.24, 11.40, 11.57; 40; about one-quarter of the way

3. 5.10, 5.20, 5.30, 5.40, 5.51, 5.61, 5.72, 5.83, 5.95, 6.06; 200; not close at all yet

7. b. 275, over estimate   **c.** 950, over estimate

9. $a = 0.05$, $b = 0.000003125$; 100, 104.97, 110.18, 115.65, 121.39

13. a. b will increase.

15. a.

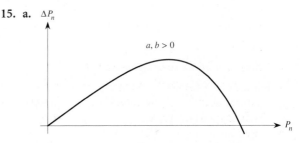

b. $L = \sqrt[3]{\dfrac{a}{b}}$

c. The population is increasing because $\Delta P_n > 0$ for $0 < P_n < L$.

d. The population is decreasing because $\Delta P_n < 0$ for $P_n > L$.

17. Logistic: $\Delta P_n = 0.1387P_n - 0.00053456P_n^2$, $r = -0.2138$; linear: $P = 3.425a + 32.5278$, $r = 0.9864$; exponential: $P = 38.3733(1.0532)^a$, $r = 0.9945$; power: $P = 29.2962a^{0.3724}$, $r = 0.9391$; the exponential model seems to be the best fit.

19. a. 255 cm, about 5 weeks   **b.** $a = 0.8772$, $b = 0.0037$

c.

t	1	2	3	4	5	6
Actual H	17.9	36.4	67.8	98.1	131.0	169.5
Predicted H	17.9	32.4	57.0	94.9	144.9	194.3

t	7	8	9	10	11	12
Actual H	205.5	228.3	247.1	250.5	253.8	254.5
Predicted H	225.0	235.1	236.8	237.0	237.077	237.080

21. If $t_0 = 0$, $t_1 = 22$, $t_2 = 46$, etc., $\Delta P_n = 1.2751P_n - 0.3149P_n^2$, with $r = -0.9107$.

23.

n	1	2	3	4	5
Actual P_n	9.6	29.0	71.1	174.6	350.7
Predicted P_n	9.6	25.3	65.5	162.7	361.0

n	6	7	8	9	10
Actual P_n	513.3	594.4	640.0	655.9	661.8
Predicted P_n	607.1	616.1	610.2	614.1	611.5

25. $\dfrac{\Delta P_n}{P_n} = 2.6480(0.9931)^{P_n}$, or $\Delta P_n = 2.6480P_n \cdot (0.9931)^{P_n}$, with $r = -0.9506$

27. a. $a = 3.792$, $b = 0.0092$
b. 303.5 gigawatts **c.** between 1980 and 1985
d. Soon after 1980

Section 5.4

1. $n = 7.9$ hours, so time of death $\approx$ 1:06 A.M.; about 18 minutes difference

3. $n = 60.7$ minutes; about 2 minutes longer

5. 36.3 minutes

7. 4.57 hours before 9 A.M. or about 4:26 A.M.

9. a. Linear: $d = -0.5567T + 34.8914$, $r = -0.8465$;
exponential: $d = 622.7097(0.8999)^T$, $r = -0.9841$;
power: $d = 1,814,899,980T^{-5.1962}$, $r = -0.9907$
b. Power function **c.** 17 days

11. b and **c** are possible; **a** and **d** are impossible

13. $t_1 < t_2$

Section 5.5

1. 1.999023438, 1.999999046, 1.999999999; the partial sums are approaching 2.

3. 4.57050327, 4.95388314, 4.99504824; the partial sums are approaching 5.

5. 170.9951, 9973.7702, 575,251.1777; the partial sums do not converge.

7. A total of 29,997 cases from 1960 through 2000; no more than 3 new cases between 2000 and 2010

9. 64.5 billion kilowatt-hours

11. 653,722.5 metric tons

13. 1,900,648 pages

15. About 45.6 years

17. 79.95 feet, 83.77 feet, 83.99 feet; 84 feet

19. 49.31 feet, 49.99 feet, 49.9998 feet; 50 feet

21. $\dfrac{25}{100}\left(\dfrac{1}{1 - 1/100}\right) = \dfrac{25}{99}$

23. The solution to $\dfrac{x_{n+1}}{x_n} = n$ is $x_n = (n - 1)!x_1$, so the solution to $\dfrac{x_{n+1}}{x_n} = n^2$, which is $x_n = [(n - 1)!]^2 x_1$, grows much faster.

Review Problems

1. a. $\{-1, 5, 11, 17, 23, \dots\}$
b. $\left\{3, \dfrac{9}{2}, 9, \dfrac{81}{4}, \dfrac{243}{5}, \dots\right\}$
c. $\{0, 0.7, 0.91, 0.973, 0.9919, \dots\}$

2. a. $\{2, 10, 18, 26, 34, \dots\}$
b. $\{12, 4, -4, -12, -20, \dots\}$
c. $\left\{5, \dfrac{5}{3}, \dfrac{5}{9}, \dfrac{5}{27}, \dfrac{5}{81}, \dots\right\}$
d. $\{10, 11, 8, 17, -10, \dots\}$

3. a. $\{2, 7, 9, 16, 25, \dots\}$ **b.** $\{3, 7, 10, 17, 27, \dots\}$

4. 167 mg **5.** 21 mg **6.** 150 pounds

7. 15,000; at year $n = 22$.

8. $b = 0.000017$; $u_{n+1} = 1.2u_n - 0.000017u_n^2$, $u_0 = 1000$

9. a. $P_{n+1} = 1.3P_n - 0.0004P_n^2$
b. $P_{n+1} = 1.3P_n - 0.0004P_n^2 - 2000$
c. $P_{n+1} = 1.3P_n - 0.0004P_n^2 + 400$
d. $P_{n+1} = 1.3P_n - 0.0004P_n^2 - 0.10P_n$

10. a. $M_{n+1} = 0.85M_n$, $M_0 = 100\%$; there will be 0.34% maple syrup content.
b. $P_{n+1} = 1.1P_n + 2000$, $P_0 = 50,000$; John will have $161,561.97.
c. $P_{n+1} = P_n - 20,000$, $P_0 = 2,000,000$; $T_{n+1} = 1.10T_n$, $T_0 = 2,000,000$; $P_5 = \$1,900,000$, $T_5 = \$3,221,020$; increasing roughly exponentially

11. a. $Q_{n+1} = 0.75Q_n + 1$, $Q_0 = 6$
b. $Q_n = 2(0.75)^n + 4$
c. 2.4 weeks

12. Increasing concave up **13.** Increasing concave up

14. Increasing concave down

15. Decreasing concave down

16. Increasing concave down

17. Increasing concave up

18. None of the above **19.** Increasing concave up

Chapter 6

Section 6.1

3. $b = 3.4409$, $c = 12.4836$

5. $a = 24$, $\theta = 36.87°$

7. The pole is 26.6 feet high.

9. a. 571.5 feet **b.** 573.7 feet

11. a. 27 feet **b.** 26.1 feet

13. 20.14 to 21.82 feet.

15. Jack is 27.3 feet from the base of the cliff.

17. a.

tan θ	0	0.5	1	1.5	2	2.5	3
θ	0°	26.6°	45°	56.3°	63.4°	68.2°	71.6°

Section 6.2

3. $a = 11.5, b = 3.3$ **5.** $\theta = 15.47°, a = 28.9$

7. $c = 135.915, b = 129.263$

9. 9659 feet

11. a. 5.71°   **c.** 20.0998 miles

13. 507.6 feet   **15.** About 36.9°

17. 41.81°

19. $\dfrac{3}{\cos 27°} \approx 3.37$ inches   **21.** 85.6 feet

23. 38.3 mph

25. a.

cos θ	1	0.8	0.6	0.4	0.2	0
θ	0°	36.9°	53.1°	66.4°	78.5°	90°

c. Inverse cosine function

Section 6.3

1. −0.707   **3.** −1   **5.** −0.5   **7.** 0.707

9. −0.5   **11.** −0.5

13. Negative; Quadrant IV   **15.** Negative; Quadrant III

17. Negative; Quadrant IV   **19.** 0

21. Negative; Quadrant III   **23.** Negative; Quadrant II

25. Positive; Quadrant III   **27.** Positive; Quadrant IV

29. Negative; Quadrant III   **31.** Positive; Quadrant III

33. a. Positive; negative; negative;
b. 0°; 90°; 180°; 270°; 360°; 450°; 540°
c.

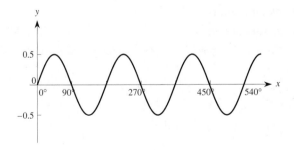

d. Yes; 180°

35. a.

cos θ	1	0.75	0.5	0.25	0	−0.25	−0.5	−0.75	−1
θ	0°	41.4°	60°	75.5°	90°	104.4°	120°	138.6°	180°

b.

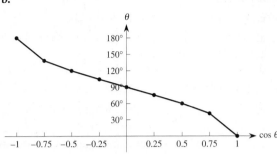

c. Inverse cosine function

Section 6.4

1. $0.954, 0.314; \theta \approx 17.5°$ **3.** $\cos \theta = 0.8, \tan \theta = 0.75$

5. $\sin \theta = \dfrac{3}{5} = 0.6, \cos \theta = \dfrac{4}{5} = 0.8$

7. −0.854, −0.609   **9.** 0.714, −0.980

11. a. (ii) Is an identity   **c.** 0° and 180° by the graphs

Exercising Your Algebra Skills

1. $\sin x$ **3.** $\sin^2 x$

5. $1 - 2 \sin \theta \cos \theta = 1 - \sin 2\theta$

7. $\dfrac{1}{\cos x}$ **9.** $-\tan^2 x$

Section 6.5

1. $C = 91°, a = 5.90,$ and $c = 13.47$

3. $C = 80°, a = 13.98,$ and $b = 22.09$

5. $B = 50.47°, C = 89.53°,$ and $c = 15.56$; or $B = 129.53°, C = 10.47°,$ and $c = 2.83$

7. There are various ways to describe the location of the third ship. The ship is 37.9 miles in the direction of 54° south of east from the northernmost ship and 29.5 miles in the direction of 41° north of east from the southernmost ship.

11. 32.2°   **13.** About 1332 miles

15. About 47.2 meters, 58°.

17. a. about 390 miles   **b.** About 307 miles

19. About 329 meters   **21.** 2.4

Review Problems

1. a. $d = \dfrac{75}{\cos \theta}$ **b.** $0° \le \theta \le 72°$
c. $-72° \le \theta \le 72°$

2. a. $x = 75 \tan \theta$ **b.** About 35 feet   **c.** $0° \le \theta \le 63°$

3. a. $y = 500 \tan \alpha$ **b.** 182 meters   **c.** 420 meters
d. 76°

4. a. 8.5°   **b.** $y = \dfrac{3}{20} x$ **c.** $y = \dfrac{5}{12} x$

5. 8.5 inches

6. a. 6.88 inches   **b.** 9.83 inches

7. a. 14,424.3 cubic inches   **b.** 17,819.1 cubic inches
    c. $13,824(1 + \cos\theta)\sin\theta$
    d. About 60°

8. a. 253.2 feet   **b.** 93.4 feet

9. 89.85°   **10.** About 6430 meters

11. a. About 5 inches
    c. From 10 to 15 minutes after, from 15 to 20 minutes after, and from 40 to 45 minutes after

12. a. $\sin\theta = 0.76822$, $\cos\theta = 0.64018$

    b. $\sin\theta = \dfrac{6}{\sqrt{61}} \approx 0.76822$, $\cos\theta = \dfrac{5}{\sqrt{61}} \approx 0.64018$

13. a. About 38 feet   **b.** 37.4 feet
    c. The tree is 31.9 feet tall, and the top of the tree is 31.5 feet above the ground.

14. The fire is 9.5 miles from tower A, and 10.8 miles from tower B.

15. a. 13 yards   **b.** 22.6°   **c.** 37 yards   **d.** 161.1°

16. a. 17.9 miles   **b.** 163.5°
    c. About 1 hour; about 2.1 hours   **d.** $18.00

17. a. 55.7°   **b.** 3268 feet

18. a. 22°   **b.** 5.2 centimeters   **c.** 49.4 cm²
    d. If the angle θ is opposite side c, the area is $\frac{1}{2}ab \sin\theta$.

Chapter 7

Section 7.1

1. The length of her fingernails is a periodic function with period of 1 week.

5. a. 135°   **b.** 144°   **c.** 120°   **d.** 85.9°
    e. 143.2°   **f.** 171.9°   **g.** 22.5°   **h.** 300°
    i. −270°   **j.** −300°

7. a. $\dfrac{\sqrt{3}}{2}$ **b.** 1   **c.** $-\dfrac{\sqrt{3}}{2}$ **d.** 1

    e. $\dfrac{\sqrt{3}}{2}$ **f.** $\dfrac{1}{2}$ **g.** $\dfrac{\sqrt{2}}{2}$ **h.** 0.9749

9. Same graph

11. a. Roughly 10 years   **b.** Roughly 10 years
    c. Maxima: 1847, 1857, 1867, 1877, 1886, 1895, 1906, 1915, 1927, 1936; minima: 1852, 1862, 1872, 1881, 1891, 1901, 1908, 1920, 1930
    d. Maxima: 1853, 1857, 1861, 1864, 1873, 1876, 1886, 1896, 1904, 1913, 1923, 1933; minima: 1847, 1855, 1859, 1862, 1868, 1882, 1890, 1899, 1908, 1918, 1928
    e. The years in which a maximum (minimum) occur are roughly 10 years apart.
    f. Clear points of inflection near: 1850, 1858, 1863, 1866, 1872, 1875, etc.; roughly halfway between a maximum and a minimum

Section 7.2

1. a. Periodic; period = 2   **b.** Periodic; period = 2
    c. Periodic; period = 2   **d.** Periodic; period = 1
    e. Periodic; period = 2   **f.** Not periodic;
    g. Not periodic
    h. A constant function can be considered periodic, but a definite period cannot be determined.
    i. Periodic; period = 2   **j.** Periodic; period = 20

3. a. 3.6   **b.** 365   **c.** 8.4 hours   **d.** 15.6 hours

5. $H(t) = 12 - 2.4 \sin\left[\dfrac{2\pi}{365}(t - 80)\right]$

7. $W(t) = 25 + 25 \sin\left(\dfrac{2\pi}{11}t\right)$

9. a. $2 \sin\left(\dfrac{\pi}{40}t\right)$ **b.** $7.5 \sin\left(\dfrac{\pi}{150}t\right)$

11. a. $T(t) = 350 + 10 \sin\left(\dfrac{\pi}{5}t\right)$

13. a. $f(x) = 51 + 21 \sin\left[\dfrac{\pi}{12}(x - 2)\right]$

    b. $f(x) = 51 + 21 \cos\left[\dfrac{\pi}{12}(x - 8)\right]$

15. $H(t) = 22 + 42 \sin\left[\dfrac{2\pi}{365}(t - 131)\right]$

17. Because the sine function is periodic there are infinitely many correct formulas for each graph.

    a. $y = 2 \sin\left(\dfrac{x}{3}\right)$ **b.** $y = 5 \sin 2x$ **c.** $y = 4 \cos 4x$

    d. $y = -3 \sin\left(\dfrac{x}{5}\right)$ **e.** $y = 1.5 \cos\left(\dfrac{x}{4}\right)$ **f.** $y = 2 - \cos 2x$

    g. $y = -5 \cos 2x$ **h.** $y = 10 + 6 \sin 8x$
    i. $y = -3 + 3 \sin 4x$ **j.** $y = 3 + 3 \cos 4x$

    k. $y = \sin\left(\dfrac{\pi}{2}x\right)$ **l.** $y = 4 \cos \pi x$

19. Vertical shift is 59; amplitude is 12; period is 24 hours; frequency is $\frac{\pi}{12}$; $T(t) = 59 + 12 \sin\left[\frac{\pi}{12}(t - 9)\right]$

23. About 12 times

25. a. For t in seconds,

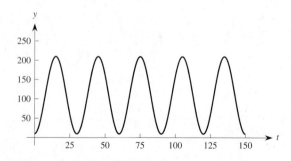

    b. $y = 110 + 100 \sin\left[\dfrac{\pi}{15}(t - 7.5)\right]$

c. $x = 100 \sin\left(\dfrac{\pi}{15}t\right)$

d. $0 < t < 7.5, 22.5 < t < 37.5, 52.5 < t < 67.5$, etc.

e. $x = -100 \sin\left(\dfrac{\pi}{15}t\right)$; $7.5 < t < 22.5, 37.5 < t < 52.5$, etc.

f. $x^2 + (y - 110)^2 = 100^2$

27. $B(t) = 4.00 + 0.35 \sin\left[\dfrac{2\pi}{5.4}(t - 1.35)\right]$

29. a. One such model might be
$T = 0.0042t + 14.67 + 0.08 \sin\left(\dfrac{\pi}{25}t\right)$.

b. Yes   **c.** 15.2°C

Section 7.3

1. Eleven hours of daylight on day 55 (February 24) and again on day 287 (October 14); ten hours of daylight on January 23 and November 16; San Diego always has more than 9 hours of daylight.

3. a. 72°   **b.** 66°

c. Reaches 70° at 1.1 minutes after 9 A.M. and every 20 minutes thereafter and also 8.9 minutes after 9 A.M. and every 20 minutes thereafter; reaches 67° at 12.3 minutes after 9 A.M. and every 20 minutes thereafter, and also 17.7 minutes after 9 A.M. and every 20 minutes thereafter.

5. a. 1.13 hours, or 1 hour 8 minutes

b. 0.35 hour, or 21 minutes

7. 99th day (April 9) and 346th day (December 12)

9. a. 1.2 and 2.8 months   **b.** 4.7 and 11.3 months

c. The average daytime high temperature never reaches 80°F.

Exercising Your Algebra Skills

1. No solution   **3.** No solution

5. $-0.85 + 2n\pi$ and $3.99 + 2n\pi$

7. $1.11 + n\pi$ and $2.03 + n\pi$

9. $\dfrac{\pi}{2} + n\pi, 0.30 + 2n\pi$, and $2.84 + 2n\pi$

Section 7.4

1. a. $y = \tan 2x$ **b.** $y = \tan\left(\dfrac{1}{2}x\right)$ **c.** $y = 3\tan x$

d. $y = -2\tan x$ **e.** $y = \tan(x - 30)$

f. $y = -10 + \tan x$

3. $y = \tan\left(\dfrac{3}{2}x\right)$

5. The building is 37.5 meters tall; the smokestack is 45.1 meters tall.

7. a. $m = \tan \alpha$, where α is the angle measured from the positive x-axis to the line

b. $m = \tan \alpha$, where α is the angle measured from the line $y = b$ to the line

11. a. $\beta = \arctan\left(\dfrac{6}{x}\right) - \arctan\left(\dfrac{1}{x}\right)$ **b.** About 2.4 feet

Exercising Your Algebra Skills

1. $\theta = 0.90 + n\pi$ **3.** $\theta = \dfrac{3\pi}{4} + n\pi$

5. $\theta = 0.67 + n\pi$ **7.** $\theta = \dfrac{3\pi}{4} + n\pi$

Review Problems

1. $\cos x = \sin\left(x + \dfrac{\pi}{2}\right)$ and $5.70 + \dfrac{\pi}{2} \approx 7.2708$.

2. a. $\theta = 120°, 420°, 480°, \dots$; $\theta = -240°, -300°, -600°, -660°, \dots$

3. a. $\theta = 315°, 405°, 675° \dots$; $\theta = -45°, -315°, -405°, -675° \dots$

b. $\theta = \dfrac{7\pi}{4}, \dfrac{9\pi}{4}, \dfrac{15\pi}{4} \dots$; $\theta = \dfrac{-\pi}{4}, \dfrac{-7\pi}{4}, \dfrac{-9\pi}{4}, \dfrac{-15\pi}{4} \dots$

	Vertical shift	Amplitude	Frequency	Period	Phase shift
4.	325	10	$\dfrac{2\pi}{9}$	9	0
5.	63	3	$\dfrac{2\pi}{25}$	25	0
6.	71	2	$\dfrac{2\pi}{15}$	15	0
7.	80	13	$\dfrac{\pi}{12}$	24	15
8.	38	8	$\dfrac{\pi}{12}$	24	5
9.	100	25	$\dfrac{\pi}{36}$	72	0
10.	100	25	$\dfrac{2\pi}{97}$	97	0
11.	145	40	$\dfrac{2\pi}{83}$	83	0

20. a. $y = 5 + 2\cos\left(\frac{2\pi}{3}t\right)$ (where t is in radians), which starts at a maximum height and maximum displacement forward of the crossbar.

b. $x = 6.93 \sin\left(\dfrac{2\pi}{3}t\right)$

21. $D = 160 + 40(0.92957)^t \cos\left(\dfrac{\pi}{3}t\right)$

22. a. $H = 1.5 + 1.5 \sin(1000\pi t)$

b. $H = 1.5 + 1.5 \sin\left[1000\,\pi\left(t - \dfrac{1}{2000}\right)\right]$

23. $H = 9 - 5 \sin\left(\dfrac{2\pi}{3}t\right)$

24. a. $y = 9 - 5 \sin\left(\dfrac{\pi}{10}x\right)$

25. a. 1000π, or 3142 miles   **b.** $\dfrac{4000\pi}{3}$, or 4189 miles

c. $\dfrac{10{,}000\pi}{3}$, or 10,472 miles

d. $\dfrac{1000\pi}{3}$, or 1047 miles

26. a. $360\pi \approx 1131$ radians **b.** $720\pi \approx 2262$ feet

28.

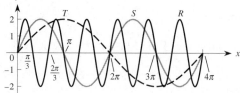

29.

	Frequency	Period	Amplitude	Phase Shift
a.	$\frac{3}{4}$	$\frac{8\pi}{3}$	2	0
b.	$\frac{3}{4}$	$\frac{8\pi}{3}$	2	$\frac{-4\pi}{3}$
c.	π	2	2	$\frac{-3}{4\pi}$
d.	$\frac{3}{4}\pi$	$\frac{8}{3}$	2	$\frac{1}{\pi}$

30. a. $\dfrac{4\pi}{9} + \dfrac{8n\pi}{3}$ and $\dfrac{20\pi}{9} + \dfrac{8n\pi}{3}$

b. $\dfrac{4\pi}{9} + \dfrac{8n\pi}{3}$ and $-\dfrac{4\pi}{9} + \dfrac{8n\pi}{3}$

c. $\dfrac{2}{3} - \dfrac{3}{4\pi} + 2n$ and $\dfrac{4}{3} - \dfrac{3}{4\pi} + 2n$

d. $\dfrac{8}{9} + \dfrac{1}{\pi} + \dfrac{8}{3}n$ and $\dfrac{16}{9} + \dfrac{1}{\pi} + \dfrac{8}{3}n$

31. a.

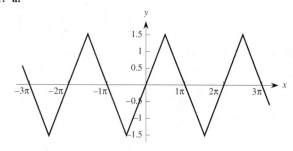

b.

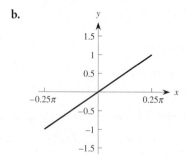

32. a. $\theta = -0.98 + n\pi$

b. $\theta = 1.11 + n\pi$ **c.** $\theta = 1.43 + n\pi$

33. a. $x = \tan 1.35 \approx 4.46$
b. $x = \sin 0.5 \approx 0.48$

Chapter 8

Section 8.1

3. Not an identity; $(0, 1)$ **5.** Identity

7. Not an identity; $(0, 0)$ **9.** Identity

11. Identity

13. Not an identity; no points of intersection

15. $\cos 3x = 4\cos^3 x - 3\cos x$

17. $\cos 5x = 16\cos^5 x - 20\cos^3 x + 5\cos x$. There are many equivalent forms.

21. $\dfrac{1 - \cos 4x}{8}$

23. a. Looks like $y = -\sin x$
b. Looks like $y = -\sin x$

25. $7.15°$ or $82.85°$

27. a. $y = 5\cos(t - 126.87°)$

31. a. Even **b.** Odd **c.** Even
d. Neither **e.** Neither **f.** Even
g. Neither **h.** Even **i.** Neither
j. Neither

33. For $x = \frac{\pi}{6}$: If $n = 5$, the sum is about 1.969. If $n = 10$, the sum is about 1.999. Values approach 2. For $x = \frac{\pi}{3}$, more terms are needed.

39. Not an identity; $(0, 0)$

41. Identity

43. Not an identity; $(-0.393, 0.172)$

45. Not an identity; $(0, 0)$

Section 8.2

1. $T_2(0) = 1$, $T_2(0.1) = 0.995$,
$T_2(0.2) = 0.98$, $T_2(0.3) = 0.955$, $T_2(0.4) = 0.92$,
$T_2(0.5) = 0.875$, $T_2(0.6) = 0.82$

3.

x	$T_3(x)$	$\sin x$	$T_3(x) - \sin x$
0	0	0	0
0.1	0.09983	0.09983	-8×10^{-8}
0.2	0.19867	0.19867	-3×10^{-6}
0.3	0.29550	0.29552	-2×10^{-5}
0.4	0.38933	0.38942	-9×10^{-5}
0.5	0.47917	0.47943	-3×10^{-4}
0.6	0.56400	0.56464	-6×10^{-4}

5.

x	$-\frac{4\pi}{25}$	$-\frac{3\pi}{25}$	$-\frac{2\pi}{25}$	$-\frac{\pi}{25}$	0	$\frac{\pi}{25}$	$\frac{2\pi}{25}$	$\frac{3\pi}{25}$	$\frac{4\pi}{25}$
$\sin x$	-0.4818	-0.3681	-0.2487	-0.1253	0	0.1253	0.2487	0.3681	0.4818

$$\sin x \approx -0.1640x^3 - (2.851 \times 10^{-14})x^2 + 0.9998x - (2.886 \times 10^{-15}); \text{ very close}$$

7. $\sin(x^2) \approx x^2 - \dfrac{1}{6}x^6 + \dfrac{1}{120}x^{10}$

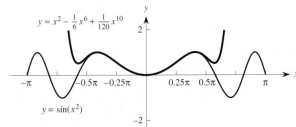

Good match for x between -0.5π and 0.5π

9. $\left(x - \dfrac{x^3}{6}\right)^2 + \left(1 - \dfrac{x^2}{2} + \dfrac{x^4}{24}\right)^2 = 1 - \dfrac{x^6}{72} + \dfrac{x^8}{576} \approx 1,$
when x is close to 0

11. $\cos x \approx 1 - \dfrac{x^2}{2} + \dfrac{x^4}{32}$

13. a. 1

  b. $1 - \dfrac{x^2}{2} - \dfrac{x\,\Delta x}{2} - \dfrac{(\Delta x)^2}{6}$

  c. $1 - \dfrac{x^2}{2}$, the quadratic approximation for $\cos x$

Section 8.3

1. $|z| = 5; \theta = -\arctan\left(\dfrac{3}{4}\right) \approx -0.64350$

3. $|z| = 13; \theta = -\arctan\left(\dfrac{5}{12}\right) \approx -0.39479$

5. $|z| = \sqrt{5392} \approx 73.43024; \theta = -\arctan\left(\dfrac{9}{16}\right) \approx -0.51239$

7. $|z| = \sqrt{74} \approx 8.60233; \theta = \pi - \arctan\left(\dfrac{7}{5}\right) \approx 2.19105$

9. $|z| = \sqrt{67} \approx 8.18535; \theta = \pi + \arctan\left(\dfrac{\sqrt{3}}{8}\right) \approx 3.35481$

11. $13\left[\cos\left(\arctan\dfrac{12}{5}\right) + i\sin\left(\arctan\dfrac{12}{5}\right)\right]$

13. $25[\cos(2.21430) + i\sin(2.21430)]$

15. $\sqrt{73}[\cos(-0.35877) + i\sin(-0.35877)]$

17. $\sqrt{17}[\cos(0.75597) + i\sin(0.75597)]$

19. $z^2 = 7 - 24i$ **21.** $z^2 = 119 - 120i$

23. $z^2 = 25\left[\cos\left(-2\arctan\dfrac{3}{4}\right) + i\sin\left(-2\arctan\dfrac{3}{4}\right)\right]$

25. $z^2 = 169\left[\cos\left(-2\arctan\dfrac{5}{12}\right) + i\sin\left(-2\arctan\dfrac{5}{12}\right)\right]$

27. $z^2 = 5392[\cos(-1.02478) + i\sin(-1.02478)]$

29. $z^2 = 74[\cos(4.38210) + i\sin(4.38210)]$

31. $z^2 = 67[\cos(6.70962) + i\sin(6.70962)]$

33. $z^3 = -2035 - 828i$ **35.** $z^3 = 14{,}625 + 5500i$

37. $z^3 = 2197\left[\cos\left(3\arctan\dfrac{12}{5}\right) + i\sin\left(3\arctan\dfrac{12}{5}\right)\right]$

39. $z^3 = 15{,}625[\cos(6.64290) + i\sin(6.64290)]$

41. $z^3 = 623.71227[\cos(-1.07631) + i\sin(-1.07631)]$

43. $z^3 = 70.09280[\cos(2.26791) + i\sin(2.26791)]$

45. $z^0 = 1, z^1 = 1 + 2i, z^2 = -3 + 4i, z^3 = -11 - 2i,$ and $z^4 = -7 - 24i$

49. a. $zw = (ac - bd) + (ad + bc)i$

  b. $zw = \sqrt{a^2 + b^2}\,\sqrt{c^2 + d^2}[\cos(\theta_1 + \theta_2) + i\sin(\theta_1 + \theta_2)],$
  where $\tan\theta_1 = b/a$ and $\tan\theta_2 = d/c$

  c. $zw = |z||w|[\cos(\theta_1 + \theta_2) + i\sin(\theta_1 + \theta_2)]$

  d. (i) $5(\cos 0 + i\sin 0) = 5$

  (ii) $1 \cdot \left(\cos\dfrac{\pi}{6} + i\sin\dfrac{\pi}{6}\right) = \dfrac{\sqrt{3}}{2} + \dfrac{1}{2}i$

51. a. $\sqrt{z} = \sqrt{|z|}\left[\cos\left(\dfrac{\theta}{2}\right) + i\sin\left(\dfrac{\theta}{2}\right)\right]$

  and $\sqrt{z} = \sqrt{|z|}\left[\cos\left(\dfrac{\theta}{2} + \pi\right) + i\sin\left(\dfrac{\theta}{2} + \pi\right)\right]$

  b. $\sqrt{\dfrac{1}{2} + \dfrac{\sqrt{3}}{2}i} = \sqrt{1}\left(\cos\dfrac{\pi}{6} + i\sin\dfrac{\pi}{6}\right) = \dfrac{\sqrt{3}}{2} + \dfrac{1}{2}i,$

  and $\sqrt{1}\left(\cos\dfrac{7\pi}{6} + i\sin\dfrac{7\pi}{6}\right) = -\dfrac{\sqrt{3}}{2} - \dfrac{1}{2}i$

  d. $z^{1/n} = |z|^{1/n}\left[\cos\left(\dfrac{\theta}{n}\right) + i\sin\left(\dfrac{\theta}{n}\right)\right],$

  $z^{1/n} = |z|^{1/n}\left[\cos\left(\dfrac{\theta + 2\pi}{n}\right) + i\sin\left(\dfrac{\theta + 2\pi}{n}\right)\right],$

  $z^{1/n} = |z|^{1/n}\left[\cos\left(\dfrac{\theta + 4\pi}{n}\right) + i\sin\left(\dfrac{\theta + 4\pi}{n}\right)\right], \ldots$

Section 8.4

1. a. The limiting value will be real whenever $C \le \dfrac{1}{4}$.

  b. $\{0.5, 0.35, 0.2225, 0.14950625, 0.12235212, 0.11497004,$
  $0.11321811, 0.11281834, 0.11272798, 0.1127076,$
  $0.112703, 0.112702, 0.112702, \ldots\}$

  c. Diverges to infinity

3. The equilibrium levels, $\dfrac{1 \pm \sqrt{1 - 4C}}{2}$, will be real when $C \le \dfrac{1}{4}$.

5. Every graph of the form $y = x^3 + C$ intersects the line $y = x$.

7. The limiting values are $\dfrac{\pi}{2} + n\pi$ for any integer n.

Review Problems

1. Not an identity

2. Identity

3. Not an identity

4. Not an identity

5. Identity

6. Identity

7. Identity

8. Identity

9. Not an identity

11. With $T_3(x) = 3x - \dfrac{9}{2}x^3$, we have $T_3(0.2) = 0.564$.

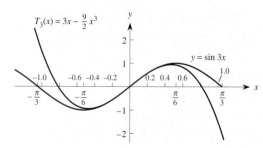

12. With $T_5(x) = 3x - \dfrac{9}{2}x^3 + \dfrac{81}{40}x^5$, we have $T_5(0.2) = 0.564648$.

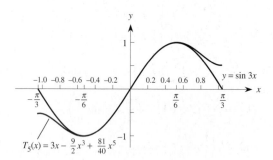

13. Within 0.055556

14. **a.** $z = 10[\cos(2.21430) + i\sin(2.21430)]$
   $w = \sqrt{29}[\cos(-0.38051) + i\sin(-0.38051)]$
   b. $z \cdot w = 53.85165[\cos(1.83379) + i\sin(1.83379)]$
   $\dfrac{z}{w} = 1.85695[\cos(2.59481) + i\sin(2.59481)]$
   $\dfrac{w}{z} = 0.53852[\cos(-2.59481) + i\sin(-2.59481)]$

15. **a.** $z = 3(\cos 52° + i\sin 52°)$
   b. $z = 1.84698 + 2.36403i$
   c. $z^5 = 243(\cos 260° + i\sin 260°)$
   d. $\sqrt{z} = 1.73205(\cos 26° + i\sin 26°)$
   and $1.73205(\cos 206° + i\sin 206°)$

16. 2π 17. 2π 18. π 19. π

20. 4π 21. 4π

Chapter 9

Section 9.2

1. 5 3. $\sqrt{41}$ 5. $\sqrt{20} \approx 4.47$

7. $(3.5, 6)$ 9. $\left(\dfrac{8}{3}, \dfrac{2}{3}\right)$

11. $(x - 5)^2 + (y - 2)^2 = 25$

13. $(x - 6)^2 + (y - 4)^2 = 16$

17. $(x + 2)^2 + (y + 3)^2 = 25$; center $(-2, -3)$, radius $= 5$

19. $(x + 5)^2 + (y - 2)^2 = 100$; center $(-5, 2)$, radius $= 10$

21. $(x - 1)^2 + (y + 3)^2 = 4$; center $(1, -3)$, radius $= 2$

23. **a.**

t	-2	-1	0	1	2	3	4	5
$x = 3 + 2t$	-1	1	3	5	7	9	11	13
$y = 4 - 5t$	14	9	4	-1	-6	-11	-16	-21

b.

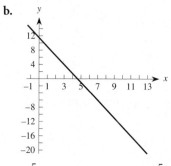

c. $-\dfrac{5}{2}$ **d.** $y + 1 = -\dfrac{5}{2}(x - 5)$

e. $\left(0, \dfrac{23}{2}\right), \left(2, \dfrac{13}{2}\right), \left(4, \dfrac{3}{2}\right), \left(6, -\dfrac{7}{2}\right), \left(8, -\dfrac{17}{2}\right),$
$\left(10, -\dfrac{27}{2}\right), \left(12, -\dfrac{37}{2}\right)$

t	-1.5	-0.5	0.5	1.5	2.5	3.5	4.5
$x = 3 + 2t$	0	2	4	6	8	10	12
$y = 4 - 5t$	11.5	6.5	1.5	-3.5	-8.5	-13.5	-18.5

25. $y - y_0 = \dfrac{y_1 - y_0}{x_1 - x_0}(x - x_0)$; the two-point form of a line

27. $(-3, 4), (5, 0)$

Exercising Your Algebra Skills

1. $(x + 4)^2 + 9$

3. $(x - 3)^2 - 4$

5. $(y + 5)^2 + 1$ 7. $(y + 2)^2 - 16$

Section 9.3

1. a. rightmost point; leftmost point
   b. point on the right   **c.** point on the left

3. The orbit will decay and the satellite spiral in.

5. Venus' orbit is the most circular and Pluto's the least circular.

7. $\dfrac{x^2}{3352.41} + \dfrac{y^2}{3210.8} = 1$

9. a. $\dfrac{x^2}{96,100} + \dfrac{y^2}{65,792.25} = 1$
   b. 348.182 feet   **c.** 402.358 feet

11. center $(0, 0)$; vertices $(-\frac{1}{2}, 0), (\frac{1}{2}, 0), (0, -\frac{1}{3})$, and $(0, \frac{1}{3})$; foci $(-\frac{\sqrt{5}}{6}, 0)$ and $(\frac{\sqrt{5}}{6}, 0)$

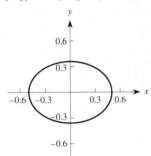

13. $\dfrac{(x + 1)^2}{4} + \dfrac{(y + 1)^2}{1} = 1$; center $(-1, -1)$; vertices $(-3, -1), (1, -1), (-1, -2)$, and $(-1, 0)$; foci $(-1 - \sqrt{3}, -1)$ and $(-1 + \sqrt{3}, -1)$

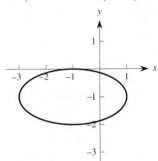

15. $\dfrac{(x + 10)^2}{100} + \dfrac{(y - 5)^2}{25} = 1$; center $(-10, 5)$; vertices $(-20, 5), (0, 5), (-10, 0)$, and $(-10, 10)$; foci $(-10 - 5\sqrt{3}, 5)$ and $(-10 + 5\sqrt{3}, 5)$

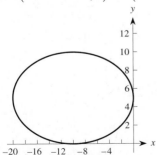

17. $\dfrac{(x - 3)^2}{1} + \dfrac{(y + 2)^2}{9} = 1$; center $(3, -2)$; vertices $(3, 1)$, $(3, -5), (2, -2), (4, -2)$; foci $(3, -2 - 2\sqrt{2})$, $(3, -2 + 2\sqrt{2})$

Section 9.4

1. a. $y = \dfrac{x^2}{24}$ **b.** 24 inches

3. b. $y = \pm\dfrac{b}{a}x$ **c.** $\pm\dfrac{b}{a}$

5. $\dfrac{(x + 1)^2}{16} + \dfrac{(y + 1)^2}{4} = 1$; ellipse with center at $(-1, -1)$, major axis parallel to x-axis, and $a = 4, b = 2$, so $c = \sqrt{12}$.

7. $\dfrac{(x + 1)^2}{16} - \dfrac{(y + 1)^2}{4} = 1$; hyperbola with center at $(-1, -1)$, axis parallel to the x-axis and $a = 4$, $b = 2$, so $c = \sqrt{20}$.

9. $\dfrac{(x - 2)^2}{\left(\frac{38}{4}\right)} - \dfrac{(y + 1)^2}{\left(\frac{38}{9}\right)} = 1$; hyperbola with center at $(2, -1)$, axis parallel to the x-axis, and $a = \dfrac{\sqrt{38}}{2}$, $b = \dfrac{\sqrt{38}}{3}$, so $c = \dfrac{\sqrt{494}}{6}$.

11. $(x - 3)^2 + (y + 2)^2 = \dfrac{9}{4}$; circle with center at $(3, -2)$ and radius $= \dfrac{3}{2}$.

13. $\dfrac{(y + 2)^2}{\left(\frac{1}{2}\right)^2} - \dfrac{(x + 1)^2}{\left(\frac{1}{3}\right)^2} = 1$; hyperbola with center at $(-1, -2)$, axis parallel to the y-axis, and $a = \dfrac{1}{2}, b = \dfrac{1}{3}$.

15. a. $-4, -\dfrac{9}{4}, -1, -\dfrac{4}{9}, -\dfrac{1}{4}$ **b.** $m = -\dfrac{1}{x^2}$

Section 9.5

1. a.

t	-2	-1	0	1	2
$x = 4 - 3t$	10	7	4	1	-2
$y = 2 - 5t$	12	7	2	-3	-8

$m = \dfrac{5}{3}$

b. $y = \dfrac{5}{3}x - \dfrac{14}{3}$

c. The ratio of the coefficients of t in y and x.

d. $y = \dfrac{5}{3}x - \dfrac{14}{3}$

e. $x = \dfrac{3}{5}y + \dfrac{14}{5}$, so that $y = \dfrac{5}{3}x - \dfrac{14}{3}$

3. a.

t	-2	-1.5	-1	-0.5	0	0.5	1	1.5	2
x	-7	-2.375	0	0.875	1	1.125	2	4.375	9
y	2	0.25	-1	-1.75	-2	-1.75	-1	0.25	2

(b), (d)

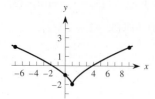

c. $y = (x - 1)^{2/3} - 2$

5. For $a = 1, b = 3$; period $= 6\pi$

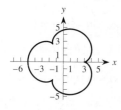

For $a = 1, b = 4$; period $= 8\pi$

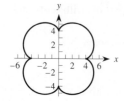

For $a = 1, b = 5$; period $= 10\pi$

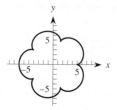

For $a = 1, b = 6$; period $= 12\pi$

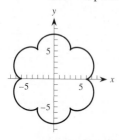

The number of loops appears to be $2b$.

7. For $a = 2, b = 3$; period $= 6\pi$

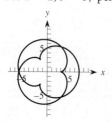

For $a = 2, b = 5$:   period $= 10\pi$

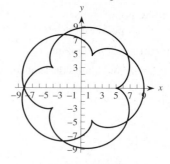

For $a = 2, b = 7$:   period $= 14\pi$

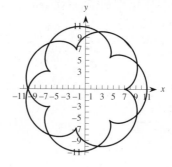

For $a = 2, b = 9$:   period $= 18\pi$

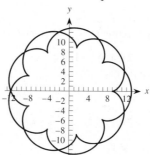

The period seems to be $2b\pi$ and the number of loops is b.

9. a. $x = 2 \cos t, y = 0$

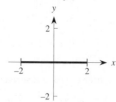

b. $x = 2 \cos t + \cos 2t, y = 2 \sin t - \sin 2t$

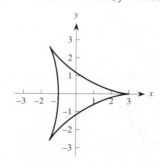

c. $x = 3 \cos t + \cos 3t, y = 3 \sin t - \sin 3t$

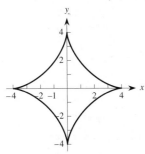

d. $x = \cos t + 2 \cos \dfrac{1}{2}t, y = \sin t - 2 \sin \dfrac{1}{2}t$

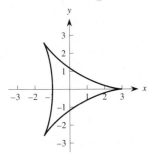

11. $x = a \arccos\left(\dfrac{a-y}{a}\right) - a \sin\left(\arccos\left(\dfrac{a-y}{a}\right)\right)$

Section 9.6

1. a. $P = \left(5, \dfrac{\pi}{6}\right)$ $Q = \left(3, \dfrac{2\pi}{3}\right)$

$R = \left(6, \dfrac{7\pi}{6}\right)$ $S = \left(2, \dfrac{5\pi}{3}\right)$

b. $P = \left(5, -\dfrac{11\pi}{6}\right)$ $Q = \left(3, -\dfrac{4\pi}{3}\right)$

$R = \left(6, -\dfrac{5\pi}{6}\right)$ $S = \left(2, -\dfrac{\pi}{3}\right)$

c. $P = \left(-5, \dfrac{7\pi}{6}\right)$ $Q = \left(-3, \dfrac{5\pi}{3}\right)$

$R = \left(-6, \dfrac{\pi}{6}\right)$ $S = \left(-2, \dfrac{2\pi}{3}\right)$

d. $P = \left(-5, -\dfrac{5\pi}{6}\right)$ $Q = \left(-3, -\dfrac{\pi}{3}\right)$

$R = \left(-6, -\dfrac{11\pi}{6}\right)$ $S = \left(-2, -\dfrac{4\pi}{3}\right)$

3. a. $r = \sqrt{32}$ and $\theta = \dfrac{\pi}{4}$

b. $r = \sqrt{32}$ and $\theta = \dfrac{3\pi}{4}$

c. $r = \sqrt{32}$ and $\theta = \dfrac{5\pi}{4}$

d. $r = \sqrt{32}$ and $\theta = \dfrac{-\pi}{4}$

e. $r = 5$ and $\theta = \arctan\left(-\dfrac{4}{3}\right)$

f. $r = 5$ and $\theta = \arctan\left(-\dfrac{4}{3}\right) + \pi$

g. $r = \sqrt{73}$ and $\theta = \arctan\left(\dfrac{3}{8}\right)$

h. $r = \sqrt{73}$ and $\theta = \arctan\left(\dfrac{8}{3}\right)$

5. a. $\left(4000, \dfrac{\pi}{4}\right)$

b. $\left(4000, \dfrac{5\pi}{8}\right)$

c. $\left(26800, -\dfrac{13\pi}{30}\right)$

Section 9.7

1.

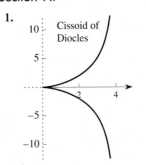

3.

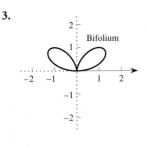

5.

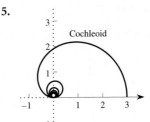

7.

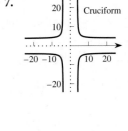

9.

Lemniscate of Bernoulli

11.

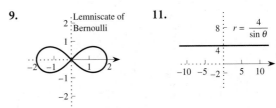

Review Problems

1. a. $\dfrac{(x-1)^2}{9} + \dfrac{(y+2)^2}{25} = 1$

b. $\dfrac{(x-1)^2}{25} + \dfrac{(y+2)^2}{9} = 1$

2. a. $\dfrac{y^2}{9} - \dfrac{(x-2)^2}{16} = 1$ **b.** $\dfrac{x^2}{4} - \dfrac{(y+1)^2}{9} = 1$

3. **a.** $y = \dfrac{(x + 4)^2}{4} + 3$ **b.** $x = \dfrac{(y - 3)^2}{8} - 1$

4. $xy = 5$ is a hyperbola with axis on the line $y = x$.

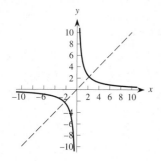

5. $y - 43 = -(x - 3)^2$ is a vertical parabola, with vertex at $(3, 43)$, focus at $(3, 42.75)$, opening downward.

6. $(x - 4)^2 + (y + 3)^2 = 16$ is a circle of radius 4 and center at $(4, -3)$.

7. $\dfrac{(x + 5)^2}{17} + \dfrac{(y - 2)^2}{\left(\frac{34}{3}\right)} = 1$ is an ellipse with center at $(-5, 2)$, major axis parallel to the x-axis and $a = \sqrt{17}$, $b = \sqrt{\frac{34}{3}}$.

8. $\dfrac{(x - 1)^2}{4} - \dfrac{(y + 3)^2}{3} = 1$ is a hyperbola with center $(1, -3)$ whose axis is parallel to the x-axis.

9. $\dfrac{(y - 2)^2}{6} - \dfrac{(x - 3)^2}{9} = 1$ is a hyperbola with center $(3, 2)$ whose axis is parallel to the y-axis.

10. $\dfrac{x^2}{100} + \dfrac{y^2}{36} = 1$

11. $\dfrac{(x + 6)^2}{40} + \dfrac{(y - 3)^2}{4} = 1$

12. $\dfrac{(y - 3)^2}{9} - \dfrac{(x - 2)^2}{7} = 1$

13. $\dfrac{y^2}{16} - \dfrac{x^2}{9} = 1$ 14. $\dfrac{x^2}{225} + \dfrac{y^2}{81} = 1$

15. $2a = 7.277$, $\dfrac{x^2}{13.24} + \dfrac{y^2}{4.24} = 1$

16. 12 ft from the point below the highest point, along the axis of the vertices.

17. $\dfrac{x^2}{\left(\frac{841}{4}\right)} + \dfrac{y^2}{78} = 1$

18. The points are $(6, -4)$, $(5, -1)$, $(4, 2)$, $(3, 5)$, $(2, 8)$; the function is $y = -3x + 14$.

19. Same as parabola $y = \dfrac{x^2}{9} + 1$, between $x = -6$ and $x = 6$.

20. **a.**

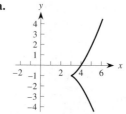

b. $x = (y + 1)^{2/3} + 3$ **c.** $x = 4$

21. **a.** Same as the graph of $y = 1 - x$, for $0 \le x \le 1$.

22. The graph has 4 "loops" and has period 4π.

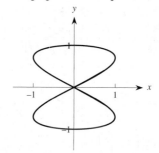

23. **a.** The graph has 2 "loops" and has period π.

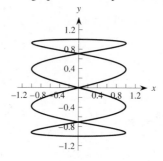

b. The graph has 3 "loops" and has period π.

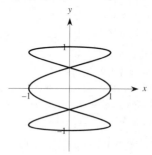

c. The graph has 6 "loops" and has period 2π.

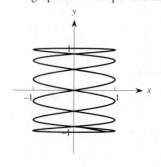

24. $x - 1 = -2y^2$

25. a. $\left(\sqrt{18}, \dfrac{\pi}{4}\right)$

   b. $(\sqrt{10}, 1.89255)$

   c. $(\sqrt{17}, -0.244979)$

   d. $\left(6, \dfrac{\pi}{2}\right)$

26. a. $(1.5, 1.5\sqrt{3})$

   b. $\left(\dfrac{3\sqrt{2}}{2}, \dfrac{3\sqrt{2}}{2}\right)$

   c. $(0, -4)$

   d. $(-2\sqrt{2}, -2\sqrt{2})$

   e. $(-2.5\sqrt{3}, 2.5)$

   f. $(-2.08073, 4.54649)$

27. Parabola $x - \dfrac{1}{2} = -\dfrac{1}{2}y^2$

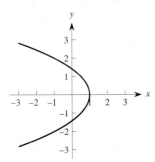

28. ellipses

29. a circle along the polar axis; a 4-petaled rose along the axes; an 8-petaled rose along the axes

30. a circle along the polar axis; a 3-petaled rose with one petal along the polar axis; a 5-petaled rose with one petal along the polar axis

31. a circle along the vertical axis; a 4-petaled rose between the axes; an 8-petaled rose between the axes

32. a circle along the positive vertical axis; a 3-petaled rose with one petal along the negative vertical axis; a 5-petaled rose with one petal along the positive vertical axis.

Chapter 10

Section 10.1

3. d. all lie on the line.

5. a. 5   **b.** $\sqrt{40}$ **c.** $\sqrt{17}$ **d.** $\sqrt{18}$
   e. $\sqrt{66}$

7. a. $\begin{bmatrix} 9 & -3 & 6 \end{bmatrix}$ **b.** $\begin{bmatrix} 30 & -10 & 20 \end{bmatrix}$
   c. $\begin{bmatrix} -21 & 7 & -14 \end{bmatrix}$ **d.** $\begin{bmatrix} 3/2 & -1/2 & 1 \end{bmatrix}$

9. a. $2\mathbf{i} + \mathbf{j}$ **b.** $-\mathbf{i} + 3\mathbf{j}$ **c.** $\frac{1}{2}\mathbf{i} + \frac{1}{2}\mathbf{j}$
   d. $3\mathbf{i}$

11. a. 27.2 pounds   **b.** 17.3 pounds   **c.** 38.6 pounds

13. a. 9.5° south of east at 608.3 mph
   b. 5.4° south of east at 679.6 mph
   c. 44.3° south of west at 349.8 mph

Section 10.2

1. $\begin{bmatrix} 64 & 73 & 86 & 85 \\ 82 & 69 & 77 & 91 \\ 82 & 84 & 81 & 83 \end{bmatrix}$, rows for people, columns for subjects

3. a. HORN   **b.** MOTHER   **c.** EARLY
   d. YESTERDAY

5. $8x_1 + 5x_2 + 3x_3 = 6200$
   $2x_1 + 5x_2 + 5x_3 = 4000$
   $3x_1 + 7x_2 + 6x_3 = 4700$

7. $2000x_1 + 500x_2 + 5000x_3 = 280{,}000$
   $5x_1 - x_2 = 0$
   $-x_1 + 2x_3 = 0$

9. a. $p_1' = 1/4,\ p_2' = 1/2,\ p_3' = 1/4$
   b. $p_1' = 3/8,\ p_2' = 3/8,\ p_3' = 1/4$
   c. $p_1' = 1/4,\ p_2' = 1/2,\ p_3' = 1/4$
   d. $p_1' = 3/8,\ p_2' = 3/8,\ p_3' = 1/4$
   e. $p_1' = 0.35,\ p_2' = 0.40,\ p_3' = 0.25$

11. a. $\begin{bmatrix} 2/3 & 1/2 \\ 1/3 & 1/2 \end{bmatrix}$ **b.** 7/12   **c.** 11/18.

13. a. $\begin{bmatrix} 1/2 & 1/4 & 0 & 0 \\ 1/2 & 1/2 & 1/4 & 0 \\ 0 & 1/4 & 1/2 & 0 \\ 0 & 0 & 1/4 & 1 \end{bmatrix}$

   b. $\begin{bmatrix} 3/8 \\ 1/2 \\ 1/8 \\ 0 \end{bmatrix},\ \begin{bmatrix} 5/16 \\ 15/32 \\ 3/16 \\ 1/32 \end{bmatrix}$

15. a. $\begin{bmatrix} 30 \\ 110 \end{bmatrix},\ \begin{bmatrix} 3 \\ 126 \end{bmatrix},\ \begin{bmatrix} -34.2 \\ 150.6 \end{bmatrix}$

   b. $\begin{bmatrix} 50 \\ 165 \end{bmatrix},\ \begin{bmatrix} 43.5 \\ 256 \end{bmatrix},\ \begin{bmatrix} 26.6 \\ 380.15 \end{bmatrix}$

Section 10.3

1. a. 10   **b.** -2 **c.** 0   **d.** -3
   e. -5 **f.** 10   **g.** 15

3. a. 3   **b.** -2 **c.** 3   **d.** 12
   e. 30   **f.** -21

5. a. $\begin{bmatrix} 14 \\ 4 \end{bmatrix}$ **b.** $\begin{bmatrix} -2 \\ 18 \end{bmatrix}$ **c.** $\begin{bmatrix} 25 \\ -3 \end{bmatrix}$ **d.** $\begin{bmatrix} 4 \\ 0 \end{bmatrix}$

   e. $\begin{bmatrix} 9 \\ 22 \\ 14 \end{bmatrix}$ **f.** $\begin{bmatrix} 11 \\ 2 \\ 14 \end{bmatrix}$

7. **a.** c **b.** d **c.** $\begin{bmatrix} 1 \\ 5 \\ 2 \end{bmatrix}$ **d.** $\begin{bmatrix} 1 \\ 3 \\ -2 \end{bmatrix}$

e. $\begin{bmatrix} 0 \\ -1 \\ 0 \end{bmatrix}$ **f.** $\begin{bmatrix} 2 \\ -5 \\ 2 \end{bmatrix}$

9. **a.** $5x_1 + x_2 = 2$
$4x_1 + 3x_2 = 5$
b. $x_1 + 4x_2 = 4$
$2x_1 - 3x_2 = 9$
c. $5x_1 + 2x_2 + x_3 = 1$
$4x_1 + x_2 + 6x_3 = 5$
$3x_1 + x_2 = 2$
d. $2x_1 - x_2 + 5x_3 = 0$
$3x_1 + x_2 + 2x_3 = 0$
$5x_1 + x_2 - 3x_3 = 0$

11. Each has the form $\mathbf{Ax} = \mathbf{b}$

a. $\mathbf{A} = \begin{bmatrix} 8 & 5 & 3 \\ 2 & 5 & 5 \\ 3 & 7 & 6 \end{bmatrix}$, $\mathbf{b} = \begin{bmatrix} 6200 \\ 4000 \\ 4700 \end{bmatrix}$

b. $\mathbf{A} = \begin{bmatrix} 10 & 50 & 200 \\ 1 & 3 & 0.2 \\ 30 & 10 & 0 \end{bmatrix}$, $\mathbf{b} = \begin{bmatrix} 600 \\ 20 \\ 200 \end{bmatrix}$

c. $\mathbf{A} = \begin{bmatrix} 2000 & 500 & 5000 \\ 5 & -1 & 0 \\ -1 & 0 & 2 \end{bmatrix}$, $\mathbf{b} = \begin{bmatrix} 280{,}000 \\ 0 \\ 0 \end{bmatrix}$

13. $\begin{bmatrix} 4 & 6 & 5 \\ 2 & 1 & 0.85 \\ 1.5 & 2 & 2.5 \\ 6 & 5 & 7 \end{bmatrix} \begin{bmatrix} 10 \\ 6 \\ 3 \\ 2 \end{bmatrix}$
b. $68.50, $82, $76.60

21. **a.** 57.53° **b.** 35.84° **c.** 81.39° **d.** 69.11°

23. State Tech because the angle is smaller.

Section 10.4

1. **a.** $\begin{bmatrix} 2 & 2 \\ 8 & 6 \end{bmatrix}$ **b.** $\begin{bmatrix} 2 & 16 \\ 1 & 6 \end{bmatrix}$ **c.** $\begin{bmatrix} 13 & 7 \\ -7 & -5 \end{bmatrix}$

d. $\begin{bmatrix} 10 & -2 \\ 2 & -2 \end{bmatrix}$ **e.** $\begin{bmatrix} -2 & -6 \\ -1 & -11 \end{bmatrix}$ **f.** $\begin{bmatrix} 2 & 8 \\ 4 & 18 \end{bmatrix}$

g. $\begin{bmatrix} 18 & 10 \\ 5 & 3 \end{bmatrix}$

3. **a.** $\begin{bmatrix} 25 & -5 \\ 5 & 7 \end{bmatrix}$ **b.** $\begin{bmatrix} 5 & 3 \\ 15 & 5 \\ 25 & 7 \end{bmatrix}$ **c.** not possible

d. $\begin{bmatrix} 5 & 25 & -10 \\ 13 & 5 & 6 \end{bmatrix}$ **e.** not possible

f. $\begin{bmatrix} 7 & 5 & 2 \\ 15 & 15 & 2 \\ 23 & 25 & 2 \end{bmatrix}$ **g.** $\begin{bmatrix} 6 & 10 \\ 13 & 18 \end{bmatrix}$

h. not possible

5. **a.** $\begin{bmatrix} 5 & 3 \\ 15 & 5 \\ 25 & 7 \end{bmatrix}$, $\begin{bmatrix} 14 & 25 & -4 \\ 30 & 75 & -20 \\ 46 & 125 & -36 \end{bmatrix}$

b. $\begin{bmatrix} 2 & 25 & -12 \\ 6 & 0 & 4 \end{bmatrix}$, $\begin{bmatrix} 14 & 25 & -4 \\ 30 & 75 & -20 \\ 46 & 125 & -36 \end{bmatrix}$

9. $\begin{bmatrix} 6.40 & 6.85 & 6.05 \\ 5.95 & 6.95 & 5.75 \end{bmatrix}$

11. **a.** $\mathbf{BA} = \begin{bmatrix} 4.70 & 4.00 \\ 3.40 & 3.65 \end{bmatrix}$

b. $\mathbf{CB} = \begin{bmatrix} 16800 & 30400 & 12000 \\ 16200 & 27600 & 12000 \end{bmatrix}$

c. $\mathbf{CBA} = \begin{bmatrix} 12250 & 10920 \\ 11730 & 10380 \end{bmatrix}$

13. **a.** $\mathbf{x}_n = \mathbf{A}^n\mathbf{x}_0$

b. $\mathbf{A}^2 = \begin{bmatrix} 4 & 2 & 1 \\ 1 & 6 & -2 \\ 3 & -1 & 4 \end{bmatrix}$, $\mathbf{A}^3 = \begin{bmatrix} 8 & 7 & 3 \\ 3 & 15 & -7 \\ 10 & -3 & 8 \end{bmatrix}$

c. $\begin{bmatrix} 8 \\ 3 \\ 10 \end{bmatrix}$, $\begin{bmatrix} 21 \\ 4 \\ 23 \end{bmatrix}$ **d.** between 4 and 5 years

15. $\begin{bmatrix} 3/4 & 1/4 \\ 1/4 & 3/4 \end{bmatrix}$

a. $\begin{bmatrix} 5/8 & 3/8 \\ 3/8 & 5/8 \end{bmatrix}$ **b.** $\begin{bmatrix} 9/16 & 7/16 \\ 7/16 & 9/16 \end{bmatrix}$, 7/16

c. 0.496.

17. Let $\mathbf{x} = \begin{bmatrix} x^3 \\ x^2 \\ x \\ 1 \end{bmatrix}$, $\mathbf{p} = \begin{bmatrix} 1 & 4 & -7 & 2 \end{bmatrix}$,

$\mathbf{q} = \begin{bmatrix} 4 & -5 & 8 & 17 \end{bmatrix}$: so $P(x) = \mathbf{p} \cdot \mathbf{x}$ and $Q(x) = \mathbf{q} \cdot \mathbf{x}$
a. $(\mathbf{p} + \mathbf{q}) \cdot \mathbf{x}$
b. $(\mathbf{p} - \mathbf{q}) \cdot \mathbf{x}$

21. The repeated application of $\mathbf{R}$ to any vector successively rotates the vector 1° at a time in the counterclockwise direction around the origin.

Section 10.5

1. **a.** $x = 28/5, y = 12/5$ **b.** $x = -21, y = -6$
c. $x = -0.7, y = -1.8$
d. $x = 19/11, y = 12/11$
e. $x = 5, y = 6$

3. a. $x_1 = 6/5, x_2 = 4/5, x_3 = 1/5$
   b. $x_1 = 7/5, x_2 = -5, x_3 = -29/5$
   c. $x_1 = -22/3, x_2 = 17/3, x_3 = 13/3$
   d. $x_1 = 10/7, x_2 = 5/7, x_3 = 6/7$
   e. $x_1 = -3, x_2 = -5, x_3 = 8$
   f. $x_1 = 12/10, x_2 = -37/10, x_3 = 29/20$

5. a. multiple solutions   **b.** inconsistent
   c. multiple solutions   **d.** multiple solutions
   e. unique solution   **f.** unique solution

7. Factory 1: 0; Factory 2: 78.57; Factory 3: 42.86

9. Gelatin 5.04 oz.; fish sticks 4.89 oz.; mystery meat 1.53 oz.

11. a. $\begin{bmatrix} 1/2 \\ 1/2 \end{bmatrix}$ **b.** $\begin{bmatrix} 1/2 \\ 1/2 \end{bmatrix}$ **c.** $\begin{bmatrix} 1 \\ 0 \end{bmatrix}$

13. $\begin{bmatrix} 2/3 & 1/2 \\ 1/3 & 1/2 \end{bmatrix}, \begin{bmatrix} 3/5 \\ 2/5 \end{bmatrix}$

15. $\begin{bmatrix} 1/2 & 1/4 & 0 & 0 \\ 1/2 & 1/2 & 1/4 & 0 \\ 0 & 1/4 & 1/2 & 0 \\ 0 & 0 & 1/4 & 1 \end{bmatrix}, \begin{bmatrix} 0 \\ 0 \\ 0 \\ 1 \end{bmatrix}$

17. $y = x^2 - 4x + 3$

19. a. $x^2 + y^2 - x - y = 0$
   b. $x^2 + y^2 - 2x + 2y - 3 = 0$
   c. $x^2 + y^2 + \dfrac{45}{19}x - \dfrac{79}{19}y - \dfrac{246}{19} = 0$
   d. $x^2 + y^2 - x - (3/2)y - 9/2 = 0$
   e. $x^2 + y^2 - 17y = 0$

21. $\mathbf{A}^{-1} = \begin{bmatrix} 1 & 0 \\ -2 & 0.5 \end{bmatrix}$

Review Problems

1. a. 5   **b.** $\sqrt{29}$ **c.** 3

2. 77.1° south of west at 172.2 mph

3.
$$4x + y + 3z = 1300$$
$$3x + 2y + 5z = 2300$$
$$2y + 4z = 1300$$

4. a. $\begin{bmatrix} 0.6 & 0.4 & 0.2 \\ 0.2 & 0.2 & 0.2 \\ 0.2 & 0.4 & 0.6 \end{bmatrix}$ **b.** $\begin{bmatrix} 0.34 \\ 0.2 \\ 0.46 \end{bmatrix}$

5. a. 2   **b.** 5   **c.** 38   **d.** 14

6. a. not defined   **b.** not defined
   c. not defined   **d.** not defined

   e. $\begin{bmatrix} 0 \\ -8 \\ 2 \end{bmatrix}$ **f.** $\begin{bmatrix} 38 \\ 18 \\ 24 \\ 29 \end{bmatrix}$

7. a. dAp   **b.** \$2210

8. a. not defined
   b. $\begin{bmatrix} -2 & -3 & -4 & -5 \\ -2 & -4 & -6 & -8 \\ 5 & 9 & 13 & 17 \end{bmatrix}$

   c. $\begin{bmatrix} 16 & 14 & 20 \\ 32 & 28 & 40 \\ 41 & 35 & 47 \end{bmatrix}$ **d.** $\begin{bmatrix} 16 & 31 & 46 & 61 \\ 7 & 12 & 17 & 22 \\ 10 & 19 & 28 & 37 \\ 11 & 19 & 27 & 35 \end{bmatrix}$

   e. $\begin{bmatrix} 13 & -7 & -4 \\ 1 & 2 & 1 \\ 7 & -3 & -2 \\ 2 & 1 & 3 \end{bmatrix}$

9. a. $\mathbf{BA} = \begin{bmatrix} \$2.30 & \$3.05 \\ \$1.65 & \$2.10 \end{bmatrix}$

   b. $\mathbf{CB} = \begin{bmatrix} 7000 & 12500 & 5500 \\ 14000 & 25000 & 11000 \end{bmatrix}$

10. a. $\mathbf{A}^2 = \begin{bmatrix} 0.48 & 0.4 & 0.32 \\ 0.2 & 0.2 & 0.2 \\ 0.32 & 0.4 & 0.48 \end{bmatrix}$ $\mathbf{A}^3 = \begin{bmatrix} 0.43 & 0.4 & 0.37 \\ 0.2 & 0.2 & 0.2 \\ 0.37 & 0.4 & 0.43 \end{bmatrix}$

   $\mathbf{A}^5 = \begin{bmatrix} 0.405 & 0.4 & 0.395 \\ 0.2 & 0.2 & 0.2 \\ 0.395 & 0.4 & 0.405 \end{bmatrix}$

   b. columns approaching $\begin{bmatrix} 0.4 \\ 0.2 \\ 0.4 \end{bmatrix}$

11. a. $x_1 = 30, x_2 = 14, x_3 = -9$
   b. $x_1 = 1/3, x_2 = -7/3, x_3 = 0$

12. 100 tables, 500 chairs, 200 sofas

13. $\begin{bmatrix} 0.4 \\ 0.2 \\ 0.4 \end{bmatrix}$

14. $y = x^2 - 2x + 2$

Appendix A

1. 21   **3.** 10   **5.** 1

7. The graph of $y = |x|$ shifted left 3 spaces.

9. 1140   **11.** $\sum_{k=5}^{40} \dfrac{1}{k}$ **13.** 5

15. a. 10   **b.** 15

17. $(x + 4)^2 + (y - 3)^2 = 36$; center $(-4, 3)$, radius 6

Appendix E

1. $-i$ **3.** -1 **5.** $2 + 4i$

7. $11 - 2i$ **9.** $5 - 5i$ **11.** 229

13. $\pm 5i$ **15.** $-\dfrac{1}{5} \pm \dfrac{2}{5}i$ **17.** $-\dfrac{2}{3} \pm \dfrac{2\sqrt{5}}{3}i$

Index

Absolute value, 722
Acceleration due to gravity, 190
Addition
    of complex numbers, 736
    of matrices, 691
AIDS, 187–188, 195, 278–279
Airplanes, wingspan vs. weight, 211–213
Amplitude, of sine function, 498
Analytic geometry. *See also* Geometry
    conic sections and, 595–618
    distance between points and,
        585–589
    ellipse and, 595–605
    equation of circle and, 589–594, 724
    explanation of, 585
    Gaussian elimination and, 711–714
    hyperbola and, 606–614
    parabola and, 614–616
Angles
    complementary, 429
    of inclination, 431
    reference, 457
Aphelion, 601, 605
Arccos functions, 522–527
Arcsin function, 517–522
Arctan function, 530–533
Arithmetic of complex numbers, 567,
    734–737
Associative law of matrices, 691
Asymptote, vertical, 105
Asymptote, horizontal, 80
Augmented coefficient matrix,
    701–702
Average rate of change, 136–139
Axis
    of ellipse, 597, 598
    explanation of, 25
    of hyperbola, 609
    polar, 630
    of symmetry, 615

Backward difference equations, 369
Base 10, 99–100
Base *e*, 109
Bases, changing, 109–111
Best fit
    curve, 181–191
    line, 166–168
Biological half-life, 363

Birds, wingspan vs. weight, 115, 120, 144
Black thread method, 65–66, 162
Bouncing ball, 423–425
Brachistochrone problem, 627

Calculators
    equation of line and, 165
    exponential functions and, 200
    graphs of polynomials on, 267–268
    linear regression, 164–166
    logarithms and, 202
    nonlinear regression, 181–191
    parametric representations on, 624
    sequences on, 360–361
    solving systems of equations on,
        730–731
    trigonometric functions on,
        456–457, 485–486, 516
Carbon dating, 120–121
Cardioid, 579, 641
Cartesian coordinate system
    explanation of, 25–26
    transforming between polar and,
        632–634
Cell phone, 341–343
Challenger disaster, 161–163, 334–336
Chaos, 574–581
    complex numbers and Julia set,
        576–579
    Mandelbrot set, 579–581
Circle
    equation of, 589–594, 724
    parametric representations of, 622
    polar coordinates, 589
    unit, 484–485, 489
Circular function. *See* Trigonometric
    function
Clothes production model, 660–662,
    675–676, 703–705
Coefficient
    correlation, 169–170
    leading, 250
    matrix, 701, 731
    of determination, 240, 273, 274
    of polynomial, 250
Column vectors, 650, 659–660
Commutative law of matrices, 691
Comparing rates of growth, 131–136
Complementary angles, 429

Complete solution of difference
    equation, 368
Completing the square, 725–727
Complex conjugates, 573, 735
Complex numbers
    arithmetic of, 567, 734–737
    chaos and, 574–575
    complex conjugates and, 573
    iteration processes applied to,
        573–575
    modulus of, 568
    powers of, 569–573
    properties of, 566–569
    trigonometric form of, 568–569
Complex plane, 567
Complex root, 573
Composite functions
    applications of, 314–315
    explanation of, 312–313
Concave down, 11
Concave down growth functions, 136
Concave up, 10–11
Concave up growth functions, 136
Concavity, 10–12
Conic sections
    ellipse, 596–605
    explanation of, 595–596, 617–618
    hyperbola, 606–614
    parabola, 614–616
Conjugate pair, 573
Constant of proportionality, 44
Contaminant model. *See* Pollutant
    model
Control systems, 290–291
Convergence, 360
Cooling model, 322–334, 408–412
Coordinate systems
    Cartesian, 25–26, 583
    coding model, 332–334, 408–412
    explanation of, 583–584
    polar, 584, 630–634
    rectangular, 25–26
    types of, 584
Coordinates, 25
Correlation coefficient
    critical values for, 171, 721
    determining significant coefficient
        and, 170–176
    explanation of, 169–170
    fit and, 222

method for finding, 170
multiple, 240–241
Cosecant function, 533
Cosine function
applications of, 450–452
approximating, 558–561
behavior of, 448, 492–493
explanation of, 448, 487–490
graph of, 462
inverse, 522–527
law of cosines, 473–474, 673
modeling periodic behavior with, 494–509
period of, 460
radian measure and, 490–492
"special" angles and, 449–450
Cotangent function, 533
Cramer's rule, 709
Cricket chirping, 27–28, 51–53, 64–65, 505–506, 521–522
Critical values for correlation coefficient, 171, 721
Cubic functions
behavior of, 263–264
characteristics of, 261–264
difference patterns, 292–294
explanation of, 250
fitting to data, 287–289
roots of, 287–289
Curve fitting procedures, 213. *See also* Fitting to data
Cycloid, 626

Data
application of fitting functions to, 228–234
capturing linear pattern in, 64–66
determining if set of data is linear, 59–68
determining if set data is exponential, 82–83
difference patterns in, 292–294
explanation of, 1–2
fitting exponential and logarithmic functions to, 196–204
fitting nonlinear functions to, 181–191
fitting polynomial functions to, 278–281
fitting power functions to, 208–218
interpreting correlation coefficient and, 222
interpreting residuals and, 222–224
interpreting scatterplot and, 225–227

interpreting sum of squares and, 227–228
linear models with several variables and, 236–243
linear regression analysis and, 164–176
linearizing, 196
measuring the center of set of, 738–739
measuring the spread in set of, 739–740
using shifting and stretching to analyze, 331–345
world population, 740–744
Data analysis
introduction to, 161–163, 737
Daylight model, 184–185, 200–202, 339–340, 349–352
Dead body model, 411
Decay factor, 88–89
Decay rate, 88
Decreasing function, 9–10
Degree of polynomial, 250, 264–266
Delta (Δ), 46
DeMoivre, Abraham, 573
DeMoivre's theorem
explanation of, 573
use of, 575, 576
Dependent variable, 19, 25
Determinant, 709
Descartes, Rene, 585
Difference, of two functions, 305
Difference, of vectors, 656–657
Difference equations. *See also* Sequences
backward, 369
behavior of logistic function and, 392–394
closed form solution and, 367
complete solution, 368
drug model and, 359–367
equilibrium, 393
explanation of, 359–360, 367–369
exponential growth and decay models and, 371–383, 416
Fibonacci model and, 382–383
first order, 383
forward, 369
geometric sequences and their sums and, 416–425
inhibited growth model. *See* logistic model
limit to growth, 388
logistic growth model and, 386–403
maximum sustainable population and, 389

Newton's laws of cooling and, 407–412
Newton's laws of heating and, 413–414
point of inflection and, 390–392
second order, 383
solution sequence to, 360–369, 374, 391
Difference identities of sine and cosine, 543–545
Dimension of a matrix, 691, 731
Directrix, of parabola, 615, 616
Discriminant, of quadratic equations, 259–260
Displacement vector, 650
Distance formula
explanation of, 585–586, 724
use of, 586, 589
Distributive law of matrices, 690
Domain, of function, 20–23, 27
Dot products. *See* Scalar products
Double-angle identities, 541–542, 545, 561
Doubling time, 76–77
Dow Jones model. *See* Stock market model
Drug doses, 357–363
Drug model
eliminating medication and, 87–88, 102, 357–358
exponential decay and, 87–88
half life, 90–93
maintenance level and, 258, 361, 363–365
repeated dosage and, 358–361

e base, 109
Earthquakes, 107–108
Earth's orbit, 601–602
Elementary row operations, 700–701
Elimination method, 728–730
Ellipse
description and graph of, 599–601
equation of, 598–599, 724–725
explanation of, 595, 596–598
foci, 596
graph of, 598–599
major and minor axes, 597
parametric representation of, 623–624
planetary orbit and, 601–605
reflection property of, 602–604
symmetry of, 604
vertex of, 597
Epicycloid, 627–628

Equation of
circle, 589–594, 724
ellipse, 598–599
hyperbola, 608–611
line, 49–51
parabola, 615–617
parametric, 622–628
quadrtic, 253
roots of, 251
systems of linear equations, 727–731
Equilibrium, 393
state, 709–710
Even function
power, 126
trigonometric, 524, 561
Even positive integer powers, 126
Excel, multivariate regression in, 241–243
Explanatory variables. *See* Independent
variables
Exponential decay
determining if set of data is exponential growth or, 94–95
half-life for, 90–91
models of, 371–372, 377–380
Exponential decay functions
decay factor, 88–89
decay rate, 88
example of, 87–88
explanation of, 88–90, 136
formula for, 89
graphs of, 88
half-life and, 90–93
radioactive decay and, 91–93
Exponential functions
behavior of, 116, 122, 131
combining sinusoidal and, 507–509
determining if a set of data is exponential, 82–83, 94–95
explanation of, 72
fit to data, 183–185, 187, 196–202
identities, 83–84, 100
logarithmic functions vs., 105–106
passing through two points, 80–82
rate of change of, 138–139
Exponential growth
applications of, 75–76
difference equation for, 372, 373
doubling time, 76–77
linear vs., 74
models of, 371–377, 382–383
population growth, 182–183
Exponential growth functions
domain of, 79
doubling time of, 76–77
explanation of, 72–73, 132
formula for, 73–74

overview of, 70–72
passing through two points, 80–82
power functions vs., 132–134, 136
prediction with, 77–80
rules for exponents and, 83–84
Exponents, rules for, 83–84
Extrapolating, 168
Extrapolation, 36

Factoral notation *n!*, 722
Factors and roots, 252, 258, 259, 260,
262, 264, 265
Falling body, 19–20, 29, 145–146,
174–175, 189–191, 273–276
Fiber optic model, 93–94
Fibonacci, 380
Fibonacci difference equation, 382
Fibonacci model
explanation of, 382, 383
for population growth, 380–383
Fibonacci sequence, 382
First order difference equations, 383
Fitting to data
correlation coefficient and, 222
examples of, 228–234
exponential functions, 196–202
interpreting scatterplot and, 225–227
interpreting sum of squares and,
227–228
linear models with several variables,
236–243
linear regression analysis, 164–176
logarithmic functions, 196, 202–204
logistic functions, 397–403
nonlinear functions, 181–191
polynomials, 278–281
power functions, 208–218
residuals and, 222–225
Florida population model, 28–29,
70–72, 76, 77–79, 101, 104, 110,
143–144
Focus
parabola, 615
ellipse, 596
hyperbola, 606
Forward difference equations, 369
Fractal, 581
Frequency, of sinusoidal function,
500–502
Fruit purchase model, 672, 678–679,
691–692
Functions. *See also specific types of
functions*
composite, 311–313
concavity of, 10–12

connecting geometric and symbolic
representations of, 25–31
cosine, 448, 462, 487–490
cubic, 250, 253, 261–266, 287–289
decreasing, 9–10
difference of, 305
domain and range of, 20–23, 27
even/odd, 126–127, 524, 556, 561
explanation of, 3, 22
exponential, 183–185, 187, 196–202
exponential decay, 87–95, 131–134
exponential growth, 70–87,
131–134, 136–137
families of, 43–44
formulas and equations of, 3
function of, 311–313
graphs of, 3–4, 26
implicit, 66–68
increasing, 9–10
inflection, point of, 11–12
inverse, 140–152
inverse cosine, 522–527
inverse sine, 517–522
inverse tangent, 530–533
linear, 44–56, 59–68, 131
logarithmic, 99–111, 132, 135–136,
189, 196, 202–204
logistic, 346–349, 392–394
mathematical models and, 34–38
monotonic, 148
multiple of, 323–328
nonlinear, 181–191
periodic, 14–15, 460
point of inflection, 11–12
polynomial, 249–269, 273–281,
285–291
power, 113–127, 131–136, 185–186,
188, 190, 208–218
products of, 305–306
quadratic, 250, 251, 255–260,
285–287
quartic, 251
quotient of, 306–307
range of, 20–23
rational, 307–310
represented symbolically, 18–23
shifting, 320–322
sine, 441–442, 485–486
sinusoidal, 496–497, 500, 501
stretching and shrinking, 323–328,
331–345
sum of, 304
surge, 349–352
tables of, 4–5
tangent, 432, 434, 527
turning point of, 9–10

Fundamental logarithmic-exponential identities, 100

Galileo, 3, 174, 273
Gateway Arch, 280–281
Gauss, Carl Friedrich, 700
Gaussian elimination
    applications of, 709–714
    clothes production model and, 703–705
    explanation of, 700–702
    geometric interpretation and, 702–703
    inverse matrix and, 708–709
    systems of linear equations with multiple solutions and, 705–706
    systems of linear equations with no solutions and, 706–708
General equation of circle, 591
Geometric sequences, 372, 417–420, 423–425
Geometry
    analytic, 585–594
    conic sections and, 595–618
    ellipse and, 595–605
    families of curves in polar coordinates and, 636–643
    hyperbola and, 606–614
    parabola and, 614–616
    parametric representations and, 619–628
    polar coordinate system and, 630–634
Golden ratio, 175, 176
Graphing calculators. See Calculators
Graphs
    of cosine function, 462
    of cubic, 263–264
    of ellipse, 598–599
    of exponential decay functions, 88–92, 131
    of exponential growth functions, 72–74, 77–79, 131
    of function, 3–4, 26
    of hyperbola, 608–611
    of inverse functions, 149–150
    of linear functions, 45–46, 131
    of logarithmic functions, 106, 132
    of logistic function, 392–394
    of polynomials, 256, 261, 262, 264, 265
    of power functions, 115–119, 131–132
    of quadratic functions, 255–257, 614
    of sine function, 461–462, 486, 487
    of tangent function, 527–530
    symmetry and, 732–734
Growth factor, 72–74
Growth rate, 72–74

Half-angle identities, 544–545
Half-life
    explanation of, 363
    for exponential decay process, 90–93
Heating model, 324–325, 413–414
Homogeneous systems, 711
Horizontal asymptote, 80
Horizontal axis, 25
Horizontal line test, 149
Horizontal shifts, 322, 340–344
Horizontal shrinking, 323
Horizontal stretching, 323, 327
Hours of daylight model. See Daylight model
Hyperbola
    applications of, 611–612, 614
    equation of, 608–611
    explanation of, 606–608
    foci of, 606
    graph of, 608–611
    parametric representations of, 625
Hyperbolic coordinate system, 584
Hyperbolic cosine, 625
Hyperbolic sine, 625

Identities. See also Trigonometric identities
    difference, 543–545
    double-angle, 541–542, 545, 561
    explanation of, 100, 464, 539, 545
    fundamental logarithmic-exponential, 100, 103
    half-angle, 544–545
    involving tangent, 548
    Pythagorean, 465, 540, 545
    reflection, 540–541, 545
    sum, 542–543, 545
Imaginary numbers, 735
Implicit functions, 66–68
Inconsistent equations, 707
Increasing functions, 9–10
Independent variables
    constant multiple of, 326
    explanation of, 19, 25, 237
    horizontal stretching and shrinking and, 323
Index of summation, 722, 723
Inflection point. See Point of inflection
Inhibiting constant, 386, 387
Inhibiting growth model. See Logistic growth model
Initial side, of angle, 457
Initial condition of difference equation, 393
Inner products. See Scalar products

Interpolation, 36, 168
Interval notation, 20
Inverse cosine function, 522–527
Inverse functions
    behavior of, 149–150
    cosine, 522–527
    determining existence of, 147–149
    examples of, 141–147
    explanation of, 140–141
    exponential, 143–145, 149
    logarithmic, 145, 149
    method for finding, 150–152
    power, 144–145, 149
    resticted domain, 518, 523
    sine, 517–522
    solving equations with, 515–517
Inverse matrix, 708–709, 732
Inverse sine function, 517–522
Inverse tangent function, 530–533
IRA account model, 374–375
Iteration, 575

Julia, Gaston, 575–576
Julia set, 576–579

Kepler, Johannes, 228
Kepler's third law of planetary motion, 228–234

Law of cosines, 473–474, 673
Law of sines, 470–473
Law of Universal Gravitation, 141–142, 310–311, 502–503
Leading coefficient, 250
Least squares line, 165
Life expectancy, 3–4, 189, 203–204
Limaçons, 642–643
Limit to growth. See Maximum sustainable population
Linear approximation, 551–558
Linear correlation coefficient. See Correlation coefficient
Linear equations, system of
    algebraic solutions to systems of, 727–731
    calculators to solve, 730–731
    matrices to solve systems of, 731–732
    with multiple solutions, 705–706
    with no solutions, 706–708
Linear functions
    behavior of, 131
    data and, 59–68
    domain of, 79

explanation of, 44
facts about lines and, 56
graphs of, 45–46
growth and, 74, 132, 136
implicit, 66–68
rate of change of, 137–138
slope and, 46–47
Linear growth, 74
Linear models
clothes production model, 660–662
explanation of, 659–660, 662
geometric view of vectors and, 660
Markov chain model for stock market, 662–665
population growth model, 665–668
with several variables, 236–242
Linear regression analysis
correlation coefficient and, 168–176
examples of, 166–168
multiple, 238–240
overview of, 164–166
Linearizing data
explanation of, 196, 199
exponential functions and, 200
power functions and, 208–209
Lines
least squares, 165
normal form of, 67
of best fit, 165
parallel, 56
parametric representations of, 619–621
perpendicular, 56
point-slope form for equation of, 50–51
slope intercept form, 49–50
slope of, 46–49
that don't pass through origin, 47–56
that pass through origin, 44–46
vertical intercept of, 47
Logarithmic functions
applications of, 106–109
behavior of, 104–105, 132
changing bases and, 109–111
explanation of, 99–100
exponential vs., 105–106
fit to data, 189, 196, 202–204
growth and, 135–136
indentities, 103
inverse of, 145–149
rate of change of, 139
Logarithms
to base 10, 99–100
to base e, 109
changing bases of, 109–111

natural, 109
nonlinear regression and, 204
properties of, 103–104
use of, 100–102
Logistic curve
explanation of, 346, 387
limiting value of, 388–389
point of inflection of, 388, 390–391
Logistic function
behavior of, 392–394
explanation of, 346–349
Logistic growth model
applications of, 394–397
difference equation for, 386–403
estimating logistic parameters, 397–403
explanation of, 386–387, 391
fitting to data, 397–403
inhibiting constant and, 386, 387
maximum sustainable population and, 388–389
point of inflection and, 390–392, 394
rabbit population and, 380–383
U.S. Population and, 400–403
LORAN system, 584, 611–612

Maintenance level of drug. *See* drug model
Mandelbrot, Benoit, 579
Mandelbrot set, 579–581
Markov chain model
explanation of, 663
matrix multiplication and, 692–697
steady state of, 709–711
for stock market, 662–665, 679–680, 709–711
Mathematical models
accuracy of, 37–38
with difference equations, 371–383
explanation of, 35, 371
expressed as functions, 35–37
linear, 659–668
for Newton's law of cooling, 409
parameters and, 37
periodic behavior with sine and cosine functions as, 494–509
Matrix
applied problems using, 660–668
augmented coefficient, 701–702
coefficient, 701, 731
dimension of, 691, 731
explanation of, 659
inverse, 708–709, 732
nonsingular powers of, 694–695
product of matrices, 686–697

product of vector and, 676–680
singular, 708–709
solving systems of linear equations with, 731–732
subtraction of, 690
transition, 662–663
Matrix algebra, 659, 691
Matrix multiplication
explanation of, 686–688
fruit purchase model and, 691–692
geometric view of powers of matrix and, 697
Markov chain transition matrices and, 692–697
methods to interpret, 686–691
population growth models and, 698–699
Matrix-vector equation, 661, 679
Matrix-vector product
explanation of, 676–680
geometric view of, 680–683
scalar product of vectors and, 686–687
Maximum of a function, 9–10
Maximum sustainable population, 388, 389
McAuliffe, Christa, 161
Mean, 738
Median, 738–739
Midline, 497
Midpoint formula, 587–588
Mile run model, 166–168, 170–172
Minimum of a function, 9–10
Models, 34. *See also* Mathematical models
Modulus, of complex number, 568, 570–572
Monotonic functions, 148
Movie industry receipts, 164, 166, 223–225
Multiple
of a function, 323–328
of a vector, 652
Multiple correlation coefficient, 240–241
Multiple regression. *See* Multivariate regression
Multiplication
of complex numbers, 736–737
matrix, 686–697
matrix and vector, 676–680
Multivariate regression
in Excel, 241–243
explanation of, 238–240

n-factorial, 722

n-vector, 659
Natural logarithm, 109
Newton, Isaac, 3, 408
Newton's law of cooling, 332–334,
    408–412
Newton's law of heating, 324–325,
    413–414
Nonhomogeneous system, 711
Nonlinear function, 181–191
Nonlinear regression
    correlation coefficient, 222
    exponential function and, 196–202
    logarithmic function and, 196,
        202–204
    logistic function and, 397–403
    power function and, 208–218
Normal form, of line, 67

Odd function, 126–127, 556
Odd positive integer powers, 126–127
Oil reserves, 420–423
Origin
    explanation of, 25
Oscillation, periodic, 483
Outliers, 227

Parabola
    axis of symmetry of, 615
    directrix of, 615, 616
    equation of, 615–617
    explanation of, 250, 614–615
    focus of, 615
    parametric representations of,
        625–626
    quadratic functions and, 255–257
    reflection property of, 616
    vertex of, 256
Parallel lines, 56
Parallel vectors, 652
Parameters, 37, 589, 619
Parametric equations of line, 589
Parametric representations
    on calculators, 624
    of circle, 622
    of ellipse, 623–624
    of hyperbola, 625
    of line, 619–621
    miscellaneous, 626–628
    of parabola, 625–626
    path of projectile and, 276–277,
        621–622
Path of projectile, 276–277, 621–622
Perihelion, 601, 605
Period, 460, 483

Periodic behavior
    of functions, 14–15
    sine and cosine functions as models
        for, 494–509
Periodic functions, 460. *See also*
    trigonometric functions
Periodic oscillatory effect, 483, 484
Perpendicular lines, 56
pH value, 106–107
Phase shift, 502–503
Planetary motion, 141–142, 228–234
Plate tectonics, 107
Point of inflection
    examples of finding, 391–392
    explanation of, 11–12, 391
    of logistic curve, 388, 390–391
    polynomials and, 263, 265
    turning point and, 263, 265
Point-slope form, 50–51, 53
Polar coordinate system
    explanation of, 630–632
    families of curves and, 636–643
    rectangular coordinates and, 632–634
Pole, 630
Pollutant model, 377–380
Polynomial functions
    approximating sine and cosine with,
        550–565
    coefficients of, 250
    cubic, 250, 253, 261–266, 287–289
    degree of, 250, 264–266
    difference patterns and, 292–294
    end behavior of, 267–269
    explanation of, 249–251
    fit to data, 278–281
    inflection points and, 263, 265
    leading coefficient of, 250
    modeling with, 273–281
    path of projectile and, 276–277
    quadratic, 250, 251, 255–260,
        285–287
    quartic, 251
    roots of, 285–291, 567
    zeros of, 251–254
Polynomial patterns
    finding, 292–294
    sums of integers and, 295–298
    sums of squares of integers and,
        298–301
Population growth, 182–183, 375–377,
    380–383, 665–668
Population models
    Florida, 28–29, 70–72, 76, 77–79,
        101, 104, 110, 143–144
    U.S. Population, 182–183, 197–198,
        344–345, 348–349, 400–402

    with harvesting, 375–377
    World, 371
Position vector, 650
Power functions
    applications of, 120–122
    behavior of, 116–119, 131–132
    even positive integer, 126
    explanation of, 113–116
    fit to data, 185–186, 188, 190,
        208–218
    fractional powers, 114–115
    graph of, 115
    growth and, 132–136
    inverse of, 145
    with integer powers, 126, 281
    negative powers, 114
    odd positive integer, 126–127
    passing through two points, 122–125
    polynomials and, 249, 250
Powers of *i*, 735–736
Principal values, 518, 522, 525
Probabilities, transition, 662–663
Product, of two functions, 305–306
    of two matrices, 686–688
Projectile motion, 314, 545–546,
    621–622
Proportionality, constant of, 44
Prozac model. *See* drug model
Pythagorean identity
    explanation of, 465, 490, 540, 545
Pythagorean theorem
    distance formula and, 585–586, 724
    explanation of, 429, 432–433, 444,
        568
    use of, 464–465
    vectors and, 651, 674

Quadrants, 25
Quadratic equation, 251–252
Quadratic formula, 252–253, 256, 267,
    285, 566
Quadratic functions
    behavior of, 255–260
    characteristics of, 260
    explanation of, 250, 251
    fitting to data, 273–281
    formula for, 252–253, 256, 267, 285
    graphs of, 255–257, 614
    linear factors of, 251–259
    turning point and, 260
    roots of, 257–260, 285–287
Quartic equations, 253
Quartic functions, 251
Quartiles, 740
Quotient, of functions, 306–307

Rabbit population, 380–383, 387
Radian measure, 490–492
Radioactive decay, 91–93, 112, 120–124
Radius, 589
Range, of function, 20–23, 27
Rate of change
    average, 136–139
    explanation of, 47
    of exponential functions, 138–139
    of linear functions, 137–138
    of logarithmic functions, 139
Rational functions
    application of, 310–311
    behavior of, 307–310
    explanation of, 307
    graph of, 300–310
    vertical asymptotes and, 308
Rectangular coordinate system, 25–26
    polar coordinates and, 632–634
Reference angles, 457
Reflection identities, 540–541, 545
Reflection property
    of ellipse, 602–604
    of parabola, 616
Regression. *See* fitting to data
Regression, multivariate, 238–243
Regression line
    equation of, 166–168, 401
    explanation of, 165
Regression plane, 238
Residual plots, 222–223, 225–227
Residuals
    explanation of, 165
    fit and, 222–225
Richter, Charles, 107
Richter scale, 107, 108
Root-mean-square (RMS), 228
Roots
    complex, 573
    of cubic equations, 262, 263
    double, 259–260
    explanation of, 251
    and factors, 252, 258, 259, 260, 262,
        264, 265
    of polynomial functions, 285–291,
        567
    of quadratic equations, 257–259,
        285–287
    real vs. complex, 259–260, 262–264,
        265
    using information on nature of,
        290–291
Rose curve, 639–641
Rotation Matrix, 681–683
Row vectors, 650, 659–660
Rumor model, 394–395

Scalar product
    clothes production model using,
        675–676
    explanation of, 649, 670–672
    geometric view of, 672–675,
        680–683
    matrix-vector product and,
        676–683, 686–687
Scatterplot
    examples of, 169
    explanation of, 162
    interpreting, 225–227
Secant function, 533
Second differences, 292
Second order difference equation, 383
Semi-log plot, 204
Sequence. *See also* Difference equations
    convergent, 416
    divergent, 416
    explanation of, 358
    exponential, 372
    Fibonacci, 382, 383
    geometric, 372, 417–420, 423–425
    limit of, 416
    notation for, 358
    solution, 360–369, 374, 391
Shifting functions
    analyzing data using, 331–344
    explanation of, 320–322
Shrinking functions, 323–328
Sigma notation, 722–723
Similar triangles, 723
Sine function. *See also* sinusoidal function
    amplitude of, 498
    applications of, 444–448
    approximating, 550–558
    behavior of, 442, 492–493
    explanation of, 441–442, 485–486
    graph of, 461–462, 486, 487
    inverse, 446, 517–522
    law of sines, 470–473
    modeling periodic behavior with,
        494–509
    period of, 460, 501
    radian measure and, 490–492
    "special" angles and, 443–444
Singular matrix, 708–709
Sinusoidal function
    amplitude of, 498
    combining exponential and, 507–509
    explanation of, 496–497
    frequency of, 500–502
    graph of, 504
    midline of, 497
    period of, 501
    phase shift of, 502

vertical shift of, 497
Skydiving model, 336–338
Slope, 46–49, 53
    parallel lines, 56
    perpendicular lines, 56
Slope-intercept form, 49–50, 53
Solution sequence, 360–369
Space shuttle. *See* Challenger
Species, number of, vs. area of habitat,
    124–125, 185–186, 192, 210–211
Square, completing the, 725–727
St. Louis Arch, 280–281
Standard deviation, 739
Standard error of the estimate, 228
Steady plate. *See* equilibrium state
Stock market model, 662–665,
    679–680, 709–711
Stretching functions
    analyzing data using, 344–345
    explanation of, 323–328
Substitution method, 728
Subtraction
    of complex numbers, 736
    matrix, 690
Sum identities, 542–543, 545
Sum of integers, 295–298
Sum of squares
    of integers, 298–301
    interpreting, 227–228
    use of, 165
Sum of two functions, 304
Sum of vectors, 653–654
Summation notation, 722–723
Surge function, 349–352
Symmetry
    axis of, 615
    explanation of, 732–734
    parabolas and, 256
Systems of linear equations
    algebraic solutions to, 727–731
    Cramer's rule and, 709
    Gaussian elimination and, 700–702
    matrices to solve, 731–732
    with multiple solutions, 705–706
    with no solutions, 706–708

3–vector, 659
Tables, representing functions with,
    4–5, 26
Tangent, of an angle, 430–439
Tangent function
    applications of, 448
    behavior of, 434–435
    explanation of, 432, 434, 527
    graph of, 527–530

identities involving, 548
inverse, 530–533
period of, 460
Tangent ratio
explanation of, 432, 527
use of, 435–437
Taylor, Brook, 556
Taylor polynomial approximations
explanation of, 556
sin *x* and cos *x* using, 564–565
Temperature oscillation model, 504, 519–520
Terminal side, of angle, 457
Terminal velocity, 336–339
Tide model, 503–504
Transformation approach
application of, 198–200
example of, 200–202
Transition matrix, 662–663
Triangle
finding angle in, 437–439
right, 429
similar, 723
Triangulation, 472
Tribbles, 385
Trigonometric functions. *See also* Cosine function; Sine function; Sinusoidal function; Tangent function.
applications from physics for, 452–455
approximating, 550–565
complex numbers and, 566–575
inverse functions, 517–522, 522–527, 530–533
Julia set and, 576–579
Mandelbrot set and, 579–581
properties of complex numbers and, 566–575
radian measure and, 490–492
reciprocals of, 533
relationships among, 463–468
Trigonometric identities
approximating sin *x* and cos *x* using, 561–563
difference, 543–545

double-angle, 541–542, 545, 561
explanation of, 539, 545
half-angle, 544–545
involving sine and cosine, 540
law of cosines, 473–474
law of sines, 470–473
Pythagorean, 465, 540, 545
reflection, 540–541, 545
sum, 542–543, 545
use of, 545–548
Trigonometry
cosine function and, 448–452
cosine of an angle, 448
finding angle in triangle and, 437–439
graph of cosine function and, 462
graph of sine function and, 461–462
law of cosines and, 473–476
law of sines and, 470–473
overview of, 429–430
sine of an angle, 441–442
sine function and, 441–448
"special" angles and, 432–434
tangent function and, 434–435
tangent of angle and, 430–432
tangent ratio and, 435–437
Turning point of function, 9–10
inflection point and, 263, 265

Unit circle, 484–485, 489
Unit vectors, 652–653
U.S. Population, 197–198, 344–345, 348–349
exponential growth model, 182–183
logistic growth model, 400–403

Variable
dependent, 19, 25
independent, 19, 25, 237, 323, 326
linear models with several, 236–243
Vector product, 671
Vectors
applications of, 654–656

angle between, 674
applied problems using, 660–668
column, 650, 659–660
constant multiple of, 652
coordinate, 654, 672
difference of two, 656–657
displacement, 649–650
explanation of, 452, 649–650
geometric view of, 660
magnitude of, 650–651
multiple of, 652
parallel, 652
position, 649
rotation of, 682–683
row, 650, 659
sum of two, 653–654
unit, 652–653
Vertex
of ellipse, 597
of hyperbola, 608–609
of parabola, 256
Vertical asymptote, 105
Vertical axis, 25
Vertical intercept
examples of, 47–50, 52
explanation of, 47, 53
Vertical line test, 30
Vertical shift, 321, 497
Vertical shrinking, 323, 324
Vertical stretching, 323, 324

Wave phenomena, 546–548
Whispering gallery, 603–604
World population, 371
Words, representing functions with, 5–6, 23

y-intercept, 47
Yeast growth model, 398–400

Zero of a function, 251–254

Function	Equation	Behavior	Graph
Sine	$y = \sin x$	Passes through the origin Periodic with period $360° = 2\pi$ radians Oscillates between -1 and $+1$	
Cosine	$y = \cos x$	Passes through $(0, 1)$ Periodic with period $360° = 2\pi$ radians Oscillates between -1 and $+1$	
Sinusoidal	$y = D + A\sin(B(x - C))$ $y = D + A\cos(B(x - C))$	Centered vertically about the midline (vertical shift) $= D$ Amplitude $= A = $ size of oscillation above and below the midline, so oscillates between $D - A$ and $D + A$ Frequency $= B = $ number of cycles between 0 and 2π radians Period $= 2\pi/$frequency Phase shift $=$ horizontal shift $= C$	
Tangent	$y = \tan x$	Passes through the origin Periodic with period $180° = \pi$ radians Vertical asymptotes at $x = \pm 90° = \pm \pi/2$ radians	